Neural Network Design

2nd Edition

Martin T. Hagan
Oklahoma State University
Stillwater, Oklahoma

Howard B. Demuth
University of Colorado
Boulder, Colorado

Mark Hudson Beale
MHB Inc.
Hayden, Idaho

Orlando De Jesús
Consultant
Frisco, Texas

MTH

To Janet, Thomas, Daniel, Mom and Dad

HBD

To Hal, Katherine, Kimberly and Mary

MHB

To Leah, Valerie, Asia, Drake, Coral and Morgan

ODJ

To: Marisela, María Victoria, Manuel, Mamá y Papá.

Neural Network Design, 2nd Edition

ISBN 978-0-9717321-1-7

OVERHEADS, DEMONSTRATION PROGRAMS and an extended electronic version of this book can be found at the following website:

hagan.okstate.edu/nnd.html

Contents

3

An Illustrative Example

4

Perceptron Learning Rule

5 Signal and Weight Vector Spaces

6 Linear Transformations for Neural Networks

7 Supervised Hebbian Learning

8 Performance Surfaces and Optimum Points

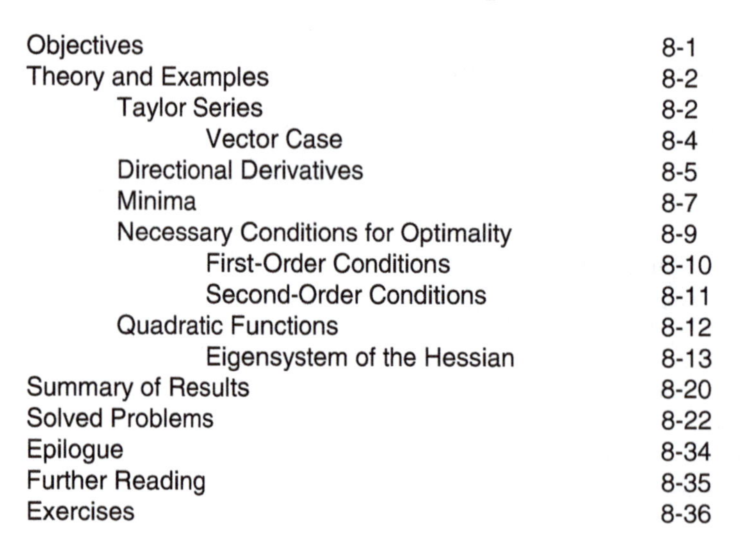

9

Performance Optimization

10

Widrow-Hoff Learning

11 Backpropagation

12 Variations on Backpropagation

13

Generalization

14

Dynamic Networks

15 Competitive Networks

16 Radial Basis Networks

17

Practical Training Issues

18

Case Study 1:Function Approximation

19

Case Study 2:Probability Estimation

20

Case Study 3:Pattern Recognition

21

Case Study 4: Clustering

Case Study 5: Prediction

Appendices

Preface

This book gives an introduction to basic neural network architectures and learning rules. Emphasis is placed on the mathematical analysis of these networks, on methods of training them and on their application to practical engineering problems in such areas as nonlinear regression, pattern recognition, signal processing, data mining and control systems.

Every effort has been made to present material in a clear and consistent manner so that it can be read and applied with ease. We have included many solved problems to illustrate each topic of discussion. We have also included a number of case studies in the final chapters to demonstrate practical issues that arise when using neural networks on real world problems.

Since this is a book on the design of neural networks, our choice of topics was guided by two principles. First, we wanted to present the most useful and practical neural network architectures, learning rules and training techniques. Second, we wanted the book to be complete in itself and to flow easily from one chapter to the next. For this reason, various introductory materials and chapters on applied mathematics are included just before they are needed for a particular subject. In summary, we have chosen some topics because of their practical importance in the application of neural networks, and other topics because of their importance in explaining how neural networks operate.

We have omitted many topics that might have been included. We have not, for instance, made this book a catalog or compendium of all known neural network architectures and learning rules, but have instead concentrated on the fundamental concepts. Second, we have not discussed neural network implementation technologies, such as VLSI, optical devices and parallel computers. Finally, we do not present the biological and psychological foundations of neural networks in any depth. These are all important topics, but we hope that we have done the reader a service by focusing on those topics that we consider to be most useful in the design of neural networks and by treating those topics in some depth.

This book has been organized for a one-semester introductory course in neural networks at the senior or first-year graduate level. (It is also suitable for short courses, self-study and reference.) The reader is expected to have some background in linear algebra, probability and differential equations.

Each chapter of the book is divided into the following sections: Objectives, Theory and Examples, Summary of Results, Solved Problems, Epilogue, Further Reading and Exercises. The *Theory and Examples* section comprises the main body of each chapter. It includes the development of fundamental ideas as well as worked examples (indicated by the icon shown here in the left margin). The *Summary of Results* section provides a convenient listing of important equations and concepts and facilitates the use of the book as an industrial reference. About a third of each chapter is devoted to the *Solved Problems* section, which provides detailed examples for all key concepts.

The following figure illustrates the dependencies among the chapters.

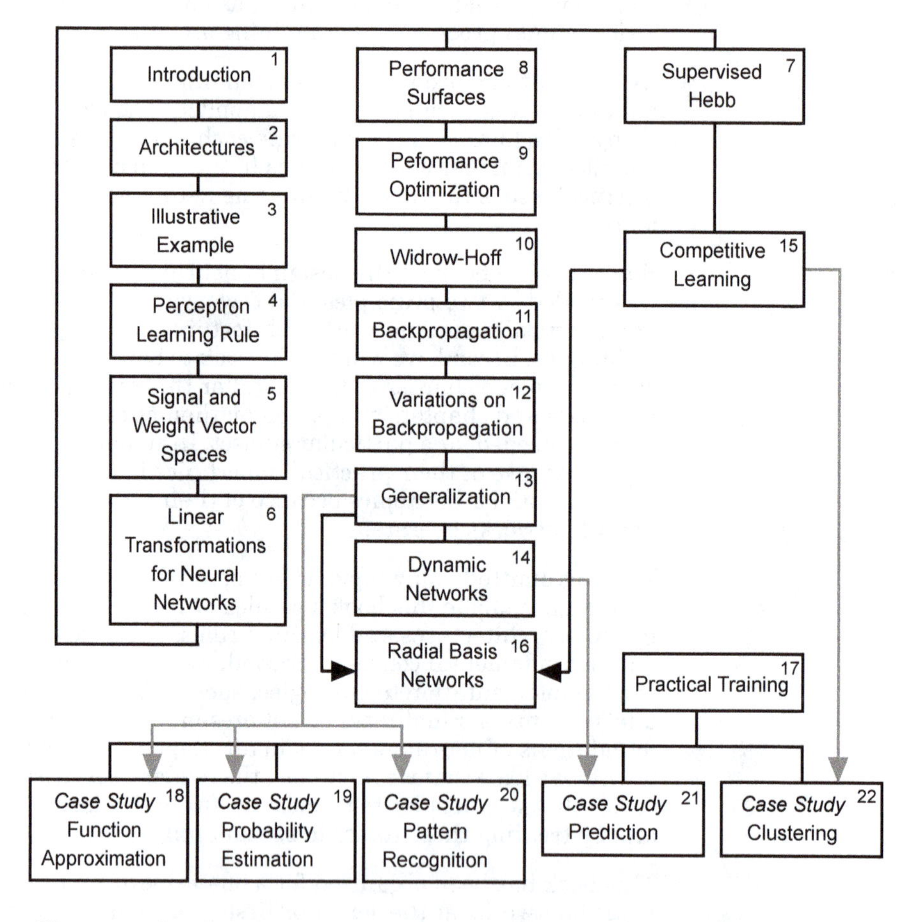

Chapter 1 through 6 cover basic concepts that are required for all of the remaining chapters. Chapter 1 is an introduction to the text, with a brief historical background and some basic biology. Chapter 2 describes the basic neural network architectures. The notation that is introduced in this chapter is used throughout the book. In Chapter 3 we present a simple pattern recognition problem and show how it can be solved using three different types of neural networks. These three networks are representative of the types of networks that are presented in the remainder of the text. In addition, the pattern recognition problem presented here provides a common thread of experience throughout the book.

Much of the focus of this book will be on methods for training neural networks to perform various tasks. In Chapter 4 we introduce learning algorithms and present the first practical algorithm: the perceptron learning rule. The perceptron network has fundamental limitations, but it is important for historical reasons and is also a useful tool for introducing key concepts that will be applied to more powerful networks in later chapters.

One of the main objectives of this book is to explain how neural networks operate. For this reason we will weave together neural network topics with important introductory material. For example, linear algebra, which is the core of the mathematics required for understanding neural networks, is reviewed in Chapter 5 and 6. The concepts discussed in these chapters will be used extensively throughout the remainder of the book.

Chapter 7 and 15 describe networks and learning rules that are heavily inspired by biology and psychology. They fall into two categories: associative networks and competitive networks. Associative networks are covered in Chapter 7 and competitive networks are covered in Chapter 15.

Chapter 8–14 and 16 develop a class of learning called performance learning, in which a network is trained to optimize its performance. Chapter 8 and 9 introduce the basic concepts of performance learning. Chapter 10–13 apply these concepts to feedforward neural networks of increasing power and complexity, Chapter 14 applies them to dynamic networks and Chapter 16 applies them to radial basis networks, which also use concepts from competitive learning.

Chapter 17–22 are different than the preceding chapters. Previous chapters focus on the fundamentals of each type of network and their learning rules. The focus is on understanding the key concepts. In Chapter 17–22, we discuss some practical issues in applying neural networks to real world problems. Chapter 17 describes many practical training tips, and Chapter 18–22 present a series of case studies, in which neural networks are applied to practical problems in function approximation, probability estimation, pattern recognition, clustering and prediction.

Software

MATLAB is not essential for using this book. The computer exercises can be performed with any available programming language, and the *Neural Network Design Demonstrations*, while helpful, are not critical to understanding the material covered in this book.

However, we have made use of the MATLAB software package to supplement the textbook. This software is widely available and, because of its matrix/vector notation and graphics, is a convenient environment in which to experiment with neural networks. We use MATLAB in two different ways. First, we have included a number of exercises for the reader to perform in MATLAB. Many of the important features of neural networks become apparent only for large-scale problems, which are computationally intensive and not feasible for hand calculations. With MATLAB, neural network algorithms can be quickly implemented, and large-scale problems can be tested conveniently. These MATLAB exercises are identified by the icon shown here to the left. (If MATLAB is not available, any other programming language can be used to perform the exercises.)

```
» 2+2
ans =
     4
```

The second way in which we use MATLAB is through the *Neural Network Design Demonstrations*, which can be downloaded from the website hagan.okstate.edu/nnd.html. These interactive demonstrations illustrate important concepts in each chapter. After the software has been loaded into the MATLAB directory on your computer (or placed on the MATLAB path), it can be invoked by typing nnd at the MATLAB prompt. All demonstrations are easily accessible from a master menu. The icon shown here to the left identifies references to these demonstrations in the text.

The demonstrations require MATLAB or the student edition of MATLAB, version 2010a or later. See Appendix C for specific information on using the demonstration software.

Overheads

As an aid to instructors who are using this text, we have prepared a companion set of overheads. Transparency masters (in Microsoft Powerpoint format or PDF) for each chapter are available on the web at hagan.okstate.edu/nnd.html.

Acknowledgments

We are deeply indebted to the reviewers who have given freely of their time to read all or parts of the drafts of this book and to test various versions of the software. In particular we are most grateful to Professor John Andreae, University of Canterbury; Dan Foresee, AT&T; Dr. Carl Latino, Oklahoma State University; Jack Hagan, MCI; Dr. Gerry Andeen, SRI; and Joan Miller and Margie Jenks, University of Idaho. We also had constructive inputs from our graduate students in ECEN 5733 at Oklahoma State University, ENEL 621 at the University of Canterbury, INSA 0506 at the Institut National des Sciences Appliquées and ECE 5120 at the University of Colorado, who read many drafts, tested the software and provided helpful suggestions for improving the book over the years. We are also grateful to the anonymous reviewers who provided several useful recommendations.

We wish to thank Dr. Peter Gough for inviting us to join the staff in the Electrical and Electronic Engineering Department at the University of Canterbury, Christchurch, New Zealand, and Dr. Andre Titli for inviting us to join the staff at the Laboratoire d'Analyse et d'Architecture des Systèms, Centre National de la Recherche Scientifique, Toulouse, France. Sabbaticals from Oklahoma State University and a year's leave from the University of Idaho gave us the time to write this book. Thanks to Texas Instruments, Halliburton, Cummins, Amgen and NSF, for their support of our neural network research. Thanks to The Mathworks for permission to use material from the *Neural Network Toolbox*.

1 Introduction

Objectives

As you read these words you are using a complex biological neural network. You have a highly interconnected set of some 10^{11} neurons to facilitate your reading, breathing, motion and thinking. Each of your biological neurons, a rich assembly of tissue and chemistry, has the complexity, if not the speed, of a microprocessor. Some of your neural structure was with you at birth. Other parts have been established by experience.

Scientists have only just begun to understand how biological neural networks operate. It is generally understood that all biological neural functions, including memory, are stored in the neurons and in the connections between them. Learning is viewed as the establishment of new connections between neurons or the modification of existing connections. This leads to the following question: Although we have only a rudimentary understanding of biological neural networks, is it possible to construct a small set of simple artificial "neurons" and perhaps train them to serve a useful function? The answer is "yes." This book, then, is about *artificial* neural networks.

The neurons that we consider here are not biological. They are extremely simple abstractions of biological neurons, realized as elements in a program or perhaps as circuits made of silicon. Networks of these artificial neurons do not have a fraction of the power of the human brain, but they can be trained to perform useful functions. This book is about such neurons, the networks that contain them and their training.

History

The history of artificial neural networks is filled with colorful, creative individuals from a variety of fields, many of whom struggled for decades to develop concepts that we now take for granted. This history has been documented by various authors. One particularly interesting book is *Neurocomputing: Foundations of Research* by John Anderson and Edward Rosenfeld. They have collected and edited a set of some 43 papers of special historical interest. Each paper is preceded by an introduction that puts the paper in historical perspective.

Histories of some of the main neural network contributors are included at the beginning of various chapters throughout this text and will not be repeated here. However, it seems appropriate to give a brief overview, a sample of the major developments.

At least two ingredients are necessary for the advancement of a technology: concept and implementation. First, one must have a concept, a way of thinking about a topic, some view of it that gives a clarity not there before. This may involve a simple idea, or it may be more specific and include a mathematical description. To illustrate this point, consider the history of the heart. It was thought to be, at various times, the center of the soul or a source of heat. In the 17th century medical practitioners finally began to view the heart as a pump, and they designed experiments to study its pumping action. These experiments revolutionized our view of the circulatory system. Without the pump concept, an understanding of the heart was out of grasp.

Concepts and their accompanying mathematics are not sufficient for a technology to mature unless there is some way to implement the system. For instance, the mathematics necessary for the reconstruction of images from computer-aided tomography (CAT) scans was known many years before the availability of high-speed computers and efficient algorithms finally made it practical to implement a useful CAT system.

The history of neural networks has progressed through both conceptual innovations and implementation developments. These advancements, however, seem to have occurred in fits and starts rather than by steady evolution.

Some of the background work for the field of neural networks occurred in the late 19th and early 20th centuries. This consisted primarily of interdisciplinary work in physics, psychology and neurophysiology by such scientists as Hermann von Helmholtz, Ernst Mach and Ivan Pavlov. This early work emphasized general theories of learning, vision, conditioning, etc., and did not include specific mathematical models of neuron operation.

The modern view of neural networks began in the 1940s with the work of Warren McCulloch and Walter Pitts [McPi43], who showed that networks of artificial neurons could, in principle, compute any arithmetic or logical function. Their work is often acknowledged as the origin of the neural network field.

McCulloch and Pitts were followed by Donald Hebb [Hebb49], who proposed that classical conditioning (as discovered by Pavlov) is present because of the properties of individual neurons. He proposed a mechanism for learning in biological neurons (see Chapter 7).

The first practical application of artificial neural networks came in the late 1950s, with the invention of the perceptron network and associated learning rule by Frank Rosenblatt [Rose58]. Rosenblatt and his colleagues built a perceptron network and demonstrated its ability to perform pattern recognition. This early success generated a great deal of interest in neural network research. Unfortunately, it was later shown that the basic perceptron network could solve only a limited class of problems. (See Chapter 4 for more on Rosenblatt and the perceptron learning rule.)

At about the same time, Bernard Widrow and Ted Hoff [WiHo60] introduced a new learning algorithm and used it to train adaptive linear neural networks, which were similar in structure and capability to Rosenblatt's perceptron. The Widrow-Hoff learning rule is still in use today. (See Chapter 10 for more on Widrow-Hoff learning.)

Unfortunately, both Rosenblatt's and Widrow's networks suffered from the same inherent limitations, which were widely publicized in a book by Marvin Minsky and Seymour Papert [MiPa69]. Rosenblatt and Widrow were aware of these limitations and proposed new networks that would overcome them. However, they were not able to successfully modify their learning algorithms to train the more complex networks.

Many people, influenced by Minsky and Papert, believed that further research on neural networks was a dead end. This, combined with the fact that there were no powerful digital computers on which to experiment, caused many researchers to leave the field. For a decade neural network research was largely suspended.

Some important work, however, did continue during the 1970s. In 1972 Teuvo Kohonen [Koho72] and James Anderson [Ande72] independently and separately developed new neural networks that could act as memories. Stephen Grossberg [Gros76] was also very active during this period in the investigation of self-organizing networks.

Interest in neural networks had faltered during the late 1960s because of the lack of new ideas and powerful computers with which to experiment. During the 1980s both of these impediments were overcome, and research in neural networks increased dramatically. New personal computers and workstations, which rapidly grew in capability, became widely available. In addition, important new concepts were introduced.

Two new concepts were most responsible for the rebirth of neural networks. The first was the use of statistical mechanics to explain the operation of a certain class of recurrent network, which could be used as an associative memory. This was described in a seminal paper by physicist John Hopfield [Hopf82].

The second key development of the 1980s was the backpropagation algorithm for training multilayer perceptron networks, which was discovered independently by several different researchers. The most influential publication of the backpropagation algorithm was by David Rumelhart and James McClelland [RuMc86]. This algorithm was the answer to the criticisms Minsky and Papert had made in the 1960s. (See Chapter 11 for a development of the backpropagation algorithm.)

These new developments reinvigorated the field of neural networks. Since the 1980s, thousands of papers have been written, neural networks have found countless applications, and the field has been buzzing with new theoretical and practical work.

The brief historical account given above is not intended to identify all of the major contributors, but is simply to give the reader some feel for how knowledge in the neural network field has progressed. As one might note, the progress has not always been "slow but sure." There have been periods of dramatic progress and periods when relatively little has been accomplished.

Many of the advances in neural networks have had to do with new concepts, such as innovative architectures and training rules. Just as important has been the availability of powerful new computers on which to test these new concepts.

Well, so much for the history of neural networks to this date. The real question is, "What will happen in the future?" Neural networks have clearly taken a permanent place as important mathematical/engineering tools. They don't provide solutions to every problem, but they are essential tools to be used in appropriate situations. In addition, remember that we still know very little about how the brain works. The most important advances in neural networks almost certainly lie in the future.

The large number and wide variety of applications of this technology are very encouraging. The next section describes some of these applications.

Applications

A newspaper article described the use of neural networks in literature research by Aston University. It stated that "the network can be taught to recognize individual writing styles, and the researchers used it to compare works attributed to Shakespeare and his contemporaries." A popular science television program documented the use of neural networks by an Italian research institute to test the purity of olive oil. Google uses neural networks for image tagging (automatically identifying an image and assigning keywords), and Microsoft has developed neural networks that can help convert spoken English speech into spoken Chinese speech. Researchers at Lund University and Skåne University Hospital in Sweden have used neural networks to improve long-term survival rates for heart transplant recipients by identifying optimal recipient and donor matches. These examples are indicative of the broad range of applications that can be found for neural networks. The applications are expanding because neural networks are good at solving problems, not just in engineering, science and mathematics, but in medicine, business, finance and literature as well. Their application to a wide variety of problems in many fields makes them very attractive. Also, faster computers and faster algorithms have made it possible to use neural networks to solve complex industrial problems that formerly required too much computation.

The following note and Table of Neural Network Applications are reproduced here from the *Neural Network Toolbox* for MATLAB with the permission of the MathWorks, Inc.

A 1988 DARPA Neural Network Study [DARP88] lists various neural network applications, beginning with the adaptive channel equalizer in about 1984. This device, which is an outstanding commercial success, is a single-neuron network used in long distance telephone systems to stabilize voice signals. The DARPA report goes on to list other commercial applications, including a small word recognizer, a process monitor, a sonar classifier and a risk analysis system.

Thousands of neural networks have been applied in hundreds of fields in the many years since the DARPA report was written. A list of some of those applications follows.

Aerospace

High performance aircraft autopilots, flight path simulations, aircraft control systems, autopilot enhancements, aircraft component simulations, aircraft component fault detectors

Automotive

Automobile automatic guidance systems, fuel injector control, automatic braking systems, misfire detection, virtual emission sensors, warranty activity analyzers

Banking

Check and other document readers, credit application evaluators, cash forecasting, firm classification, exchange rate forecasting, predicting loan recovery rates, measuring credit risk

Defense

Weapon steering, target tracking, object discrimination, facial recognition, new kinds of sensors, sonar, radar and image signal processing including data compression, feature extraction and noise suppression, signal/image identification

Electronics

Code sequence prediction, integrated circuit chip layout, process control, chip failure analysis, machine vision, voice synthesis, nonlinear modeling

Entertainment

Animation, special effects, market forecasting

Financial

Real estate appraisal, loan advisor, mortgage screening, corporate bond rating, credit line use analysis, portfolio trading program, corporate financial analysis, currency price prediction

Insurance

Policy application evaluation, product optimization

Manufacturing

Manufacturing process control, product design and analysis, process and machine diagnosis, real-time particle identification, visual quality inspection systems, beer testing, welding quality analysis, paper quality prediction, computer chip quality analysis, analysis of grinding operations, chemical product design analysis, machine maintenance analysis, project bidding, planning and management, dynamic modeling of chemical process systems

Medical

Breast cancer cell analysis, EEG and ECG analysis, prosthesis design, optimization of transplant times, hospital expense reduction, hospital quality improvement, emergency room test advisement

Oil and Gas

Exploration, smart sensors, reservoir modeling, well treatment decisions, seismic interpretation

Robotics

Trajectory control, forklift robot, manipulator controllers, vision systems, autonomous vehicles

Speech

Speech recognition, speech compression, vowel classification, text to speech synthesis

Securities

Market analysis, automatic bond rating, stock trading advisory systems

Telecommunications

Image and data compression, automated information services, real-time translation of spoken language, customer payment processing systems

Transportation

Truck brake diagnosis systems, vehicle scheduling, routing systems

Conclusion

The number of neural network applications, the money that has been invested in neural network software and hardware, and the depth and breadth of interest in these devices is enormous.

Biological Inspiration

The artificial neural networks discussed in this text are only remotely related to their biological counterparts. In this section we will briefly describe those characteristics of brain function that have inspired the development of artificial neural networks.

The brain consists of a large number (approximately 10^{11}) of highly connected elements (approximately 10^4 connections per element) called neurons. For our purposes these neurons have three principal components: the dendrites, the cell body and the axon. The dendrites are tree-like receptive networks of nerve fibers that carry electrical signals into the cell body. The cell body effectively sums and thresholds these incoming signals. The axon is a single long fiber that carries the signal from the cell body out to other neurons. The point of contact between an axon of one cell and a dendrite of another cell is called a synapse. It is the arrangement of neurons and the strengths of the individual synapses, determined by a complex chemical process, that establishes the function of the neural network. Figure 1.1 is a simplified schematic diagram of two biological neurons.

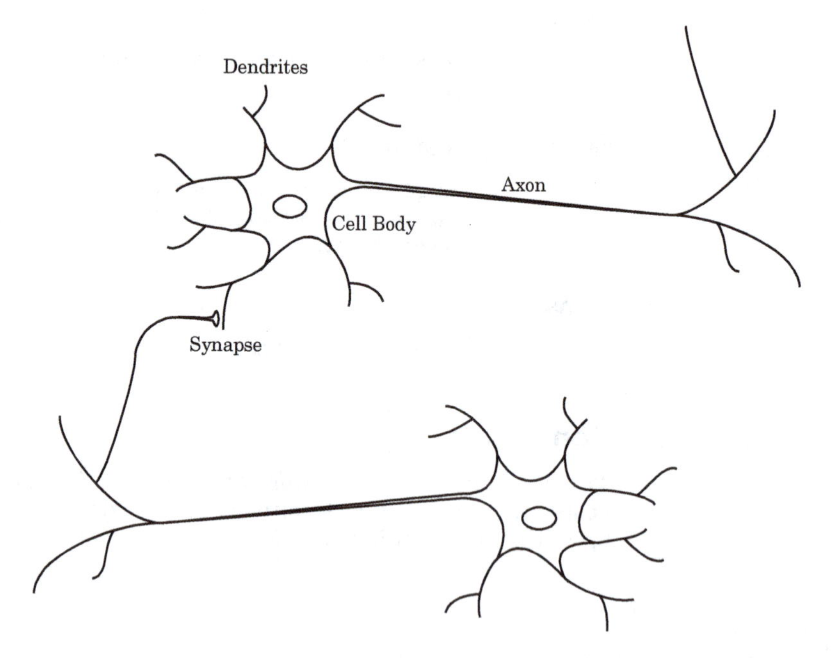

Figure 1.1 Schematic Drawing of Biological Neurons

Some of the neural structure is defined at birth. Other parts are developed through learning, as new connections are made and others waste away. This development is most noticeable in the early stages of life. For example, it has been shown that if a young cat is denied use of one eye during a critical window of time, it will never develop normal vision in that eye. Linguists have discovered that infants over six months of age can no longer discriminate certain speech sounds, unless they were exposed to them earlier in life [WeTe84].

Neural structures continue to change throughout life. These later changes tend to consist mainly of strengthening or weakening of synaptic junctions. For instance, it is believed that new memories are formed by modification of these synaptic strengths. Thus, the process of learning a new friend's face consists of altering various synapses. Neuroscientists have discovered [MaGa2000], for example, that the hippocampi of London taxi drivers are significantly larger than average. This is because they must memorize a large amount of navigational information—a process that takes more than two years.

Artificial neural networks do not approach the complexity of the brain. There are, however, two key similarities between biological and artificial neural networks. First, the building blocks of both networks are simple computational devices (although artificial neurons are much simpler than biological neurons) that are highly interconnected. Second, the connections between neurons determine the function of the network. The primary objective of this book will be to determine the appropriate connections to solve particular problems.

It is worth noting that even though biological neurons are very slow when compared to electrical circuits (10^{-3} s compared to 10^{-10} s), the brain is able to perform many tasks much faster than any conventional computer. This is in part because of the massively parallel structure of biological neural networks; all of the neurons are operating at the same time. Artificial neural networks share this parallel structure. Even though most artificial neural networks are currently implemented on conventional digital computers, their parallel structure makes them ideally suited to implementation using VLSI, optical devices and parallel processors.

In the following chapter we will introduce our basic artificial neuron and will explain how we can combine such neurons to form networks. This will provide a background for Chapter 3, where we take our first look at neural networks in action.

Further Reading

[Ande72] J. A. Anderson, "A simple neural network generating an interactive memory," *Mathematical Biosciences*, Vol. 14, pp. 197–220, 1972.

Anderson proposed a "linear associator" model for associative memory. The model was trained, using a generalization of the Hebb postulate, to learn an association between input and output vectors. The physiological plausibility of the network was emphasized. Kohonen published a closely related paper at the same time [Koho72], although the two researchers were working independently.

[AnRo88] J. A. Anderson and E. Rosenfeld, *Neurocomputing: Foundations of Research*, Cambridge, MA: MIT Press, 1989.

Neurocomputing is a fundamental reference book. It contains over forty of the most important neurocomputing writings. Each paper is accompanied by an introduction that summarizes its results and gives a perspective on the position of the paper in the history of the field.

[DARP88] *DARPA Neural Network Study*, Lexington, MA: MIT Lincoln Laboratory, 1988.

This study is a compendium of knowledge of neural networks as they were known to 1988. It presents the theoretical foundations of neural networks and discusses their current applications. It contains sections on associative memories, recurrent networks, vision, speech recognition, and robotics. Finally, it discusses simulation tools and implementation technology.

[Gros76] S. Grossberg, "Adaptive pattern classification and universal recoding: I. Parallel development and coding of neural feature detectors," *Biological Cybernetics*, Vol. 23, pp. 121–134, 1976.

Grossberg describes a self-organizing neural network based on the visual system. The network, which consists of short-term and long-term memory mechanisms, is a continuous-time competitive network. It forms a basis for the adaptive resonance theory (ART) networks.

[Gros80] S. Grossberg, "How does the brain build a cognitive code?" *Psychological Review*, Vol. 88, pp. 375–407, 1980.

Grossberg's 1980 paper proposes neural structures and mechanisms that can explain many physiological behaviors including spatial frequency adaptation, binocular rivalry, etc. His systems perform error correction by themselves, without outside help.

[Hebb 49] D. O. Hebb, *The Organization of Behavior*. New York: Wiley, 1949.

The main premise of this seminal book is that behavior can be explained by the action of neurons. In it, Hebb proposed one of the first learning laws, which postulated a mechanism for learning at the cellular level.

Hebb proposes that classical conditioning in biology is present because of the properties of individual neurons.

[Hopf82] J. J. Hopfield, "Neural networks and physical systems with emergent collective computational abilities," *Proceedings of the National Academy of Sciences*, Vol. 79, pp. 2554–2558, 1982.

Hopfield describes a content-addressable neural network. He also presents a clear picture of how his neural network operates, and of what it can do.

[Koho72] T. Kohonen, "Correlation matrix memories," *IEEE Transactions on Computers*, vol. 21, pp. 353–359, 1972.

Kohonen proposed a correlation matrix model for associative memory. The model was trained, using the outer product rule (also known as the Hebb rule), to learn an association between input and output vectors. The mathematical structure of the network was emphasized. Anderson published a closely related paper at the same time [Ande72], although the two researchers were working independently.

[MaGa00] E. A. Maguire, D. G. Gadian, I. S. Johnsrude, C. D. Good, J. Ashburner, R. S. J. Frackowiak, and C. D. Frith, "Navigation-related structural change in the hippocampi of taxi drivers," Proceedings of the National Academy of Sciences, Vol. 97, No. 8, pp. 4398-4403, 2000.

Taxi drivers in London must undergo extensive training, learning how to navigate between thousands of places in the city. This training is colloquially known as "being on The Knowledge" and takes about 2 years to acquire on average. This study demonstrated that the posterior hippocampi of London taxi drivers were significantly larger relative to those of control subjects.

[McPi43] W. McCulloch and W. Pitts, "A logical calculus of the ideas immanent in nervous activity," *Bulletin of Mathematical Biophysics.*, Vol. 5, pp. 115–133, 1943.

This article introduces the first mathematical model of a neuron, in which a weighted sum of input signals is compared to a threshold to determine whether or not the neuron fires. This was the first attempt to describe what the brain does, based on computing elements known at the time. It shows that simple neural networks can compute any arithmetic or logical function.

[MiPa69] M. Minsky and S. Papert, *Perceptrons*, Cambridge, MA: MIT Press, 1969.

A landmark book that contains the first rigorous study devoted to determining what a perceptron network is capable of learning. A formal treatment of the perceptron was needed both to explain the perceptron's limitations and to indicate directions for overcoming them. Unfortunately, the book pessimistically predicted that the limitations of perceptrons indicated that the field of neural networks was a dead end. Although this was not true it temporarily cooled research and funding for research for several years.

[Rose58] F. Rosenblatt, "The perceptron: A probabilistic model for information storage and organization in the brain," *Psychological Review*, Vol. 65, pp. 386–408, 1958.

Rosenblatt presents the first practical artificial neural network — the perceptron.

[RuMc86] D. E. Rumelhart and J. L. McClelland, eds., *Parallel Distributed Processing: Explorations in the Microstructure of Cognition*, Vol. 1, Cambridge, MA: MIT Press, 1986.

One of the two key influences in the resurgence of interest in the neural network field during the 1980s. Among other topics, it presents the backpropagation algorithm for training multilayer networks.

[WeTe84] J. F. Werker and R. C. Tees, "Cross-language speech perception: Evidence for perceptual reorganization during the first year of life," Infant Behavior and Development, Vol. 7, pp. 49-63, 1984.

This work describes an experiment in which infants from the Interior Salish ethnic group in British Columbia, and other infants outside that group, were tested on their ability to discriminate two different sounds from the Thompson language, which is spoken by the Interior Salish. The researchers discovered that infants less than 6 or 8 months of age were generally able to distinguish the sounds, whether or not they were Interior Salish. By 10 to 12 months of age, only the Interior Salish children were able to distinguish the two sounds.

[WiHo60] B. Widrow and M. E. Hoff, "Adaptive switching circuits," *1960 IRE WESCON Convention Record*, New York: IRE Part 4, pp. 96–104, 1960.

This seminal paper describes an adaptive perceptron-like network that can learn quickly and accurately. The authors assume that the system has inputs and a desired output classification for each input, and that the system can calculate the error between the actual and desired output. The weights are adjusted, using a gradient descent method, so as to minimize the mean square error. (Least Mean Square error or LMS algorithm.)

This paper is reprinted in [AnRo88].

2 Neuron Model and Network Architectures

Objectives

In Chapter 1 we presented a simplified description of biological neurons and neural networks. Now we will introduce our simplified mathematical model of the neuron and will explain how these artificial neurons can be interconnected to form a variety of network architectures. We will also illustrate the basic operation of these networks through some simple examples. The concepts and notation introduced in this chapter will be used throughout this book.

This chapter does not cover all of the architectures that will be used in this book, but it does present the basic building blocks. More complex architectures will be introduced and discussed as they are needed in later chapters. Even so, a lot of detail is presented here. Please note that it is not necessary for the reader to memorize all of the material in this chapter on a first reading. Instead, treat it as a sample to get you started and a resource to which you can return.

Theory and Examples

Notation

Unfortunately, there is no single neural network notation that is universally accepted. Papers and books on neural networks have come from many diverse fields, including engineering, physics, psychology and mathematics, and many authors tend to use vocabulary peculiar to their specialty. As a result, many books and papers in this field are difficult to read, and concepts are made to seem more complex than they actually are. This is a shame, as it has prevented the spread of important new ideas. It has also led to more than one "reinvention of the wheel."

In this book we have tried to use standard notation where possible, to be clear and to keep matters simple without sacrificing rigor. In particular, we have tried to define practical conventions and use them consistently.

Figures, mathematical equations and text discussing both figures and mathematical equations will use the following notation:

Scalars — small *italic* letters: a,b,c

Vectors — small **bold** nonitalic letters: **a,b,c**

Matrices — capital **BOLD** nonitalic letters: **A,B,C**

Additional notation concerning the network architectures will be introduced as you read this chapter. A complete list of the notation that we use throughout the book is given in Appendix B, so you can look there if you have a question.

Neuron Model

Single-Input Neuron

Weight
Bias
Net Input
Transfer Function

A single-input neuron is shown in Figure 2.1. The scalar input p is multiplied by the scalar *weight* w to form wp, one of the terms that is sent to the summer. The other input, 1, is multiplied by a *bias* b and then passed to the summer. The summer output n, often referred to as the *net input*, goes into a *transfer function* f, which produces the scalar neuron output a. (Some authors use the term "activation function" rather than *transfer function* and "offset" rather than *bias*.)

If we relate this simple model back to the biological neuron that we discussed in Chapter 1, the weight w corresponds to the strength of a synapse, the cell body is represented by the summation and the transfer function, and the neuron output a represents the signal on the axon.

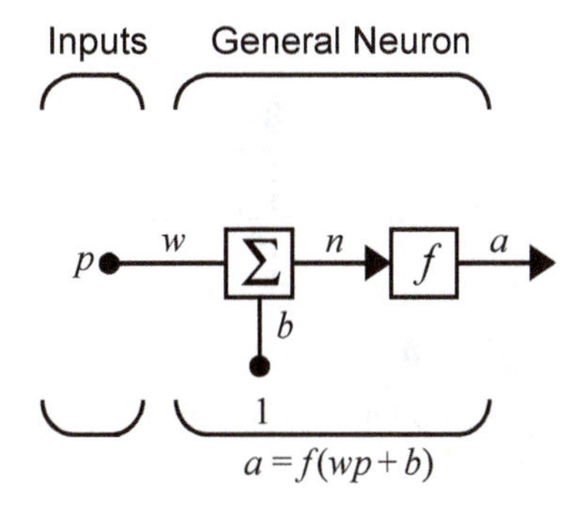

Figure 2.1 Single-Input Neuron

The neuron output is calculated as

$$a = f(wp + b).$$

If, for instance, $w = 3$, $p = 2$ and $b = -1.5$, then

$$a = f(3(2) - 1.5) = f(4.5)$$

The actual output depends on the particular transfer function that is chosen. We will discuss transfer functions in the next section.

The bias is much like a weight, except that it has a constant input of 1. However, if you do not want to have a bias in a particular neuron, it can be omitted. We will see examples of this in Chapter 3, 7 and 15.

Note that w and b are both *adjustable* scalar parameters of the neuron. Typically the transfer function is chosen by the designer and then the parameters w and b will be adjusted by some learning rule so that the neuron input/output relationship meets some specific goal (see Chapter 4 for an introduction to learning rules). As described in the following section, we have different transfer functions for different purposes.

Transfer Functions

The transfer function in Figure 2.1 may be a linear or a nonlinear function of n. A particular transfer function is chosen to satisfy some specification of the problem that the neuron is attempting to solve.

A variety of transfer functions have been included in this book. Three of the most commonly used functions are discussed below.

Hard Limit Transfer Function

The *hard limit transfer function*, shown on the left side of Figure 2.2, sets the output of the neuron to 0 if the function argument is less than 0, or 1 if its argument is greater than or equal to 0. We will use this function to create neurons that classify inputs into two distinct categories. It will be used extensively in Chapter 4.

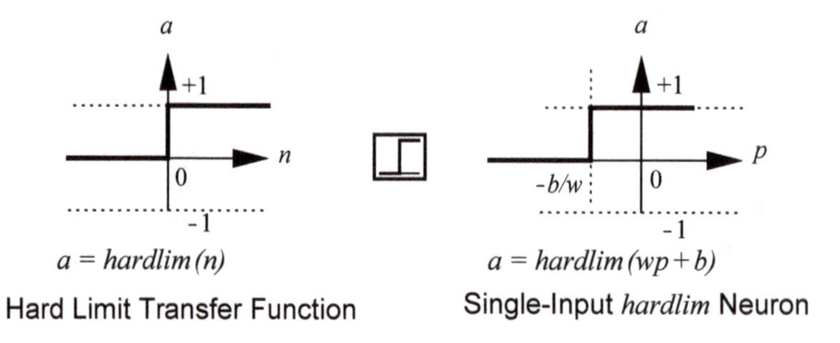

Figure 2.2 Hard Limit Transfer Function

The graph on the right side of Figure 2.2 illustrates the input/output characteristic of a single-input neuron that uses a hard limit transfer function. Here we can see the effect of the weight and the bias. Note that an icon for the hard limit transfer function is shown between the two figures. Such icons will replace the general f in network diagrams to show the particular transfer function that is being used.

Linear Transfer Function

The output of a *linear transfer function* is equal to its input:

$$a = n, \tag{2.1}$$

as illustrated in Figure 2.3.

Neurons with this transfer function are used in the ADALINE networks, which are discussed in Chapter 10.

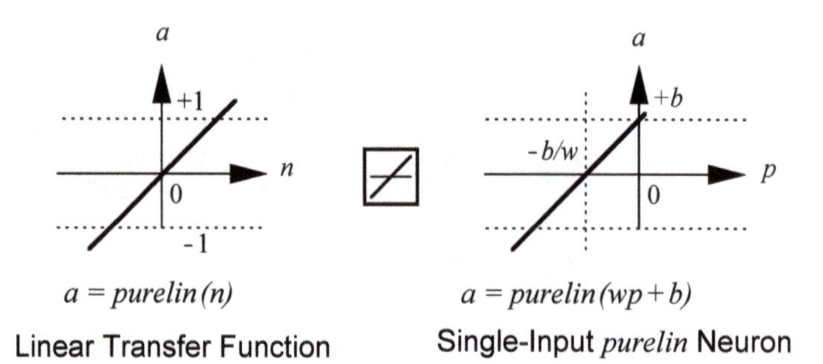

Figure 2.3 Linear Transfer Function

The output (a) versus input (p) characteristic of a single-input linear neuron with a bias is shown on the right of Figure 2.3.

Log-Sigmoid Transfer Function

The *log-sigmoid transfer function* is shown in Figure 2.4.

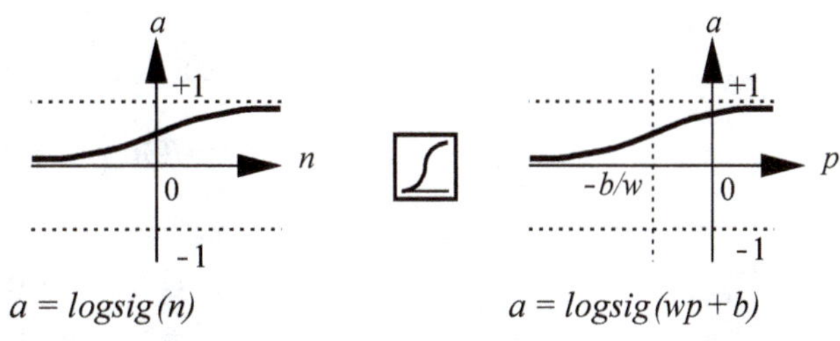

Figure 2.4 Log-Sigmoid Transfer Function

This transfer function takes the input (which may have any value between plus and minus infinity) and squashes the output into the range 0 to 1, according to the expression:

$$a = \frac{1}{1 + e^{-n}}. \tag{2.2}$$

The log-sigmoid transfer function is commonly used in multilayer networks that are trained using the backpropagation algorithm, in part because this function is differentiable (see Chapter 11).

Most of the transfer functions used in this book are summarized in Table 2.1. Of course, you can define other transfer functions in addition to those shown in Table 2.1 if you wish.

To experiment with a single-input neuron, use the Neural Network Design Demonstration One-Input Neuron **nnd2n1**.

Name	Input/Output Relation	Icon	MATLAB Function
Hard Limit	$a = 0 \quad n < 0$ $a = 1 \quad n \geq 0$		hardlim
Symmetrical Hard Limit	$a = -1 \quad n < 0$ $a = +1 \quad n \geq 0$		hardlims
Linear	$a = n$		purelin
Saturating Linear	$a = 0 \quad n < 0$ $a = n \quad 0 \leq n \leq 1$ $a = 1 \quad n > 1$		satlin
Symmetric Saturating Linear	$a = -1 \quad n < -1$ $a = n \quad -1 \leq n \leq 1$ $a = 1 \quad n > 1$		satlins
Log-Sigmoid	$a = \frac{1}{1 + e^{-n}}$		logsig
Hyperbolic Tangent Sigmoid	$a = \frac{e^{n} - e^{-n}}{e^{n} + e^{-n}}$		tansig
Positive Linear	$a = 0 \quad n < 0$ $a = n \quad 0 \leq n$		poslin
Competitive	$a = 1$ neuron with max n $a = 0$ all other neurons	C	compet

Table 2.1 Transfer Functions

Multiple-Input Neuron

Typically, a neuron has more than one input. A neuron with R inputs is shown in Figure 2.5. The individual inputs $p_1, p_2, ..., p_R$ are each weighted by corresponding elements $w_{1,1}, w_{1,2}, ..., w_{1,R}$ of the *weight matrix* $\mathbf{W}$.

Weight Matrix

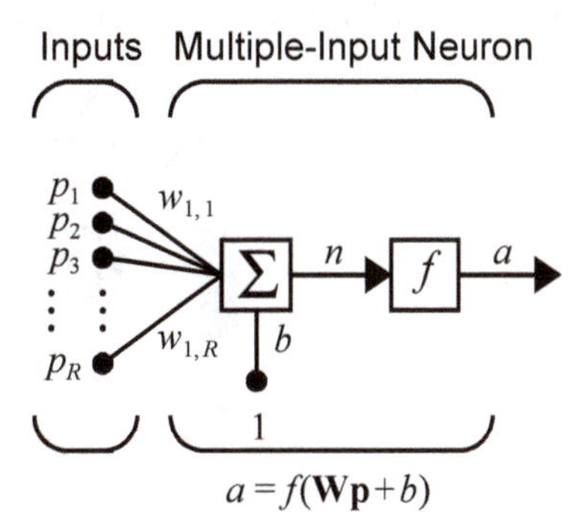

Figure 2.5 Multiple-Input Neuron

The neuron has a bias b, which is summed with the weighted inputs to form the net input n:

$$n = w_{1,1}p_1 + w_{1,2}p_2 + \cdots + w_{1,R}p_R + b. \quad (2.3)$$

This expression can be written in matrix form:

$$n = \mathbf{Wp} + b, \quad (2.4)$$

where the matrix $\mathbf{W}$ for the single neuron case has only one row.

Now the neuron output can be written as

$$a = f(\mathbf{Wp} + b). \quad (2.5)$$

Fortunately, neural networks can often be described with matrices. This kind of matrix expression will be used throughout the book. Don't be concerned if you are rusty with matrix and vector operations. We will review these topics in Chapter 5 and Chapter 6, and we will provide many examples and solved problems that will spell out the procedures.

Weight Indices

We have adopted a particular convention in assigning the indices of the elements of the weight matrix. The first index indicates the particular neuron destination for that weight. The second index indicates the source of the signal fed to the neuron. Thus, the indices in $w_{1,2}$ say that this weight represents the connection *to* the first (and only) neuron *from* the second source. Of course, this convention is more useful if there is more than one neuron, as will be the case later in this chapter.

We would like to draw networks with several neurons, each having several inputs. Further, we would like to have more than one layer of neurons. You can imagine how complex such a network might appear if all the lines were drawn. It would take a lot of ink, could hardly be read, and the mass of detail might obscure the main features. Thus, we will use an *abbreviated notation*. A multiple-input neuron using this notation is shown in Figure 2.6.

Abbreviated Notation

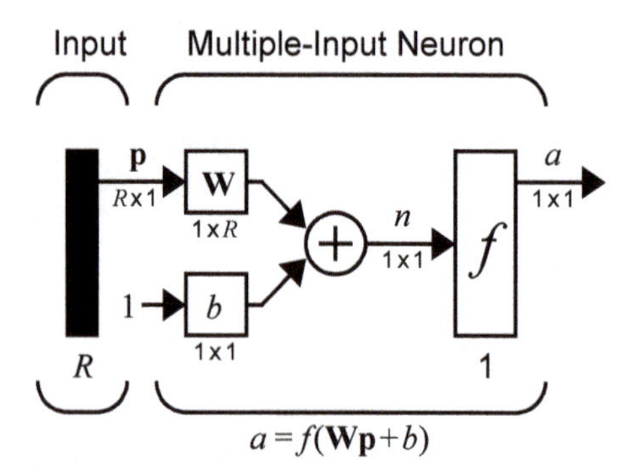

Figure 2.6 Neuron with R Inputs, Abbreviated Notation

As shown in Figure 2.6, the input vector $\mathbf{p}$ is represented by the solid vertical bar at the left. The dimensions of $\mathbf{p}$ are displayed below the variable as $R \times 1$, indicating that the input is a single vector of R elements. These inputs go to the weight matrix $\mathbf{W}$, which has R columns but only one row in this single neuron case. A constant 1 enters the neuron as an input and is multiplied by a scalar bias b. The net input to the transfer function f is n, which is the sum of the bias b and the product $\mathbf{Wp}$. The neuron's output a is a scalar in this case. If we had more than one neuron, the network output would be a vector.

The dimensions of the variables in these abbreviated notation figures will always be included, so that you can tell immediately if we are talking about a scalar, a vector or a matrix. You will not have to guess the kind of variable or its dimensions.

Note that the number of inputs to a network is set by the external specifications of the problem. If, for instance, you want to design a neural network that is to predict kite-flying conditions and the inputs are air temperature, wind velocity and humidity, then there would be three inputs to the network.

To experiment with a two-input neuron, use the Neural Network Design Demonstration Two-Input Neuron (nnd2n2).

Network Architectures

Commonly one neuron, even with many inputs, may not be sufficient. We might need five or ten, operating in parallel, in what we will call a "layer." This concept of a layer is discussed below.

A Layer of Neurons

Layer

A single-*layer* network of S neurons is shown in Figure 2.7. Note that each of the R inputs is connected to each of the neurons and that the weight matrix now has S rows.

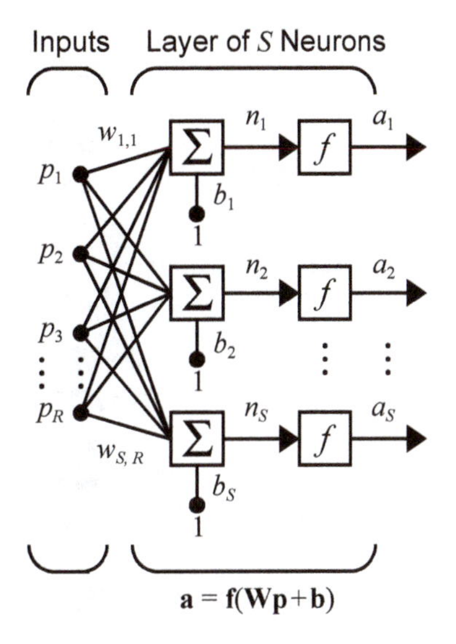

Figure 2.7 Layer of S Neurons

The layer includes the weight matrix, the summers, the bias vector $\mathbf{b}$, the transfer function boxes and the output vector $\mathbf{a}$. Some authors refer to the inputs as another layer, but we will not do that here.

Each element of the input vector $\mathbf{p}$ is connected to each neuron through the weight matrix $\mathbf{W}$. Each neuron has a bias b_i, a summer, a transfer function f and an output a_i. Taken together, the outputs form the output vector $\mathbf{a}$.

It is common for the number of inputs to a layer to be different from the number of neurons (i.e., $R \neq S$).

You might ask if all the neurons in a layer must have the same transfer function. The answer is no; you can define a single (composite) layer of neurons having different transfer functions by combining two of the networks shown above in parallel. Both networks would have the same inputs, and each network would create some of the outputs.

The input vector elements enter the network through the weight matrix $\mathbf{W}$:

$$\mathbf{W} = \begin{bmatrix} w_{1,1} & w_{1,2} & \dots & w_{1,R} \\ w_{2,1} & w_{2,2} & \dots & w_{2,R} \\ \vdots & \vdots & & \vdots \\ w_{S,1} & w_{S,2} & \dots & w_{S,R} \end{bmatrix}. \tag{2.6}$$

As noted previously, the row indices of the elements of matrix $\mathbf{W}$ indicate the destination neuron associated with that weight, while the column indices indicate the source of the input for that weight. Thus, the indices in $w_{3,2}$ say that this weight represents the connection *to* the third neuron *from* the second source.

Fortunately, the S-neuron, R-input, one-layer network also can be drawn in abbreviated notation, as shown in Figure 2.8.

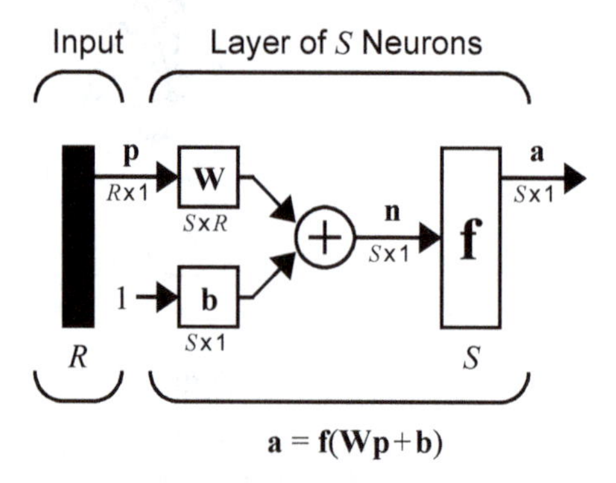

Figure 2.8 Layer of S Neurons, Abbreviated Notation

Here again, the symbols below the variables tell you that for this layer, $\mathbf{p}$ is a vector of length R, $\mathbf{W}$ is an $S \times R$ matrix, and $\mathbf{a}$ and $\mathbf{b}$ are vectors of length S. As defined previously, the layer includes the weight matrix, the summation and multiplication operations, the bias vector $\mathbf{b}$, the transfer function boxes and the output vector.

Multiple Layers of Neurons

Now consider a network with several layers. Each layer has its own weight matrix $\mathbf{W}$, its own bias vector $\mathbf{b}$, a net input vector $\mathbf{n}$ and an output vector $\mathbf{a}$. We need to introduce some additional notation to distinguish between

Layer Superscript

these layers. We will use superscripts to identify the layers. Specifically, we append the number of the layer as a *superscript* to the names for each of these variables. Thus, the weight matrix for the first layer is written as $\mathbf{W}^1$, and the weight matrix for the second layer is written as $\mathbf{W}^2$. This notation is used in the three-layer network shown in Figure 2.9.

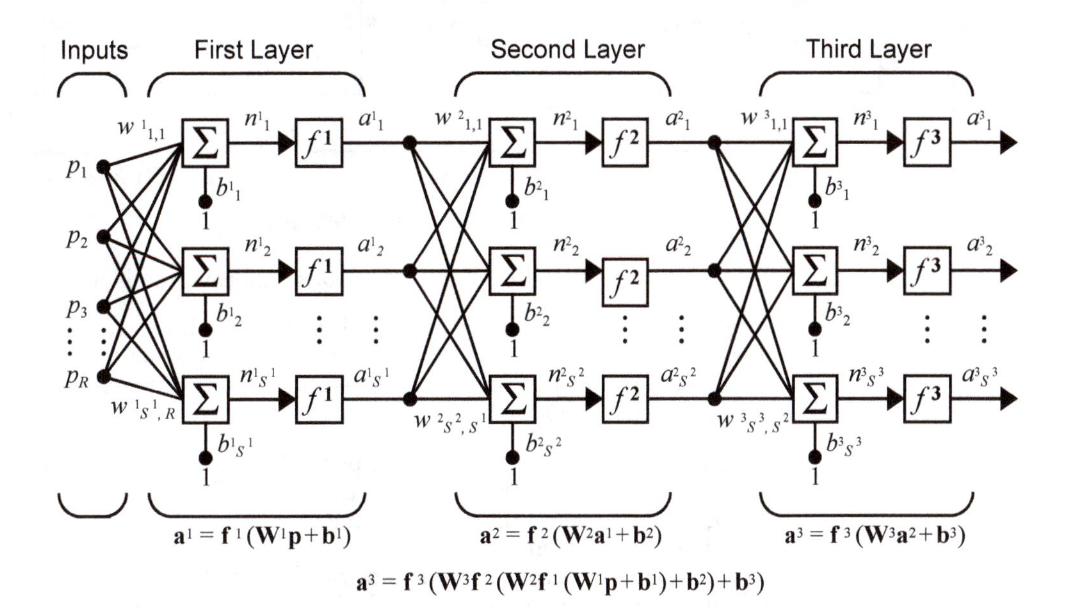

Figure 2.9 Three-Layer Network

As shown, there are R inputs, S^1 neurons in the first layer, S^2 neurons in the second layer, etc. As noted, different layers can have different numbers of neurons.

The outputs of layers one and two are the inputs for layers two and three. Thus layer 2 can be viewed as a one-layer network with $R = S^1$ inputs, $S = S^2$ neurons, and an $S^2 \times S^1$ weight matrix $\mathbf{W}^2$. The input to layer 2 is $\mathbf{a}^1$, and the output is $\mathbf{a}^2$.

Output Layer
Hidden Layers

A layer whose output is the network output is called an *output layer*. The other layers are called *hidden layers*. The network shown above has an output layer (layer 3) and two hidden layers (layers 1 and 2).

The same three-layer network discussed previously also can be drawn using our abbreviated notation, as shown in Figure 2.10.

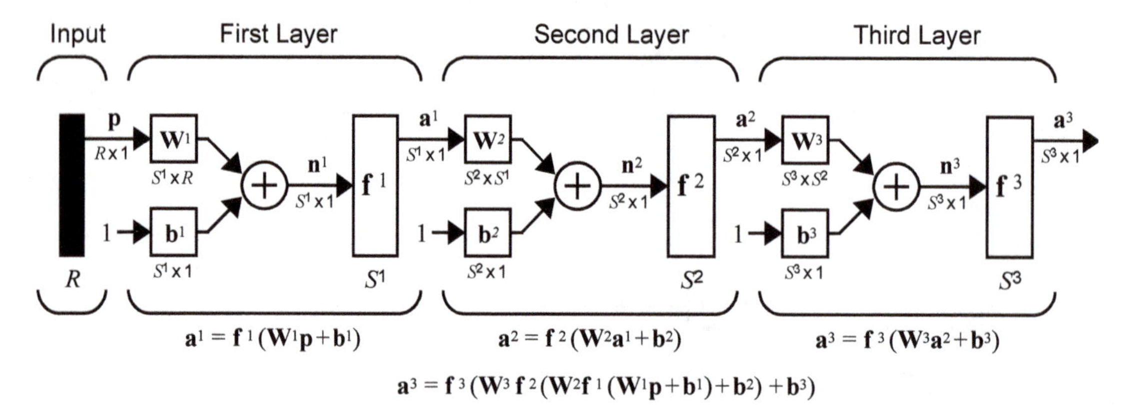

Figure 2.10 Three-Layer Network, Abbreviated Notation

Multilayer networks are more powerful than single-layer networks. For instance, a two-layer network having a sigmoid first layer and a linear second layer can be trained to approximate most functions arbitrarily well. Single-layer networks cannot do this.

At this point the number of choices to be made in specifying a network may look overwhelming, so let us consider this topic. The problem is not as bad as it looks. First, recall that the number of inputs to the network and the number of outputs from the network are defined by external problem specifications. So if there are four external variables to be used as inputs, there are four inputs to the network. Similarly, if there are to be seven outputs from the network, there must be seven neurons in the output layer. Finally, the desired characteristics of the output signal also help to select the transfer function for the output layer. If an output is to be either -1 or 1, then a symmetrical hard limit transfer function should be used. Thus, the architecture of a single-layer network is almost completely determined by problem specifications, including the specific number of inputs and outputs and the particular output signal characteristic.

Now, what if we have more than two layers? Here the external problem does not tell you directly the number of neurons required in the hidden layers. In fact, there are few problems for which one can predict the optimal number of neurons needed in a hidden layer. This problem is an active area of research. We will develop some feeling on this matter as we proceed to Chapter 11, Backpropagation.

As for the number of layers, most practical neural networks have just two or three layers. Four or more layers are used rarely.

We should say something about the use of biases. One can choose neurons with or without biases. The bias gives the network an extra variable, and so you might expect that networks with biases would be more powerful than those without, and that is true. Note, for instance, that a neuron without a bias will always have a net input n of zero when the network inputs $\mathbf{p}$ are zero. This may not be desirable and can be avoided by the use of a bias. The effect of the bias is discussed more fully in Chapter 3, 4 and 5.

In later chapters we will omit a bias in some examples or demonstrations. In some cases this is done simply to reduce the number of network parameters. With just two variables, we can plot system convergence in a two-dimensional plane. Three or more variables are difficult to display.

Recurrent Networks

Delay

Before we discuss recurrent networks, we need to introduce some simple building blocks. The first is the *delay* block, which is illustrated in Figure 2.11.

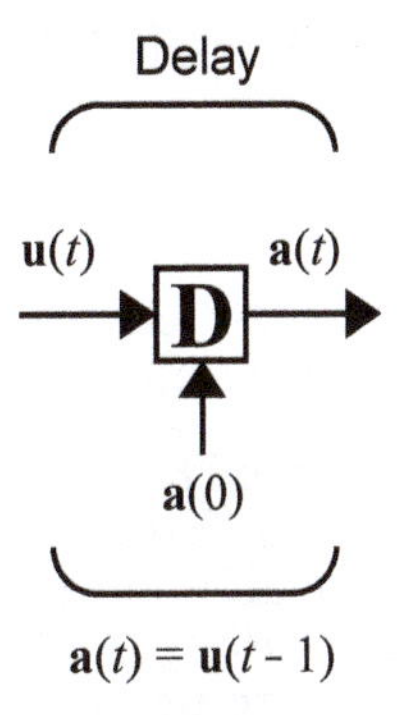

Figure 2.11 Delay Block

The delay output $\mathbf{a}(t)$ is computed from its input $\mathbf{u}(t)$ according to

$$\mathbf{a}(t) = \mathbf{u}(t-1). \tag{2.7}$$

Thus the output is the input delayed by one time step. (This assumes that time is updated in discrete steps and takes on only integer values.) Eq. (2.7) requires that the output be initialized at time $t = 0$. This initial condition is indicated in Figure 2.11 by the arrow coming into the bottom of the delay block.

Integrator

Another related building block, which we can use for continuous-time recurrent networks, is the *integrator*, which is shown in Figure 2.12.

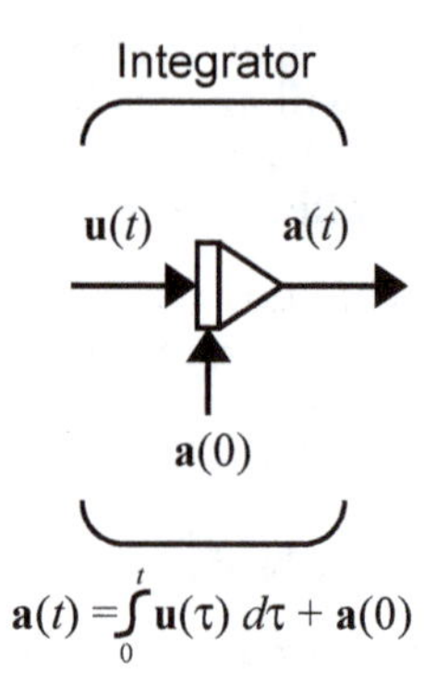

Figure 2.12 Integrator Block

The integrator output $\mathbf{a}(t)$ is computed from its input $\mathbf{u}(t)$ according to

$$\mathbf{a}(t) = \int_0^t \mathbf{u}(\tau)d\tau + \mathbf{a}(0). \tag{2.8}$$

The initial condition $\mathbf{a}(0)$ is indicated by the arrow coming into the bottom of the integrator block.

Recurrent Network

We are now ready to introduce recurrent networks. A *recurrent network* is a network with feedback; some of its outputs are connected to its inputs. This is quite different from the networks that we have studied thus far, which were strictly feedforward with no backward connections. One type of discrete-time recurrent network is shown in Figure 2.13.

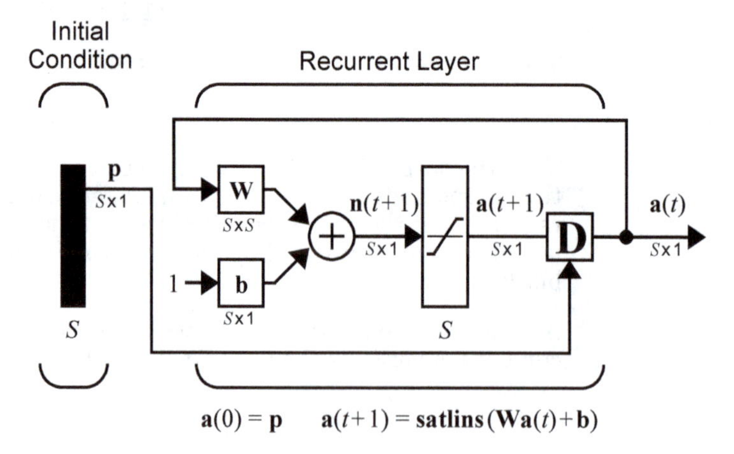

Figure 2.13 Recurrent Network

In this particular network the vector $\mathbf{p}$ supplies the initial conditions (i.e., $\mathbf{a}(0) = \mathbf{p}$). Then future outputs of the network are computed from previous outputs:

$$\mathbf{a}(1) = \mathbf{satlins}(\mathbf{Wa}(0) + \mathbf{b}),\ \mathbf{a}(2) = \mathbf{satlins}(\mathbf{Wa}(1) + \mathbf{b}),\ \ldots$$

Recurrent networks are potentially more powerful than feedforward networks and can exhibit temporal behavior. These types of networks are discussed in Chapter 3 and Chapter 14.

Summary of Results

Single-Input Neuron

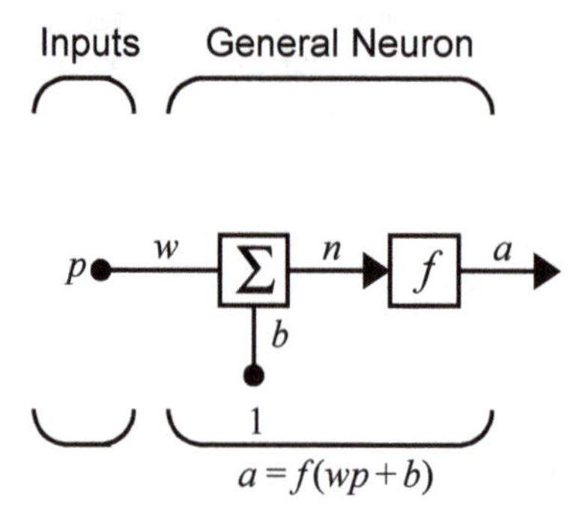

Multiple-Input Neuron

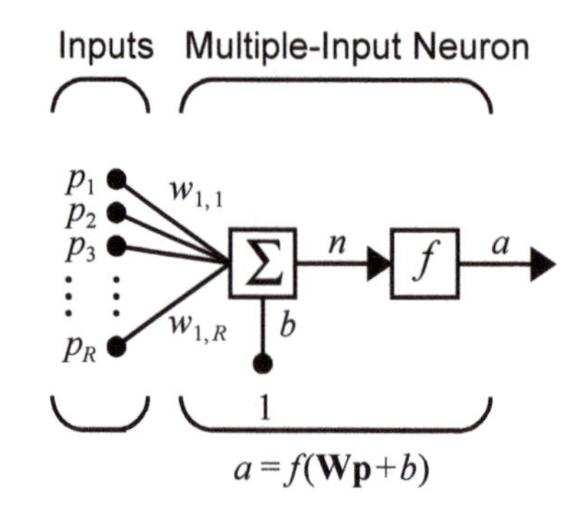

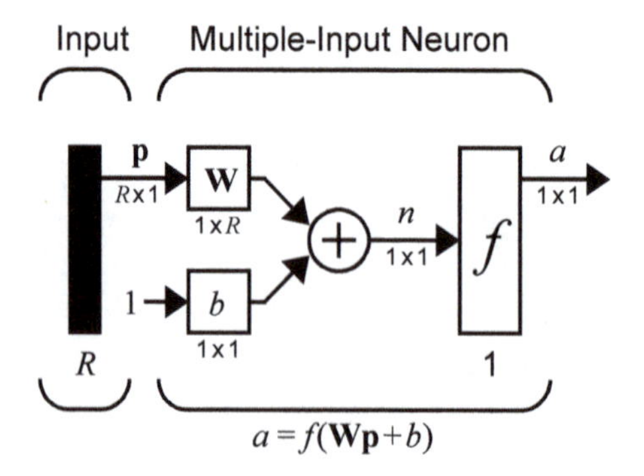

Transfer Functions

Name	Input/Output Relation	Icon	MATLAB Function
Hard Limit	$a = 0 \quad n < 0$ $a = 1 \quad n \geq 0$		hardlim
Symmetrical Hard Limit	$a = -1 \quad n < 0$ $a = +1 \quad n \geq 0$		hardlims
Linear	$a = n$		purelin
Saturating Linear	$a = 0 \quad n < 0$ $a = n \quad 0 \leq n \leq 1$ $a = 1 \quad n > 1$		satlin
Symmetric Saturating Linear	$a = -1 \quad n < -1$ $a = n \quad -1 \leq n \leq 1$ $a = 1 \quad n > 1$		satlins
Log-Sigmoid	$a = \frac{1}{1 + e^{-n}}$		logsig
Hyperbolic Tangent Sigmoid	$a = \frac{e^{n} - e^{-n}}{e^{n} + e^{-n}}$		tansig
Positive Linear	$a = 0 \quad n < 0$ $a = n \quad 0 \leq n$		poslin
Competitive	$a = 1$ neuron with max n $a = 0$ all other neurons	C	compet

Layer of Neurons

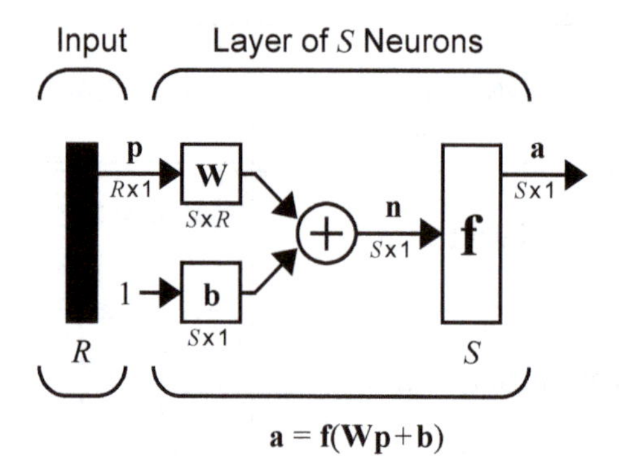

Three Layers of Neurons

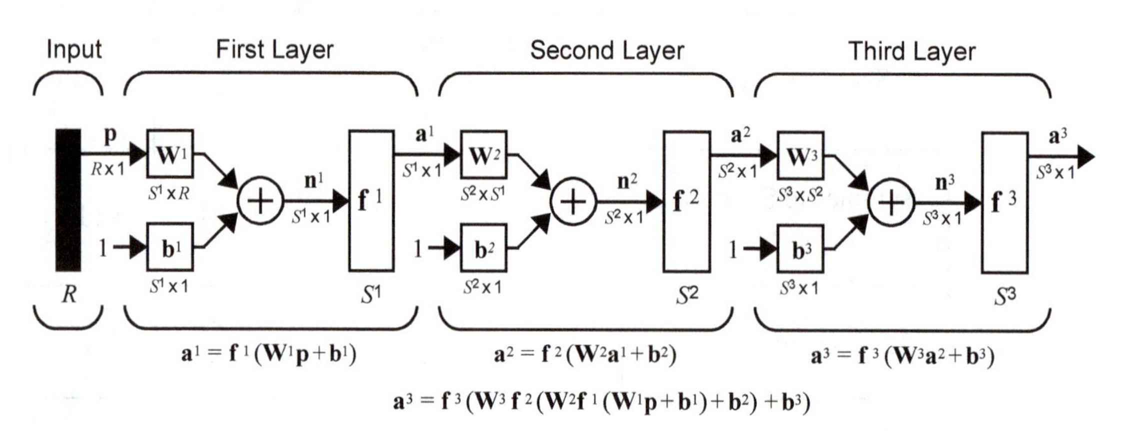

Delay

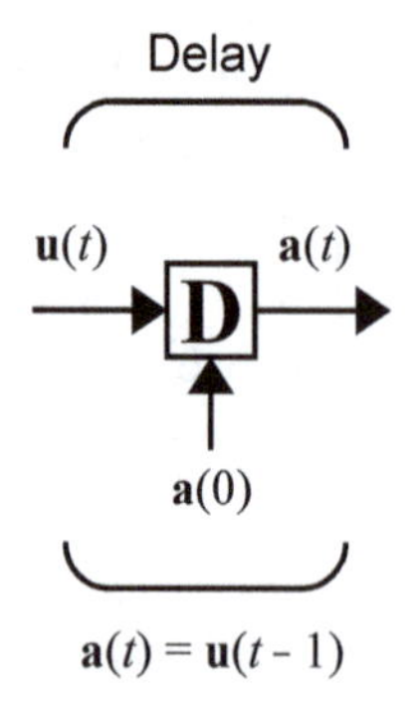

Integrator

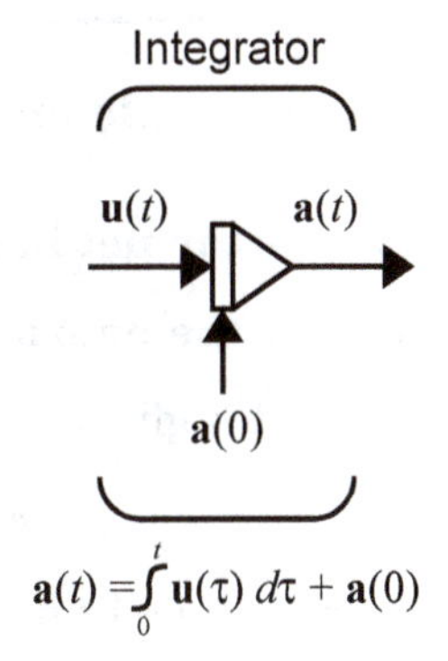

Recurrent Network

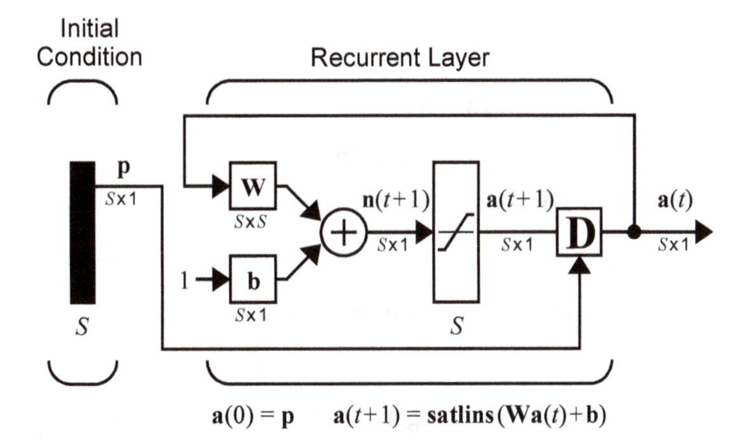

How to Pick an Architecture

Problem specifications help define the network in the following ways:

1. Number of network inputs = number of problem inputs
2. Number of neurons in output layer = number of problem outputs
3. Output layer transfer function choice at least partly determined by problem specification of the outputs

Solved Problems

P2.1 The input to a single-input neuron is 2.0, its weight is 2.3 and its bias is -3.

i. What is the net input to the transfer function?

ii. What is the neuron output?

i. The net input is given by:

$$n = wp + b = (2.3)(2) + (-3) = 1.6$$

ii. The output cannot be determined because the transfer function is not specified.

P2.2 What is the output of the neuron of P2.1 if it has the following transfer functions?

i. Hard limit

ii. Linear

iii. Log-sigmoid

i. For the hard limit transfer function:

$$a = hardlim(1.6) = 1.0$$

ii. For the linear transfer function:

$$a = purelin(1.6) = 1.6$$

iii. For the log-sigmoid transfer function:

$$a = logsig(1.6) = \frac{1}{1 + e^{-1.6}} = 0.8320$$

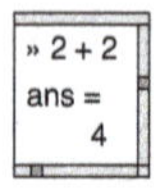

Verify this result using MATLAB and the function `logsig`, which is in the MININNET directory (see Appendix B).

P2.3 Given a two-input neuron with the following parameters: $b = 1.2$, $\mathbf{W} = \begin{bmatrix} 3 & 2 \end{bmatrix}$ **and** $\mathbf{p} = \begin{bmatrix} -5 & 6 \end{bmatrix}^T$**, calculate the neuron output for the following transfer functions:**

i. A symmetrical hard limit transfer function

ii. A saturating linear transfer function

iii. A hyperbolic tangent sigmoid (tansig) transfer function

First calculate the net input n:

$$n = \mathbf{W}\mathbf{p} + b = \begin{bmatrix} 3 & 2 \end{bmatrix} \begin{bmatrix} -5 \\ 6 \end{bmatrix} + (1.2) = -1.8 .$$

Now find the outputs for each of the transfer functions.

i. $a = hardlims(-1.8) = -1$

ii. $a = satlin(-1.8) = 0$

iii. $a = tansig(-1.8) = -0.9468$

P2.4 A single-layer neural network is to have six inputs and two outputs. The outputs are to be limited to and continuous over the range 0 to 1. What can you tell about the network architecture? Specifically:

i. How many neurons are required?

ii. What are the dimensions of the weight matrix?

iii. What kind of transfer functions could be used?

iv. Is a bias required?

The problem specifications allow you to say the following about the network.

i. Two neurons, one for each output, are required.

ii. The weight matrix has two rows corresponding to the two neurons and six columns corresponding to the six inputs. (The product $\mathbf{W}\mathbf{p}$ is a two-element vector.)

iii. Of the transfer functions we have discussed, the *logsig* transfer function would be most appropriate.

iv. Not enough information is given to determine if a bias is required.

Epilogue

This chapter has introduced a simple artificial neuron and has illustrated how different neural networks can be created by connecting groups of neurons in various ways. One of the main objectives of this chapter has been to introduce our basic notation. As the networks are discussed in more detail in later chapters, you may wish to return to Chapter 2 to refresh your memory of the appropriate notation.

This chapter was not meant to be a complete presentation of the networks we have discussed here. That will be done in the chapters that follow. We will begin in Chapter 3, which will present a simple example that uses some of the networks described in this chapter, and will give you an opportunity to see these networks in action. The networks demonstrated in Chapter 3 are representative of the types of networks that are covered in the remainder of this text.

Exercises

E2.1 A single input neuron has a weight of 1.3 and a bias of 3.0. What possible kinds of transfer functions, from Table 2.1, could this neuron have, if its output is given below. In each case, give the value of the input that would produce these outputs.

i. 1.6

ii. 1.0

iii. 0.9963

iv. -1.0

E2.2 Consider a single-input neuron with a bias. We would like the output to be -1 for inputs less than 3 and +1 for inputs greater than or equal to 3.

i. What kind of a transfer function is required?

ii. What bias would you suggest? Is your bias in any way related to the input weight? If yes, how?

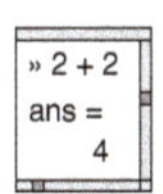

iii. Summarize your network by naming the transfer function and stating the bias and the weight. Draw a diagram of the network. Verify the network performance using MATLAB.

E2.3 Given a two-input neuron with the following weight matrix and input vector: $\mathbf{W} = \begin{bmatrix} 3 & 2 \end{bmatrix}$ and $\mathbf{p} = \begin{bmatrix} -5 & 7 \end{bmatrix}^T$, we would like to have an output of 0.5. Do you suppose that there is a combination of bias and transfer function that might allow this?

i. Is there a transfer function from Table 2.1 that will do the job if the bias is zero?

ii. Is there a bias that will do the job if the linear transfer function is used? If yes, what is it?

iii. Is there a bias that will do the job if a log-sigmoid transfer function is used? Again, if yes, what is it?

iv. Is there a bias that will do the job if a symmetrical hard limit transfer function is used? Again, if yes, what is it?

E2.4 A two-layer neural network is to have four inputs and six outputs. The range of the outputs is to be continuous between 0 and 1. What can you tell about the network architecture? Specifically:

i. How many neurons are required in each layer?

ii. What are the dimensions of the first-layer and second-layer weight matrices?

iii. What kinds of transfer functions can be used in each layer?

iv. Are biases required in either layer?

E2.5 Consider the following neuron.

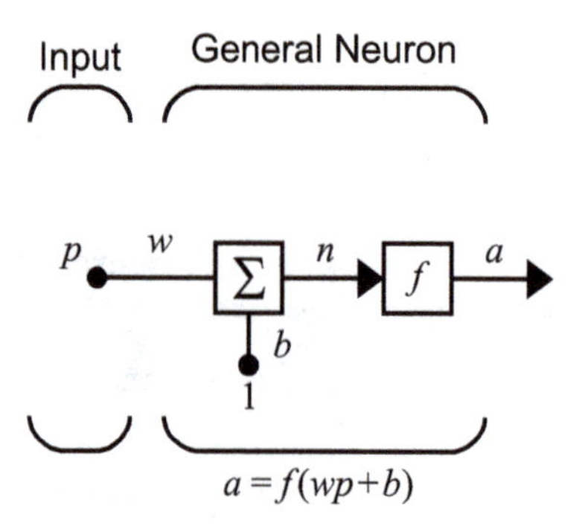

Figure P15.1 General Neuron

Sketch the neuron response (plot a versus p for $-2<p<2$) for the following cases.

i. $w = 1$, $b = 1$, $f = hardlims$.

ii. $w = -1$, $b = 1$, $f = hardlims$.

iii. $w = 2$, $b = 3$, $f = purelin$.

iv. $w = 2$, $b = 3$, $f = satlins$.

v. $w = -2$, $b = -1$, $f = poslin$.

E2.6 Consider the following neural network.

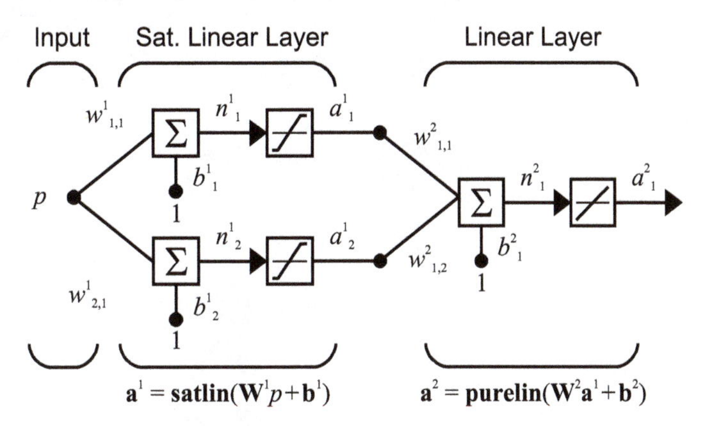

$$w_{1,1}^{1} = 2, \; w_{2,1}^{1} = 1, \; b_1^1 = 2, \; b_2^1 = -1, \; w_{1,1}^{2} = 1, \; w_{1,2}^{2} = -1, \; b_1^2 = 0$$

Sketch the following responses (plot the indicated variable versus p for $-3 < p < 3$).

i. n_1^1.

ii. a_1^1.

iii. n_2^1

iv. a_2^1.

v. n_1^2.

vi. a_1^2.

3 An Illustrative Example

Objectives

Think of this chapter as a preview of coming attractions. We will take a simple pattern recognition problem and show how it can be solved using three different neural network architectures. It will be an opportunity to see how the architectures described in the previous chapter can be used to solve a practical (although extremely oversimplified) problem. Do not expect to completely understand these three networks after reading this chapter. We present them simply to give you a taste of what can be done with neural networks, and to demonstrate that there are many different types of networks that can be used to solve a given problem.

The three networks presented in this chapter are representative of the types of networks discussed in the remaining chapters: feedforward networks (represented here by the perceptron), competitive networks (represented here by the Hamming network) and recurrent associative memory networks (represented here by the Hopfield network).

Theory and Examples

Problem Statement

A produce dealer has a warehouse that stores a variety of fruits and vegetables. When fruit is brought to the warehouse, various types of fruit may be mixed together. The dealer wants a machine that will sort the fruit according to type. There is a conveyer belt on which the fruit is loaded. This conveyer passes through a set of sensors, which measure three properties of the fruit: *shape*, *texture* and *weight*. These sensors are somewhat primitive. The shape sensor will output a 1 if the fruit is approximately round and a -1 if it is more elliptical. The texture sensor will output a 1 if the surface of the fruit is smooth and a -1 if it is rough. The weight sensor will output a 1 if the fruit is more than one pound and a -1 if it is less than one pound.

The three sensor outputs will then be input to a neural network. The purpose of the network is to decide which kind of fruit is on the conveyor, so that the fruit can be directed to the correct storage bin. To make the problem even simpler, let's assume that there are only two kinds of fruit on the conveyor: apples and oranges.

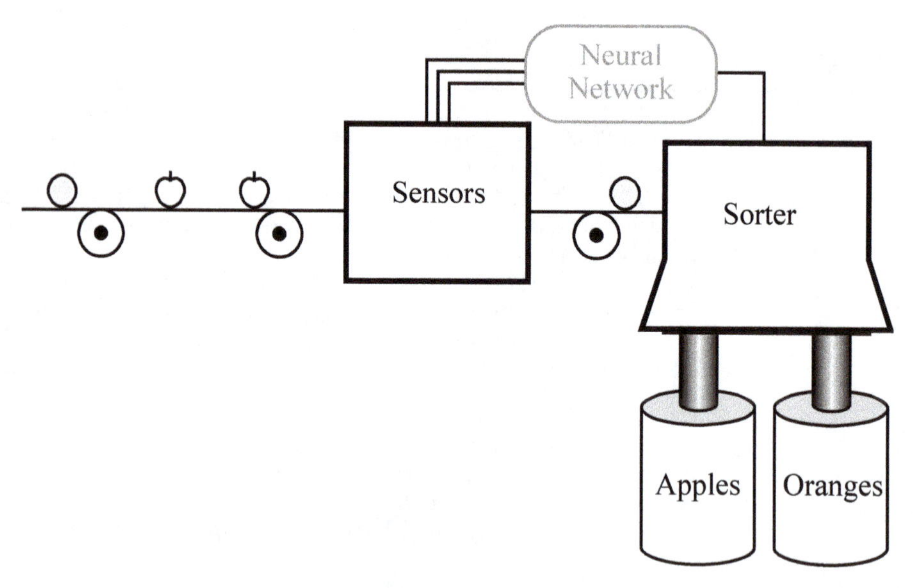

As each fruit passes through the sensors it can be represented by a three-dimensional vector. The first element of the vector will represent shape, the second element will represent texture and the third element will represent weight:

$$\mathbf{p} = \begin{bmatrix} shape \\ texture \\ weight \end{bmatrix}. \quad (3.1)$$

Therefore, a prototype orange would be represented by

$$\mathbf{p}_1 = \begin{bmatrix} 1 \\ -1 \\ -1 \end{bmatrix}, \quad (3.2)$$

and a prototype apple would be represented by

$$\mathbf{p}_2 = \begin{bmatrix} 1 \\ 1 \\ -1 \end{bmatrix}. \quad (3.3)$$

The neural network will receive one three-dimensional input vector for each fruit on the conveyer and must make a decision as to whether the fruit is an *orange* $(\mathbf{p}_1)$ or an *apple* $(\mathbf{p}_2)$.

Now that we have defined this simple (trivial?) pattern recognition problem, let's look briefly at three different neural networks that could be used to solve it. The simplicity of our problem will facilitate our understanding of the operation of the networks.

Perceptron

The first network we will discuss is the perceptron. Figure 3.1 illustrates a single-layer perceptron with a symmetric hard limit transfer function *hardlims*.

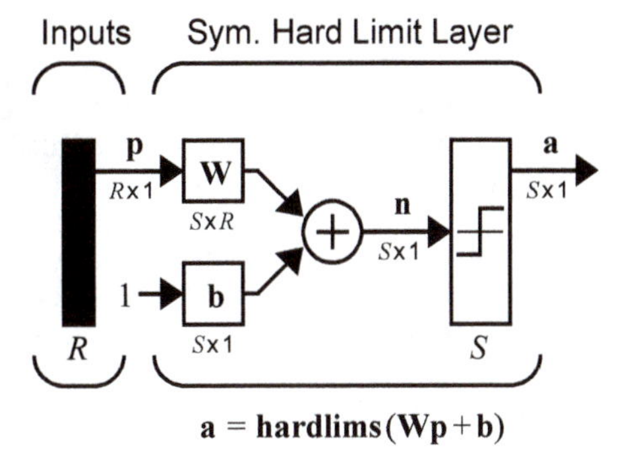

Figure 3.1 Single-Layer Perceptron

Two-Input Case

Before we use the perceptron to solve the orange and apple recognition problem (which will require a three-input perceptron, i.e., $R = 3$), it is useful to investigate the capabilities of a two-input/single-neuron perceptron ($R = 2$), which can be easily analyzed graphically. The two-input perceptron is shown in Figure 3.2.

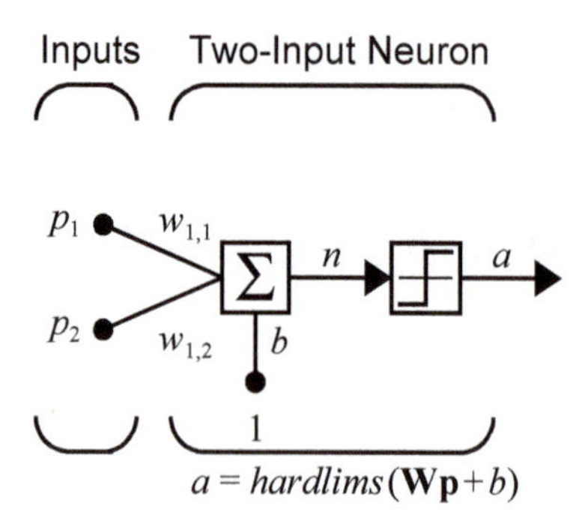

Figure 3.2 Two-Input/Single-Neuron Perceptron

Single-neuron perceptrons can classify input vectors into two categories. For example, for a two-input perceptron, if $w_{1,1} = -1$ and $w_{1,2} = 1$ then

$$a = hardlims(n) = hardlims(\begin{bmatrix} -1 & 1 \end{bmatrix}\mathbf{p} + b). \tag{3.4}$$

Therefore, if the inner product of the weight matrix (a single row vector in this case) with the input vector is greater than or equal to $-b$, the output will be 1. If the inner product of the weight vector and the input is less than $-b$, the output will be -1. This divides the input space into two parts. Figure 3.3 illustrates this for the case where $b = -1$. The blue line in the figure represents all points for which the net input n is equal to 0:

$$n = \begin{bmatrix} -1 & 1 \end{bmatrix}\mathbf{p} - 1 = 0. \tag{3.5}$$

Notice that this decision boundary will always be orthogonal to the weight matrix, and the position of the boundary can be shifted by changing b. (In the general case, $\mathbf{W}$ is a matrix consisting of a number of row vectors, each of which will be used in an equation like Eq. (3.5). There will be one boundary for each row of $\mathbf{W}$. See Chapter 4 for more on this topic.) The shaded region contains all input vectors for which the output of the network will be 1. The output will be -1 for all other input vectors.

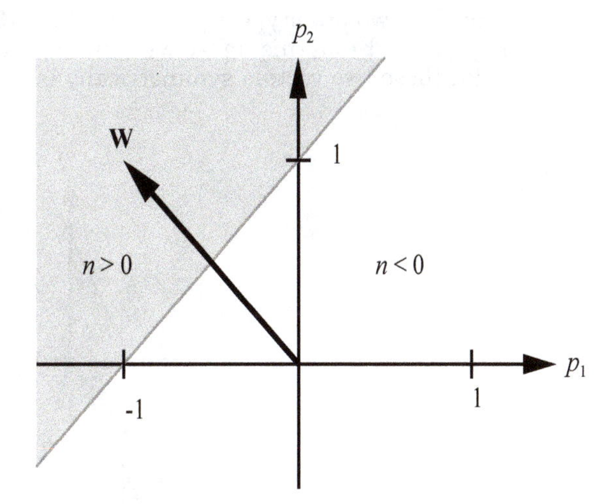

Figure 3.3 Perceptron Decision Boundary

The key property of the single-neuron perceptron, therefore, is that it can separate input vectors into two categories. The decision boundary between the categories is determined by the equation

$$\mathbf{W}\mathbf{p} + b = 0 . \tag{3.6}$$

Because the boundary must be linear, the single-layer perceptron can only be used to recognize patterns that are linearly separable (can be separated by a linear boundary). These concepts will be discussed in more detail in Chapter 4.

Pattern Recognition Example

Now consider the apple and orange pattern recognition problem. Because there are only two categories, we can use a single-neuron perceptron. The vector inputs are three-dimensional ($R = 3$), therefore the perceptron equation will be

$$a = hardlims\left(\begin{bmatrix} w_{1,1} & w_{1,2} & w_{1,3} \end{bmatrix} \begin{bmatrix} p_1 \\ p_2 \\ p_3 \end{bmatrix} + b \right) . \tag{3.7}$$

We want to choose the bias b and the elements of the weight matrix so that the perceptron will be able to distinguish between apples and oranges. For example, we may want the output of the perceptron to be 1 when an apple is input and -1 when an orange is input. Using the concept illustrated in Figure 3.3, let's find a linear boundary that can separate oranges and apples. The two prototype vectors (recall Eq. (3.2) and Eq. (3.3)) are shown in Figure 3.4. From this figure we can see that the linear boundary that divides these two vectors symmetrically is the p_1, p_3 plane.

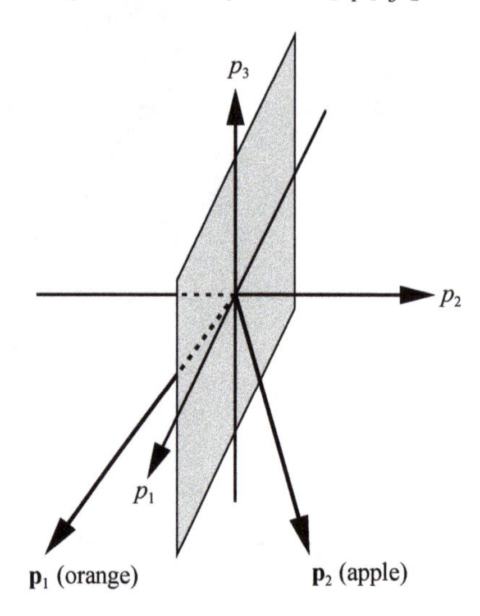

Figure 3.4 Prototype Vectors

The p_1, p_3 plane, which will be our decision boundary, can be described by the equation

$$p_2 = 0, \tag{3.8}$$

or

$$\begin{bmatrix} 0 & 1 & 0 \end{bmatrix} \begin{bmatrix} p_1 \\ p_2 \\ p_3 \end{bmatrix} + 0 = 0. \tag{3.9}$$

Therefore the weight matrix and bias will be

$$\mathbf{W} = \begin{bmatrix} 0 & 1 & 0 \end{bmatrix}, b = 0. \tag{3.10}$$

The weight matrix is orthogonal to the decision boundary and points toward the region that contains the prototype pattern $\mathbf{p}_2$ *(apple)* for which we want the perceptron to produce an output of 1. The bias is 0 because the decision boundary passes through the origin.

Now let's test the operation of our perceptron pattern classifier. It classifies perfect apples and oranges correctly since

Orange:

$$a = hardlims\left(\begin{bmatrix} 0 & 1 & 0 \end{bmatrix} \begin{bmatrix} 1 \\ -1 \\ -1 \end{bmatrix} + 0 \right) = -1(orange), \tag{3.11}$$

Apple:

$$a = hardlims\left(\begin{bmatrix} 0 & 1 & 0 \end{bmatrix} \begin{bmatrix} 1 \\ 1 \\ -1 \end{bmatrix} + 0 \right) = 1(apple). \tag{3.12}$$

But what happens if we put a not-so-perfect orange into the classifier? Let's say that an orange with an elliptical shape is passed through the sensors. The input vector would then be

$$\mathbf{p} = \begin{bmatrix} -1 \\ -1 \\ -1 \end{bmatrix}. \tag{3.13}$$

The response of the network would be

$$a = hardlims\left(\begin{bmatrix} 0 & 1 & 0 \end{bmatrix} \begin{bmatrix} -1 \\ -1 \\ -1 \end{bmatrix} + 0 \right) = -1(orange). \tag{3.14}$$

In fact, any input vector that is closer to the orange prototype vector than to the apple prototype vector (in Euclidean distance) will be classified as an orange (and vice versa).

To experiment with the perceptron network and the apple/orange classification problem, use the Neural Network Design Demonstration Perceptron Classification (**nnd3pc**).

This example has demonstrated some of the features of the perceptron network, but by no means have we exhausted our investigation of perceptrons. This network, and variations on it, will be examined in Chapter 4 through Chapter 13. Let's consider some of these future topics.

In the apple/orange example we were able to design a network graphically, by choosing a decision boundary that clearly separated the patterns. What about practical problems, with high dimensional input spaces? In Chapter 4, 7, 10 and 11 we will introduce learning algorithms that can be used to train networks to solve complex problems by using a set of examples of proper network behavior.

The key characteristic of the single-layer perceptron is that it creates linear decision boundaries to separate categories of input vector. What if we have categories that cannot be separated by linear boundaries? This question will be addressed in Chapter 11, where we will introduce the multilayer perceptron. The multilayer networks are able to solve classification problems of arbitrary complexity.

Hamming Network

The next network we will consider is the Hamming network [Lipp87]. It was designed explicitly to solve binary pattern recognition problems (where each element of the input vector has only two possible values — in our example 1 or -1). This is an interesting network, because it uses both feedforward and recurrent (feedback) layers, which were both described in Chapter 2. Figure 3.5 shows the standard Hamming network. Note that the number of neurons in the first layer is the same as the number of neurons in the second layer.

The objective of the Hamming network is to decide which prototype vector is closest to the input vector. This decision is indicated by the output of the recurrent layer. There is one neuron in the recurrent layer for each prototype pattern. When the recurrent layer converges, there will be only one neuron with nonzero output. This neuron indicates the prototype pattern that is closest to the input vector. Now let's investigate the two layers of the Hamming network in detail.

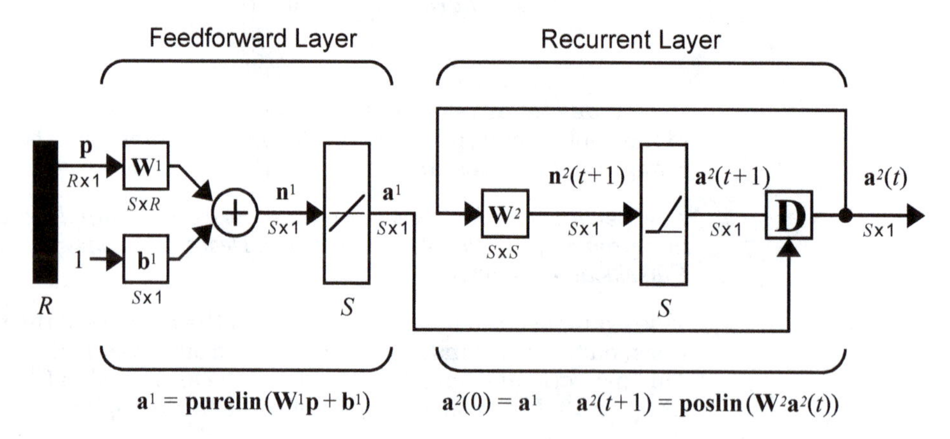

Figure 3.5 Hamming Network

Feedforward Layer

The feedforward layer performs a correlation, or inner product, between each of the prototype patterns and the input pattern (as we will see in Eq. (3.17)). In order for the feedforward layer to perform this correlation, the rows of the weight matrix in the feedforward layer, represented by the connection matrix $\mathbf{W}^1$, are set to the prototype patterns. For our apple and orange example this would mean

$$\mathbf{W}^1 = \begin{bmatrix} \mathbf{p}_1^T \\ \mathbf{p}_2^T \end{bmatrix} = \begin{bmatrix} 1 & -1 & -1 \\ 1 & 1 & -1 \end{bmatrix}. \tag{3.15}$$

The feedforward layer uses a linear transfer function, and each element of the bias vector is equal to R, where R is the number of elements in the input vector. For our example the bias vector would be

$$\mathbf{b}^1 = \begin{bmatrix} 3 \\ 3 \end{bmatrix}. \tag{3.16}$$

With these choices for the weight matrix and bias vector, the output of the feedforward layer is

$$\mathbf{a}^1 = \mathbf{W}^1\mathbf{p} + \mathbf{b}^1 = \begin{bmatrix} \mathbf{p}_1^T \\ \mathbf{p}_2^T \end{bmatrix}\mathbf{p} + \begin{bmatrix} 3 \\ 3 \end{bmatrix} = \begin{bmatrix} \mathbf{p}_1^T\mathbf{p} + 3 \\ \mathbf{p}_2^T\mathbf{p} + 3 \end{bmatrix}. \tag{3.17}$$

Note that the outputs of the feedforward layer are equal to the inner products of each prototype pattern with the input, plus R. For two vectors of equal length (norm), their inner product will be largest when the vectors point in the same direction, and will be smallest when they point in opposite directions. (We will discuss this concept in more depth in Chapter 5, 8 and 9.) By adding R to the inner product we guarantee that the outputs of the feedforward layer can never be negative. This is required for proper operation of the recurrent layer.

This network is called the Hamming network because the neuron in the feedforward layer with the largest output will correspond to the prototype pattern that is closest in Hamming distance to the input pattern. (The Hamming distance between two vectors is equal to the number of elements that are different. It is defined only for binary vectors.) We leave it to the reader to show that the outputs of the feedforward layer are equal to $2R$ minus twice the Hamming distances from the prototype patterns to the input pattern.

Recurrent Layer

The recurrent layer of the Hamming network is what is known as a "competitive" layer. The neurons in this layer are initialized with the outputs of the feedforward layer, which indicate the correlation between the prototype patterns and the input vector. Then the neurons compete with each other to determine a winner. After the competition, only one neuron will have a nonzero output. The winning neuron indicates which category of input was presented to the network (for our example the two categories are *apples* and *oranges*). The equations that describe the competition are:

$$\mathbf{a}^2(0) = \mathbf{a}^1 \quad \text{(Initial Condition)}, \tag{3.18}$$

and

$$\mathbf{a}^2(t+1) = \mathbf{poslin}(\mathbf{W}^2\mathbf{a}^2(t)) . \tag{3.19}$$

(Don't forget that the superscripts here indicate the layer number, not a power of 2.) The *poslin* transfer function is linear for positive values and zero for negative values. The weight matrix $\mathbf{W}^2$ has the form

$$\mathbf{W}^2 = \begin{bmatrix} 1 & -\varepsilon \\ -\varepsilon & 1 \end{bmatrix}, \tag{3.20}$$

where ε is some number less than $1/(S-1)$, and S is the number of neurons in the recurrent layer. (Can you show why ε must be less than $1/(S-1)$?)

An iteration of the recurrent layer proceeds as follows:

$$\mathbf{a}^2(t+1) = \mathbf{poslin}\left(\begin{bmatrix} 1 & -\varepsilon \\ -\varepsilon & 1 \end{bmatrix}\mathbf{a}^2(t)\right) = \mathbf{poslin}\left(\begin{bmatrix} a_1^2(t) - \varepsilon a_2^2(t) \\ a_2^2(t) - \varepsilon a_1^2(t) \end{bmatrix}\right). \tag{3.21}$$

Each element is reduced by the same fraction of the other. The larger element will be reduced by less, and the smaller element will be reduced by more, therefore the difference between large and small will be increased. The effect of the recurrent layer is to zero out all neuron outputs, except the one with the largest initial value (which corresponds to the prototype pattern that is closest in Hamming distance to the input).

To illustrate the operation of the Hamming network, consider again the oblong orange that we used to test the perceptron:

$$\mathbf{p} = \begin{bmatrix} -1 \\ -1 \\ -1 \end{bmatrix}. \quad (3.22)$$

The output of the feedforward layer will be

$$\mathbf{a}^1 = \begin{bmatrix} 1 & -1 & -1 \\ 1 & 1 & -1 \end{bmatrix} \begin{bmatrix} -1 \\ -1 \\ -1 \end{bmatrix} + \begin{bmatrix} 3 \\ 3 \end{bmatrix} = \begin{bmatrix} (1+3) \\ (-1+3) \end{bmatrix} = \begin{bmatrix} 4 \\ 2 \end{bmatrix}, \quad (3.23)$$

which will then become the initial condition for the recurrent layer.

The weight matrix for the recurrent layer will be given by Eq. (3.20) with $\varepsilon = 1/2$ (any number less than 1 would work). The first iteration of the recurrent layer produces

$$\mathbf{a}^2(1) = \mathbf{poslin}(\mathbf{W}^2\mathbf{a}^2(0)) = \begin{cases} \mathbf{poslin}\left(\begin{bmatrix} 1 & -0.5 \\ -0.5 & 1 \end{bmatrix} \begin{bmatrix} 4 \\ 2 \end{bmatrix}\right) \\ \mathbf{poslin}\left(\begin{bmatrix} 3 \\ 0 \end{bmatrix}\right) = \begin{bmatrix} 3 \\ 0 \end{bmatrix} \end{cases}. \quad (3.24)$$

The second iteration produces

$$\mathbf{a}^2(2) = \mathbf{poslin}(\mathbf{W}^2\mathbf{a}^2(1)) = \begin{cases} \mathbf{poslin}\left(\begin{bmatrix} 1 & -0.5 \\ -0.5 & 1 \end{bmatrix} \begin{bmatrix} 3 \\ 0 \end{bmatrix}\right) \\ \mathbf{poslin}\left(\begin{bmatrix} 3 \\ -1.5 \end{bmatrix}\right) = \begin{bmatrix} 3 \\ 0 \end{bmatrix} \end{cases}. \quad (3.25)$$

Since the outputs of successive iterations produce the same result, the network has converged. Prototype pattern number one, the *orange*, is chosen as the correct match, since neuron number one has the only nonzero output. (Recall that the first element of $\mathbf{a}^1$ was $(\mathbf{p}_1^T\mathbf{p} + 3)$.) This is the correct choice, since the Hamming distance from the *orange* prototype to the input pattern is 1, and the Hamming distance from the *apple* prototype to the input pattern is 2.

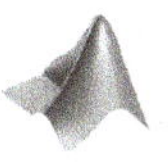

To experiment with the Hamming network and the apple/orange classification problem, use the Neural Network Design Demonstration Hamming Classification (nnd3hamc).

There are a number of networks whose operation is based on the same principles as the Hamming network; that is, where an inner product operation (feedforward layer) is followed by a competitive dynamic layer. These competitive networks will be discussed in Chapter 15. They are *self-organizing* networks, which can learn to adjust their prototype vectors based on the inputs that have been presented.

Hopfield Network

The final network we will discuss in this brief preview is the Hopfield network. This is a recurrent network that is similar in some respects to the recurrent layer of the Hamming network, but which can effectively perform the operations of both layers of the Hamming network. A diagram of the Hopfield network is shown in Figure 3.6. (This figure is actually a slight variation of the standard Hopfield network. We use this variation because it is somewhat simpler to describe and yet demonstrates the basic concepts.)

The neurons in this network are initialized with the input vector, then the network iterates until the output converges. When the network is operating correctly, the resulting output should be one of the prototype vectors. Therefore, whereas in the Hamming network the nonzero neuron indicates which prototype pattern is chosen, the Hopfield network actually produces the selected prototype pattern at its output.

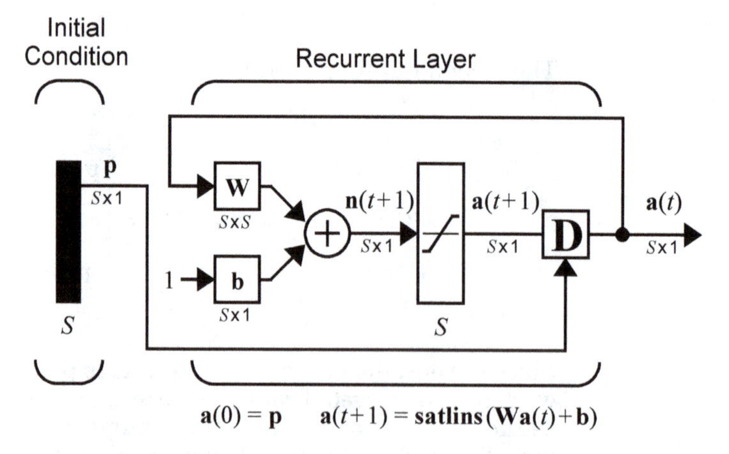

Figure 3.6 Hopfield Network

The equations that describe the network operation are

$$\mathbf{a}(0) = \mathbf{p} \tag{3.26}$$

and

$$\mathbf{a}(t+1) = \mathbf{satlins}(\mathbf{W}\mathbf{a}(t) + \mathbf{b}), \tag{3.27}$$

where *satlins* is the transfer function that is linear in the range [-1, 1] and saturates at 1 for inputs greater than 1 and at -1 for inputs less than -1.

The design of the weight matrix and the bias vector for the Hopfield network is a more complex procedure than it is for the Hamming network, where the weights in the feedforward layer are the prototype patterns.

To illustrate the operation of the network, we have determined a weight matrix and a bias vector that can solve our orange and apple pattern recognition problem. They are given in Eq. (3.28).

$$\mathbf{W} = \begin{bmatrix} 0.2 & 0 & 0 \\ 0 & 1.2 & 0 \\ 0 & 0 & 0.2 \end{bmatrix}, \mathbf{b} = \begin{bmatrix} 0.9 \\ 0 \\ -0.9 \end{bmatrix} \tag{3.28}$$

Although the procedure for computing the weights and biases for the Hopfield network is beyond the scope of this chapter, we can say a few things about why the parameters in Eq. (3.28) work for the apple and orange example.

We want the network output to converge to either the orange pattern, $\mathbf{p}_1$, or the apple pattern, $\mathbf{p}_2$. In both patterns, the first element is 1, and the third element is -1. The difference between the patterns occurs in the second element. Therefore, no matter what pattern is input to the network, we want the first element of the output pattern to converge to 1, the third element to converge to -1, and the second element to go to either 1 or -1, whichever is closer to the second element of the input vector.

The equations of operation of the Hopfield network, using the parameters given in Eq. (3.28), are

$$a_1(t+1) = satlins(0.2a_1(t) + 0.9)$$

$$a_2(t+1) = satlins(1.2a_2(t))$$

$$a_3(t+1) = satlins(0.2a_3(t) - 0.9) \tag{3.29}$$

Regardless of the initial values, $a_i(0)$, the first element will be increased until it saturates at 1, and the third element will be decreased until it saturates at -1. The second element is multiplied by a number larger than 1. Therefore, if it is initially negative, it will eventually saturate at -1; if it is initially positive it will saturate at 1.

(It should be noted that this is not the only $(\mathbf{W}, \mathbf{b})$ pair that could be used. You might want to try some others. See if you can discover what makes these work.)

Let's again take our oblong orange to test the Hopfield network. The outputs of the Hopfield network for the first three iterations would be

$$\mathbf{a}(0) = \begin{bmatrix} -1 \\ -1 \\ -1 \end{bmatrix}, \mathbf{a}(1) = \begin{bmatrix} 0.7 \\ -1 \\ -1 \end{bmatrix}, \mathbf{a}(2) = \begin{bmatrix} 1 \\ -1 \\ -1 \end{bmatrix}, \mathbf{a}(3) = \begin{bmatrix} 1 \\ -1 \\ -1 \end{bmatrix} \qquad (3.30)$$

The network has converged to the *orange* pattern, as did both the Hamming network and the perceptron, although each network operated in a different way. The perceptron had a single output, which could take on values of -1 (*orange*) or 1 (*apple*). In the Hamming network the single nonzero neuron indicated which prototype pattern had the closest match. If the first neuron was nonzero, that indicated *orange*, and if the second neuron was nonzero, that indicated *apple*. In the Hopfield network the prototype pattern itself appears at the output of the network.

To experiment with the Hopfield network and the apple/orange classification problem, use the Neural Network Design Demonstration Hopfield Classification (`nnd3hopc`).

As with the other networks demonstrated in this chapter, do not expect to feel completely comfortable with the Hopfield network at this point. There are a number of questions that we have not discussed. For example, "How do we know that the network will eventually converge?" It is possible for recurrent networks to oscillate or to have chaotic behavior. In addition, we have not discussed general procedures for designing the weight matrix and the bias vector.

Epilogue

The three networks that we have introduced in this chapter demonstrate many of the characteristics that are found in the architectures which are discussed throughout this book.

Feedforward networks, of which the perceptron is one example, are presented in Chapter 4, 7, 11, 12, Chapter 13 and 16. In these networks, the output is computed directly from the input in one pass; no feedback is involved. Feedforward networks are used for pattern recognition, as in the apple and orange example, and also for function approximation (see Chapter 11). Function approximation applications are found in such areas as adaptive filtering (see Chapter 10) and automatic control.

Competitive networks, represented here by the Hamming network, are characterized by two properties. First, they compute some measure of distance between stored prototype patterns and the input pattern. Second, they perform a competition to determine which neuron represents the prototype pattern closest to the input. In the competitive networks that are discussed in Chapter 15, the prototype patterns are adjusted as new inputs are applied to the network. These adaptive networks learn to cluster the inputs into different categories.

Recurrent networks, like the Hopfield network, were originally inspired by statistical mechanics. They have been used as associative memories, in which stored data is recalled by association with input data, rather than by an address. They have also been used to solve a variety of optimization problems.

We hope this chapter has piqued your curiosity about the capabilities of neural networks and has raised some questions. A few of the questions we will answer in later chapters are:

1. How do we determine the weight matrix and bias for perceptron networks with many inputs, where it is impossible to visualize the decision boundary? (Chapter 4 and 10)
2. If the categories to be recognized are not linearly separable, can we extend the standard perceptron to solve the problem? (Chapter 11, 12 and Chapter 13)
3. Can we learn the weights and biases of the Hamming network when we don't know the prototype patterns? (Chapter 15)

Exercises

E3.1 In this chapter we have designed three different neural networks to distinguish between apples and oranges, based on three sensor measurements (shape, texture and weight). Suppose that we want to distinguish between bananas and pineapples:

$$\mathbf{p}_1 = \begin{bmatrix} -1 \\ 1 \\ -1 \end{bmatrix} \text{ (Banana)}$$

$$\mathbf{p}_2 = \begin{bmatrix} -1 \\ -1 \\ 1 \end{bmatrix} \text{ (Pineapple)}$$

i. Design a perceptron to recognize these patterns.

ii. Design a Hamming network to recognize these patterns.

iii. Design a Hopfield network to recognize these patterns.

iv. Test the operation of your networks by applying several different input patterns. Discuss the advantages and disadvantages of each network.

E3.2 Consider the following prototype patterns.

$$\mathbf{p}_1 = \begin{bmatrix} 1 \\ 0.5 \end{bmatrix}, \mathbf{p}_2 = \begin{bmatrix} 2 \\ 1 \end{bmatrix}$$

i. Find and sketch a decision boundary for a perceptron network that will recognize these two vectors.

ii. Find weights and bias which will produce the decision boundary you found in part i, and sketch the network diagram.

iii. Calculate the network output for the following input. Is the network response (decision) reasonable? Explain.

$$\mathbf{p} = \begin{bmatrix} 1 \\ 0 \end{bmatrix}$$

iv. Design a Hamming network to recognize the two prototype vectors above.

v. Calculate the network output for the Hamming network with the input vector given in part iii, showing all steps. Does the Hamming network produce the same decision as the perceptron? Explain why or why not. Which network is better suited to this problem? Explain.

E3.3 Consider a Hopfield network, with the following weight and bias.

$$\mathbf{W} = \begin{bmatrix} 1 & -1 \\ -1 & 1 \end{bmatrix}, \quad \mathbf{b} = \begin{bmatrix} 0 \\ 0 \end{bmatrix}$$

i. The following input (initial condition) is applied to the network. Find the network response (show the network output at each iteration until the network converges).

$$\mathbf{p} = \begin{bmatrix} 0.9 \\ 1 \end{bmatrix}$$

ii. Draw a sketch indicating what region of the input space will converge to the same final output that you found in part i. (In other words, for what other **p** vectors will the network converge to the same final output?) Explain how you obtained your answer.

iii. What other prototypes will this network converge to, and what regions of the input space correspond to each prototype (sketch the regions). Explain how you obtained your answer.

E3.4 Consider the following perceptron network.

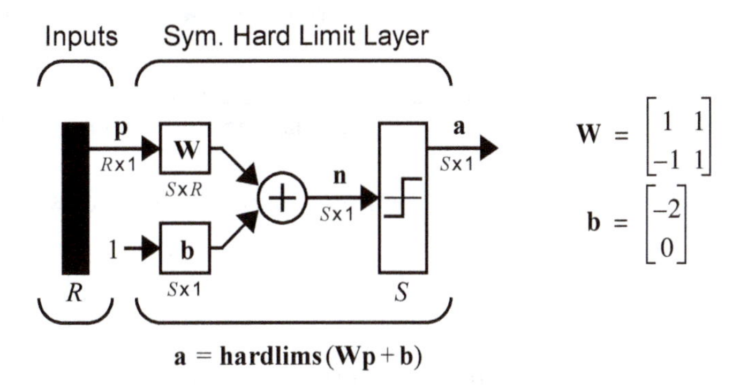

$$\mathbf{W} = \begin{bmatrix} 1 & 1 \\ -1 & 1 \end{bmatrix}$$

$$\mathbf{b} = \begin{bmatrix} -2 \\ 0 \end{bmatrix}$$

i. How many different classes can this network classify?

ii. Draw a diagram illustrating the regions corresponding to each class. Label each region with the corresponding network output.

iii. Calculate the network output for the following input.

$$\mathbf{p} = \begin{bmatrix} 1 \\ -1 \end{bmatrix}$$

iv. Plot the input from part iii in your diagram from part ii, and verify that it falls in the correctly labeled region.

E3.5 We want to design a perceptron network to output a 1 when either of these two vectors are input to the network:

$$\left\{ \begin{bmatrix} -1 \\ 0 \end{bmatrix}, \begin{bmatrix} 1 \\ 2 \end{bmatrix} \right\},$$

and to output a -1 when either of the following vectors are input to the network:

$$\left\{ \begin{bmatrix} -1 \\ 1 \end{bmatrix}, \begin{bmatrix} 0 \\ 2 \end{bmatrix} \right\}.$$

i. Find and sketch a decision boundary for a network that will solve this problem.

ii. Find weights and biases that will produce the decision boundary you found in part i. Show all work.

iii. Draw the network diagram using abreviated notation.

iv. For each of the four vectors given above, calculate the net input, **n**, and the network output, **a**, for the network you have designed. Verify that your network solves the problem.

v. Are there other weights and biases that would solve the problem? If so, would you consider your weights best? Explain.

E3.6 We have the folowing two prototype vectors:

$$\left\{ \begin{bmatrix} -1 \\ 1 \end{bmatrix}, \begin{bmatrix} 1 \\ 1 \end{bmatrix} \right\}.$$

i. Find and sketch a decision boundary for a perceptron network that will recognize these two vectors.

ii. Find weights and bias that will produce the decision boundary you found in part i.

iii. Draw the network diagram using abreviated notation.

iv. For the vector given below, calculate the net input, n, and the network output, a, for the network you have designed. Does the network produce a good output? Explain.

$$\begin{bmatrix} 0.5 \\ -0.5 \end{bmatrix}$$

v. Design a Hamming network to recognize the two vectors used in part i.

vi. Calculate the network output for the Hamming network for the input vector given in part iv. Does the network produce a good output? Explain.

vii. Design a Hopfield network to recognize the two vectors used in part i.

viii. Calculate the network output for the Hopfield network for the input vector given in part iv. Does the network produce a good output? Explain.

E3.7 We want to design a Hamming network to recognize the following prototype vectors:

$$\left\{ \begin{bmatrix} 1 \\ 1 \end{bmatrix}, \begin{bmatrix} -1 \\ -1 \end{bmatrix}, \begin{bmatrix} -1 \\ 1 \end{bmatrix} \right\}.$$

i. Find the weight matrices and bias vectors for the Hamming network.

ii. Draw the network diagram.

iii. Apply the following input vector and calculate the total network response (iterating the second layer to convergence). Explain the meaning of the final network output.

$$\mathbf{p} = \begin{bmatrix} 1 \\ 0 \end{bmatrix}$$

iv. Sketch the decision boundaries for this network. Explain how you determined the boundaries.

4 Perceptron Learning Rule

Objectives

One of the questions we raised in Chapter 3 was: "How do we determine the weight matrix and bias for perceptron networks with many inputs, where it is impossible to visualize the decision boundaries?" In this chapter we will describe an algorithm for *training* perceptron networks, so that they can *learn* to solve classification problems. We will begin by explaining what a learning rule is and will then develop the perceptron learning rule. We will conclude by discussing the advantages and limitations of the single-layer perceptron network. This discussion will lead us into future chapters.

Theory and Examples

In 1943, Warren McCulloch and Walter Pitts introduced one of the first artificial neurons [McPi43]. The main feature of their neuron model is that a weighted sum of input signals is compared to a threshold to determine the neuron output. When the sum is greater than or equal to the threshold, the output is 1. When the sum is less than the threshold, the output is 0. They went on to show that networks of these neurons could, in principle, compute any arithmetic or logical function. Unlike biological networks, the parameters of their networks had to be designed, as no training method was available. However, the perceived connection between biology and digital computers generated a great deal of interest.

In the late 1950s, Frank Rosenblatt and several other researchers developed a class of neural networks called perceptrons. The neurons in these networks were similar to those of McCulloch and Pitts. Rosenblatt's key contribution was the introduction of a learning rule for training perceptron networks to solve pattern recognition problems [Rose58]. He proved that his learning rule will always converge to the correct network weights, if weights exist that solve the problem. Learning was simple and automatic. Examples of proper behavior were presented to the network, which learned from its mistakes. The perceptron could even learn when initialized with random values for its weights and biases.

Unfortunately, the perceptron network is inherently limited. These limitations were widely publicized in the book *Perceptrons* [MiPa69] by Marvin Minsky and Seymour Papert. They demonstrated that the perceptron networks were incapable of implementing certain elementary functions. It was not until the 1980s that these limitations were overcome with improved (multilayer) perceptron networks and associated learning rules. We will discuss these improvements in Chapter 11 and 12.

Today the perceptron is still viewed as an important network. It remains a fast and reliable network for the class of problems that it can solve. In addition, an understanding of the operations of the perceptron provides a good basis for understanding more complex networks. Thus, the perceptron network, and its associated learning rule, are well worth discussing here.

In the remainder of this chapter we will define what we mean by a learning rule, explain the perceptron network and learning rule, and discuss the limitations of the perceptron network.

Learning Rules

Learning Rule

As we begin our presentation of the perceptron learning rule, we want to discuss learning rules in general. By *learning rule* we mean a procedure for modifying the weights and biases of a network. (This procedure may also be referred to as a training algorithm.) The purpose of the learning rule is to train the network to perform some task. There are many types of neural network learning rules. They fall into three broad categories: supervised learning, unsupervised learning and reinforcement (or graded) learning.

Supervised Learning
Training Set

In *supervised learning*, the learning rule is provided with a set of examples (the *training set*) of proper network behavior:

$$\{\mathbf{p}_1,\mathbf{t}_1\}, \{\mathbf{p}_2,\mathbf{t}_2\}, \ldots, \{\mathbf{p}_Q,\mathbf{t}_Q\}, \tag{4.1}$$

Target

where $\mathbf{p}_q$ is an input to the network and $\mathbf{t}_q$ is the corresponding correct (*target*) output. As the inputs are applied to the network, the network outputs are compared to the targets. The learning rule is then used to adjust the weights and biases of the network in order to move the network outputs closer to the targets. The perceptron learning rule falls in this supervised learning category. We will also investigate supervised learning algorithms in Chapter 7–14.

Reinforcement Learning

Reinforcement learning is similar to supervised learning, except that, instead of being provided with the correct output for each network input, the algorithm is only given a grade. The grade (or score) is a measure of the network performance over some sequence of inputs. This type of learning is currently much less common than supervised learning. It appears to be most suited to control system applications (see [BaSu83], [WhSo92]).

Unsupervised Learning

In *unsupervised learning*, the weights and biases are modified in response to network inputs only. There are no target outputs available. At first glance this might seem to be impractical. How can you train a network if you don't know what it is supposed to do? Most of these algorithms perform some kind of clustering operation. They learn to categorize the input patterns into a finite number of classes. This is especially useful in such applications as vector quantization. We will see in Chapter 15 that there are a number of unsupervised learning algorithms.

Perceptron Architecture

Before we present the perceptron learning rule, let's expand our investigation of the perceptron network, which we began in Chapter 3. The general perceptron network is shown in Figure 4.1.

The output of the network is given by

$$\mathbf{a} = \mathbf{hardlim}(\mathbf{Wp} + \mathbf{b}). \tag{4.2}$$

(Note that in Chapter 3 we used the *hardlims* transfer function, instead of *hardlim*. This does not affect the capabilities of the network. See Exercise E4.10.)

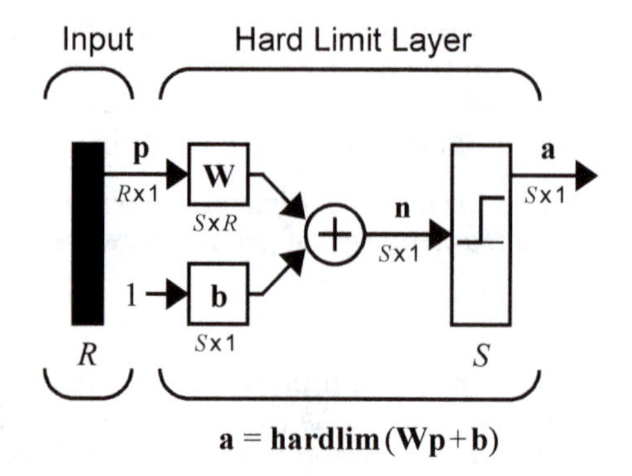

Figure 4.1 Perceptron Network

It will be useful in our development of the perceptron learning rule to be able to conveniently reference individual elements of the network output. Let's see how this can be done. First, consider the network weight matrix:

$$\mathbf{W} = \begin{bmatrix} w_{1,1} & w_{1,2} & \cdots & w_{1,R} \\ w_{2,1} & w_{2,2} & \cdots & w_{2,R} \\ \vdots & \vdots & & \vdots \\ w_{S,1} & w_{S,2} & \cdots & w_{S,R} \end{bmatrix}. \tag{4.3}$$

We will define a vector composed of the elements of the ith row of $\mathbf{W}$:

$$_{i}\mathbf{w} = \begin{bmatrix} w_{i,1} \\ w_{i,2} \\ \vdots \\ w_{i,R} \end{bmatrix}. \tag{4.4}$$

Now we can partition the weight matrix:

$$\mathbf{W} = \begin{bmatrix} {}_{1}\mathbf{w}^{T} \\ {}_{2}\mathbf{w}^{T} \\ \vdots \\ {}_{S}\mathbf{w}^{T} \end{bmatrix}. \tag{4.5}$$

This allows us to write the ith element of the network output vector as

$$a_i = hardlim(n_i) = hardlim({}_{i}\mathbf{w}^{T}\mathbf{p} + b_i). \tag{4.6}$$

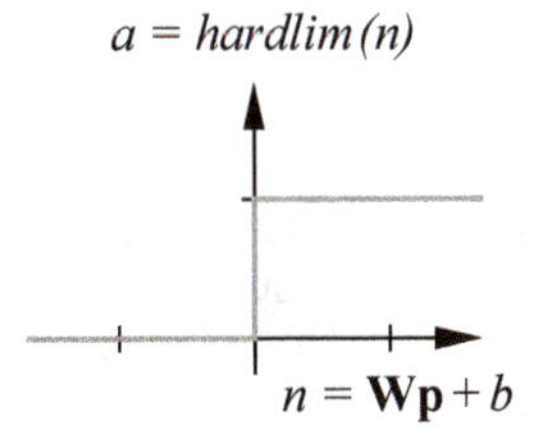

Recall that the *hardlim* transfer function (shown at left) is defined as:

$$a = hardlim(n) = \begin{cases} 1 & if \;\; n \geq 0 \\ 0 & otherwise. \end{cases} \tag{4.7}$$

Therefore, if the inner product of the ith row of the weight matrix with the input vector is greater than or equal to $-b_i$, the output will be 1, otherwise the output will be 0. *Thus each neuron in the network divides the input space into two regions.* It is useful to investigate the boundaries between these regions. We will begin with the simple case of a single-neuron perceptron with two inputs.

Single-Neuron Perceptron

Let's consider a two-input perceptron with one neuron, as shown in Figure 4.2.

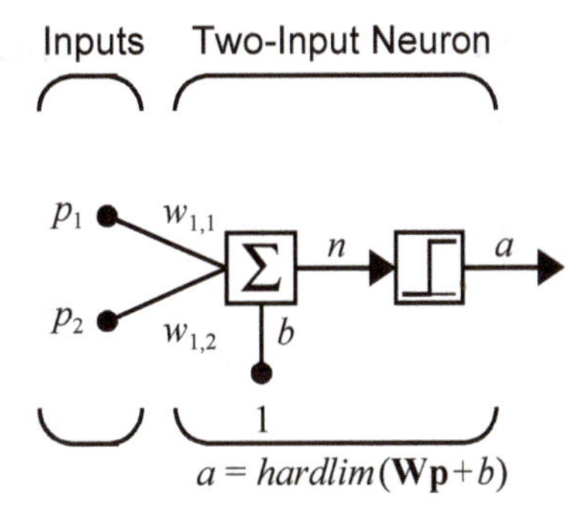

Figure 4.2 Two-Input/Single-Output Perceptron

The output of this network is determined by

$$a = hardlim(n) = hardlim(\mathbf{Wp} + b)$$

$$= hardlim({}_1\mathbf{w}^T\mathbf{p} + b) = hardlim(w_{1,1}p_1 + w_{1,2}p_2 + b) \tag{4.8}$$

Decision Boundary

The *decision boundary* is determined by the input vectors for which the net input n is zero:

$$n = {}_1\mathbf{w}^T\mathbf{p} + b = w_{1,1}p_1 + w_{1,2}p_2 + b = 0\,. \tag{4.9}$$

To make the example more concrete, let's assign the following values for the weights and bias:

$$w_{1,1} = 1\,,\; w_{1,2} = 1\,,\; b = -1\,. \tag{4.10}$$

The decision boundary is then

$$n = {}_1\mathbf{w}^T\mathbf{p} + b = w_{1,1}p_1 + w_{1,2}p_2 + b = p_1 + p_2 - 1 = 0 . \tag{4.11}$$

This defines a line in the input space. On one side of the line the network output will be 0; on the line and on the other side of the line the output will be 1. To draw the line, we can find the points where it intersects the p_1 and p_2 axes. To find the p_2 intercept set $p_1 = 0$:

$$p_2 = -\frac{b}{w_{1,2}} = -\frac{-1}{1} = 1 \quad \text{if } p_1 = 0 . \tag{4.12}$$

To find the p_1 intercept, set $p_2 = 0$:

$$p_1 = -\frac{b}{w_{1,1}} = -\frac{-1}{1} = 1 \quad \text{if } p_2 = 0 . \tag{4.13}$$

The resulting decision boundary is illustrated in Figure 4.3.

To find out which side of the boundary corresponds to an output of 1, we just need to test one point. For the input $\mathbf{p} = \begin{bmatrix} 2 & 0 \end{bmatrix}^T$, the network output will be

$$a = hardlim({}_1\mathbf{w}^T\mathbf{p} + b) = hardlim\left(\begin{bmatrix} 1 & 1 \end{bmatrix} \begin{bmatrix} 2 \\ 0 \end{bmatrix} - 1 \right) = 1 . \tag{4.14}$$

Therefore, the network output will be 1 for the region above and to the right of the decision boundary. This region is indicated by the shaded area in Figure 4.3.

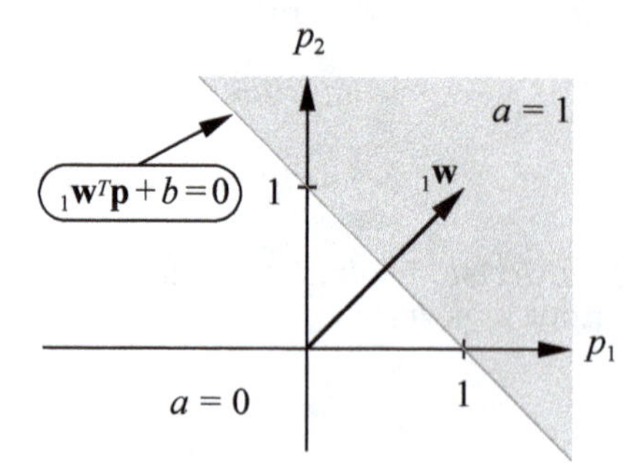

Figure 4.3 Decision Boundary for Two-Input Perceptron

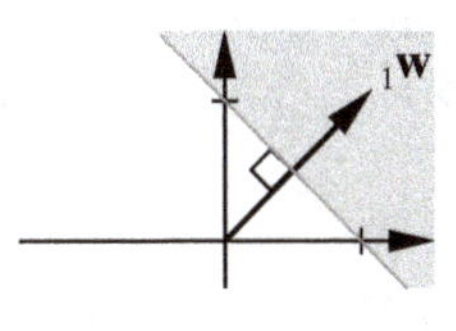

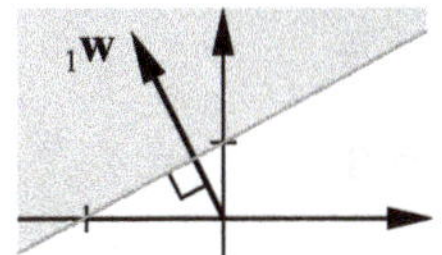

We can also find the decision boundary graphically. The first step is to note that the boundary is always orthogonal to $_1\mathbf{w}$, as illustrated in the adjacent figures. The boundary is defined by

$$_1\mathbf{w}^T\mathbf{p} + b = 0. \tag{4.15}$$

For all points on the boundary, the inner product of the input vector with the weight vector is the same. This implies that these input vectors will all have the same projection onto the weight vector, so they must lie on a line orthogonal to the weight vector. (These concepts will be covered in more detail in Chapter 5.) In addition, any vector in the shaded region of Figure 4.3 will have an inner product greater than $-b$, and vectors in the unshaded region will have inner products less than $-b$. Therefore the weight vector $_1\mathbf{w}$ will always point toward the region where the neuron output is 1.

After we have selected a weight vector with the correct angular orientation, the bias value can be computed by selecting a point on the boundary and satisfying Eq. (4.15).

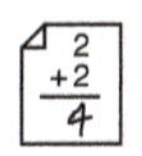

Let's apply some of these concepts to the design of a perceptron network to implement a simple logic function: the AND gate. The input/target pairs for the AND gate are

$$\left\{\mathbf{p}_1 = \begin{bmatrix}0\\0\end{bmatrix}, t_1 = 0\right\} \left\{\mathbf{p}_2 = \begin{bmatrix}0\\1\end{bmatrix}, t_2 = 0\right\} \left\{\mathbf{p}_3 = \begin{bmatrix}1\\0\end{bmatrix}, t_3 = 0\right\} \left\{\mathbf{p}_4 = \begin{bmatrix}1\\1\end{bmatrix}, t_4 = 1\right\}.$$

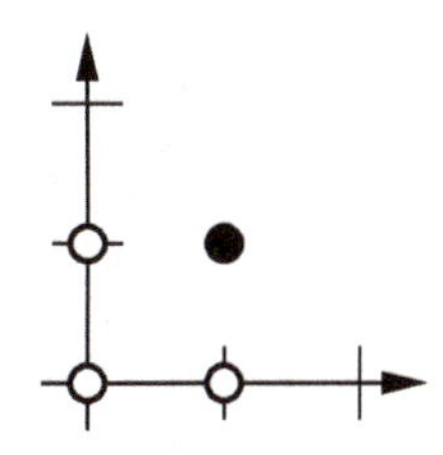

The figure to the left illustrates the problem graphically. It displays the input space, with each input vector labeled according to its target. The dark circles ● indicate that the target is 1, and the light circles ○ indicate that the target is 0.

The first step of the design is to select a decision boundary. We want to have a line that separates the dark circles and the light circles. There are an infinite number of solutions to this problem. It seems reasonable to choose the line that falls "halfway" between the two categories of inputs, as shown in the adjacent figure.

Next we want to choose a weight vector that is orthogonal to the decision boundary. The weight vector can be any length, so there are infinite possibilities. One choice is

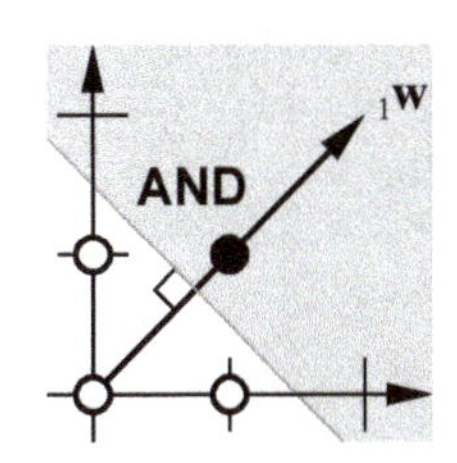

$$_1\mathbf{w} = \begin{bmatrix}2\\2\end{bmatrix}, \tag{4.16}$$

as displayed in the figure to the left.

Finally, we need to find the bias, b. We can do this by picking a point on the decision boundary and satisfying Eq. (4.15). If we use $\mathbf{p} = \begin{bmatrix} 1.5 & 0 \end{bmatrix}^T$ we find

$$ {}_1\mathbf{w}^T\mathbf{p} + b = \begin{bmatrix} 2 & 2 \end{bmatrix} \begin{bmatrix} 1.5 \\ 0 \end{bmatrix} + b = 3 + b = 0 \quad \Rightarrow \quad b = -3. \tag{4.17} $$

We can now test the network on one of the input/target pairs. If we apply $\mathbf{p}_2$ to the network, the output will be

$$ a = hardlim({}_1\mathbf{w}^T\mathbf{p}_2 + b) = hardlim\left(\begin{bmatrix} 2 & 2 \end{bmatrix} \begin{bmatrix} 0 \\ 1 \end{bmatrix} - 3 \right) $$

$$ a = hardlim(-1) = 0, \tag{4.18} $$

which is equal to the target output t_2. Verify for yourself that all inputs are correctly classified.

To experiment with decision boundaries, use the Neural Network Design Demonstration Decision Boundaries (nnd4db).

Multiple-Neuron Perceptron

Note that for perceptrons with multiple neurons, as in Figure 4.1, there will be one decision boundary for each neuron. The decision boundary for neuron i will be defined by

$$ {}_i\mathbf{w}^T\mathbf{p} + b_i = 0. \tag{4.19} $$

A single-neuron perceptron can classify input vectors into two categories, since its output can be either 0 or 1. A multiple-neuron perceptron can classify inputs into many categories. Each category is represented by a different output vector. Since each element of the output vector can be either 0 or 1, there are a total of 2^S possible categories, where S is the number of neurons.

Perceptron Learning Rule

Now that we have examined the performance of perceptron networks, we are in a position to introduce the perceptron learning rule. This learning rule is an example of supervised training, in which the learning rule is provided with a set of examples of proper network behavior:

$$ \{\mathbf{p}_1, \mathbf{t}_1\}, \{\mathbf{p}_2, \mathbf{t}_2\}, \ldots, \{\mathbf{p}_Q, \mathbf{t}_Q\}, \tag{4.20} $$

where $\mathbf{p}_q$ is an input to the network and $\mathbf{t}_q$ is the corresponding target output. As each input is applied to the network, the network output is compared to the target. The learning rule then adjusts the weights and biases of the network in order to move the network output closer to the target.

Test Problem

In our presentation of the perceptron learning rule we will begin with a simple test problem and will experiment with possible rules to develop some intuition about how the rule should work. The input/target pairs for our test problem are

$$\left\{\mathbf{p}_1 = \begin{bmatrix}1\\2\end{bmatrix}, t_1 = 1\right\}\left\{\mathbf{p}_2 = \begin{bmatrix}-1\\2\end{bmatrix}, t_2 = 0\right\}\left\{\mathbf{p}_3 = \begin{bmatrix}0\\-1\end{bmatrix}, t_3 = 0\right\}.$$

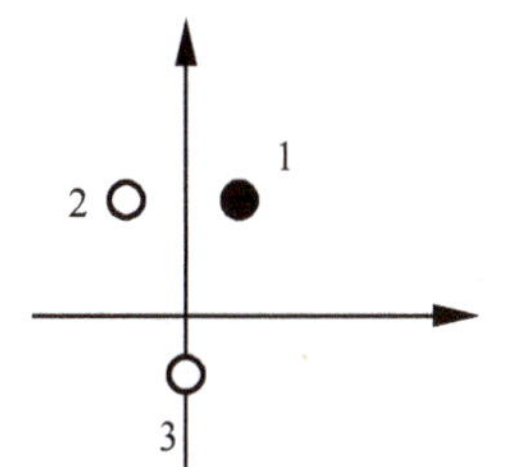

The problem is displayed graphically in the adjacent figure, where the two input vectors whose target is 0 are represented with a light circle ○, and the vector whose target is 1 is represented with a dark circle ●. This is a very simple problem, and we could almost obtain a solution by inspection. This simplicity will help us gain some intuitive understanding of the basic concepts of the perceptron learning rule.

The network for this problem should have two-inputs and one output. To simplify our development of the learning rule, we will begin with a network without a bias. The network will then have just two parameters, $w_{1,1}$ and $w_{1,2}$, as shown in Figure 4.4.

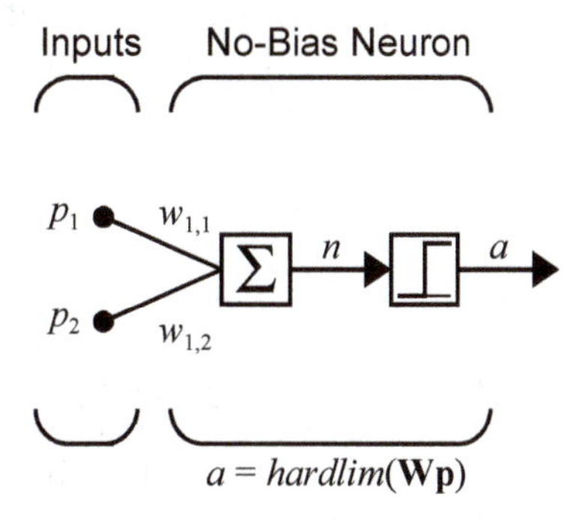

Figure 4.4 Test Problem Network

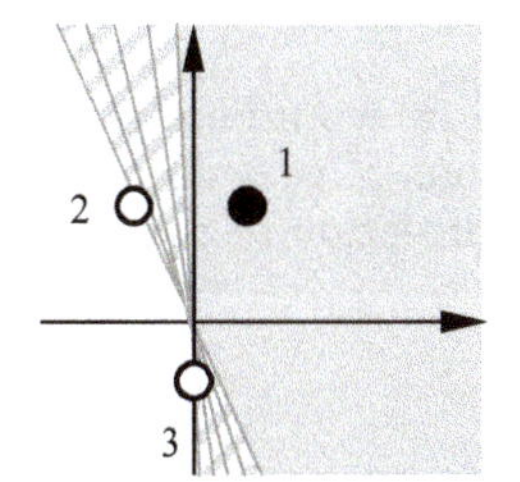

By removing the bias we are left with a network whose decision boundary must pass through the origin. We need to be sure that this network is still able to solve the test problem. There must be an allowable decision boundary that can separate the vectors $\mathbf{p}_2$ and $\mathbf{p}_3$ from the vector $\mathbf{p}_1$. The figure to the left illustrates that there are indeed an infinite number of such boundaries.

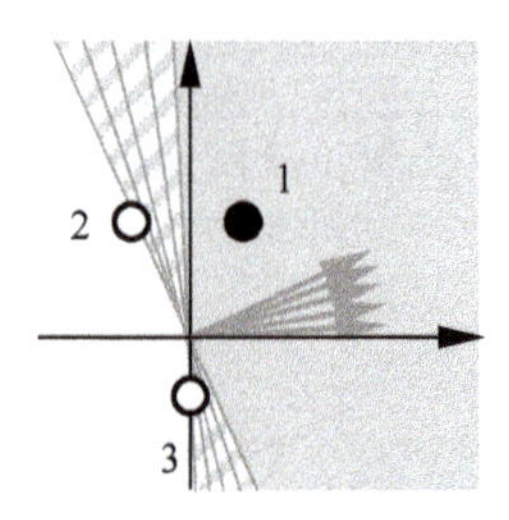

The adjacent figure shows the weight vectors that correspond to the allowable decision boundaries. (Recall that the weight vector is orthogonal to the decision boundary.) We would like a learning rule that will find a weight vector that points in one of these directions. Remember that the length of the weight vector does not matter; only its direction is important.

Constructing Learning Rules

Training begins by assigning some initial values for the network parameters. In this case we are training a two-input/single-output network without a bias, so we only have to initialize its two weights. Here we set the elements of the weight vector, ${}_1\mathbf{w}$, to the following randomly generated values:

$$ {}_1\mathbf{w}^T = \begin{bmatrix} 1.0 & -0.8 \end{bmatrix} . \tag{4.21} $$

We will now begin presenting the input vectors to the network. We begin with $\mathbf{p}_1$:

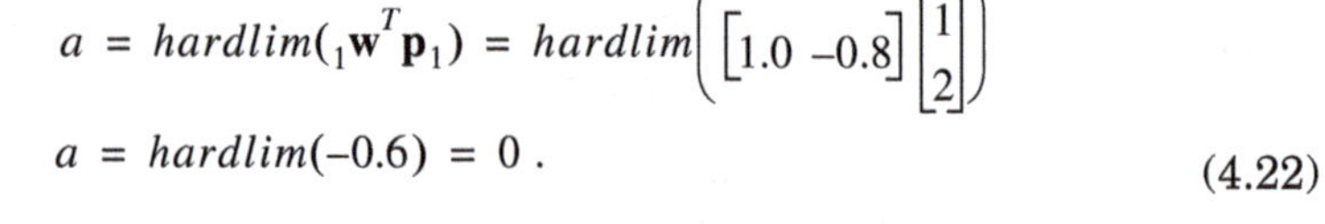

$$ a = hardlim({}_1\mathbf{w}^T\mathbf{p}_1) = hardlim\left(\begin{bmatrix} 1.0 & -0.8 \end{bmatrix}\begin{bmatrix} 1 \\ 2 \end{bmatrix}\right) $$
$$ a = hardlim(-0.6) = 0 . \tag{4.22} $$

The network has not returned the correct value. The network output is 0, while the target response, t_1, is 1.

We can see what happened by looking at the adjacent diagram. The initial weight vector results in a decision boundary that incorrectly classifies the vector $\mathbf{p}_1$. We need to alter the weight vector so that it points more toward $\mathbf{p}_1$, so that in the future it has a better chance of classifying it correctly.

One approach would be to set ${}_1\mathbf{w}$ equal to $\mathbf{p}_1$. This is simple and would ensure that $\mathbf{p}_1$ was classified properly in the future. Unfortunately, it is easy to construct a problem for which this rule cannot find a solution. The diagram to the lower left shows a problem that cannot be solved with the weight vector pointing directly at either of the two class 1 vectors. If we apply the rule ${}_1\mathbf{w} = \mathbf{p}$ every time one of these vectors is misclassified, the network's weights will simply oscillate back and forth and will never find a solution.

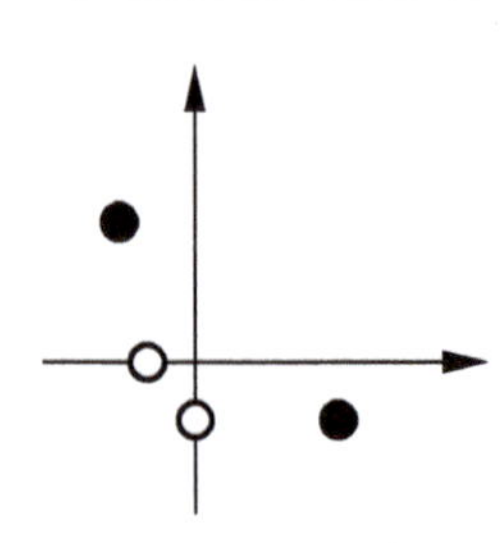

Another possibility would be to add $\mathbf{p}_1$ to ${}_1\mathbf{w}$. Adding $\mathbf{p}_1$ to ${}_1\mathbf{w}$ would make ${}_1\mathbf{w}$ point more in the direction of $\mathbf{p}_1$. Repeated presentations of $\mathbf{p}_1$ would cause the direction of ${}_1\mathbf{w}$ to asymptotically approach the direction of $\mathbf{p}_1$. This rule can be stated:

$$ \text{If } t = 1 \text{ and } a = 0, \text{ then } {}_1\mathbf{w}^{new} = {}_1\mathbf{w}^{old} + \mathbf{p} . \tag{4.23} $$

Applying this rule to our test problem results in new values for ${}_1\mathbf{w}$:

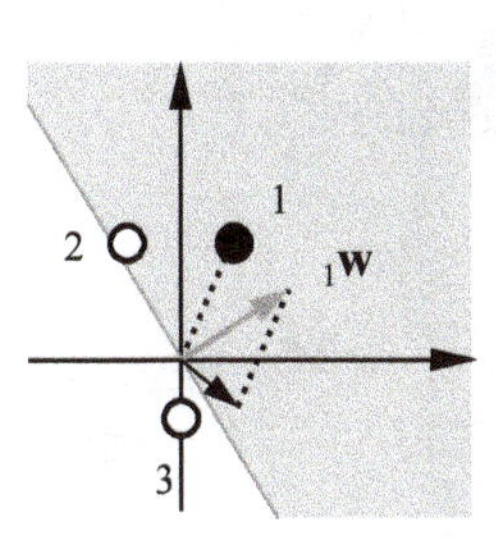

$$ {}_1\mathbf{w}^{new} = {}_1\mathbf{w}^{old} + \mathbf{p}_1 = \begin{bmatrix} 1.0 \\ -0.8 \end{bmatrix} + \begin{bmatrix} 1 \\ 2 \end{bmatrix} = \begin{bmatrix} 2.0 \\ 1.2 \end{bmatrix}. \quad (4.24) $$

This operation is illustrated in the adjacent figure.

We now move on to the next input vector and will continue making changes to the weights and cycling through the inputs until they are all classified correctly.

The next input vector is $\mathbf{p}_2$. When it is presented to the network we find:

$$ a = hardlim({}_1\mathbf{w}^T\mathbf{p}_2) = hardlim\left(\begin{bmatrix} 2.0 & 1.2 \end{bmatrix}\begin{bmatrix} -1 \\ 2 \end{bmatrix}\right) $$

$$ = hardlim(0.4) = 1 . \quad (4.25) $$

The target t_2 associated with $\mathbf{p}_2$ is 0 and the output a is 1. A class 0 vector was misclassified as a 1.

Since we would now like to move the weight vector ${}_1\mathbf{w}$ away from the input, we can simply change the addition in Eq. (4.23) to subtraction:

$$ \text{If } t = 0 \text{ and } a = 1, \text{ then } {}_1\mathbf{w}^{new} = {}_1\mathbf{w}^{old} - \mathbf{p}. \quad (4.26) $$

If we apply this to the test problem we find:

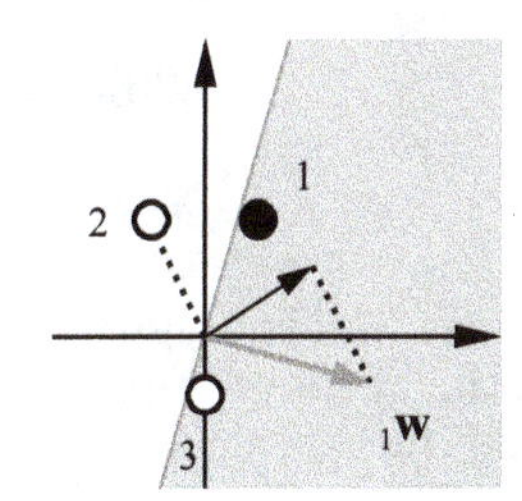

$$ {}_1\mathbf{w}^{new} = {}_1\mathbf{w}^{old} - \mathbf{p}_2 = \begin{bmatrix} 2.0 \\ 1.2 \end{bmatrix} - \begin{bmatrix} -1 \\ 2 \end{bmatrix} = \begin{bmatrix} 3.0 \\ -0.8 \end{bmatrix}, \quad (4.27) $$

which is illustrated in the adjacent figure.

Now we present the third vector $\mathbf{p}_3$:

$$ a = hardlim({}_1\mathbf{w}^T\mathbf{p}_3) = hardlim\left(\begin{bmatrix} 3.0 & -0.8 \end{bmatrix}\begin{bmatrix} 0 \\ -1 \end{bmatrix}\right) $$

$$ = hardlim(0.8) = 1 . \quad (4.28) $$

The current ${}_1\mathbf{w}$ results in a decision boundary that misclassifies $\mathbf{p}_3$. This is a situation for which we already have a rule, so ${}_1\mathbf{w}$ will be updated again, according to Eq. (4.26):

$$ {}_1\mathbf{w}^{new} = {}_1\mathbf{w}^{old} - \mathbf{p}_3 = \begin{bmatrix} 3.0 \\ -0.8 \end{bmatrix} - \begin{bmatrix} 0 \\ -1 \end{bmatrix} = \begin{bmatrix} 3.0 \\ 0.2 \end{bmatrix}. \quad (4.29) $$

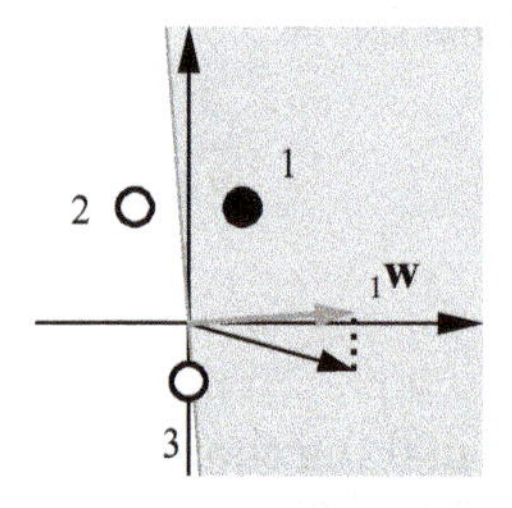

The diagram to the left shows that the perceptron has finally learned to classify the three vectors properly. If we present any of the input vectors to the neuron, it will output the correct class for that input vector.

This brings us to our third and final rule: if it works, don't fix it.

$$\text{If } t = a, \text{ then } {}_1\mathbf{w}^{new} = {}_1\mathbf{w}^{old}. \tag{4.30}$$

Here are the three rules, which cover all possible combinations of output and target values:

$$\begin{aligned} &\text{If } t = 1 \text{ and } a = 0, \text{ then } {}_1\mathbf{w}^{new} = {}_1\mathbf{w}^{old} + \mathbf{p}. \\ &\text{If } t = 0 \text{ and } a = 1, \text{ then } {}_1\mathbf{w}^{new} = {}_1\mathbf{w}^{old} - \mathbf{p}. \\ &\text{If } t = a, \text{ then } {}_1\mathbf{w}^{new} = {}_1\mathbf{w}^{old}. \end{aligned} \tag{4.31}$$

Unified Learning Rule

The three rules in Eq. (4.31) can be rewritten as a single expression. First we will define a new variable, the perceptron error e:

$$e = t - a. \tag{4.32}$$

We can now rewrite the three rules of Eq. (4.31) as:

$$\begin{aligned} &\text{If } e = 1, \text{ then } {}_1\mathbf{w}^{new} = {}_1\mathbf{w}^{old} + \mathbf{p}. \\ &\text{If } e = -1, \text{ then } {}_1\mathbf{w}^{new} = {}_1\mathbf{w}^{old} - \mathbf{p}. \\ &\text{If } e = 0, \text{ then } {}_1\mathbf{w}^{new} = {}_1\mathbf{w}^{old}. \end{aligned} \tag{4.33}$$

Looking carefully at the first two rules in Eq. (4.33) we can see that the sign of $\mathbf{p}$ is the same as the sign on the error, e. Furthermore, the absence of $\mathbf{p}$ in the third rule corresponds to an e of 0. Thus, we can unify the three rules into a single expression:

$${}_1\mathbf{w}^{new} = {}_1\mathbf{w}^{old} + e\mathbf{p} = {}_1\mathbf{w}^{old} + (t - a)\mathbf{p}. \tag{4.34}$$

This rule can be extended to train the bias by noting that a bias is simply a weight whose input is always 1. We can thus replace the input $\mathbf{p}$ in Eq. (4.34) with the input to the bias, which is 1. The result is the perceptron rule for a bias:

$$b^{new} = b^{old} + e. \tag{4.35}$$

Training Multiple-Neuron Perceptrons

The perceptron rule, as given by Eq. (4.34) and Eq. (4.35), updates the weight vector of a single neuron perceptron. We can generalize this rule for the multiple-neuron perceptron of Figure 4.1 as follows. To update the *i*th row of the weight matrix use:

$$ {}_i\mathbf{w}^{new} = {}_i\mathbf{w}^{old} + e_i\mathbf{p} . \tag{4.36}$$

To update the *i*th element of the bias vector use:

$$ b_i^{new} = b_i^{old} + e_i . \tag{4.37}$$

Perceptron Rule

The *perceptron rule* can be written conveniently in matrix notation:

$$ \mathbf{W}^{new} = \mathbf{W}^{old} + \mathbf{e}\mathbf{p}^T , \tag{4.38}$$

and

$$ \mathbf{b}^{new} = \mathbf{b}^{old} + \mathbf{e} . \tag{4.39}$$

2 +2 4

To test the perceptron learning rule, consider again the apple/orange recognition problem of Chapter 3. The input/output prototype vectors will be

$$ \left\{ \mathbf{p}_1 = \begin{bmatrix} 1 \\ -1 \\ -1 \end{bmatrix}, t_1 = \begin{bmatrix} 0 \end{bmatrix} \right\} \quad \left\{ \mathbf{p}_2 = \begin{bmatrix} 1 \\ 1 \\ -1 \end{bmatrix}, t_2 = \begin{bmatrix} 1 \end{bmatrix} \right\} . \tag{4.40}$$

(Note that we are using 0 as the target output for the orange pattern, $\mathbf{p}_1$, instead of -1, as was used in Chapter 3. This is because we are using the *hardlim* transfer function, instead of *hardlims*.)

Typically the weights and biases are initialized to small random numbers. Suppose that here we start with the initial weight matrix and bias:

$$ \mathbf{W} = \begin{bmatrix} 0.5 & -1 & -0.5 \end{bmatrix} , b = 0.5 . \tag{4.41}$$

The first step is to apply the first input vector, $\mathbf{p}_1$, to the network:

$$ a = hardlim(\mathbf{W}\mathbf{p}_1 + b) = hardlim\left(\begin{bmatrix} 0.5 & -1 & -0.5 \end{bmatrix} \begin{bmatrix} 1 \\ -1 \\ -1 \end{bmatrix} + 0.5 \right) $$

$$ = hardlim(2.5) = 1 \tag{4.42}$$

Then we calculate the error:

$$e = t_1 - a = 0 - 1 = -1 . \quad (4.43)$$

The weight update is

$$\mathbf{W}^{new} = \mathbf{W}^{old} + e\mathbf{p}^T = \begin{bmatrix} 0.5 & -1 & -0.5 \end{bmatrix} + (-1)\begin{bmatrix} 1 & -1 & -1 \end{bmatrix} \quad (4.44)$$
$$= \begin{bmatrix} -0.5 & 0 & 0.5 \end{bmatrix} .$$

The bias update is

$$b^{new} = b^{old} + e = 0.5 + (-1) = -0.5 . \quad (4.45)$$

This completes the first iteration.

The second iteration of the perceptron rule is:

$$a = hardlim\,(\mathbf{W}\mathbf{p}_2 + b) = hardlim\,\left(\begin{bmatrix} -0.5 & 0 & 0.5 \end{bmatrix}\begin{bmatrix} 1 \\ 1 \\ -1 \end{bmatrix} + (-0.5)\right)$$

$$= hardlim\,(-0.5) = 0 \quad (4.46)$$

$$e = t_2 - a = 1 - 0 = 1 \quad (4.47)$$

$$\mathbf{W}^{new} = \mathbf{W}^{old} + e\mathbf{p}^T = \begin{bmatrix} -0.5 & 0 & 0.5 \end{bmatrix} + 1\begin{bmatrix} 1 & 1 & -1 \end{bmatrix} = \begin{bmatrix} 0.5 & 1 & -0.5 \end{bmatrix} \quad (4.48)$$

$$b^{new} = b^{old} + e = -0.5 + 1 = 0.5 \quad (4.49)$$

The third iteration begins again with the first input vector:

$$a = hardlim\,(\mathbf{W}\mathbf{p}_1 + b) = hardlim\,\left(\begin{bmatrix} 0.5 & 1 & -0.5 \end{bmatrix}\begin{bmatrix} 1 \\ -1 \\ -1 \end{bmatrix} + 0.5\right)$$

$$= hardlim\,(0.5) = 1 \quad (4.50)$$

$$e = t_1 - a = 0 - 1 = -1 \quad (4.51)$$

$$\mathbf{W}^{new} = \mathbf{W}^{old} + e\mathbf{p}^T = \begin{bmatrix} 0.5 & 1 & -0.5 \end{bmatrix} + (-1)\begin{bmatrix} 1 & -1 & -1 \end{bmatrix} \quad (4.52)$$
$$= \begin{bmatrix} -0.5 & 2 & 0.5 \end{bmatrix}$$

$$b^{new} = b^{old} + e = 0.5 + (-1) = -0.5 . \tag{4.53}$$

If you continue with the iterations you will find that both input vectors will now be correctly classified. The algorithm has converged to a solution. Note that the final decision boundary is not the same as the one we developed in Chapter 3, although both boundaries correctly classify the two input vectors.

To experiment with the perceptron learning rule, use the Neural Network Design Demonstration Perceptron Rule (**nnd4pr**).

Proof of Convergence

Although the perceptron learning rule is simple, it is quite powerful. In fact, it can be shown that the rule will always converge to weights that accomplish the desired classification (assuming that such weights exist). In this section we will present a proof of convergence for the perceptron learning rule for the single-neuron perceptron shown in Figure 4.5.

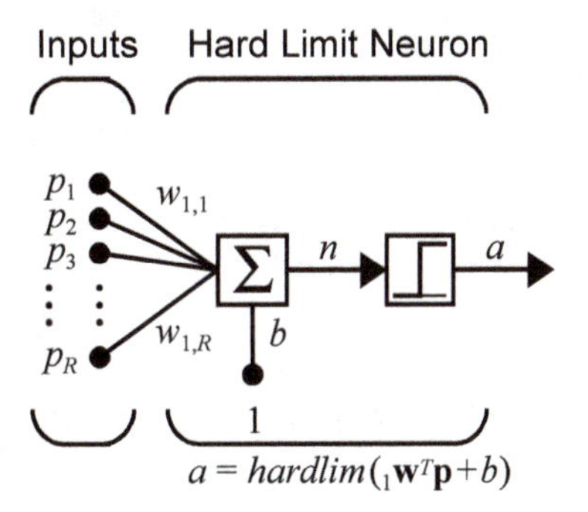

Figure 4.5 Single-Neuron Perceptron

The output of this perceptron is obtained from

$$a = hardlim({}_1\mathbf{w}^T\mathbf{p} + b) . \tag{4.54}$$

The network is provided with the following examples of proper network behavior:

$$\{\mathbf{p}_1, t_1\} , \{\mathbf{p}_2, t_2\} , \ldots, \{\mathbf{p}_Q, t_Q\} . \tag{4.55}$$

where each target output, t_q, is either 0 or 1.

Notation

To conveniently present the proof we will first introduce some new notation. We will combine the weight matrix and the bias into a single vector:

$$\mathbf{x} = \begin{bmatrix} {}_1\mathbf{w} \\ b \end{bmatrix}. \quad (4.56)$$

We will also augment the input vectors with a 1, corresponding to the bias input:

$$\mathbf{z}_q = \begin{bmatrix} \mathbf{p}_q \\ 1 \end{bmatrix}. \quad (4.57)$$

Now we can express the net input to the neuron as follows:

$$n = {}_1\mathbf{w}^T\mathbf{p} + b = \mathbf{x}^T\mathbf{z}. \quad (4.58)$$

The perceptron learning rule for a single-neuron perceptron (Eq. (4.34) and Eq. (4.35)) can now be written

$$\mathbf{x}^{new} = \mathbf{x}^{old} + e\mathbf{z}. \quad (4.59)$$

The error e can be either 1, -1 or 0. If $e = 0$, then no change is made to the weights. If $e = 1$, then the input vector is added to the weight vector. If $e = -1$, then the negative of the input vector is added to the weight vector. If we count only those iterations for which the weight vector is changed, the learning rule becomes

$$\mathbf{x}(k) = \mathbf{x}(k-1) + \mathbf{z}'(k-1), \quad (4.60)$$

where $\mathbf{z}'(k-1)$ is the appropriate member of the set

$$\{\mathbf{z}_1, \mathbf{z}_2, \ldots, \mathbf{z}_Q, -\mathbf{z}_1, -\mathbf{z}_2, \ldots, -\mathbf{z}_Q\}. \quad (4.61)$$

We will assume that a weight vector exists that can correctly categorize all Q input vectors. This solution will be denoted $\mathbf{x}^*$. For this weight vector we will assume that

$$\mathbf{x}^{*T}\mathbf{z}_q > \delta > 0 \text{ if } t_q = 1, \quad (4.62)$$

and

$$\mathbf{x}^{*T}\mathbf{z}_q < -\delta < 0 \text{ if } t_q = 0. \quad (4.63)$$

Proof

We are now ready to begin the proof of the perceptron convergence theorem. The objective of the proof is to find upper and lower bounds on the length of the weight vector at each stage of the algorithm.

Assume that the algorithm is initialized with the zero weight vector: $\mathbf{x}(0) = \mathbf{0}$. (This does not affect the generality of our argument.) Then, after k iterations (changes to the weight vector), we find from Eq. (4.60):

$$\mathbf{x}(k) = \mathbf{z}'(0) + \mathbf{z}'(1) + \cdots + \mathbf{z}'(k-1). \tag{4.64}$$

If we take the inner product of the solution weight vector with the weight vector at iteration k we obtain

$$\mathbf{x}^{*T}\mathbf{x}(k) = \mathbf{x}^{*T}\mathbf{z}'(0) + \mathbf{x}^{*T}\mathbf{z}'(1) + \cdots + \mathbf{x}^{*T}\mathbf{z}'(k-1). \tag{4.65}$$

From Eq. (4.61)–Eq. (4.63) we can show that

$$\mathbf{x}^{*T}\mathbf{z}'(i) > \delta. \tag{4.66}$$

Therefore

$$\mathbf{x}^{*T}\mathbf{x}(k) > k\delta. \tag{4.67}$$

From the Cauchy-Schwartz inequality (see [Brog91])

$$(\mathbf{x}^{*T}\mathbf{x}(k))^2 \le \|\mathbf{x}^*\|^2\|\mathbf{x}(k)\|^2, \tag{4.68}$$

where

$$\|\mathbf{x}\|^2 = \mathbf{x}^T\mathbf{x}. \tag{4.69}$$

If we combine Eq. (4.67) and Eq. (4.68) we can put a lower bound on the squared length of the weight vector at iteration k:

$$\|\mathbf{x}(k)\|^2 \ge \frac{(\mathbf{x}^{*T}\mathbf{x}(k))^2}{\|\mathbf{x}^*\|^2} > \frac{(k\delta)^2}{\|\mathbf{x}^*\|^2}. \tag{4.70}$$

Next we want to find an upper bound for the length of the weight vector. We begin by finding the change in the length at iteration k:

$$\begin{aligned}\|\mathbf{x}(k)\|^2 &= \mathbf{x}^T(k)\mathbf{x}(k)\\ &= [\mathbf{x}(k-1) + \mathbf{z}'(k-1)]^T[\mathbf{x}(k-1) + \mathbf{z}'(k-1)]\\ &= \mathbf{x}^T(k-1)\mathbf{x}(k-1) + 2\mathbf{x}^T(k-1)\mathbf{z}'(k-1)\\ &\quad + \mathbf{z}'^T(k-1)\mathbf{z}'(k-1)\end{aligned} \tag{4.71}$$

Note that

$$\mathbf{x}^T(k-1)\mathbf{z}'(k-1) \le 0, \tag{4.72}$$

since the weights would not be updated unless the previous input vector had been misclassified. Now Eq. (4.71) can be simplified to

$$\|\mathbf{x}(k)\|^2 \le \|\mathbf{x}(k-1)\|^2 + \|\mathbf{z}'(k-1)\|^2 . \quad (4.73)$$

We can repeat this process for $\|\mathbf{x}(k-1)\|^2$, $\|\mathbf{x}(k-2)\|^2$, etc., to obtain

$$\|\mathbf{x}(k)\|^2 \le \|\mathbf{z}'(0)\|^2 + \cdots + \|\mathbf{z}'(k-1)\|^2 . \quad (4.74)$$

If $\Pi = max\{\|\mathbf{z}'(i)\|^2\}$, this upper bound can be simplified to

$$\|\mathbf{x}(k)\|^2 \le k\Pi . \quad (4.75)$$

We now have an upper bound (Eq. (4.75)) and a lower bound (Eq. (4.70)) on the squared length of the weight vector at iteration k. If we combine the two inequalities we find

$$k\Pi \ge \|\mathbf{x}(k)\|^2 > \frac{(k\delta)^2}{\|\mathbf{x}^*\|^2} \text{ or } k < \frac{\Pi\|\mathbf{x}^*\|^2}{\delta^2} . \quad (4.76)$$

Because k has an upper bound, this means that the weights will only be changed a finite number of times. Therefore, the perceptron learning rule will converge in a finite number of iterations.

The maximum number of iterations (changes to the weight vector) is inversely related to the square of δ. This parameter is a measure of how close the solution decision boundary is to the input patterns. This means that if the input classes are difficult to separate (are close to the decision boundary) it will take many iterations for the algorithm to converge.

Note that there are only three key assumptions required for the proof:

1. A solution to the problem exists, so that Eq. (4.66) is satisfied.
2. The weights are only updated when the input vector is misclassified, therefore Eq. (4.72) is satisfied.
3. An upper bound, Π, exists for the length of the input vectors.

Because of the generality of the proof, there are many variations of the perceptron learning rule that can also be shown to converge. (See Exercise E4.13.)

Limitations

The perceptron learning rule is guaranteed to converge to a solution in a finite number of steps, so long as a solution exists. This brings us to an important question. What problems can a perceptron solve? Recall that a single-neuron perceptron is able to divide the input space into two regions. The boundary between the regions is defined by the equation

$$ {}_1\mathbf{w}^T\mathbf{p} + b = 0 . \qquad (4.77)$$

Linear Separability

This is a linear boundary (hyperplane). The perceptron can be used to classify input vectors that can be separated by a linear boundary. We call such vectors *linearly separable*. The logical AND gate example on page 4-7 illustrates a two-dimensional example of a linearly separable problem. The apple/orange recognition problem of Chapter 3 was a three-dimensional example.

Unfortunately, many problems are not linearly separable. The classic example is the XOR gate. The input/target pairs for the XOR gate are

$$\left\{\mathbf{p}_1 = \begin{bmatrix}0\\0\end{bmatrix}, t_1 = 0\right\}\left\{\mathbf{p}_2 = \begin{bmatrix}0\\1\end{bmatrix}, t_2 = 1\right\}\left\{\mathbf{p}_3 = \begin{bmatrix}1\\0\end{bmatrix}, t_3 = 1\right\}\left\{\mathbf{p}_4 = \begin{bmatrix}1\\1\end{bmatrix}, t_4 = 0\right\}.$$

This problem is illustrated graphically on the left side of Figure 4.6, which also shows two other linearly inseparable problems. Try drawing a straight line between the vectors with targets of 1 and those with targets of 0 in any of the diagrams of Figure 4.6.

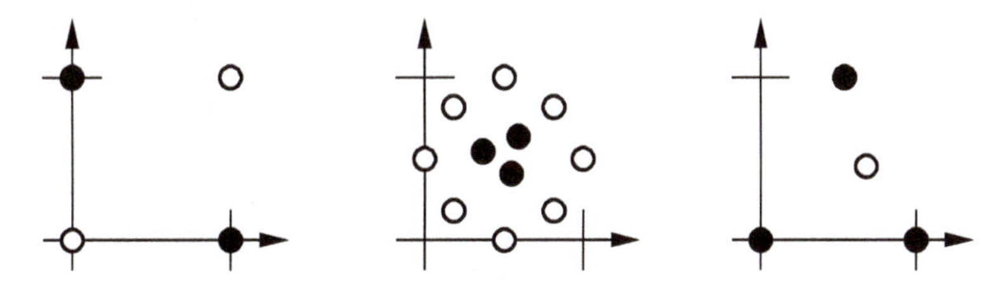

Figure 4.6 Linearly Inseparable Problems

It was the inability of the basic perceptron to solve such simple problems that led, in part, to a reduction in interest in neural network research during the 1970s. Rosenblatt had investigated more complex networks, which he felt would overcome the limitations of the basic perceptron, but he was never able to effectively extend the perceptron rule to such networks. In Chapter 11 we will introduce multilayer perceptrons, which can solve arbitrary classification problems, and will describe the backpropagation algorithm, which can be used to train them.

Summary of Results

Perceptron Architecture

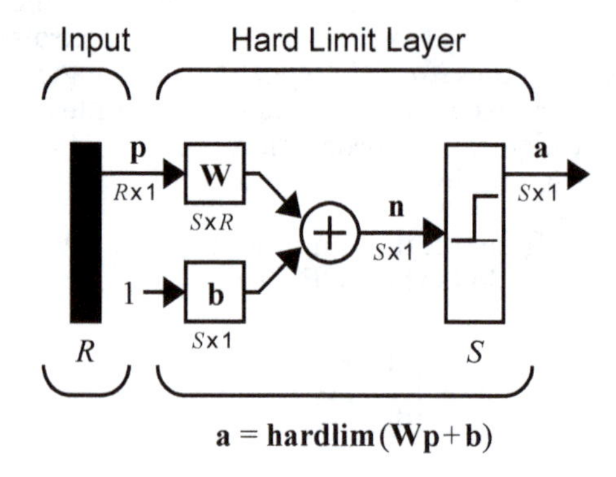

$$\mathbf{W} = \begin{bmatrix} {}_1\mathbf{w}^T \\ {}_2\mathbf{w}^T \\ \vdots \\ {}_S\mathbf{w}^T \end{bmatrix}$$

$$\mathbf{a} = \mathbf{hardlim}(\mathbf{Wp} + \mathbf{b})$$

$$a_i = hardlim(n_i) = hardlim({}_i\mathbf{w}^T\mathbf{p} + b_i)$$

Decision Boundary

$${}_i\mathbf{w}^T\mathbf{p} + b_i = 0 .$$

The decision boundary is always orthogonal to the weight vector.

Single-layer perceptrons can only classify linearly separable vectors.

Perceptron Learning Rule

$$\mathbf{W}^{new} = \mathbf{W}^{old} + \mathbf{e}\mathbf{p}^T$$

$$\mathbf{b}^{new} = \mathbf{b}^{old} + \mathbf{e}$$

where $\mathbf{e} = \mathbf{t} - \mathbf{a}$.

Solved Problems

P4.1 Solve the three simple classification problems shown in Figure P4.1 by drawing a decision boundary. Find weight and bias values that result in single-neuron perceptrons with the chosen decision boundaries.

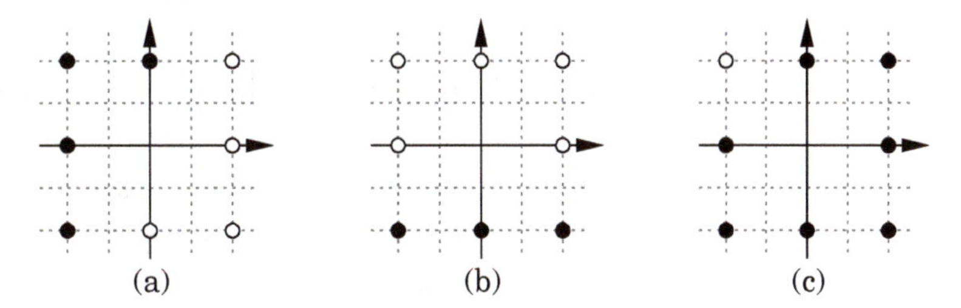

Figure P4.1 Simple Classification Problems

First we draw a line between each set of dark and light data points.

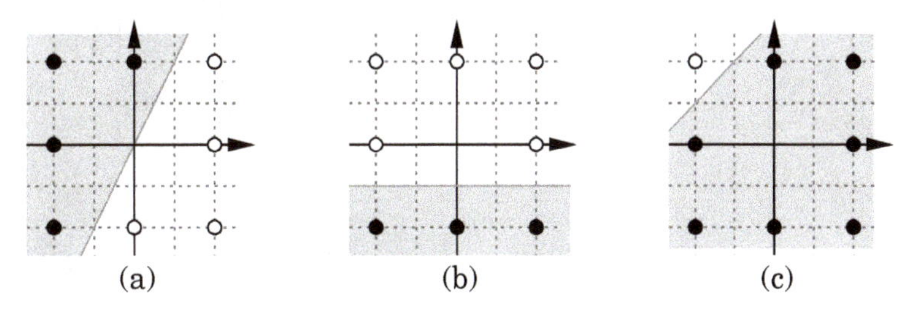

The next step is to find the weights and biases. The weight vectors must be orthogonal to the decision boundaries, and pointing in the direction of points to be classified as 1 (the dark points). The weight vectors can have any length we like.

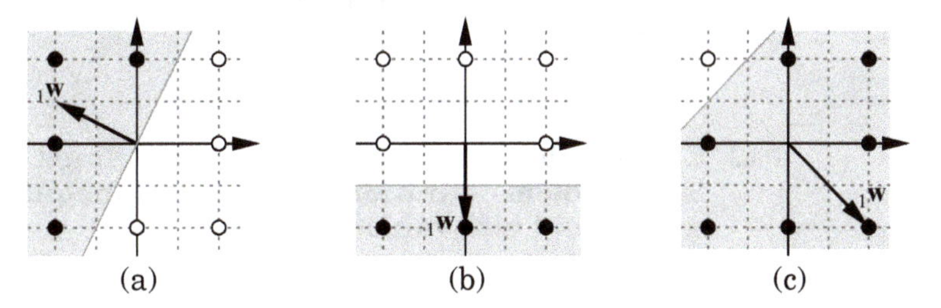

Here is one set of choices for the weight vectors:

$$(a)\ {}_1\mathbf{w}^T = \begin{bmatrix} -2 & 1 \end{bmatrix},\quad (b)\ {}_1\mathbf{w}^T = \begin{bmatrix} 0 & -2 \end{bmatrix},\quad (c)\ {}_1\mathbf{w}^T = \begin{bmatrix} 2 & -2 \end{bmatrix}.$$

Now we find the bias values for each perceptron by picking a point on the decision boundary and satisfying Eq. (4.15).

$$ {}_1\mathbf{w}^T\mathbf{p} + b = 0 $$
$$ b = -{}_1\mathbf{w}^T\mathbf{p} $$

This gives us the following three biases:

$$ \text{(a) } b = -\begin{bmatrix} -2 & 1 \end{bmatrix}\begin{bmatrix} 0 \\ 0 \end{bmatrix} = 0 \text{, (b) } b = -\begin{bmatrix} 0 & -2 \end{bmatrix}\begin{bmatrix} 0 \\ -1 \end{bmatrix} = -2 \text{, (c) } b = -\begin{bmatrix} 2 & -2 \end{bmatrix}\begin{bmatrix} -2 \\ 1 \end{bmatrix} = 6 $$

We can now check our solution against the original points. Here we test the first network on the input vector $\mathbf{p} = \begin{bmatrix} -2 & 2 \end{bmatrix}^T$.

$$ a = hardlim({}_1\mathbf{w}^T\mathbf{p} + b) $$
$$ = hardlim\left(\begin{bmatrix} -2 & 1 \end{bmatrix}\begin{bmatrix} -2 \\ 2 \end{bmatrix} + 0 \right) $$
$$ = hardlim(6) $$
$$ = 1 $$

» 2 + 2
ans =
4

We can use MATLAB to automate the testing process and to try new points. Here the first network is used to classify a point that was not in the original problem.

```
w=[-2 1]; b = 0;
a = hardlim(w*[1;1]+b)
a =
        0
```

P4.2 Convert the classification problem defined below into an equivalent problem definition consisting of inequalities constraining weight and bias values.

$$ \left\{ \mathbf{p}_1 = \begin{bmatrix} 0 \\ 2 \end{bmatrix}, t_1 = 1 \right\} \left\{ \mathbf{p}_2 = \begin{bmatrix} 1 \\ 0 \end{bmatrix}, t_2 = 1 \right\} \left\{ \mathbf{p}_3 = \begin{bmatrix} 0 \\ -2 \end{bmatrix}, t_3 = 0 \right\} \left\{ \mathbf{p}_4 = \begin{bmatrix} 2 \\ 0 \end{bmatrix}, t_4 = 0 \right\} $$

Each target t_i indicates whether or not the net input in response to $\mathbf{p}_i$ must be less than 0, or greater than or equal to 0. For example, since t_1 is 1, we know that the net input corresponding to $\mathbf{p}_1$ must be greater than or equal to 0. Thus we get the following inequality:

$$\begin{aligned} \mathbf{W}\mathbf{p}_1 + b &\geq 0 \\ 0w_{1,1} + 2w_{1,2} + b &\geq 0 \\ 2w_{1,2} + b &\geq 0. \end{aligned}$$

Applying the same procedure to the input/target pairs for $\{\mathbf{p}_2, t_2\}$, $\{\mathbf{p}_3, t_3\}$ and $\{\mathbf{p}_4, t_4\}$ results in the following set of inequalities.

$$\begin{aligned} 2w_{1,2} + b &\geq 0 \quad (i) \\ w_{1,1} + b &\geq 0 \quad (ii) \\ -2w_{1,2} + b &< 0 \quad (iii) \\ 2w_{1,1} + b &< 0 \quad (iv) \end{aligned}$$

Solving a set of inequalities is more difficult than solving a set of equalities. One added complexity is that there are often an infinite number of solutions (just as there are often an infinite number of linear decision boundaries that can solve a linearly separable classification problem).

However, because of the simplicity of this problem, we can solve it by graphing the solution spaces defined by the inequalities. Note that $w_{1,1}$ only appears in inequalities (*ii*) and (*iv*), and $w_{1,2}$ only appears in inequalities (*i*) and (*iii*). We can plot each pair of inequalities with two graphs.

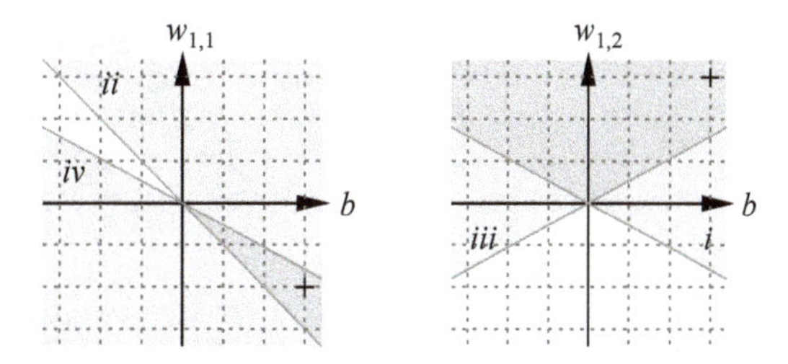

Any weight and bias values that fall in both dark gray regions will solve the classification problem.

Here is one such solution:

$$\mathbf{W} = \begin{bmatrix} -2 & 3 \end{bmatrix} \qquad b = 3.$$

P4.3 We have a classification problem with four classes of input vector. The four classes are

$$\text{class 1:}\left\{\mathbf{p}_1 = \begin{bmatrix}1\\1\end{bmatrix}, \mathbf{p}_2 = \begin{bmatrix}1\\2\end{bmatrix}\right\}, \text{ class 2: }\left\{\mathbf{p}_3 = \begin{bmatrix}2\\-1\end{bmatrix}, \mathbf{p}_4 = \begin{bmatrix}2\\0\end{bmatrix}\right\},$$

$$\text{class 3:}\left\{\mathbf{p}_5 = \begin{bmatrix}-1\\2\end{bmatrix}, \mathbf{p}_6 = \begin{bmatrix}-2\\1\end{bmatrix}\right\}, \text{ class 4: }\left\{\mathbf{p}_7 = \begin{bmatrix}-1\\-1\end{bmatrix}, \mathbf{p}_8 = \begin{bmatrix}-2\\-2\end{bmatrix}\right\}.$$

Design a perceptron network to solve this problem.

To solve a problem with four classes of input vector we will need a perceptron with at least two neurons, since an S-neuron perceptron can categorize 2^S classes. The two-neuron perceptron is shown in Figure P4.2.

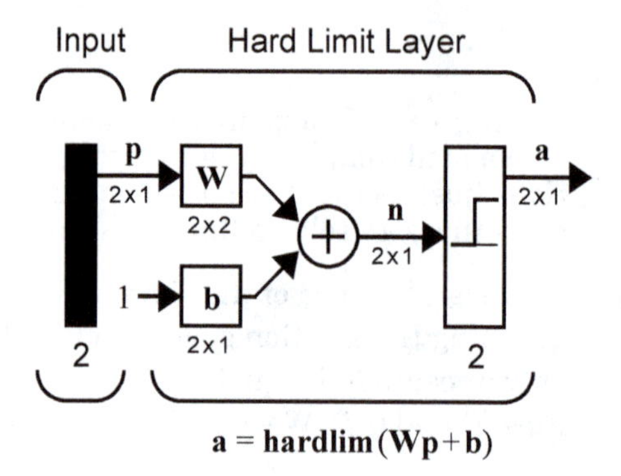

Figure P4.2 Two-Neuron Perceptron

Let's begin by displaying the input vectors, as in Figure P4.3. The light circles ○ indicate class 1 vectors, the light squares □ indicate class 2 vectors, the dark circles ● indicate class 3 vectors, and the dark squares ■ indicate class 4 vectors.

A two-neuron perceptron creates two decision boundaries. Therefore, to divide the input space into the four categories, we need to have one decision boundary divide the four classes into two sets of two. The remaining boundary must then isolate each class. Two such boundaries are illustrated in Figure P4.4. We now know that our patterns are linearly separable.

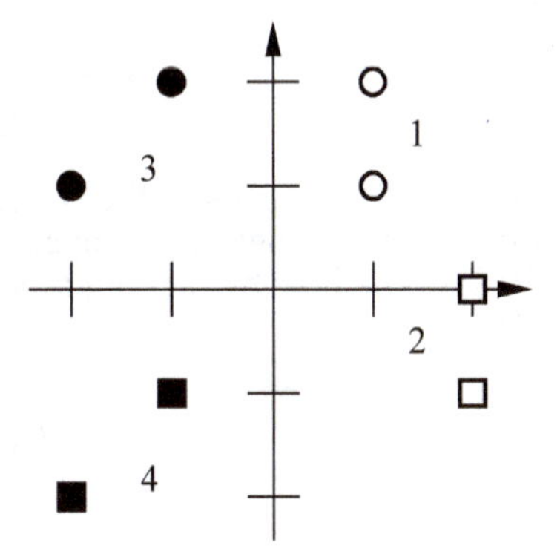

Figure P4.3 Input Vectors for Problem P4.3

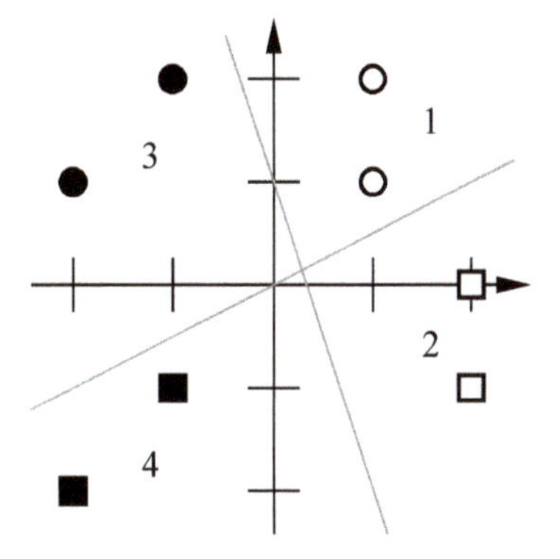

Figure P4.4 Tentative Decision Boundaries for Problem P4.3

The weight vectors should be orthogonal to the decision boundaries and should point toward the regions where the neuron outputs are 1. The next step is to decide which side of each boundary should produce a 1. One choice is illustrated in Figure P4.5, where the shaded areas represent outputs of 1. The darkest shading indicates that both neuron outputs are 1. Note that this solution corresponds to target values of

$$\text{class 1:}\left\{\mathbf{t}_1 = \begin{bmatrix}0\\0\end{bmatrix}, \mathbf{t}_2 = \begin{bmatrix}0\\0\end{bmatrix}\right\}, \text{ class 2: } \left\{\mathbf{t}_3 = \begin{bmatrix}0\\1\end{bmatrix}, \mathbf{t}_4 = \begin{bmatrix}0\\1\end{bmatrix}\right\},$$

$$\text{class 3:}\left\{\mathbf{t}_5 = \begin{bmatrix}1\\0\end{bmatrix}, \mathbf{t}_6 = \begin{bmatrix}1\\0\end{bmatrix}\right\}, \text{ class 4: } \left\{\mathbf{t}_7 = \begin{bmatrix}1\\1\end{bmatrix}, \mathbf{t}_8 = \begin{bmatrix}1\\1\end{bmatrix}\right\}.$$

We can now select the weight vectors:

$$_1\mathbf{w} = \begin{bmatrix} -3 \\ -1 \end{bmatrix} \text{ and } _2\mathbf{w} = \begin{bmatrix} 1 \\ -2 \end{bmatrix}.$$

Note that the lengths of the weight vectors is not important, only their directions. They must be orthogonal to the decision boundaries. Now we can calculate the bias by picking a point on a boundary and satisfying Eq. (4.15):

$$b_1 = -{}_1\mathbf{w}^T\mathbf{p} = -\begin{bmatrix} -3 & -1 \end{bmatrix} \begin{bmatrix} 0 \\ 1 \end{bmatrix} = 1,$$

$$b_2 = -{}_2\mathbf{w}^T\mathbf{p} = -\begin{bmatrix} 1 & -2 \end{bmatrix} \begin{bmatrix} 0 \\ 0 \end{bmatrix} = 0.$$

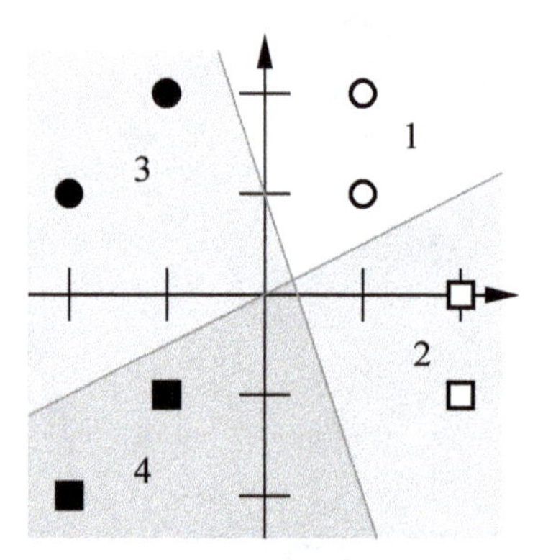

Figure P4.5 Decision Regions for Problem P4.3

In matrix form we have

$$\mathbf{W} = \begin{bmatrix} {}_1\mathbf{w}^T \\ {}_2\mathbf{w}^T \end{bmatrix} = \begin{bmatrix} -3 & -1 \\ 1 & -2 \end{bmatrix} \text{ and } \mathbf{b} = \begin{bmatrix} 1 \\ 0 \end{bmatrix},$$

which completes our design.

P4.4 Solve the following classification problem with the perceptron rule. Apply each input vector in order, for as many repetitions as it takes to ensure that the problem is solved. Draw a graph of the problem only after you have found a solution.

$$\left\{\mathbf{p}_1 = \begin{bmatrix} 2 \\ 2 \end{bmatrix}, t_1 = 0\right\} \left\{\mathbf{p}_2 = \begin{bmatrix} 1 \\ -2 \end{bmatrix}, t_2 = 1\right\} \left\{\mathbf{p}_3 = \begin{bmatrix} -2 \\ 2 \end{bmatrix}, t_3 = 0\right\} \left\{\mathbf{p}_4 = \begin{bmatrix} -1 \\ 1 \end{bmatrix}, t_4 = 1\right\}$$

Use the initial weights and bias:

$$\mathbf{W}(0) = \begin{bmatrix} 0 & 0 \end{bmatrix} \qquad b(0) = 0 .$$

We start by calculating the perceptron's output a for the first input vector $\mathbf{p}_1$, using the initial weights and bias.

$$a = hardlim(\mathbf{W}(0)\mathbf{p}_1 + b(0))$$

$$= hardlim\left(\begin{bmatrix} 0 & 0 \end{bmatrix} \begin{bmatrix} 2 \\ 2 \end{bmatrix} + 0\right) = hardlim(0) = 1$$

The output a does not equal the target value t_1, so we use the perceptron rule to find new weights and biases based on the error.

$$e = t_1 - a = 0 - 1 = -1$$

$$\mathbf{W}(1) = \mathbf{W}(0) + e\mathbf{p}_1^T = \begin{bmatrix} 0 & 0 \end{bmatrix} + (-1)\begin{bmatrix} 2 & 2 \end{bmatrix} = \begin{bmatrix} -2 & -2 \end{bmatrix}$$

$$b(1) = b(0) + e = 0 + (-1) = -1$$

We now apply the second input vector $\mathbf{p}_2$, using the updated weights and bias.

$$a = hardlim(\mathbf{W}(1)\mathbf{p}_2 + b(1))$$

$$= hardlim\left(\begin{bmatrix} -2 & -2 \end{bmatrix} \begin{bmatrix} 1 \\ -2 \end{bmatrix} - 1\right) = hardlim(1) = 1$$

This time the output a is equal to the target t_2. Application of the perceptron rule will not result in any changes.

$$\mathbf{W}(2) = \mathbf{W}(1)$$

$$b(2) = b(1)$$

We now apply the third input vector.

$$a = hardlim(\mathbf{W}(2)\mathbf{p}_3 + b(2))$$

$$= hardlim\left(\begin{bmatrix}-2 & -2\end{bmatrix}\begin{bmatrix}-2\\2\end{bmatrix} - 1\right) = hardlim(-1) = 0$$

The output in response to input vector $\mathbf{p}_3$ is equal to the target t_3, so there will be no changes.

$$\mathbf{W}(3) = \mathbf{W}(2)$$

$$b(3) = b(2)$$

We now move on to the last input vector $\mathbf{p}_4$.

$$a = hardlim(\mathbf{W}(3)\mathbf{p}_4 + b(3))$$

$$= hardlim\left(\begin{bmatrix}-2 & -2\end{bmatrix}\begin{bmatrix}-1\\1\end{bmatrix} - 1\right) = hardlim(-1) = 0$$

This time the output a does not equal the appropriate target t_4. The perceptron rule will result in a new set of values for $\mathbf{W}$ and b.

$$e = t_4 - a = 1 - 0 = 1$$

$$\mathbf{W}(4) = \mathbf{W}(3) + e\mathbf{p}_4^T = \begin{bmatrix}-2 & -2\end{bmatrix} + (1)\begin{bmatrix}-1 & 1\end{bmatrix} = \begin{bmatrix}-3 & -1\end{bmatrix}$$

$$b(4) = b(3) + e = -1 + 1 = 0$$

We now must check the first vector $\mathbf{p}_1$ again. This time the output a is equal to the associated target t_1.

$$a = hardlim(\mathbf{W}(4)\mathbf{p}_1 + b(4))$$

$$= hardlim\left(\begin{bmatrix}-3 & -1\end{bmatrix}\begin{bmatrix}2\\2\end{bmatrix} + 0\right) = hardlim(-8) = 0$$

Therefore there are no changes.

$$\mathbf{W}(5) = \mathbf{W}(4)$$

$$b(5) = b(4)$$

The second presentation of $\mathbf{p}_2$ results in an error and therefore a new set of weight and bias values.

$$a = hardlim(\mathbf{W}(5)\mathbf{p}_2 + b(5))$$

$$= hardlim\left(\begin{bmatrix}-3 & -1\end{bmatrix}\begin{bmatrix}1 \\ -2\end{bmatrix} + 0\right) = hardlim(-1) = 0$$

Here are those new values:

$$e = t_2 - a = 1 - 0 = 1$$

$$\mathbf{W}(6) = \mathbf{W}(5) + e\mathbf{p}_2^T = \begin{bmatrix}-3 & -1\end{bmatrix} + (1)\begin{bmatrix}1 & -2\end{bmatrix} = \begin{bmatrix}-2 & -3\end{bmatrix}$$

$$b(6) = b(5) + e = 0 + 1 = 1.$$

Cycling through each input vector once more results in no errors.

$$a = hardlim(\mathbf{W}(6)\mathbf{p}_3 + b(6)) = hardlim\left(\begin{bmatrix}-2 & -3\end{bmatrix}\begin{bmatrix}-2 \\ 2\end{bmatrix} + 1\right) = 0 = t_3$$

$$a = hardlim(\mathbf{W}(6)\mathbf{p}_4 + b(6)) = hardlim\left(\begin{bmatrix}-2 & -3\end{bmatrix}\begin{bmatrix}-1 \\ 1\end{bmatrix} + 1\right) = 1 = t_4$$

$$a = hardlim(\mathbf{W}(6)\mathbf{p}_1 + b(6)) = hardlim\left(\begin{bmatrix}-2 & -3\end{bmatrix}\begin{bmatrix}2 \\ 2\end{bmatrix} + 1\right) = 0 = t_1$$

$$a = hardlim(\mathbf{W}(6)\mathbf{p}_2 + b(6)) = hardlim\left(\begin{bmatrix}-2 & -3\end{bmatrix}\begin{bmatrix}1 \\ -2\end{bmatrix} + 1\right) = 1 = t_2$$

Therefore the algorithm has converged. The final solution is:

$$\mathbf{W} = \begin{bmatrix}-2 & -3\end{bmatrix} \qquad b = 1.$$

Now we can graph the training data and the decision boundary of the solution. The decision boundary is given by

$$n = \mathbf{W}\mathbf{p} + b = w_{1,1}p_1 + w_{1,2}p_2 + b = -2p_1 - 3p_2 + 1 = 0.$$

To find the p_2 intercept of the decision boundary, set $p_1 = 0$:

$$p_2 = -\frac{b}{w_{1,2}} = -\frac{1}{-3} = \frac{1}{3} \qquad \text{if } p_1 = 0$$

To find the p_1 intercept, set $p_2 = 0$:

$$p_1 = -\frac{b}{w_{1,1}} = -\frac{1}{-2} = \frac{1}{2} \qquad \text{if } p_2 = 0$$

The resulting decision boundary is illustrated in Figure P4.6.

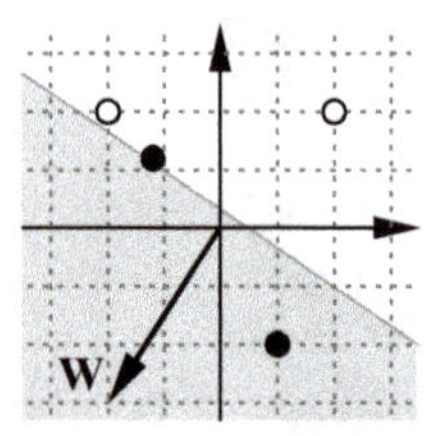

Figure P4.6 Decision Boundary for Problem P4.4

Note that the decision boundary falls across one of the training vectors. This is acceptable, given the problem definition, since the hard limit function returns 1 when given an input of 0, and the target for the vector in question is indeed 1.

P4.5 Consider again the four-class decision problem that we introduced in Problem P4.3. Train a perceptron network to solve this problem using the perceptron learning rule.

If we use the same target vectors that we introduced in Problem P4.3, the training set will be:

$$\left\{\mathbf{p}_1 = \begin{bmatrix}1\\1\end{bmatrix}, \mathbf{t}_1 = \begin{bmatrix}0\\0\end{bmatrix}\right\} \left\{\mathbf{p}_2 = \begin{bmatrix}1\\2\end{bmatrix}, \mathbf{t}_2 = \begin{bmatrix}0\\0\end{bmatrix}\right\} \left\{\mathbf{p}_3 = \begin{bmatrix}2\\-1\end{bmatrix}, \mathbf{t}_3 = \begin{bmatrix}0\\1\end{bmatrix}\right\}$$

$$\left\{\mathbf{p}_4 = \begin{bmatrix}2\\0\end{bmatrix}, \mathbf{t}_4 = \begin{bmatrix}0\\1\end{bmatrix}\right\} \left\{\mathbf{p}_5 = \begin{bmatrix}-1\\2\end{bmatrix}, \mathbf{t}_5 = \begin{bmatrix}1\\0\end{bmatrix}\right\} \left\{\mathbf{p}_6 = \begin{bmatrix}-2\\1\end{bmatrix}, \mathbf{t}_6 = \begin{bmatrix}1\\0\end{bmatrix}\right\}$$

$$\left\{\mathbf{p}_7 = \begin{bmatrix}-1\\-1\end{bmatrix}, \mathbf{t}_7 = \begin{bmatrix}1\\1\end{bmatrix}\right\} \left\{\mathbf{p}_8 = \begin{bmatrix}-2\\-2\end{bmatrix}, \mathbf{t}_8 = \begin{bmatrix}1\\1\end{bmatrix}\right\}.$$

Let's begin the algorithm with the following initial weights and biases:

$$\mathbf{W}(0) = \begin{bmatrix}1 & 0\\0 & 1\end{bmatrix}, \mathbf{b}(0) = \begin{bmatrix}1\\1\end{bmatrix}.$$

The first iteration is

$$\mathbf{a} = hardlim\,(\mathbf{W}(0)\mathbf{p}_1 + \mathbf{b}(0)) = hardlim\,(\begin{bmatrix} 1 & 0 \\ 0 & 1 \end{bmatrix}\begin{bmatrix} 1 \\ 1 \end{bmatrix} + \begin{bmatrix} 1 \\ 1 \end{bmatrix}) = \begin{bmatrix} 1 \\ 1 \end{bmatrix},$$

$$\mathbf{e} = \mathbf{t}_1 - \mathbf{a} = \begin{bmatrix} 0 \\ 0 \end{bmatrix} - \begin{bmatrix} 1 \\ 1 \end{bmatrix} = \begin{bmatrix} -1 \\ -1 \end{bmatrix},$$

$$\mathbf{W}(1) = \mathbf{W}(0) + \mathbf{e}\mathbf{p}_1^T = \begin{bmatrix} 1 & 0 \\ 0 & 1 \end{bmatrix} + \begin{bmatrix} -1 \\ -1 \end{bmatrix}\begin{bmatrix} 1 & 1 \end{bmatrix} = \begin{bmatrix} 0 & -1 \\ -1 & 0 \end{bmatrix},$$

$$\mathbf{b}(1) = \mathbf{b}(0) + \mathbf{e} = \begin{bmatrix} 1 \\ 1 \end{bmatrix} + \begin{bmatrix} -1 \\ -1 \end{bmatrix} = \begin{bmatrix} 0 \\ 0 \end{bmatrix}.$$

The second iteration is

$$\mathbf{a} = hardlim\,(\mathbf{W}(1)\mathbf{p}_2 + \mathbf{b}(1)) = hardlim\,(\begin{bmatrix} 0 & -1 \\ -1 & 0 \end{bmatrix}\begin{bmatrix} 1 \\ 2 \end{bmatrix} + \begin{bmatrix} 0 \\ 0 \end{bmatrix}) = \begin{bmatrix} 0 \\ 0 \end{bmatrix},$$

$$\mathbf{e} = \mathbf{t}_2 - \mathbf{a} = \begin{bmatrix} 0 \\ 0 \end{bmatrix} - \begin{bmatrix} 0 \\ 0 \end{bmatrix} = \begin{bmatrix} 0 \\ 0 \end{bmatrix},$$

$$\mathbf{W}(2) = \mathbf{W}(1) + \mathbf{e}\mathbf{p}_2^T = \begin{bmatrix} 0 & -1 \\ -1 & 0 \end{bmatrix} + \begin{bmatrix} 0 \\ 0 \end{bmatrix}\begin{bmatrix} 1 & 2 \end{bmatrix} = \begin{bmatrix} 0 & -1 \\ -1 & 0 \end{bmatrix},$$

$$\mathbf{b}(2) = \mathbf{b}(1) + \mathbf{e} = \begin{bmatrix} 0 \\ 0 \end{bmatrix} + \begin{bmatrix} 0 \\ 0 \end{bmatrix} = \begin{bmatrix} 0 \\ 0 \end{bmatrix}.$$

The third iteration is

$$\mathbf{a} = hardlim\,(\mathbf{W}(2)\mathbf{p}_3 + \mathbf{b}(2)) = hardlim\,(\begin{bmatrix} 0 & -1 \\ -1 & 0 \end{bmatrix}\begin{bmatrix} 2 \\ -1 \end{bmatrix} + \begin{bmatrix} 0 \\ 0 \end{bmatrix}) = \begin{bmatrix} 1 \\ 0 \end{bmatrix},$$

$$\mathbf{e} = \mathbf{t}_3 - \mathbf{a} = \begin{bmatrix} 0 \\ 1 \end{bmatrix} - \begin{bmatrix} 1 \\ 0 \end{bmatrix} = \begin{bmatrix} -1 \\ 1 \end{bmatrix},$$

$$\mathbf{W}(3) = \mathbf{W}(2) + \mathbf{e}\mathbf{p}_3^T = \begin{bmatrix} 0 & -1 \\ -1 & 0 \end{bmatrix} + \begin{bmatrix} -1 \\ 1 \end{bmatrix} \begin{bmatrix} 2 & -1 \end{bmatrix} = \begin{bmatrix} -2 & 0 \\ 1 & -1 \end{bmatrix},$$

$$\mathbf{b}(3) = \mathbf{b}(2) + \mathbf{e} = \begin{bmatrix} 0 \\ 0 \end{bmatrix} + \begin{bmatrix} -1 \\ 1 \end{bmatrix} = \begin{bmatrix} -1 \\ 1 \end{bmatrix}.$$

Iterations four through eight produce no changes in the weights.

$$\mathbf{W}(8) = \mathbf{W}(7) = \mathbf{W}(6) = \mathbf{W}(5) = \mathbf{W}(4) = \mathbf{W}(3)$$

$$\mathbf{b}(8) = \mathbf{b}(7) = \mathbf{b}(6) = \mathbf{b}(5) = \mathbf{b}(4) = \mathbf{b}(3)$$

The ninth iteration produces

$$\mathbf{a} = hardlim\,(\mathbf{W}(8)\mathbf{p}_1 + \mathbf{b}(8)) = hardlim\,(\begin{bmatrix} -2 & 0 \\ 1 & -1 \end{bmatrix} \begin{bmatrix} 1 \\ 1 \end{bmatrix} + \begin{bmatrix} -1 \\ 1 \end{bmatrix}) = \begin{bmatrix} 0 \\ 1 \end{bmatrix},$$

$$\mathbf{e} = \mathbf{t}_1 - \mathbf{a} = \begin{bmatrix} 0 \\ 0 \end{bmatrix} - \begin{bmatrix} 0 \\ 1 \end{bmatrix} = \begin{bmatrix} 0 \\ -1 \end{bmatrix},$$

$$\mathbf{W}(9) = \mathbf{W}(8) + \mathbf{e}\mathbf{p}_1^T = \begin{bmatrix} -2 & 0 \\ 1 & -1 \end{bmatrix} + \begin{bmatrix} 0 \\ -1 \end{bmatrix} \begin{bmatrix} 1 & 1 \end{bmatrix} = \begin{bmatrix} -2 & 0 \\ 0 & -2 \end{bmatrix},$$

$$\mathbf{b}(9) = \mathbf{b}(8) + \mathbf{e} = \begin{bmatrix} -1 \\ 1 \end{bmatrix} + \begin{bmatrix} 0 \\ -1 \end{bmatrix} = \begin{bmatrix} -1 \\ 0 \end{bmatrix}.$$

At this point the algorithm has converged, since all input patterns will be correctly classified. The final decision boundaries are displayed in Figure P4.7. Compare this result with the network we designed in Problem P4.3.

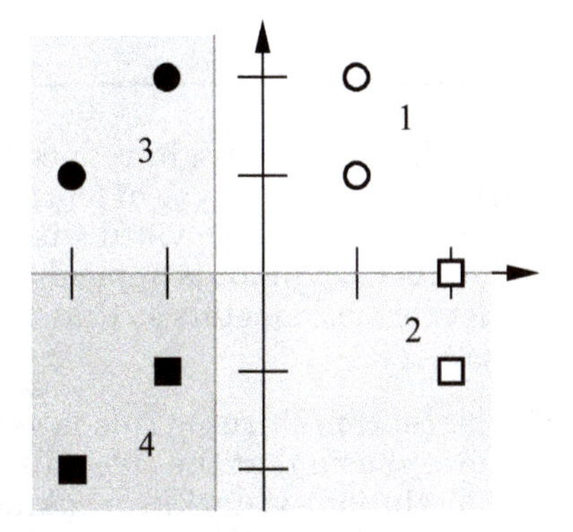

Figure P4.7 Final Decision Boundaries for Problem P4.5

Epilogue

In this chapter we have introduced our first learning rule — the perceptron learning rule. It is a type of learning called *supervised learning*, in which the learning rule is provided with a set of examples of proper network behavior. As each input is applied to the network, the learning rule adjusts the network parameters so that the network output will move closer to the target.

The perceptron learning rule is very simple, but it is also quite powerful. We have shown that the rule will always converge to a correct solution, if such a solution exists. The weakness of the perceptron network lies not with the learning rule, but with the structure of the network. The standard perceptron is only able to classify vectors that are linearly separable. We will see in Chapter 11 that the perceptron architecture can be generalized to multilayer perceptrons, which can solve arbitrary classification problems. The backpropagation learning rule, which is introduced in Chapter 11, can be used to train these networks.

In Chapter 3 and 4 we have used many concepts from the field of linear algebra, such as inner product, projection, distance (norm), etc. We will find in later chapters that a good foundation in linear algebra is essential to our understanding of all neural networks. In Chapter 5 and 6 we will review some of the key concepts from linear algebra that will be most important in our study of neural networks. Our objective will be to obtain a fundamental understanding of how neural networks work.

Further Reading

[BaSu83] A. Barto, R. Sutton and C. Anderson, "Neuron-like adaptive elements can solve difficult learning control problems," *IEEE Transactions on Systems, Man and Cybernetics*, Vol. 13, No. 5, pp. 834–846, 1983.

A classic paper in which a reinforcement learning algorithm is used to train a neural network to balance an inverted pendulum.

[Brog91] W. L. Brogan, *Modern Control Theory*, 3rd Ed., Englewood Cliffs, NJ: Prentice-Hall, 1991.

A well-written book on the subject of linear systems. The first half of the book is devoted to linear algebra. It also has good sections on the solution of linear differential equations and the stability of linear and nonlinear systems. It has many worked problems.

[McPi43] W. McCulloch and W. Pitts, "A logical calculus of the ideas immanent in nervous activity," *Bulletin of Mathematical Biophysics*, Vol. 5, pp. 115–133, 1943.

This article introduces the first mathematical model of a neuron, in which a weighted sum of input signals is compared to a threshold to determine whether or not the neuron fires.

[MiPa69] M. Minsky and S. Papert, *Perceptrons*, Cambridge, MA: MIT Press, 1969.

A landmark book that contains the first rigorous study devoted to determining what a perceptron network is capable of learning. A formal treatment of the perceptron was needed both to explain the perceptron's limitations and to indicate directions for overcoming them. Unfortunately, the book pessimistically predicted that the limitations of perceptrons indicated that the field of neural networks was a dead end. Although this was not true, it temporarily cooled research and funding for research for several years.

[Rose58] F. Rosenblatt, "The perceptron: A probabilistic model for information storage and organization in the brain," *Psychological Review*, Vol. 65, pp. 386–408, 1958.

This paper presents the first practical artificial neural network — the perceptron.

[Rose61] F. Rosenblatt, *Principles of Neurodynamics*, Washington DC: Spartan Press, 1961.

One of the first books on neurocomputing.

[WhSo92] D. White and D. Sofge (Eds.), *Handbook of Intelligent Control*, New York: Van Nostrand Reinhold, 1992.

Collection of articles describing current research and applications of neural networks and fuzzy logic to control systems.

Exercises

E4.1 Consider the classification problem defined below:

$$\left\{\mathbf{p}_1 = \begin{bmatrix} -1 \\ 1 \end{bmatrix}, t_1 = 1\right\} \left\{\mathbf{p}_2 = \begin{bmatrix} 0 \\ 0 \end{bmatrix}, t_2 = 1\right\} \left\{\mathbf{p}_3 = \begin{bmatrix} 1 \\ -1 \end{bmatrix}, t_3 = 1\right\} \left\{\mathbf{p}_4 = \begin{bmatrix} 1 \\ 0 \end{bmatrix}, t_4 = 0\right\}$$

$$\left\{\mathbf{p}_5 = \begin{bmatrix} 0 \\ 1 \end{bmatrix}, t_5 = 0\right\}.$$

i. Draw a diagram of the single-neuron perceptron you would use to solve this problem. How many inputs are required?

ii. Draw a graph of the data points, labeled according to their targets. Is this problem solvable with the network you defined in part (i)? Why or why not?

E4.2 Consider the classification problem defined below.

$$\left\{\mathbf{p}_1 = \begin{bmatrix} -1 \\ 1 \end{bmatrix}, t_1 = 1\right\} \left\{\mathbf{p}_2 = \begin{bmatrix} -1 \\ -1 \end{bmatrix}, t_2 = 1\right\} \left\{\mathbf{p}_3 = \begin{bmatrix} 0 \\ 0 \end{bmatrix}, t_3 = 0\right\} \left\{\mathbf{p}_4 = \begin{bmatrix} 1 \\ 0 \end{bmatrix}, t_4 = 0\right\}.$$

i. Design a single-neuron perceptron to solve this problem. Design the network graphically, by choosing weight vectors that are orthogonal to the decision boundaries.

ii. Test your solution with all four input vectors.

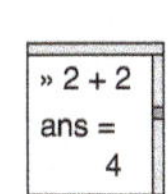

iii. Classify the following input vectors with your solution. You can either perform the calculations manually or with MATLAB.

$$\mathbf{p}_5 = \begin{bmatrix} -2 \\ 0 \end{bmatrix} \quad \mathbf{p}_6 = \begin{bmatrix} 1 \\ 1 \end{bmatrix} \quad \mathbf{p}_7 = \begin{bmatrix} 0 \\ 1 \end{bmatrix} \quad \mathbf{p}_8 = \begin{bmatrix} -1 \\ -2 \end{bmatrix}$$

iv. Which of the vectors in part (iii) will always be classified the same way, regardless of the solution values for $\mathbf{W}$ and b? Which may vary depending on the solution? Why?

E4.3 Solve the classification problem in Exercise E4.2 by solving inequalities (as in Problem P4.2), and repeat parts (ii) and (iii) with the new solution. (The solution is more difficult than Problem P4.2, since you can't isolate the weights and biases in a pairwise manner.)

E4.4 Solve the classification problem in Exercise E4.2 by applying the perceptron rule to the following initial parameters, and repeat parts (ii) and (iii) with the new solution.

$$\mathbf{W}(0) = \begin{bmatrix} 0 & 0 \end{bmatrix} \qquad b(0) = 0$$

E4.5 Prove mathematically (not graphically) that the following problem is unsolvable for a two-input/single-neuron perceptron.

$$\left\{\mathbf{p}_1 = \begin{bmatrix} -1 \\ 1 \end{bmatrix}, t_1 = 1\right\} \left\{\mathbf{p}_2 = \begin{bmatrix} -1 \\ -1 \end{bmatrix}, t_2 = 0\right\} \left\{\mathbf{p}_3 = \begin{bmatrix} 1 \\ -1 \end{bmatrix}, t_3 = 1\right\} \left\{\mathbf{p}_4 = \begin{bmatrix} 1 \\ 1 \end{bmatrix}, t_4 = 0\right\}$$

(Hint: start by rewriting the input/target requirements as inequalities that constrain the weight and bias values.)

E4.6 We have four categories of vectors.

$$\text{Category I: } \left\{ \begin{bmatrix} -1 \\ 1 \end{bmatrix}, \begin{bmatrix} -1 \\ 0 \end{bmatrix} \right\}, \text{ Category II: } \left\{ \begin{bmatrix} 0 \\ 2 \end{bmatrix}, \begin{bmatrix} 1 \\ 2 \end{bmatrix} \right\}$$

$$\text{Category III: } \left\{ \begin{bmatrix} 2 \\ 0 \end{bmatrix}, \begin{bmatrix} 2 \\ 1 \end{bmatrix} \right\}, \text{ Category IV: } \left\{ \begin{bmatrix} 1 \\ -1 \end{bmatrix}, \begin{bmatrix} 0 \\ -1 \end{bmatrix} \right\}$$

i. Design a two-neuron perceptron network (single layer) to recognize these four categories of vectors. Sketch the decision boundaries.

ii. Draw the network diagram.

iii. Suppose the following vector is to be added to Category I.

$$\begin{bmatrix} -1 \\ -3 \end{bmatrix}$$

Perform one iteration of the perceptron learning rule with this vector. (Start with the weights you determined in part i.) Draw the new decision boundaries.

E4.7 We have two categories of vectors. Category I consists of

$$\left\{ \begin{bmatrix} 0 \\ 0 \end{bmatrix}, \begin{bmatrix} -1 \\ 0 \end{bmatrix}, \begin{bmatrix} 0 \\ 1 \end{bmatrix} \right\}.$$

Category II consists of

$$\left\{\begin{bmatrix}-1\\1\end{bmatrix}, \begin{bmatrix}0\\2\end{bmatrix}, \begin{bmatrix}-2\\0\end{bmatrix}\right\}.$$

i. Design a single-neuron perceptron network to recognize these two categories of vectors.

ii. Draw the network diagram.

iii. Sketch the decision boundary.

iv. If we add the following vector to Category I, will your network classify it correctly? Demonstrate by computing the network response.

$$\begin{bmatrix}-3\\0\end{bmatrix}$$

v. Can your weight matrix and bias be modified so your network can classify this new vector correctly (while continuing to classify the other vectors correctly)? Explain.

E4.8 We want to train a perceptron network with the following training set:

$$\left\{\mathbf{p}_1 = \begin{bmatrix}-1\\-1\end{bmatrix}, t_1 = 0\right\}\left\{\mathbf{p}_2 = \begin{bmatrix}0\\0\end{bmatrix}, t_2 = 0\right\}\left\{\mathbf{p}_3 = \begin{bmatrix}-1\\1\end{bmatrix}, t_3 = 1\right\}.$$

The initial weight matrix and bias are

$$\mathbf{W}(0) = \begin{bmatrix}1 & 0\end{bmatrix}, \; b(0) = 0.5\,.$$

i. Plot the initial decision boundary, weight vector and input patterns. Which patterns are correctly classified using the initial weight and bias?

ii. Train the network with the perceptron rule. Present each input vector once, in the order shown.

iii. Plot the final decision boundary, and demonstrate graphically which patterns are correctly classified.

iv. Will the perceptron rule (given enough iterations) always learn to correctly classify the patterns in this training set, no matter what initial weights we use? Explain.

E4.9 We want to train a perceptron network using the following training set:

$$\left\{\mathbf{p}_1 = \begin{bmatrix}1\\0\end{bmatrix}, t_1 = 0\right\}\left\{\mathbf{p}_2 = \begin{bmatrix}-1\\2\end{bmatrix}, t_2 = 0\right\}\left\{\mathbf{p}_3 = \begin{bmatrix}1\\2\end{bmatrix}, t_3 = 1\right\},$$

starting from the initial conditions

$$\mathbf{W}(0) = \begin{bmatrix}0 & 1\end{bmatrix}, \; b(0) = \begin{bmatrix}1\end{bmatrix}.$$

i. Sketch the initial decision boundary, and show the weight vector and the three training input vectors, $\mathbf{p}_1, \mathbf{p}_2, \mathbf{p}_3$. Indicate the class of each input vector, and show which ones are correctly classified by the initial decision boundary.

ii. Present the input $\mathbf{p}_1$ to the network, and perform one iteration of the perceptron learning rule.

iii. Sketch the new decision boundary and weight vector, and again indicate which of the three input vectors are correctly classified.

iv. Present the input $\mathbf{p}_2$ to the network, and perform one more iteration of the perceptron learning rule.

v. Sketch the new decision boundary and weight vector, and again indicate which of the three input vectors are correctly classified.

vi. If you continued to use the perceptron learning rule, and presented all of the patterns many times, would the network eventually learn to correctly classify the patterns? Explain your answer. (This part does not require any calculations.)

E4.10 The symmetric hard limit function is sometimes used in perceptron networks, instead of the hard limit function. Target values are then taken from the set [-1, 1] instead of [0, 1].

$a = hardlims\,(n)$

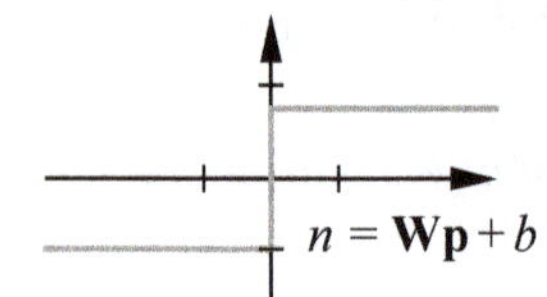

i. Write a simple expression that maps numbers in the ordered set [0, 1] into the ordered set [-1, 1]. Write the expression that performs the inverse mapping.

ii. Consider two single-neuron perceptrons with the same weight and bias values. The first network uses the hard limit function ([0, 1] values), and the second network uses the symmetric hard limit function. If the two networks are given the same input $\mathbf{p}$, and updated with the perceptron learning rule, will their weights continue to have the same value?

iii. If the changes to the weights of the two neurons are different, how do they differ? Why?

iv. Given initial weight and bias values for a standard hard limit perceptron, create a method for initializing a symmetric hard limit perceptron so that the two neurons will always respond identically when trained on identical data.

E4.11 The vectors in the ordered set defined below were obtained by measuring the weight and ear lengths of toy rabbits and bears in the Fuzzy Wuzzy Animal Factory. The target values indicate whether the respective input vector was taken from a rabbit (0) or a bear (1). The first element of the input vector is the weight of the toy, and the second element is the ear length.

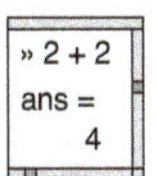

$$\left\{\mathbf{p}_1 = \begin{bmatrix}1\\4\end{bmatrix}, t_1 = 0\right\}\left\{\mathbf{p}_2 = \begin{bmatrix}1\\5\end{bmatrix}, t_2 = 0\right\}\left\{\mathbf{p}_3 = \begin{bmatrix}2\\4\end{bmatrix}, t_3 = 0\right\}\left\{\mathbf{p}_4 = \begin{bmatrix}2\\5\end{bmatrix}, t_4 = 0\right\}$$

$$\left\{\mathbf{p}_5 = \begin{bmatrix}3\\1\end{bmatrix}, t_5 = 1\right\}\left\{\mathbf{p}_6 = \begin{bmatrix}3\\2\end{bmatrix}, t_6 = 1\right\}\left\{\mathbf{p}_7 = \begin{bmatrix}4\\1\end{bmatrix}, t_7 = 1\right\}\left\{\mathbf{p}_8 = \begin{bmatrix}4\\2\end{bmatrix}, t_8 = 1\right\}$$

i. Use MATLAB to initialize and train a network to solve this "practical" problem.

ii. Use MATLAB to test the resulting weight and bias values against the input vectors.

iii. Add input vectors to the training set to ensure that the decision boundary of any solution will not intersect one of the original input vectors (i.e., to ensure only robust solutions are found). Then retrain the network. Your method for adding the input vectors should be general purpose (not designed specifically for this problem).

E4.12 Consider again the four-category classification problem described in Problems P4.3 and P4.5. Suppose that we change the input vector $\mathbf{p}_3$ to

$$\mathbf{p}_3 = \begin{bmatrix}2\\2\end{bmatrix}.$$

i. Is the problem still linearly separable? Demonstrate your answer graphically.

ii. Use MATLAB to initialize and train a network to solve this problem. Explain your results.

iii. If $\mathbf{p}_3$ is changed to

$$\mathbf{p}_3 = \begin{bmatrix}2\\1.5\end{bmatrix}$$

is the problem linearly separable?

iv. With the $\mathbf{p}_3$ from (iii), use MATLAB to initialize and train a network to solve this problem. Explain your results.

E4.13 One variation of the perceptron learning rule is

$$\mathbf{W}^{new} = \mathbf{W}^{old} + \alpha \mathbf{e} \mathbf{p}^T$$

$$\mathbf{b}^{new} = \mathbf{b}^{old} + \alpha \mathbf{e}$$

where α is called the learning rate. Prove convergence of this algorithm. Does the proof require a limit on the learning rate? Explain.

5 Signal and Weight Vector Spaces

Objectives

It is clear from Chapter 3 and 4 that it is very useful to think of the inputs and outputs of a neural network, and the rows of a weight matrix, as vectors. In this chapter we want to examine these vector spaces in detail and to review those properties of vector spaces that are most helpful when analyzing neural networks. We will begin with general definitions and then apply these definitions to specific neural network problems. The concepts that are discussed in this chapter and in Chapter 6 will be used extensively throughout the remaining chapters of this book. They are critical to our understanding of why neural networks work.

Theory and Examples

Linear algebra is the core of the mathematics required for understanding neural networks. In Chapter 3 and 4 we saw the utility of representing the inputs and outputs of neural networks as vectors. In addition, we saw that it is often useful to think of the rows of a weight matrix as vectors in the same vector space as the input vectors.

Recall from Chapter 3 that in the Hamming network the rows of the weight matrix of the feedforward layer were equal to the prototype vectors. In fact, the purpose of the feedforward layer was to calculate the inner products between the prototype vectors and the input vector.

In the single neuron perceptron network we noted that the decision boundary was always orthogonal to the weight matrix (a row vector).

In this chapter we want to review the basic concepts of vector spaces (e.g., inner products, orthogonality) in the context of neural networks. We will begin with a general definition of vector spaces. Then we will present the basic properties of vectors that are most useful for neural network applications.

One comment about notation before we begin. All of the vectors we have discussed so far have been ordered n-tuples (columns) of real numbers and are represented by bold small letters, e.g.,

$$\mathbf{x} = \begin{bmatrix} x_1 & x_2 & \dots & x_n \end{bmatrix}^T . \qquad (5.1)$$

These are vectors in $\Re^n$, the standard n-dimensional Euclidean space. In this chapter we will also be talking about more general vector spaces than $\Re^n$. These more general vectors will be represented with a script typeface, as in χ. We will show in this chapter how these general vectors can often be represented by columns of numbers.

Linear Vector Spaces

What do we mean by a vector space? We will begin with a very general definition. While this definition may seem abstract, we will provide many concrete examples. By using a general definition we can solve a larger class of problems, and we can impart a deeper understanding of the concepts.

Vector Space

Definition. A linear *vector space*, X, is a set of elements (vectors) defined over a scalar field, F, that satisfies the following conditions:

1. An operation called vector addition is defined such that if $\chi \in X$ (χ is an element of X) and $y \in X$, then $\chi + y \in X$.

2. $\chi + y = y + \chi$.

3. $(\chi + y) + z = \chi + (y + z)$.

4. There is a unique vector $0 \in X$, called the zero vector, such that $\chi + 0 = \chi$ for all $\chi \in X$.

5. For each vector $\chi \in X$ there is a unique vector in X, to be called $-\chi$, such that $\chi + (-\chi) = 0$.

6. An operation, called multiplication, is defined such that for all scalars $a \in F$, and all vectors $\chi \in X$, $a\chi \in X$.

7. For any $\chi \in X$, $1\chi = \chi$ (for scalar 1).

8. For any two scalars $a \in F$ and $b \in F$, and any $\chi \in X$, $a(b\chi) = (ab)\chi$.

9. $(a + b)\chi = a\chi + b\chi$.

10. $a(\chi + y) = a\chi + ay$.

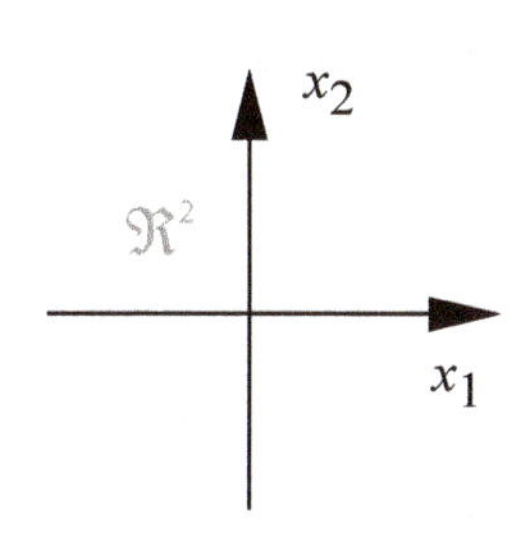

To illustrate these conditions, let's investigate a few sample sets and determine whether or not they are vector spaces. First consider the standard two-dimensional Euclidean space, $\Re^2$, shown in the upper left figure. This is clearly a vector space, and all ten conditions are satisfied for the standard definitions of vector addition and scalar multiplication.

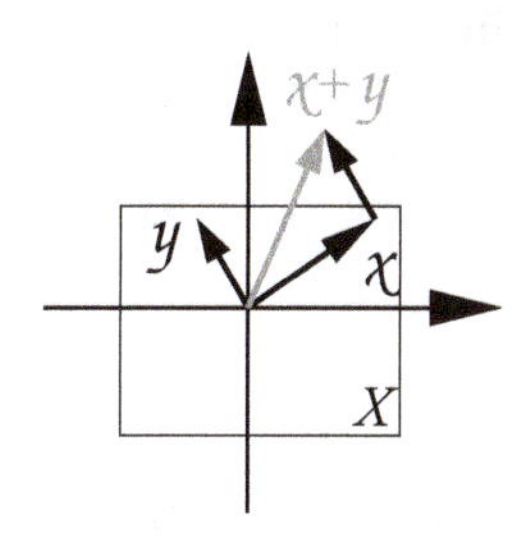

What about subsets of $\Re^2$? What subsets of $\Re^2$ are also vector spaces (subspaces)? Consider the boxed area (X) in the center left figure. Does it satisfy all ten conditions? No. Clearly even condition 1 is not satisfied. The vectors χ and y shown in the figure are in X, but $\chi + y$ is not. From this example it is clear that no bounded sets can be vector spaces.

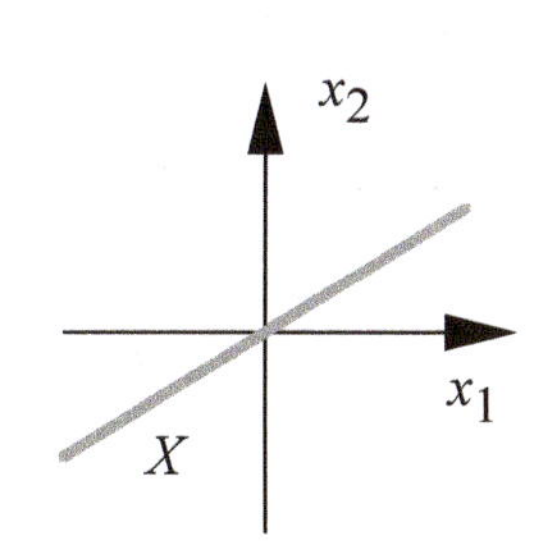

Are there any subsets of $\Re^2$ that are vector spaces? Consider the line (X) shown in the bottom left figure. (Assume that the line extends to infinity in both directions.) Is this line a vector space? We leave it to you to show that indeed all ten conditions are satisfied. Will any such infinite line satisfy the ten conditions? Well, any line that passes through the origin will work. If it does not pass through the origin then condition 4, for instance, would not be satisfied.

In addition to the standard Euclidean spaces, there are other sets that also satisfy the ten conditions of a vector space. Consider, for example, the set P^2 of all polynomials of degree less than or equal to 2. Two members of this set would be

$$\chi = 2 + t + 4t^2$$

$$y = 1 + 5t \,. \tag{5.2}$$

If you are used to thinking of vectors only as columns of numbers, these may seem to be strange vectors indeed. However, recall that to be a vector space, a set need only satisfy the ten conditions we presented. Are these conditions satisfied for the set P^2? If we add two polynomials of degree less than or equal to 2, the result will also be a polynomial of degree less than or equal to 2. Therefore condition 1 is satisfied. We can also multiply a polynomial by a scalar without changing the order of the polynomial. Therefore condition 6 is satisfied. It is not difficult to show that all ten conditions are satisfied, showing that P^2 is a vector space.

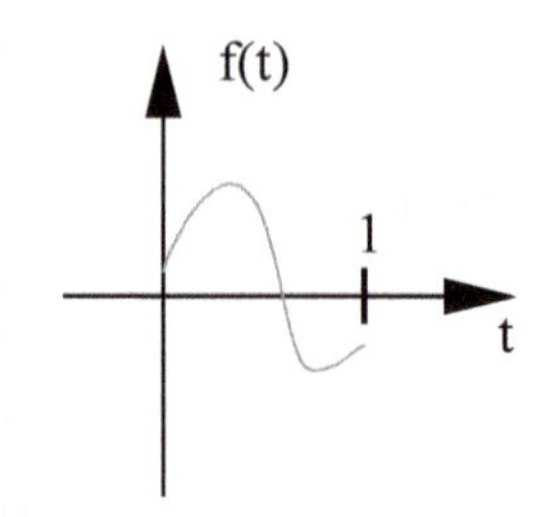

Consider the set $C_{[0,\,1]}$ of all continuous functions defined on the interval [0, 1]. Two members of this set would be

$$\chi = \sin(t)$$

$$y = e^{-2t} \,. \tag{5.3}$$

Another member of the set is shown in the figure to the left.

The sum of two continuous functions is also a continuous function, and a scalar times a continuous function is a continuous function. The set $C_{[0,\,1]}$ is also a vector space. This set is different than the other vector spaces we have discussed; it is infinite dimensional. We will define what we mean by dimension later in this chapter.

Linear Independence

Now that we have defined what we mean by a vector space, we will investigate some of the properties of vectors. The first properties are linear dependence and linear independence.

Consider n vectors $\{\chi_1, \chi_2, \ldots, \chi_n\}$. If there exist n scalars $a_1, a_2, \ldots, a_n$, at least one of which is nonzero, such that

$$a_1\chi_1 + a_2\chi_2 + \cdots + a_n\chi_n = 0, \tag{5.4}$$

then the $\{\chi_i\}$ are linearly dependent.

Linear Independence

The converse statement would be: If $a_1\chi_1 + a_2\chi_2 + \cdots + a_n\chi_n = 0$ implies that each $a_i = 0$, then $\{\chi_i\}$ is a set of *linearly independent* vectors.

Note that these definitions are equivalent to saying that if a set of vectors is independent then no vector in the set can be written as a linear combination of the other vectors.

2 +2 4

As an example of independence, consider the pattern recognition problem of Chapter 3. The two prototype patterns (*orange* and *apple*) were given by:

$$\mathbf{p}_1 = \begin{bmatrix} 1 \\ -1 \\ -1 \end{bmatrix}, \mathbf{p}_2 = \begin{bmatrix} 1 \\ 1 \\ -1 \end{bmatrix}. \tag{5.5}$$

Let $a_1\mathbf{p}_1 + a_2\mathbf{p}_2 = \mathbf{0}$, then

$$\begin{bmatrix} a_1 + a_2 \\ -a_1 + a_2 \\ -a_1 + (-a_2) \end{bmatrix} = \begin{bmatrix} 0 \\ 0 \\ 0 \end{bmatrix}, \tag{5.6}$$

but this can only be true if $a_1 = a_2 = 0$. Therefore $\mathbf{p}_1$ and $\mathbf{p}_2$ are linearly independent.

2 +2 4

Consider vectors from the space P^2 of polynomials of degree less than or equal to 2. Three vectors from this space would be

$$\chi_1 = 1 + t + t^2, \; \chi_2 = 2 + 2t + t^2, \; \chi_3 = 1 + t. \tag{5.7}$$

Note that if we let $a_1 = 1$, $a_2 = -1$ and $a_3 = 1$, then

$$a_1\chi_1 + a_2\chi_2 + a_3\chi_3 = 0. \tag{5.8}$$

Therefore these three vectors are linearly dependent.

Spanning a Space

Next we want to define what we mean by the dimension (size) of a vector space. To do so we must first define the concept of a spanning set.

Let X be a linear vector space and let $\{u_1, u_2, \ldots, u_m\}$ be a subset of general vectors in X. This subset spans X if and only if for every vector $\chi \in X$ there exist scalars $x_1, x_2, \ldots, x_n$ such that $\chi = x_1 u_1 + x_2 u_2 + \cdots + x_m u_m$. In other words, a subset spans a space if every vector in the space can be written as a linear combination of the vectors in the subset.

The dimension of a vector space is determined by the minimum number of vectors it takes to span the space. This leads to the definition of a basis set.

Basis Set

A *basis set* for X is a set of linearly independent vectors that spans X. Any basis set contains the minimum number of vectors required to span the space. The dimension of X is therefore equal to the number of elements in the basis set. Any vector space can have many basis sets, but each one must contain the same number of elements. (See [Stra80] for a proof of this fact.)

Take, for example, the linear vector space P^2. One possible basis for this space is

$$u_1 = 1 , u_2 = t , u_3 = t^2 . \tag{5.9}$$

Clearly any polynomial of degree two or less can be created by taking a linear combination of these three vectors. Note, however, that *any* three independent vectors from P^2 would form a basis for this space. One such alternate basis is:

$$u_1 = 1 , u_2 = 1 + t , u_3 = 1 + t + t^2 . \tag{5.10}$$

Inner Product

From our brief encounter with neural networks in Chapter 3 and 4, it is clear that the inner product is fundamental to the operation of many neural networks. Here we will introduce a general definition for inner products and then give several examples.

Inner Product

Any scalar function of χ and y can be defined as an *inner product*, (χ, y), provided that the following properties are satisfied:

1. $(\chi, y) = (y, \chi)$.
2. $(\chi, a y_1 + b y_2) = a(\chi, y_1) + b(\chi, y_2)$.
3. $(\chi, \chi) \geq 0$, where equality holds if and only if χ is the zero vector.

The standard inner product for vectors in R^n is

$$\mathbf{x}^T\mathbf{y} = x_1y_1 + x_2y_2 + \cdots + x_ny_n , \tag{5.11}$$

but this is not the only possible inner product. Consider again the set $C_{[0, 1]}$ of all continuous functions defined on the interval [0, 1]. Show that the following scalar function is man inner product (see Problem P5.6).

$$(\chi, y) = \int_0^1 \chi(t)y(t)dt \tag{5.12}$$

Norm

The next operation we need to define is the norm, which is based on the concept of vector length.

Norm A scalar function $\|\chi\|$ is called a *norm* if it satisfies the following properties:

1. $\|\chi\| \geq 0$.
2. $\|\chi\| = 0$ if and only if $\chi = 0$.
3. $\|a\chi\| = |a|\|\chi\|$ for scalar a.
4. $\|\chi + y\| \leq \|\chi\| + \|y\|$.

There are many functions that would satisfy these conditions. One common norm is based on the inner product:

$$\|\chi\| = (\chi,\chi)^{1/2}. \tag{5.13}$$

For Euclidean spaces, $\Re^n$, this yields the norm with which we are most familiar:

$$\|\mathbf{x}\| = (\mathbf{x}^T\mathbf{x})^{1/2} = \sqrt{x_1^2 + x_2^2 + \cdots + x_n^2}. \tag{5.14}$$

In neural network applications it is often useful to normalize the input vectors. This means that $\|\mathbf{p}_i\| = 1$ for each input vector.

Using the norm and the inner product we can generalize the concept of angle for vector spaces of dimension greater than two. The *angle* θ between two vectors χ and y is defined by

Angle

$$\cos\theta = \frac{(\chi,y)}{\|\chi\|\|y\|}. \tag{5.15}$$

Orthogonality

Now that we have defined the inner product operation, we can introduce the important concept of orthogonality.

Orthogonality Two vectors $\chi, y \in X$ are said to be *orthogonal* if $(\chi,y) = 0$.

Orthogonality is an important concept in neural networks. We will see in Chapter 7 that when the prototype vectors of a pattern recognition problem are orthogonal and normalized, a linear associator neural network can be trained, using the Hebb rule, to achieve perfect recognition.

In addition to orthogonal vectors, we can also have orthogonal spaces. A vector $\chi \in X$ is orthogonal to a subspace X_1 if χ is orthogonal to every vec-

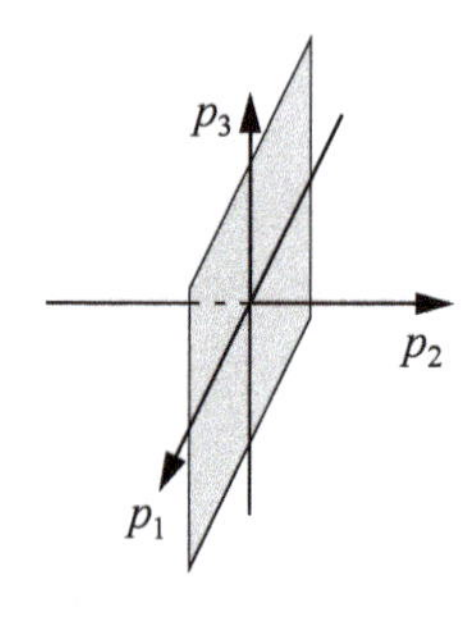

tor in X_1. This is typically represented as $\chi \perp X_1$. A subspace X_1 is orthogonal to a subspace X_2 if every vector in X_1 is orthogonal to every vector in X_2. This is represented by $X_1 \perp X_2$.

The figure to the left illustrates the two orthogonal spaces that were used in the perceptron example of Chapter 3. (See Figure 3.4.) The p_1, p_3 plane is a subspace of $\Re^3$, which is orthogonal to the p_2 axis (which is another subspace of $\Re^3$). The p_1, p_3 plane was the decision boundary of a perceptron network. In Solved Problem P5.1 we will show that the perceptron decision boundary will be a vector space whenever the bias value is zero.

Gram-Schmidt Orthogonalization

There is a relationship between orthogonality and independence. It is possible to convert a set of independent vectors into a set of orthogonal vectors that spans the same vector space. The standard procedure to accomplish this is called Gram-Schmidt orthogonalization.

Assume that we have n independent vectors $y_1, y_2, \ldots, y_n$. From these vectors we want to obtain n orthogonal vectors $v_1, v_2, \ldots, v_n$. The first orthogonal vector is chosen to be the first independent vector:

$$v_1 = y_1 . \tag{5.16}$$

To obtain the second orthogonal vector we use y_2, but subtract off the portion of y_2 that is in the direction of v_1. This leads to the equation

$$v_2 = y_2 - a v_1 , \tag{5.17}$$

where a is chosen so that v_2 is orthogonal to v_1. This requires that

$$(v_1, v_2) = (v_1, y_2 - a v_1) = (v_1, y_2) - a(v_1, v_1) = 0 , \tag{5.18}$$

or

$$a = \frac{(v_1, y_2)}{(v_1, v_1)} . \tag{5.19}$$

Therefore to find the component of y_2 in the direction of v_1, $a v_1$, we need to find the inner product between the two vectors. We call $a v_1$ the *projection* of y_2 on the vector v_1.

Projection

If we continue this process, the kth step will be

$$v_k = y_k - \sum_{i=1}^{k-1} \frac{(v_i, y_k)}{(v_i, v_i)} v_i . \tag{5.20}$$

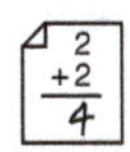

To illustrate this process, we consider the following independent vectors in $\Re^2$:

$$\mathbf{y}_1 = \begin{bmatrix} 2 \\ 1 \end{bmatrix}, \mathbf{y}_2 = \begin{bmatrix} 1 \\ 2 \end{bmatrix}. \tag{5.21}$$

The first orthogonal vector would be

$$\mathbf{v}_1 = \mathbf{y}_1 = \begin{bmatrix} 2 \\ 1 \end{bmatrix}. \tag{5.22}$$

The second orthogonal vector is calculated as follows:

$$\mathbf{v}_2 = \mathbf{y}_2 - \frac{\mathbf{v}_1^T \mathbf{y}_2}{\mathbf{v}_1^T \mathbf{v}_1} \mathbf{v}_1 = \begin{bmatrix} 1 \\ 2 \end{bmatrix} - \frac{\begin{bmatrix} 2 & 1 \end{bmatrix} \begin{bmatrix} 1 \\ 2 \end{bmatrix}}{\begin{bmatrix} 2 & 1 \end{bmatrix} \begin{bmatrix} 2 \\ 1 \end{bmatrix}} \begin{bmatrix} 2 \\ 1 \end{bmatrix} = \begin{bmatrix} 1 \\ 2 \end{bmatrix} - \begin{bmatrix} 1.6 \\ 0.8 \end{bmatrix} = \begin{bmatrix} -0.6 \\ 1.2 \end{bmatrix}. \tag{5.23}$$

See Figure 5.1 for a graphical representation of this process.

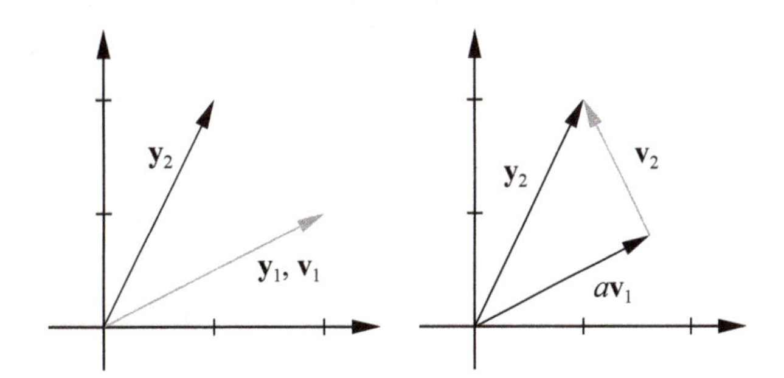

Figure 5.1 Gram-Schmidt Orthogonalization Example

Orthonormal

We could convert $\mathbf{v}_1$ and $\mathbf{v}_2$ to a set of *orthonormal* (orthogonal and normalized) vectors by dividing each vector by its norm.

To experiment with this orthogonalization process, use the Neural Network Design Demonstration Gram-Schmidt (**nnd5gs**).

Vector Expansions

Note that we have been using a script font (χ) to represent general vectors and bold type ($\mathbf{x}$) to represent vectors in $\Re^n$, which can be written as columns of numbers. In this section we will show that general vectors in finite dimensional vector spaces can also be written as columns of numbers and therefore are in some ways equivalent to vectors in $\Re^n$.

Vector Expansion

If a vector space X has a basis set $\{v_1, v_2, \ldots, v_n\}$, then any $\chi \in X$ has a unique *vector expansion*:

$$\chi = \sum_{i=1}^{n} x_i v_i = x_1 v_1 + x_2 v_2 + \cdots + x_n v_n . \tag{5.24}$$

Therefore any vector in a finite dimensional vector space can be represented by a column of numbers:

$$\mathbf{x} = \begin{bmatrix} x_1 & x_2 & \ldots & x_n \end{bmatrix}^T . \tag{5.25}$$

This $\mathbf{x}$ is a representation of the general vector χ. Of course in order to interpret the meaning of $\mathbf{x}$ we need to know the basis set. If the basis set changes, $\mathbf{x}$ will change, even though it still represents the same general vector χ. We will discuss this in more detail in the next subsection.

If the vectors in the basis set are orthogonal ($(v_i, v_j) = 0$, $i \neq j$) it is very easy to compute the coefficients in the expansion. We simply take the inner product of v_j with both sides of Eq. (5.24):

$$(v_j, \chi) = (v_j, \sum_{i=1}^{n} x_i v_i) = \sum_{i=1}^{n} x_i (v_j, v_i) = x_j (v_j, v_j) . \tag{5.26}$$

Therefore the coefficients of the expansion are given by

$$x_j = \frac{(v_j, \chi)}{(v_j, v_j)} . \tag{5.27}$$

When the vectors in the basis set are not orthogonal, the computation of the coefficients in the vector expansion is more complex. This case is covered in the following subsection.

Reciprocal Basis Vectors

If a vector expansion is required and the basis set is not orthogonal, the reciprocal basis vectors are introduced. These are defined by the following equations:

$$\begin{aligned} (r_i, v_j) &= 0 \qquad i \neq j \\ &= 1 \qquad i = j , \end{aligned} \tag{5.28}$$

Reciprocal Basis Vectors

where the basis vectors are $\{ v_1, v_2, \ldots, v_n \}$ and the *reciprocal basis vectors* are $\{ r_1, r_2, \ldots, r_n \}$.

If the vectors have been represented by columns of numbers (through vector expansion), and the standard inner product is used

$$(r_i,v_j) = \mathbf{r}_i^T \mathbf{v}_j , \tag{5.29}$$

then Eq. (5.28) can be represented in matrix form as

$$\mathbf{R}^T\mathbf{B} = \mathbf{I} , \tag{5.30}$$

where

$$\mathbf{B} = \begin{bmatrix} \mathbf{v}_1 & \mathbf{v}_2 & \ldots & \mathbf{v}_n \end{bmatrix} , \tag{5.31}$$

$$\mathbf{R} = \begin{bmatrix} \mathbf{r}_1 & \mathbf{r}_2 & \ldots & \mathbf{r}_n \end{bmatrix} . \tag{5.32}$$

Therefore **R** can be found from

$$\mathbf{R}^T = \mathbf{B}^{-1} , \tag{5.33}$$

and the reciprocal basis vectors can be obtained from the columns of **R**.

Now consider again the vector expansion

$$\chi = x_1 v_1 + x_2 v_2 + \cdots + x_n v_n . \tag{5.34}$$

Taking the inner product of r_1 with both sides of Eq. (5.34) we obtain

$$(r_1,\chi) = x_1(r_1,v_1) + x_2(r_1,v_2) + \cdots + x_n(r_1,v_n) . \tag{5.35}$$

By definition

$$(r_1,v_2) = (r_1,v_3) = \cdots = (r_1,v_n) = 0$$

$$(r_1,v_1) = 1 . \tag{5.36}$$

Therefore the first coefficient of the expansion is

$$x_1 = (r_1,\chi) , \tag{5.37}$$

and in general

$$x_j = (r_j,\chi) . \tag{5.38}$$

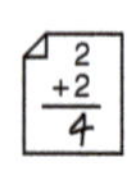

As an example, consider the two basis vectors

$$\mathbf{v}_1^s = \begin{bmatrix} 2 \\ 1 \end{bmatrix},\ \mathbf{v}_2^s = \begin{bmatrix} 1 \\ 2 \end{bmatrix}. \tag{5.39}$$

Suppose that we want to expand the vector

$$\mathbf{x}^s = \begin{bmatrix} 0 \\ \frac{3}{2} \end{bmatrix} \tag{5.40}$$

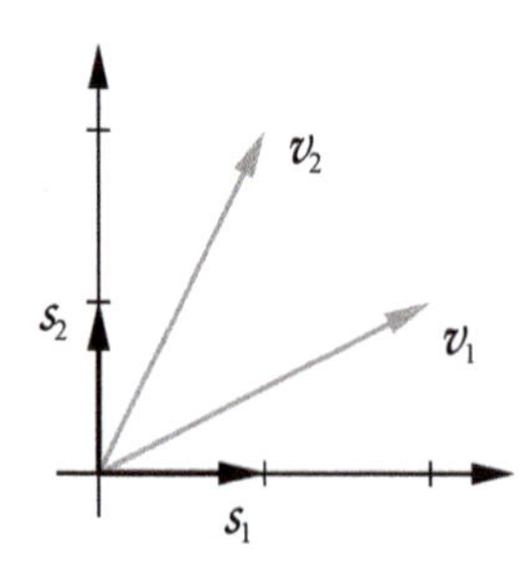

in terms of the two basis vectors. (We are using the superscript s to indicate that these columns of numbers represent expansions of the vectors in terms of the standard basis in $\Re^2$. The elements of the standard basis are indicated in the adjacent figure as the vectors s_1 and s_2. We need to use this explicit notation in this example because we will be expanding the vectors in terms of two different basis sets.)

The first step in the vector expansion is to find the reciprocal basis vectors.

$$\mathbf{R}^T = \begin{bmatrix} 2 & 1 \\ 1 & 2 \end{bmatrix}^{-1} = \begin{bmatrix} \frac{2}{3} & -\frac{1}{3} \\ -\frac{1}{3} & \frac{2}{3} \end{bmatrix} \qquad \mathbf{r}_1 = \begin{bmatrix} \frac{2}{3} \\ -\frac{1}{3} \end{bmatrix} \qquad \mathbf{r}_2 = \begin{bmatrix} -\frac{1}{3} \\ \frac{2}{3} \end{bmatrix}. \tag{5.41}$$

Now we can find the coefficients in the expansion.

$$x_1^v = \mathbf{r}_1^T \mathbf{x}^s = \begin{bmatrix} \frac{2}{3} & -\frac{1}{3} \end{bmatrix} \begin{bmatrix} 0 \\ \frac{3}{2} \end{bmatrix} = -\frac{1}{2}$$

$$x_2^v = \mathbf{r}_2^T \mathbf{x}^s = \begin{bmatrix} -\frac{1}{3} & \frac{2}{3} \end{bmatrix} \begin{bmatrix} 0 \\ \frac{3}{2} \end{bmatrix} = 1 \tag{5.42}$$

or, in matrix form,

$$\mathbf{x}^v = \mathbf{R}^T \mathbf{x}^s = \mathbf{B}^{-1} \mathbf{x}^s = \begin{bmatrix} \frac{2}{3} & -\frac{1}{3} \\ -\frac{1}{3} & \frac{2}{3} \end{bmatrix} \begin{bmatrix} 0 \\ \frac{3}{2} \end{bmatrix} = \begin{bmatrix} -\frac{1}{2} \\ 1 \end{bmatrix}. \tag{5.43}$$

So that

$$\chi = -\frac{1}{2}v_1 + 1v_2, \tag{5.44}$$

as indicated in Figure 5.2.

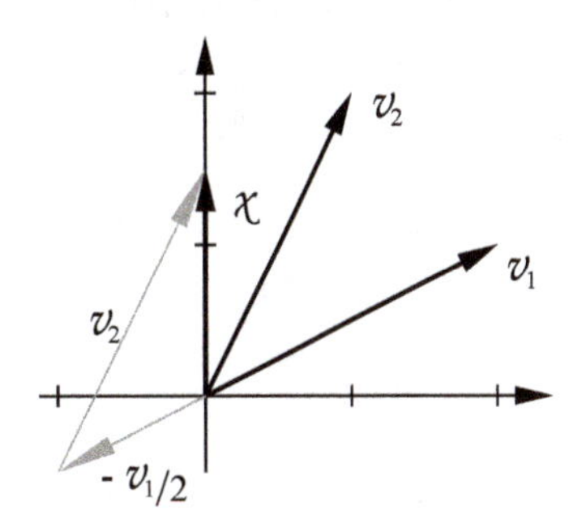

Figure 5.2 Vector Expansion

Note that we now have two different vector expansions for χ, represented by $\mathbf{x}^s$ and $\mathbf{x}^v$. In other words,

$$\chi = 0s_1 + \frac{3}{2}s_2 = -\frac{1}{2}v_1 + 1v_2. \tag{5.45}$$

When we represent a general vector as a column of numbers we need to know what basis set was used for the expansion. In this text, unless otherwise stated, assume the standard basis set was used.

Eq. (5.43) shows the relationship between the two different representations of χ, $\mathbf{x}^v = \mathbf{B}^{-1}\mathbf{x}^s$. This operation, called a change of basis, will become very important in later chapters for the performance analysis of certain neural networks.

To experiment with the vector expansion process, use the Neural Network Design Demonstration Reciprocal Basis (`nnd5rb`).

Summary of Results

Linear Vector Spaces

Definition. A linear vector space, X, is a set of elements (vectors) defined over a scalar field, F, that satisfies the following conditions:

1. An operation called vector addition is defined such that if $\chi \in X$ and $y \in X$, then $\chi + y \in X$.
2. $\chi + y = y + \chi$.
3. $(\chi + y) + z = \chi + (y + z)$.
4. There is a unique vector $0 \in X$, called the zero vector, such that $\chi + 0 = \chi$ for all $\chi \in X$.
5. For each vector $\chi \in X$ there is a unique vector in X, to be called $-\chi$, such that $\chi + (-\chi) = 0$.
6. An operation, called multiplication, is defined such that for all scalars $a \in F$, and all vectors $\chi \in X$, $a\chi \in X$.
7. For any $\chi \in X$, $1\chi = \chi$ (for scalar 1).
8. For any two scalars $a \in F$ and $b \in F$, and any $\chi \in X$, $a(b\chi) = (ab)\chi$.
9. $(a + b)\chi = a\chi + b\chi$.
10. $a(\chi + y) = a\chi + ay$.

Linear Independence

Consider n vectors $\{\chi_1, \chi_2, \dots, \chi_n\}$. If there exist n scalars $a_1, a_2, \dots, a_n$, at least one of which is nonzero, such that

$$a_1\chi_1 + a_2\chi_2 + \cdots + a_n\chi_n = 0,$$

then the $\{\chi_i\}$ are linearly dependent.

Spanning a Space

Let X be a linear vector space and let $\{u_1, u_2, \ldots, u_m\}$ be a subset of vectors in X. This subset spans X if and only if for every vector $\chi \in X$ there exist scalars $x_1, x_2, \ldots, x_n$ such that $\chi = x_1 u_1 + x_2 u_2 + \cdots + x_m u_m$.

Inner Product

Any scalar function of χ and y can be defined as an inner product, (χ, y), provided that the following properties are satisfied.

1. $(\chi, y) = (y, \chi)$.
2. $(\chi, a y_1 + b y_2) = a(\chi, y_1) + b(\chi, y_2)$.
3. $(\chi, \chi) \geq 0$, where equality holds if and only if χ is the zero vector.

Norm

A scalar function $\|\chi\|$ is called a norm if it satisfies the following properties:

1. $\|\chi\| \geq 0$.
2. $\|\chi\| = 0$ if and only if $\chi = 0$.
3. $\|a\chi\| = |a|\|\chi\|$ for scalar a.
4. $\|\chi + y\| \leq \|\chi\| + \|y\|$.

Angle

The angle θ between two vectors χ and y is defined by

$$\cos\theta = \frac{(\chi, y)}{\|\chi\|\|y\|}.$$

Orthogonality

Two vectors $\chi, y \in X$ are said to be orthogonal if $(\chi, y) = 0$.

Gram-Schmidt Orthogonalization

Assume that we have n independent vectors $y_1, y_2, \ldots, y_n$. From these vectors we will obtain n orthogonal vectors $v_1, v_2, \ldots, v_n$.

$$v_1 = y_1$$

$$v_k = y_k - \sum_{i=1}^{k-1} \frac{(v_i, y_k)}{(v_i, v_i)} v_i,$$

where

$$\frac{(v_i, y_k)}{(v_i, v_i)} v_i$$

is the projection of y_k on v_i.

Vector Expansions

$$\chi = \sum_{i=1}^{n} x_i v_i = x_1 v_1 + x_2 v_2 + \cdots + x_n v_n .$$

For orthogonal vectors,

$$x_j = \frac{(v_j, \chi)}{(v_j, v_j)}$$

Reciprocal Basis Vectors

$$(r_i, v_j) = 0 \qquad i \neq j$$

$$= 1 \qquad i = j$$

$$x_j = (r_j, \chi) .$$

To compute the reciprocal basis vectors:

$$\mathbf{B} = \begin{bmatrix} \mathbf{v}_1 & \mathbf{v}_2 & \dots & \mathbf{v}_n \end{bmatrix},$$

$$\mathbf{R} = \begin{bmatrix} \mathbf{r}_1 & \mathbf{r}_2 & \dots & \mathbf{r}_n \end{bmatrix},$$

$$\mathbf{R}^T = \mathbf{B}^{-1} .$$

In matrix form:

$$\mathbf{x}^v = \mathbf{B}^{-1} \mathbf{x}^s .$$

Solved Problems

P5.1 Consider the single-neuron perceptron network shown in Figure P5.1. Recall from Chapter 8 (see Eq. (3.6)) that the decision boundary for this network is given by $\mathbf{W}\mathbf{p} + b = 0$. Show that the decision boundary is a vector space if $b = 0$.

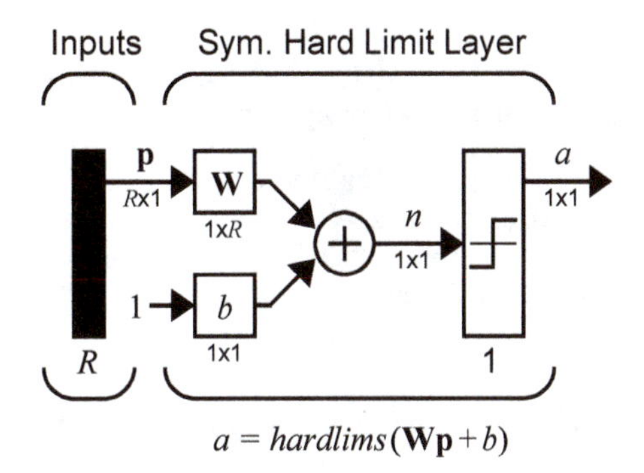

Figure P5.1 Single-Neuron Perceptron

To be a vector space the boundary must satisfy the ten conditions given at the beginning of this chapter. Condition 1 requires that when we add two vectors together the sum remains in the vector space. Let $\mathbf{p}_1$ and $\mathbf{p}_2$ be two vectors on the decision boundary. To be on the boundary they must satisfy

$$\mathbf{W}\mathbf{p}_1 = 0 \qquad \mathbf{W}\mathbf{p}_2 = 0.$$

If we add these two equations together we find

$$\mathbf{W}(\mathbf{p}_1 + \mathbf{p}_2) = 0.$$

Therefore the sum is also on the decision boundary.

Conditions 2 and 3 are clearly satisfied. Condition 4 requires that the zero vector be on the boundary. Since $\mathbf{W}\mathbf{0} = 0$, the zero vector is on the decision boundary. Condition 5 implies that if $\mathbf{p}$ is on the boundary, then $-\mathbf{p}$ must also be on the boundary. If $\mathbf{p}$ is on the boundary, then

$$\mathbf{W}\mathbf{p} = 0.$$

If we multiply both sides of this equation by -1 we find

$$\mathbf{W}(-\mathbf{p}) = 0.$$

Therefore condition 5 is satisfied.

Condition 6 will be satisfied if for any $\mathbf{p}$ on the boundary $a\mathbf{p}$ is also on the boundary. This can be shown in the same way as condition 5. Just multiply both sides of the equation by a instead of by 1.

$$\mathbf{W}(a\mathbf{p}) = 0$$

Conditions 7 through 10 are clearly satisfied. Therefore the perceptron decision boundary is a vector space.

P5.2 Show that the set Y of nonnegative ($f(t) \geq 0$) continuous functions is not a vector space.

This set violates several of the conditions required of a vector space. For example, there are no negative vectors, so condition 5 cannot be satisfied. Also, consider condition 6. The function $f(t) = |t|$ is a member of Y. Let $a = -2$. Then

$$af(2) = -2|2| = -4 < 0 .$$

Therefore $af(t)$ is not a member of Y, and condition 6 is not satisfied.

P5.3 Which of the following sets of vectors are independent? Find the dimension of the vector space spanned by each set.

i. $\begin{bmatrix} 1 \\ 1 \\ 1 \end{bmatrix} \quad \begin{bmatrix} 1 \\ 0 \\ 1 \end{bmatrix} \quad \begin{bmatrix} 1 \\ 2 \\ 1 \end{bmatrix}$

ii. $\sin t \quad \cos t \quad 2\cos\left(t + \frac{\pi}{4}\right)$

iii. $\begin{bmatrix} 1 \\ 1 \\ 1 \\ 1 \end{bmatrix} \quad \begin{bmatrix} 1 \\ 0 \\ 1 \\ 1 \end{bmatrix} \quad \begin{bmatrix} 1 \\ 2 \\ 1 \\ 1 \end{bmatrix}$

i. We can solve this problem several ways. First, let's assume that the vectors are dependent. Then we can write

$$a_1 \begin{bmatrix} 1 \\ 1 \\ 1 \end{bmatrix} + a_2 \begin{bmatrix} 1 \\ 0 \\ 1 \end{bmatrix} + a_3 \begin{bmatrix} 1 \\ 2 \\ 1 \end{bmatrix} = \begin{bmatrix} 0 \\ 0 \\ 0 \end{bmatrix} .$$

If we can solve for the coefficients and they are not all zero, then the vectors are dependent. By inspection we can see that if we let $a_1 = 2$, $a_2 = -1$ and $a_3 = -1$, then the equation is satisfied. Therefore the vectors are dependent.

Another approach, when we have n vectors in $\Re^n$, is to write the above equation in matrix form:

$$\begin{bmatrix} 1 & 1 & 1 \\ 1 & 0 & 2 \\ 1 & 1 & 1 \end{bmatrix} \begin{bmatrix} a_1 \\ a_2 \\ a_3 \end{bmatrix} = \begin{bmatrix} 0 \\ 0 \\ 0 \end{bmatrix}$$

If the matrix in this equation has an inverse, then the solution will require that all coefficients be zero; therefore the vectors are independent. If the matrix is singular (has no inverse), then a nonzero set of coefficients will work, and the vectors are dependent. The test, then, is to create a matrix using the vectors as columns. If the determinant of the matrix is zero (singular matrix), then the vectors are dependent; otherwise they are independent. Using the Laplace expansion [Brog91] on the first column, the determinant of this matrix is

$$\begin{vmatrix} 1 & 1 & 1 \\ 1 & 0 & 2 \\ 1 & 1 & 1 \end{vmatrix} = 1\begin{vmatrix} 0 & 2 \\ 1 & 1 \end{vmatrix} + (-1)\begin{vmatrix} 1 & 1 \\ 1 & 1 \end{vmatrix} + 1\begin{vmatrix} 1 & 1 \\ 0 & 2 \end{vmatrix} = -2 + 0 + 2 = 0$$

Therefore the vectors are dependent.

The dimension of the space spanned by the vectors is two, since any two of the vectors can be shown to be independent.

ii. By using some trigonometric identities we can write

$$\cos\left(t + \frac{\pi}{4}\right) = \frac{-1}{\sqrt{2}}\sin t + \frac{1}{\sqrt{2}}\cos t .$$

Therefore the vectors are dependent. The dimension of the space spanned by the vectors is two, since no linear combination of $\sin t$ and $\cos t$ is identically zero.

iii. This is similar to part (i), except that the number of vectors is less than the size of the vector space they are drawn from (three vectors in $\Re^4$). In this case the matrix made up of the vectors will not be square, so we will not be able to compute a determinant. However, we can use something called the Gramian [Brog91]. It is the determinant of a matrix whose i, j element is the inner product of vector i and vector j. The vectors are dependent if and only if the Gramian is zero.

For our problem the Gramian would be

$$G = \begin{vmatrix} (\mathbf{x}_1,\mathbf{x}_1) & (\mathbf{x}_1,\mathbf{x}_2) & (\mathbf{x}_1,\mathbf{x}_3) \\ (\mathbf{x}_2,\mathbf{x}_1) & (\mathbf{x}_2,\mathbf{x}_2) & (\mathbf{x}_2,\mathbf{x}_3) \\ (\mathbf{x}_3,\mathbf{x}_1) & (\mathbf{x}_3,\mathbf{x}_2) & (\mathbf{x}_3,\mathbf{x}_3) \end{vmatrix},$$

where

$$\mathbf{x}_1 = \begin{bmatrix} 1 \\ 1 \\ 1 \\ 1 \end{bmatrix} \quad \mathbf{x}_2 = \begin{bmatrix} 1 \\ 0 \\ 1 \\ 1 \end{bmatrix} \quad \mathbf{x}_3 = \begin{bmatrix} 1 \\ 2 \\ 1 \\ 1 \end{bmatrix}.$$

Therefore

$$G = \begin{vmatrix} 4 & 3 & 5 \\ 3 & 3 & 3 \\ 5 & 3 & 7 \end{vmatrix} = 4\begin{vmatrix} 3 & 3 \\ 3 & 7 \end{vmatrix} + (-3)\begin{vmatrix} 3 & 5 \\ 3 & 7 \end{vmatrix} + 5\begin{vmatrix} 3 & 5 \\ 3 & 3 \end{vmatrix} = 48 - 18 - 30 = 0.$$

We can also show that these vectors are dependent by noting

$$2\begin{bmatrix} 1 \\ 1 \\ 1 \\ 1 \end{bmatrix} - 1\begin{bmatrix} 1 \\ 0 \\ 1 \\ 1 \end{bmatrix} - 1\begin{bmatrix} 1 \\ 2 \\ 1 \\ 1 \end{bmatrix} = \begin{bmatrix} 0 \\ 0 \\ 0 \\ 0 \end{bmatrix}.$$

The dimension of the space must therefore be less than 3. We can show that $\mathbf{x}_1$ and $\mathbf{x}_2$ are independent, since

$$G = \begin{vmatrix} 4 & 3 \\ 3 & 3 \end{vmatrix} = 4 \neq 0.$$

Therefore the dimension of the space is 2.

P5.4 Recall from Chapter 3 and 4 that one-layer perceptrons can only be used to recognize patterns that are linearly separable (can be separated by a linear boundary — see Figure 3.3). If two patterns are linearly separable, are they always linearly independent?

No, these are two unrelated concepts. Take the following simple example. Consider the two input perceptron shown in Figure P5.2.

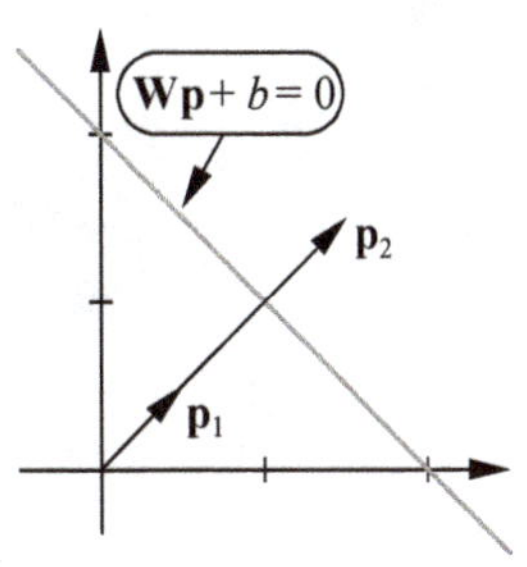

Suppose that we want to separate the two vectors

$$\mathbf{p}_1 = \begin{bmatrix} 0.5 \\ 0.5 \end{bmatrix} \qquad \mathbf{p}_2 = \begin{bmatrix} 1.5 \\ 1.5 \end{bmatrix}.$$

If we choose the weights and offsets to be $w_{11} = 1$, $w_{12} = 1$ and $b = -2$, then the decision boundary ($\mathbf{Wp} + b = 0$) is shown in the figure to the left. Clearly these two vectors are linearly separable. However, they are not linearly independent since $\mathbf{p}_2 = 3\mathbf{p}_1$.

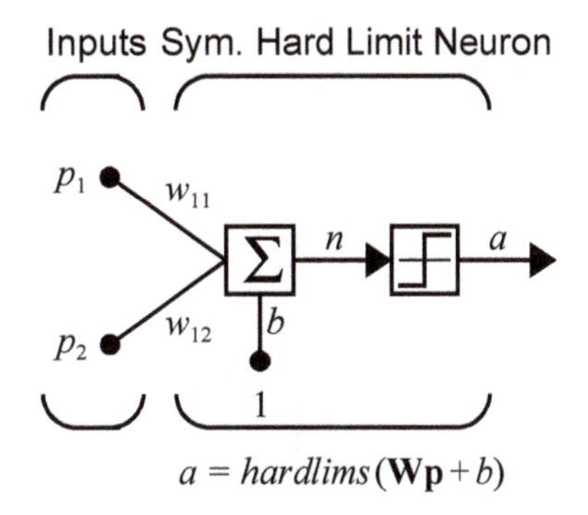

Figure P5.2 Two-Input Perceptron

P5.5 Using the following basis vectors, find an orthogonal set using Gram-Schmidt orthogonalization.

$$\mathbf{y}_1 = \begin{bmatrix} 1 \\ 1 \\ 1 \end{bmatrix} \qquad \mathbf{y}_2 = \begin{bmatrix} 1 \\ 0 \\ 0 \end{bmatrix} \qquad \mathbf{y}_3 = \begin{bmatrix} 0 \\ 1 \\ 0 \end{bmatrix}$$

Step 1.

$$\mathbf{v}_1 = \mathbf{y}_1 = \begin{bmatrix} 1 \\ 1 \\ 1 \end{bmatrix}$$

Step 2.

$$\mathbf{v}_2 = \mathbf{y}_2 - \frac{\mathbf{v}_1^T \mathbf{y}_2}{\mathbf{v}_1^T \mathbf{v}_1}\mathbf{v}_1 = \begin{bmatrix}1\\0\\0\end{bmatrix} - \frac{\begin{bmatrix}1 & 1 & 1\end{bmatrix}\begin{bmatrix}1\\0\\0\end{bmatrix}}{\begin{bmatrix}1 & 1 & 1\end{bmatrix}\begin{bmatrix}1\\1\\1\end{bmatrix}}\begin{bmatrix}1\\1\\1\end{bmatrix} = \begin{bmatrix}1\\0\\0\end{bmatrix} - \begin{bmatrix}1/3\\1/3\\1/3\end{bmatrix} = \begin{bmatrix}2/3\\-1/3\\-1/3\end{bmatrix}$$

Step 3.

$$\mathbf{v}_3 = \mathbf{y}_3 - \frac{\mathbf{v}_1^T \mathbf{y}_3}{\mathbf{v}_1^T \mathbf{v}_1}\mathbf{v}_1 - \frac{\mathbf{v}_2^T \mathbf{y}_3}{\mathbf{v}_2^T \mathbf{v}_2}\mathbf{v}_2$$

$$\mathbf{v}_3 = \begin{bmatrix}0\\1\\0\end{bmatrix} - \frac{\begin{bmatrix}1 & 1 & 1\end{bmatrix}\begin{bmatrix}0\\1\\0\end{bmatrix}}{\begin{bmatrix}1 & 1 & 1\end{bmatrix}\begin{bmatrix}1\\1\\1\end{bmatrix}}\begin{bmatrix}1\\1\\1\end{bmatrix} - \frac{\begin{bmatrix}2/3 & -1/3 & -1/3\end{bmatrix}\begin{bmatrix}0\\1\\0\end{bmatrix}}{\begin{bmatrix}2/3 & -1/3 & -1/3\end{bmatrix}\begin{bmatrix}2/3\\-1/3\\-1/3\end{bmatrix}}\begin{bmatrix}2/3\\-1/3\\-1/3\end{bmatrix}$$

$$\mathbf{v}_3 = \begin{bmatrix}0\\1\\0\end{bmatrix} - \begin{bmatrix}1/3\\1/3\\1/3\end{bmatrix} - \begin{bmatrix}-1/3\\1/6\\1/6\end{bmatrix} = \begin{bmatrix}0\\1/2\\-1/2\end{bmatrix}$$

P5.6 Consider the vector space of all polynomials defined on the interval [-1, 1]. Show that $(\chi,y) = \int_{-1}^{1} \chi(t)y(t)dt$ is a valid inner product.

An inner product must satisfy the following properties.

1. $(\chi,y) = (y,\chi)$

$$(\chi,y) = \int_{-1}^{1} \chi(t)y(t)dt = \int_{-1}^{1} y(t)\chi(t)dt = (y,\chi)$$

2. $(\chi, a y_1 + b y_2) = a(\chi,y_1) + b(\chi,y_2)$

$$(x, ay_1 + by_2) = \int_{-1}^{1} x(t)(ay_1(t) + by_2(t))dt = a\int_{-1}^{1} x(t)y_1(t)dt + b\int_{-1}^{1} x(t)y_2(t)dt$$
$$= a(x,y_1) + b(x,y_2)$$

3. $(x,x) \geq 0$, where equality holds if and only if x is the zero vector.

$$(x,x) = \int_{-1}^{1} x(t)x(t)dt = \int_{-1}^{1} x^2(t)\, dt \geq 0$$

Equality holds here only if $x(t) = 0$ for $-1 \leq t \leq 1$, which is the zero vector.

P5.7 Two vectors from the vector space described in the previous problem (polynomials defined on the interval [-1, 1]) are $1 + t$ and $1 - t$. Find an orthogonal set of vectors based on these two vectors.

Step 1.

$$v_1 = y_1 = 1 + t$$

Step 2.

$$v_2 = y_2 - \frac{(v_1,y_2)}{(v_1,v_1)}v_1$$

where

$$(v_1,y_2) = \int_{-1}^{1} (1+t)(1-t)dt = \left(t - \frac{t^3}{3}\right)\Bigg|_{-1}^{1} = \left(\frac{2}{3}\right) - \left(-\frac{2}{3}\right) = \frac{4}{3}$$

$$(v_1,v_1) = \int_{-1}^{1} (1+t)^2 dt = \frac{(1+t)^3}{3}\Bigg|_{-1}^{1} = \left(\frac{8}{3}\right) - (0) = \frac{8}{3}.$$

Therefore

$$v_2 = (1-t) - \frac{4/3}{8/3}(1+t) = \frac{1}{2} - \frac{3}{2}t.$$

P5.8 Expand $\mathbf{x} = \begin{bmatrix} 6 & 9 & 9 \end{bmatrix}^T$ in terms of the following basis set.

$$\mathbf{v}_1 = \begin{bmatrix} 1 \\ 1 \\ 1 \end{bmatrix} \qquad \mathbf{v}_2 = \begin{bmatrix} 1 \\ 2 \\ 3 \end{bmatrix} \qquad \mathbf{v}_3 = \begin{bmatrix} 1 \\ 3 \\ 2 \end{bmatrix}$$

The first step is to calculate the reciprocal basis vectors.

$$\mathbf{B} = \begin{bmatrix} 1 & 1 & 1 \\ 1 & 2 & 3 \\ 1 & 3 & 2 \end{bmatrix} \qquad \mathbf{B}^{-1} = \begin{bmatrix} \frac{5}{3} & -\frac{1}{3} & -\frac{1}{3} \\ -\frac{1}{3} & -\frac{1}{3} & \frac{2}{3} \\ -\frac{1}{3} & \frac{2}{3} & -\frac{1}{3} \end{bmatrix}$$

Therefore taking the rows of $\mathbf{B}^{-1}$,

$$\mathbf{r}_1 = \begin{bmatrix} 5/3 \\ -1/3 \\ -1/3 \end{bmatrix} \qquad \mathbf{r}_2 = \begin{bmatrix} -1/3 \\ -1/3 \\ 2/3 \end{bmatrix} \qquad \mathbf{r}_3 = \begin{bmatrix} -1/3 \\ 2/3 \\ -1/3 \end{bmatrix}.$$

The coefficients in the expansion are calculated

$$x_1^v = \mathbf{r}_1^T \mathbf{x} = \begin{bmatrix} \frac{5}{3} & \frac{-1}{3} & \frac{-1}{3} \end{bmatrix} \begin{bmatrix} 6 \\ 9 \\ 9 \end{bmatrix} = 4$$

$$x_2^v = \mathbf{r}_2^T \mathbf{x} = \begin{bmatrix} \frac{-1}{3} & \frac{-1}{3} & \frac{2}{3} \end{bmatrix} \begin{bmatrix} 6 \\ 9 \\ 9 \end{bmatrix} = 1$$

$$x_3^v = \mathbf{r}_3^T \mathbf{x} = \begin{bmatrix} \frac{-1}{3} & \frac{2}{3} & \frac{-1}{3} \end{bmatrix} \begin{bmatrix} 6 \\ 9 \\ 9 \end{bmatrix} = 1 ,$$

and the expansion is written

$$\mathbf{x} = x_1^v \mathbf{v}_1 + x_2^v \mathbf{v}_2 + x_3^v \mathbf{v}_3 = 4 \begin{bmatrix} 1 \\ 1 \\ 1 \end{bmatrix} + 1 \begin{bmatrix} 1 \\ 2 \\ 3 \end{bmatrix} + 1 \begin{bmatrix} 1 \\ 3 \\ 2 \end{bmatrix} .$$

We can represent the process in matrix form:

$$\mathbf{x}^v = \mathbf{B}^{-1} \mathbf{x} = \begin{bmatrix} \frac{5}{3} & -\frac{1}{3} & -\frac{1}{3} \\ -\frac{1}{3} & -\frac{1}{3} & \frac{2}{3} \\ -\frac{1}{3} & \frac{2}{3} & -\frac{1}{3} \end{bmatrix} \begin{bmatrix} 6 \\ 9 \\ 9 \end{bmatrix} = \begin{bmatrix} 4 \\ 1 \\ 1 \end{bmatrix} .$$

Recall that both $\mathbf{x}^v$ and $\mathbf{x}$ are representations of the same vector, but are expanded in terms of different basis sets. (It is assumed that $\mathbf{x}$ uses the standard basis set, unless otherwise indicated.)

Epilogue

This chapter has presented a few of the basic concepts of vector spaces, material that is critical to the understanding of how neural networks work. This subject of vector spaces is very large, and we have made no attempt to cover all its aspects. Instead, we have presented those concepts that we feel are most relevant to neural networks. The topics covered here will be revisited in almost every chapter that follows.

The next chapter will continue our investigation of the topics of linear algebra most relevant to neural networks. There we will concentrate on linear transformations and matrices.

Further Reading

[Brog91] W. L. Brogan, *Modern Control Theory*, 3rd Ed., Englewood Cliffs, NJ: Prentice-Hall, 1991.

This is a well-written book on the subject of linear systems. The first half of the book is devoted to linear algebra. It also has good sections on the solution of linear differential equations and the stability of linear and nonlinear systems. It has many worked problems.

[Stra76] G. Strang, *Linear Algebra and Its Applications*, New York: Academic Press, 1980.

Strang has written a good basic text on linear algebra. Many applications of linear algebra are integrated into the text.

Exercises

E5.1 Consider again the perceptron described in Problem P5.1. If $b \neq 0$, show that the decision boundary is not a vector space.

E5.2 What is the dimension of the vector space described in Problem P5.1?

E5.3 Consider the set of all continuous functions that satisfy the condition $f(0) = 0$. Show that this is a vector space.

E5.4 Show that the set of 2×2 matrices is a vector space.

E5.5 Consider a perceptron network, with the following weights and bias.

$$\mathbf{W} = \begin{bmatrix} 1 & 0 & -1 \end{bmatrix}, \ b = 0.$$

i. Write out the equation for the decision boundary.

ii. Show that the decision boundary is a vector space. (Demonstrate that the 10 criteria are satisfied for any point on the boundary.)

iii. What is the dimension of the vector space?

iv. Find a basis set for the vector space.

E5.6 The three parts to this question refer to subsets of the set of real-valued continuous functions defined on the interval [0,1]. Tell which of these subsets are vector spaces. If the subset is not a vector space, identify which of the 10 criteria are not satisfied.

i. All functions such that $f(0.5) = 2$.

ii. All functions such that $f(0.75) = 0$.

iii. All functions such that $f(0.5) = -f(0.75) - 3$.

E5.7 The next three questions refer to subsets of the set of real polynomials defined over the real line (e.g., $3 + 2t + 6t^2$). Tell which of these subsets are vector spaces. If the subset is not a vector space, identify which of the 10 criteria are not satisfied.

i. Polynomials of degree 5 or less.

ii. Polynomials that are positive for positive t.

iii. Polynomials that go to zero as t goes to zero.

E5.8 Which of the following sets of vectors are independent? Find the dimension of the vector space spanned by each set. (Verify your answers to parts (i) and (iv) using the MATLAB function `rank`.)

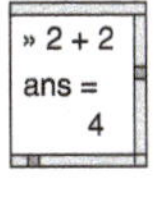

i. $\begin{bmatrix} 1 \\ 2 \\ 3 \end{bmatrix}$ $\begin{bmatrix} 1 \\ 0 \\ 1 \end{bmatrix}$ $\begin{bmatrix} 1 \\ 2 \\ 1 \end{bmatrix}$

ii. $\sin t$ $\cos t$ $\cos(2t)$

iii. $1 + t$ $1 - t$

iv. $\begin{bmatrix} 1 \\ 2 \\ 2 \\ 1 \end{bmatrix}$ $\begin{bmatrix} 1 \\ 0 \\ 0 \\ 1 \end{bmatrix}$ $\begin{bmatrix} 3 \\ 4 \\ 4 \\ 3 \end{bmatrix}$

E5.9 Recall the apple and orange pattern recognition problem of Chapter 3. Find the angles between each of the prototype patterns (*orange* and *apple*) and the test input pattern (*oblong orange*). Verify that the angles make intuitive sense.

$$\mathbf{p}_1 = \begin{bmatrix} 1 \\ -1 \\ -1 \end{bmatrix} (orange) \qquad \mathbf{p}_2 = \begin{bmatrix} 1 \\ 1 \\ -1 \end{bmatrix} (apple) \qquad \mathbf{p} = \begin{bmatrix} -1 \\ -1 \\ -1 \end{bmatrix}$$

E5.10 Using the following basis vectors, find an orthogonal set using Gram-Schmidt orthogonalization. (Check your answer using MATLAB.)

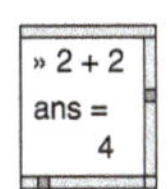

$$\mathbf{y}_1 = \begin{bmatrix} 1 \\ 0 \\ 0 \end{bmatrix} \qquad \mathbf{y}_2 = \begin{bmatrix} 1 \\ 1 \\ 0 \end{bmatrix} \qquad \mathbf{y}_3 = \begin{bmatrix} 1 \\ 1 \\ 1 \end{bmatrix}$$

E5.11 Consider the vector space of all piecewise continuous functions on the interval [0, 1]. The set $\{f_1, f_2, f_3\}$, which is defined in Figure E15.1, contains three vectors from this vector space.

i. Show that this set is linearly independent.

ii. Generate an orthogonal set using the Gram-Schmidt procedure. The inner product is defined to be

$$(f,g) = \int_0^1 f(t)g(t)dt.$$

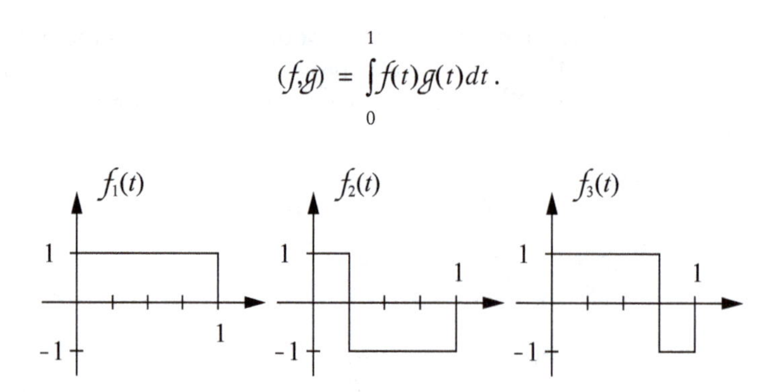

Figure E15.1 Basis Set for Exercise E5.11

E5.12 Consider the vector space of all piece wise continuous functions on the interval [0,1]. The set $\{f_1, f_2\}$, which is defined in Figure E15.2, contains two vectors from this vector space.

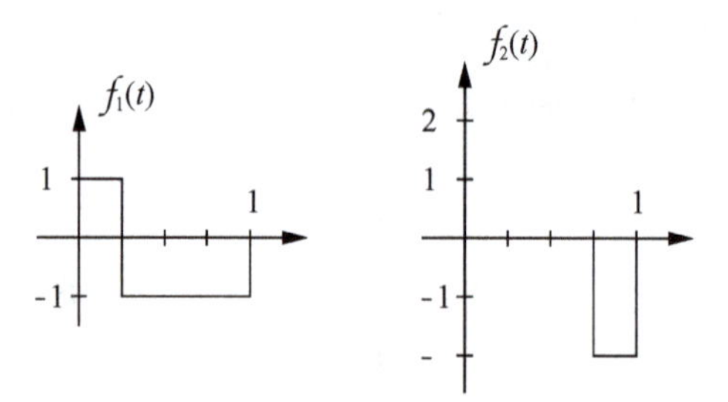

Figure E15.2 Basis Set for Exercise E5.12

i. Generate an orthogonal set using the Gram-Schmidt procedure. The inner product is defined to be

$$(f,g) = \int_0^1 f(t)g(t)dt.$$

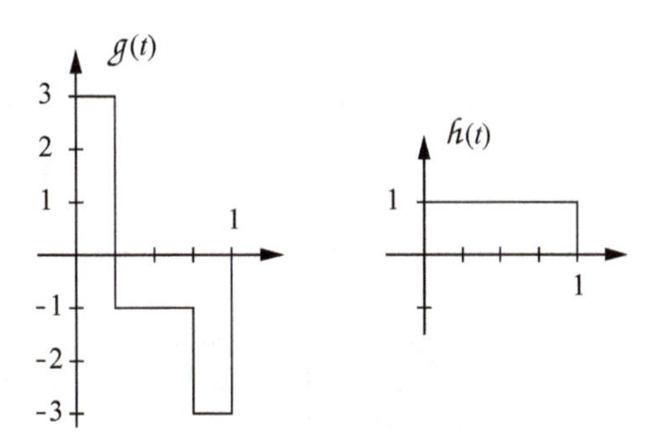

Figure E15.3 Vectors vectors g and h for Exercise E5.12 part ii.

ii. Expand the vectors g and h in Figure E15.3 in terms of the orthogonal set you created in Part 1. Explain any problems you find.

E5.13 Consider the set of polynomials of degree 1 or less. This is a linear vector space. One basis set for this space is

$$\{u_1 = 1, u_2 = t\}$$

Using this basis set, the polynomial $y = 2 + 4t$ can be represented as

$$\mathbf{y}^u = \begin{bmatrix} 2 \\ 4 \end{bmatrix}$$

Consider the new basis set

$$\{v_1 = 1 + t, v_2 = 1 - t\}$$

Use reciprocal basis vectors to find the representation of y in terms of this new basis set.

E5.14 A vector χ can be expanded in terms of the basis vectors $\{v_1, v_2\}$ as

$$\chi = 1v_1 + 1v_2$$

The vectors v_1 and v_2 can be expanded in terms of the basis vectors $\{s_1, s_2\}$ as

$$v_1 = 1s_1 - 1s_2$$

$$v_2 = 1s_1 + 1s_2$$

i. Find the expansion for x in terms of the basis vectors $\{s_1, s_2\}$.

ii. A vector y can be expanded in terms of the basis vectors $\{s_1, s_2\}$ as

$$y = 1s_1 + 1s_2 .$$

Find the expansion of y in terms of the basis vectors $\{v_1, v_2\}$.

E5.15 Consider the vector space of all continuous functions on the interval [0,1]. The set $\{f_1, f_2\}$, which is defined in the figure below, contains two vectors from this vector space.

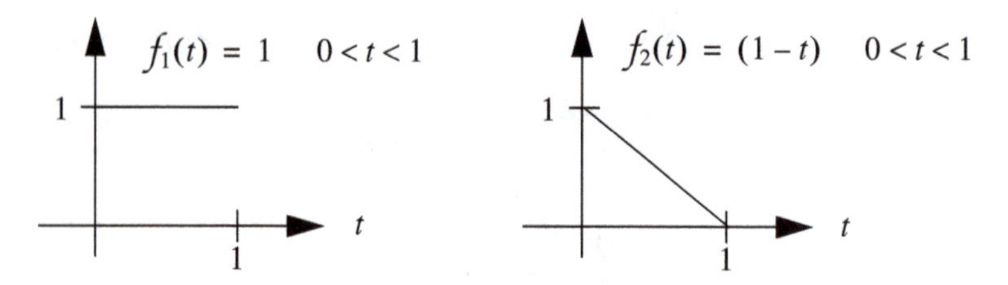

Figure E15.4 Independent Vectors for Exercise E5.15

i. From these two vectors, generate an orthogonal set $\{g_1, g_2\}$ using the Gram-Schmidt procedure. The inner product is defined to be

$$(f,g) = \int_0^1 f(t)g(t)dt .$$

Plot the two orthogonal vectors g_1 and g_2 as functions of time.

ii. Expand the following vector h in terms of the orthogonal set you created in part i., using Eq. (5.27). Demonstrate that the expansion is correct by reproducing h as a combination of g_1 and g_2 .

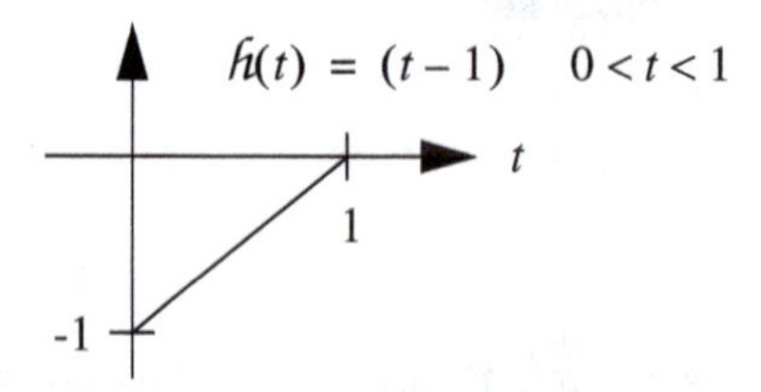

Figure E15.5 Vector h for Exercise E5.15

E5.16 Consider the set of all complex numbers. This can be considered a vector space, because it satisfies the ten defining properties. We can also define an inner product for this vector space $(\chi, y) = Re(\chi)Re(y) + Im(\chi)Im(y)$, where $Re(\chi)$ is the real part of χ, and $Im(\chi)$ is the imaginary part of χ. This leads to the following definition for norm: $\|\chi\| = \sqrt{(\chi, y)}$.

i. Consider the following basis set for the vector space described above: $v_1 = 1 + 2j$, $v_2 = 2 + j$. Using the Gram-Schmidt method, find an orthogonal basis set.

ii. Using your orthogonal basis set from part i., find vector expansions for $u_1 = 1 - j$, $u_2 = 1 + j$, and $\chi = 3 + j$. This will allow you to write χ, u_1, and u_2 as a columns of numbers $\mathbf{x}$, $\mathbf{u}_1$ and $\mathbf{u}_2$.

iii. We now want to represent the vector χ using the basis set $\{u_1, u_2\}$. Use reciprocal basis vectors to find the expansion for χ in terms of the basis vectors $\{u_1, u_2\}$. This will allow you to write χ as a new column of numbers $\mathbf{x}^u$.

iv. Show that the representations for χ that you found in parts ii. and iii. are equivalent (the two columns of numbers $\mathbf{x}$ and $\mathbf{x}^u$ both represent the same vector χ).

E5.17 Consider the vectors defined in Figure E15.6. The set $\{s_1, s_2\}$ is the standard basis set. The set $\{u_1, u_2\}$ is an alternate basis set. The vector χ is a vector that we wish to represent with respect to the two basis sets.

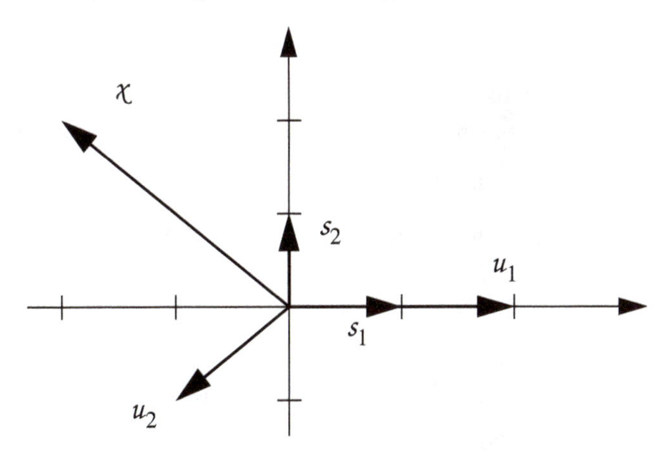

Figure E15.6 Vector Definitions for Exercise E5.17

i. Write the expansion for χ in terms of the standard basis $\{s_1, s_2\}$.

ii. Write the expansions for u_1 and u_2 in terms of the standard basis $\{s_1, s_2\}$

iii. Using reciprocal basis vectors, write the expansion for χ in terms of the basis $\{u_1, u_2\}$.

iv. Draw sketches, similar to Figure 5.2, that demonstrate that the expansions of part i. and part iii. are equivalent.

E5.18 Consider the set of all functions that can be written in the form $A\sin(t+\theta)$. This set can be considered a vector space, because it satisfies the ten defining properties.

i. Consider the following basis set for the vector space described above: $v_1 = \sin(t)$, $v_2 = \cos(t)$. Represent the vector $\chi = 2\sin(t) + 4\cos(t)$ as a column of numbers $\mathbf{x}^v$ (find the vector expansion), using this basis set.

ii. Using your basis set from part i., find vector expansions for $u_1 = 2\sin(t) + \cos(t)$, $u_2 = 3\sin(t)$.

iii. We now want to represent the vector χ of part i., using the basis set $\{u_1, u_2\}$. Use reciprocal basis vectors to find the expansion for χ in terms of the basis vectors $\{u_1, u_2\}$. This will allow you to write χ as a new column of numbers $\mathbf{x}^u$.

iv. Show that the representations for χ that you found in parts i. and iii. are equivalent (the two columns of numbers $\mathbf{x}^v$ and $\mathbf{x}^u$ both represent the same vector χ).

E5.19 Suppose that we have three vectors: $\chi, y, z \in X$. We want to add some multiple of y to χ, so that the resulting vector is orthogonal to z.

i. How would you determine the appropriate multiple of y to add to χ?

ii. Verify your results in part i. using the following vectors.

$$\mathbf{x} = \begin{bmatrix} 1 \\ 0 \end{bmatrix} \quad \mathbf{y} = \begin{bmatrix} 1 \\ 0.5 \end{bmatrix} \quad \mathbf{z} = \begin{bmatrix} 0.5 \\ 1 \end{bmatrix}$$

iii. Use a sketch to illustrate your results from part ii.

E5.20 Expand $\mathbf{x} = \begin{bmatrix} 1 & 2 & 2 \end{bmatrix}^T$ in terms of the following basis set. (Verify your answer using MATLAB.)

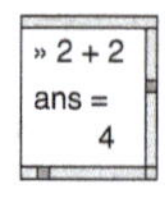

$$\mathbf{v}_1 = \begin{bmatrix} -1 \\ 1 \\ 0 \end{bmatrix} \quad \mathbf{v}_2 = \begin{bmatrix} 1 \\ 1 \\ -2 \end{bmatrix} \quad \mathbf{v}_3 = \begin{bmatrix} 1 \\ 1 \\ 0 \end{bmatrix}$$

E5.21 Find the value of a that makes $\|\chi - a y\|$ a minimum. (Use $\|\chi\| = (\chi,\chi)^{1/2}$.) Show that for this value of a the vector $z = \chi - ay$ is orthogonal to y and that

$$\|\chi - ay\|^2 + \|ay\|^2 = \|\chi\|^2 .$$

(The vector ay is the projection of χ on y.) Draw a diagram for the case where χ and y are two-dimensional. Explain how this concept is related to Gram-Schmidt orthogonalization.

6 Linear Transformations for Neural Networks

Objectives

This chapter will continue the work of Chapter 5 in laying out the mathematical foundations for our analysis of neural networks. In Chapter 5 we reviewed vector spaces; in this chapter we investigate linear transformations as they apply to neural networks.

As we have seen in previous chapters, the multiplication of an input vector by a weight matrix is one of the key operations that is performed by neural networks. This operation is an example of a linear transformation. We want to investigate general linear transformations and determine their fundamental characteristics. The concepts covered in this chapter, such as eigenvalues, eigenvectors and change of basis, will be critical to our understanding of such key neural network topics as performance learning (including the Widrow-Hoff rule and backpropagation) and Hopfield network convergence.

Theory and Examples

Recall the Hopfield network that was discussed in Chapter 3. (See Figure 6.1.) The output of the network is updated synchronously according to the equation

$$\mathbf{a}(t+1) = satlin(\mathbf{W}\mathbf{a}(t) + \mathbf{b}) . \tag{6.1}$$

Notice that at each iteration the output of the network is again multiplied by the weight matrix $\mathbf{W}$. What is the effect of this repeated operation? Can we determine whether or not the output of the network will converge to some steady state value, go to infinity, or oscillate? In this chapter we will lay the foundation for answering these questions, along with many other questions about neural networks discussed in this book.

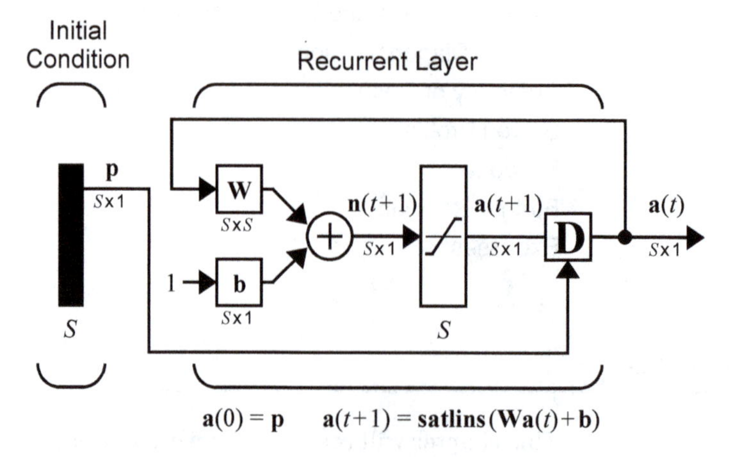

Figure 6.1 Hopfield Network

Linear Transformations

We begin with some general definitions.

Transformation

A *transformation* consists of three parts:

1. a set of elements $X = \{\chi_i\}$, called the domain,
2. a set of elements $Y = \{y_i\}$, called the range, and
3. a rule relating each $\chi_i \in X$ to an element $y_i \in Y$.

Linear Transformation

A transformation $\mathcal{A}$ is *linear* if:

1. for all $\chi_1, \chi_2 \in X$, $\mathcal{A}(\chi_1 + \chi_2) = \mathcal{A}(\chi_1) + \mathcal{A}(\chi_2)$,

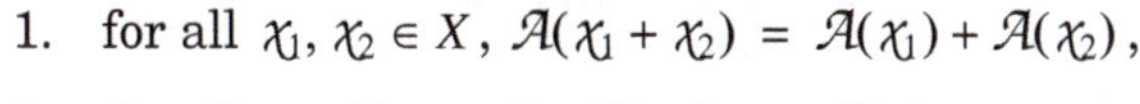

2. for all $\chi \in X$, $a \in R$, $\mathcal{A}(a\chi) = a\mathcal{A}(\chi)$.

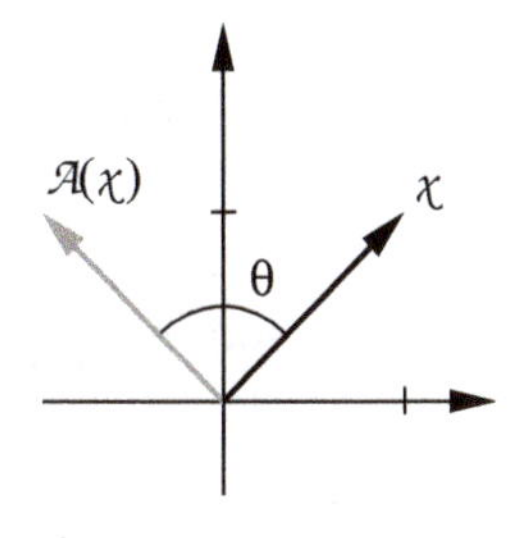

Consider, for example, the transformation obtained by rotating vectors in $\Re^2$ by an angle θ, as shown in the figure to the left. The next two figures illustrate that property 1 is satisfied for rotation. They show that if you want to rotate a sum of two vectors, you can rotate each vector first and then sum them. The fourth figure illustrates property 2. If you want to rotate a scaled vector, you can rotate it first and then scale it. Therefore rotation is a linear operation.

Matrix Representations

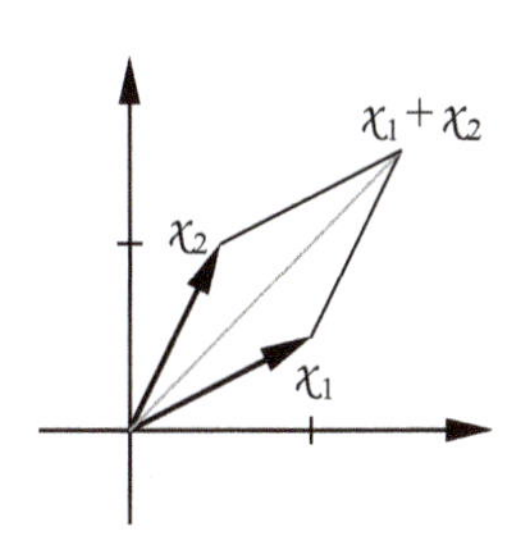

As we mentioned at the beginning of this chapter, matrix multiplication is an example of a linear transformation. We can also show that any linear transformation between two finite-dimensional vector spaces can be represented by a matrix (just as in the last chapter we showed that any general vector in a finite-dimensional vector space can be represented by a column of numbers). To show this we will use most of the concepts covered in the previous chapter.

Let $\{v_1, v_2, \ldots, v_n\}$ be a basis for vector space X, and let $\{u_1, u_2, \ldots, u_m\}$ be a basis for vector space Y. This means that for any two vectors $\chi \in X$ and $y \in Y$

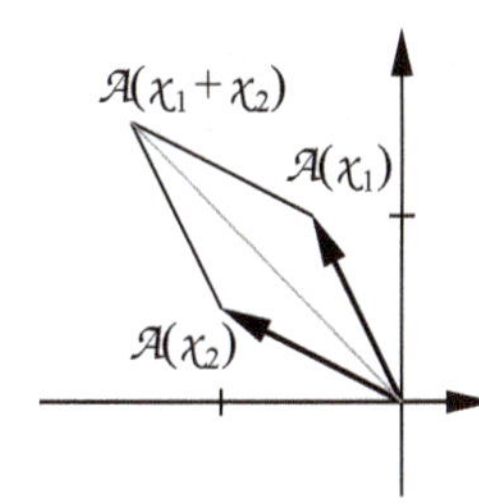

$$\chi = \sum_{i=1}^{n} x_i v_i \text{ and } y = \sum_{i=1}^{m} y_i u_i . \tag{6.2}$$

Let $\mathcal{A}$ be a linear transformation with domain X and range Y ($\mathcal{A}: X \to Y$). Then

$$\mathcal{A}(\chi) = y \tag{6.3}$$

can be written

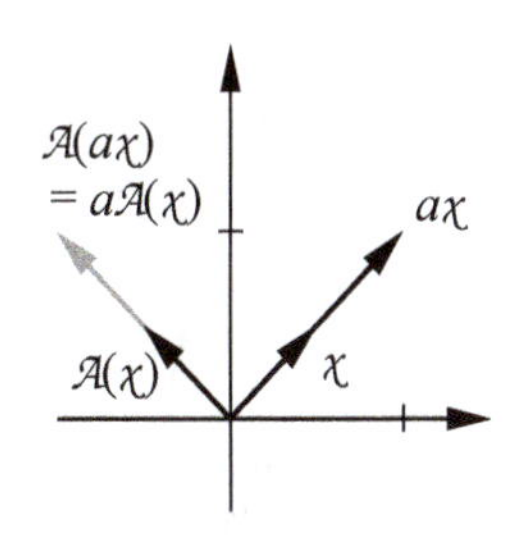

$$\mathcal{A}\left(\sum_{j=1}^{n} x_j v_j\right) = \sum_{i=1}^{m} y_i u_i . \tag{6.4}$$

Since $\mathcal{A}$ is a linear operator, Eq. (6.4) can be written

$$\sum_{j=1}^{n} x_j \mathcal{A}(v_j) = \sum_{i=1}^{m} y_i u_i . \tag{6.5}$$

Since the vectors $\mathcal{A}(v_j)$ are elements of Y, they can be written as linear combinations of the basis vectors for Y:

$$\mathcal{A}(v_j) = \sum_{i=1}^{m} a_{ij} u_i . \tag{6.6}$$

(Note that the notation used for the coefficients of this expansion, a_{ij}, was not chosen by accident.) If we substitute Eq. (6.6) into Eq. (6.5) we obtain

$$\sum_{j=1}^{n} x_j \sum_{i=1}^{m} a_{ij} u_i = \sum_{i=1}^{m} y_i u_i . \tag{6.7}$$

The order of the summations can be reversed, to produce

$$\sum_{i=1}^{m} u_i \sum_{j=1}^{n} a_{ij} x_j = \sum_{i=1}^{m} y_i u_i . \tag{6.8}$$

This equation can be rearranged, to obtain

$$\sum_{i=1}^{m} u_i \left(\sum_{j=1}^{n} a_{ij} x_j - y_i \right) = 0. \tag{6.9}$$

Recall that since the u_i form a basis set they must be independent. This means that each coefficient that multiplies u_i in Eq. (6.9) must be identically zero (see Eq. (5.4)), therefore

$$\sum_{j=1}^{n} a_{ij} x_j = y_i . \tag{6.10}$$

This is just matrix multiplication, as in

$$\begin{bmatrix} a_{11} & a_{12} & \dots & a_{1n} \\ a_{21} & a_{22} & \dots & a_{2n} \\ \vdots & \vdots & & \vdots \\ a_{m1} & a_{m2} & \dots & a_{mn} \end{bmatrix} \begin{bmatrix} x_1 \\ x_2 \\ \vdots \\ x_n \end{bmatrix} = \begin{bmatrix} y_1 \\ y_2 \\ \vdots \\ y_m \end{bmatrix} \tag{6.11}$$

We can summarize these results: *For any linear transformation between two finite-dimensional vector spaces there is a matrix representation. When we multiply the matrix times the vector expansion for the domain vector x, we obtain the vector expansion for the transformed vector y.*

Keep in mind that the matrix representation is not unique (just as the representation of a general vector by a column of numbers is not unique — see Chapter 5). If we change the basis set for the domain or for the range, the matrix representation will also change. We will use this fact to our advantage in later chapters.

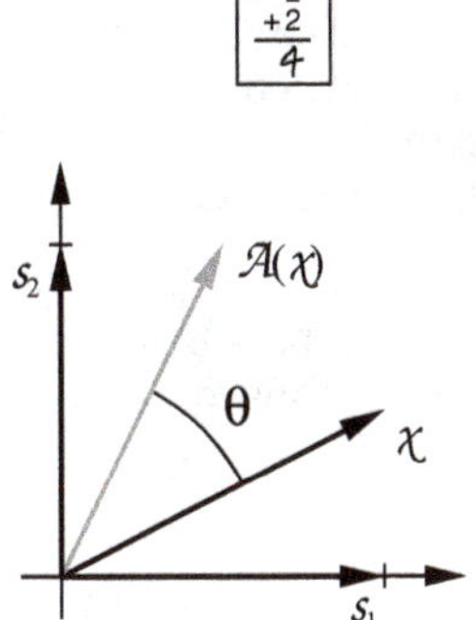

As an example of a matrix representation, consider the rotation transformation. Let's find a matrix representation for that transformation. The key step is given in Eq. (6.6). We must transform each basis vector for the domain and then expand it in terms of the basis vectors of the range. In this example the domain and the range are the same ($X = Y = \Re^2$), so to keep things simple we will use the standard basis for both ($u_i = v_i = s_i$), as shown in the adjacent figure.

The first step is to transform the first basis vector and expand the resulting transformed vector in terms of the basis vectors. If we rotate s_1 counterclockwise by the angle θ we obtain

$$\mathcal{A}(s_1) = \cos(\theta)s_1 + \sin(\theta)s_2 = \sum_{i=1}^{2} a_{i1}s_i = a_{11}s_1 + a_{21}s_2, \tag{6.12}$$

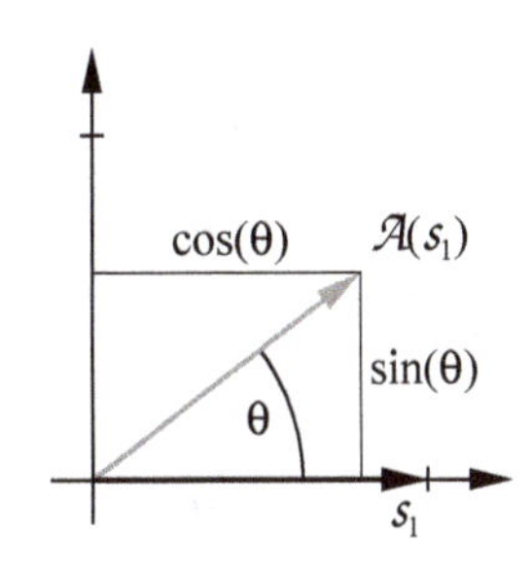

as can be seen in the middle left figure. The two coefficients in this expansion make up the first column of the matrix representation.

The next step is to transform the second basis vector. If we rotate s_2 counterclockwise by the angle θ we obtain

$$\mathcal{A}(s_2) = -\sin(\theta)s_1 + \cos(\theta)s_2 = \sum_{i=1}^{2} a_{i2}s_i = a_{12}s_1 + a_{22}s_2, \tag{6.13}$$

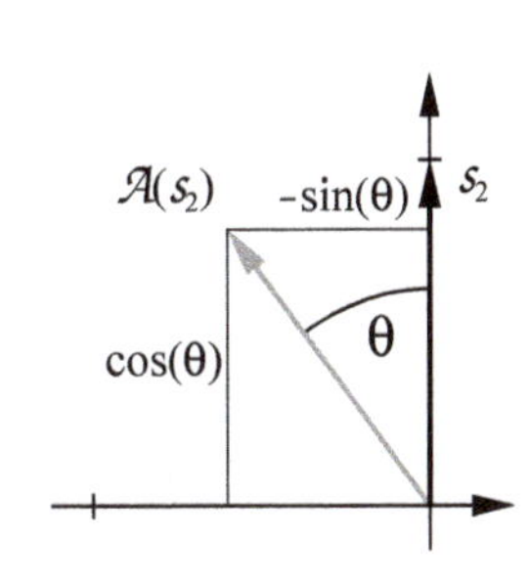

as can be seen in the lower left figure. From this expansion we obtain the second column of the matrix representation. The complete matrix representation is thus given by

$$\mathbf{A} = \begin{bmatrix} \cos(\theta) & -\sin(\theta) \\ \sin(\theta) & \cos(\theta) \end{bmatrix}. \tag{6.14}$$

Verify for yourself that when you multiply a vector by the matrix of Eq. (6.14), the vector is rotated by an angle θ.

In summary, to obtain the matrix representation of a transformation we use Eq. (6.6). We transform each basis vector for the domain and expand it in terms of the basis vectors of the range. The coefficients of each expansion produce one column of the matrix.

To graphically investigate the process of creating a matrix representation, use the Neural Network Design Demonstration Linear Transformations (**nnd6lt**).

Change of Basis

We notice from the previous section that the matrix representation of a linear transformation is not unique. The representation will depend on what basis sets are used for the domain and the range of the transformation. In this section we will illustrate exactly how a matrix representation changes as the basis sets are changed.

Consider a linear transformation $\mathcal{A}:X \rightarrow Y$. Let $\{v_1, v_2, \ldots, v_n\}$ be a basis for vector space X, and let $\{u_1, u_2, \ldots, u_m\}$ be a basis for vector space Y. Therefore, any vector $\chi \in X$ can be written

$$\chi = \sum_{i=1}^{n} x_i v_i, \tag{6.15}$$

and any vector $y \in Y$ can be written

$$y = \sum_{i=1}^{m} y_i u_i. \tag{6.16}$$

So if

$$\mathcal{A}(\chi) = y \tag{6.17}$$

the matrix representation will be

$$\begin{bmatrix} a_{11} & a_{12} & \ldots & a_{1n} \\ a_{21} & a_{22} & \ldots & a_{2n} \\ \vdots & \vdots & & \vdots \\ a_{m1} & a_{m2} & \ldots & a_{mn} \end{bmatrix} \begin{bmatrix} x_1 \\ x_2 \\ \vdots \\ x_n \end{bmatrix} = \begin{bmatrix} y_1 \\ y_2 \\ \vdots \\ y_m \end{bmatrix}, \tag{6.18}$$

or

$$\mathbf{Ax} = \mathbf{y}. \tag{6.19}$$

Now suppose that we use different basis sets for X and Y. Let $\{t_1, t_2, \ldots, t_n\}$ be the new basis for X, and let $\{w_1, w_2, \ldots, w_m\}$ be the new basis for Y. With the new basis sets, the vector $\chi \in X$ is written

$$x = \sum_{i=1}^{n} x'_i t_i , \tag{6.20}$$

and the vector $y \in Y$ is written

$$y = \sum_{i=1}^{m} y'_i w_i . \tag{6.21}$$

This produces a new matrix representation:

$$\begin{bmatrix} a'_{11} & a'_{12} & \dots & a'_{1n} \\ a'_{21} & a'_{22} & \dots & a'_{2n} \\ \vdots & \vdots & & \vdots \\ a'_{m1} & a'_{m2} & \dots & a'_{mn} \end{bmatrix} \begin{bmatrix} x'_1 \\ x'_2 \\ \vdots \\ x'_n \end{bmatrix} = \begin{bmatrix} y'_1 \\ y'_2 \\ \vdots \\ y'_m \end{bmatrix} , \tag{6.22}$$

or

$$\mathbf{A}'\mathbf{x}' = \mathbf{y}' . \tag{6.23}$$

What is the relationship between $\mathbf{A}$ and $\mathbf{A}'$? To find out, we need to find the relationship between the two basis sets. First, since each t_i is an element of X, they can be expanded in terms of the original basis for X:

$$t_i = \sum_{j=1}^{n} t_{ji} v_j . \tag{6.24}$$

Next, since each w_i is an element of Y, they can be expanded in terms of the original basis for Y:

$$w_i = \sum_{j=1}^{m} w_{ji} u_j . \tag{6.25}$$

Therefore, the basis vectors can be written as columns of numbers:

$$\mathbf{t}_i = \begin{bmatrix} t_{1i} \\ t_{2i} \\ \vdots \\ t_{ni} \end{bmatrix} \qquad \mathbf{w}_i = \begin{bmatrix} w_{1i} \\ w_{2i} \\ \vdots \\ w_{mi} \end{bmatrix} . \tag{6.26}$$

Define a matrix whose columns are the $\mathbf{t}_i$:

$$\mathbf{B}_t = \begin{bmatrix} \mathbf{t}_1 \ \mathbf{t}_2 \ \dots \ \mathbf{t}_n \end{bmatrix} . \quad (6.27)$$

Then we can write Eq. (6.20) in matrix form:

$$\mathbf{x} = x'_1\mathbf{t}_1 + x'_2\mathbf{t}_2 + \cdots + x'_n\mathbf{t}_n = \mathbf{B}_t\mathbf{x}' . \quad (6.28)$$

This equation demonstrates the relationships between the two different representations for the vector χ. (Note that this is effectively the same as Eq. (5.43). You may want to revisit our discussion of reciprocal basis vectors in Chapter 5.)

Now define a matrix whose columns are the $\mathbf{w}_i$:

$$\mathbf{B}_w = \begin{bmatrix} \mathbf{w}_1 \ \mathbf{w}_2 \ \dots \ \mathbf{w}_m \end{bmatrix} . \quad (6.29)$$

This allows us to write Eq. (6.21) in matrix form,

$$\mathbf{y} = \mathbf{B}_w\mathbf{y}' , \quad (6.30)$$

which then demonstrates the relationships between the two different representations for the vector y.

Now substitute Eq. (6.28) and Eq. (6.30) into Eq. (6.19):

$$\mathbf{A}\mathbf{B}_t\mathbf{x}' = \mathbf{B}_w\mathbf{y}' . \quad (6.31)$$

If we multiply both sides of this equation by $\mathbf{B}_w^{-1}$ we obtain

$$[\mathbf{B}_w^{-1}\mathbf{A}\mathbf{B}_t]\mathbf{x}' = \mathbf{y}' . \quad (6.32)$$

Change of Basis

A comparison of Eq. (6.32) and Eq. (6.23) yields the following operation for a *change of basis*:

$$\mathbf{A}' = [\mathbf{B}_w^{-1}\mathbf{A}\mathbf{B}_t] . \quad (6.33)$$

Similarity Transform

This key result, which describes the relationship between any two matrix representations of a given linear transformation, is called a *similarity transform* [Brog91]. It will be of great use to us in later chapters. It turns out that with the right choice of basis vectors we can obtain a matrix representation that reveals the key characteristics of the linear transformation it represents. This will be discussed in the next section.

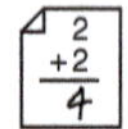

As an example of changing basis sets, let's revisit the vector rotation example of the previous section. In that section a matrix representation was developed using the standard basis set $\{s_1, s_2\}$. Now let's find a new representation using the basis $\{t_1, t_2\}$, which is shown in the adjacent fig-

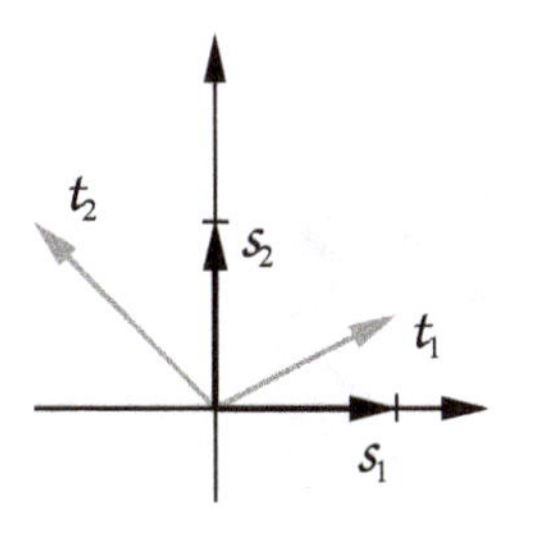

ure. (Note that in this example the same basis set is used for both the domain and the range.)

The first step is to expand t_1 and t_2 in terms of the standard basis set, as in Eq. (6.24) and Eq. (6.25). By inspection of the adjacent figure we find:

$$t_1 = s_1 + 0.5 s_2 , \quad (6.34)$$

$$t_2 = -s_1 + s_2 . \quad (6.35)$$

Therefore we can write

$$\mathbf{t}_1 = \begin{bmatrix} 1 \\ 0.5 \end{bmatrix} \qquad \mathbf{t}_2 = \begin{bmatrix} -1 \\ 1 \end{bmatrix} . \quad (6.36)$$

Now we can form the matrix

$$\mathbf{B}_t = \begin{bmatrix} \mathbf{t}_1 & \mathbf{t}_2 \end{bmatrix} = \begin{bmatrix} 1 & -1 \\ 0.5 & 1 \end{bmatrix} , \quad (6.37)$$

and, because we are using the same basis set for both the domain and the range of the transformation,

$$\mathbf{B}_w = \mathbf{B}_t = \begin{bmatrix} 1 & -1 \\ 0.5 & 1 \end{bmatrix} . \quad (6.38)$$

We can now compute the new matrix representation from Eq. (6.33):

$$\mathbf{A}' = [\mathbf{B}_w^{-1} \mathbf{A} \mathbf{B}_t] = \begin{bmatrix} 2/3 & 2/3 \\ -1/3 & 2/3 \end{bmatrix} \begin{bmatrix} \cos\theta & -\sin\theta \\ \sin\theta & \cos\theta \end{bmatrix} \begin{bmatrix} 1 & -1 \\ 0.5 & 1 \end{bmatrix}$$
$$= \begin{bmatrix} 1/3\sin\theta + \cos\theta & -4/3\sin\theta \\ \frac{5}{6}\sin\theta & -1/3\sin\theta + \cos\theta \end{bmatrix} . \quad (6.39)$$

Take, for example, the case where $\theta = 30°$.

$$\mathbf{A}' = \begin{bmatrix} 1.033 & -0.667 \\ 0.417 & 0.699 \end{bmatrix} , \quad (6.40)$$

and

$$\mathbf{A} = \begin{bmatrix} 0.866 & -0.5 \\ 0.5 & 0.866 \end{bmatrix}. \tag{6.41}$$

To check that these matrices are correct, let's try a test vector

$$\mathbf{x} = \begin{bmatrix} 1 \\ 0.5 \end{bmatrix}, \text{ which corresponds to } \mathbf{x}' = \begin{bmatrix} 1 \\ 0 \end{bmatrix}. \tag{6.42}$$

(Note that the vector represented by $\mathbf{x}$ and $\mathbf{x}'$ is t_1, a member of the second basis set.) The transformed test vector would be

$$\mathbf{y} = \mathbf{A}\mathbf{x} = \begin{bmatrix} 0.866 & -0.5 \\ 0.5 & 0.866 \end{bmatrix} \begin{bmatrix} 1 \\ 0.5 \end{bmatrix} = \begin{bmatrix} 0.616 \\ 0.933 \end{bmatrix}, \tag{6.43}$$

which should correspond to

$$\mathbf{y}' = \mathbf{A}'\mathbf{x}' = \begin{bmatrix} 1.033 & -0.667 \\ 0.416 & 0.699 \end{bmatrix} \begin{bmatrix} 1 \\ 0 \end{bmatrix} = \begin{bmatrix} 1.033 \\ 0.416 \end{bmatrix}. \tag{6.44}$$

How can we test to see if $\mathbf{y}'$ does correspond to $\mathbf{y}$? Both should be representations of the same vector, y, in terms of two different basis sets; $\mathbf{y}$ uses the basis $\{s_1, s_2\}$ and $\mathbf{y}'$ uses the basis $\{t_1, t_2\}$. In Chapter 5 we used the reciprocal basis vectors to transform from one representation to another (see Eq. (5.43)). Using that concept we have

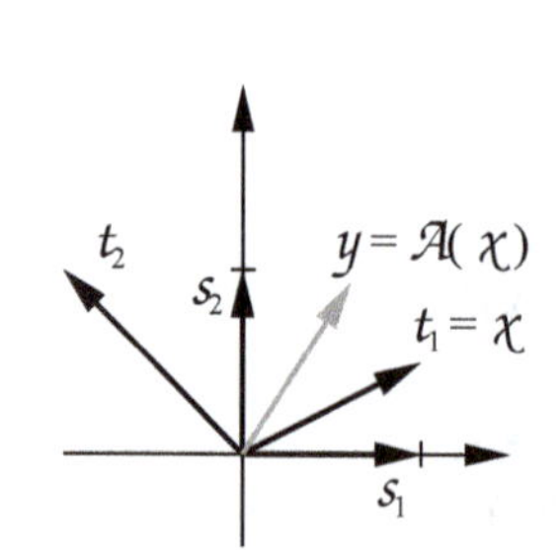

$$\mathbf{y}' = \mathbf{B}^{-1}\mathbf{y} = \begin{bmatrix} 1 & -1 \\ 0.5 & 1 \end{bmatrix}^{-1} \begin{bmatrix} 0.616 \\ 0.933 \end{bmatrix} = \begin{bmatrix} 2/3 & 2/3 \\ -1/3 & 2/3 \end{bmatrix} \begin{bmatrix} 0.616 \\ 0.933 \end{bmatrix} = \begin{bmatrix} 1.033 \\ 0.416 \end{bmatrix}, \tag{6.45}$$

which verifies our previous result. The vectors are displayed in the figure to the left. Verify graphically that the two representations, $\mathbf{y}$ and $\mathbf{y}'$, given by Eq. (6.43) and Eq. (6.44), are reasonable.

Eigenvalues and Eigenvectors

In this final section we want to discuss two key properties of linear transformations: eigenvalues and eigenvectors. Knowledge of these properties will allow us to answer some key questions about neural network performance, such as the question we posed at the beginning of this chapter, concerning the stability of Hopfield networks.

Eigenvalues
Eigenvectors

Let's first define what we mean by *eigenvalues* and *eigenvectors*. Consider a linear transformation $\mathcal{A}: X \rightarrow X$. (The domain is the same as the range.) Those vectors $z \in X$ that are not equal to zero and those scalars λ that satisfy

$$\mathcal{A}(z) = \lambda z \tag{6.46}$$

are called eigenvectors (z) and eigenvalues (λ), respectively. Notice that the term eigenvector is a little misleading, since it is not really a vector but a vector space, since if z satisfies Eq. (6.46), then az will also satisfy it.

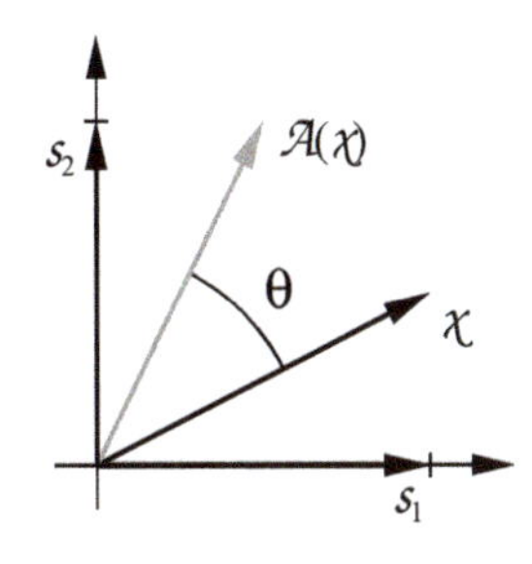

Therefore an eigenvector of a given transformation represents a direction, such that any vector in that direction, when transformed, will continue to point in the same direction, but will be scaled by the eigenvalue. As an example, consider again the rotation example used in the previous sections. Is there any vector that, when rotated by 30°, continues to point in the same direction? No; this is a case where there are no real eigenvalues. (If we allow complex scalars, then two eigenvalues exist, as we will see later.)

How can we compute the eigenvalues and eigenvectors? Suppose that a basis has been chosen for the n-dimensional vector space X. Then the matrix representation for Eq. (6.46) can be written

$$\mathbf{Az} = \lambda\mathbf{z}, \tag{6.47}$$

or

$$[\mathbf{A} - \lambda\mathbf{I}]\mathbf{z} = \mathbf{0}. \tag{6.48}$$

This means that the columns of $[\mathbf{A} - \lambda\mathbf{I}]$ are dependent, and therefore the determinant of this matrix must be zero:

$$|[\mathbf{A} - \lambda\mathbf{I}]| = 0. \tag{6.49}$$

This determinant is an nth-order polynomial. Therefore Eq. (6.49) always has n roots, some of which may be complex and some of which may be repeated.

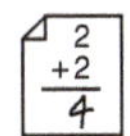

As an example, let's revisit the rotation example. If we use the standard basis set, the matrix of the transformation is

$$\mathbf{A} = \begin{bmatrix} \cos\theta & -\sin\theta \\ \sin\theta & \cos\theta \end{bmatrix}. \tag{6.50}$$

We can then write Eq. (6.49) as

$$\left|\begin{bmatrix} \cos\theta - \lambda & -\sin\theta \\ \sin\theta & \cos\theta - \lambda \end{bmatrix}\right| = 0, \tag{6.51}$$

or

$$\lambda^2 - 2\lambda\cos\theta + ((\cos\theta)^2 + (\sin\theta)^2) = \lambda^2 - 2\lambda\cos\theta + 1 = 0. \tag{6.52}$$

The roots of this equation are

$$\lambda_1 = \cos\theta + j\sin\theta \qquad \lambda_2 = \cos\theta - j\sin\theta . \tag{6.53}$$

Therefore, as we predicted, this transformation has no real eigenvalues (if $\sin\theta \neq 0$). This means that when any real vector is transformed, it will point in a new direction.

2 +2 4

Consider another matrix:

$$\mathbf{A} = \begin{bmatrix} -1 & 1 \\ 0 & -2 \end{bmatrix} . \tag{6.54}$$

To find the eigenvalues we must solve

$$\left| \begin{bmatrix} -1-\lambda & 1 \\ 0 & -2-\lambda \end{bmatrix} \right| = 0 , \tag{6.55}$$

or

$$\lambda^2 + 3\lambda + 2 = (\lambda + 1)(\lambda + 2) = 0 , \tag{6.56}$$

and the eigenvalues are

$$\lambda_1 = -1 \qquad \lambda_2 = -2 . \tag{6.57}$$

To find the eigenvectors we must solve Eq. (6.48), which in this example becomes

$$\begin{bmatrix} -1-\lambda & 1 \\ 0 & -2-\lambda \end{bmatrix} \mathbf{z} = \begin{bmatrix} 0 \\ 0 \end{bmatrix} . \tag{6.58}$$

We will solve this equation twice, once using λ_1 and once using λ_2. Beginning with λ_1 we have

$$\begin{bmatrix} 0 & 1 \\ 0 & -1 \end{bmatrix} \mathbf{z}_1 = \begin{bmatrix} 0 & 1 \\ 0 & -1 \end{bmatrix} \begin{bmatrix} z_{11} \\ z_{21} \end{bmatrix} = \begin{bmatrix} 0 \\ 0 \end{bmatrix} \tag{6.59}$$

or

$$z_{21} = 0 \text{, no constraint on } z_{11} . \tag{6.60}$$

Therefore the first eigenvector will be

$$\mathbf{z}_1 = \begin{bmatrix} 1 \\ 0 \end{bmatrix}, \qquad (6.61)$$

or any scalar multiple. For the second eigenvector we use λ_2:

$$\begin{bmatrix} 1 & 1 \\ 0 & 0 \end{bmatrix} \mathbf{z}_2 = \begin{bmatrix} 1 & 1 \\ 0 & 0 \end{bmatrix} \begin{bmatrix} z_{12} \\ z_{22} \end{bmatrix} = \begin{bmatrix} 0 \\ 0 \end{bmatrix}, \qquad (6.62)$$

or

$$z_{22} = -z_{12} . \qquad (6.63)$$

Therefore the second eigenvector will be

$$\mathbf{z}_2 = \begin{bmatrix} 1 \\ -1 \end{bmatrix}, \qquad (6.64)$$

or any scalar multiple.

To verify our results we consider the following:

$$\mathbf{A}\mathbf{z}_1 = \begin{bmatrix} -1 & 1 \\ 0 & -2 \end{bmatrix} \begin{bmatrix} 1 \\ 0 \end{bmatrix} = \begin{bmatrix} -1 \\ 0 \end{bmatrix} = (-1) \begin{bmatrix} 1 \\ 0 \end{bmatrix} = \lambda_1 \mathbf{z}_1 , \qquad (6.65)$$

$$\mathbf{A}\mathbf{z}_2 = \begin{bmatrix} -1 & 1 \\ 0 & -2 \end{bmatrix} \begin{bmatrix} 1 \\ -1 \end{bmatrix} = \begin{bmatrix} -2 \\ 2 \end{bmatrix} = (-2) \begin{bmatrix} 1 \\ -1 \end{bmatrix} = \lambda_2 \mathbf{z}_2 . \qquad (6.66)$$

To test your understanding of eigenvectors, use the Neural Network Design Demonstration Eigenvector Game (**nnd6eg**).

Diagonalization

Whenever we have n distinct eigenvalues we are guaranteed that we can find n independent eigenvectors [Brog91]. Therefore the eigenvectors make up a basis set for the vector space of the transformation. Let's find the matrix of the previous transformation (Eq. (6.54)) using the eigenvectors as the basis vectors. From Eq. (6.33) we have

$$\mathbf{A}' = [\mathbf{B}^{-1}\mathbf{A}\mathbf{B}] = \begin{bmatrix} 1 & 1 \\ 0 & -1 \end{bmatrix} \begin{bmatrix} -1 & 1 \\ 0 & -2 \end{bmatrix} \begin{bmatrix} 1 & 1 \\ 0 & -1 \end{bmatrix} = \begin{bmatrix} -1 & 0 \\ 0 & -2 \end{bmatrix} . \qquad (6.67)$$

Note that this is a diagonal matrix, with the eigenvalues on the diagonal. This is not a coincidence. Whenever we have distinct eigenvalues we can diagonalize the matrix representation by using the eigenvectors as the ba-

Diagonalization

sis vectors. This *diagonalization* process is summarized in the following. Let

$$\mathbf{B} = \begin{bmatrix} \mathbf{z}_1 & \mathbf{z}_2 & \ldots & \mathbf{z}_n \end{bmatrix}, \tag{6.68}$$

where $\{\mathbf{z}_1, \mathbf{z}_2, \ldots, \mathbf{z}_n\}$ are the eigenvectors of a matrix **A**. Then

$$[\mathbf{B}^{-1}\mathbf{A}\mathbf{B}] = \begin{bmatrix} \lambda_1 & 0 & \ldots & 0 \\ 0 & \lambda_2 & \ldots & 0 \\ \vdots & \vdots & & \vdots \\ 0 & 0 & \ldots & \lambda_n \end{bmatrix}, \tag{6.69}$$

where $\{\lambda_1, \lambda_2, \ldots, \lambda_n\}$ are the eigenvalues of the matrix **A**.

This result will be very helpful as we analyze the performance of several neural networks in later chapters.

Summary of Results

Transformations

A *transformation* consists of three parts:

1. a set of elements $X = \{\chi_i\}$, called the domain,
2. a set of elements $Y = \{y_i\}$, called the range, and
3. a rule relating each $\chi_i \in X$ to an element $y_i \in Y$.

Linear Transformations

A transformation $\mathcal{A}$ is *linear* if:

1. for all $\chi_1, \chi_2 \in X$, $\mathcal{A}(\chi_1 + \chi_2) = \mathcal{A}(\chi_1) + \mathcal{A}(\chi_2)$,
2. for all $\chi \in X$, $a \in R$, $\mathcal{A}(a\chi) = a\mathcal{A}(\chi)$.

Matrix Representations

Let $\{v_1, v_2, \ldots, v_n\}$ be a basis for vector space X, and let $\{u_1, u_2, \ldots, u_m\}$ be a basis for vector space Y. Let $\mathcal{A}$ be a linear transformation with domain X and range Y:

$$\mathcal{A}(\chi) = y.$$

The coefficients of the matrix representation are obtained from

$$\mathcal{A}(v_j) = \sum_{i=1}^{m} a_{ij} u_i .$$

Change of Basis

$$\mathbf{B}_t = \begin{bmatrix} \mathbf{t}_1 & \mathbf{t}_2 & \ldots & \mathbf{t}_n \end{bmatrix}$$

$$\mathbf{B}_w = \begin{bmatrix} \mathbf{w}_1 & \mathbf{w}_2 & \ldots & \mathbf{w}_m \end{bmatrix}$$

$$\mathbf{A}' = [\mathbf{B}_w^{-1} \mathbf{A} \mathbf{B}_t]$$

Eigenvalues and Eigenvectors

$$\mathbf{A}\mathbf{z} = \lambda\mathbf{z}$$

$$|[\mathbf{A} - \lambda\mathbf{I}]| = 0$$

Diagonalization

$$\mathbf{B} = \begin{bmatrix} \mathbf{z}_1 & \mathbf{z}_2 & \dots & \mathbf{z}_n \end{bmatrix},$$

where $\{\mathbf{z}_1, \mathbf{z}_2, \dots, \mathbf{z}_n\}$ are the eigenvectors of a square matrix $\mathbf{A}$.

$$[\mathbf{B}^{-1}\mathbf{A}\mathbf{B}] = \begin{bmatrix} \lambda_1 & 0 & \dots & 0 \\ 0 & \lambda_2 & \dots & 0 \\ \vdots & \vdots & & \vdots \\ 0 & 0 & \dots & \lambda_n \end{bmatrix}$$

Solved Problems

P6.1 Consider the single-layer network shown in Figure P6.1, which has a linear transfer function. Is the transformation from the input vector to the output vector a linear transformation?

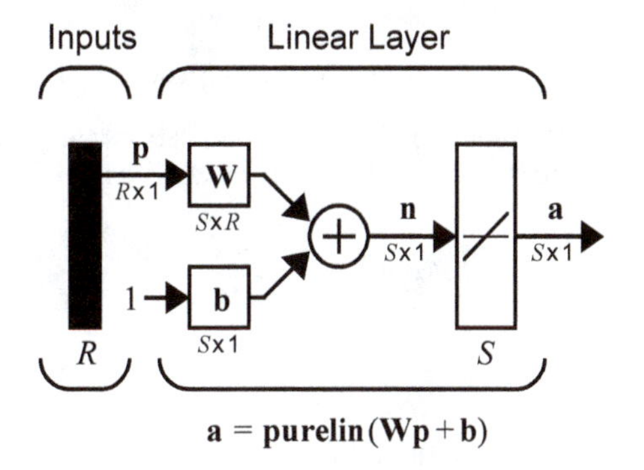

Figure P6.1 Single-Neuron Perceptron

The network equation is

$$\mathbf{a} = \mathcal{A}(\mathbf{p}) = \mathbf{W}\mathbf{p} + \mathbf{b}.$$

In order for this transformation to be linear it must satisfy

1. $\mathcal{A}(\mathbf{p}_1 + \mathbf{p}_2) = \mathcal{A}(\mathbf{p}_1) + \mathcal{A}(\mathbf{p}_2)$,
2. $\mathcal{A}(a\mathbf{p}) = a\mathcal{A}(\mathbf{p})$.

Let's test condition 1 first.

$$\mathcal{A}(\mathbf{p}_1 + \mathbf{p}_2) = \mathbf{W}(\mathbf{p}_1 + \mathbf{p}_2) + \mathbf{b} = \mathbf{W}\mathbf{p}_1 + \mathbf{W}\mathbf{p}_2 + \mathbf{b}.$$

Compare this with

$$\mathcal{A}(\mathbf{p}_1) + \mathcal{A}(\mathbf{p}_2) = \mathbf{W}\mathbf{p}_1 + \mathbf{b} + \mathbf{W}\mathbf{p}_2 + \mathbf{b} = \mathbf{W}\mathbf{p}_1 + \mathbf{W}\mathbf{p}_2 + 2\mathbf{b}.$$

Clearly these two expressions will be equal only if $\mathbf{b} = \mathbf{0}$. Therefore this network performs a nonlinear transformation, even though it has a linear transfer function. This particular type of nonlinearity is called an affine transformation.

P6.2 We discussed projections in Chapter 5. Is a projection a linear transformation?

The projection of a vector χ onto a vector v is computed as

$$y = \mathcal{A}(\chi) = \frac{(\chi, v)}{(v, v)} v,$$

where (χ, v) is the inner product of χ with v.

We need to check to see if this transformation satisfies the two conditions for linearity. Let's start with condition 1:

$$\mathcal{A}(\chi_1 + \chi_2) = \frac{(\chi_1 + \chi_2, v)}{(v, v)} v = \frac{(\chi_1, v) + (\chi_2, v)}{(v, v)} v = \frac{(\chi_1, v)}{(v, v)} v + \frac{(\chi_2, v)}{(v, v)} v$$

$$= \mathcal{A}(\chi_1) + \mathcal{A}(\chi_2).$$

(Here we used linearity properties of inner products.) Checking condition 2:

$$\mathcal{A}(a\chi) = \frac{(a\chi, v)}{(v, v)} v = \frac{a(\chi, v)}{(v, v)} v = a\mathcal{A}(\chi).$$

Therefore projection is a linear operation.

P6.3 Consider the transformation $\mathcal{A}$ created by reflecting a vector χ in $\Re^2$ about the line $x_1 + x_2 = 0$, as shown in Figure P6.2. Find the matrix of this transformation relative to the standard basis in $\Re^2$.

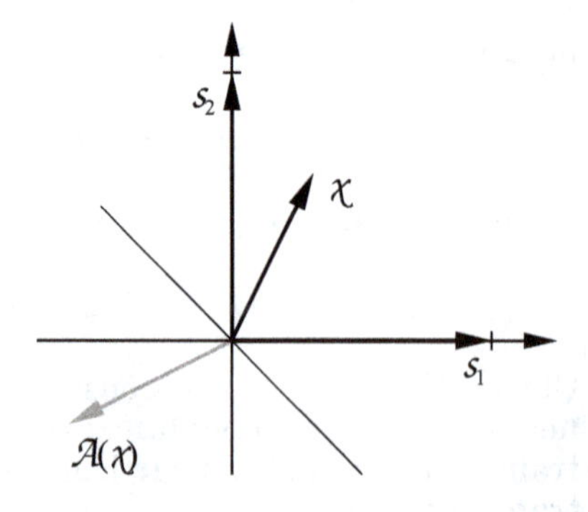

Figure P6.2 Reflection Transformation

The key to finding the matrix of a transformation is given in Eq. (6.6):

$$\mathcal{A}(v_j) = \sum_{i=1}^{m} a_{ij} u_i .$$

We need to transform each basis vector of the domain and then expand the result in terms of the basis vectors for the range. Each time we do the expansion we get one column of the matrix representation. In this case the basis set for both the domain and the range is $\{s_1, s_2\}$. So let's transform s_1 first. If we reflect s_1 about the line $x_1 + x_2 = 0$, we find

$$\mathcal{A}(s_1) = -s_2 = \sum_{i=1}^{2} a_{i1} s_i = a_{11} s_1 + a_{21} s_2 = 0 s_1 + (-1) s_2$$

(as shown in the top left figure), which gives us the first column of the matrix. Next we transform s_2:

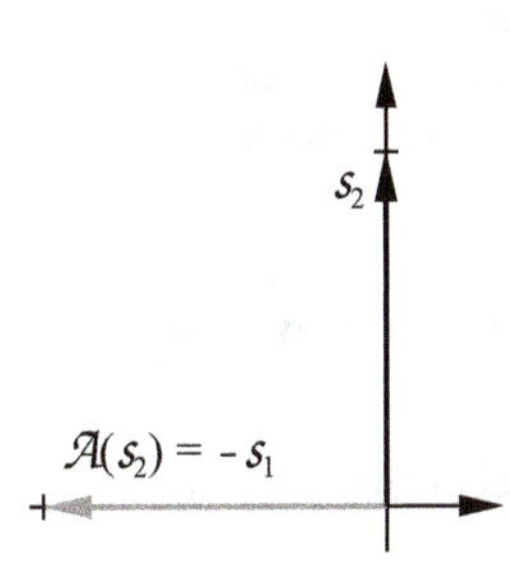

$$\mathcal{A}(s_2) = -s_1 = \sum_{i=1}^{2} a_{i2} s_i = a_{12} s_1 + a_{22} s_2 = (-1) s_1 + 0 s_2$$

(as shown in the second figure on the left), which gives us the second column of the matrix. The final result is

$$\begin{bmatrix} 0 & -1 \\ -1 & 0 \end{bmatrix} .$$

Let's test our result by transforming the vector $\mathbf{x} = \begin{bmatrix} 1 & 1 \end{bmatrix}^T$:

$$\mathbf{Ax} = \begin{bmatrix} 0 & -1 \\ -1 & 0 \end{bmatrix} \begin{bmatrix} 1 \\ 1 \end{bmatrix} = \begin{bmatrix} -1 \\ -1 \end{bmatrix} .$$

This is indeed the reflection of $\mathbf{x}$ about the line $x_1 + x_2 = 0$, as we can see in Figure P6.3.

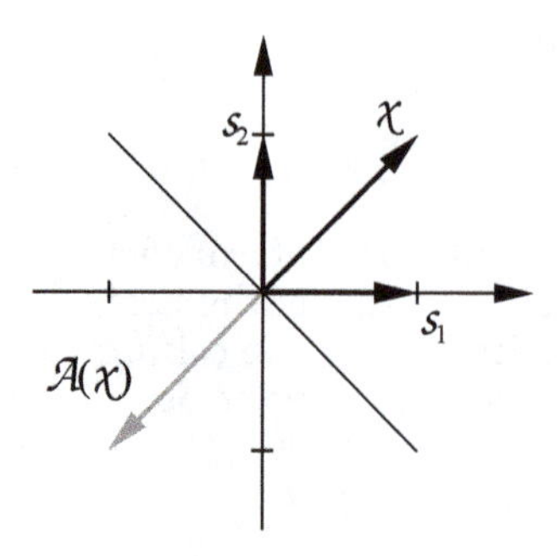

Figure P6.3 Test of Reflection Operation

(Can you guess the eigenvalues and eigenvectors of this transformation? Use the *Neural Network Design Demonstration* *Linear Transformations* (**nnd6lt**) to investigate this graphically. Compute the eigenvalues and eigenvectors, using the MATLAB function **eig**, and check your guess.)

P6.4 Consider the space of complex numbers. Let this be the vector space X, and let the basis for X be $\{1+j, 1-j\}$. Let $\mathcal{A}: X \rightarrow X$ be the conjugation operator (i.e., $\mathcal{A}(\chi) = \chi^*$).

i. Find the matrix of the transformation $\mathcal{A}$ relative to the basis set given above.

ii. Find the eigenvalues and eigenvectors of the transformation.

iii. Find the matrix representation for $\mathcal{A}$ relative to the eigenvectors as the basis vectors.

i. To find the matrix of the transformation, transform each of the basis vectors (by finding their conjugate):

$$\mathcal{A}(v_1) = \mathcal{A}(1+j) = 1-j = v_2 = a_{11}v_1 + a_{21}v_2 = 0v_1 + 1v_2,$$

$$\mathcal{A}(v_2) = \mathcal{A}(1-j) = 1+j = v_1 = a_{12}v_1 + a_{22}v_2 = 1v_1 + 0v_2.$$

This gives us the matrix representation

$$\mathbf{A} = \begin{bmatrix} 0 & 1 \\ 1 & 0 \end{bmatrix}.$$

ii. To find the eigenvalues, we need to use Eq. (6.49):

$$|[\mathbf{A} - \lambda\mathbf{I}]| = \left|\begin{bmatrix} -\lambda & 1 \\ 1 & -\lambda \end{bmatrix}\right| = \lambda^2 - 1 = (\lambda - 1)(\lambda + 1) = 0 .$$

So the eigenvalues are: $\lambda_1 = 1$, $\lambda_2 = -1$. To find the eigenvectors, use Eq. (6.48):

$$[\mathbf{A} - \lambda\mathbf{I}]\mathbf{z} = \begin{bmatrix} -\lambda & 1 \\ 1 & -\lambda \end{bmatrix}\mathbf{z} = \begin{bmatrix} 0 \\ 0 \end{bmatrix} .$$

For $\lambda = \lambda_1 = 1$ this gives us

$$\begin{bmatrix} -1 & 1 \\ 1 & -1 \end{bmatrix}\mathbf{z}_1 = \begin{bmatrix} -1 & 1 \\ 1 & -1 \end{bmatrix}\begin{bmatrix} z_{11} \\ z_{21} \end{bmatrix} = \begin{bmatrix} 0 \\ 0 \end{bmatrix},$$

or

$$z_{11} = z_{21} .$$

Therefore the first eigenvector will be

$$\mathbf{z}_1 = \begin{bmatrix} 1 \\ 1 \end{bmatrix},$$

or any scalar multiple. For the second eigenvector we use $\lambda = \lambda_2 = -1$:

$$\begin{bmatrix} 1 & 1 \\ 1 & 1 \end{bmatrix}\mathbf{z}_1 = \begin{bmatrix} 1 & 1 \\ 1 & 1 \end{bmatrix}\begin{bmatrix} z_{12} \\ z_{22} \end{bmatrix} = \begin{bmatrix} 0 \\ 0 \end{bmatrix},$$

or

$$z_{12} = -z_{22} .$$

Therefore the second eigenvector is

$$\mathbf{z}_2 = \begin{bmatrix} 1 \\ -1 \end{bmatrix},$$

or any scalar multiple.

Note that while these eigenvectors can be represented as columns of numbers, in reality they are complex numbers. For example:

$$z_1 = 1v_1 + 1v_2 = (1 + j) + (1 - j) = 2 ,$$

$$z_2 = 1v_1 + (-1)v_2 = (1+j) - (1-j) = 2j .$$

Checking that these are indeed eigenvectors:

$$\mathcal{A}(z_1) = (2)^* = 2 = \lambda_1 z_1 ,$$

$$\mathcal{A}(z_2) = (2j)^* = -2j = \lambda_2 z_2 .$$

iii. To perform a change of basis we need to use Eq. (6.33):

$$\mathbf{A}' = [\mathbf{B}_w^{-1}\mathbf{A}\mathbf{B}_t] = [\mathbf{B}^{-1}\mathbf{A}\mathbf{B}] ,$$

where

$$\mathbf{B} = \begin{bmatrix} \mathbf{z}_1 & \mathbf{z}_2 \end{bmatrix} = \begin{bmatrix} 1 & 1 \\ 1 & -1 \end{bmatrix} .$$

(We are using the same basis set for the range and the domain.) Therefore we have

$$\mathbf{A}' = \begin{bmatrix} 0.5 & 0.5 \\ 0.5 & -0.5 \end{bmatrix} \begin{bmatrix} 0 & 1 \\ 1 & 0 \end{bmatrix} \begin{bmatrix} 1 & 1 \\ 1 & -1 \end{bmatrix} = \begin{bmatrix} 1 & 0 \\ 0 & -1 \end{bmatrix} = \begin{bmatrix} \lambda_1 & 0 \\ 0 & \lambda_2 \end{bmatrix} .$$

As expected from Eq. (6.69), we have diagonalized the matrix representation.

P6.5 Diagonalize the following matrix:

$$\mathbf{A} = \begin{bmatrix} 2 & -2 \\ -1 & 3 \end{bmatrix} .$$

The first step is to find the eigenvalues:

$$|[\mathbf{A} - \lambda\mathbf{I}]| = \left| \begin{bmatrix} 2-\lambda & -2 \\ -1 & 3-\lambda \end{bmatrix} \right| = \lambda^2 - 5\lambda + 4 = (\lambda - 1)(\lambda - 4) = 0 ,$$

so the eigenvalues are $\lambda_1 = 1$, $\lambda_2 = 4$. To find the eigenvectors,

$$[\mathbf{A} - \lambda\mathbf{I}]\mathbf{z} = \begin{bmatrix} 2-\lambda & -2 \\ -1 & 3-\lambda \end{bmatrix} \mathbf{z} = \begin{bmatrix} 0 \\ 0 \end{bmatrix} .$$

For $\lambda = \lambda_1 = 1$

$$\begin{bmatrix} 1 & -2 \\ -1 & 2 \end{bmatrix} \mathbf{z}_1 = \begin{bmatrix} 1 & -2 \\ -1 & 2 \end{bmatrix} \begin{bmatrix} z_{11} \\ z_{21} \end{bmatrix} = \begin{bmatrix} 0 \\ 0 \end{bmatrix},$$

or

$$z_{11} = 2z_{21}.$$

Therefore the first eigenvector will be

$$\mathbf{z}_1 = \begin{bmatrix} 2 \\ 1 \end{bmatrix},$$

or any scalar multiple.

For $\lambda = \lambda_2 = 4$

$$\begin{bmatrix} -2 & -2 \\ -1 & -1 \end{bmatrix} \mathbf{z}_1 = \begin{bmatrix} -2 & -2 \\ -1 & -1 \end{bmatrix} \begin{bmatrix} z_{12} \\ z_{22} \end{bmatrix} = \begin{bmatrix} 0 \\ 0 \end{bmatrix},$$

or

$$z_{12} = -z_{22}.$$

Therefore the second eigenvector will be

$$\mathbf{z}_2 = \begin{bmatrix} 1 \\ -1 \end{bmatrix},$$

or any scalar multiple.

To diagonalize the matrix we use Eq. (6.69):

$$\mathbf{A}' = [\mathbf{B}^{-1}\mathbf{A}\mathbf{B}],$$

where

$$\mathbf{B} = \begin{bmatrix} \mathbf{z}_1 & \mathbf{z}_2 \end{bmatrix} = \begin{bmatrix} 2 & 1 \\ 1 & -1 \end{bmatrix}.$$

Therefore we have

$$\mathbf{A}' = \begin{bmatrix} \frac{1}{3} & \frac{1}{3} \\ \frac{1}{3} & -\frac{2}{3} \end{bmatrix} \begin{bmatrix} 2 & -2 \\ -1 & 3 \end{bmatrix} \begin{bmatrix} 2 & 1 \\ 1 & -1 \end{bmatrix} = \begin{bmatrix} 1 & 0 \\ 0 & 4 \end{bmatrix} = \begin{bmatrix} \lambda_1 & 0 \\ 0 & \lambda_2 \end{bmatrix}.$$

P6.6 Consider a transformation $\mathcal{A}: R^3 \to R^2$ whose matrix representation relative to the standard basis sets is

$$\mathbf{A} = \begin{bmatrix} 3 & -1 & 0 \\ 0 & 0 & 1 \end{bmatrix}.$$

Find the matrix for this transformation relative to the basis sets:

$$T = \left\{ \begin{bmatrix} 2 \\ 0 \\ 1 \end{bmatrix}, \begin{bmatrix} 0 \\ -1 \\ 0 \end{bmatrix}, \begin{bmatrix} 0 \\ -2 \\ 3 \end{bmatrix} \right\} \qquad W = \left\{ \begin{bmatrix} 1 \\ 0 \end{bmatrix}, \begin{bmatrix} 0 \\ -2 \end{bmatrix} \right\}.$$

The first step is to form the matrices

$$\mathbf{B}_t = \begin{bmatrix} 2 & 0 & 0 \\ 0 & -1 & -2 \\ 1 & 0 & 3 \end{bmatrix} \qquad \mathbf{B}_w = \begin{bmatrix} 1 & 0 \\ 0 & -2 \end{bmatrix}.$$

Now we use Eq. (6.33) to form the new matrix representation:

$$\mathbf{A}' = [\mathbf{B}_w^{-1} \mathbf{A} \mathbf{B}_t],$$

$$\mathbf{A}' = \begin{bmatrix} 1 & 0 \\ 0 & -\frac{1}{2} \end{bmatrix} \begin{bmatrix} 3 & -1 & 0 \\ 0 & 0 & 1 \end{bmatrix} \begin{bmatrix} 2 & 0 & 0 \\ 0 & -1 & -2 \\ 1 & 0 & 3 \end{bmatrix} = \begin{bmatrix} 6 & 1 & 2 \\ -\frac{1}{2} & 0 & -\frac{3}{2} \end{bmatrix}.$$

Therefore this is the matrix of the transformation with respect to the basis sets T and W.

P6.7 Consider a transformation $\mathcal{A}: \Re^2 \to \Re^2$. One basis set for $\Re^2$ is given as $V = \{v_1, v_2\}$.

i. Find the matrix of the transformation $\mathcal{A}$ relative to the basis set V if it is given that

$$\mathcal{A}(v_1) = v_1 + 2v_2,$$

$$\mathcal{A}(v_2) = v_1 + v_2.$$

ii. **Consider a new basis set $W = \{w_1, w_2\}$. Find the matrix of the transformation $\mathcal{A}$ relative to the basis set W if it is given that**

$$w_1 = v_1 + v_2,$$

$$w_2 = v_1 - v_2.$$

i. Each of the two equations gives us one column of the matrix, as defined in Eq. (6.6). Therefore the matrix is

$$\mathbf{A} = \begin{bmatrix} 1 & 1 \\ 2 & 1 \end{bmatrix}.$$

ii. We can represent the W basis vectors as columns of numbers in terms of the V basis vectors:

$$\mathbf{w}_1 = \begin{bmatrix} 1 \\ 1 \end{bmatrix} \qquad \mathbf{w}_2 = \begin{bmatrix} 1 \\ -1 \end{bmatrix}.$$

We can now form the basis matrix that we need to perform the similarity transform:

$$\mathbf{B}_w = \begin{bmatrix} 1 & 1 \\ 1 & -1 \end{bmatrix}.$$

The new matrix representation can then be obtained from Eq. (6.33):

$$\mathbf{A}' = [\mathbf{B}_w^{-1}\mathbf{A}\mathbf{B}_w],$$

$$\mathbf{A}' = \begin{bmatrix} \frac{1}{2} & \frac{1}{2} \\ \frac{1}{2} & -\frac{1}{2} \end{bmatrix} \begin{bmatrix} 1 & 1 \\ 2 & 1 \end{bmatrix} \begin{bmatrix} 1 & 1 \\ 1 & -1 \end{bmatrix} = \begin{bmatrix} \frac{5}{2} & \frac{1}{2} \\ -\frac{1}{2} & -\frac{1}{2} \end{bmatrix}.$$

P6.8 Consider the vector space P^2 of all polynomials of degree less than or equal to 2. One basis for this vector space is $V = \{1, t, t^2\}$. Consider the differentiation transformation $\mathcal{D}$.

i. Find the matrix of this transformation relative to the basis set V.

ii. Find the eigenvalues and eigenvectors of the transformation.

i. The first step is to transform each of the basis vectors:

$$\mathcal{D}(1) = 0 = (0)1 + (0)t + (0)t^2,$$

$$\mathcal{D}(t) = 1 = (1)1 + (0)t + (0)t^2,$$

$$\mathcal{D}(t^2) = 2t = (0)1 + (2)t + (0)t^2.$$

The matrix of the transformation is then given by

$$\mathbf{D} = \begin{bmatrix} 0 & 1 & 0 \\ 0 & 0 & 2 \\ 0 & 0 & 0 \end{bmatrix}.$$

ii. To find the eigenvalues we must solve

$$|[\mathbf{D} - \lambda\mathbf{I}]| = \left| \begin{bmatrix} -\lambda & 1 & 0 \\ 0 & -\lambda & 2 \\ 0 & 0 & -\lambda \end{bmatrix} \right| = -\lambda^3 = 0.$$

Therefore all three eigenvalues are zero. To find the eigenvectors we need to solve

$$[\mathbf{D} - \lambda\mathbf{I}]\mathbf{z} = \begin{bmatrix} -\lambda & 1 & 0 \\ 0 & -\lambda & 2 \\ 0 & 0 & -\lambda \end{bmatrix} \mathbf{z} = \begin{bmatrix} 0 \\ 0 \\ 0 \end{bmatrix}.$$

For $\lambda = 0$ we have

$$\begin{bmatrix} 0 & 1 & 0 \\ 0 & 0 & 2 \\ 0 & 0 & 0 \end{bmatrix} \begin{bmatrix} z_1 \\ z_2 \\ z_3 \end{bmatrix} = \begin{bmatrix} 0 \\ 0 \\ 0 \end{bmatrix}.$$

This means that

$$z_2 = z_3 = 0 .$$

Therefore we have a single eigenvector:

$$\mathbf{z} = \begin{bmatrix} 1 \\ 0 \\ 0 \end{bmatrix} .$$

Therefore the only polynomial whose derivative is a scaled version of itself is a constant (a zeroth-order polynomial).

P6.9 Consider a transformation $\mathcal{A}:R^2 \rightarrow R^2$. Two examples of transformed vectors are given in Figure P6.4. Find the matrix representation of this transformation relative to the standard basis set.

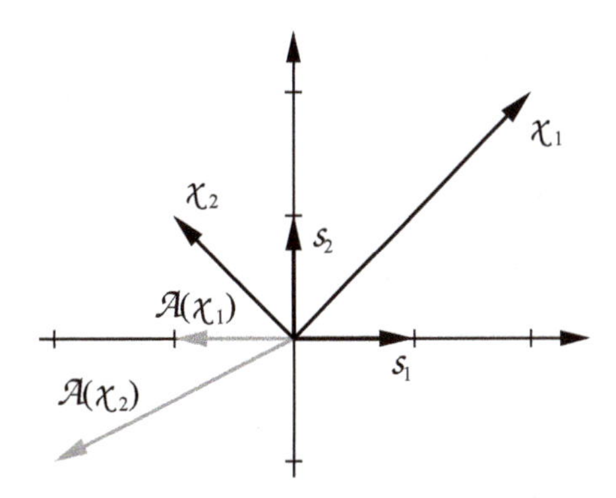

Figure P6.4 Transformation for Problem P6.9

For this problem we do not know how the basis vectors are transformed, so we cannot use Eq. (6.6) to find the matrix representation. However, we do know how two vectors are transformed, and we do know how those vectors can be represented in terms of the standard basis set. From Figure P6.4 we can write the following equations:

$$\mathbf{A}\begin{bmatrix} 2 \\ 2 \end{bmatrix} = \begin{bmatrix} -1 \\ 0 \end{bmatrix} , \mathbf{A}\begin{bmatrix} -1 \\ 1 \end{bmatrix} = \begin{bmatrix} -2 \\ -1 \end{bmatrix} .$$

We then put these two equations together to form

$$\mathbf{A}\begin{bmatrix} 2 & -1 \\ 2 & 1 \end{bmatrix} = \begin{bmatrix} -1 & -2 \\ 0 & -1 \end{bmatrix} .$$

So that

$$\mathbf{A} = \begin{bmatrix} -1 & -2 \\ 0 & -1 \end{bmatrix} \begin{bmatrix} 2 & -1 \\ 2 & 1 \end{bmatrix}^{-1} = \begin{bmatrix} -1 & -2 \\ 0 & -1 \end{bmatrix} \begin{bmatrix} \frac{1}{4} & \frac{1}{4} \\ -\frac{1}{2} & \frac{1}{2} \end{bmatrix} = \begin{bmatrix} \frac{3}{4} & -\frac{5}{4} \\ \frac{1}{2} & -\frac{1}{2} \end{bmatrix}.$$

This is the matrix representation of the transformation with respect to the standard basis set.

This procedure is used in the *Neural Network Design Demonstration Linear Transformations* (`nnd6lt`).

Epilogue

In this chapter we have reviewed those properties of linear transformations and matrices that are most important to our study of neural networks. The concepts of eigenvalues, eigenvectors, change of basis (similarity transformation) and diagonalization will be used again and again throughout the remainder of this text. Without this linear algebra background our study of neural networks could only be superficial.

In the next chapter we will use linear algebra to analyze the operation of one of the first neural network training algorithms — the Hebb rule.

Further Reading

[Brog91] W. L. Brogan, *Modern Control Theory*, 3rd Ed., Englewood Cliffs, NJ: Prentice-Hall, 1991.

This is a well-written book on the subject of linear systems. The first half of the book is devoted to linear algebra. It also has good sections on the solution of linear differential equations and the stability of linear and nonlinear systems. It has many worked problems.

[Stra76] G. Strang, *Linear Algebra and Its Applications*, New York: Academic Press, 1980.

Strang has written a good basic text on linear algebra. Many applications of linear algebra are integrated into the text.

Exercises

E6.1 Is the operation of transposing a matrix a linear transformation?

E6.2 Consider again the neural network shown in Figure P6.1. Show that if the bias vector $\mathbf{b}$ is equal to zero then the network performs a linear operation.

E6.3 Consider the linear transformation illustrated in Figure E6.1.

i. Find the matrix representation of this transformation relative to the standard basis set.

ii. Find the matrix of this transformation relative to the basis set $\{v_1, v_2\}$.

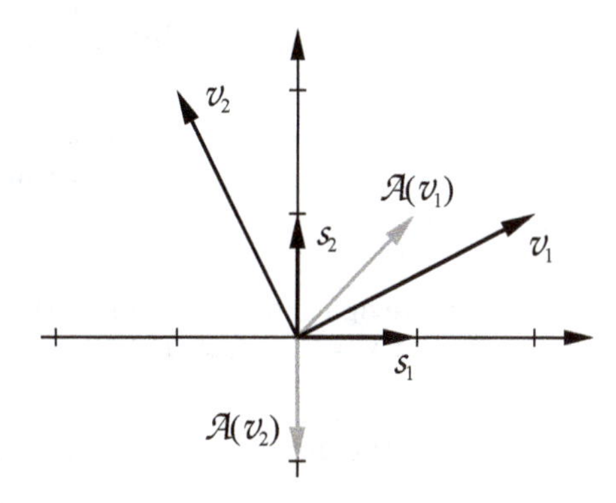

Figure E6.1 Example Transformation for Exercise E6.3

E6.4 Consider the space of complex numbers. Let this be the vector space X, and let the basis for X be $\{1+j, 1-j\}$. Let $\mathcal{A}: X \to X$ be the operation of multiplication by $(1+j)$ (i.e., $\mathcal{A}(\chi) = (1+j)\chi$).

i. Find the matrix of the transformation $\mathcal{A}$ relative to the basis set given above.

ii. Find the eigenvalues and eigenvectors of the transformation.

iii. Find the matrix representation for $\mathcal{A}$ relative to the eigenvectors as the basis vectors.

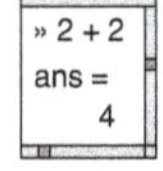

iv. Check your answers to parts (ii) and (iii) using MATLAB.

E6.5 Consider a transformation $\mathcal{A}: P^2 \rightarrow P^3$, from the space of second-order polynomials to the space of third-order polynomials, which is defined by the following:

$$\chi = a_0 + a_1 t + a_2 t^2 ,$$

$$\mathcal{A}(\chi) = a_0(t+1) + a_1(t+1)^2 + a_2(t+1)^3 .$$

Find the matrix representation of this transformation relative to the basis sets $V^2 = \{1, t, t^2\}$, $V^3 = \{1, t, t^2, t^3\}$.

E6.6 Consider the vector space of polynomials of degree two or less. These polynomials have the form $f(t) = a_0 + a_1 t + a_2 t^2$. Now consider the transformation in which the variable t is replaced by $t+1$. (for example, $t^2 + 2t + 3 \Rightarrow (t+1)^2 + 2(t+1) + 3 = t^2 + 4t + 6$)

i. Find the matrix of this transformation with respect to the basis set $\{1, t-1, t^2\}$.

ii. Find the eigenvalues and eigenvectors of the transformation. Show the eigenvectors as columns of numbers and as functions of time (polynomials).

E6.7 Consider the space of functions of the form $\alpha \sin(t + \phi)$. One basis set for this space is $V = \{\sin t, \cos t\}$. Consider the differentiation transformation $\mathcal{D}$.

i. Find the matrix of the transformation $\mathcal{D}$ relative to the basis set V.

ii. Find the eigenvalues and eigenvectors of the transformation. Show the eigenvectors as columns of numbers and as functions of t.

iii. Find the matrix of the transformation relative to the eigenvectors as basis vectors.

E6.8 Consider the vector space of functions of the form $\alpha + \beta e^{2t}$. One basis set for this vector space is $V = \{1 + e^{2t}, 1 - e^{2t}\}$. Consider the differentiation transformation $\mathcal{D}$.

i. Find the matrix of the transformation $\mathcal{D}$ relative to the basis set V, using Eq. (6.6).

ii. Verify the operation of the matrix on the function $2e^{2t}$.

iii. Find the eigenvalues and eigenvectors of the transformation. Show the eigenvectors as columns of numbers (with respect to the basis set V) and as functions of t.

iv. Find the matrix of the transformation relative to the eigenvectors as basis vectors.

E6.9 Consider the set of all 2x2 matrices. This set is a vector space, which we will call X (yes, matrices can be vectors). If $\mathbf{M}$ is an element of this vector space, define the transformation $\mathcal{A}:X \to X$, such that $\mathcal{A}(\mathbf{M}) = \mathbf{M} + \mathbf{M}^T$. Consider the following basis set for the vector space X.

$$v_1 = \begin{bmatrix} 1 & 0 \\ 0 & 0 \end{bmatrix}, \; v_2 = \begin{bmatrix} 0 & 1 \\ 0 & 0 \end{bmatrix}, \; v_3 = \begin{bmatrix} 0 & 0 \\ 1 & 0 \end{bmatrix}, \; v_4 = \begin{bmatrix} 0 & 0 \\ 0 & 1 \end{bmatrix}$$

i. Find the matrix representation of the transformation $\mathcal{A}$ relative to the basis set $\{v_1, v_2, v_3, v_4\}$ (for both domain and range) (using Eq. (6.6)).

ii. Verify the operation of the matrix representation from part i. on the element of X given below. (Verify that the matrix multiplication produces the same result as the transformation.)

$$\begin{bmatrix} 1 & 2 \\ 0 & 1 \end{bmatrix}$$

iii. Find the eigenvalues and eigenvectors of the transformation. You do not need to use the matrix representation that you found in part i. You can find the eigenvalues and eigenvectors directly from the definition of the transformation. Your eigenvectors should be 2x2 matrices (elements of the vector space X). This does not require much computation. Use the definition of eigenvector in Eq. (6.46).

E6.10 Consider a transformation $\mathcal{A}:P^1 \to P^2$, from the space of first degree polynomials into the space of second degree polynomials. The transformation is defined as follows

$$\mathcal{A}(a + bt) = at + \frac{b}{2}t^2$$

(e.g., $\mathcal{A}(2 + 6t) = 2t + 3t^2$). One basis set for P^1 is $U = \{1, t\}$. One basis for P^2 is $V = \{1, t, t^2\}$.

i. Find the matrix representation of the transformation A relative to the basis sets U and V, using Eq. (6.6).

ii. Verify the operation of the matrix on the polynomial $6 + 8t$. (Verify that the matrix multiplication produces the same result as the transformation.)

iii. Using a similarity transform, find the matrix of the transformation with respect to the basis sets $S = \{1+t, 1-t\}$ and V.

E6.11 Let $\mathcal{D}$ be the differentiation operator ($\mathcal{D}(f) = df/dt$), and use the basis set

$$\{u_1, u_2\} = \{e^{5t}, te^{5t}\}$$

for both the domain and the range of the transformation $\mathcal{D}$.

i. Show that the transformation $\mathcal{D}$ is linear.,

ii. Find the matrix of this transformation relative to the basis shown above.

iii. Find the eigenvalues and eigenvectors of the transformation $\mathcal{D}$.

E6.12 A certain linear transformation has the following eigenvalues and eigenvectors (represented in terms of the standard basis set):

$$\left\{\mathbf{z}_1 = \begin{bmatrix}1\\2\end{bmatrix}, \lambda_1 = 1\right\}, \left\{\mathbf{z}_2 = \begin{bmatrix}-1\\2\end{bmatrix}, \lambda_2 = 2\right\}$$

i. Find the matrix representation of the transformation, relative to the standard basis set.

ii. Find the matrix representation of the transformation relative to the eigenvectors as the basis vectors.

E6.13 Consider a transformation $\mathcal{A}:\Re^2 \to \Re^2$. In the figure below, we show a set of basis vectors $V = \{v_1, v_2\}$ and the transformed basis vectors.

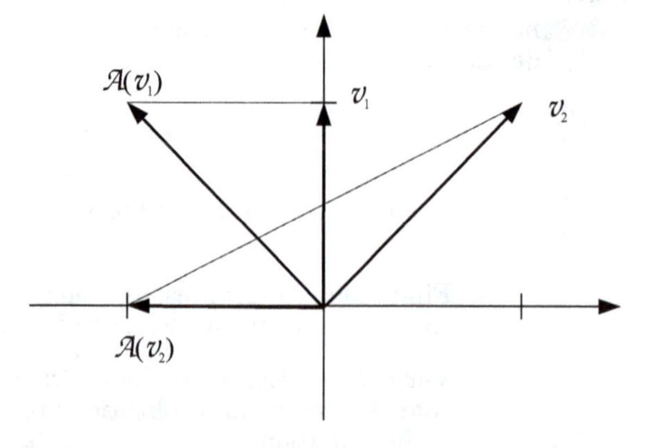

Figure E6.2 Definition of Transformation for Exercise E6.13

i. Find the matrix representation of this transformation with respect to the basis vectors $V = \{v_1, v_2\}$.

ii. Find the matrix representation of this transformation with respect to the standard basis vectors.

iii. Find the eigenvalues and eigenvectors of this transformation. Sketch the eigenvectors and their transformations.

iv. Find the matrix representation of this transformation with respect to the eigenvectors as the basis vectors.

E6.14 Consider the vector spaces P^2 and P^3 of second-order and third-order polynomials. Find the matrix representation of the integration transformation $I: P^2 \to P^3$, relative to the basis sets $V^2 = \{1, t, t^2\}$, $V^3 = \{1, t, t^2, t^3\}$.

E6.15 A certain linear transformation $\mathcal{A}: \Re^2 \to \Re^2$ has a matrix representation relative to the standard basis set of

$$\mathbf{A} = \begin{bmatrix} 1 & 2 \\ 3 & 4 \end{bmatrix}.$$

Find the matrix representation of this transformation relative to the new basis set:

$$V = \left\{ \begin{bmatrix} 1 \\ 3 \end{bmatrix}, \begin{bmatrix} 2 \\ 5 \end{bmatrix} \right\}.$$

E6.16 We know that a certain linear transformation $\mathcal{A}: R^2 \to R^2$ has eigenvalues and eigenvectors given by

$$\lambda_1 = 1 \qquad \mathbf{z}_1 = \begin{bmatrix} 1 \\ 1 \end{bmatrix} \qquad \lambda_2 = 2 \qquad \mathbf{z}_2 = \begin{bmatrix} 1 \\ 2 \end{bmatrix}.$$

(The eigenvectors are represented relative to the standard basis set.)

i. Find the matrix representation of the transformation $\mathcal{A}$ relative to the standard basis set.

ii. Find the matrix representation relative to the new basis

$$V = \left\{ \begin{bmatrix} 1 \\ 1 \end{bmatrix}, \begin{bmatrix} -1 \\ 1 \end{bmatrix} \right\}.$$

E6.17 Consider the transformation $\mathcal{A}$ created by projecting a vector χ onto the line shown in Figure E6.3. An example of the transformation is shown in the figure.

i. Using Eq. (6.6), find the matrix representation of this transformation relative to the standard basis set $\{s_1, s_2\}$.

ii. Using your answer to part i, find the matrix representation of this transformation relative to the basis set $\{v_1, v_2\}$ shown in Figure E6.3.

iii. What are the eigenvalues and eigenvectors of this transformation? Sketch the eigenvectors and their transformations.

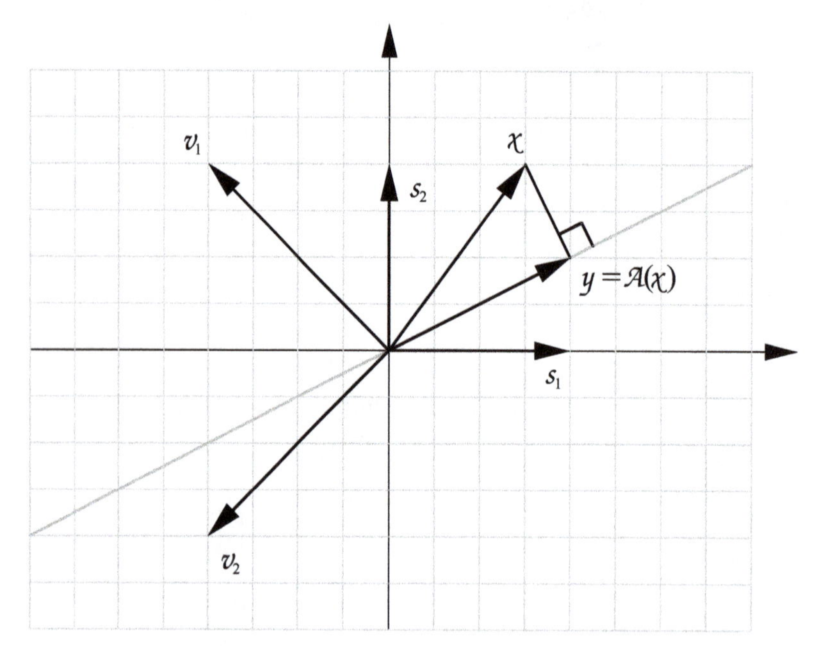

Figure E6.3 Definition of Transformation for Exercise E6.17

E6.18 Consider the following basis set for $\Re^2$:

$$V = \{\mathbf{v}_1, \mathbf{v}_2\} = \left\{ \begin{bmatrix} 1 \\ -1 \end{bmatrix}, \begin{bmatrix} 1 \\ -2 \end{bmatrix} \right\}.$$

(The basis vectors are represented relative to the standard basis set.)

i. Find the reciprocal basis vectors for this basis set.

ii. Consider a transformation $\mathcal{A}: \Re^2 \to \Re^2$. The matrix representation for $\mathcal{A}$ relative to the standard basis in $\Re^2$ is

$$\mathbf{A} = \begin{bmatrix} 0 & 1 \\ -2 & -3 \end{bmatrix}.$$

Find the expansion of $\mathbf{A}\mathbf{v}_1$ in terms of the basis set V. (Use the reciprocal basis vectors.)

iii. Find the expansion of $\mathbf{A}\mathbf{v}_2$ in terms of the basis set V.

iv. Find the matrix representation for $\mathcal{A}$ relative to the basis V. (This step should require no further computation.)

7 Supervised Hebbian Learning

Objectives

The Hebb rule was one of the first neural network learning laws. It was proposed by Donald Hebb in 1949 as a possible mechanism for synaptic modification in the brain and since then has been used to train artificial neural networks.

In this chapter we will use the linear algebra concepts of the previous two chapters to explain why Hebbian learning works. We will also show how the Hebb rule can be used to train neural networks for pattern recognition.

Theory and Examples

Donald O. Hebb was born in Chester, Nova Scotia, just after the turn of the century. He originally planned to become a novelist, and obtained a degree in English from Dalhousie University in Halifax in 1925. Since every first-rate novelist needs to have a good understanding of human nature, he began to study Freud after graduation and became interested in psychology. He then pursued a master's degree in psychology at McGill University, where he wrote a thesis on Pavlovian conditioning. He received his Ph.D. from Harvard in 1936, where his dissertation investigated the effects of early experience on the vision of rats. Later he joined the Montreal Neurological Institute, where he studied the extent of intellectual changes in brain surgery patients. In 1942 he moved to the Yerkes Laboratories of Primate Biology in Florida, where he studied chimpanzee behavior.

In 1949 Hebb summarized his two decades of research in *The Organization of Behavior* [Hebb49]. The main premise of this book was that behavior could be explained by the action of neurons. This was in marked contrast to the behaviorist school of psychology (with proponents such as B. F. Skinner), which emphasized the correlation between stimulus and response and discouraged the use of any physiological hypotheses. It was a confrontation between a top-down philosophy and a bottom-up philosophy. Hebb stated his approach: "The method then calls for learning as much as one can about what the parts of the brain do (primarily the physiologist's field), and relating the behavior as far as possible to this knowledge (primarily for the psychologist); then seeing what further information is to be had about how the total brain works, from the discrepancy between (1) actual behavior and (2) the behavior that would be predicted from adding up what is known about the action of the various parts."

The most famous idea contained in *The Organization of Behavior* was the postulate that came to be known as Hebbian learning:

Hebb's Postulate

"When an axon of cell A is near enough to excite a cell B and repeatedly or persistently takes part in firing it, some growth process or metabolic change takes place in one or both cells such that A's efficiency, as one of the cells firing B, is increased."

This postulate suggested a physical mechanism for learning at the cellular level. Although Hebb never claimed to have firm physiological evidence for his theory, subsequent research has shown that some cells do exhibit Hebbian learning. Hebb's theories continue to influence current research in neuroscience.

As with most historic ideas, Hebb's postulate was not completely new, as he himself emphasized. It had been foreshadowed by several others, including Freud. Consider, for example, the following principle of association stated by psychologist and philosopher William James in 1890: "When two brain processes are active together or in immediate succession, one of them, on reoccurring tends to propagate its excitement into the other."

Linear Associator

Hebb's learning law can be used in combination with a variety of neural network architectures. We will use a very simple architecture for our initial presentation of Hebbian learning. In this way we can concentrate on the learning law rather than the architecture. The network we will use is the *linear associator*, which is shown in Figure 7.1. (This network was introduced independently by James Anderson [Ande72] and Teuvo Kohonen [Koho72].)

Linear Associator

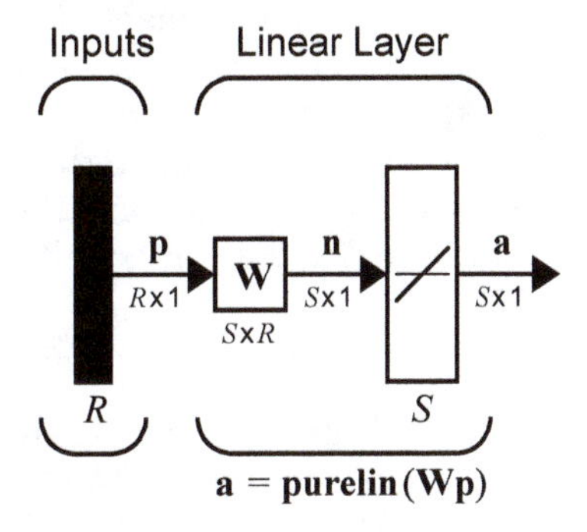

Figure 7.1 Linear Associator

The output vector $\mathbf{a}$ is determined from the input vector $\mathbf{p}$ according to:

$$\mathbf{a} = \mathbf{W}\mathbf{p}, \tag{7.1}$$

or

$$a_i = \sum_{j=1}^{R} w_{ij} p_j . \tag{7.2}$$

The linear associator is an example of a type of neural network called an *associative memory*. The task of an associative memory is to learn Q pairs of prototype input/output vectors:

Associative Memory

$$\{\mathbf{p}_1, \mathbf{t}_1\}, \{\mathbf{p}_2, \mathbf{t}_2\}, \dots, \{\mathbf{p}_Q, \mathbf{t}_Q\} . \tag{7.3}$$

In other words, if the network receives an input $\mathbf{p} = \mathbf{p}_q$ then it should produce an output $\mathbf{a} = \mathbf{t}_q$, for $q = 1, 2, \dots, Q$. In addition, if the input is changed slightly (i.e., $\mathbf{p} = \mathbf{p}_q + \delta$) then the output should only be changed slightly (i.e., $\mathbf{a} = \mathbf{t}_q + \varepsilon$).

The Hebb Rule

How can we interpret Hebb's postulate mathematically, so that we can use it to train the weight matrix of the linear associator? First, let's rephrase the postulate: If two neurons on either side of a synapse are activated simultaneously, the strength of the synapse will increase. Notice from Eq. (7.2) that the connection (synapse) between input p_j and output a_i is the weight w_{ij}. Therefore Hebb's postulate would imply that if a positive p_j produces a positive a_i then w_{ij} should increase. This suggests that one mathematical interpretation of the postulate could be

The Hebb Rule

$$w_{ij}^{new} = w_{ij}^{old} + \alpha f_i(a_{iq}) g_j(p_{jq}), \tag{7.4}$$

where p_{jq} is the jth element of the qth input vector $\mathbf{p}_q$; a_{iq} is the ith element of the network output when the qth input vector is presented to the network; and α is a positive constant, called the learning rate. This equation says that the change in the weight w_{ij} is proportional to a product of functions of the activities on either side of the synapse. For this chapter we will simplify Eq. (7.4) to the following form

$$w_{ij}^{new} = w_{ij}^{old} + \alpha a_{iq} p_{jq}. \tag{7.5}$$

Note that this expression actually extends Hebb's postulate beyond its strict interpretation. The change in the weight is proportional to a product of the activity on either side of the synapse. Therefore, not only do we increase the weight when both p_j and a_i are positive, but we also increase the weight when they are both negative. In addition, this implementation of the Hebb rule will decrease the weight whenever p_j and a_i have opposite sign.

The Hebb rule defined in Eq. (7.5) is an *unsupervised* learning rule. It does not require any information concerning the target output. In this chapter we are interested in using the Hebb rule for supervised learning, in which the target output is known for each input vector. For the *supervised* Hebb rule we substitute the target output for the actual output. In this way, we are telling the algorithm what the network *should* do, rather than what it is currently doing. The resulting equation is

$$w_{ij}^{new} = w_{ij}^{old} + t_{iq} p_{jq}, \tag{7.6}$$

where t_{iq} is the ith element of the qth target vector $\mathbf{t}_q$. (We have set the learning rate α to one, for simplicity.)

Notice that Eq. (7.6) can be written in vector notation:

$$\mathbf{W}^{new} = \mathbf{W}^{old} + \mathbf{t}_q \mathbf{p}_q^T. \tag{7.7}$$

If we assume that the weight matrix is initialized to zero and then each of the Q input/output pairs are applied once to Eq. (7.7), we can write

$$\mathbf{W} = \mathbf{t}_1\mathbf{p}_1^T + \mathbf{t}_2\mathbf{p}_2^T + \cdots + \mathbf{t}_Q\mathbf{p}_Q^T = \sum_{q=1}^{Q} \mathbf{t}_q\mathbf{p}_q^T. \tag{7.8}$$

This can be represented in matrix form:

$$\mathbf{W} = \begin{bmatrix} \mathbf{t}_1 & \mathbf{t}_2 & \ldots & \mathbf{t}_Q \end{bmatrix} \begin{bmatrix} \mathbf{p}_1^T \\ \mathbf{p}_2^T \\ \vdots \\ \mathbf{p}_Q^T \end{bmatrix} = \mathbf{T}\mathbf{P}^T, \tag{7.9}$$

where

$$\mathbf{T} = \begin{bmatrix} \mathbf{t}_1 & \mathbf{t}_2 & \ldots & \mathbf{t}_Q \end{bmatrix}, \quad \mathbf{P} = \begin{bmatrix} \mathbf{p}_1 & \mathbf{p}_2 & \ldots & \mathbf{p}_Q \end{bmatrix}. \tag{7.10}$$

Performance Analysis

Let's analyze the performance of Hebbian learning for the linear associator. First consider the case where the $\mathbf{p}_q$ vectors are orthonormal (orthogonal and unit length). If $\mathbf{p}_k$ is input to the network, then the network output can be computed

$$\mathbf{a} = \mathbf{W}\mathbf{p}_k = \left(\sum_{q=1}^{Q} \mathbf{t}_q\mathbf{p}_q^T \right) \mathbf{p}_k = \sum_{q=1}^{Q} \mathbf{t}_q(\mathbf{p}_q^T\mathbf{p}_k). \tag{7.11}$$

Since the $\mathbf{p}_q$ are orthonormal,

$$\begin{aligned} (\mathbf{p}_q^T\mathbf{p}_k) &= 1 \qquad q = k \\ &= 0 \qquad q \neq k. \end{aligned} \tag{7.12}$$

Therefore Eq. (7.11) can be rewritten

$$\mathbf{a} = \mathbf{W}\mathbf{p}_k = \mathbf{t}_k. \tag{7.13}$$

The output of the network is equal to the target output. This shows that, if the input prototype vectors are orthonormal, the Hebb rule will produce the correct output for each input.

But what about non-orthogonal prototype vectors? Let's assume that each $\mathbf{p}_q$ vector is unit length, but that they are not orthogonal. Then Eq. (7.11) becomes

$$\mathbf{a} = \mathbf{W}\mathbf{p}_k = \mathbf{t}_k + \underbrace{\sum_{q \neq k} \mathbf{t}_q(\mathbf{p}_q^T \mathbf{p}_k)}_{Error}. \quad (7.14)$$

Because the vectors are not orthogonal, the network will not produce the correct output. The magnitude of the error will depend on the amount of correlation between the prototype input patterns.

As an example, suppose that the prototype input/output vectors are

$$\left\{ \mathbf{p}_1 = \begin{bmatrix} 0.5 \\ -0.5 \\ 0.5 \\ -0.5 \end{bmatrix}, \mathbf{t}_1 = \begin{bmatrix} 1 \\ -1 \end{bmatrix} \right\} \quad \left\{ \mathbf{p}_2 = \begin{bmatrix} 0.5 \\ 0.5 \\ -0.5 \\ -0.5 \end{bmatrix}, \mathbf{t}_2 = \begin{bmatrix} 1 \\ 1 \end{bmatrix} \right\}. \quad (7.15)$$

(Check that the two input vectors are orthonormal.)

The weight matrix would be

$$\mathbf{W} = \mathbf{T}\mathbf{P}^T = \begin{bmatrix} 1 & 1 \\ -1 & 1 \end{bmatrix} \begin{bmatrix} 0.5 & -0.5 & 0.5 & -0.5 \\ 0.5 & 0.5 & -0.5 & -0.5 \end{bmatrix} = \begin{bmatrix} 1 & 0 & 0 & -1 \\ 0 & 1 & -1 & 0 \end{bmatrix}. \quad (7.16)$$

If we test this weight matrix on the two prototype inputs we find

$$\mathbf{W}\mathbf{p}_1 = \begin{bmatrix} 1 & 0 & 0 & -1 \\ 0 & 1 & -1 & 0 \end{bmatrix} \begin{bmatrix} 0.5 \\ -0.5 \\ 0.5 \\ -0.5 \end{bmatrix} = \begin{bmatrix} 1 \\ -1 \end{bmatrix}, \quad (7.17)$$

and

$$\mathbf{W}\mathbf{p}_2 = \begin{bmatrix} 1 & 0 & 0 & -1 \\ 0 & 1 & -1 & 0 \end{bmatrix} \begin{bmatrix} 0.5 \\ 0.5 \\ -0.5 \\ -0.5 \end{bmatrix} = \begin{bmatrix} 1 \\ 1 \end{bmatrix}. \quad (7.18)$$

Success!! The outputs of the network are equal to the targets.

2 +2 4

Now let's revisit the *apple* and *orange* recognition problem described in Chapter 3. Recall that the prototype inputs were

$$\mathbf{p}_1 = \begin{bmatrix} 1 \\ -1 \\ -1 \end{bmatrix} (orange) \qquad \mathbf{p}_2 = \begin{bmatrix} 1 \\ 1 \\ -1 \end{bmatrix} (apple). \tag{7.19}$$

(Note that they are not orthogonal.) If we normalize these inputs and choose as desired outputs -1 and 1, we obtain

$$\left\{ \mathbf{p}_1 = \begin{bmatrix} 0.5774 \\ -0.5774 \\ -0.5774 \end{bmatrix}, \mathbf{t}_1 = \begin{bmatrix} -1 \end{bmatrix} \right\} \qquad \left\{ \mathbf{p}_2 = \begin{bmatrix} 0.5774 \\ 0.5774 \\ -0.5774 \end{bmatrix}, \mathbf{t}_2 = \begin{bmatrix} 1 \end{bmatrix} \right\}. \tag{7.20}$$

Our weight matrix becomes

$$\mathbf{W} = \mathbf{T}\mathbf{P}^T = \begin{bmatrix} -1 & 1 \end{bmatrix} \begin{bmatrix} 0.5774 & -0.5774 & -0.5774 \\ 0.5774 & 0.5774 & -0.5774 \end{bmatrix} = \begin{bmatrix} 0 & 1.1548 & 0 \end{bmatrix}. \tag{7.21}$$

So, if we use our two prototype patterns,

$$\mathbf{W}\mathbf{p}_1 = \begin{bmatrix} 0 & 1.1548 & 0 \end{bmatrix} \begin{bmatrix} 0.5774 \\ -0.5774 \\ -0.5774 \end{bmatrix} = \begin{bmatrix} -0.6668 \end{bmatrix}, \tag{7.22}$$

$$\mathbf{W}\mathbf{p}_2 = \begin{bmatrix} 0 & 1.1548 & 0 \end{bmatrix} \begin{bmatrix} 0.5774 \\ 0.5774 \\ -0.5774 \end{bmatrix} = \begin{bmatrix} 0.6668 \end{bmatrix}. \tag{7.23}$$

The outputs are close, but do not quite match the target outputs.

Pseudoinverse Rule

When the prototype input patterns are not orthogonal, the Hebb rule produces some errors. There are several procedures that can be used to reduce these errors. In this section we will discuss one of those procedures, the pseudoinverse rule.

Recall that the task of the linear associator was to produce an output of $\mathbf{t}_q$ for an input of $\mathbf{p}_q$. In other words,

$$\mathbf{W}\mathbf{p}_q = \mathbf{t}_q \qquad q = 1, 2, \ldots, Q. \tag{7.24}$$

If it is not possible to choose a weight matrix so that these equations are exactly satisfied, then we want them to be approximately satisfied. One approach would be to choose the weight matrix to minimize the following performance index:

$$F(\mathbf{W}) = \sum_{q=1}^{Q} \|\mathbf{t}_q - \mathbf{W}\mathbf{p}_q\|^2 . \tag{7.25}$$

If the prototype input vectors $\mathbf{p}_q$ are orthonormal and we use the Hebb rule to find $\mathbf{W}$, then $F(\mathbf{W})$ will be zero. When the input vectors are not orthogonal and we use the Hebb rule, then $F(\mathbf{W})$ will be not be zero, and it is not clear that $F(\mathbf{W})$ will be minimized. It turns out that the weight matrix that will minimize $F(\mathbf{W})$ is obtained by using the pseudoinverse matrix, which we will define next.

First, let's rewrite Eq. (7.24) in matrix form:

$$\mathbf{W}\mathbf{P} = \mathbf{T} , \tag{7.26}$$

where

$$\mathbf{T} = \begin{bmatrix} \mathbf{t}_1 & \mathbf{t}_2 & \dots & \mathbf{t}_Q \end{bmatrix} , \mathbf{P} = \begin{bmatrix} \mathbf{p}_1 & \mathbf{p}_2 & \dots & \mathbf{p}_Q \end{bmatrix} . \tag{7.27}$$

Then Eq. (7.25) can be written

$$F(\mathbf{W}) = \|\mathbf{T} - \mathbf{W}\mathbf{P}\|^2 = \|\mathbf{E}\|^2 , \tag{7.28}$$

where

$$\mathbf{E} = \mathbf{T} - \mathbf{W}\mathbf{P} , \tag{7.29}$$

and

$$\|\mathbf{E}\|^2 = \sum_i \sum_j e_{ij}^2 . \tag{7.30}$$

Note that $F(\mathbf{W})$ can be made zero if we can solve Eq. (7.26). If the $\mathbf{P}$ matrix has an inverse, the solution is

$$\mathbf{W} = \mathbf{T}\mathbf{P}^{-1} . \tag{7.31}$$

However, this is rarely possible. Normally the $\mathbf{p}_q$ vectors (the columns of $\mathbf{P}$) will be independent, but R (the dimension of $\mathbf{p}_q$) will be larger than Q (the number of $\mathbf{p}_q$ vectors). Therefore, $\mathbf{P}$ will not be a square matrix, and no exact inverse will exist.

Pseudoinverse Rule

It has been shown [Albe72] that the weight matrix that minimizes Eq. (7.25) is given by the *pseudoinverse rule*:

$$\mathbf{W} = \mathbf{T}\mathbf{P}^{+}, \tag{7.32}$$

where $\mathbf{P}^{+}$ is the Moore-Penrose pseudoinverse. The pseudoinverse of a real matrix $\mathbf{P}$ is the unique matrix that satisfies

$$\begin{aligned} \mathbf{P}\mathbf{P}^{+}\mathbf{P} &= \mathbf{P}, \\ \mathbf{P}^{+}\mathbf{P}\mathbf{P}^{+} &= \mathbf{P}^{+}, \\ \mathbf{P}^{+}\mathbf{P} &= (\mathbf{P}^{+}\mathbf{P})^{T}, \\ \mathbf{P}\mathbf{P}^{+} &= (\mathbf{P}\mathbf{P}^{+})^{T}. \end{aligned} \tag{7.33}$$

When the number, R, of rows of $\mathbf{P}$ is greater than the number of columns, Q, of $\mathbf{P}$, and the columns of $\mathbf{P}$ are independent, then the pseudoinverse can be computed by

$$\mathbf{P}^{+} = (\mathbf{P}^{T}\mathbf{P})^{-1}\mathbf{P}^{T}. \tag{7.34}$$

2 +2 4

To test the pseudoinverse rule (Eq. (7.32)), consider again the apple and orange recognition problem. Recall that the input/output prototype vectors are

$$\left\{\mathbf{p}_1 = \begin{bmatrix} 1 \\ -1 \\ -1 \end{bmatrix}, \mathbf{t}_1 = \begin{bmatrix} -1 \end{bmatrix}\right\} \quad \left\{\mathbf{p}_2 = \begin{bmatrix} 1 \\ 1 \\ -1 \end{bmatrix}, \mathbf{t}_2 = \begin{bmatrix} 1 \end{bmatrix}\right\}. \tag{7.35}$$

(Note that we do not need to normalize the input vectors when using the pseudoinverse rule.)

The weight matrix is calculated from Eq. (7.32):

$$\mathbf{W} = \mathbf{T}\mathbf{P}^{+} = \begin{bmatrix} -1 & 1 \end{bmatrix} \left(\begin{bmatrix} 1 & 1 \\ -1 & 1 \\ -1 & -1 \end{bmatrix} \right)^{+}, \tag{7.36}$$

where the pseudoinverse is computed from Eq. (7.34):

$$\mathbf{P}^{+} = (\mathbf{P}^{T}\mathbf{P})^{-1}\mathbf{P}^{T} = \begin{bmatrix} 3 & 1 \\ 1 & 3 \end{bmatrix}^{-1} \begin{bmatrix} 1 & -1 & -1 \\ 1 & 1 & -1 \end{bmatrix} = \begin{bmatrix} 0.25 & -0.5 & -0.25 \\ 0.25 & 0.5 & -0.25 \end{bmatrix}. \tag{7.37}$$

This produces the following weight matrix:

$$\mathbf{W} = \mathbf{T}\mathbf{P}^{+} = \begin{bmatrix} -1 & 1 \end{bmatrix} \begin{bmatrix} 0.25 & -0.5 & -0.25 \\ 0.25 & 0.5 & -0.25 \end{bmatrix} = \begin{bmatrix} 0 & 1 & 0 \end{bmatrix}. \quad (7.38)$$

Let's try this matrix on our two prototype patterns.

$$\mathbf{W}\mathbf{p}_1 = \begin{bmatrix} 0 & 1 & 0 \end{bmatrix} \begin{bmatrix} 1 \\ -1 \\ -1 \end{bmatrix} = \begin{bmatrix} -1 \end{bmatrix} \quad (7.39)$$

$$\mathbf{W}\mathbf{p}_2 = \begin{bmatrix} 0 & 1 & 0 \end{bmatrix} \begin{bmatrix} 1 \\ 1 \\ -1 \end{bmatrix} = \begin{bmatrix} 1 \end{bmatrix} \quad (7.40)$$

The network outputs exactly match the desired outputs. Compare this result with the performance of the Hebb rule. As you can see from Eq. (7.22) and Eq. (7.23), the Hebbian outputs are only close, while the pseudoinverse rule produces exact results.

Application

Now let's see how we might use the Hebb rule on a practical, although greatly oversimplified, pattern recognition problem. For this problem we will use a special type of associative memory — the autoassociative memory. In an *autoassociative memory* the desired output vector is equal to the input vector (i.e., $\mathbf{t}_q = \mathbf{p}_q$). We will use an autoassociative memory to store a set of patterns and then to recall these patterns, even when corrupted patterns are provided as input.

Autoassociative Memory

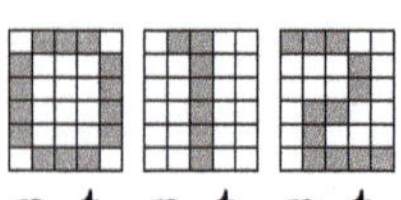
$\mathbf{p}_1,\mathbf{t}_1$ $\mathbf{p}_2,\mathbf{t}_2$ $\mathbf{p}_3,\mathbf{t}_3$

The patterns we want to store are shown to the left. (Since we are designing an autoassociative memory, these patterns represent the input vectors and the targets.) They represent the digits {0, 1, 2} displayed in a 6X5 grid. We need to convert these digits to vectors, which will become the prototype patterns for our network. Each white square will be represented by a "-1", and each dark square will be represented by a "1". Then, to create the input vectors, we will scan each 6X5 grid one column at a time. For example, the first prototype pattern will be

$$\mathbf{p}_1 = \begin{bmatrix} -1 & 1 & 1 & 1 & 1 & -1 & 1 & -1 & -1 & -1 & -1 & 1 & 1 & -1 & \ldots & 1 & -1 \end{bmatrix}^T. \quad (7.41)$$

The vector $\mathbf{p}_1$ corresponds to the digit "0", $\mathbf{p}_2$ to the digit "1", and $\mathbf{p}_3$ to the digit "2". Using the Hebb rule, the weight matrix is computed

$$\mathbf{W} = \mathbf{p}_1\mathbf{p}_1^T + \mathbf{p}_2\mathbf{p}_2^T + \mathbf{p}_3\mathbf{p}_3^T . \tag{7.42}$$

(Note that $\mathbf{p}_q$ replaces $\mathbf{t}_q$ in Eq. (7.8), since this is autoassociative memory.)

Because there are only two allowable values for the elements of the prototype vectors, we will modify the linear associator so that its output elements can only take on values of "-1" or "1". We can do this by replacing the linear transfer function with a symmetrical hard limit transfer function. The resulting network is displayed in Figure 7.2.

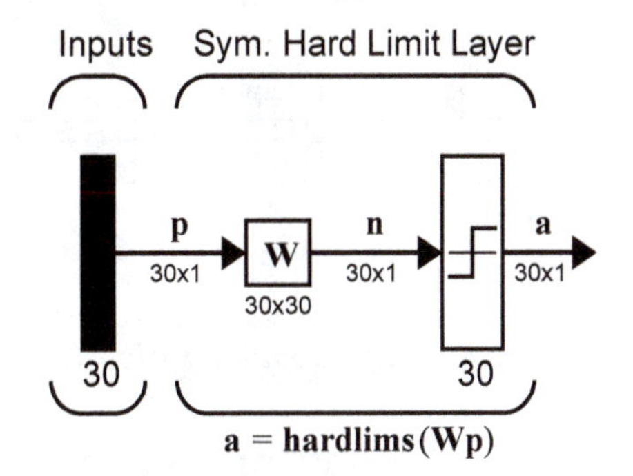

Figure 7.2 Autoassociative Network for Digit Recognition

Now let's investigate the operation of this network. We will provide the network with corrupted versions of the prototype patterns and then check the network output. In the first test, which is shown in Figure 7.3, the network is presented with a prototype pattern in which the lower half of the pattern is occluded. In each case the correct pattern is produced by the network.

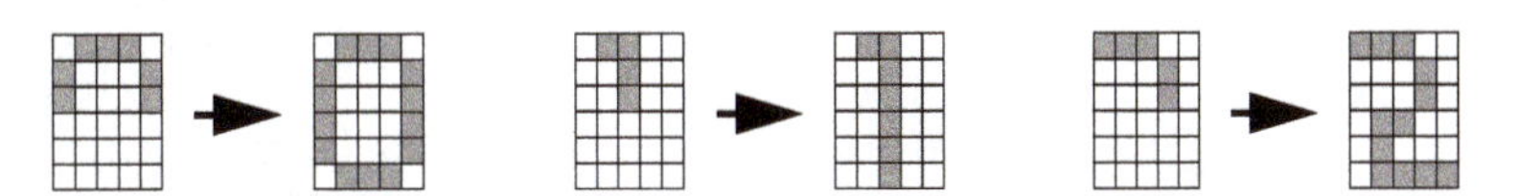

Figure 7.3 Recovery of 50% Occluded Patterns

In the next test we remove even more of the prototype patterns. Figure 7.4 illustrates the result of removing the lower two-thirds of each pattern. In this case only the digit "1" is recovered correctly. The other two patterns produce results that do not correspond to any of the prototype patterns. This is a common problem in associative memories. We would like to design networks so that the number of such spurious patterns would be minimized.

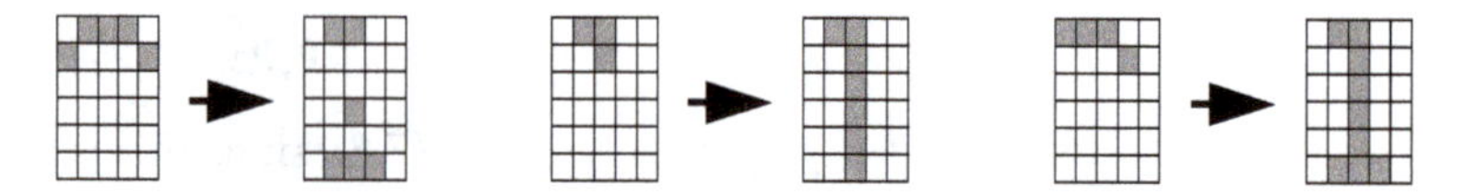

Figure 7.4 Recovery of 67% Occluded Patterns

In our final test we will present the autoassociative network with noisy versions of the prototype pattern. To create the noisy patterns we will randomly change seven elements of each pattern. The results are shown in Figure 7.5. For these examples all of the patterns were correctly recovered.

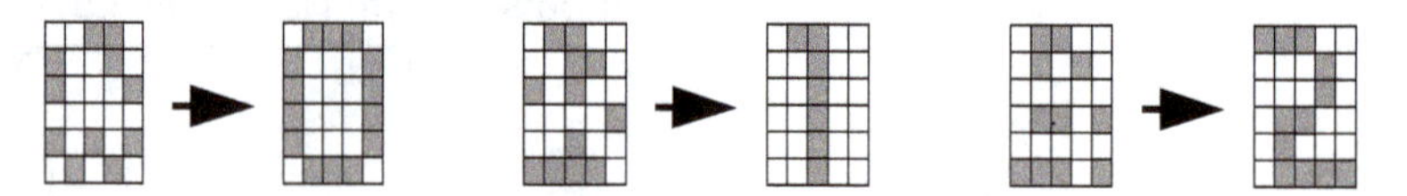

Figure 7.5 Recovery of Noisy Patterns

To experiment with this type of pattern recognition problem, use the Neural Network Design Demonstration Supervised Hebb (**nnd7sh**).

Variations of Hebbian Learning

There have been a number of variations on the basic Hebb rule. In fact, many of the learning laws that will be discussed in the remainder of this text have some relationship to the Hebb rule.

One of the problems of the Hebb rule is that it can lead to weight matrices having very large elements if there are many prototype patterns in the training set. Consider again the basic rule:

$$\mathbf{W}^{new} = \mathbf{W}^{old} + \mathbf{t}_q \mathbf{p}_q^T . \tag{7.43}$$

A positive parameter α, called the learning rate, can be used to limit the amount of increase in the weight matrix elements, if the learning rate is less than one, as in:

$$\mathbf{W}^{new} = \mathbf{W}^{old} + \alpha \mathbf{t}_q \mathbf{p}_q^T . \tag{7.44}$$

We can also add a decay term, so that the learning rule behaves like a smoothing filter, remembering the most recent inputs more clearly:

$$\mathbf{W}^{new} = \mathbf{W}^{old} + \alpha \mathbf{t}_q \mathbf{p}_q^T - \gamma \mathbf{W}^{old} = (1-\gamma)\mathbf{W}^{old} + \alpha \mathbf{t}_q \mathbf{p}_q^T , \tag{7.45}$$

where γ is a positive constant less than one. As γ approaches zero, the learning law becomes the standard rule. As γ approaches one, the learning law quickly forgets old inputs and remembers only the most recent patterns. This keeps the weight matrix from growing without bound.

The idea of filtering the weight changes and of having an adjustable learning rate are important ones, and we will discuss them again in Chapter 10 and Chapter 12.

If we modify Eq. (7.44) by replacing the desired output with the difference between the desired output and the actual output, we get another important learning rule:

$$\mathbf{W}^{new} = \mathbf{W}^{old} + \alpha(\mathbf{t}_q - \mathbf{a}_q)\mathbf{p}_q^T . \tag{7.46}$$

This is sometimes known as the delta rule, since it uses the difference between desired and actual output. It is also known as the Widrow-Hoff algorithm, after the researchers who introduced it. The delta rule adjusts the weights so as to minimize the mean square error (see Chapter 10). For this reason it will produce the same results as the pseudoinverse rule, which minimizes the sum of squares of errors (see Eq. (7.25)). The advantage of the delta rule is that it can update the weights after each new input pattern is presented, whereas the pseudoinverse rule computes the weights in one step, after all of the input/target pairs are known. This sequential updating allows the delta rule to adapt to a changing environment. The delta rule will be discussed in detail in Chapter 10.

In the present chapter we have used a supervised form of the Hebb rule. We have assumed that the desired outputs of the network, $\mathbf{t}_q$, are known, and can be used in the learning rule. In the unsupervised Hebb rule, the *actual* network output is used instead of the *desired* network output, as in:

$$\mathbf{W}^{new} = \mathbf{W}^{old} + \alpha\mathbf{a}_q\mathbf{p}_q^T , \tag{7.47}$$

where $\mathbf{a}_q$ is the output of the network when $\mathbf{p}_q$ is given as the input (see also Eq. (7.5)). This unsupervised form of the Hebb rule, which does not require knowledge of the desired output, is actually a more direct interpretation of Hebb's postulate than is the supervised form discussed in this chapter.

Summary of Results

Hebb's Postulate

"When an axon of cell A is near enough to excite a cell B and repeatedly or persistently takes part in firing it, some growth process or metabolic change takes place in one or both cells such that A's efficiency, as one of the cells firing B, is increased."

Linear Associator

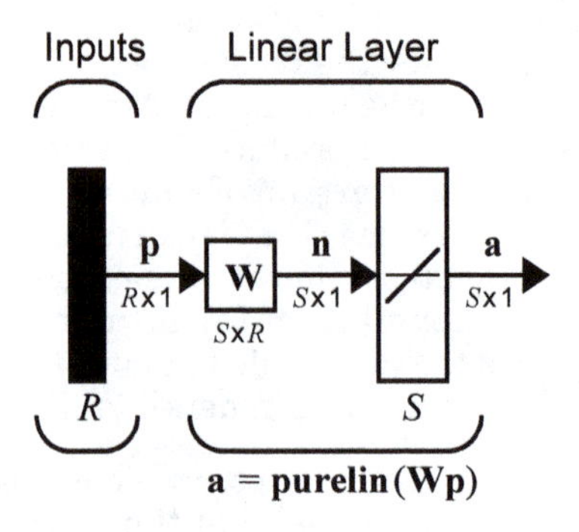

The Hebb Rule

$$w_{ij}^{new} = w_{ij}^{old} + t_{qi}p_{qj}$$

$$\mathbf{W} = \mathbf{t}_1\mathbf{p}_1^T + \mathbf{t}_2\mathbf{p}_2^T + \cdots + \mathbf{t}_Q\mathbf{p}_Q^T$$

$$\mathbf{W} = \begin{bmatrix} \mathbf{t}_1 & \mathbf{t}_2 & \ldots & \mathbf{t}_Q \end{bmatrix} \begin{bmatrix} \mathbf{p}_1^T \\ \mathbf{p}_2^T \\ \vdots \\ \mathbf{p}_Q^T \end{bmatrix} = \mathbf{T}\mathbf{P}^T$$

Pseudoinverse Rule

$$\mathbf{W} = \mathbf{T}\mathbf{P}^+$$

When the number, R, of rows of $\mathbf{P}$ is greater than the number of columns, Q, of $\mathbf{P}$ and the columns of $\mathbf{P}$ are independent, then the pseudoinverse can be computed by

$$\mathbf{P}^{+} = (\mathbf{P}^{T}\mathbf{P})^{-1}\mathbf{P}^{T}.$$

Variations of Hebbian Learning

Filtered Learning

(See Chapter 15)

$$\mathbf{W}^{new} = (1-\gamma)\mathbf{W}^{old} + \alpha\mathbf{t}_q\mathbf{p}_q^T$$

Delta Rule

(See Chapter 10)

$$\mathbf{W}^{new} = \mathbf{W}^{old} + \alpha(\mathbf{t}_q - \mathbf{a}_q)\mathbf{p}_q^T$$

Unsupervised Hebb

$$\mathbf{W}^{new} = \mathbf{W}^{old} + \alpha\mathbf{a}_q\mathbf{p}_q^T$$

Solved Problems

P7.1 Consider the linear associator shown in Figure P7.1.

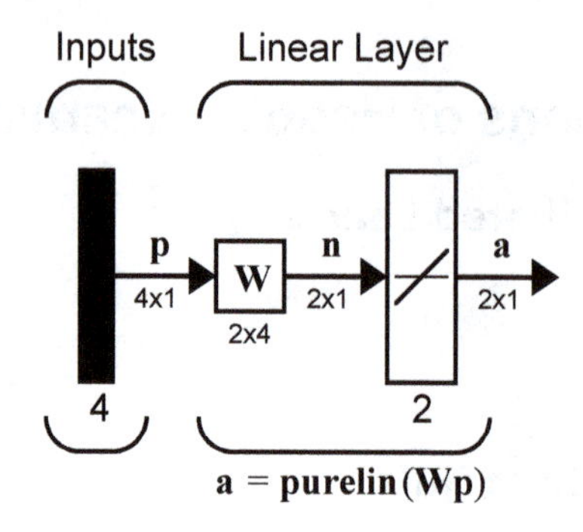

Figure P7.1 Single-Neuron Perceptron

Let the input/output prototype vectors be

$$\left\{ \mathbf{p}_1 = \begin{bmatrix} 1 \\ -1 \\ 1 \\ -1 \end{bmatrix}, \mathbf{t}_1 = \begin{bmatrix} 1 \\ -1 \end{bmatrix} \right\} \quad \left\{ \mathbf{p}_2 = \begin{bmatrix} 1 \\ 1 \\ -1 \\ -1 \end{bmatrix}, \mathbf{t}_2 = \begin{bmatrix} 1 \\ 1 \end{bmatrix} \right\}.$$

- i. **Use the Hebb rule to find the appropriate weight matrix for this linear associator.**
- ii. **Repeat part (i) using the pseudoinverse rule.**
- iii. **Apply the input $\mathbf{p}_1$ to the linear associator using the weight matrix of part (i), then using the weight matrix of part (ii).**

i. The first step is to create the **P** and **T** matrices of Eq. (7.10):

$$\mathbf{P} = \begin{bmatrix} 1 & 1 \\ -1 & 1 \\ 1 & -1 \\ -1 & -1 \end{bmatrix}, \quad \mathbf{T} = \begin{bmatrix} 1 & 1 \\ -1 & 1 \end{bmatrix}.$$

Then the weight matrix can be computed using Eq. (7.9):

$$\mathbf{W}^h = \mathbf{T}\mathbf{P}^T = \begin{bmatrix} 1 & 1 \\ -1 & 1 \end{bmatrix}\begin{bmatrix} 1 & -1 & 1 & -1 \\ 1 & 1 & -1 & -1 \end{bmatrix} = \begin{bmatrix} 2 & 0 & 0 & -2 \\ 0 & 2 & -2 & 0 \end{bmatrix}.$$

ii. For the pseudoinverse rule we use Eq. (7.32):

$$\mathbf{W} = \mathbf{T}\mathbf{P}^+.$$

Since the number of rows of $\mathbf{P}$, four, is greater than the number of columns of $\mathbf{P}$, two, and the columns of $\mathbf{P}$ are independent, then the pseudoinverse can be computed by Eq. (7.34):

$$\mathbf{P}^+ = (\mathbf{P}^T\mathbf{P})^{-1}\mathbf{P}^T.$$

$$\mathbf{P}^+ = \left(\begin{bmatrix} 1 & -1 & 1 & -1 \\ 1 & 1 & -1 & -1 \end{bmatrix}\begin{bmatrix} 1 & 1 \\ -1 & 1 \\ 1 & -1 \\ -1 & -1 \end{bmatrix}\right)^{-1}\begin{bmatrix} 1 & -1 & 1 & -1 \\ 1 & 1 & -1 & -1 \end{bmatrix} = \left(\begin{bmatrix} 4 & 0 \\ 0 & 4 \end{bmatrix}\right)^{-1}\begin{bmatrix} 1 & -1 & 1 & -1 \\ 1 & 1 & -1 & -1 \end{bmatrix}$$

$$= \begin{bmatrix} \frac{1}{4} & 0 \\ 0 & \frac{1}{4} \end{bmatrix}\begin{bmatrix} 1 & -1 & 1 & -1 \\ 1 & 1 & -1 & -1 \end{bmatrix} = \begin{bmatrix} \frac{1}{4} & -\frac{1}{4} & \frac{1}{4} & -\frac{1}{4} \\ \frac{1}{4} & \frac{1}{4} & -\frac{1}{4} & -\frac{1}{4} \end{bmatrix}$$

The weight matrix can now be computed:

$$\mathbf{W}^p = \mathbf{T}\mathbf{P}^+ = \begin{bmatrix} 1 & 1 \\ -1 & 1 \end{bmatrix}\begin{bmatrix} \frac{1}{4} & -\frac{1}{4} & \frac{1}{4} & -\frac{1}{4} \\ \frac{1}{4} & \frac{1}{4} & -\frac{1}{4} & -\frac{1}{4} \end{bmatrix} = \begin{bmatrix} \frac{1}{2} & 0 & 0 & -\frac{1}{2} \\ 0 & \frac{1}{2} & -\frac{1}{2} & 0 \end{bmatrix}.$$

iii. We now test the two weight matrices.

$$\mathbf{W}^h\mathbf{p}_1 = \begin{bmatrix} 2 & 0 & 0 & -2 \\ 0 & 2 & -2 & 0 \end{bmatrix}\begin{bmatrix} 1 \\ -1 \\ 1 \\ -1 \end{bmatrix} = \begin{bmatrix} 4 \\ -4 \end{bmatrix} \neq \mathbf{t}_1$$

$$\mathbf{W}^p \mathbf{p}_1 = \begin{bmatrix} \frac{1}{2} & 0 & 0 & -\frac{1}{2} \\ 0 & \frac{1}{2} & -\frac{1}{2} & 0 \end{bmatrix} \begin{bmatrix} 1 \\ -1 \\ 1 \\ -1 \end{bmatrix} = \begin{bmatrix} 1 \\ -1 \end{bmatrix} = \mathbf{t}_1$$

Why didn't the Hebb rule produce the correct results? Well, consider again Eq. (7.11). Since $\mathbf{p}_1$ and $\mathbf{p}_2$ are orthogonal (check that they are) this equation can be written

$$\mathbf{W}^h \mathbf{p}_1 = \mathbf{t}_1(\mathbf{p}_1^T \mathbf{p}_1),$$

but the $\mathbf{p}_1$ vector is not normalized, so $(\mathbf{p}_1^T \mathbf{p}_1) \neq 1$. Therefore the output of the network will not be $\mathbf{t}_1$.

The pseudoinverse rule, on the other hand, is guaranteed to minimize

$$\sum_{q=1}^{2} \left\| \mathbf{t}_q - \mathbf{W}\mathbf{p}_q \right\|^2,$$

which in this case can be made equal to zero.

$\mathbf{p}_1$ $\mathbf{p}_2$

P7.2 Consider the prototype patterns shown to the left.

i. Are these patterns orthogonal?

ii. Design an autoassociator for these patterns. Use the Hebb rule.

$\mathbf{p}_t$

iii. What response does the network give to the test input pattern, $\mathbf{p}_t$, shown to the left?

i. The first thing we need to do is to convert the patterns into vectors. Let's assign any solid square the value 1 and any open square the value -1. Then to convert from the two-dimensional pattern to a vector we will scan the pattern column by column. (We could use rows if we wished.) The two prototype vectors then become:

$$\mathbf{p}_1 = \begin{bmatrix} 1 & 1 & -1 & 1 & -1 & -1 \end{bmatrix}^T \qquad \mathbf{p}_2 = \begin{bmatrix} -1 & 1 & 1 & 1 & 1 & -1 \end{bmatrix}^T.$$

To test orthogonality we take the inner product of $\mathbf{p}_1$ and $\mathbf{p}_2$:

$$\mathbf{p}_1^T\mathbf{p}_2 = \begin{bmatrix}1 & 1 & -1 & 1 & -1 & -1\end{bmatrix}\begin{bmatrix}-1\\1\\1\\1\\1\\-1\end{bmatrix} = 0 \cdot$$

Therefore they are orthogonal. (Although they are not normalized since

$$\mathbf{p}_1^T\mathbf{p}_1 = \mathbf{p}_2^T\mathbf{p}_2 = 6.)$$

ii. We will use an autoassociator like the one in Figure 7.2, except that the number of inputs and outputs to the network will be six. To find the weight matrix we use the Hebb rule:

$$\mathbf{W} = \mathbf{T}\mathbf{P}^T,$$

where

$$\mathbf{P} = \mathbf{T} = \begin{bmatrix}1 & -1\\1 & 1\\-1 & 1\\1 & 1\\-1 & 1\\-1 & -1\end{bmatrix}.$$

Therefore the weight matrix is

$$\mathbf{W} = \mathbf{T}\mathbf{P}^T = \begin{bmatrix}1 & -1\\1 & 1\\-1 & 1\\1 & 1\\-1 & 1\\-1 & -1\end{bmatrix}\begin{bmatrix}1 & 1 & -1 & 1 & -1 & -1\\-1 & 1 & 1 & 1 & 1 & -1\end{bmatrix} = \begin{bmatrix}2 & 0 & -2 & 0 & -2 & 0\\0 & 2 & 0 & 2 & 0 & -2\\-2 & 0 & 2 & 0 & 2 & 0\\0 & 2 & 0 & 2 & 0 & -2\\-2 & 0 & 2 & 0 & 2 & 0\\0 & -2 & 0 & -2 & 0 & 2\end{bmatrix}.$$

iii. To apply the test pattern to the network we convert it to a vector:

$$\mathbf{p}_t = \begin{bmatrix}1 & 1 & 1 & 1 & 1 & -1\end{bmatrix}^T.$$

The network response is then

$$\mathbf{a} = \mathbf{hardlims}(\mathbf{W}\mathbf{p}_t) = \mathbf{hardlims}\left(\begin{bmatrix} 2 & 0 & -2 & 0 & -2 & 0 \\ 0 & 2 & 0 & 2 & 0 & -2 \\ -2 & 0 & 2 & 0 & 2 & 0 \\ 0 & 2 & 0 & 2 & 0 & -2 \\ -2 & 0 & 2 & 0 & 2 & 0 \\ 0 & -2 & 0 & -2 & 0 & 2 \end{bmatrix}\begin{bmatrix} 1 \\ 1 \\ 1 \\ 1 \\ 1 \\ -1 \end{bmatrix}\right)$$

$$\mathbf{a} = \mathbf{hardlims}\left(\begin{bmatrix} -2 \\ 6 \\ 2 \\ 6 \\ 2 \\ -6 \end{bmatrix}\right) = \begin{bmatrix} -1 \\ 1 \\ 1 \\ 1 \\ 1 \\ -1 \end{bmatrix} = \mathbf{p}_2 .$$

Is this a satisfactory response? How would we want the network to respond to this input pattern? The network should produce the prototype pattern that is closest to the input pattern. In this case the test input pattern, $\mathbf{p}_t$, has a Hamming distance of 1 from $\mathbf{p}_2$, and a distance of 2 from $\mathbf{p}_1$. Therefore the network did produce the correct response. (See Chapter 3 for a discussion of Hamming distance.)

Note that in this example the prototype vectors were not normalized. This did not cause the same problem with network performance that we saw in Problem P7.1, because of the *hardlims* nonlinearity. It forces the network output to be either 1 or -1. In fact, most of the interesting and useful properties of neural networks are due to the effects of nonlinearities.

P7.3 Consider an autoassociation problem in which there are three prototype patterns (shown below as $\mathbf{p}_1$, $\mathbf{p}_2$, $\mathbf{p}_3$). Design autoassociative networks to recognize these patterns, using both the Hebb rule and the pseudoinverse rule. Check their performance on the test pattern $\mathbf{p}_t$ shown below.

$$\mathbf{p}_1 = \begin{bmatrix} 1 \\ 1 \\ -1 \\ -1 \\ 1 \\ 1 \\ 1 \end{bmatrix} \quad \mathbf{p}_2 = \begin{bmatrix} 1 \\ 1 \\ 1 \\ -1 \\ 1 \\ -1 \\ 1 \end{bmatrix} \quad \mathbf{p}_3 = \begin{bmatrix} -1 \\ 1 \\ -1 \\ 1 \\ 1 \\ -1 \\ 1 \end{bmatrix} \quad \mathbf{p}_t = \begin{bmatrix} -1 \\ 1 \\ -1 \\ -1 \\ 1 \\ -1 \\ 1 \end{bmatrix}$$

```
» 2 + 2
ans =
     4
```

This problem is a little tedious to work out by hand, so let's use MATLAB. First we create the prototype vectors.

```
p1=[ 1  1 -1 -1  1  1  1]';
p2=[ 1  1  1 -1  1 -1  1]';
p3=[-1  1 -1  1  1 -1  1]';
P=[p1 p2 p3];
```

Now we can compute the weight matrix using the Hebb rule.

```
wh=P*P';
```

To check the network we create the test vector.

```
pt=[-1  1 -1 -1  1 -1  1]';
```

The network response is then calculated.

```
ah=hardlims(wh*pt);
ah'
ans =
       1     1    -1    -1     1    -1     1
```

Notice that this response does not match any of the prototype vectors. This is not surprising since the prototype patterns are not orthogonal. Let's try the pseudoinverse rule.

```
pseu=inv(P'*P)*P';
wp=P*pseu;
ap=hardlims(wp*pt);
ap'
ans =
      -1     1    -1     1     1    -1     1
```

Note that the network response is equal to $\mathbf{p}_3$. Is this the correct response? As usual, we want the response to be the prototype pattern closest to the input pattern. In this case $\mathbf{p}_t$ is a Hamming distance of 2 from both $\mathbf{p}_1$ and $\mathbf{p}_2$, but only a distance of 1 from $\mathbf{p}_3$. Therefore the pseudoinverse rule produces the correct response.

Try other test inputs to see if there are additional cases where the pseudoinverse rule produces better results than the Hebb rule.

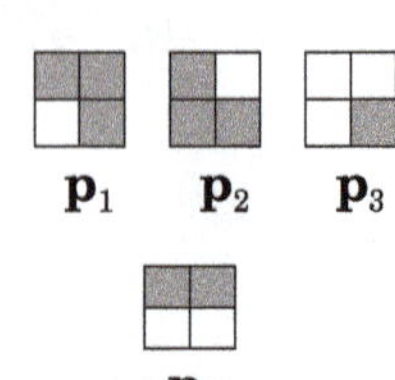

P7.4 Consider the three prototype patterns shown to the left.

i. Use the Hebb rule to design a perceptron network that will recognize these three patterns.

ii. Find the response of the network to the pattern $\mathbf{p}_t$ shown to the left. Is the response correct?

i. We can convert the patterns to vectors, as we did in previous problems, to obtain:

$$\mathbf{p}_1 = \begin{bmatrix} 1 \\ -1 \\ 1 \\ 1 \end{bmatrix} \quad \mathbf{p}_2 = \begin{bmatrix} 1 \\ 1 \\ -1 \\ 1 \end{bmatrix} \quad \mathbf{p}_3 = \begin{bmatrix} -1 \\ -1 \\ -1 \\ 1 \end{bmatrix} \quad \mathbf{p}_t = \begin{bmatrix} 1 \\ -1 \\ 1 \\ -1 \end{bmatrix}.$$

We now need to choose the desired output vectors for each prototype input vector. Since there are three prototype vectors that we need to distinguish, we will need two elements in the output vector. We can choose the three desired outputs to be:

$$\mathbf{t}_1 = \begin{bmatrix} -1 \\ -1 \end{bmatrix} \quad \mathbf{t}_2 = \begin{bmatrix} -1 \\ 1 \end{bmatrix} \quad \mathbf{t}_3 = \begin{bmatrix} 1 \\ -1 \end{bmatrix}.$$

(Note that this choice was arbitrary. Any distinct combination of 1 and -1 could have been chosen for each vector.)

The resulting network is shown in Figure P7.2.

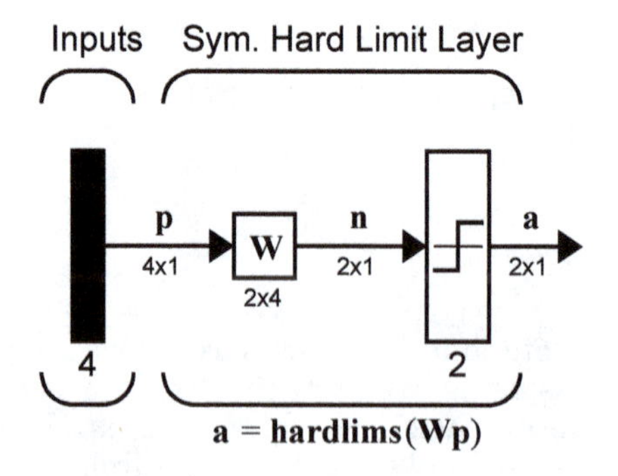

Figure P7.2 Perceptron Network for Problem P7.4

The next step is to determine the weight matrix using the Hebb rule.

$$\mathbf{W} = \mathbf{T}\mathbf{P}^T = \begin{bmatrix} -1 & -1 & 1 \\ -1 & 1 & -1 \end{bmatrix} \begin{bmatrix} 1 & -1 & 1 & 1 \\ 1 & 1 & -1 & 1 \\ -1 & -1 & -1 & 1 \end{bmatrix} = \begin{bmatrix} -3 & -1 & -1 & -1 \\ 1 & 3 & -1 & -1 \end{bmatrix}$$

ii. The response of the network to the test input pattern is calculated as follows.

$$\mathbf{a} = \mathbf{hardlims}(\mathbf{W}\mathbf{p}_t) = \mathbf{hardlims}\left(\begin{bmatrix} -3 & -1 & -1 & -1 \\ 1 & 3 & -1 & -1 \end{bmatrix} \begin{bmatrix} 1 \\ -1 \\ 1 \\ -1 \end{bmatrix} \right)$$

$$= \mathbf{hardlims}\left(\begin{bmatrix} -2 \\ -2 \end{bmatrix} \right) = \begin{bmatrix} -1 \\ -1 \end{bmatrix} \rightarrow \mathbf{p}_1 .$$

So the response of the network indicates that the test input pattern is closest to $\mathbf{p}_1$. Is this correct? Yes, the Hamming distance to $\mathbf{p}_1$ is 1, while the distance to $\mathbf{p}_2$ and $\mathbf{p}_3$ is 3.

P7.5 Suppose that we have a linear autoassociator that has been designed for Q orthogonal prototype vectors of length R using the Hebb rule. The vector elements are either 1 or -1.

i. Show that the Q prototype patterns are eigenvectors of the weight matrix.

ii. What are the other $(R - Q)$ eigenvectors of the weight matrix?

i. Suppose the prototype vectors are:

$$\mathbf{p}_1, \mathbf{p}_2, \dots, \mathbf{p}_Q .$$

Since this is an autoassociator, these are both the input vectors and the desired output vectors. Therefore

$$\mathbf{T} = \begin{bmatrix} \mathbf{p}_1 & \mathbf{p}_2 & \dots & \mathbf{p}_Q \end{bmatrix} \qquad \mathbf{P} = \begin{bmatrix} \mathbf{p}_1 & \mathbf{p}_2 & \dots & \mathbf{p}_Q \end{bmatrix} .$$

If we then use the Hebb rule to calculate the weight matrix we find

$$\mathbf{W} = \mathbf{T}\mathbf{P}^T = \sum_{q=1}^{Q} \mathbf{p}_q \mathbf{p}_q^T ,$$

from Eq. (7.8). Now, if we apply one of the prototype vectors as input to the network we obtain

$$\mathbf{a} = \mathbf{W}\mathbf{p}_k = \left(\sum_{q=1}^{Q} \mathbf{p}_q\mathbf{p}_q^T\right)\mathbf{p}_k = \sum_{q=1}^{Q} \mathbf{p}_q(\mathbf{p}_q^T\mathbf{p}_k) .$$

Because the patterns are orthogonal, this reduces to

$$\mathbf{a} = \mathbf{p}_k(\mathbf{p}_k^T\mathbf{p}_k) .$$

And since every element of $\mathbf{p}_k$ must be either -1 or 1, we find that

$$\mathbf{a} = \mathbf{p}_k R .$$

To summarize the results:

$$\mathbf{W}\mathbf{p}_k = R\mathbf{p}_k ,$$

which implies that $\mathbf{p}_k$ is an eigenvector of $\mathbf{W}$ and R is the corresponding eigenvalue. Each prototype vector is an eigenvector with the same eigenvalue.

ii. Note that the repeated eigenvalue R has a Q-dimensional eigenspace associated with it: the subspace spanned by the Q prototype vectors. Now consider the subspace that is orthogonal to this eigenspace. Every vector in this subspace should be orthogonal to each prototype vector. The dimension of the orthogonal subspace will be $R - Q$. Consider the following arbitrary basis set for this orthogonal space:

$$\mathbf{z}_1, \mathbf{z}_2, \ldots , \mathbf{z}_{R-Q} .$$

If we apply any one of these basis vectors to the network we obtain:

$$\mathbf{a} = \mathbf{W}\mathbf{z}_k = \left(\sum_{q=1}^{Q} \mathbf{p}_q\mathbf{p}_q^T\right)\mathbf{z}_k = \sum_{q=1}^{Q} \mathbf{p}_q(\mathbf{p}_q^T\mathbf{z}_k) = 0 ,$$

since each $\mathbf{z}_k$ is orthogonal to every $\mathbf{p}_q$. This implies that each $\mathbf{z}_k$ is an eigenvector of $\mathbf{W}$ with eigenvalue 0.

To summarize, the weight matrix $\mathbf{W}$ has two eigenvalues, R and 0. This means that any vector in the space spanned by the prototype vectors will be amplified by R, whereas any vector that is orthogonal to the prototype vectors will be set to 0.

P7.6 The networks we have used so far in this chapter have not included a bias vector. Consider the problem of designing a perceptron network (Figure P7.3) to recognize the following patterns:

$$\mathbf{p}_1 = \begin{bmatrix} 1 \\ 1 \end{bmatrix} \qquad \mathbf{p}_2 = \begin{bmatrix} 2 \\ 2 \end{bmatrix}.$$

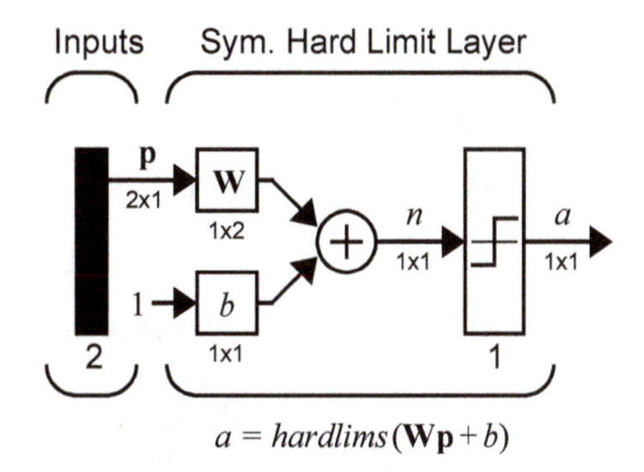

Figure P7.3 Single-Neuron Perceptron

i. **Why is a bias required to solve this problem?**

ii. **Use the pseudoinverse rule to design a network with bias to solve this problem.**

i. Recall from Chapter 3 and Chapter 4 that the decision boundary for the perceptron network is the line defined by:

$$\mathbf{W}\mathbf{p} + b = 0 .$$

If there is no bias, then $b = 0$ and the boundary is defined by:

$$\mathbf{W}\mathbf{p} = 0 ,$$

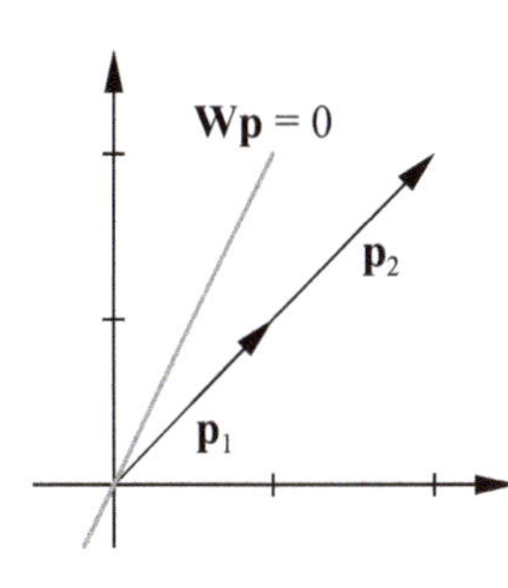

which is a line that must pass through the origin. Now consider the two vectors, $\mathbf{p}_1$ and $\mathbf{p}_2$, which are given in this problem. They are shown graphically in the figure to the left, along with an arbitrary decision boundary that passes through the origin. It is clear that no decision boundary that passes through the origin could separate these two vectors. Therefore a bias is required to solve this problem.

ii. To use the pseudoinverse rule (or the Hebb rule) when there is a bias term, we should treat the bias as another weight, with an input of 1 (as is shown in all of the network figures). We then augment the input vectors with a 1 as the last element:

$$\mathbf{p'}_1 = \begin{bmatrix} 1 \\ 1 \\ 1 \end{bmatrix} \qquad \mathbf{p'}_2 = \begin{bmatrix} 2 \\ 2 \\ 1 \end{bmatrix}.$$

Let's choose the desired outputs to be

$$t_1 = 1 \qquad t_2 = -1,$$

so that

$$\mathbf{P} = \begin{bmatrix} 1 & 2 \\ 1 & 2 \\ 1 & 1 \end{bmatrix}, \mathbf{T} = \begin{bmatrix} 1 & -1 \end{bmatrix}.$$

We now form the pseudoinverse matrix:

$$\mathbf{P}^{+} = \left(\begin{bmatrix} 1 & 1 & 1 \\ 2 & 2 & 1 \end{bmatrix} \begin{bmatrix} 1 & 2 \\ 1 & 2 \\ 1 & 1 \end{bmatrix} \right)^{-1} \begin{bmatrix} 1 & 1 & 1 \\ 2 & 2 & 1 \end{bmatrix} = \begin{bmatrix} 3 & 5 \\ 5 & 9 \end{bmatrix}^{-1} \begin{bmatrix} 1 & 1 & 1 \\ 2 & 2 & 1 \end{bmatrix} = \begin{bmatrix} -0.5 & -0.5 & 2 \\ 0.5 & 0.5 & -1 \end{bmatrix}.$$

The augmented weight matrix is then computed:

$$\mathbf{W'} = \mathbf{T}\mathbf{P}^{+} = \begin{bmatrix} 1 & -1 \end{bmatrix} \begin{bmatrix} -0.5 & -0.5 & 2 \\ 0.5 & 0.5 & -1 \end{bmatrix} = \begin{bmatrix} -1 & -1 & 3 \end{bmatrix}.$$

We can then pull out the standard weight matrix and bias:

$$\mathbf{W} = \begin{bmatrix} -1 & -1 \end{bmatrix} \qquad b = 3.$$

The decision boundary for this weight and bias is shown in the Figure P7.4. This boundary does separate the two prototype vectors.

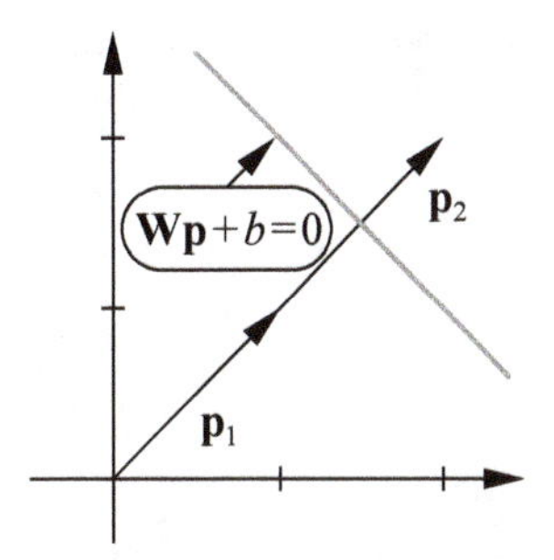

Figure P7.4 Decision Boundary for Solved Problem P7.6

P7.7 In all of our pattern recognition examples thus far, we have represented patterns as vectors by using "1" and "-1" to represent dark and light pixels (picture elements), respectively. What if we were to use "1" and "0" instead? How should the Hebb rule be changed?

First, let's introduce some notation to distinguish the two different representations (usually referred to as the bipolar {-1, 1} representation and the binary {0, 1} representation). The bipolar representation of the prototype input/output vectors will be denoted

$$\{\mathbf{p}_1, \mathbf{t}_1\}, \{\mathbf{p}_2, \mathbf{t}_2\}, \dots, \{\mathbf{p}_Q, \mathbf{t}_Q\},$$

and the binary representation will be denoted

$$\{\mathbf{p}'_1, \mathbf{t}'_1\}, \{\mathbf{p}'_2, \mathbf{t}'_2\}, \dots, \{\mathbf{p}'_Q, \mathbf{t}'_Q\}.$$

The relationship between the two representations is given by:

$$\mathbf{p}'_q = \frac{1}{2}\mathbf{p}_q + \frac{1}{2}\mathbf{1} \qquad \mathbf{p}_q = 2\mathbf{p}'_q - \mathbf{1},$$

where $\mathbf{1}$ is a vector of ones.

Next, we determine the form of the binary associative network. We will use the network shown in Figure P7.5. It is different than the bipolar associative network, as shown in Figure 7.2, in two ways. First, it uses the *hardlim* nonlinearity rather than *hardlims*, since the output should be either 0 or 1. Secondly, it uses a bias vector. It requires a bias vector because all binary vectors will fall into one quadrant of the vector space, so a boundary that passes through the origin will not always be able to divide the patterns. (See Problem P7.6.)

The next step is to determine the weight matrix and the bias vector for this network. If we want the binary network of Figure P7.5 to have the same effective response as a bipolar network (as in Figure 7.2), then the net input, $\mathbf{n}$, should be the same for both networks:

$$\mathbf{W}'\mathbf{p}' + \mathbf{b} = \mathbf{W}\mathbf{p}.$$

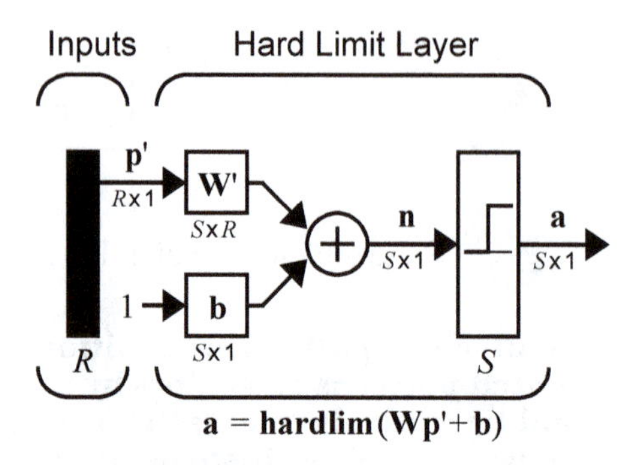

Figure P7.5 Binary Associative Network

This will guarantee that whenever the bipolar network produces a "1" the binary network will produce a "1", and whenever the bipolar network produces a "-1" the binary network will produce a "0".

If we then substitute for $\mathbf{p}'$ as a function of $\mathbf{p}$ we find:

$$\mathbf{W}'\left(\frac{1}{2}\mathbf{p} + \frac{1}{2}\mathbf{1}\right) + \mathbf{b} = \frac{1}{2}\mathbf{W}'\mathbf{p} + \frac{1}{2}\mathbf{W}'\mathbf{1} + \mathbf{b} = \mathbf{W}\mathbf{p}.$$

Therefore, to produce the same results as the bipolar network, we should choose

$$\mathbf{W}' = 2\mathbf{W} \qquad \mathbf{b} = -\mathbf{W}\mathbf{1},$$

where $\mathbf{W}$ is the bipolar weight matrix.

Epilogue

We had two main objectives for this chapter. First, we wanted to introduce one of the most influential neural network learning rules: the Hebb rule. This was one of the first neural learning rules ever proposed, and yet it continues to influence even the most recent developments in network learning theory. Second, we wanted to show how the performance of this learning rule could be explained using the linear algebra concepts discussed in the two preceding chapters. This is one of the key objectives of this text. We want to show how certain important mathematical concepts underlie the operation of all artificial neural networks. We plan to continue to weave together the mathematical ideas with the neural network applications, and hope in the process to increase our understanding of both.

The next two chapters introduce some mathematics that are critical to our understanding of the two learning laws covered in Chapter 10 and Chapter 11. Those learning laws fall under a subheading called *performance* learning, because they attempt to optimize the performance of the network. In order to understand these performance learning laws, we need to introduce some basic concepts in optimization. As with the material on the Hebb rule, our understanding of these topics in optimization will be greatly aided by our previous work in linear algebra.

Further Reading

[Albe72] A. Albert, *Regression and the Moore-Penrose Pseudoinverse*, New York: Academic Press, 1972.

Albert's text is the major reference for the theory and basic properties of the pseudoinverse. Proofs are included for all major pseudoinverse theorems.

[Ande72] J. Anderson, "A simple neural network generating an interactive memory," *Mathematical Biosciences*, vol. 14, pp. 197–220, 1972.

Anderson proposed a "linear associator" model for associative memory. The model was trained, using a generalization of the Hebb postulate, to learn an association between input and output vectors. The physiological plausibility of the network was emphasized. Kohonen published a closely related paper at the same time [Koho72], although the two researchers were working independently.

[Hebb49] D. O. Hebb, *The Organization of Behavior*, New York: Wiley, 1949.

The main premise of this seminal book is that behavior can be explained by the action of neurons. In it, Hebb proposes one of the first learning laws, which postulated a mechanism for learning at the cellular level.

[Koho72] T. Kohonen, "Correlation matrix memories," *IEEE Transactions on Computers*, vol. 21, pp. 353–359, 1972.

Kohonen proposed a correlation matrix model for associative memory. The model was trained, using the outer product rule (also known as the Hebb rule), to learn an association between input and output vectors. The mathematical structure of the network was emphasized. Anderson published a closely related paper at the same time [Ande72], although the two researchers were working independently.

Exercises

E7.1 Consider the prototype patterns given to the left.

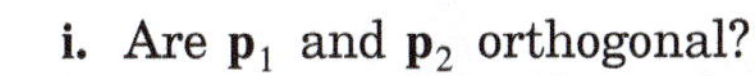

i. Are $\mathbf{p}_1$ and $\mathbf{p}_2$ orthogonal?

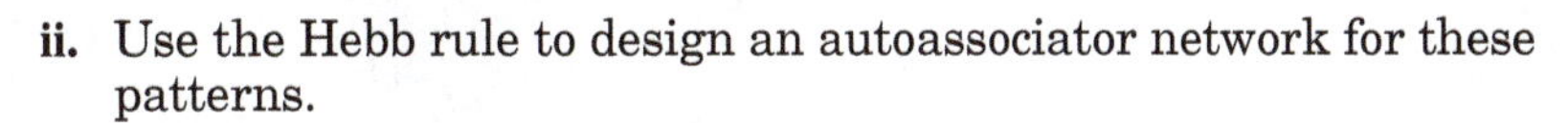

ii. Use the Hebb rule to design an autoassociator network for these patterns.

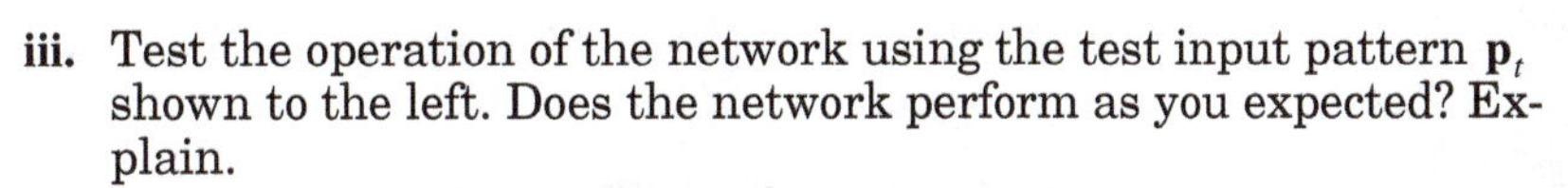

iii. Test the operation of the network using the test input pattern $\mathbf{p}_t$ shown to the left. Does the network perform as you expected? Explain.

E7.2 Repeat Exercise E7.1 using the pseudoinverse rule.

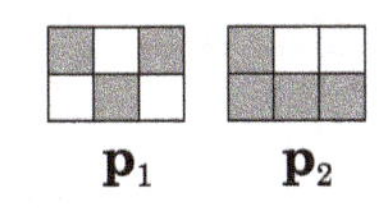

E7.3 Use the Hebb rule to determine the weight matrix for a perceptron network (shown in Figure E7.1) to recognize the patterns shown to the left.

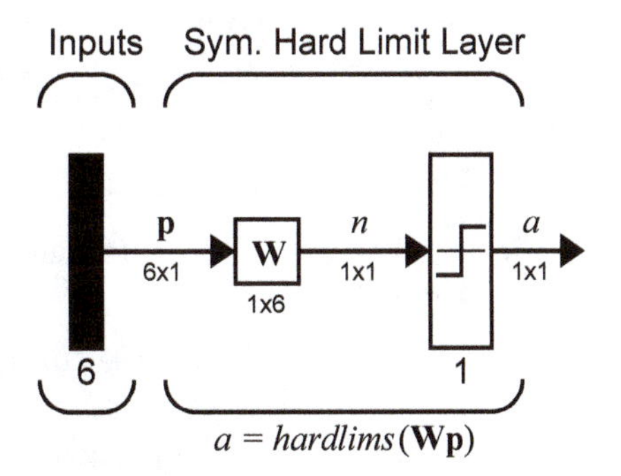

Figure E7.1 Perceptron Network for Exercise E7.3

E7.4 In Problem P7.7 we demonstrated how networks can be trained using the Hebb rule when the prototype vectors are given in binary (as opposed to bipolar) form. Repeat Exercise E7.1 using the binary representation for the prototype vectors. Show that the response of this binary network is equivalent to the response of the original bipolar network.

E7.5 Show that an autoassociator network will continue to perform if we zero the diagonal elements of a weight matrix that has been determined by the Hebb rule. In other words, suppose that the weight matrix is determined from:

$$\mathbf{W} = \mathbf{P}\mathbf{P}^T - Q\mathbf{I} ,$$

where Q is the number of prototype vectors. (Hint: show that the prototype vectors continue to be eigenvectors of the new weight matrix.)

E7.6 We have three input/output prototype vector pairs:

$$\left\{\mathbf{p}_1 = \begin{bmatrix}1\\0\end{bmatrix}, t_1 = 1\right\}, \left\{\mathbf{p}_2 = \begin{bmatrix}1\\1\end{bmatrix}, t_2 = -1\right\}, \left\{\mathbf{p}_3 = \begin{bmatrix}0\\1\end{bmatrix}, t_3 = 1\right\}.$$

i. Show that this problem cannot be solved unless the network uses a bias.

ii. Use the pseudoinverse rule to design a network for these prototype vectors. Verify that the network correctly transforms the prototype vectors.

E7.7 Consider the reference patterns and targets given below. We want to use these data to train a linear associator network.

$$\left\{\mathbf{p}_1 = \begin{bmatrix}2\\4\end{bmatrix}, t_1 = \begin{bmatrix}26\end{bmatrix}\right\} \quad \left\{\mathbf{p}_2 = \begin{bmatrix}4\\2\end{bmatrix}, t_2 = \begin{bmatrix}26\end{bmatrix}\right\} \quad \left\{\mathbf{p}_3 = \begin{bmatrix}-2\\-2\end{bmatrix}, t_3 = \begin{bmatrix}-26\end{bmatrix}\right\}$$

i. Use the Hebb rule to find the weights of the network.

ii. Find and sketch the decision boundary for the network with the Hebb rule weights.

iii. Use the pseudo-inverse rule to find the weights of the network. Because the number, R, of rows of $\mathbf{P}$ is less than the number of columns, Q, of $\mathbf{P}$, the pseudoinverse can be computed by $\mathbf{P}^+ = \mathbf{P}^T(\mathbf{P}\mathbf{P}^T)^{-1}$.

iv. Find and sketch the decision boundary for the network with the pseudo-inverse rule weights.

v. Compare (discuss) the decision boundaries and weights for each of the methods (Hebb and pseudo-inverse).

E7.8 Consider the three prototype patterns shown in Figure E7.2.

i. Are these patterns orthogonal? Demonstrate.

ii. Use the Hebb rule to determine the weight matrix for a linear autoassociator to recognize these patterns.

iii. Draw the network diagram.

iv. Find the eigenvalues and eigenvectors of the weight matrix. (Do **not** solve the equation $|\mathbf{W} - \lambda\mathbf{I}| = 0$. Use an analysis of the Hebb rule.)

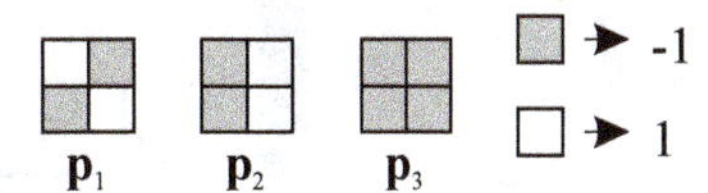

Figure E7.2 Prototype Patterns for Exercise E7.8

E7.9 Suppose that we have the following three reference patterns and their targets.

$$\left\{\mathbf{p}_1 = \begin{bmatrix}3\\6\end{bmatrix}, t_1 = \begin{bmatrix}75\end{bmatrix}\right\} \quad \left\{\mathbf{p}_2 = \begin{bmatrix}6\\3\end{bmatrix}, t_2 = \begin{bmatrix}75\end{bmatrix}\right\} \quad \left\{\mathbf{p}_3 = \begin{bmatrix}-6\\3\end{bmatrix}, t_3 = \begin{bmatrix}-75\end{bmatrix}\right\}$$

i. Draw the network diagram for a linear associator network that could be trained on these patterns.

ii. Use the Hebb rule to find the weights of the network.

iii. Find and sketch the decision boundary for the network with the Hebb rule weights. Does the boundary separate the patterns? Demonstrate.

iv. Use the pseudo-inverse rule to find the weights of the network. Describe the difference between this boundary and the Hebb rule boundary.

E7.10 We have the following input/output pairs:

$$\left\{\mathbf{p}_1 = \begin{bmatrix}1\\1\end{bmatrix}, t_1 = \begin{bmatrix}1\end{bmatrix}\right\} \quad \left\{\mathbf{p}_2 = \begin{bmatrix}1\\-1\end{bmatrix}, t_2 = \begin{bmatrix}-1\end{bmatrix}\right\}$$

i. Use the Hebb rule to determine the weight matrix for the perceptron network shown in Figure E7.3.

ii. Plot the resulting decision boundary. Is this a "good" decision boundary? Explain.

iii. Repeat part i. using the Pseudoinverse rule.

iv. Will there be any difference in the operation of the network if the Pseudoinverse weight matrix is used? Explain.

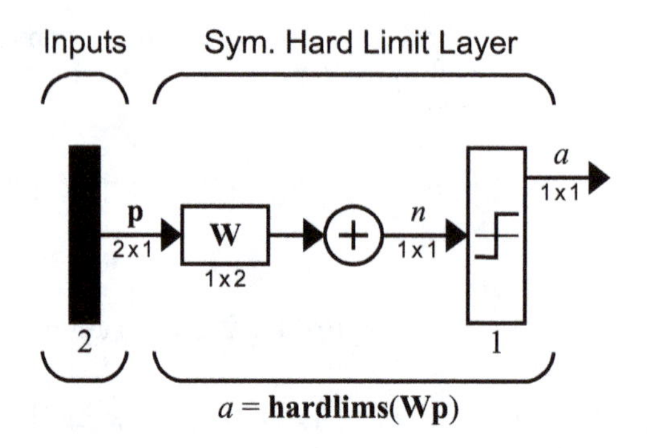

Figure E7.3 Network for Exercise E7.10

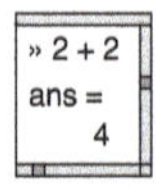

E7.11 One question we might ask about the Hebb and pseudoinverse rules is: How many prototype patterns can be stored in one weight matrix? Test this experimentally using the digit recognition problem that was discussed on page 7-10. Begin with the digits "0" and "1". Add one digit at a time up to "6", and test how often the correct digit is reconstructed after randomly changing 2, 4 and 6 pixels.

i. First use the Hebb rule to create the weight matrix for the digits "0" and "1". Then randomly change 2 pixels of each digit and apply the noisy digits to the network. Repeat this process 10 times, and record the percentage of times in which the correct pattern (without noise) is produced at the output of the network. Repeat as 4 and 6 pixels of each digit are modified. The entire process is then repeated when the digits "0", "1" and "2" are used. This continues, one digit at a time, until you test the network when all of the digits "0" through "6" are used. When you have completed all of the tests, you will be able to plot three curves showing percentage error versus number of digits stored, one curve each for 2, 4 and 6 pixel errors.

ii. Repeat part (i) using the pseudoinverse rule, and compare the results of the two rules.

8 Performance Surfaces and Optimum Points

Objectives

This chapter lays the foundation for a type of neural network training technique called performance learning. There are several different classes of network learning laws, including associative learning (as in the Hebbian learning of Chapter 7) and competitive learning (which we will discuss in Chapter 15). Performance learning is another important class of learning law, in which the network parameters are adjusted to optimize the performance of the network. In the next two chapters we will lay the groundwork for the development of performance learning, which will then be presented in detail in Chapter 10–14. The main objective of the present chapter is to investigate performance surfaces and to determine conditions for the existence of minima and maxima of the performance surface. Chapter 9 will follow this up with a discussion of procedures to locate the minima or maxima.

Theory and Examples

Performance Learning

There are several different learning laws that fall under the category of *performance learning*. Two of these will be presented in this text. These learning laws are distinguished by the fact that during training the network parameters (weights and biases) are adjusted in an effort to optimize the "performance" of the network.

Performance Index

There are two steps involved in this optimization process. The first step is to define what we mean by "performance." In other words, we must find a quantitative measure of network performance, called the *performance index*, which is small when the network performs well and large when the network performs poorly. In this chapter, and in Chapter 9, we will assume that the performance index is given. In Chapter 10, 11 and 13 we will discuss the choice of performance index.

The second step of the optimization process is to search the parameter space (adjust the network weights and biases) in order to reduce the performance index. In this chapter we will investigate the characteristics of performance surfaces and set some conditions that will guarantee that a surface does have a minimum point (the optimum we are searching for). Thus, in this chapter we will obtain some understanding of what performance surfaces look like. Then, in Chapter 9 we will develop procedures for locating the optimum points.

Taylor Series

Taylor Series Expansion

Let us say that the performance index that we want to minimize is represented by $F(x)$, where x is the scalar parameter we are adjusting. We will assume that the performance index is an analytic function, so that all of its derivatives exist. Then it can be represented by its *Taylor series expansion* about some nominal point x^*:

$$\begin{aligned} F(x) = {} & F(x^*) + \frac{d}{dx}F(x)\Big|_{x=x^*}(x - x^*) \\ & + \frac{1}{2}\frac{d^2}{dx^2}F(x)\Big|_{x=x^*}(x - x^*)^2 + \cdots \\ & + \frac{1}{n!}\frac{d^n}{dx^n}F(x)\Big|_{x=x^*}(x - x^*)^n + \cdots \end{aligned} \tag{8.1}$$

We will use the Taylor series expansion to approximate the performance index, by limiting the expansion to a finite number of terms. For example, let

$$F(x) = \cos(x) . \tag{8.2}$$

The Taylor series expansion for $F(x)$ about the point $x^* = 0$ is

$$F(x) = \cos(x) = \cos(0) - \sin(0)(x-0) - \frac{1}{2}\cos(0)(x-0)^2 + \frac{1}{6}\sin(0)(x-0)^3 + \cdots$$

$$= 1 - \frac{1}{2}x^2 + \frac{1}{24}x^4 + \cdots \tag{8.3}$$

The zeroth-order approximation of $F(x)$ (using only the zeroth power of x) is

$$F(x) \approx F_0(x) = 1 . \tag{8.4}$$

The second-order approximation is

$$F(x) \approx F_2(x) = 1 - \frac{1}{2}x^2 . \tag{8.5}$$

(Note that in this case the first-order approximation is the same as the zeroth-order approximation, since the first derivative is zero.)

The fourth-order approximation is

$$F(x) \approx F_4(x) = 1 - \frac{1}{2}x^2 + \frac{1}{24}x^4 . \tag{8.6}$$

A graph showing $F(x)$ and these three approximations is shown in Figure 8.1.

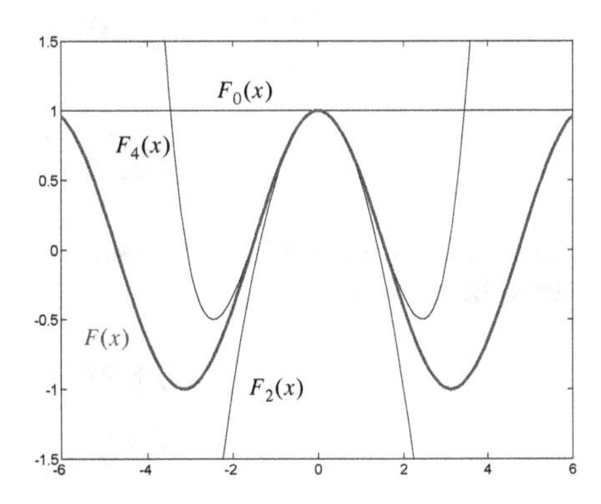

Figure 8.1 Cosine Function and Taylor Series Approximations

From the figure we can see that all three approximations are accurate if x is very close to $x^* = 0$. However, as x moves farther away from x^* only the higher-order approximations are accurate. The second-order approximation is accurate over a wider range than the zeroth-order approximation, and the fourth-order approximation is accurate over a wider range than the second-order approximation. An investigation of Eq. (8.1) explains this behavior. Each succeeding term in the series involves a higher power of $(x - x^*)$. As x gets closer to x^*, these terms will become geometrically smaller.

We will use the Taylor series approximations of the performance index to investigate the shape of the performance index in the neighborhood of possible optimum points.

To experiment with Taylor series expansions of the cosine function, use the MATLAB® Neural Network Design Demonstration Taylor Series (**nnd8ts1**).

Vector Case

Of course the neural network performance index will not be a function of a scalar x. It will be a function of all of the network parameters (weights and biases), of which there may be a very large number. Therefore, we need to extend the Taylor series expansion to functions of many variables. Consider the following function of n variables:

$$F(\mathbf{x}) = F(x_1, x_2, \dots, x_n). \tag{8.7}$$

The Taylor series expansion for this function, about the point x^*, will be

$$\begin{aligned} F(\mathbf{x}) = {} & F(\mathbf{x}^*) + \frac{\partial}{\partial x_1}F(\mathbf{x})\Big|_{\mathbf{x}=\mathbf{x}^*}(x_1 - x_1^*) + \frac{\partial}{\partial x_2}F(\mathbf{x})\Big|_{\mathbf{x}=\mathbf{x}^*}(x_2 - x_2^*) \\ & + \dots + \frac{\partial}{\partial x_n}F(\mathbf{x})\Big|_{\mathbf{x}=\mathbf{x}^*}(x_n - x_n^*) + \frac{1}{2}\frac{\partial^2}{\partial x_1^2}F(\mathbf{x})\Big|_{\mathbf{x}=\mathbf{x}^*}(x_1 - x_1^*)^2 \\ & + \frac{1}{2}\frac{\partial^2}{\partial x_1 \partial x_2}F(\mathbf{x})\Big|_{\mathbf{x}=\mathbf{x}^*}(x_1 - x_1^*)(x_2 - x_2^*) + \dots \end{aligned} \tag{8.8}$$

This notation is a bit cumbersome. It is more convenient to write it in matrix form, as in:

$$\begin{aligned} F(\mathbf{x}) = {} & F(\mathbf{x}^*) + \nabla F(\mathbf{x})^T\Big|_{\mathbf{x}=\mathbf{x}^*}(\mathbf{x} - \mathbf{x}^*) \\ & + \frac{1}{2}(\mathbf{x} - \mathbf{x}^*)^T \nabla^2 F(\mathbf{x})\Big|_{\mathbf{x}=\mathbf{x}^*}(\mathbf{x} - \mathbf{x}^*) + \dots \end{aligned} \tag{8.9}$$

Gradient

where $\nabla F(\mathbf{x})$ is the *gradient*, and is defined as

$$\nabla F(\mathbf{x}) = \left[\frac{\partial}{\partial x_1}F(\mathbf{x}) \; \frac{\partial}{\partial x_2}F(\mathbf{x}) \; \dots \; \frac{\partial}{\partial x_n}F(\mathbf{x})\right]^T, \tag{8.10}$$

Hessian

and $\nabla^2 F(\mathbf{x})$ is the *Hessian*, and is defined as:

$$\nabla^2 F(\mathbf{x}) = \begin{bmatrix} \frac{\partial^2}{\partial x_1^2}F(\mathbf{x}) & \frac{\partial^2}{\partial x_1 \partial x_2}F(\mathbf{x}) & \dots & \frac{\partial^2}{\partial x_1 \partial x_n}F(\mathbf{x}) \\ \frac{\partial^2}{\partial x_2 \partial x_1}F(\mathbf{x}) & \frac{\partial^2}{\partial x_2^2}F(\mathbf{x}) & \dots & \frac{\partial^2}{\partial x_2 \partial x_n}F(\mathbf{x}) \\ \vdots & \vdots & & \vdots \\ \frac{\partial^2}{\partial x_n \partial x_1}F(\mathbf{x}) & \frac{\partial^2}{\partial x_n \partial x_2}F(\mathbf{x}) & \dots & \frac{\partial^2}{\partial x_n^2}F(\mathbf{x}) \end{bmatrix}. \tag{8.11}$$

The gradient and the Hessian are very important to our understanding of performance surfaces. In the next section we discuss the practical meaning of these two concepts.

To experiment with Taylor series expansions of a function of two variables, use the MATLAB® Neural Network Design Demonstration Vector Taylor Series(`nnd8ts2`).

Directional Derivatives

The ith element of the gradient, $\partial F(\mathbf{x})/\partial x_i$, is the first derivative of the performance index F along the x_i axis. The ith element of the diagonal of the Hessian matrix, $\partial^2 F(\mathbf{x})/\partial x_i^2$, is the second derivative of the performance index F along the x_i axis. What if we want to know the derivative of the function in an arbitrary direction? We let $\mathbf{p}$ be a vector in the direction along which we wish to know the derivative. This *directional derivative* can be computed from

Directional Derivative

$$\frac{\mathbf{p}^T \nabla F(\mathbf{x})}{\|\mathbf{p}\|}. \tag{8.12}$$

The second derivative along $\mathbf{p}$ can also be computed:

$$\frac{\mathbf{p}^T \nabla^2 F(\mathbf{x}) \mathbf{p}}{\|\mathbf{p}\|^2}. \tag{8.13}$$

To illustrate these concepts, consider the function

$$F(\mathbf{x}) = x_1^2 + 2x_2^2 . \tag{8.14}$$

Suppose that we want to know the derivative of the function at the point $\mathbf{x}^* = \begin{bmatrix} 0.5 & 0.5 \end{bmatrix}^T$ in the direction $\mathbf{p} = \begin{bmatrix} 2 & -1 \end{bmatrix}^T$. First we evaluate the gradient at $\mathbf{x}^*$:

$$\nabla F(\mathbf{x})\Big|_{\mathbf{x} = \mathbf{x}^*} = \left.\begin{bmatrix} \frac{\partial}{\partial x_1} F(\mathbf{x}) \\ \frac{\partial}{\partial x_2} F(\mathbf{x}) \end{bmatrix}\right|_{\mathbf{x} = \mathbf{x}^*} = \left.\begin{bmatrix} 2x_1 \\ 4x_2 \end{bmatrix}\right|_{\mathbf{x} = \mathbf{x}^*} = \begin{bmatrix} 1 \\ 2 \end{bmatrix} . \tag{8.15}$$

The derivative in the direction $\mathbf{p}$ can then be computed:

$$\frac{\mathbf{p}^T \nabla F(\mathbf{x})}{\|\mathbf{p}\|} = \frac{\begin{bmatrix} 2 & -1 \end{bmatrix} \begin{bmatrix} 1 \\ 2 \end{bmatrix}}{\left\| \begin{bmatrix} 2 \\ -1 \end{bmatrix} \right\|} = \frac{\begin{bmatrix} 0 \end{bmatrix}}{\sqrt{5}} = 0 . \tag{8.16}$$

Therefore the function has zero slope in the direction $\mathbf{p}$ from the point $\mathbf{x}^*$. Why did this happen? What can we say about those directions that have zero slope? If we consider the definition of directional derivative in Eq. (8.12), we can see that the numerator is an inner product between the direction vector and the gradient. Therefore any direction that is orthogonal to the gradient will have zero slope.

Which direction has the greatest slope? The maximum slope will occur when the inner product of the direction vector and the gradient is a maximum. This happens when the direction vector is the same as the gradient. (Notice that the magnitude of the direction vector has no effect, since we normalize by its magnitude.) This effect is illustrated in Figure 8.2, which shows a contour plot and a 3-D plot of $F(\mathbf{x})$. On the contour plot we see five vectors starting from our nominal point $\mathbf{x}^*$ and pointing in different directions. At the end of each vector the first directional derivative is displayed. The maximum derivative occurs in the direction of the gradient. The zero derivative is in the direction orthogonal to the gradient (tangent to the contour line).

To experiment with directional derivatives, use the MATLAB® Neural Network Design Demonstration Directional Derivatives (nnd8dd).

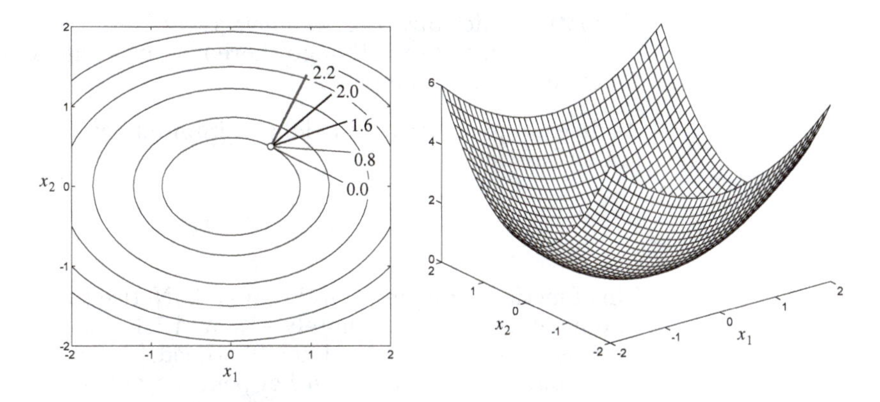

Figure 8.2 Quadratic Function and Directional Derivatives

Minima

Recall that the objective of performance learning will be to optimize the network performance index. In this section we want to define what we mean by an optimum point. We will assume that the optimum point is a minimum of the performance index. The definitions can be easily modified for maximization problems.

Strong Minimum

Strong Minimum

The point $\mathbf{x}^*$ is a strong minimum of $F(\mathbf{x})$ if a scalar $\delta > 0$ exists, such that $F(\mathbf{x}^*) < F(\mathbf{x}^* + \Delta\mathbf{x})$ for all $\Delta\mathbf{x}$ such that $\delta > \|\Delta\mathbf{x}\| > 0$.

In other words, if we move away from a strong minimum a small distance in *any* direction the function will increase.

Global Minimum

Global Minimum

The point $\mathbf{x}^*$ is a unique global minimum of $F(\mathbf{x})$ if $F(\mathbf{x}^*) < F(\mathbf{x}^* + \Delta\mathbf{x})$ for all $\Delta\mathbf{x} \neq \mathbf{0}$.

For a simple strong minimum, $\mathbf{x}^*$, the function may be smaller than $F(\mathbf{x}^*)$ at some points outside a small neighborhood of $\mathbf{x}^*$. Therefore this is sometimes called a local minimum. For a global minimum the function will be larger than the minimum point at every other point in the parameter space.

Weak Minimum

Weak Minimum

The point $\mathbf{x}^*$ is a weak minimum of $F(\mathbf{x})$ if it is not a strong minimum, and a scalar $\delta > 0$ exists, such that $F(\mathbf{x}^*) \leq F(\mathbf{x}^* + \Delta\mathbf{x})$ for all $\Delta\mathbf{x}$ such that $\delta > \|\Delta\mathbf{x}\| > 0$.

No matter which direction we move away from a weak minimum, the function cannot decrease, although there may be some directions in which the function does not change.

As an example of local and global minimum points, consider the following scalar function:

$$F(x) = 3x^4 - 7x^2 - \frac{1}{2}x + 6 . \tag{8.17}$$

This function is displayed in Figure 8.3. Notice that it has two strong minimum points: at approximately -1.1 and 1.1. For both of these points the function increases in a local neighborhood. The minimum at 1.1 is a global minimum, since there is no other point for which the function is as small.

There is no weak minimum for this function. We will show a two-dimensional example of a weak minimum later.

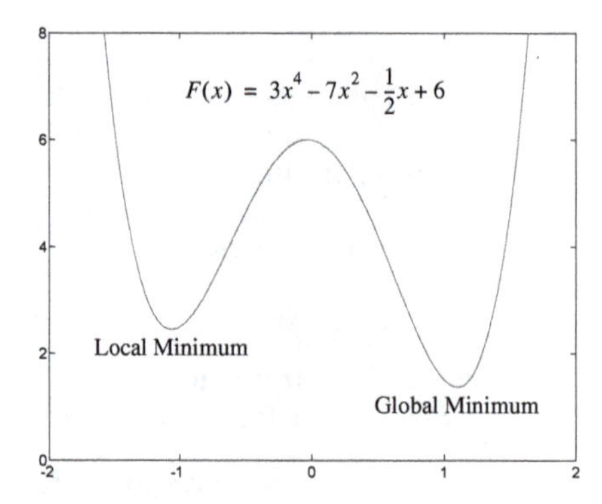

Figure 8.3 Scalar Example of Local and Global Minima

Now let's consider some vector cases. First, consider the following function:

$$F(\mathbf{x}) = (x_2 - x_1)^4 + 8x_1x_2 - x_1 + x_2 + 3 . \tag{8.18}$$

Contour Plot

In Figure 8.4 we have a *contour plot* (a series of curves along which the function value remains constant) and a 3-D surface plot for this function (for function values less than 12). We can see that the function has two strong local minimum points: one at $(-0.42, 0.42)$, and the other at $(0.55, -0.55)$. The global minimum point is at $(0.55, -0.55)$.

Saddle Point

There is also another interesting feature of this function at $(-0.13, 0.13)$. It is called a *saddle point* because of the shape of the surface in the neighborhood of the point. It is characterized by the fact that along the line $x_1 = -x_2$ the saddle point is a local maximum, but along a line orthogonal to that line it is a local minimum. We will investigate this example in more detail in Problems P8.2 and P8.5.

This function is used in the MATLAB® Neural Network Design Demonstration Vector Taylor Series (`nnd8ts2`).

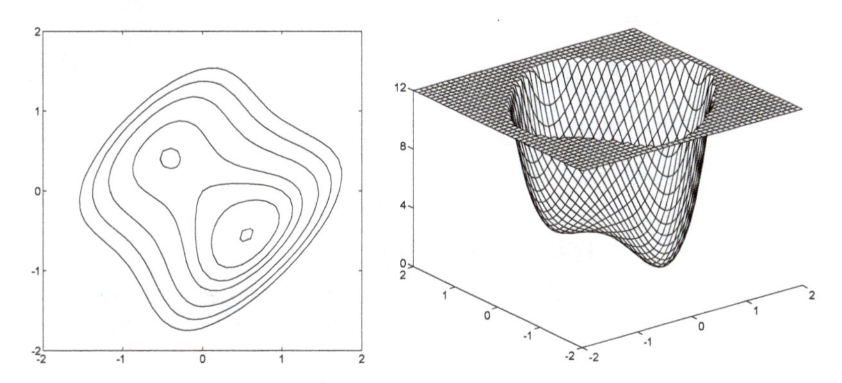

Figure 8.4 Vector Example of Minima and Saddle Point

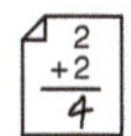

As a final example, consider the function defined in Eq. (8.19):

$$F(\mathbf{x}) = (x_1^2 - 1.5x_1x_2 + 2x_2^2)x_1^2 \quad (8.19)$$

The contour and 3-D plots of this function are given in Figure 8.5. Here we can see that any point along the line $x_1 = 0$ is a weak minimum.

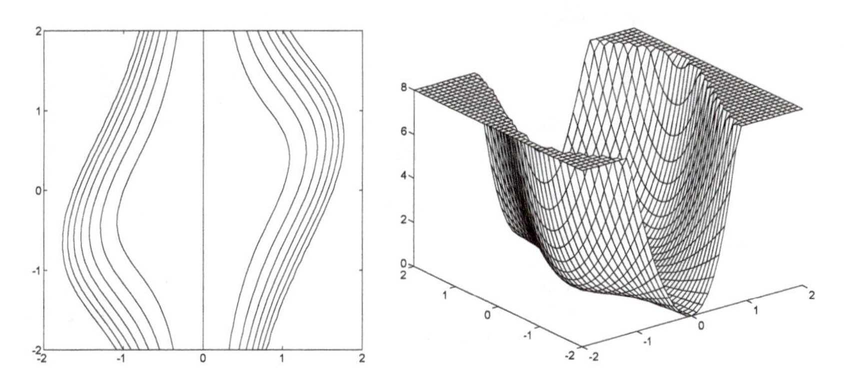

Figure 8.5 Weak Minimum Example

Necessary Conditions for Optimality

Now that we have defined what we mean by an optimum (minimum) point, let's identify some conditions that would have to be satisfied by such a point. We will again use the Taylor series expansion to derive these conditions:

$$F(\mathbf{x}) = F(\mathbf{x}^* + \Delta\mathbf{x}) = F(\mathbf{x}^*) + \nabla F(\mathbf{x})^T\big|_{\mathbf{x} = \mathbf{x}^*}\Delta\mathbf{x} + \frac{1}{2}\Delta\mathbf{x}^T\nabla^2 F(\mathbf{x})\big|_{\mathbf{x} = \mathbf{x}^*}\Delta\mathbf{x} + \cdots , \quad (8.20)$$

where

$$\Delta\mathbf{x} = \mathbf{x} - \mathbf{x}^* . \quad (8.21)$$

First-Order Conditions

If $\|\Delta\mathbf{x}\|$ is very small then the higher order terms in Eq. (8.20) will be negligible and we can approximate the function as

$$F(\mathbf{x}^* + \Delta\mathbf{x}) \cong F(\mathbf{x}^*) + \nabla F(\mathbf{x})^T\big|_{\mathbf{x} = \mathbf{x}^*}\Delta\mathbf{x} . \quad (8.22)$$

The point $\mathbf{x}^*$ is a candidate minimum point, which means that the function should go up (or at least not go down) if $\Delta\mathbf{x}$ is not zero. For this to happen the second term in Eq. (8.22) should not be negative. In other words

$$\nabla F(\mathbf{x})^T\big|_{\mathbf{x} = \mathbf{x}^*}\Delta\mathbf{x} \geq 0 . \quad (8.23)$$

However, if this term is positive,

$$\nabla F(\mathbf{x})^T\big|_{\mathbf{x} = \mathbf{x}^*}\Delta\mathbf{x} > 0 , \quad (8.24)$$

then this would imply that

$$F(\mathbf{x}^* - \Delta\mathbf{x}) \cong F(\mathbf{x}^*) - \nabla F(\mathbf{x})^T\big|_{\mathbf{x} = \mathbf{x}^*}\Delta\mathbf{x} < F(\mathbf{x}^*) . \quad (8.25)$$

But this is a contradiction, since $\mathbf{x}^*$ should be a minimum point. Therefore, since Eq. (8.23) must be true, and Eq. (8.24) must be false, the only alternative must be that

$$\nabla F(\mathbf{x})^T\big|_{\mathbf{x} = \mathbf{x}^*}\Delta\mathbf{x} = 0 . \quad (8.26)$$

Since this must be true for any $\Delta\mathbf{x}$, we have

$$\nabla F(\mathbf{x})\big|_{\mathbf{x} = \mathbf{x}^*} = \mathbf{0} . \quad (8.27)$$

Stationary Points

Therefore the gradient must be zero at a minimum point. This is a first-order, necessary (but not sufficient) condition for $\mathbf{x}^*$ to be a local minimum point. Any points that satisfy Eq. (8.27) are called *stationary points*.

Second-Order Conditions

Assume that we have a stationary point $\mathbf{x}^*$. Since the gradient of $F(\mathbf{x})$ is zero at all stationary points, the Taylor series expansion will be

$$F(\mathbf{x}^* + \Delta\mathbf{x}) = F(\mathbf{x}^*) + \frac{1}{2}\Delta\mathbf{x}^T \nabla^2 F(\mathbf{x})\big|_{\mathbf{x} = \mathbf{x}^*} \Delta\mathbf{x} + \cdots . \tag{8.28}$$

As before, we will consider only those points in a small neighborhood of $\mathbf{x}^*$, so that $\|\Delta\mathbf{x}\|$ is small and $F(\mathbf{x})$ can be approximated by the first two terms in Eq. (8.28). Therefore a strong minimum will exist at $\mathbf{x}^*$ if

$$\Delta\mathbf{x}^T \nabla^2 F(\mathbf{x})\big|_{\mathbf{x} = \mathbf{x}^*} \Delta\mathbf{x} > 0. \tag{8.29}$$

Positive Definite

For this to be true for arbitrary $\Delta\mathbf{x} \neq \mathbf{0}$ requires that the Hessian matrix be positive definite. (By definition, a matrix $\mathbf{A}$ is *positive definite* if

$$\mathbf{z}^T \mathbf{A} \mathbf{z} > 0 \tag{8.30}$$

Positive Semidefinite

for any vector $\mathbf{z} \neq \mathbf{0}$. It is *positive semidefinite* if

$$\mathbf{z}^T \mathbf{A} \mathbf{z} \geq 0 \tag{8.31}$$

for any vector $\mathbf{z}$. We can check these conditions by testing the eigenvalues of the matrix. If all eigenvalues are positive, then the matrix is positive definite. If all eigenvalues are nonnegative, then the matrix is positive semidefinite.)

Sufficient Condition

A positive definite Hessian matrix is a second-order, *sufficient* condition for a strong minimum to exist. It is not a necessary condition. A minimum can still be strong if the second-order term of the Taylor series is zero, but the third-order term is positive. Therefore the second-order, *necessary* condition for a strong minimum is that the Hessian matrix be positive semi-definite.

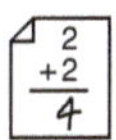

To illustrate these conditions, consider the following function of two variables:

$$F(\mathbf{x}) = x_1^4 + x_2^2 . \tag{8.32}$$

First, we want to locate any stationary points, so we need to evaluate the gradient:

$$\nabla F(\mathbf{x}) = \begin{bmatrix} 4x_1^3 \\ 2x_2 \end{bmatrix} = \mathbf{0} . \tag{8.33}$$

Therefore the only stationary point is the point $\mathbf{x}^* = \mathbf{0}$. We now need to test the second-order condition, which requires the Hessian matrix:

$$\nabla^2 F(\mathbf{x})\big|_{\mathbf{x}=\mathbf{0}} = \begin{bmatrix} 12x_1^2 & 0 \\ 0 & 2 \end{bmatrix}\Bigg|_{\mathbf{x}=\mathbf{0}} = \begin{bmatrix} 0 & 0 \\ 0 & 2 \end{bmatrix}. \tag{8.34}$$

This matrix is positive semidefinite, which is a necessary condition for $\mathbf{x}^* = \mathbf{0}$ to be a strong minimum point. We cannot guarantee from first-order and second-order conditions that it is a minimum point, but we have not eliminated it as a possibility. Actually, even though the Hessian matrix is only positive semidefinite, $\mathbf{x}^* = \mathbf{0}$ is a strong minimum point, but we cannot prove it from the conditions we have discussed.

Just to summarize, the necessary conditions for $\mathbf{x}^*$ to be a minimum, strong or weak, of $F(\mathbf{x})$ are:

$$\nabla F(\mathbf{x})\big|_{\mathbf{x}=\mathbf{x}^*} = \mathbf{0} \text{ and } \nabla^2 F(\mathbf{x})\big|_{\mathbf{x}=\mathbf{x}^*} \text{ positive semidefinite.}$$

The sufficient conditions for $\mathbf{x}^*$ to be a strong minimum point of $F(\mathbf{x})$ are:

$$\nabla F(\mathbf{x})\big|_{\mathbf{x}=\mathbf{x}^*} = \mathbf{0} \text{ and } \nabla^2 F(\mathbf{x})\big|_{\mathbf{x}=\mathbf{x}^*} \text{ positive definite.}$$

Quadratic Functions

We will find throughout this text that one type of performance index is universal — the quadratic function. This is true because there are many applications in which the quadratic function appears, but also because many functions can be approximated by quadratic functions in small neighborhoods, especially near local minimum points. For this reason we want to spend a little time investigating the characteristics of the quadratic function.

Quadratic Function

The general form of a *quadratic function* is

$$F(\mathbf{x}) = \frac{1}{2}\mathbf{x}^T\mathbf{A}\mathbf{x} + \mathbf{d}^T\mathbf{x} + c, \tag{8.35}$$

where the matrix $\mathbf{A}$ is symmetric. (If the matrix is not symmetric it can be replaced by a symmetric matrix that produces the same $F(\mathbf{x})$. Try it!)

To find the gradient for this function, we will use the following useful properties of the gradient:

$$\nabla(\mathbf{h}^T\mathbf{x}) = \nabla(\mathbf{x}^T\mathbf{h}) = \mathbf{h}, \tag{8.36}$$

where $\mathbf{h}$ is a constant vector, and

$$\nabla \mathbf{x}^T\mathbf{Q}\mathbf{x} = \mathbf{Q}\mathbf{x} + \mathbf{Q}^T\mathbf{x} = 2\mathbf{Q}\mathbf{x} \text{ (for symmetric } \mathbf{Q}) . \quad (8.37)$$

We can now compute the gradient of $F(\mathbf{x})$:

$$\nabla F(\mathbf{x}) = \mathbf{A}\mathbf{x} + \mathbf{d} , \quad (8.38)$$

and in a similar way we can find the Hessian:

$$\nabla^2 F(\mathbf{x}) = \mathbf{A} . \quad (8.39)$$

All higher derivatives of the quadratic function are zero. Therefore the first three terms of the Taylor series expansion (as in Eq. (8.20)) give an exact representation of the function. We can also say that all analytic functions behave like quadratics over a small neighborhood (i.e., when $\|\Delta\mathbf{x}\|$ is small).

Eigensystem of the Hessian

We now want to investigate the general shape of the quadratic function. It turns out that we can tell a lot about the shape by looking at the eigenvalues and eigenvectors of the Hessian matrix. Consider a quadratic function that has a stationary point at the origin, and whose value there is zero:

$$F(\mathbf{x}) = \frac{1}{2}\mathbf{x}^T\mathbf{A}\mathbf{x} . \quad (8.40)$$

The shape of this function can be seen more clearly if we perform a change of basis (see Chapter 6). We want to use the eigenvectors of the Hessian matrix, $\mathbf{A}$, as the new basis vectors. Since $\mathbf{A}$ is symmetric, its eigenvectors will be mutually orthogonal. (See [Brog91].) This means that if we make up a matrix with the eigenvectors as the columns, as in Eq. (6.68):

$$\mathbf{B} = \begin{bmatrix} \mathbf{z}_1 & \mathbf{z}_2 & \dots & \mathbf{z}_n \end{bmatrix} , \quad (8.41)$$

the inverse of the matrix will be the same as the transpose:

$$\mathbf{B}^{-1} = \mathbf{B}^T . \quad (8.42)$$

(This assumes that we have normalized the eigenvectors.)

If we now perform a change of basis, so that the eigenvectors are the basis vectors (as in Eq. (6.69)), the new $\mathbf{A}$ matrix will be

$$\mathbf{A}' = [\mathbf{B}^T\mathbf{A}\mathbf{B}] = \begin{bmatrix} \lambda_1 & 0 & \dots & 0 \\ 0 & \lambda_2 & \dots & 0 \\ \vdots & \vdots & & \vdots \\ 0 & 0 & \dots & \lambda_n \end{bmatrix} = \Lambda , \quad (8.43)$$

where the λ_i are the eigenvalues of $\mathbf{A}$. We can also write this equation as

$$\mathbf{A} = \mathbf{B}\Lambda\mathbf{B}^T. \tag{8.44}$$

We will now use the concept of the directional derivative to explain the physical meaning of the eigenvalues and eigenvectors of $\mathbf{A}$, and to explain how they determine the shape of the surface of the quadratic function.

Recall from Eq. (8.13) that the second derivative of a function $F(\mathbf{x})$ in the direction of a vector $\mathbf{p}$ is given by

$$\frac{\mathbf{p}^T\nabla^2 F(\mathbf{x})\mathbf{p}}{\|\mathbf{p}\|^2} = \frac{\mathbf{p}^T\mathbf{A}\mathbf{p}}{\|\mathbf{p}\|^2}. \tag{8.45}$$

Now define

$$\mathbf{p} = \mathbf{B}\mathbf{c}, \tag{8.46}$$

where $\mathbf{c}$ is the representation of the vector $\mathbf{p}$ with respect to the eigenvectors of $\mathbf{A}$. (See Eq. (6.28) and the discussion that follows.) With this definition, and Eq. (8.44), we can rewrite Eq. (8.45):

$$\frac{\mathbf{p}^T\mathbf{A}\mathbf{p}}{\|\mathbf{p}\|^2} = \frac{\mathbf{c}^T\mathbf{B}^T(\mathbf{B}\Lambda\mathbf{B}^T)\mathbf{B}\mathbf{c}}{\mathbf{c}^T\mathbf{B}^T\mathbf{B}\mathbf{c}} = \frac{\mathbf{c}^T\Lambda\mathbf{c}}{\mathbf{c}^T\mathbf{c}} = \frac{\sum_{i=1}^{n}\lambda_i c_i^2}{\sum_{i=1}^{n} c_i^2}. \tag{8.47}$$

This result tells us several useful things. First, note that this second derivative is just a weighted average of the eigenvalues. Therefore it can never be larger than the largest eigenvalue, or smaller than the smallest eigenvalue. In other words,

$$\lambda_{min} \le \frac{\mathbf{p}^T\mathbf{A}\mathbf{p}}{\|\mathbf{p}\|^2} \le \lambda_{max}. \tag{8.48}$$

Under what condition, if any, will this second derivative be equal to the largest eigenvalue? What if we choose

$$\mathbf{p} = \mathbf{z}_{max}, \tag{8.49}$$

where $\mathbf{z}_{max}$ is the eigenvector associated with the largest eigenvalue, λ_{max}? For this case the $\mathbf{c}$ vector will be

$$\mathbf{c} = \mathbf{B}^T\mathbf{p} = \mathbf{B}^T\mathbf{z}_{max} = \begin{bmatrix}0 & 0 & \dots & 0 & 1 & 0 & \dots & 0\end{bmatrix}^T, \tag{8.50}$$

where the one occurs only in the position that corresponds to the largest eigenvalue (i.e., $c_{max} = 1$). This is because the eigenvectors are orthonormal.

If we now substitute $\mathbf{z}_{max}$ for $\mathbf{p}$ in Eq. (8.47) we obtain

$$\frac{\mathbf{z}_{max}{}^T \mathbf{A} \mathbf{z}_{max}}{\|\mathbf{z}_{max}\|^2} = \frac{\sum_{i=1}^{n} \lambda_i c_i^2}{\sum_{i=1}^{n} c_i^2} = \lambda_{max} . \tag{8.51}$$

So the maximum second derivative occurs in the direction of the eigenvector that corresponds to the largest eigenvalue. In fact, in each of the eigenvector directions the second derivatives will be equal to the corresponding eigenvalue. In other directions the second derivative will be a weighted average of the eigenvalues. The eigenvalues are the second derivatives in the directions of the eigenvectors.

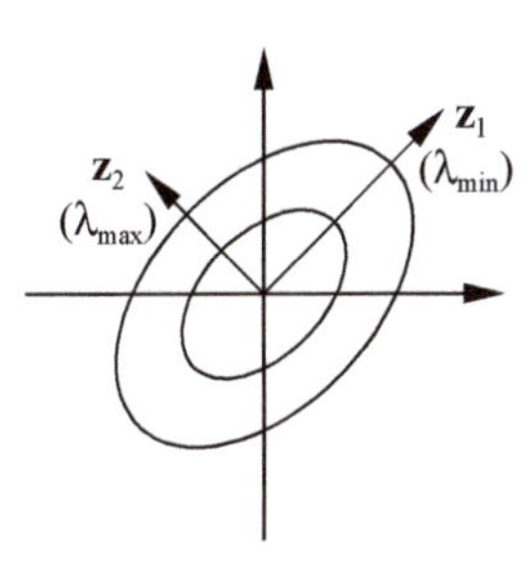

The eigenvectors define a new coordinate system in which the quadratic cross terms vanish. The eigenvectors are known as the principal axes of the function contours. The figure to the left illustrates these concepts in two dimensions. This figure illustrates the case where the first eigenvalue is smaller than the second eigenvalue. Therefore the minimum curvature (second derivative) will occur in the direction of the first eigenvector. This means that we will cross contour lines more slowly in this direction. The maximum curvature will occur in the direction of the second eigenvector, therefore we will cross contour lines more quickly in that direction.

One caveat about this figure: it is only valid when both eigenvalues have the same sign, so that we have either a strong minimum or a strong maximum. For these cases the contour lines are always elliptical. We will provide examples later where the eigenvalues have opposite signs and where one of the eigenvalues is zero.

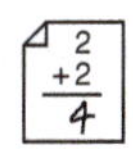

For our first example, consider the following function:

$$F(\mathbf{x}) = x_1^2 + x_2^2 = \frac{1}{2}\mathbf{x}^T \begin{bmatrix} 2 & 0 \\ 0 & 2 \end{bmatrix} \mathbf{x} . \tag{8.52}$$

The Hessian matrix and its eigenvalues and eigenvectors are

$$\nabla^2 F(\mathbf{x}) = \begin{bmatrix} 2 & 0 \\ 0 & 2 \end{bmatrix}, \lambda_1 = 2, \mathbf{z}_1 = \begin{bmatrix} 1 \\ 0 \end{bmatrix}, \lambda_2 = 2, \mathbf{z}_2 = \begin{bmatrix} 0 \\ 1 \end{bmatrix}. \tag{8.53}$$

(Actually, any two independent vectors could be the eigenvectors in this case. There is a repeated eigenvalue, and its eigenvector is the plane.) Since all the eigenvalues are equal, the curvature should be the same in all directions, and therefore the function should have circular contours. Figure 8.6 shows the contour and 3-D plots for this function, a circular hollow.

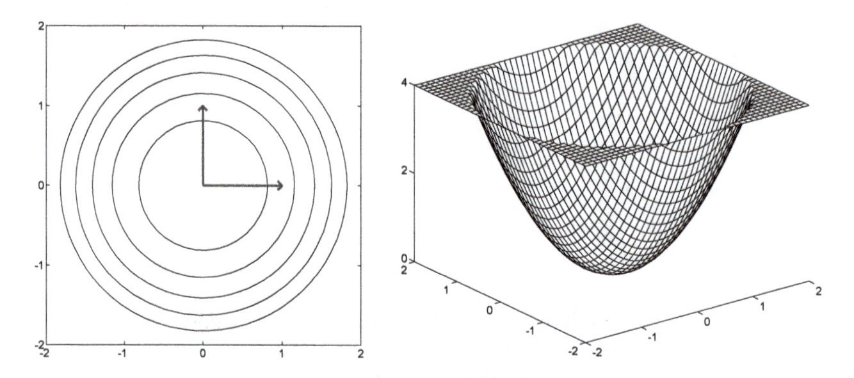

Figure 8.6 Circular Hollow

Let's try an example with distinct eigenvalues. Consider the following quadratic function:

$$F(\mathbf{x}) = x_1^2 + x_1 x_2 + x_2^2 = \frac{1}{2}\mathbf{x}^T \begin{bmatrix} 2 & 1 \\ 1 & 2 \end{bmatrix} \mathbf{x} \tag{8.54}$$

The Hessian matrix and its eigenvalues and eigenvectors are

$$\nabla^2 F(\mathbf{x}) = \begin{bmatrix} 2 & 1 \\ 1 & 2 \end{bmatrix}, \lambda_1 = 1, \mathbf{z}_1 = \begin{bmatrix} 1 \\ -1 \end{bmatrix}, \lambda_2 = 3, \mathbf{z}_2 = \begin{bmatrix} 1 \\ 1 \end{bmatrix}. \tag{8.55}$$

(As we discussed in Chapter 6, the eigenvectors are not unique, they can be multiplied by any scalar.) In this case the maximum curvature is in the direction of $\mathbf{z}_2$ so we should cross contour lines more quickly in that direction. Figure 8.7 shows the contour and 3-D plots for this function, an elliptical hollow.

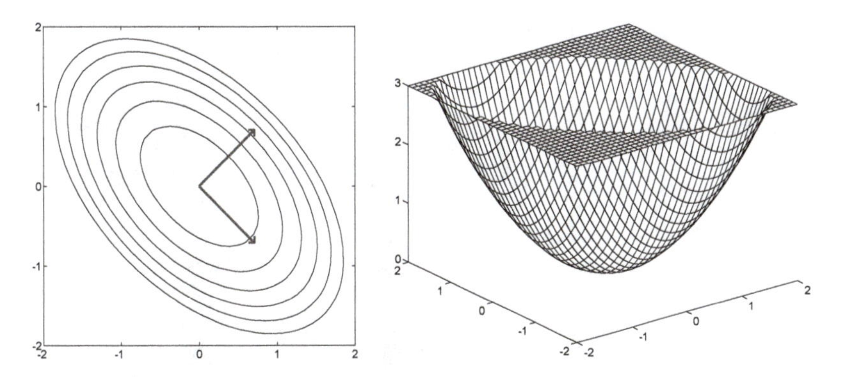

Figure 8.7 Elliptical Hollow

What happens when the eigenvalues have opposite signs? Consider the following function:

$$F(\mathbf{x}) = -\frac{1}{4}x_1^2 - \frac{3}{2}x_1x_2 - \frac{1}{4}x_2^2 = \frac{1}{2}\mathbf{x}^T\begin{bmatrix} -0.5 & -1.5 \\ -1.5 & -0.5 \end{bmatrix}\mathbf{x}. \tag{8.56}$$

The Hessian matrix and its eigenvalues and eigenvectors are

$$\nabla^2 F(\mathbf{x}) = \begin{bmatrix} -0.5 & -1.5 \\ -1.5 & -0.5 \end{bmatrix},\ \lambda_1 = 1\ ,\ \mathbf{z}_1 = \begin{bmatrix} -1 \\ 1 \end{bmatrix},\ \lambda_2 = -2\ ,\ \mathbf{z}_2 = \begin{bmatrix} -1 \\ -1 \end{bmatrix}. \tag{8.57}$$

The first eigenvalue is positive, so there is positive curvature in the direction of $\mathbf{z}_1$. The second eigenvalue is negative, so there is negative curvature in the direction of $\mathbf{z}_2$. Also, since the magnitude of the second eigenvalue is greater than the magnitude of the first eigenvalue, we will cross contour lines faster in the direction of $\mathbf{z}_2$.

Figure 8.8 shows the contour and 3-D plots for this function, an elongated saddle. Note that the stationary point,

$$\mathbf{x}^* = \begin{bmatrix} 0 \\ 0 \end{bmatrix}, \tag{8.58}$$

is no longer a strong minimum point, since the Hessian matrix is not positive definite. Since the eigenvalues are of opposite sign, we know that the Hessian is indefinite (see [Brog91]). The stationary point is therefore a saddle point. It is a minimum of the function along the first eigenvector (positive eigenvalue), but it is a maximum of the function along the second eigenvector (negative eigenvalue).

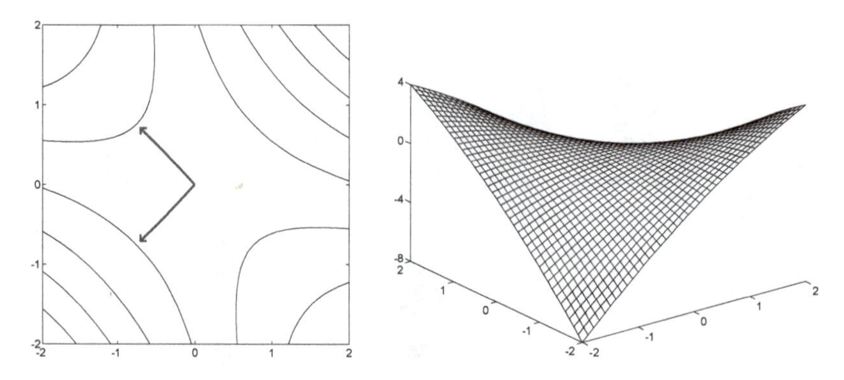

Figure 8.8 Elongated Saddle

As a final example, let's try a case where one of the eigenvalues is zero. An example of this is given by the following function:

$$F(\mathbf{x}) = \frac{1}{2}x_1^2 - x_1 x_2 + \frac{1}{2}x_2^2 = \frac{1}{2}\mathbf{x}^T\begin{bmatrix} 1 & -1 \\ -1 & 1 \end{bmatrix}\mathbf{x}. \tag{8.59}$$

The Hessian matrix and its eigenvalues and eigenvectors are

$$\nabla^2 F(\mathbf{x}) = \begin{bmatrix} 1 & -1 \\ -1 & 1 \end{bmatrix}, \lambda_1 = 2, \mathbf{z}_1 = \begin{bmatrix} -1 \\ 1 \end{bmatrix}, \lambda_2 = 0, \mathbf{z}_2 = \begin{bmatrix} -1 \\ -1 \end{bmatrix}. \tag{8.60}$$

The second eigenvalue is zero, so we would expect to have zero curvature along $\mathbf{z}_2$. Figure 8.9 shows the contour and 3-D plots for this function, a stationary valley. In this case the Hessian matrix is positive semidefinite, and we have a weak minimum along the line

$$x_1 = x_2, \tag{8.61}$$

corresponding to the second eigenvector.

For quadratic functions the Hessian matrix must be positive definite in order for a strong minimum to exist. For higher-order functions it is possible to have a strong minimum with a positive semidefinite Hessian matrix, as we discussed previously in the section on minima.

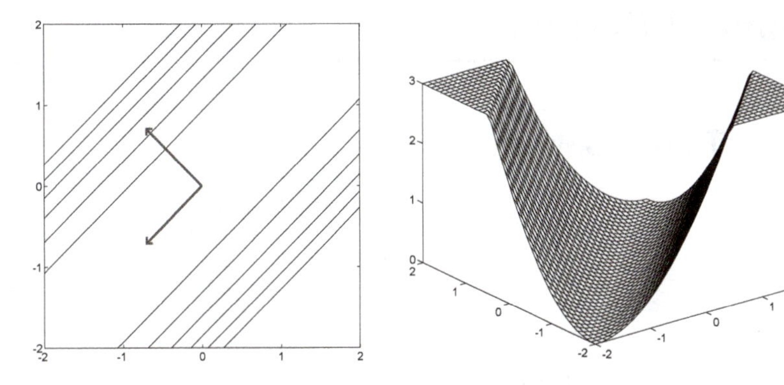

Figure 8.9 Stationary Valley

To experiment with other quadratic functions, use the MATLAB® Neural Network Design Demonstration Quadratic Function (`nnd8qf`).

At this point we can summarize some characteristics of the quadratic function.

1. If the eigenvalues of the Hessian matrix are all positive, the function will have a single strong minimum.
2. If the eigenvalues are all negative, the function will have a single strong maximum.
3. If some eigenvalues are positive and other eigenvalues are negative, the function will have a single saddle point.
4. If the eigenvalues are all nonnegative, but some eigenvalues are zero, then the function will either have a weak minimum (as in Figure 8.9) or will have no stationary point (see Solved Problem P8.7).
5. If the eigenvalues are all nonpositive, but some eigenvalues are zero, then the function will either have a weak maximum or will have no stationary point.

We should note that in this discussion we have assumed, for simplicity, that the stationary point of the quadratic function was at the origin, and that it had a zero value there. This requires that the terms $\mathbf{d}$ and c in Eq. (8.35) both be zero. If c is nonzero then the function is simply increased in magnitude by c at every point. The shape of the contours do not change. When $\mathbf{d}$ is nonzero, and $\mathbf{A}$ is invertible, the shape of the contours are not changed, but the stationary point of the function moves to

$$\mathbf{x}^* = -\mathbf{A}^{-1}\mathbf{d}. \tag{8.62}$$

If $\mathbf{A}$ is not invertible (has some zero eigenvalues) and $\mathbf{d}$ is nonzero then stationary points may not exist (see Solved Problem P8.7).

Summary of Results

Taylor Series

$$F(\mathbf{x}) = F(\mathbf{x}^*) + \nabla F(\mathbf{x})^T\big|_{\mathbf{x}=\mathbf{x}^*}(\mathbf{x}-\mathbf{x}^*) + \frac{1}{2}(\mathbf{x}-\mathbf{x}^*)^T \nabla^2 F(\mathbf{x})\big|_{\mathbf{x}=\mathbf{x}^*}(\mathbf{x}-\mathbf{x}^*) + \cdots$$

Gradient

$$\nabla F(\mathbf{x}) = \begin{bmatrix} \frac{\partial}{\partial x_1}F(\mathbf{x}) & \frac{\partial}{\partial x_2}F(\mathbf{x}) & \ldots & \frac{\partial}{\partial x_n}F(\mathbf{x}) \end{bmatrix}^T$$

Hessian Matrix

$$\nabla^2 F(\mathbf{x}) = \begin{bmatrix} \frac{\partial^2}{\partial x_1^2}F(\mathbf{x}) & \frac{\partial^2}{\partial x_1 \partial x_2}F(\mathbf{x}) & \ldots & \frac{\partial^2}{\partial x_1 \partial x_n}F(\mathbf{x}) \\ \frac{\partial^2}{\partial x_2 \partial x_1}F(\mathbf{x}) & \frac{\partial^2}{\partial x_2^2}F(\mathbf{x}) & \ldots & \frac{\partial^2}{\partial x_2 \partial x_n}F(\mathbf{x}) \\ \vdots & \vdots & & \vdots \\ \frac{\partial^2}{\partial x_n \partial x_1}F(\mathbf{x}) & \frac{\partial^2}{\partial x_n \partial x_2}F(\mathbf{x}) & \ldots & \frac{\partial^2}{\partial x_n^2}F(\mathbf{x}) \end{bmatrix}$$

Directional Derivatives

First Directional Derivative

$$\frac{\mathbf{p}^T \nabla F(\mathbf{x})}{\|\mathbf{p}\|}$$

Second Directional Derivative

$$\frac{\mathbf{p}^T \nabla^2 F(\mathbf{x}) \mathbf{p}}{\|\mathbf{p}\|^2}$$

Minima

Strong Minimum

The point $\mathbf{x}^*$ is a strong minimum of $F(\mathbf{x})$ if a scalar $\delta > 0$ exists, such that $F(\mathbf{x}) < F(\mathbf{x} + \Delta\mathbf{x})$ for all $\Delta\mathbf{x}$ such that $\delta > \|\Delta\mathbf{x}\| > 0$.

Global Minimum

The point $\mathbf{x}^*$ is a unique global minimum of $F(\mathbf{x})$ if $F(\mathbf{x}) < F(\mathbf{x} + \Delta\mathbf{x})$ for all $\Delta\mathbf{x} \neq \mathbf{0}$.

Weak Minimum

The point $\mathbf{x}^*$ is a weak minimum of $F(\mathbf{x})$ if it is not a strong minimum, and a scalar $\delta > 0$ exists, such that $F(\mathbf{x}) \le F(\mathbf{x} + \Delta\mathbf{x})$ for all $\Delta\mathbf{x}$ such that $\delta > \|\Delta\mathbf{x}\| > 0$.

Necessary Conditions for Optimality

First-Order Condition

$$\nabla F(\mathbf{x})\big|_{\mathbf{x} = \mathbf{x}^*} = \mathbf{0} \text{ (Stationary Points)}$$

Second-Order Condition

$$\nabla^2 F(\mathbf{x})\big|_{\mathbf{x} = \mathbf{x}^*} \ge 0 \text{ (Positive Semidefinite Hessian Matrix)}$$

Quadratic Functions

$$F(\mathbf{x}) = \frac{1}{2}\mathbf{x}^T\mathbf{A}\mathbf{x} + \mathbf{d}^T\mathbf{x} + c$$

Gradient

$$\nabla F(\mathbf{x}) = \mathbf{A}\mathbf{x} + \mathbf{d}$$

Hessian

$$\nabla^2 F(\mathbf{x}) = \mathbf{A}$$

Directional Derivatives

$$\lambda_{min} \le \frac{\mathbf{p}^T\mathbf{A}\mathbf{p}}{\|\mathbf{p}\|^2} \le \lambda_{max}$$

Solved Problems

P8.1 In Figure 8.1 we illustrated 3 approximations to the cosine function about the point $x^* = 0$. Repeat that procedure about the point $x^* = \pi/2$.

The function we want to approximate is

$$F(x) = \cos(x).$$

The Taylor series expansion for $F(x)$ about the point $x^* = \pi/2$ is

$$F(x) = \cos(x) = \cos\left(\frac{\pi}{2}\right) - \sin\left(\frac{\pi}{2}\right)\left(x - \frac{\pi}{2}\right) - \frac{1}{2}\cos\left(\frac{\pi}{2}\right)\left(x - \frac{\pi}{2}\right)^2 + \frac{1}{6}\sin\left(\frac{\pi}{2}\right)\left(x - \frac{\pi}{2}\right)^3 + \cdots$$

$$= -\left(x - \frac{\pi}{2}\right) + \frac{1}{6}\left(x - \frac{\pi}{2}\right)^3 - \frac{1}{120}\left(x - \frac{\pi}{2}\right)^5 + \cdots$$

The zeroth-order approximation of $F(x)$ is

$$F(x) \approx F_0(x) = 0.$$

The first-order approximation is

$$F(\mathbf{x}) \approx F_1(x) = -\left(x - \frac{\pi}{2}\right) = \frac{\pi}{2} - x.$$

(Note that in this case the second-order approximation is the same as the first-order approximation, since the second derivative is zero.)

The third-order approximation is

$$F(\mathbf{x}) \approx F_3(x) = -\left(x - \frac{\pi}{2}\right) + \frac{1}{6}\left(x - \frac{\pi}{2}\right)^3.$$

A graph showing $F(x)$ and these three approximations is shown in Figure P8.1. Note that in this case the zeroth-order approximation is very poor, while the first-order approximation is accurate over a reasonably wide range. Compare this result with Figure 8.1. In that case we were expanding about a local maximum point, $x^* = 0$, so the first derivative was zero.

Check the Taylor series expansions at other points using the Neural Network Design Demonstration Taylor Series (`nnd8ts1`).

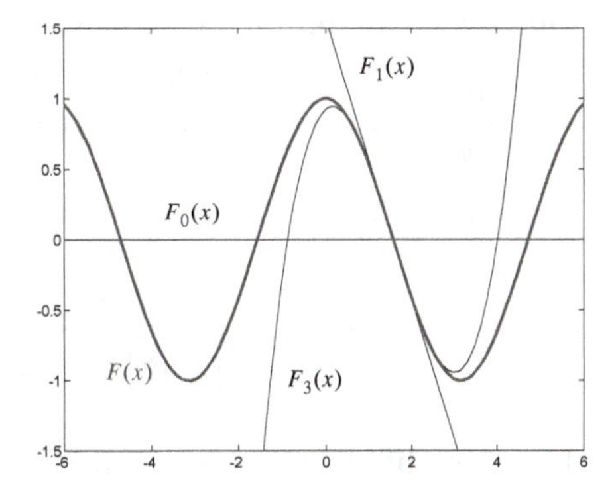

Figure P8.1 Cosine Approximation About $x = \pi/2$

P8.2 Recall the function that is displayed in Figure 8.4, on page 8-9. We know that this function has two strong minima. Find the second-order Taylor series expansions for this function about the two minima.

The equation for this function is

$$F(\mathbf{x}) = (x_2 - x_1)^4 + 8x_1x_2 - x_1 + x_2 + 3 .$$

To find the second-order Taylor series expansion, we need to find the gradient and the Hessian for $F(\mathbf{x})$. For the gradient we have

$$\nabla F(\mathbf{x}) = \begin{bmatrix} \frac{\partial}{\partial x_1} F(\mathbf{x}) \\ \frac{\partial}{\partial x_2} F(\mathbf{x}) \end{bmatrix} = \begin{bmatrix} -4(x_2 - x_1)^3 + 8x_2 - 1 \\ 4(x_2 - x_1)^3 + 8x_1 + 1 \end{bmatrix},$$

and the Hessian matrix is

$$\nabla^2 F(\mathbf{x}) = \begin{bmatrix} \frac{\partial^2}{\partial x_1^2} F(\mathbf{x}) & \frac{\partial^2}{\partial x_1 \partial x_2} F(\mathbf{x}) \\ \frac{\partial^2}{\partial x_2 \partial x_1} F(\mathbf{x}) & \frac{\partial^2}{\partial x_2^2} F(\mathbf{x}) \end{bmatrix}$$

$$= \begin{bmatrix} 12(x_2 - x_1)^2 & -12(x_2 - x_1)^2 + 8 \\ -12(x_2 - x_1)^2 + 8 & 12(x_2 - x_1)^2 \end{bmatrix}$$

One strong minimum occurs at $\mathbf{x}^1 = \begin{bmatrix} -0.42 & 0.42 \end{bmatrix}^T$, and the other at $\mathbf{x}^2 = \begin{bmatrix} 0.55 & -0.55 \end{bmatrix}^T$. If we perform the second-order Taylor series expansion of $F(\mathbf{x})$ about these two points we obtain:

$$F^1(\mathbf{x}) = F(\mathbf{x}^1) + \nabla F(\mathbf{x})^T\Big|_{\mathbf{x}=\mathbf{x}^1}(\mathbf{x}-\mathbf{x}^1) + \frac{1}{2}(\mathbf{x}-\mathbf{x}^1)^T \nabla^2 F(\mathbf{x})\Big|_{\mathbf{x}=\mathbf{x}^1}(\mathbf{x}-\mathbf{x}^1)$$

$$= 2.93 + \frac{1}{2}\left(\mathbf{x} - \begin{bmatrix} -0.42 \\ 0.42 \end{bmatrix}\right)^T \begin{bmatrix} 8.42 & -0.42 \\ -0.42 & 8.42 \end{bmatrix}\left(\mathbf{x} - \begin{bmatrix} -0.42 \\ 0.42 \end{bmatrix}\right).$$

If we simplify this expression we find

$$F^1(\mathbf{x}) = 4.49 - \begin{bmatrix} -3.7128 & 3.7128 \end{bmatrix}\mathbf{x} + \frac{1}{2}\mathbf{x}^T \begin{bmatrix} 8.42 & -0.42 \\ -0.42 & 8.42 \end{bmatrix}\mathbf{x}.$$

Repeating this process for $\mathbf{x}^2$ results in

$$F^2(\mathbf{x}) = 7.41 - \begin{bmatrix} 11.781 & -11.781 \end{bmatrix}\mathbf{x} + \frac{1}{2}\mathbf{x}^T \begin{bmatrix} 14.71 & -6.71 \\ -6.71 & 14.71 \end{bmatrix}\mathbf{x}.$$

The original function and the two approximations are plotted in the following figures.

Check the Taylor series expansions at other points using the Neural Network Design Demonstration Vector Taylor Series (**nnd8ts2**).

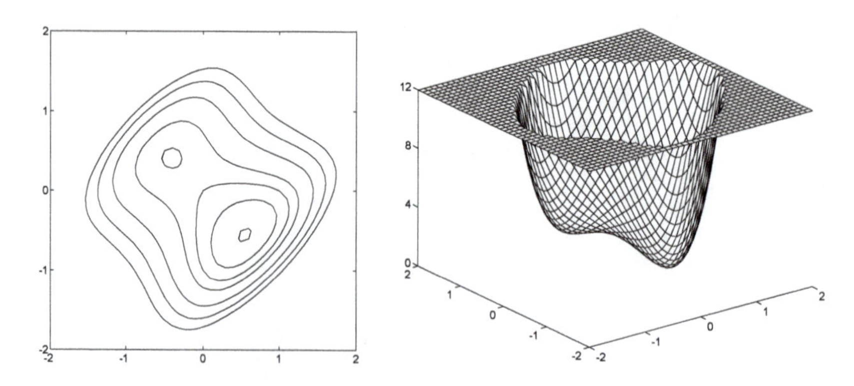

Figure P8.2 Function $F(\mathbf{x})$ for Problem P8.2

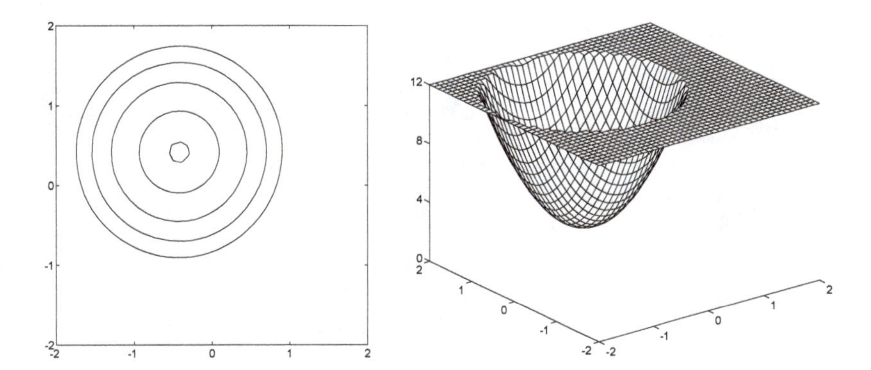

Figure P8.3 Function $F^1(\mathbf{x})$ for Problem P8.2

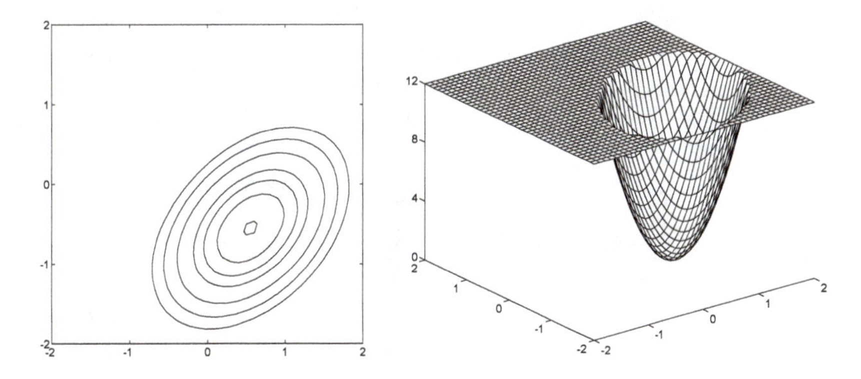

Figure P8.4 Function $F^2(\mathbf{x})$ for Problem P8.2

P8.3 For the function $F(\mathbf{x})$ given below, find the equation for the line that is tangent to the contour line at $\mathbf{x} = \begin{bmatrix} 0 & 0 \end{bmatrix}^T$.

$$F(\mathbf{x}) = (2 + x_1)^2 + 5(1 - x_1 - x_2^2)^2$$

To solve this problem we can use the directional derivative. What is the derivative of $F(\mathbf{x})$ along a line that is tangent to a contour line? Since the contour is a line along which the function does not change, the derivative of $F(\mathbf{x})$ should be zero in the direction of the contour. So we can get the equation for the tangent to the contour line by setting the directional derivative equal to zero.

First we need to find the gradient:

$$\nabla F(\mathbf{x}) = \begin{bmatrix} 2(2+x_1) + 10(1-x_1-x_2^2)(-1) \\ 10(1-x_1-x_2^2)(-2x_2) \end{bmatrix} = \begin{bmatrix} -6+12x_1+10x_2^2 \\ -20x_2+20x_1x_2+20x_2^3 \end{bmatrix}.$$

If we evaluate this at $\mathbf{x}^* = \begin{bmatrix} 0 & 0 \end{bmatrix}^T$, we obtain

$$\nabla F(\mathbf{x}^*) = \begin{bmatrix} -6 \\ 0 \end{bmatrix}.$$

Now recall that the equation for the derivative of $F(\mathbf{x})$ in the direction of a vector $\mathbf{p}$ is

$$\frac{\mathbf{p}^T \nabla F(\mathbf{x})}{\|\mathbf{p}\|}.$$

Therefore if we want the equation for the line that passes through $\mathbf{x}^* = \begin{bmatrix} 0 & 0 \end{bmatrix}^T$ and along which the derivative is zero, we can set the numerator of the directional derivative in the direction of $\Delta\mathbf{x}$ to zero:

$$\Delta\mathbf{x}^T \nabla F(\mathbf{x}^*) = 0,$$

where $\Delta\mathbf{x} = \mathbf{x} - \mathbf{x}^*$. For this case we have

$$\mathbf{x}^T \begin{bmatrix} -6 \\ 0 \end{bmatrix} = 0, \text{ or } x_1 = 0.$$

This result is illustrated in Figure P8.5.

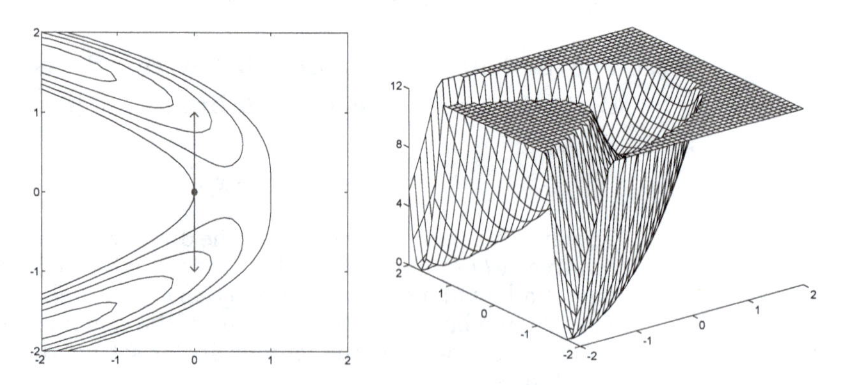

Figure P8.5 Plot of $F(\mathbf{x})$ for Problem P8.3

P8.4 Consider the following fourth-order polynomial:

$$F(x) = x^4 - \frac{2}{3}x^3 - 2x^2 + 2x + 4 .$$

Find any stationary points and test them to see if they are minima.

To find the stationary points we set the derivative of $F(x)$ to zero:

$$\frac{d}{dx}F(x) = 4x^3 - 2x^2 - 4x + 2 = 0 .$$

We can use MATLAB to find the roots of this polynomial:

```
coef=[4 -2 -4 2];

stapoints=roots(coef);

stapoints'

ans =

    1.0000    -1.0000     0.5000
```

Now we need to check the second derivative at each of these points. The second derivative of $F(x)$ is

$$\frac{d^2}{dx^2}F(x) = 12x^2 - 4x - 4 .$$

If we evaluate this at each of the stationary points we find

$$\left(\frac{d^2}{dx^2}F(1) = 4\right), \left(\frac{d^2}{dx^2}F(-1) = 12\right), \left(\frac{d^2}{dx^2}F(0.5) = -3\right) .$$

Therefore we should have strong local minima at 1 and -1 (since the second derivatives were positive), and a strong local maximum at 0.5 (since the second derivative was negative). To find the global minimum we would have to evaluate the function at the two local minima:

$$(F(1) = 4.333), (F(-1) = 1.667) .$$

Therefore the global minimum occurs at -1. But are we sure that this is a global minimum? What happens to the function as $x \to \infty$ or $x \to -\infty$? In this case, because the highest power of x has a positive coefficient and is an even power (x^4), the function goes to ∞ at both limits. So we can safely say that the global minimum occurs at -1. The function is plotted in Figure P8.6.

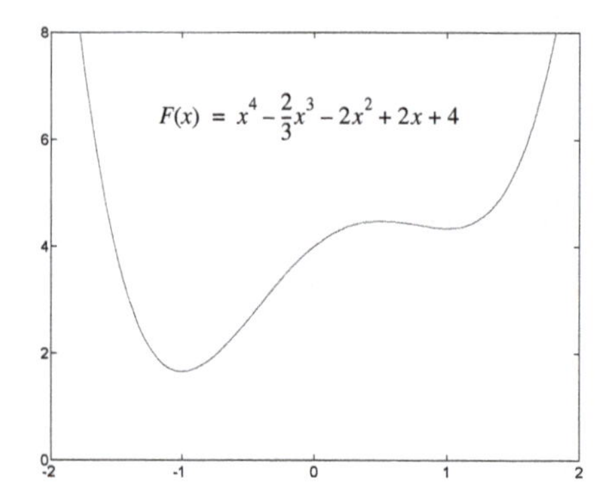

Figure P8.6 Graph of $F(x)$ for Problem P8.4

P8.5 Look back to the function of Problem P8.2. This function has three stationary points:

$$\mathbf{x}^1 = \begin{bmatrix} -0.41878 \\ 0.41878 \end{bmatrix}, \mathbf{x}^2 = \begin{bmatrix} -0.134797 \\ 0.134797 \end{bmatrix}, \mathbf{x}^3 = \begin{bmatrix} 0.55358 \\ -0.55358 \end{bmatrix}.$$

Test whether or not any of these points could be local minima.

From Problem P8.2 we know that the Hessian matrix for the function is

$$\nabla^2 F(\mathbf{x}) = \begin{bmatrix} 12(x_2 - x_1)^2 & -12(x_2 - x_1)^2 + 8 \\ -12(x_2 - x_1)^2 + 8 & 12(x_2 - x_1)^2 \end{bmatrix}.$$

To test the definiteness of this matrix we can check the eigenvalues. If the eigenvalues are all positive, the Hessian is positive definite, which guarantees a strong minimum. If the eigenvalues are nonnegative, the Hessian is positive semidefinite, which is consistent with either a strong or a weak minimum. If one eigenvalue is positive and the other eigenvalue is negative, the Hessian is indefinite, which would signal a saddle point.

If we evaluate the Hessian at $\mathbf{x}^1$, we find

$$\nabla^2 F(\mathbf{x}^1) = \begin{bmatrix} 8.42 & -0.42 \\ -0.42 & 8.42 \end{bmatrix}.$$

The eigenvalues of this matrix are

$$\lambda_1 = 8.84, \lambda_2 = 8.0,$$

therefore $\mathbf{x}^1$ must be a strong minimum point.

If we evaluate the Hessian at $\mathbf{x}^2$, we find

$$\nabla^2 F(\mathbf{x}^2) = \begin{bmatrix} 0.87 & 7.13 \\ 7.13 & 0.87 \end{bmatrix}.$$

The eigenvalues of this matrix are

$$\lambda_1 = -6.26, \lambda_2 = 8.0,$$

therefore $\mathbf{x}^2$ must be a saddle point. In one direction the curvature is negative, and in another direction the curvature is positive. The negative curvature is in the direction of the first eigenvector, and the positive curvature is in the direction of the second eigenvector. The eigenvectors are

$$\mathbf{z}_1 = \begin{bmatrix} 1 \\ -1 \end{bmatrix} \text{ and } \mathbf{z}_2 = \begin{bmatrix} 1 \\ 1 \end{bmatrix}.$$

(Note that this is consistent with our previous discussion of this function on page 8-8.)

If we evaluate the Hessian at $\mathbf{x}^3$, we find

$$\nabla^2 F(\mathbf{x}^3) = \begin{bmatrix} 14.7 & -6.71 \\ -6.71 & 14.7 \end{bmatrix}.$$

The eigenvalues of this matrix are

$$\lambda_1 = 21.42, \lambda_2 = 8.0,$$

therefore $\mathbf{x}^3$ must be a strong minimum point.

Check these results using the Neural Network Design Demonstration Vector Taylor Series (`nnd8ts2`).

P8.6 Let's apply the concepts in this chapter to a neural network problem. Consider the linear network shown in Figure P8.7. Suppose that the desired inputs/outputs for the network are

$$\{(p_1 = 2), (t_1 = 0.5)\}, \{(p_2 = -1), (t_2 = 0)\}.$$

Sketch the following performance index for this network:

$$F(\mathbf{x}) = (t_1 - a_1(\mathbf{x}))^2 + (t_2 - a_2(\mathbf{x}))^2.$$

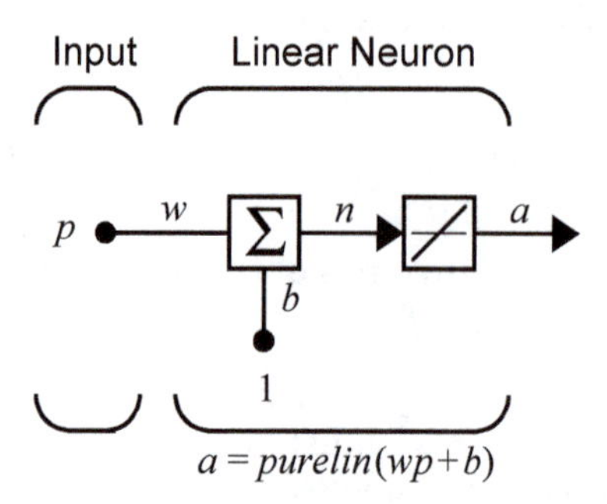

Figure P8.7 Linear Network for Problem P8.6

The parameters of this network are w and b, which make up the parameter vector

$$\mathbf{x} = \begin{bmatrix} w \\ b \end{bmatrix}.$$

We want to sketch the performance index $F(\mathbf{x})$. First we will show that the performance index is a quadratic function. Then we will find the eigenvectors and eigenvalues of the Hessian matrix and use them to sketch the contour plot of the function.

Begin by writing $F(\mathbf{x})$ as an explicit function of the parameter vector $\mathbf{x}$:

$$F(\mathbf{x}) = e_1^2 + e_2^2,$$

where

$$(e_1 = t_1 - (wp_1 + b)), (e_2 = t_2 - (wp_2 + b)).$$

This can be written in matrix form:

$$F(\mathbf{x}) = \mathbf{e}^T\mathbf{e},$$

where

$$\mathbf{e} = \mathbf{t} - \begin{bmatrix} p_1 & 1 \\ p_2 & 1 \end{bmatrix}\mathbf{x} = \mathbf{t} - \mathbf{G}\mathbf{x}.$$

The performance index can now be rewritten:

$$F(\mathbf{x}) = [\mathbf{t} - \mathbf{G}\mathbf{x}]^T[\mathbf{t} - \mathbf{G}\mathbf{x}] = \mathbf{t}^T\mathbf{t} - 2\mathbf{t}^T\mathbf{G}\mathbf{x} + \mathbf{x}^T\mathbf{G}^T\mathbf{G}\mathbf{x}.$$

If we compare this with Eq. (8.35):

$$F(\mathbf{x}) = \frac{1}{2}\mathbf{x}^T\mathbf{A}\mathbf{x} + \mathbf{d}^T\mathbf{x} + c,$$

we can see that the performance index for this linear network is a quadratic function, with

$$c = \mathbf{t}^T\mathbf{t},\ \mathbf{d} = -2\mathbf{G}^T\mathbf{t},\ \text{and}\ \mathbf{A} = 2\mathbf{G}^T\mathbf{G}.$$

The gradient of the quadratic function is given in Eq. (8.38):

$$\nabla F(\mathbf{x}) = \mathbf{A}\mathbf{x} + \mathbf{d} = 2\mathbf{G}^T\mathbf{G}\mathbf{x} - 2\mathbf{G}^T\mathbf{t}.$$

The stationary point (also the center of the function contours) will occur where the gradient is equal to zero:

$$\mathbf{x}^* = -\mathbf{A}^{-1}\mathbf{d} = [\mathbf{G}^T\mathbf{G}]^{-1}\mathbf{G}^T\mathbf{t}.$$

For

$$\mathbf{G} = \begin{bmatrix} p_1 & 1 \\ p_2 & 1 \end{bmatrix} = \begin{bmatrix} 2 & 1 \\ -1 & 1 \end{bmatrix} \text{ and } \mathbf{t} = \begin{bmatrix} 0.5 \\ 0 \end{bmatrix}$$

we have

$$\mathbf{x}^* = [\mathbf{G}^T\mathbf{G}]^{-1}\mathbf{G}^T\mathbf{t} = \begin{bmatrix} 5 & 1 \\ 1 & 2 \end{bmatrix}^{-1}\begin{bmatrix} 1 \\ 0.5 \end{bmatrix} = \begin{bmatrix} 0.167 \\ 0.167 \end{bmatrix}.$$

(Therefore the optimal network parameters are $w = 0.167$ and $b = 0.167$.)

The Hessian matrix of the quadratic function is given by Eq. (8.39):

$$\nabla^2 F(\mathbf{x}) = \mathbf{A} = 2\mathbf{G}^T\mathbf{G} = \begin{bmatrix} 10 & 2 \\ 2 & 4 \end{bmatrix}.$$

To sketch the contour plot we need the eigenvectors and eigenvalues of the Hessian. For this case we find

$$\left\{(\lambda_1 = 10.6), \left(\mathbf{z}_1 = \begin{bmatrix} 1 \\ 0.3 \end{bmatrix}\right)\right\}, \left\{(\lambda_2 = 3.4), \left(\mathbf{z}_2 = \begin{bmatrix} 0.3 \\ -1 \end{bmatrix}\right)\right\}.$$

Therefore we know that $\mathbf{x}^*$ is a strong minimum. Also, since the first eigenvalue is larger than the second, we know that the contours will be elliptical and that the long axis of the ellipses will be in the direction of the second eigenvector. The contours will be centered at $\mathbf{x}^*$. This is demonstrated in Figure P8.8.

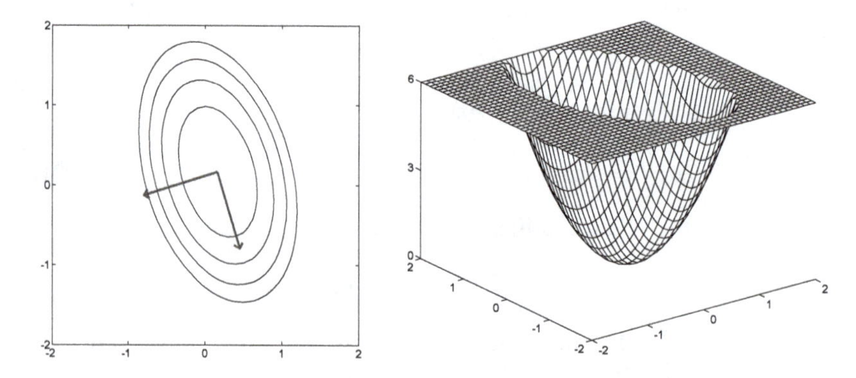

Figure P8.8 Graph of Function for Problem P8.6

P8.7 There are quadratic functions that do not have stationary points. This problem illustrates one such case. Consider the following function:

$$F(\mathbf{x}) = \begin{bmatrix} 1 & -1 \end{bmatrix} \mathbf{x} + \frac{1}{2} \mathbf{x}^T \begin{bmatrix} 1 & 1 \\ 1 & 1 \end{bmatrix} \mathbf{x} .$$

Sketch the contour plot of this function.

As with Problem P8.6, we need to find the eigenvalues and eigenvectors of the Hessian matrix. By inspection of the quadratic function we see that the Hessian matrix is

$$\nabla^2 F(\mathbf{x}) = \mathbf{A} = \begin{bmatrix} 1 & 1 \\ 1 & 1 \end{bmatrix} . \tag{8.63}$$

The eigenvalues and eigenvectors are

$$\left\{ (\lambda_1 = 0), \left(\mathbf{z}_1 = \begin{bmatrix} 1 \\ -1 \end{bmatrix} \right) \right\}, \left\{ (\lambda_2 = 2), \left(\mathbf{z}_2 = \begin{bmatrix} 1 \\ 1 \end{bmatrix} \right) \right\} .$$

Notice that the first eigenvalue is zero, so there is no curvature along the first eigenvector. The second eigenvalue is positive, so there is positive curvature along the second eigenvector. If we had no linear term in $F(\mathbf{x})$, the plot of the function would show a stationary valley, as in Figure 8.9. In this case we must find out if the linear term creates a slope in the direction of the valley (the direction of the first eigenvector).

The linear term is

$$F_{lin}(\mathbf{x}) = \begin{bmatrix} 1 & -1 \end{bmatrix} \mathbf{x} .$$

From Eq. (8.36) we know that the gradient of this term is

$$\nabla F_{lin}(\mathbf{x}) = \begin{bmatrix} 1 \\ -1 \end{bmatrix} ,$$

which means that the linear term is increasing most rapidly in the direction of this gradient. Since the quadratic term has no curvature in this direction, the overall function will have a linear slope in this direction. Therefore $F(\mathbf{x})$ will have positive curvature in the direction of the second eigenvector and a linear slope in the direction of the first eigenvector. The contour plot and the 3-D plot for this function are given in Figure P8.9.

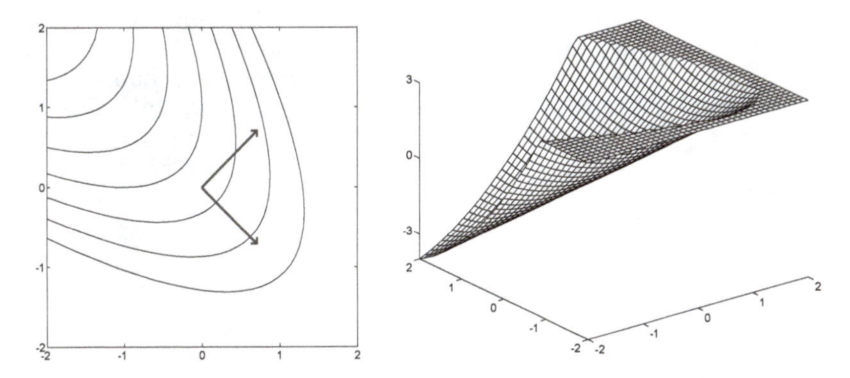

Figure P8.9 Falling Valley Function for Problem P8.7

Whenever any of the eigenvalues of the Hessian matrix are zero it is impossible to solve for the stationary point of the quadratic function using

$$\mathbf{x}^* = -\mathbf{A}^{-1}\mathbf{d} ,$$

since the Hessian matrix does not have an inverse. This lack of an inverse could mean that we have a weak minimum point, as illustrated in Figure 8.9, or that there is no stationary point, as this example shows.

Epilogue

Performance learning is one of the most important classes of neural network learning rules. With performance learning, network parameters are adjusted to optimize network performance. In this chapter we have introduced tools that we will need to understand performance learning rules. After reading this chapter and solving the exercises, you should be able to:

i. Perform a Taylor series expansion and use it to approximate a function.

ii. Calculate a directional derivative.

iii. Find stationary points and test whether they could be minima.

iv. Sketch contour plots of quadratic functions.

We will be using these concepts in a number of succeeding chapters, including the chapters on performance learning (9–14) and the radial basis network chapter (Chapter 16). In the next chapter we will build on the concepts we have covered here, to design algorithms that will optimize performance functions. Then, in succeeding chapters, we will apply these algorithms to the training of neural networks.

Further Reading

[Brog91] W. L. Brogan, *Modern Control Theory,* 3rd Ed., Englewood Cliffs, NJ: Prentice-Hall, 1991.

This is a well-written book on the subject of linear systems. The first half of the book is devoted to linear algebra. It also has good sections on the solution of linear differential equations and the stability of linear and nonlinear systems. It has many worked problems.

[Gill81] P. E. Gill, W. Murray, and M. H. Wright, *Practical Optimization*, New York: Academic Press, 1981.

As the title implies, this text emphasizes the practical implementation of optimization algorithms. It provides motivation for the optimization methods, as well as details of implementation that affect algorithm performance.

[Himm72] D. M. Himmelblau, *Applied Nonlinear Programming*, New York: McGraw-Hill, 1972.

This is a comprehensive text on nonlinear optimization. It covers both constrained and unconstrained optimization problems. The text is very complete, with many examples worked out in detail.

[Scal85] L. E. Scales, *Introduction to Non-Linear Optimization*, New York: Springer-Verlag, 1985.

A very readable text describing the major optimization algorithms, this text emphasizes methods of optimization rather than existence theorems and proofs of convergence. Algorithms are presented with intuitive explanations, along with illustrative figures and examples. Pseudo-code is presented for most algorithms.

Exercises

E8.1 Consider the following scalar function:

$$F(x) = \frac{1}{x^3 - \frac{3}{4}x - \frac{1}{2}}.$$

i. Find the second-order Taylor series approximation for $F(x)$ about the point $x = -0.5$.

ii. Find the second-order Taylor series approximation for $F(x)$ about the point $x = 1.1$.

iii. Plot $F(x)$ and the two approximations and discuss their accuracy.

E8.2 Consider the following function of two variables:

$$F(\mathbf{x}) = e^{(2x_1^2 + 2x_2^2 + x_1 - 5x_2 + 10)}.$$

i. Find the second-order Taylor series approximation for $F(\mathbf{x})$ about the point $\mathbf{x} = \begin{bmatrix} 0 & 0 \end{bmatrix}^T$.

ii. Find the stationary point for this approximation.

iii. Find the stationary point for $F(\mathbf{x})$. (Note that the exponent of $F(\mathbf{x})$ is simply a quadratic function.)

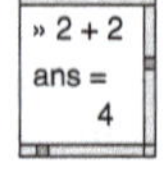

iv. Explain the difference between the two stationary points. (Use MATLAB to plot the two functions.)

E8.3 For the following functions find the first and second directional derivatives from the point $\mathbf{x} = \begin{bmatrix} 1 & 1 \end{bmatrix}^T$ in the direction $\mathbf{p} = \begin{bmatrix} -1 & 1 \end{bmatrix}^T$.

i. $F(\mathbf{x}) = \frac{7}{2}x_1^2 - 6x_1x_2 - x_2^2$

ii. $F(\mathbf{x}) = 5x_1^2 - 6x_1x_2 + 5x_2^2 + 4x_1 + 4x_2$

iii. $F(\mathbf{x}) = \frac{9}{2}x_1^2 - 2x_1x_2 + 3x_2^2 + 2x_1 - x_2$

iv. $F(\mathbf{x}) = -\frac{1}{2}(7x_1^2 + 12x_1x_2 - 2x_2^2)$

v. $F(\mathbf{x}) = x_1^2 + x_1x_2 + x_2^2 + 3x_1 + 3x_2$

vi. $F(\mathbf{x}) = \frac{1}{2}x_1^2 - 3x_1x_2 + \frac{1}{2}x_2^2 - 4x_1 + 4x_2$

vii. $F(\mathbf{x}) = \frac{1}{2}x_1^2 - 2x_1x_2 + 2x_2^2 + x_1 - 2x_2$

viii. $F(\mathbf{x}) = \frac{3}{2}x_1^2 + 2x_1x_2 + 4x_1 + 4x_2$

ix. $F(\mathbf{x}) = -\frac{3}{2}x_1^2 + 4x_1x_2 + \frac{3}{2}x_2^2 + 5x_1$

x. $F(\mathbf{x}) = 2x_1^2 - 2x_1x_2 + \frac{1}{2}x_2^2 + x_1 + x_2$

E8.4 For the following function,

$$F(x) = x^4 - \frac{1}{2}x^2 + 1,$$

i. find the stationary points,

ii. test the stationary points to find minimum and maximum points, and

iii. plot the function using MATLAB to verify your answers.

E8.5 Consider the following function of two variables:

$$F(\mathbf{x}) = (x_1 + x_2)^4 - 12x_1x_2 + x_1 + x_2 + 1.$$

i. Verify that the function has three stationary points at

$$\mathbf{x}^1 = \begin{bmatrix} -0.6504 \\ -0.6504 \end{bmatrix}, \mathbf{x}^2 = \begin{bmatrix} 0.085 \\ 0.085 \end{bmatrix}, \mathbf{x}^3 = \begin{bmatrix} 0.5655 \\ 0.5655 \end{bmatrix}.$$

ii. Test the stationary points to find any minima, maxima or saddle points.

iii. Find the second-order Taylor series approximations for the function at each of the stationary points.

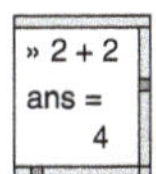

iv. Plot the function and the approximations using MATLAB.

E8.6 For the functions of Exercise E8.3:

i. find the stationary points,

ii. test the stationary points to find minima, maxima or saddle points,

iii. provide rough sketches of the contour plots, using the eigenvalues

» 2 + 2
ans =
4

and eigenvectors of the Hessian matrices, and

iv. plot the functions using MATLAB to verify your answers.

E8.7 Consider the following quadratic function:

$$F(\mathbf{x}) = \frac{1}{2}\mathbf{x}^T\begin{bmatrix} 1 & -3 \\ -3 & 1 \end{bmatrix}\mathbf{x} + \begin{bmatrix} 4 & -4 \end{bmatrix}\mathbf{x} + 2.$$

i. Find the gradient and Hessian matrix for $F(\mathbf{x})$.

ii. Sketch the contour plot for $F(\mathbf{x})$.

iii. Find the directional derivative of $F(\mathbf{x})$ at the point $\mathbf{x}_0 = \begin{bmatrix} 0 & 0 \end{bmatrix}^T$ in the direction $\mathbf{p} = \begin{bmatrix} 1 & 1 \end{bmatrix}^T$.

iv. Is your answer to part iii. consistent with your contour plot of part ii.? Explain.

E8.8 Repeat Exercise E8.7 with the following quadratic function:

$$F(\mathbf{x}) = \frac{1}{2}\mathbf{x}^T\begin{bmatrix} 3 & -2 \\ -2 & 0 \end{bmatrix}\mathbf{x} + \begin{bmatrix} 4 & 4 \end{bmatrix}\mathbf{x} + 2.$$

E8.9 Consider the following function:

$$F(\mathbf{x}) = (1 + x_1 + x_2)^2 + \frac{1}{4}x_1^4.$$

i. Find the quadratic approximation to $F(\mathbf{x})$ about the point $\mathbf{x}_0 = \begin{bmatrix} 1 & 0 \end{bmatrix}^T$

ii. Sketch the contour plot of the quadratic approximation in part i.

E8.10 Consider the following function:

$$F(\mathbf{x}) = \frac{3}{2}x_1^2 + 2x_1x_2 + x_2^3 + 4x_1 + 4x_2.$$

i. Find the quadratic approximation to $F(\mathbf{x})$ about the point $\mathbf{x}_0 = \begin{bmatrix} 1 & 0 \end{bmatrix}^T$.

ii. Locate the stationary point of the quadratic approximation you found in part i.

iii. Is the answer to part ii a minimum of $F(\mathbf{x})$?

E8.11 Consider the following function:

$$F(\mathbf{x}) = x_1 x_2 - x_1 + 2x_2 .$$

i. Locate any stationary points.

ii. For each answer to part i., determine, if possible, whether the stationary point is a minimum point, a maximum point, or a saddle point.

iii. Find the directional derivative of the function at the point $\mathbf{x}_0 = \begin{bmatrix} -1 & 1 \end{bmatrix}^T$ in the direction $\mathbf{p} = \begin{bmatrix} -1 & 1 \end{bmatrix}^T$.

E8.12 Consider the following function:

$$F(\mathbf{x}) = x_1^2 + 2x_1 x_2 + x_2^2 + (x_1 - x_2)^3 .$$

i. Find the quadratic approximation to $F(\mathbf{x})$ about the point $\mathbf{x}_0 = \begin{bmatrix} 2 & 1 \end{bmatrix}^T$.

ii. Sketch the contour plot of the quadratic approximation.

E8.13 Recall the function in Problem P8.7. For that function there was no stationary point. It is possible to modify the function, by changing only the $\mathbf{d}$ vector, so that a stationary point will exist. Find a new nonzero $\mathbf{d}$ vector that will create a weak minimum.

9 Performance Optimization

Objectives

We initiated our discussion of performance optimization in Chapter 8. There we introduced the Taylor series expansion as a tool for analyzing the performance surface, and then used it to determine conditions that must be satisfied by optimum points. In this chapter we will again use the Taylor series expansion, in this case to develop algorithms to locate the optimum points. We will discuss three different categories of optimization algorithm: steepest descent, Newton's method and conjugate gradient. In Chapters 10–14 we will apply all of these algorithms to the training of neural networks.

Theory and Examples

In the previous chapter we began our investigation of performance surfaces. Now we are in a position to develop algorithms to search the parameter space and locate minimum points of the surface (find the optimum weights and biases for a given neural network).

It is interesting to note that most of the algorithms presented in this chapter were developed hundreds of years ago. The basic principles of optimization were discovered during the 17th century, by such scientists and mathematicians as Kepler, Fermat, Newton and Leibniz. From 1950 on, these principles were rediscovered to be implemented on "high speed" (in comparison to the pen and paper available to Newton) digital computers. The success of these efforts stimulated significant research on new algorithms, and the field of optimization theory became recognized as a major branch of mathematics. Now neural network researchers have access to a vast storehouse of optimization theory and practice that can be applied to the training of neural networks.

The objective of this chapter, then, is to develop algorithms to optimize a performance index $F(\mathbf{x})$. For our purposes the word "optimize" will mean to find the value of $\mathbf{x}$ that minimizes $F(\mathbf{x})$. All of the optimization algorithms we will discuss are iterative. We begin from some initial guess, $\mathbf{x}_0$, and then update our guess in stages according to an equation of the form

$$\mathbf{x}_{k+1} = \mathbf{x}_k + \alpha_k \mathbf{p}_k , \tag{9.1}$$

or

$$\Delta\mathbf{x}_k = (\mathbf{x}_{k+1} - \mathbf{x}_k) = \alpha_k \mathbf{p}_k , \tag{9.2}$$

where the vector $\mathbf{p}_k$ represents a search direction, and the positive scalar α_k is the learning rate, which determines the length of the step.

The algorithms we will discuss in this chapter are distinguished by the choice of the search direction, $\mathbf{p}_k$. We will discuss three different possibilities. There are also a variety of ways to select the learning rate, α_k, and we will discuss several of these.

Steepest Descent

When we update our guess of the optimum (minimum) point using Eq. (9.1), we would like to have the function decrease at each iteration. In other words,

$$F(\mathbf{x}_{k+1}) < F(\mathbf{x}_k) . \tag{9.3}$$

How can we choose a direction, $\mathbf{p}_k$, so that for sufficiently small learning rate, α_k, we will move "downhill" in this way? Consider the first-order Taylor series expansion (see Eq. (8.9)) of $F(\mathbf{x})$ about the old guess $\mathbf{x}_k$:

$$F(\mathbf{x}_{k+1}) = F(\mathbf{x}_k + \Delta\mathbf{x}_k) \approx F(\mathbf{x}_k) + \mathbf{g}_k^T \Delta\mathbf{x}_k, \quad (9.4)$$

where $\mathbf{g}_k$ is the gradient evaluated at the old guess $\mathbf{x}_k$:

$$\mathbf{g}_k \equiv \nabla F(\mathbf{x})\big|_{\mathbf{x} = \mathbf{x}_k}. \quad (9.5)$$

For $F(\mathbf{x}_{k+1})$ to be less than $F(\mathbf{x}_k)$, the second term on the right-hand side of Eq. (9.4) must be negative:

$$\mathbf{g}_k^T \Delta\mathbf{x}_k = \alpha_k \mathbf{g}_k^T \mathbf{p}_k < 0. \quad (9.6)$$

We will select an α_k that is small, but greater than zero. This implies:

$$\mathbf{g}_k^T \mathbf{p}_k < 0. \quad (9.7)$$

Descent Direction

Any vector $\mathbf{p}_k$ that satisfies this equation is called a *descent direction*. The function must go down if we take a small enough step in this direction. This brings up another question. What is the direction of steepest descent? (In what direction will the function decrease most rapidly?) This will occur when

$$\mathbf{g}_k^T \mathbf{p}_k \quad (9.8)$$

is most negative. (We assume that the length of $\mathbf{p}_k$ does not change, only the direction.) This is an inner product between the gradient and the direction vector. It will be most negative when the direction vector is the negative of the gradient. (Review our discussion of directional derivatives on page 8-6.) Therefore a vector that points in the steepest descent direction is

$$\mathbf{p}_k = -\mathbf{g}_k. \quad (9.9)$$

Steepest Descent

Using this in the iteration of Eq. (9.1) produces the method of *steepest descent*:

$$\mathbf{x}_{k+1} = \mathbf{x}_k - \alpha_k \mathbf{g}_k. \quad (9.10)$$

Learning Rate

For steepest descent there are two general methods for determining the *learning rate*, α_k. One approach is to minimize the performance index $F(\mathbf{x})$ with respect to α_k at each iteration. In this case we are minimizing along the line

$$\mathbf{x}_k - \alpha_k \mathbf{g}_k. \quad (9.11)$$

The other method for selecting α_k is to use a fixed value (e.g., $\alpha_k = 0.02$), or to use variable, but predetermined, values (e.g., $\alpha_k = 1/k$). We will discuss the choice of α_k in more detail in the following examples.

Let's apply the steepest descent algorithm to the following function,

$$F(\mathbf{x}) = x_1^2 + 25x_2^2, \tag{9.12}$$

starting from the initial guess

$$\mathbf{x}_0 = \begin{bmatrix} 0.5 \\ 0.5 \end{bmatrix}. \tag{9.13}$$

The first step is to find the gradient:

$$\nabla F(\mathbf{x}) = \begin{bmatrix} \frac{\partial}{\partial x_1} F(\mathbf{x}) \\ \frac{\partial}{\partial x_2} F(\mathbf{x}) \end{bmatrix} = \begin{bmatrix} 2x_1 \\ 50x_2 \end{bmatrix}. \tag{9.14}$$

If we evaluate the gradient at the initial guess we find

$$\mathbf{g}_0 = \nabla F(\mathbf{x})\big|_{\mathbf{x} = \mathbf{x}_0} = \begin{bmatrix} 1 \\ 25 \end{bmatrix}. \tag{9.15}$$

Assume that we use a fixed learning rate of $\alpha = 0.01$. The first iteration of the steepest descent algorithm would be

$$\mathbf{x}_1 = \mathbf{x}_0 - \alpha \mathbf{g}_0 = \begin{bmatrix} 0.5 \\ 0.5 \end{bmatrix} - 0.01 \begin{bmatrix} 1 \\ 25 \end{bmatrix} = \begin{bmatrix} 0.49 \\ 0.25 \end{bmatrix}. \tag{9.16}$$

The second iteration of steepest descent produces

$$\mathbf{x}_2 = \mathbf{x}_1 - \alpha \mathbf{g}_1 = \begin{bmatrix} 0.49 \\ 0.25 \end{bmatrix} - 0.01 \begin{bmatrix} 0.98 \\ 12.5 \end{bmatrix} = \begin{bmatrix} 0.4802 \\ 0.125 \end{bmatrix}. \tag{9.17}$$

If we continue the iterations we obtain the trajectory illustrated in Figure 9.1.

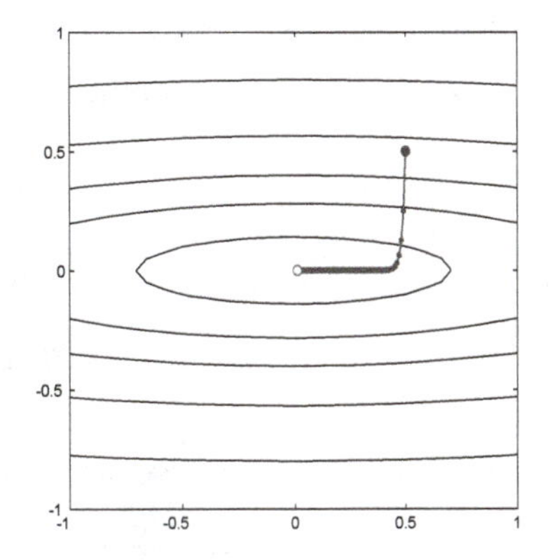

Figure 9.1 Trajectory for Steepest Descent with $\alpha = 0.01$

Note that the steepest descent trajectory, for small learning rate, follows a path that is always orthogonal to the contour lines. This is because the gradient is orthogonal to the contour lines. (See the discussion on page 8-6.)

How would a change in the learning rate change the performance of the algorithm? If we increase the learning rate to $\alpha = 0.035$, we obtain the trajectory illustrated in Figure 9.2. Note that the trajectory now oscillates. If we make the learning rate too large the algorithm will become unstable; the oscillations will increase instead of decaying.

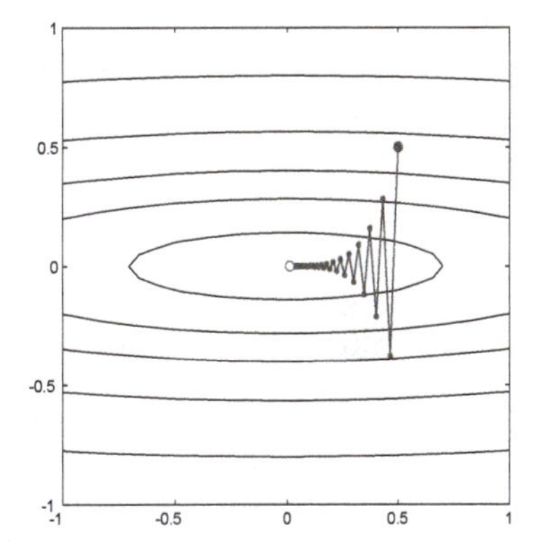

Figure 9.2 Trajectory for Steepest Descent with $\alpha = 0.035$

We would like to make the learning rate large, since then we will be taking large steps and would expect to converge faster. However, as we can see from this example, if we make the learning rate too large the algorithm will become unstable. Is there some way to predict the maximum allowable learning rate? This is not possible for arbitrary functions, but for quadratic functions we can set an upper limit.

Stable Learning Rates

Suppose that the performance index is a quadratic function:

$$F(\mathbf{x}) = \frac{1}{2}\mathbf{x}^T\mathbf{A}\mathbf{x} + \mathbf{d}^T\mathbf{x} + c. \tag{9.18}$$

From Eq. (8.38) the gradient of the quadratic function is

$$\nabla F(\mathbf{x}) = \mathbf{A}\mathbf{x} + \mathbf{d}. \tag{9.19}$$

If we now insert this expression into our expression for the steepest descent algorithm (assuming a constant learning rate), we obtain

$$\mathbf{x}_{k+1} = \mathbf{x}_k - \alpha\mathbf{g}_k = \mathbf{x}_k - \alpha(\mathbf{A}\mathbf{x}_k + \mathbf{d}) \tag{9.20}$$

or

$$\mathbf{x}_{k+1} = [\mathbf{I} - \alpha\mathbf{A}]\mathbf{x}_k - \alpha\mathbf{d}. \tag{9.21}$$

This is a linear dynamic system, which will be stable if the eigenvalues of the matrix $[\mathbf{I} - \alpha\mathbf{A}]$ are less than one in magnitude (see [Brog91]). We can express the eigenvalues of this matrix in terms of the eigenvalues of the Hessian matrix $\mathbf{A}$. Let $\{\lambda_1, \lambda_2, \dots, \lambda_n\}$ and $\{\mathbf{z}_1, \mathbf{z}_2, \dots, \mathbf{z}_n\}$ be the eigenvalues and eigenvectors of the Hessian matrix. Then

$$[\mathbf{I} - \alpha\mathbf{A}]\mathbf{z}_i = \mathbf{z}_i - \alpha\mathbf{A}\mathbf{z}_i = \mathbf{z}_i - \alpha\lambda_i\mathbf{z}_i = (1 - \alpha\lambda_i)\mathbf{z}_i. \tag{9.22}$$

Therefore the eigenvectors of $[\mathbf{I} - \alpha\mathbf{A}]$ are the same as the eigenvectors of $\mathbf{A}$, and the eigenvalues of $[\mathbf{I} - \alpha\mathbf{A}]$ are $(1 - \alpha\lambda_i)$. Our condition for the stability of the steepest descent algorithm is then

$$|(1 - \alpha\lambda_i)| < 1. \tag{9.23}$$

If we assume that the quadratic function has a strong minimum point, then its eigenvalues must be positive numbers. Eq. (9.23) then reduces to

$$\alpha < \frac{2}{\lambda_i}. \tag{9.24}$$

Since this must be true for all the eigenvalues of the Hessian matrix we have

$$\alpha < \frac{2}{\lambda_{max}}. \tag{9.25}$$

The maximum stable learning rate is inversely proportional to the maximum curvature of the quadratic function. The curvature tells us how fast the gradient is changing. If the gradient is changing too fast we may jump past the minimum point so far that the gradient at the new location will be larger in magnitude (but opposite direction) than the gradient at the old location. This will cause the steps to increase in size at each iteration.

Let's apply this result to our previous example. The Hessian matrix for that quadratic function is

$$\mathbf{A} = \begin{bmatrix} 2 & 0 \\ 0 & 50 \end{bmatrix}. \quad (9.26)$$

The eigenvalues and eigenvectors of $\mathbf{A}$ are

$$\left\{(\lambda_1 = 2), \left(\mathbf{z}_1 = \begin{bmatrix} 1 \\ 0 \end{bmatrix}\right)\right\}, \left\{(\lambda_2 = 50), \left(\mathbf{z}_2 = \begin{bmatrix} 0 \\ 1 \end{bmatrix}\right)\right\}. \quad (9.27)$$

Therefore the maximum allowable learning rate is

$$\alpha < \frac{2}{\lambda_{max}} = \frac{2}{50} = 0.04. \quad (9.28)$$

This result is illustrated experimentally in Figure 9.3, which shows the steepest descent trajectories when the learning rate is just below ($\alpha = 0.039$) and just above ($\alpha = 0.041$), the maximum stable value.

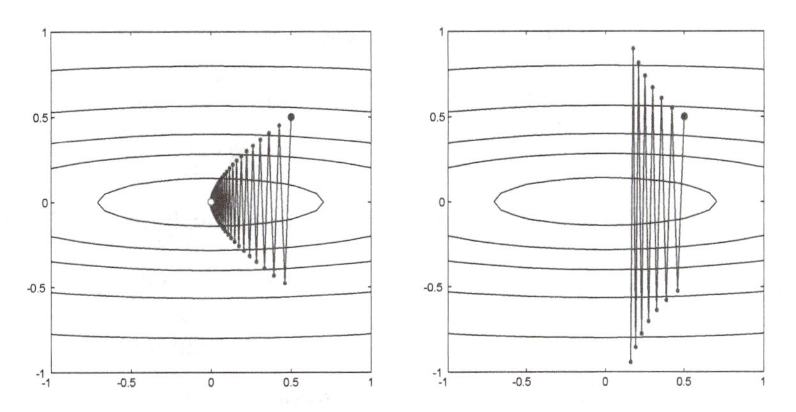

Figure 9.3 Trajectories for $\alpha = 0.039$ (left) and $\alpha = 0.041$ (right).

This example has illustrated several points. The learning rate is limited by the largest eigenvalue (second derivative) of the Hessian matrix. The algorithm tends to converge most quickly in the direction of the eigenvector corresponding to this largest eigenvalue, and we don't want to overshoot the minimum point by too far in that direction. (Note that in our examples the initial step is almost parallel to the x_2 axis, which is $\mathbf{z}_2$.) However, the algorithm will tend to converge most slowly in the direction of the eigenvector that corresponds to the smallest eigenvalue ($\mathbf{z}_1$ for our example). In the end it is the smallest eigenvalue, in combination with the learning rate, that determines how quickly the algorithm will converge. When there is a great difference in magnitude between the largest and smallest eigenvalues, the steepest descent algorithm will converge slowly.

To experiment with steepest descent on this quadratic function, use the Neural Network Design Demonstration Steepest Descent for a Quadratic (nnd9sdq).

Minimizing Along a Line

Another approach for selecting the learning rate is to minimize the performance index with respect to α_k at each iteration. In other words, choose α_k to minimize

$$F(\mathbf{x}_k + \alpha_k \mathbf{p}_k) . \tag{9.29}$$

To do this for arbitrary functions requires a line search, which we will discuss in Chapter 12. For quadratic functions it is possible to perform the linear minimization analytically. The derivative of Eq. (9.29) with respect to α_k, for quadratic $F(\mathbf{x})$, can be shown to be

$$\frac{d}{d\alpha_k} F(\mathbf{x}_k + \alpha_k \mathbf{p}_k) = \nabla F(\mathbf{x})^T \Big|_{\mathbf{x} = \mathbf{x}_k} \mathbf{p}_k + \alpha_k \mathbf{p}_k^T \nabla^2 F(\mathbf{x}) \Big|_{\mathbf{x} = \mathbf{x}_k} \mathbf{p}_k . \tag{9.30}$$

If we set this derivative equal to zero and solve for α_k, we obtain

$$\alpha_k = - \frac{\nabla F(\mathbf{x})^T \Big|_{\mathbf{x} = \mathbf{x}_k} \mathbf{p}_k}{\mathbf{p}_k^T \nabla^2 F(\mathbf{x}) \Big|_{\mathbf{x} = \mathbf{x}_k} \mathbf{p}_k} = - \frac{\mathbf{g}_k^T \mathbf{p}_k}{\mathbf{p}_k^T \mathbf{A}_k \mathbf{p}_k} , \tag{9.31}$$

where $\mathbf{A}_k$ is the Hessian matrix evaluated at the old guess $\mathbf{x}_k$:

$$\mathbf{A}_k \equiv \nabla^2 F(\mathbf{x}) \Big|_{\mathbf{x} = \mathbf{x}_k} . \tag{9.32}$$

(For quadratic functions the Hessian matrix is not a function of k.)

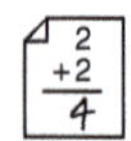

Let's apply steepest descent with line minimization to the following quadratic function:

$$F(\mathbf{x}) = \frac{1}{2} \mathbf{x}^T \begin{bmatrix} 2 & 1 \\ 1 & 2 \end{bmatrix} \mathbf{x} , \tag{9.33}$$

starting from the initial guess

$$\mathbf{x}_0 = \begin{bmatrix} 0.8 \\ -0.25 \end{bmatrix}. \tag{9.34}$$

The gradient of this function is

$$\nabla F(\mathbf{x}) = \begin{bmatrix} 2x_1 + x_2 \\ x_1 + 2x_2 \end{bmatrix}. \tag{9.35}$$

The search direction for steepest descent is the negative of the gradient. For the first iteration this will be

$$\mathbf{p}_0 = -\mathbf{g}_0 = -\nabla F(\mathbf{x})\big|_{\mathbf{x} = \mathbf{x}_0} = \begin{bmatrix} -1.35 \\ -0.3 \end{bmatrix}. \tag{9.36}$$

From Eq. (9.31), the learning rate for the first iteration will be

$$\alpha_0 = -\frac{\begin{bmatrix} 1.35 & 0.3 \end{bmatrix} \begin{bmatrix} -1.35 \\ -0.3 \end{bmatrix}}{\begin{bmatrix} -1.35 & -0.3 \end{bmatrix} \begin{bmatrix} 2 & 1 \\ 1 & 2 \end{bmatrix} \begin{bmatrix} -1.35 \\ -0.3 \end{bmatrix}} = 0.413. \tag{9.37}$$

The first step of steepest descent will then produce

$$\mathbf{x}_1 = \mathbf{x}_0 - \alpha_0 \mathbf{g}_0 = \begin{bmatrix} 0.8 \\ -0.25 \end{bmatrix} - 0.413 \begin{bmatrix} 1.35 \\ 0.3 \end{bmatrix} = \begin{bmatrix} 0.24 \\ -0.37 \end{bmatrix}. \tag{9.38}$$

The first five iterations of the algorithm are illustrated in Figure 9.4.

Note that the successive steps of the algorithm are orthogonal. Why does this happen? First, when we minimize along a line we will always stop at a point that is tangent to a contour line. Then, since the gradient is orthogonal to the contour line, the next step, which is along the negative of the gradient, will be orthogonal to the previous step.

We can show this analytically by using the chain rule on Eq. (9.30):

$$\frac{d}{d\alpha_k} F(\mathbf{x}_k + \alpha_k \mathbf{p}_k) = \frac{d}{d\alpha_k} F(\mathbf{x}_{k+1}) = \nabla F(\mathbf{x})^T \big|_{\mathbf{x} = \mathbf{x}_{k+1}} \frac{d}{d\alpha_k} [\mathbf{x}_k + \alpha_k \mathbf{p}_k]$$

$$= \nabla F(\mathbf{x})^T \big|_{\mathbf{x} = \mathbf{x}_{k+1}} \mathbf{p}_k = \mathbf{g}_{k+1}^T \mathbf{p}_k. \tag{9.39}$$

Therefore at the minimum point, where this derivative is zero, the gradient is orthogonal to the previous search direction. Since the next search direction is the negative of this gradient, the consecutive search directions must be orthogonal. (Note that this result implies that when minimizing in any direction, the gradient at the minimum point will be orthogonal to the search direction, even if we are not using steepest descent. We will use this result in our discussion of conjugate directions.)

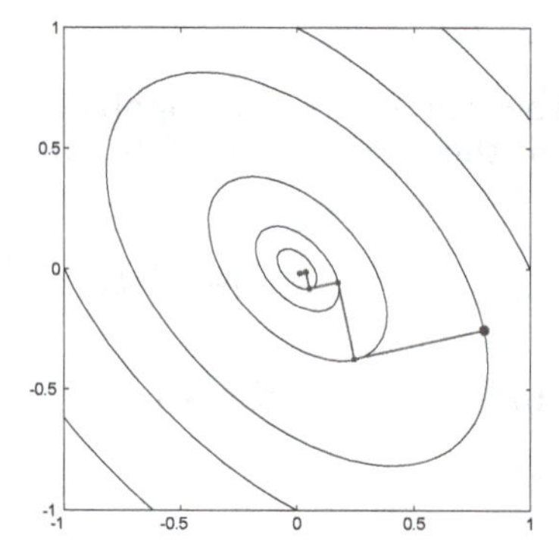

Figure 9.4 Steepest Descent with Minimization Along a Line

To experiment with steepest descent with minimization along a line, use the Neural Network Design Demonstration Method Comparison (**nnd9mc**).

Later in this chapter we will find that we can improve performance if we adjust the search directions, so that instead of being orthogonal they are *conjugate*. (We will define this term later.) If conjugate directions are used the function can be exactly minimized in at most n steps, where n is the dimension of $\mathbf{x}$. (There are certain types of quadratic functions that are minimized in one step by the steepest descent algorithm. Can you think of such a function? How is its Hessian matrix characterized?)

Newton's Method

The derivation of the steepest descent algorithm was based on the first-order Taylor series expansion (Eq. (9.4)). Newton's method is based on the second-order Taylor series:

$$F(\mathbf{x}_{k+1}) = F(\mathbf{x}_k + \Delta\mathbf{x}_k) \approx F(\mathbf{x}_k) + \mathbf{g}_k^T \Delta\mathbf{x}_k + \frac{1}{2}\Delta\mathbf{x}_k^T \mathbf{A}_k \Delta\mathbf{x}_k . \qquad (9.40)$$

The principle behind Newton's method is to locate the stationary point of this quadratic approximation to $F(\mathbf{x})$. If we use Eq. (8.38) to take the gradient of this quadratic function with respect to $\Delta\mathbf{x}_k$ and set it equal to zero, we find

$$\mathbf{g}_k + \mathbf{A}_k \Delta\mathbf{x}_k = \mathbf{0}. \quad (9.41)$$

Solving for $\Delta\mathbf{x}_k$ produces

$$\Delta\mathbf{x}_k = -\mathbf{A}_k^{-1}\mathbf{g}_k. \quad (9.42)$$

Newton's Method

Newton's method is then defined:

$$\mathbf{x}_{k+1} = \mathbf{x}_k - \mathbf{A}_k^{-1}\mathbf{g}_k. \quad (9.43)$$

To illustrate the operation of Newton's method, let's apply it to our previous example function of Eq. (9.12):

$$F(\mathbf{x}) = x_1^2 + 25x_2^2. \quad (9.44)$$

The gradient and Hessian matrices are

$$\nabla F(\mathbf{x}) = \begin{bmatrix} \frac{\partial}{\partial x_1}F(\mathbf{x}) \\ \frac{\partial}{\partial x_2}F(\mathbf{x}) \end{bmatrix} = \begin{bmatrix} 2x_1 \\ 50x_2 \end{bmatrix}, \quad \nabla^2 F(\mathbf{x}) = \begin{bmatrix} 2 & 0 \\ 0 & 50 \end{bmatrix}. \quad (9.45)$$

If we start from the same initial guess

$$\mathbf{x}_0 = \begin{bmatrix} 0.5 \\ 0.5 \end{bmatrix}, \quad (9.46)$$

the first step of Newton's method would be

$$\mathbf{x}_1 = \begin{bmatrix} 0.5 \\ 0.5 \end{bmatrix} - \begin{bmatrix} 2 & 0 \\ 0 & 50 \end{bmatrix}^{-1} \begin{bmatrix} 1 \\ 25 \end{bmatrix} = \begin{bmatrix} 0.5 \\ 0.5 \end{bmatrix} - \begin{bmatrix} 0.5 \\ 0.5 \end{bmatrix} = \begin{bmatrix} 0 \\ 0 \end{bmatrix}. \quad (9.47)$$

This method will always find the minimum of a quadratic function in one step. This is because Newton's method is designed to approximate a function as quadratic and then locate the stationary point of the quadratic approximation. If the original function is quadratic (with a strong minimum) it will be minimized in one step. The trajectory of Newton's method for this problem is given in Figure 9.5.

If the function $F(\mathbf{x})$ is not quadratic, then Newton's method will not generally converge in one step. In fact, we cannot be sure that it will converge at all, since this will depend on the function and the initial guess.

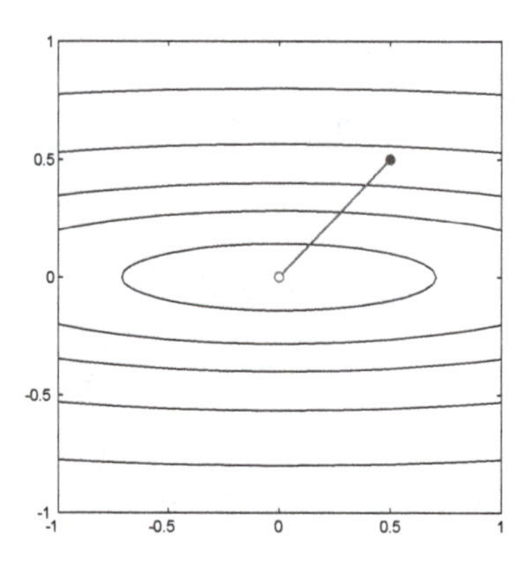

Figure 9.5 Trajectory for Newton's Method

Recall the function given by Eq. (8.18):

$$F(\mathbf{x}) = (x_2 - x_1)^4 + 8x_1x_2 - x_1 + x_2 + 3\,. \tag{9.48}$$

We know from Chapter 8 (see Problem P8.5) that this function has three stationary points:

$$\mathbf{x}^1 = \begin{bmatrix} -0.41878 \\ 0.41878 \end{bmatrix},\ \mathbf{x}^2 = \begin{bmatrix} -0.134797 \\ 0.134797 \end{bmatrix},\ \mathbf{x}^3 = \begin{bmatrix} 0.55358 \\ -0.55358 \end{bmatrix}. \tag{9.49}$$

The first point is a strong local minimum, the second point is a saddle point, and the third point is a strong global minimum.

If we apply Newton's method to this problem, starting from the initial guess $\mathbf{x}_0 = \begin{bmatrix} 1.5 & 0 \end{bmatrix}^T$, our first iteration will be as shown in Figure 9.6. The graph on the left-hand side of the figure is a contour plot of the original function. On the right we see the quadratic approximation to the function at the initial guess.

The function is not minimized in one step, which is not surprising since the function is not quadratic. However, we do take a step toward the global minimum, and if we continue for two more iterations the algorithm will converge to within 0.01 of the global minimum. Newton's method converges quickly in many applications because analytic functions can be accurately approximated by quadratic functions in a small neighborhood of a strong minimum. So as we move closer to the minimum point, Newton's method will more accurately predict its location. In this case we can see that the contour plot of the quadratic approximation is similar to the contour plot of the original function near the initial guess.

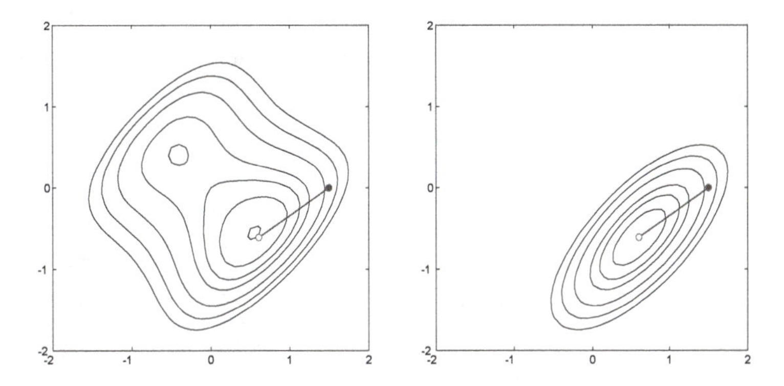

Figure 9.6 One Iteration of Newton's Method from $\mathbf{x}_0 = \begin{bmatrix} 1.5 & 0 \end{bmatrix}^T$

In Figure 9.7 we see one iteration of Newton's method from the initial guess $\mathbf{x}_0 = \begin{bmatrix} -1.5 & 0 \end{bmatrix}^T$. In this case we are converging to the local minimum. Clearly Newton's method cannot distinguish between a local minimum and a global minimum, since it approximates the function as a quadratic, and the quadratic function can have only one minimum. Newton's method, like steepest descent, relies on the local features of the surface (the first and second derivatives). It cannot know the global character of the function.

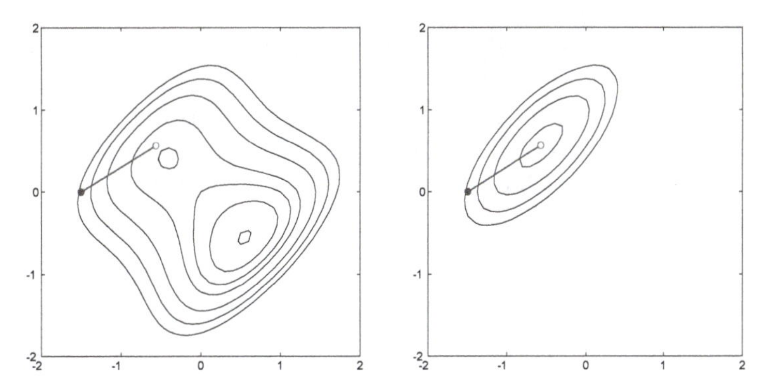

Figure 9.7 One Iteration of Newton's Method from $\mathbf{x}_0 = \begin{bmatrix} -1.5 & 0 \end{bmatrix}^T$

In Figure 9.8 we see one iteration of Newton's method from the initial guess $\mathbf{x}_0 = \begin{bmatrix} 0.75 & 0.75 \end{bmatrix}^T$. Now we are converging toward the saddle point of the function. Note that Newton's method locates the stationary point of the quadratic approximation to the function at the current guess. It does not distinguish between minima, maxima and saddle points. For this problem the quadratic approximation has a saddle point (indefinite Hessian matrix), which is near the saddle point of the original function. If we continue the iterations, the algorithm does converge to the saddle point of $F(\mathbf{x})$.

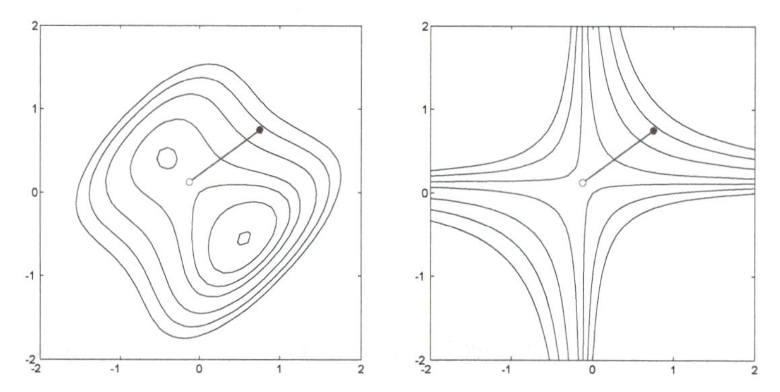

Figure 9.8 One Iteration of Newton's Method from $\mathbf{x}_0 = \begin{bmatrix} 0.75 & 0.75 \end{bmatrix}^T$

In each of the cases we have looked at so far the stationary point of the quadratic approximation has been close to a corresponding stationary point of $F(\mathbf{x})$. This is not always the case. In fact, Newton's method can produce very unpredictable results.

In Figure 9.9 we see one iteration of Newton's method from the initial guess $\mathbf{x}_0 = \begin{bmatrix} 1.15 & 0.75 \end{bmatrix}^T$. In this case the quadratic approximation predicts a saddle point, however, the saddle point is located very close to the local minimum of $F(\mathbf{x})$. If we continue the iterations, the algorithm will converge to the local minimum. Notice that the initial guess was actually farther away from the local minimum than it was for the previous case, in which the algorithm converged to the saddle point.

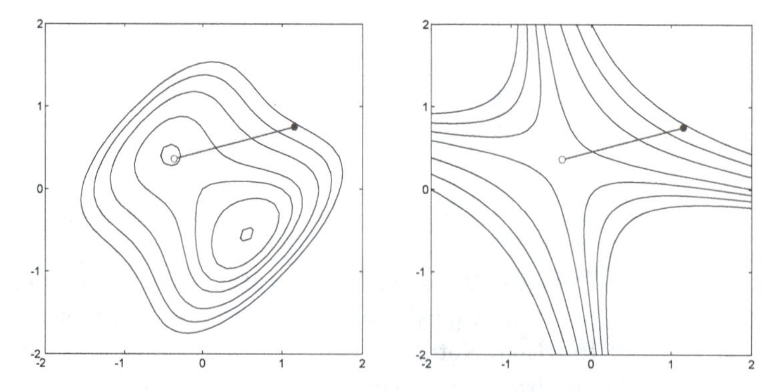

Figure 9.9 One Iteration of Newton's Method from $\mathbf{x}_0 = \begin{bmatrix} 1.15 & 0.75 \end{bmatrix}^T$

To experiment with Newton's method and steepest descent on this function, use the Neural Network Design Demonstrations Newton's Method (nnd9nm) *and Steepest Descent* (nnd9sd).

This is a good place to summarize some of the properties of Newton's method that we have observed.

While Newton's method usually produces faster convergence than steepest descent, the behavior of Newton's method can be quite complex. In addition to the problem of convergence to saddle points (which is very unlikely with steepest descent), it is possible for the algorithm to oscillate or diverge. Steepest descent is guaranteed to converge, if the learning rate is not too large or if we perform a linear minimization at each stage.

In Chapter 12 we will discuss a variation of Newton's method that is well suited to neural network training. It eliminates the divergence problem by using steepest descent steps whenever divergence begins to occur.

Another problem with Newton's method is that it requires the computation and storage of the Hessian matrix, as well as its inverse. If we compare steepest descent, Eq. (9.10), with Newton's method, Eq. (9.43), we see that their search directions will be identical when

$$\mathbf{A}_k = \mathbf{A}_k^{-1} = \mathbf{I}. \tag{9.50}$$

This observation has lead to a class of optimization algorithms know as quasi-Newton or one-step-secant methods. These methods replace $\mathbf{A}_k^{-1}$ with a positive definite matrix, $\mathbf{H}_k$, which is updated at each iteration without matrix inversion. The algorithms are typically designed so that for quadratic functions $\mathbf{H}_k$ will converge to $\mathbf{A}^{-1}$. (The Hessian is constant for quadratic functions.) See [Gill81], [Scal85] or [Batt92] for a discussion of these methods.

Conjugate Gradient

Quadratic Termination

Newton's method has a property called *quadratic termination*, which means that it minimizes a quadratic function exactly in a finite number of iterations. Unfortunately, it requires calculation and storage of the second derivatives. When the number of parameters, n, is large, it may be impractical to compute all of the second derivatives. (Note that the gradient has n elements, while the Hessian has n^2 elements.) This is especially true with neural networks, where practical applications can require several hundred to many thousand weights. For these cases we would like to have methods that require only first derivatives but still have quadratic termination.

Recall the performance of the steepest descent algorithm, with linear searches at each iteration. The search directions at consecutive iterations were orthogonal (see Figure 9.4). For quadratic functions with elliptical contours this produces a zig-zag trajectory of short steps. Perhaps quadratic search directions are not the best choice. Is there a set of search directions that will guarantee quadratic termination? One possibility is conjugate directions.

Suppose that we wish to locate the minimum of the following quadratic function:

$$F(\mathbf{x}) = \frac{1}{2}\mathbf{x}^T\mathbf{A}\mathbf{x} + \mathbf{d}^T\mathbf{x} + c. \tag{9.51}$$

Conjugate

A set of vectors $\{\mathbf{p}_k\}$ is mutually *conjugate* with respect to a positive definite Hessian matrix $\mathbf{A}$ if and only if

$$\mathbf{p}_k^T\mathbf{A}\mathbf{p}_j = 0 \qquad k \neq j. \tag{9.52}$$

As with orthogonal vectors, there are an infinite number of mutually conjugate sets of vectors that span a given n-dimensional space. One set of conjugate vectors consists of the eigenvectors of $\mathbf{A}$. Let $\{\lambda_1, \lambda_2, \ldots, \lambda_n\}$ and $\{\mathbf{z}_1, \mathbf{z}_2, \ldots, \mathbf{z}_n\}$ be the eigenvalues and eigenvectors of the Hessian matrix. To see that the eigenvectors are conjugate, replace $\mathbf{p}_k$ with $\mathbf{z}_k$ in Eq. (9.52):

$$\mathbf{z}_k^T\mathbf{A}\mathbf{z}_j = \lambda_j\mathbf{z}_k^T\mathbf{z}_j = 0 \qquad k \neq j, \tag{9.53}$$

where the last equality holds because the eigenvectors of a symmetric matrix are mutually orthogonal. Therefore the eigenvectors are both conjugate and orthogonal. (Can you find a quadratic function where all orthogonal vectors are also conjugate?)

It is not surprising that we can minimize a quadratic function exactly by searching along the eigenvectors of the Hessian matrix, since they form the principal axes of the function contours. (See the discussion on pages 8-13 through 8-19.) Unfortunately this is not of much practical help, since to find the eigenvectors we must first find the Hessian matrix. We want to find an algorithm that does not require the computation of second derivatives.

It can be shown (see [Scal85] or [Gill81]) that if we make a sequence of exact linear searches along any set of conjugate directions $\{\mathbf{p}_1, \mathbf{p}_2, \ldots, \mathbf{p}_n\}$, then the exact minimum of any quadratic function, with n parameters, will be reached in at most n searches. The question is “How can we construct these conjugate search directions?” First, we want to restate the conjugacy condition, which is given in Eq. (9.52), without use of the Hessian matrix. Recall that for quadratic functions

$$\nabla F(\mathbf{x}) = \mathbf{A}\mathbf{x} + \mathbf{d}, \tag{9.54}$$

$$\nabla^2 F(\mathbf{x}) = \mathbf{A}. \tag{9.55}$$

By combining these equations we find that the change in the gradient at iteration $k+1$ is

$$\Delta \mathbf{g}_k = \mathbf{g}_{k+1} - \mathbf{g}_k = (\mathbf{A}\mathbf{x}_{k+1} + \mathbf{d}) - (\mathbf{A}\mathbf{x}_k + \mathbf{d}) = \mathbf{A}\Delta\mathbf{x}_k , \tag{9.56}$$

where, from Eq. (9.2), we have

$$\Delta \mathbf{x}_k = (\mathbf{x}_{k+1} - \mathbf{x}_k) = \alpha_k \mathbf{p}_k , \tag{9.57}$$

and α_k is chosen to minimize $F(\mathbf{x})$ in the direction $\mathbf{p}_k$.

We can now restate the conjugacy conditions (Eq. (9.52)):

$$\alpha_k \mathbf{p}_k^T \mathbf{A} \mathbf{p}_j = \Delta \mathbf{x}_k^T \mathbf{A} \mathbf{p}_j = \Delta \mathbf{g}_k^T \mathbf{p}_j = 0 \qquad k \neq j. \tag{9.58}$$

Note that we no longer need to know the Hessian matrix. We have restated the conjugacy conditions in terms of the changes in the gradient at successive iterations of the algorithm. The search directions will be conjugate if they are orthogonal to the changes in the gradient.

Note that the first search direction, $\mathbf{p}_0$, is arbitrary, and $\mathbf{p}_1$ can be any vector that is orthogonal to $\Delta \mathbf{g}_0$. Therefore there are an infinite number of sets of conjugate vectors. It is common to begin the search in the steepest descent direction:

$$\mathbf{p}_0 = -\mathbf{g}_0 . \tag{9.59}$$

Then, at each iteration we need to construct a vector $\mathbf{p}_k$ that is orthogonal to $\{\Delta \mathbf{g}_0, \Delta \mathbf{g}_1, \ldots, \Delta \mathbf{g}_{k-1}\}$. It is a procedure similar to Gram-Schmidt orthogonalization, which we discussed in Chapter 5. It can be simplified (see [Scal85]) to iterations of the form

$$\mathbf{p}_k = -\mathbf{g}_k + \beta_k \mathbf{p}_{k-1} . \tag{9.60}$$

The scalars β_k can be chosen by several different methods, which produce equivalent results for quadratic functions. The most common choices (see [Scal85]) are

$$\beta_k = \frac{\Delta \mathbf{g}_{k-1}^T \mathbf{g}_k}{\Delta \mathbf{g}_{k-1}^T \mathbf{p}_{k-1}} , \tag{9.61}$$

due to Hestenes and Stiefel,

$$\beta_k = \frac{\mathbf{g}_k^T \mathbf{g}_k}{\mathbf{g}_{k-1}^T \mathbf{g}_{k-1}} \tag{9.62}$$

due to Fletcher and Reeves, and

$$\beta_k = \frac{\Delta\mathbf{g}_{k-1}^T \mathbf{g}_k}{\mathbf{g}_{k-1}^T \mathbf{g}_{k-1}} \tag{9.63}$$

due to Polak and Ribiére.

Conjugate Gradient

To summarize our discussion, the *conjugate gradient* method consists of the following steps:

1. Select the first search direction to be the negative of the gradient, as in Eq. (9.59).
2. Take a step according to Eq. (9.57), selecting the learning rate α_k to minimize the function along the search direction. We will discuss general linear minimization techniques in Chapter 12. For quadratic functions we can use Eq. (9.31).
3. Select the next search direction according to Eq. (9.60), using Eq. (9.61), Eq. (9.62), or Eq. (9.63) to calculate β_k.
4. If the algorithm has not converged, return to step 2.

To illustrate the performance of the algorithm, recall the example we used to demonstrate steepest descent with linear minimization:

$$F(\mathbf{x}) = \frac{1}{2}\mathbf{x}^T \begin{bmatrix} 2 & 1 \\ 1 & 2 \end{bmatrix} \mathbf{x}, \tag{9.64}$$

with initial guess

$$\mathbf{x}_0 = \begin{bmatrix} 0.8 \\ -0.25 \end{bmatrix}. \tag{9.65}$$

The gradient of this function is

$$\nabla F(\mathbf{x}) = \begin{bmatrix} 2x_1 + x_2 \\ x_1 + 2x_2 \end{bmatrix}. \tag{9.66}$$

As with steepest descent, the first search direction is the negative of the gradient:

$$\mathbf{p}_0 = -\mathbf{g}_0 = -\nabla F(\mathbf{x})^T \Big|_{\mathbf{x} = \mathbf{x}_0} = \begin{bmatrix} -1.35 \\ -0.3 \end{bmatrix}. \tag{9.67}$$

From Eq. (9.31), the learning rate for the first iteration will be

$$\alpha_0 = -\frac{\begin{bmatrix} 1.35 & 0.3 \end{bmatrix} \begin{bmatrix} -1.35 \\ -0.3 \end{bmatrix}}{\begin{bmatrix} -1.35 & -0.3 \end{bmatrix} \begin{bmatrix} 2 & 1 \\ 1 & 2 \end{bmatrix} \begin{bmatrix} -1.35 \\ -0.3 \end{bmatrix}} = 0.413 . \quad (9.68)$$

The first step of conjugate gradient is therefore:

$$\mathbf{x}_1 = \mathbf{x}_0 + \alpha_0 \mathbf{p}_0 = \begin{bmatrix} 0.8 \\ -0.25 \end{bmatrix} + 0.413 \begin{bmatrix} -1.35 \\ -0.3 \end{bmatrix} = \begin{bmatrix} 0.24 \\ -0.37 \end{bmatrix} , \quad (9.69)$$

which is equivalent to the first step of steepest descent with minimization along a line.

Now we need to find the second search direction from Eq. (9.60). This requires the gradient at $\mathbf{x}_1$:

$$\mathbf{g}_1 = \nabla F(\mathbf{x})\big|_{\mathbf{x} = \mathbf{x}_1} = \begin{bmatrix} 2 & 1 \\ 1 & 2 \end{bmatrix} \begin{bmatrix} 0.24 \\ -0.37 \end{bmatrix} = \begin{bmatrix} 0.11 \\ -0.5 \end{bmatrix} . \quad (9.70)$$

We can now find β_1:

$$\beta_1 = \frac{\mathbf{g}_1^T \mathbf{g}_1}{\mathbf{g}_0^T \mathbf{g}_0} = \frac{\begin{bmatrix} 0.11 & -0.5 \end{bmatrix} \begin{bmatrix} 0.11 \\ -0.5 \end{bmatrix}}{\begin{bmatrix} 1.35 & 0.3 \end{bmatrix} \begin{bmatrix} 1.35 \\ 0.3 \end{bmatrix}} = \frac{0.2621}{1.9125} = 0.137 , \quad (9.71)$$

using the method of Fletcher and Reeves (Eq. (9.62)). The second search direction is then computed from Eq. (9.60):

$$\mathbf{p}_1 = -\mathbf{g}_1 + \beta_1 \mathbf{p}_0 = \begin{bmatrix} -0.11 \\ 0.5 \end{bmatrix} + 0.137 \begin{bmatrix} -1.35 \\ -0.3 \end{bmatrix} = \begin{bmatrix} -0.295 \\ 0.459 \end{bmatrix} . \quad (9.72)$$

From Eq. (9.31), the learning rate for the second iteration will be

$$\alpha_1 = -\frac{\begin{bmatrix} 0.11 & -0.5 \end{bmatrix} \begin{bmatrix} -0.295 \\ 0.459 \end{bmatrix}}{\begin{bmatrix} -0.295 & 0.459 \end{bmatrix} \begin{bmatrix} 2 & 1 \\ 1 & 2 \end{bmatrix} \begin{bmatrix} -0.295 \\ 0.459 \end{bmatrix}} = \frac{0.262}{0.325} = 0.807 . \quad (9.73)$$

The second step of conjugate gradient is therefore

$$\mathbf{x}_2 = \mathbf{x}_1 + \alpha_1 \mathbf{p}_1 = \begin{bmatrix} 0.24 \\ -0.37 \end{bmatrix} + 0.807 \begin{bmatrix} -0.295 \\ 0.459 \end{bmatrix} = \begin{bmatrix} 0 \\ 0 \end{bmatrix}. \qquad (9.74)$$

As predicted, the algorithm converges exactly to the minimum in two iterations (since this is a two-dimensional quadratic function), as illustrated in Figure 9.10. Compare this result with the steepest descent algorithm, as shown in Figure 9.4. The conjugate gradient algorithm adjusts the second search direction so that it will pass through the minimum of the function (center of the function contours), instead of using an orthogonal search direction, as in steepest descent.

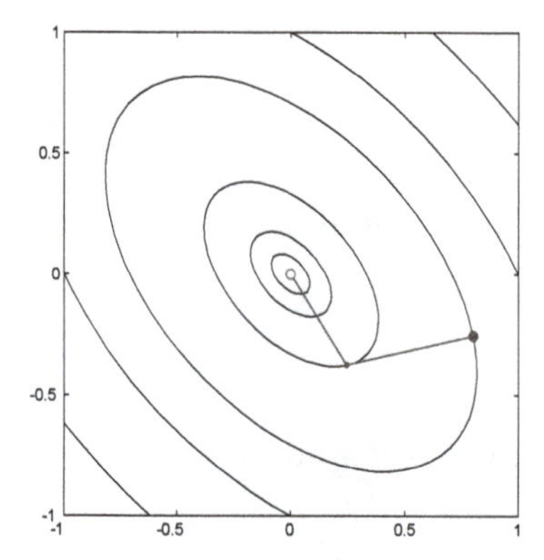

Figure 9.10 Conjugate Gradient Algorithm

We will return to the conjugate gradient algorithm in Chapter 12. In that chapter we will discuss how the algorithm should be adjusted for non-quadratic functions.

To experiment with the conjugate gradient algorithm and compare it with steepest descent, use the Neural Network Design Demonstration Method Comparison (`nnd9mc`).

Summary of Results

General Minimization Algorithm

$$\mathbf{x}_{k+1} = \mathbf{x}_k + \alpha_k \mathbf{p}_k$$

or

$$\Delta\mathbf{x}_k = (\mathbf{x}_{k+1} - \mathbf{x}_k) = \alpha_k \mathbf{p}_k$$

Steepest Descent Algorithm

$$\mathbf{x}_{k+1} = \mathbf{x}_k - \alpha_k \mathbf{g}_k$$

Where $\mathbf{g}_k \equiv \nabla F(\mathbf{x})\big|_{\mathbf{x} = \mathbf{x}_k}$

Stable Learning Rate ($\alpha_k = \alpha$, constant)

$$\alpha < \frac{2}{\lambda_{max}}$$

$\{\lambda_1, \lambda_2, \ldots, \lambda_n\}$ Eigenvalues of Hessian matrix $\mathbf{A}$

Learning Rate to Minimize Along the Line $\mathbf{x}_{k+1} = \mathbf{x}_k + \alpha_k \mathbf{p}_k$

$$\alpha_k = -\frac{\mathbf{g}_k^T \mathbf{p}_k}{\mathbf{p}_k^T \mathbf{A} \mathbf{p}_k} \qquad \text{(For quadratic functions)}$$

After Minimizing Along the Line $\mathbf{x}_{k+1} = \mathbf{x}_k + \alpha_k \mathbf{p}_k$

$$\mathbf{g}_{k+1}^T \mathbf{p}_k = 0$$

Newton's Method

$$\mathbf{x}_{k+1} = \mathbf{x}_k - \mathbf{A}_k^{-1} \mathbf{g}_k$$

Where $\mathbf{A}_k \equiv \nabla^2 F(\mathbf{x})\big|_{\mathbf{x} = \mathbf{x}_k}$

Conjugate Gradient Algorithm

$$\Delta \mathbf{x}_k = \alpha_k \mathbf{p}_k$$

Learning rate α_k is chosen to minimize along the line $\mathbf{x}_{k+1} = \mathbf{x}_k + \alpha_k \mathbf{p}_k$.

$$\mathbf{p}_0 = -\mathbf{g}_0$$

$$\mathbf{p}_k = -\mathbf{g}_k + \beta_k \mathbf{p}_{k-1}$$

$$\beta_k = \frac{\Delta \mathbf{g}_{k-1}^T \mathbf{g}_k}{\Delta \mathbf{g}_{k-1}^T \mathbf{p}_{k-1}} \quad \text{or} \quad \beta_k = \frac{\mathbf{g}_k^T \mathbf{g}_k}{\mathbf{g}_{k-1}^T \mathbf{g}_{k-1}} \quad \text{or} \quad \beta_k = \frac{\Delta \mathbf{g}_{k-1}^T \mathbf{g}_k}{\mathbf{g}_{k-1}^T \mathbf{g}_{k-1}}$$

Where $\mathbf{g}_k \equiv \nabla F(\mathbf{x})\big|_{\mathbf{x} = \mathbf{x}_k}$ and $\Delta \mathbf{g}_k = \mathbf{g}_{k+1} - \mathbf{g}_k$.

Solved Problems

P9.1 We want to find the minimum of the following function:

$$F(\mathbf{x}) = 5x_1^2 - 6x_1x_2 + 5x_2^2 + 4x_1 + 4x_2 .$$

i. **Sketch a contour plot of this function.**

ii. **Sketch the trajectory of the steepest descent algorithm on the contour plot of part (i) if the initial guess is $\mathbf{x}_0 = \begin{bmatrix} -1 & -2.5 \end{bmatrix}^T$. Assume a very small learning rate is used.**

iii. **What is the maximum stable learning rate?**

i. To sketch the contour plot we first need to find the Hessian matrix. For quadratic functions we can do this by putting the function into the standard form (see Eq. (8.35)):

$$F(\mathbf{x}) = \frac{1}{2}\mathbf{x}^T\mathbf{A}\mathbf{x} + \mathbf{d}^T\mathbf{x} + c = \frac{1}{2}\mathbf{x}^T\begin{bmatrix} 10 & -6 \\ -6 & 10 \end{bmatrix}\mathbf{x} + \begin{bmatrix} 4 & 4 \end{bmatrix}\mathbf{x} .$$

From Eq. (8.39) the Hessian matrix is

$$\nabla^2 F(\mathbf{x}) = \mathbf{A} = \begin{bmatrix} 10 & -6 \\ -6 & 10 \end{bmatrix} .$$

The eigenvalues and eigenvectors of this matrix are

$$\lambda_1 = 4 ,\ \mathbf{z}_1 = \begin{bmatrix} 1 \\ 1 \end{bmatrix} ,\ \lambda_2 = 16 ,\ \mathbf{z}_2 = \begin{bmatrix} 1 \\ -1 \end{bmatrix} .$$

From the discussion on quadratic functions in Chapter 8 (see page 8-15) we know that the function contours are elliptical. The maximum curvature of $F(\mathbf{x})$ is in the direction of $\mathbf{z}_2$, since λ_2 is larger than λ_1, and the minimum curvature is in the direction of $\mathbf{z}_1$ (the long axis of the ellipses).

Next we need to find the center of the contours (the stationary point). This occurs when the gradient is equal to zero. From Eq. (8.38) we find

$$\nabla F(\mathbf{x}) = \mathbf{A}\mathbf{x} + \mathbf{d} = \begin{bmatrix} 10 & -6 \\ -6 & 10 \end{bmatrix}\mathbf{x} + \begin{bmatrix} 4 \\ 4 \end{bmatrix} = \begin{bmatrix} 0 \\ 0 \end{bmatrix} .$$

Therefore

$$\mathbf{x}^* = -\begin{bmatrix} 10 & -6 \\ -6 & 10 \end{bmatrix}^{-1} \begin{bmatrix} 4 \\ 4 \end{bmatrix} = \begin{bmatrix} -1 \\ -1 \end{bmatrix}.$$

The contours will be elliptical, centered at $\mathbf{x}^*$, with long axis in the direction of $\mathbf{z}_1$. The contour plot is shown in Figure P9.1.

ii. We know that the gradient is always orthogonal to the contour line, therefore the steepest descent trajectory, if we take small enough steps, will follow a path that is orthogonal to each contour line it intersects. We can therefore trace the trajectory without performing any computations. The result is shown in Figure P9.1.

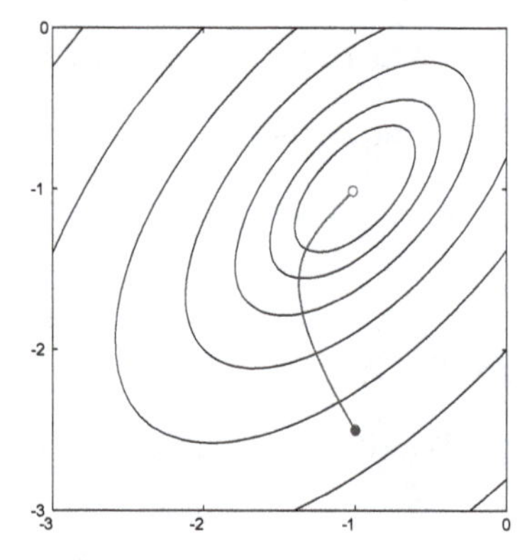

Figure P9.1 Contour Plot and Steep. Desc. Trajectory for Problem P9.1

iii. From Eq. (9.25) we know that the maximum stable learning rate for a quadratic function is determined by the maximum eigenvalue of the Hessian matrix:

$$\alpha < \frac{2}{\lambda_{max}}.$$

The maximum eigenvalue for this problem is $\lambda_2 = 16$, therefore for stability

$$\alpha < \frac{2}{16} = 0.125.$$

This result is verified experimentally in Figure P9.2, which shows the steepest descent trajectories when the learning rate is just below ($\alpha = 0.12$) and just above ($\alpha = 0.13$) the maximum stable value.

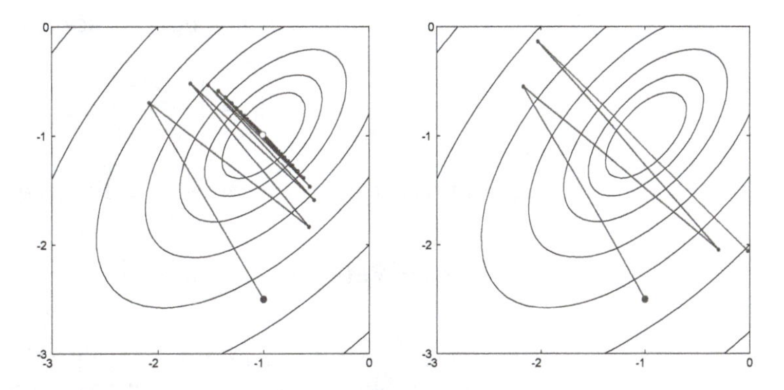

Figure P9.2 Trajectories for $\alpha = 0.12$ (left) and $\alpha = 0.13$ (right)

P9.2 Consider again the quadratic function of Problem P9.1. Take two steps of the steepest descent algorithm, minimizing along a line at each step. Use the following initial condition:

$$\mathbf{x}_0 = \begin{bmatrix} 0 & -2 \end{bmatrix}^T .$$

In Problem P9.1 we found the gradient of the function to be

$$\nabla F(\mathbf{x}) = \mathbf{A}\mathbf{x} + \mathbf{d} = \begin{bmatrix} 10 & -6 \\ -6 & 10 \end{bmatrix} \mathbf{x} + \begin{bmatrix} 4 \\ 4 \end{bmatrix} .$$

If we evaluate this at $\mathbf{x}_0$, we find

$$\mathbf{g}_0 = \nabla F(\mathbf{x}_0) = \mathbf{A}\mathbf{x}_0 + \mathbf{d} = \begin{bmatrix} 10 & -6 \\ -6 & 10 \end{bmatrix} \begin{bmatrix} 0 \\ -2 \end{bmatrix} + \begin{bmatrix} 4 \\ 4 \end{bmatrix} = \begin{bmatrix} 16 \\ -16 \end{bmatrix} .$$

Therefore the first search direction is

$$\mathbf{p}_0 = -\mathbf{g}_0 = \begin{bmatrix} -16 \\ 16 \end{bmatrix} .$$

To minimize along a line, for a quadratic function, we can use Eq. (9.31):

$$\alpha_0 = -\frac{\mathbf{g}_0^T \mathbf{p}_0}{\mathbf{p}_0^T \mathbf{A} \mathbf{p}_0} = -\frac{\begin{bmatrix} 16 & -16 \end{bmatrix} \begin{bmatrix} -16 \\ 16 \end{bmatrix}}{\begin{bmatrix} -16 & 16 \end{bmatrix} \begin{bmatrix} 10 & -6 \\ -6 & 10 \end{bmatrix} \begin{bmatrix} -16 \\ 16 \end{bmatrix}} = -\frac{-512}{8192} = 0.0625 .$$

Therefore the first iteration of steepest descent will be

$$\mathbf{x}_1 = \mathbf{x}_0 - \alpha_0 \mathbf{g}_0 = \begin{bmatrix} 0 \\ -2 \end{bmatrix} - 0.0625 \begin{bmatrix} 16 \\ -16 \end{bmatrix} = \begin{bmatrix} -1 \\ -1 \end{bmatrix}.$$

To begin the second iteration we need to find the gradient at $\mathbf{x}_1$:

$$\mathbf{g}_1 = \nabla F(\mathbf{x}_1) = \mathbf{A}\mathbf{x}_1 + \mathbf{d} = \begin{bmatrix} 10 & -6 \\ -6 & 10 \end{bmatrix} \begin{bmatrix} -1 \\ -1 \end{bmatrix} + \begin{bmatrix} 4 \\ 4 \end{bmatrix} = \begin{bmatrix} 0 \\ 0 \end{bmatrix}.$$

Therefore we have reached a stationary point; the algorithm has converged. From Problem P9.1 we know that $\mathbf{x}_1$ is indeed the minimum point of this quadratic function. The trajectory is shown in Figure P9.3.

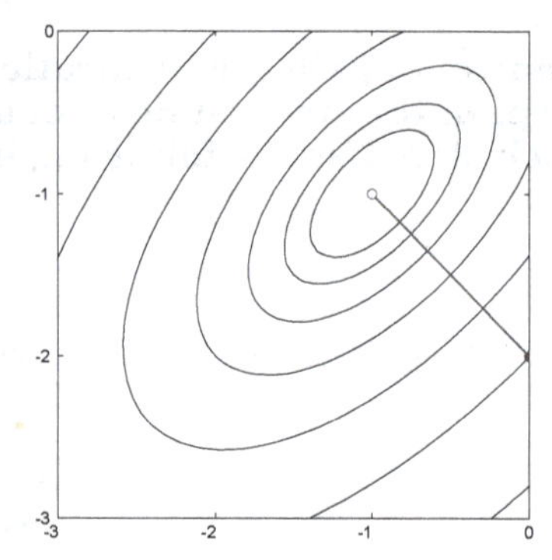

Figure P9.3 Steepest Descent with Linear Minimization for Problem P9.2

This is an unusual case, where the steepest descent algorithm located the minimum in one iteration. Notice that this occurred because the initial guess was located in the direction of one of the eigenvectors of the Hessian matrix, with respect to the minimum point. For those cases where every direction is an eigenvector, the steepest descent algorithm will always locate the minimum in one iteration. What would this imply about the eigenvalues of the Hessian matrix?

P9.3 Recall Problem P8.6, in which we derived a performance index for a linear neural network. The network, which is displayed again in Figure P9.4, was to be trained for the following input/output pairs:

$$\{(p_1 = 2), (t_1 = 0.5)\}, \{(p_2 = -1), (t_2 = 0)\}$$

The performance index for the network was defined to be

$$F(\mathbf{x}) = (t_1 - a_1(\mathbf{x}))^2 + (t_2 - a_2(\mathbf{x}))^2,$$

which was displayed in Figure P8.8.

i. Use the steepest descent algorithm to locate the optimal parameters for this network (recall that $\mathbf{x} = \begin{bmatrix} w & b \end{bmatrix}^T$), starting from the initial guess $\mathbf{x}_0 = \begin{bmatrix} 1 & 1 \end{bmatrix}^T$. Use a learning rate of $\alpha = 0.05$.

ii. What is the maximum stable learning rate?

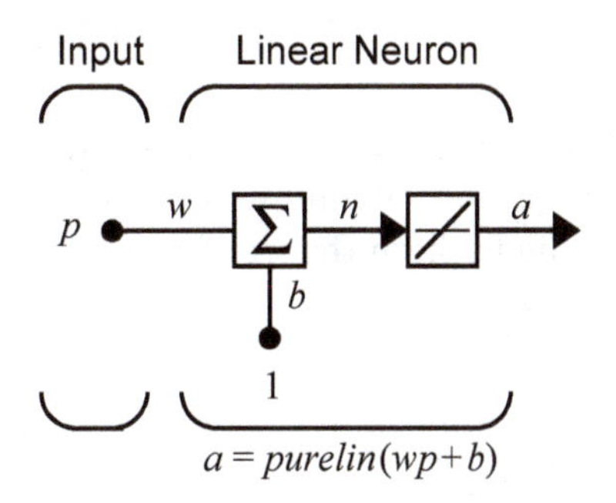

Figure P9.4 Linear Network for Problems P9.3 and P8.6

i. In Problem P8.6 we found that the performance index could be written in quadratic form:

$$F(\mathbf{x}) = \frac{1}{2}\mathbf{x}^T\mathbf{A}\mathbf{x} + \mathbf{d}^T\mathbf{x} + c,$$

where

$$c = \mathbf{t}^T\mathbf{t} = \begin{bmatrix} 0.5 & 0 \end{bmatrix}\begin{bmatrix} 0.5 \\ 0 \end{bmatrix} = 0.25,$$

$$\mathbf{d} = -2\mathbf{G}^T\mathbf{t} = -2\begin{bmatrix} 2 & -1 \\ 1 & 1 \end{bmatrix}\begin{bmatrix} 0.5 \\ 0 \end{bmatrix} = \begin{bmatrix} -2 \\ -1 \end{bmatrix},$$

$$\mathbf{A} = 2\mathbf{G}^T\mathbf{G} = \begin{bmatrix} 10 & 2 \\ 2 & 4 \end{bmatrix}.$$

The gradient at $\mathbf{x}_0$ is

$$\mathbf{g}_0 = \nabla F(\mathbf{x}_0) = \mathbf{A}\mathbf{x}_0 + \mathbf{d} = \begin{bmatrix} 10 & 2 \\ 2 & 4 \end{bmatrix}\begin{bmatrix} 1 \\ 1 \end{bmatrix} + \begin{bmatrix} -2 \\ -1 \end{bmatrix} = \begin{bmatrix} 10 \\ 5 \end{bmatrix}.$$

The first iteration of steepest descent will be

$$\mathbf{x}_1 = \mathbf{x}_0 - \alpha \mathbf{g}_0 = \begin{bmatrix} 1 \\ 1 \end{bmatrix} - 0.05 \begin{bmatrix} 10 \\ 5 \end{bmatrix} = \begin{bmatrix} 0.5 \\ 0.75 \end{bmatrix}.$$

The second iteration will be

$$\mathbf{x}_2 = \mathbf{x}_1 - \alpha \mathbf{g}_1 = \begin{bmatrix} 0.5 \\ 0.75 \end{bmatrix} - 0.05 \begin{bmatrix} 4.5 \\ 3 \end{bmatrix} = \begin{bmatrix} 0.275 \\ 0.6 \end{bmatrix}.$$

The remaining iterations are displayed in Figure P9.5. The algorithm converges to the minimum point $\mathbf{x}^* = \begin{bmatrix} 0.167 & 0.167 \end{bmatrix}^T$. Therefore the optimal value for both the weight and the bias of this network is 0.167.

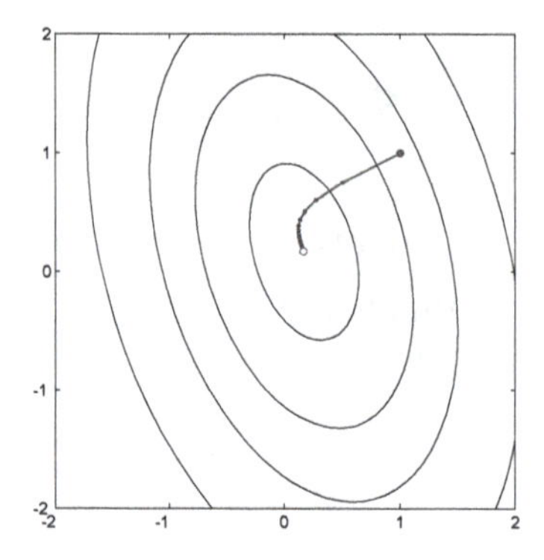

Figure P9.5 Steepest Descent Trajectory for Problem P9.3 with $\alpha = 0.05$

Note that in order to train this network we needed to know all of the input/output pairs. We then performed iterations of the steepest descent algorithm until convergence was achieved. In Chapter 10 we will introduce an adaptive algorithm, based on steepest descent, for training linear networks. With this adaptive algorithm the network parameters are updated after each input/output pair is presented. We will show how this allows the network to adapt to a changing environment.

ii. The maximum eigenvalue of the Hessian matrix for this problem is $\lambda_1 = 10.6$ (see Problem P8.6), therefore for stability

$$\alpha < \frac{2}{10.6} = 0.1887.$$

P9.4 Consider the function

$$F(\mathbf{x}) = e^{(x_1^2 - x_1 + 2x_2^2 + 4)}.$$

Take one iteration of Newton's method from the initial guess $\mathbf{x}_0 = \begin{bmatrix} 1 & -2 \end{bmatrix}^T$. How close is this result to the minimum point of $F(\mathbf{x})$? Explain.

The first step is to find the gradient and the Hessian matrix. The gradient is given by

$$\nabla F(\mathbf{x}) = \begin{bmatrix} \frac{\partial}{\partial x_1} F(\mathbf{x}) \\ \frac{\partial}{\partial x_2} F(\mathbf{x}) \end{bmatrix} = e^{(x_1^2 - x_1 + 2x_2^2 + 4)} \begin{bmatrix} (2x_1 - 1) \\ (4x_2) \end{bmatrix},$$

and the Hessian matrix is given by

$$\nabla^2 F(\mathbf{x}) = \begin{bmatrix} \frac{\partial^2}{\partial x_1^2} F(\mathbf{x}) & \frac{\partial^2}{\partial x_1 \partial x_2} F(\mathbf{x}) \\ \frac{\partial^2}{\partial x_2 \partial x_1} F(\mathbf{x}) & \frac{\partial^2}{\partial x_2^2} F(\mathbf{x}) \end{bmatrix}$$

$$= e^{(x_1^2 - x_1 + 2x_2^2 + 4)} \begin{bmatrix} 4x_1^2 - 4x_1 + 3 & (2x_1 - 1)(4x_2) \\ (2x_1 - 1)(4x_2) & 16x_2^2 + 4 \end{bmatrix}$$

If we evaluate these at the initial guess we find

$$\mathbf{g}_0 = \nabla F(\mathbf{x})\big|_{\mathbf{x} = \mathbf{x}_0} = \begin{bmatrix} 0.163 \times 10^6 \\ -1.302 \times 10^6 \end{bmatrix},$$

and

$$\mathbf{A}_0 = \nabla^2 F(\mathbf{x})\big|_{\mathbf{x} = \mathbf{x}_0} = \begin{bmatrix} 0.049 \times 10^7 & -0.130 \times 10^7 \\ -0.130 \times 10^7 & 1.107 \times 10^7 \end{bmatrix}.$$

Therefore the first iteration of Newton's method, from Eq. (9.43), will be

$$\mathbf{x}_1 = \mathbf{x}_0 - \mathbf{A}_0^{-1}\mathbf{g}_0 = \begin{bmatrix} 1 \\ -2 \end{bmatrix} - \begin{bmatrix} 0.049\times10^7 & -0.130\times10^7 \\ -0.130\times10^7 & 1.107\times10^7 \end{bmatrix}^{-1} \begin{bmatrix} 0.163\times10^6 \\ -1.302\times10^6 \end{bmatrix} = \begin{bmatrix} 0.971 \\ -1.886 \end{bmatrix}$$

How close is this to the true minimum point of $F(\mathbf{x})$? First, note that the exponent of $F(\mathbf{x})$ is a quadratic function:

$$x_1^2 - x_1 + 2x_2^2 + 4 = \frac{1}{2}\mathbf{x}^T\mathbf{A}\mathbf{x} + \mathbf{d}^T\mathbf{x} + c = \frac{1}{2}\mathbf{x}^T\begin{bmatrix} 2 & 0 \\ 0 & 4 \end{bmatrix}\mathbf{x} + \begin{bmatrix} -1 & 0 \end{bmatrix}\mathbf{x} + 4 .$$

The minimum point of $F(\mathbf{x})$ will be the same as the minimum point of the exponent, which is

$$\mathbf{x}^* = -\mathbf{A}^{-1}\mathbf{d} = -\begin{bmatrix} 2 & 0 \\ 0 & 4 \end{bmatrix}^{-1}\begin{bmatrix} -1 \\ 0 \end{bmatrix} = \begin{bmatrix} 0.5 \\ 0 \end{bmatrix} .$$

Therefore Newton's method has taken only a very small step toward the true minimum point. This is because $F(\mathbf{x})$ cannot be accurately approximated by a quadratic function in the neighborhood of $\mathbf{x}_0 = \begin{bmatrix} 1 & -2 \end{bmatrix}^T$.

For this problem Newton's method will converge to the true minimum point, but it will take many iterations. The trajectory for Newton's method is illustrated in Figure P9.6.

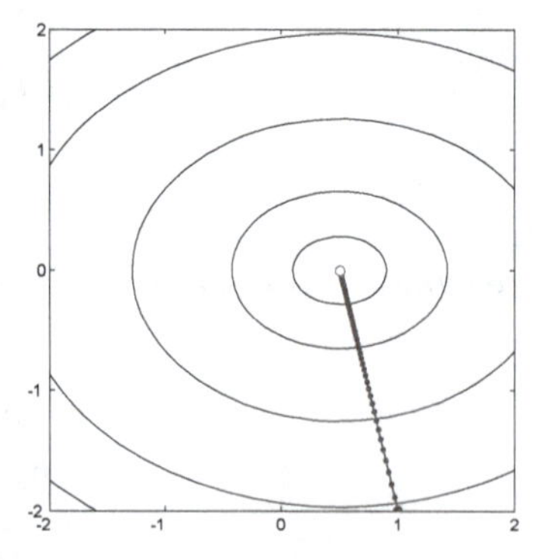

Figure P9.6 Newton's Method Trajectory for Problem P9.4

P9.5 Compare the performance of Newton's method and steepest descent on the following function:

$$F(\mathbf{x}) = \frac{1}{2}\mathbf{x}^T\begin{bmatrix} 1 & -1 \\ -1 & 1 \end{bmatrix}\mathbf{x}.$$

Start from the initial guess

$$\mathbf{x}_0 = \begin{bmatrix} 1 \\ 0 \end{bmatrix}.$$

Recall that this function is an example of a stationary valley (see Eq. (8.59) and Figure 8.9). The gradient is

$$\nabla F(\mathbf{x}) = \mathbf{A}\mathbf{x} + \mathbf{d} = \begin{bmatrix} 1 & -1 \\ -1 & 1 \end{bmatrix}\mathbf{x}$$

and the Hessian matrix is

$$\nabla^2 F(\mathbf{x}) = \mathbf{A} = \begin{bmatrix} 1 & -1 \\ -1 & 1 \end{bmatrix}.$$

Newton's method is given by

$$\mathbf{x}_{k+1} = \mathbf{x}_k - \mathbf{A}_k^{-1}\mathbf{g}_k.$$

Note, however, that we cannot actually perform this algorithm, because the Hessian matrix is singular. We know from our discussion of this function in Chapter 8 that this function does not have a strong minimum, but it does have a weak minimum along the line $x_1 = x_2$.

What about steepest descent? If we start from the initial guess, with learning rate $\alpha = 0.1$, the first two iterations will be

$$\mathbf{x}_1 = \mathbf{x}_0 - \alpha\mathbf{g}_0 = \begin{bmatrix} 1 \\ 0 \end{bmatrix} - 0.1\begin{bmatrix} 1 \\ -1 \end{bmatrix} = \begin{bmatrix} 0.9 \\ 0.1 \end{bmatrix},$$

$$\mathbf{x}_2 = \mathbf{x}_1 - \alpha\mathbf{g}_1 = \begin{bmatrix} 0.9 \\ 0.1 \end{bmatrix} - 0.1\begin{bmatrix} 0.8 \\ -0.8 \end{bmatrix} = \begin{bmatrix} 0.82 \\ 0.18 \end{bmatrix}.$$

The complete trajectory is shown in Figure P9.7. This is a case where the steepest descent algorithm performs better than Newton's method. Steepest descent converges to a minimum point (weak minimum), while Newton's method fails to converge. In Chapter 12 we will discuss a technique that combines steepest descent with Newton's method, to overcome the problem of singular (or almost singular) Hessian matrices.

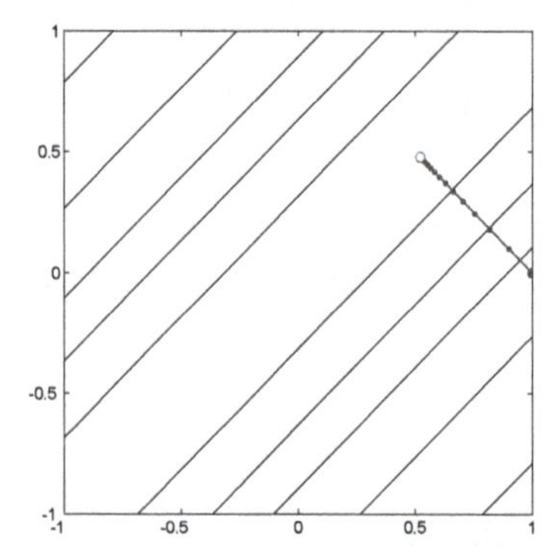

Figure P9.7 Steepest Descent Trajectory for Problem P9.5 with $\alpha = 0.1$

P9.6 Consider the following function:

$$F(\mathbf{x}) = x_1^3 + x_1 x_2 - x_1^2 x_2^2$$

i. **Perform one iteration of Newton's method from the initial guess $\mathbf{x}_0 = \begin{bmatrix} 1 & 1 \end{bmatrix}^T$.**

ii. **Find the second-order Taylor series expansion of $F(\mathbf{x})$ about $\mathbf{x}_0$. Is this quadratic function minimized at the point $\mathbf{x}_1$ found in part (i)? Explain.**

i. The gradient of $F(\mathbf{x})$ is

$$\nabla F(\mathbf{x}) = \begin{bmatrix} \frac{\partial}{\partial x_1} F(\mathbf{x}) \\ \frac{\partial}{\partial x_2} F(\mathbf{x}) \end{bmatrix} = \begin{bmatrix} 3x_1^2 + x_2 - 2x_1 x_2^2 \\ x_1 - 2x_1^2 x_2 \end{bmatrix},$$

and the Hessian matrix is

$$\nabla^2 F(\mathbf{x}) = \begin{bmatrix} 6x_1 - 2x_2^2 & 1 - 4x_1x_2 \\ 1 - 4x_1x_2 & -2x_1^2 \end{bmatrix}.$$

If we evaluate these at the initial guess we find

$$\mathbf{g}_0 = \nabla F(\mathbf{x})\big|_{\mathbf{x} = \mathbf{x}_0} = \begin{bmatrix} 2 \\ -1 \end{bmatrix},$$

$$\mathbf{A}_0 = \nabla^2 F(\mathbf{x})\big|_{\mathbf{x} = \mathbf{x}_0} = \begin{bmatrix} 4 & -3 \\ -3 & -2 \end{bmatrix}.$$

The first iteration of Newton's method is then

$$\mathbf{x}_1 = \mathbf{x}_0 - \mathbf{A}_0^{-1}\mathbf{g}_0 = \begin{bmatrix} 1 \\ 1 \end{bmatrix} - \begin{bmatrix} 4 & -3 \\ -3 & -2 \end{bmatrix}^{-1} \begin{bmatrix} 2 \\ -1 \end{bmatrix} = \begin{bmatrix} 0.5882 \\ 1.1176 \end{bmatrix}.$$

ii. From Eq. (9.40), the second-order Taylor series expansion of $F(\mathbf{x})$ about $\mathbf{x}_0$ is

$$F(\mathbf{x}) = F(\mathbf{x}_0 + \Delta\mathbf{x}_0) \approx F(\mathbf{x}_0) + \mathbf{g}_0^T \Delta\mathbf{x}_0 + \frac{1}{2}\Delta\mathbf{x}_0^T \mathbf{A}_0 \Delta\mathbf{x}_0 .$$

If we substitute the values for $\mathbf{x}_0$, $\mathbf{g}_0$ and $\mathbf{A}_0$, we find

$$F(\mathbf{x}) \approx 1 + \begin{bmatrix} 2 & -1 \end{bmatrix} \left\{ \mathbf{x} - \begin{bmatrix} 1 \\ 1 \end{bmatrix} \right\} + \frac{1}{2} \left\{ \mathbf{x} - \begin{bmatrix} 1 \\ 1 \end{bmatrix} \right\}^T \begin{bmatrix} 4 & -3 \\ -3 & -2 \end{bmatrix} \left\{ \mathbf{x} - \begin{bmatrix} 1 \\ 1 \end{bmatrix} \right\}.$$

This can be reduced to

$$F(\mathbf{x}) \approx -2 + \begin{bmatrix} 1 & 4 \end{bmatrix} \mathbf{x} + \frac{1}{2}\mathbf{x}^T \begin{bmatrix} 4 & -3 \\ -3 & -2 \end{bmatrix} \mathbf{x} .$$

This function has a stationary point at $\mathbf{x}_1$. The question is whether or not the stationary point is a strong minimum. This can be determined from the eigenvalues of the Hessian matrix. If both eigenvalues are positive, it is a strong minimum. If both eigenvalues are negative, it is a strong maximum. If the two eigenvalues have opposite signs, it is a saddle point. In this case the eigenvalues of $\mathbf{A}_0$ are

$$\lambda_1 = 5.24 \text{ and } \lambda_2 = -3.24 .$$

Therefore the quadratic approximation to $F(\mathbf{x})$ at $\mathbf{x}_0$ is not minimized at $\mathbf{x}_1$, since it is a saddle point. Figure P9.8 displays the contour plots of $F(\mathbf{x})$ and its quadratic approximation.

This sort of problem was also illustrated in Figure 9.8 and Figure 9.9. Newton's method does locate the stationary point of the quadratic approximation of the function at the current guess. It does not distinguish between minima, maxima and saddle points.

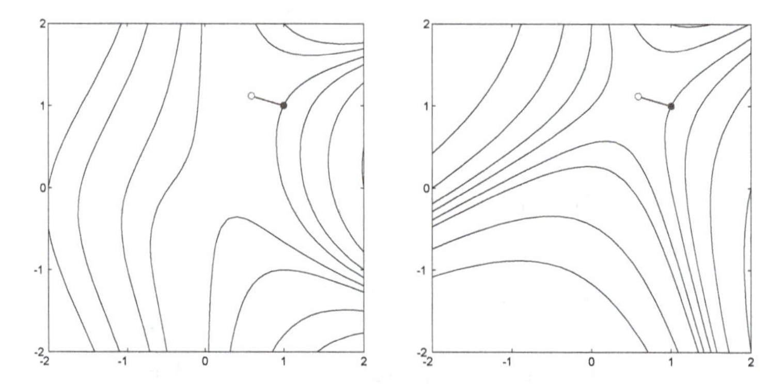

Figure P9.8 One Iteration of Newton's Method from $\mathbf{x}_0 = \begin{bmatrix} 1 & 1 \end{bmatrix}^T$

P9.7 Repeat Problem P9.3 (i) using the conjugate gradient algorithm.

Recall that the function to be minimized was

$$F(\mathbf{x}) = 0.25 + \begin{bmatrix} -2 & -1 \end{bmatrix}\mathbf{x} + \frac{1}{2}\mathbf{x}^T\begin{bmatrix} 10 & 2 \\ 2 & 4 \end{bmatrix}\mathbf{x}.$$

The gradient at $\mathbf{x}_0$ is

$$\mathbf{g}_0 = \nabla F(\mathbf{x}_0) = \mathbf{A}\mathbf{x}_0 + \mathbf{d} = \begin{bmatrix} 10 & 2 \\ 2 & 4 \end{bmatrix}\begin{bmatrix} 1 \\ 1 \end{bmatrix} + \begin{bmatrix} -2 \\ -1 \end{bmatrix} = \begin{bmatrix} 10 \\ 5 \end{bmatrix}.$$

The first search direction is then

$$\mathbf{p}_0 = -\mathbf{g}_0 = \begin{bmatrix} -10 \\ -5 \end{bmatrix}.$$

To minimize along a line, for a quadratic function, we can use Eq. (9.31):

$$\alpha_0 = -\frac{\mathbf{g}_0^T \mathbf{p}_0}{\mathbf{p}_0^T \mathbf{A} \mathbf{p}_0} = -\frac{\begin{bmatrix} 10 & 5 \end{bmatrix} \begin{bmatrix} -10 \\ -5 \end{bmatrix}}{\begin{bmatrix} -10 & -5 \end{bmatrix} \begin{bmatrix} 10 & 2 \\ 2 & 4 \end{bmatrix} \begin{bmatrix} -10 \\ -5 \end{bmatrix}} = -\frac{-125}{1300} = 0.0962 .$$

Therefore the first iteration of conjugate gradient will be

$$\mathbf{x}_1 = \mathbf{x}_0 + \alpha_0 \mathbf{p}_0 = \begin{bmatrix} 1 \\ 1 \end{bmatrix} + 0.0962 \begin{bmatrix} -10 \\ -5 \end{bmatrix} = \begin{bmatrix} 0.038 \\ 0.519 \end{bmatrix} .$$

Now we need to find the second search direction from Eq. (9.60). This requires the gradient at $\mathbf{x}_1$:

$$\mathbf{g}_1 = \nabla F(\mathbf{x})\big|_{\mathbf{x} = \mathbf{x}_1} = \begin{bmatrix} 10 & 2 \\ 2 & 4 \end{bmatrix} \begin{bmatrix} 0.038 \\ 0.519 \end{bmatrix} + \begin{bmatrix} -2 \\ -1 \end{bmatrix} = \begin{bmatrix} -0.577 \\ 1.154 \end{bmatrix} .$$

We can now find β_1:

$$\beta_1 = \frac{\Delta \mathbf{g}_0^T \mathbf{g}_1}{\mathbf{g}_0^T \mathbf{g}_0} = \frac{\begin{bmatrix} -10.577 & -3.846 \end{bmatrix} \begin{bmatrix} -0.577 \\ 1.154 \end{bmatrix}}{\begin{bmatrix} 10 & 5 \end{bmatrix} \begin{bmatrix} 10 \\ 5 \end{bmatrix}} = \frac{1.665}{125} = 0.0133 ,$$

using the method of Polak and Ribiére (Eq. (9.63)). (The other two methods for computing β_1 will produce the same results for a quadratic function. You may want to try them.) The second search direction is then computed from Eq. (9.60):

$$\mathbf{p}_1 = -\mathbf{g}_1 + \beta_1 \mathbf{p}_0 = \begin{bmatrix} 0.577 \\ -1.154 \end{bmatrix} + 0.0133 \begin{bmatrix} -10 \\ -5 \end{bmatrix} = \begin{bmatrix} 0.444 \\ -1.220 \end{bmatrix} .$$

From Eq. (9.31), the learning rate for the second iteration will be

$$\alpha_1 = -\frac{\begin{bmatrix} -0.577 & 1.154 \end{bmatrix} \begin{bmatrix} 0.444 \\ -1.220 \end{bmatrix}}{\begin{bmatrix} 0.444 & -1.220 \end{bmatrix} \begin{bmatrix} 10 & 2 \\ 2 & 4 \end{bmatrix} \begin{bmatrix} 0.444 \\ -1.220 \end{bmatrix}} = -\frac{-1.664}{5.758} = 0.2889 .$$

The second step of conjugate gradient is therefore

$$\mathbf{x}_2 = \mathbf{x}_1 + \alpha_1 \mathbf{p}_1 = \begin{bmatrix} 0.038 \\ 0.519 \end{bmatrix} + 0.2889 \begin{bmatrix} 0.444 \\ -1.220 \end{bmatrix} = \begin{bmatrix} 0.1667 \\ 0.1667 \end{bmatrix}.$$

As expected, the minimum is reached in two iterations. The trajectory is illustrated in Figure P9.9.

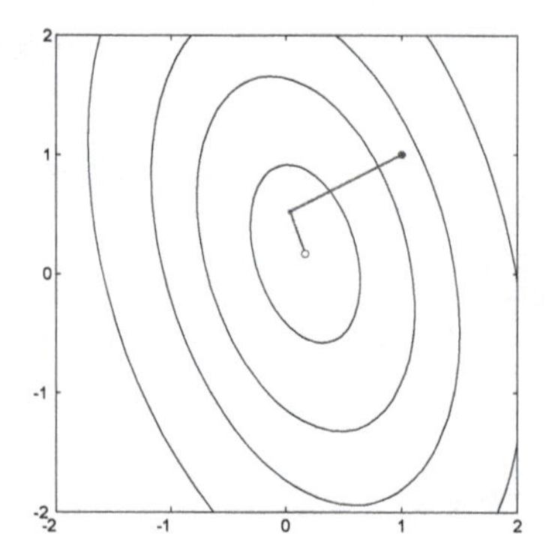

Figure P9.9 Conjugate Gradient Trajectory for Problem P9.7

P9.8 Show that conjugate vectors are independent.

Suppose that we have a set of vectors, $\{\mathbf{p}_0, \mathbf{p}_1, \ldots, \mathbf{p}_{n-1}\}$, which are conjugate with respect to the Hessian matrix $\mathbf{A}$. If these vectors are dependent, then, from Eq. (5.4), it must be true that

$$\sum_{j=0}^{n-1} a_j \mathbf{p}_j = \mathbf{0},$$

for some set of constants $a_0, a_1, \ldots, a_{n-1}$, at least one of which is nonzero. If we multiply both sides of this equation by $\mathbf{p}_k^T \mathbf{A}$, we obtain

$$\mathbf{p}_k^T \mathbf{A} \sum_{j=0}^{n-1} a_j \mathbf{p}_j = \sum_{j=0}^{n-1} a_j \mathbf{p}_k^T \mathbf{A} \mathbf{p}_j = a_k \mathbf{p}_k^T \mathbf{A} \mathbf{p}_k = 0,$$

where the second equality comes from the definition of conjugate vectors in Eq. (9.52). If $\mathbf{A}$ is positive definite (a unique strong minimum exists), then $\mathbf{p}_k^T \mathbf{A} \mathbf{p}_k$ must be strictly positive. This implies that a_k must be zero for all k. Therefore conjugate directions must be independent.

Epilogue

In this chapter we have introduced three different optimization algorithms: steepest descent, Newton's method and conjugate gradient. The basis for these algorithms is the Taylor series expansion. Steepest descent is derived by using a first-order expansion, whereas Newton's method and conjugate gradient are designed for second-order (quadratic) functions.

Steepest descent has the advantage that it is very simple, requiring calculation only of the gradient. It is also guaranteed to converge to a stationary point if the learning rate is small enough. The disadvantage of steepest descent is that training times are generally longer than for other algorithms. This is especially true when the eigenvalues of the Hessian matrix, for quadratic functions, have a wide range of magnitudes.

Newton's method is generally much faster than steepest descent. For quadratic functions it will locate a stationary point in one iteration. One disadvantage is that it requires calculation and storage of the Hessian matrix, as well as its inverse. In addition, the convergence properties of Newton's method are quite complex. In Chapter 12 we will introduce a modification of Newton's method that overcomes some of the disadvantages of the standard algorithm.

The conjugate gradient algorithm is something of a compromise between steepest descent and Newton's method. It will locate the minimum of a quadratic function in a finite number of iterations, but it does not require calculation and storage of the Hessian matrix. It is well suited to problems with large numbers of parameters, where it is impractical to compute and store the Hessian.

In later chapters we will apply each of these optimization algorithms to the training of neural networks. In Chapter 10 we will demonstrate how an approximate steepest descent algorithm, Widrow-Hoff learning, can be used to train linear networks. In Chapter 11 we generalize Widrow-Hoff learning to train multilayer networks. In Chapter 12 the conjugate gradient algorithm, and a variation of Newton's method, are used to speed up the training of multilayer networks.

Further Reading

[Batt92] R. Battiti, "First and Second Order Methods for Learning: Between Steepest Descent and Newton's Method," *Neural Computation*, Vol. 4, No. 2, pp. 141-166, 1992.

This article reviews the latest developments in unconstrained optimization using first and second derivatives. The techniques discussed are those that are most suitable for neural network applications.

[Brog91] W. L. Brogan, *Modern Control Theory,* 3rd Ed., Englewood Cliffs, NJ: Prentice-Hall, 1991.

This is a well-written book on the subject of linear systems. The first half of the book is devoted to linear algebra. It also has good sections on the solution of linear differential equations and the stability of linear and nonlinear systems. It has many worked problems.

[Gill81] P. E. Gill, W. Murray and M. H. Wright, *Practical Optimization*, New York: Academic Press, 1981.

As the title implies, this text emphasizes the practical implementation of optimization algorithms. It provides motivation for the optimization methods, as well as details of implementation that affect algorithm performance.

[Himm72] D. M. Himmelblau, *Applied Nonlinear Programming*, New York: McGraw-Hill, 1972.

This is a comprehensive text on nonlinear optimization. It covers both constrained and unconstrained optimization problems. The text is very complete, with many examples worked out in detail.

[Scal85] L. E. Scales, *Introduction to Non-Linear Optimization*, New York: Springer-Verlag, 1985.

A very readable text describing the major optimization algorithms, this text emphasizes methods of optimization rather than existence theorems and proofs of convergence. Algorithms are presented with intuitive explanations, along with illustrative figures and examples. Pseudo-code is presented for most algorithms.

Exercises

E9.1 In Problem P9.1 we found the maximum stable learning rate for the steepest descent algorithm when applied to a particular quadratic function. Will the algorithm always diverge when a larger learning rate is used, or are there any conditions for which the algorithm will still converge?

E9.2 We want to find the minimum of the following function:

$$F(\mathbf{x}) = \frac{1}{2}\mathbf{x}^T \begin{bmatrix} 6 & -2 \\ -2 & 6 \end{bmatrix} \mathbf{x} + \begin{bmatrix} -1 & -1 \end{bmatrix} \mathbf{x} .$$

i. Sketch a contour plot of this function.

ii. Sketch the trajectory of the steepest descent algorithm on the contour plot of part (i), if the initial guess is $\mathbf{x}_0 = \begin{bmatrix} 0 & 0 \end{bmatrix}^T$. Assume a very small learning rate is used.

iii. Perform two iterations of steepest descent with learning rate $\alpha = 0.1$.

iv. What is the maximum stable learning rate?

v. What is the maximum stable learning rate for the initial guess given in part (ii)? (See Exercise E9.1.)

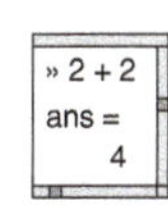

vi. Write a MATLAB M-file to implement the steepest descent algorithm for this problem, and use it to check your answers to parts (i). through (v).

E9.3 For the quadratic function

$$F(\mathbf{x}) = x_1^2 + 2x_2^2 ,$$

i. Find the minimum of the function along the line

$$\mathbf{x} = \begin{bmatrix} 1 \\ 1 \end{bmatrix} + \alpha \begin{bmatrix} -1 \\ -2 \end{bmatrix} .$$

ii. Verify that the gradient of $F(\mathbf{x})$ at the minimum point from part (i) is orthogonal to the line along which the minimization occurred.

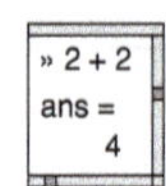

E9.4 For the functions given in Exercise E8.3 perform two iterations of the steepest descent algorithm with linear minimization, starting from the initial guess $\mathbf{x}_0 = \begin{bmatrix} 1 & 1 \end{bmatrix}^T$. Write MATLAB M-files to check your answer.

E9.5 Consider the following function:

$$F(\mathbf{x}) = [1 + (x_1 + x_2 - 5)^2][1 + (3x_1 - 2x_2)^2].$$

i. Perform one iteration of Newton's method, starting from the initial guess $\mathbf{x}_0 = \begin{bmatrix} 10 & 10 \end{bmatrix}^T$.

ii. Repeat part (i), starting from the initial guess $\mathbf{x}_0 = \begin{bmatrix} 2 & 2 \end{bmatrix}^T$.

iii. Find the minimum of the function, and compare with your results from the previous two parts.

E9.6 Consider the following quadratic function

$$F(\mathbf{x}) = \frac{1}{2}\mathbf{x}^T \begin{bmatrix} 3 & 2 \\ 2 & 0 \end{bmatrix} \mathbf{x} + \begin{bmatrix} 4 & 4 \end{bmatrix} \mathbf{x}$$

i. Sketch the contour plot for $F(\mathbf{x})$. Show all work.

ii. Take one iteration of Newton's method from the initial guess $\mathbf{x}_0 = \begin{bmatrix} 0 & 0 \end{bmatrix}^T$.

iii. In part (ii), did you reach the minimum of $F(\mathbf{x})$? Explain.

E9.7 Consider the function

$$F(\mathbf{x}) = (x_1 + x_2)^4 + 2(x_2 - 1)^2$$

i. Find the second-order Taylor series approximation of this function about the point $\mathbf{x}_0 = \begin{bmatrix} -1 & 1 \end{bmatrix}^T$.

ii. Is this point a minimum point? Does it satisfy the first and second order conditions?

iii. Perform one iteration of Newton's method from the initial guess $\mathbf{x}_0 = \begin{bmatrix} 0.5 & 0 \end{bmatrix}^T$.

E9.8 Consider the following quadratic function:

$$F(\mathbf{x}) = \frac{1}{2}\mathbf{x}^T \begin{bmatrix} 7 & -9 \\ -9 & -17 \end{bmatrix} \mathbf{x} + \begin{bmatrix} 16 & 8 \end{bmatrix} \mathbf{x}$$

i. Sketch the contour plot for this function.

ii. Take one step of Newton's method from the initial guess $\mathbf{x}_0 = \begin{bmatrix} 2 & 2 \end{bmatrix}^T$.

iii. Did you reach the minimum of the function after the Newton step of part (ii)? Explain.

iv. From the initial guess in part ii, trace the path of steepest descent, with very small learning rate, on your contour plot from part (i). Explain how you determined the path. Will steepest descent eventually converge to the same result you found in part (ii)? Explain.

E9.9 Consider the following function:

$$F(\mathbf{x}) = (1 + x_1 + x_2)^2 + \frac{1}{4}x_1^4.$$

i. Find the quadratic approximation to $F(\mathbf{x})$ about the point $\mathbf{x}_0 = \begin{bmatrix} 2 & 2 \end{bmatrix}^T$.

ii. Sketch the contour plot of the quadratic approximation in part i.

iii. Perform one iteration of Newton's method on the function $F(\mathbf{x})$ from the initial condition $\mathbf{x}_0$ given in part (i). Sketch the path from $\mathbf{x}_0$ to $\mathbf{x}_1$ on your contour plot from part (ii).

iv. Is the $\mathbf{x}_1$ in part iii. a strong minimum of the quadratic approximation? Is it a strong minimum of the original function $F(\mathbf{x})$? Explain.

v. Will Newton's method always converge to a strong minimum of $F(\mathbf{x})$, given enough iterations? Will it always converge to a strong minimum of the quadratic approximation of $F(\mathbf{x})$? Explain your answers in detail.

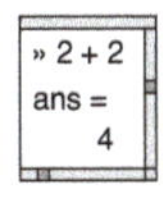

E9.10 Recall the function presented in Exercise E8.5. Write MATLAB M-files to implement the steepest descent algorithm and Newton's method for that function. Test the performance of the algorithms for various initial guesses.

E9.11 Repeat Exercise E9.4 using the conjugate gradient algorithm. Use each of the three methods (Eq. (9.61)–Eq. (9.63)) at least once.

E9.12 Prove or disprove the following statement:

$$\text{If } \mathbf{p}_1 \text{ is conjugate to } \mathbf{p}_2 \text{ and } \mathbf{p}_2 \text{ is conjugate to } \mathbf{p}_3,$$
$$\text{then } \mathbf{p}_1 \text{ is conjugate to } \mathbf{p}_3.$$

10 Widrow-Hoff Learning

Objectives

In the previous two chapters we laid the foundation for *performance learning*, in which a network is trained to optimize its performance. In this chapter we apply the principles of performance learning to a single-layer linear neural network.

Widrow-Hoff learning is an approximate steepest descent algorithm, in which the performance index is mean square error. This algorithm is important to our discussion for two reasons. First, it is widely used today in many signal processing applications, several of which we will discuss in this chapter. In addition, it is the precursor to the backpropagation algorithm for multilayer networks, which is presented in Chapter 11.

Theory and Examples

Bernard Widrow began working in neural networks in the late 1950s, at about the same time that Frank Rosenblatt developed the perceptron learning rule. In 1960 Widrow, and his graduate student Marcian Hoff, introduced the ADALINE (ADAptive LInear NEuron) network, and a learning rule which they called the LMS (Least Mean Square) algorithm [WiHo60].

Their ADALINE network is very similar to the perceptron, except that its transfer function is linear, instead of hard-limiting. Both the ADALINE and the perceptron suffer from the same inherent limitation: they can only solve linearly separable problems (recall our discussion in Chapter 3 and 4). The LMS algorithm, however, is more powerful than the perceptron learning rule. While the perceptron rule is guaranteed to converge to a solution that correctly categorizes the training patterns, the resulting network can be sensitive to noise, since patterns often lie close to the decision boundaries. The LMS algorithm minimizes mean square error, and therefore tries to move the decision boundaries as far from the training patterns as possible.

The LMS algorithm has found many more practical uses than the perceptron learning rule. This is especially true in the area of digital signal processing. For example, most long distance phone lines use ADALINE networks for echo cancellation. We will discuss these applications in detail later in the chapter.

Because of the great success of the LMS algorithm in signal processing applications, and because of the lack of success in adapting the algorithm to multilayer networks, Widrow stopped work on neural networks in the early 1960s and began to work full time on adaptive signal processing. He returned to the neural network field in the 1980s and began research on the use of neural networks in adaptive control, using temporal backpropagation, a descendant of his original LMS algorithm.

ADALINE Network

The ADALINE network is shown in Figure 10.1. Notice that it has the same basic structure as the perceptron network we discussed in Chapter 4. The only difference is that it has a linear transfer function.

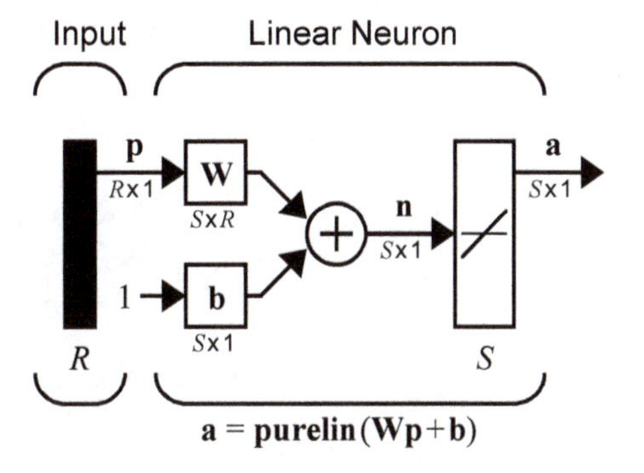

Figure 10.1 ADALINE Network

The output of the network is given by

$$\mathbf{a} = \mathbf{purelin}(\mathbf{Wp} + \mathbf{b}) = \mathbf{Wp} + \mathbf{b}. \tag{10.1}$$

Recall from our discussion of the perceptron network that the ith element of the network output vector can be written

$$a_i = purelin(n_i) = purelin({}_i\mathbf{w}^T\mathbf{p} + b_i) = {}_i\mathbf{w}^T\mathbf{p} + b_i, \tag{10.2}$$

where ${}_i\mathbf{w}$ is made up of the elements of the ith row of $\mathbf{W}$:

$${}_i\mathbf{w} = \begin{bmatrix} w_{i,1} \\ w_{i,2} \\ \vdots \\ w_{i,R} \end{bmatrix}. \tag{10.3}$$

Single ADALINE

To simplify our discussion, let's consider a single ADALINE with two inputs. The diagram for this network is shown in Figure 10.2.

The output of the network is given by

$$\begin{aligned} a &= purelin(n) = purelin({}_1\mathbf{w}^T\mathbf{p} + b) = {}_1\mathbf{w}^T\mathbf{p} + b \\ &= {}_1\mathbf{w}^T\mathbf{p} + b = w_{1,1}p_1 + w_{1,2}p_2 + b. \end{aligned} \tag{10.4}$$

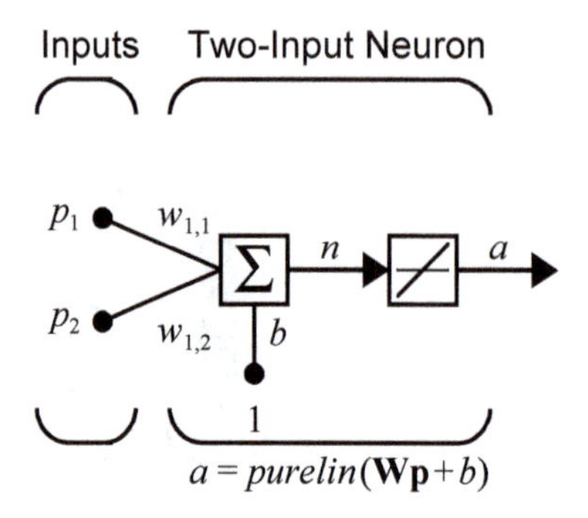

Figure 10.2 Two-Input Linear Neuron

You may recall from Chapter 4 that the perceptron has a *decision boundary*, which is determined by the input vectors for which the net input n is zero. Now, does the ADALINE also have such a boundary? Clearly it does. If we set $n = 0$ then ${}_1\mathbf{w}^T\mathbf{p} + b = 0$ specifies such a line, as shown in Figure 10.3.

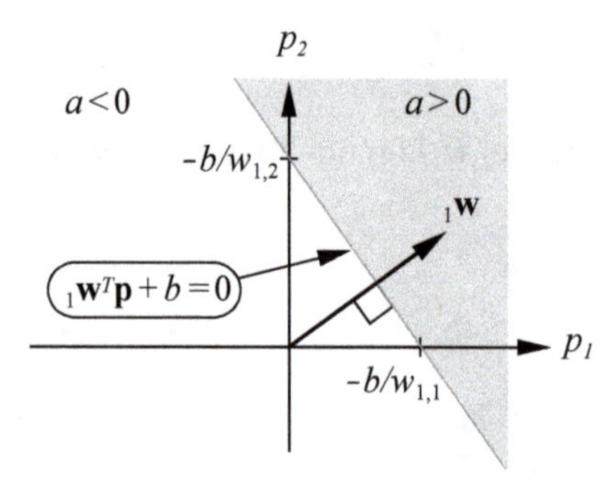

Figure 10.3 Decision Boundary for Two-Input ADALINE

The neuron output is greater than 0 in the gray area. In the white area the output is less than zero. Now, what does this imply about the ADALINE? It says that the ADALINE can be used to classify objects into two categories. However, it can do so only if the objects are linearly separable. Thus, in this respect, the ADALINE has the same limitation as the perceptron.

Mean Square Error

Now that we have examined the characteristics of the ADALINE network, we are ready to begin our development of the LMS algorithm. As with the perceptron rule, the LMS algorithm is an example of supervised training, in which the learning rule is provided with a set of examples of proper network behavior:

$$\{\mathbf{p}_1, \mathbf{t}_1\}, \{\mathbf{p}_2, \mathbf{t}_2\}, \dots, \{\mathbf{p}_Q, \mathbf{t}_Q\}, \tag{10.5}$$

where $\mathbf{p}_q$ is an input to the network, and $\mathbf{t}_q$ is the corresponding target output. As each input is applied to the network, the network output is compared to the target.

The LMS algorithm will adjust the weights and biases of the ADALINE in order to minimize the mean square error, where the error is the difference between the target output and the network output. In this section we want to discuss this performance index. We will consider first the single-neuron case.

To simply our development, we will lump all of the parameters we are adjusting, including the bias, into one vector:

$$\mathbf{x} = \begin{bmatrix} {}_1\mathbf{w} \\ b \end{bmatrix}. \tag{10.6}$$

Similarly, we include the bias input "1" as a component of the input vector

$$\mathbf{z} = \begin{bmatrix} \mathbf{p} \\ 1 \end{bmatrix}. \tag{10.7}$$

Now the network output, which we usually write in the form

$$a = {}_1\mathbf{w}^T\mathbf{p} + b, \tag{10.8}$$

can be written as

$$a = \mathbf{x}^T\mathbf{z}. \tag{10.9}$$

Mean Square Error

This allows us to conveniently write out an expression for the ADALINE network *mean square error*:

$$F(\mathbf{x}) = E[e^2] = E[(t-a)^2] = E[(t-\mathbf{x}^T\mathbf{z})^2], \tag{10.10}$$

where the expectation is taken over all sets of input/target pairs. (Here we use $E[\ ]$ to denote expected value. We use a generalized definition of expectation, which becomes a time-average for deterministic signals. See [WiSt85].) We can expand this expression as follows:

$$F(\mathbf{x}) = E[t^2 - 2t\mathbf{x}^T\mathbf{z} + \mathbf{x}^T\mathbf{z}\mathbf{z}^T\mathbf{x}]$$
$$= E[t^2] - 2\mathbf{x}^T E[t\mathbf{z}] + \mathbf{x}^T E[\mathbf{z}\mathbf{z}^T]\mathbf{x}. \tag{10.11}$$

This can be written in the following convenient form:

$$F(\mathbf{x}) = c - 2\mathbf{x}^T\mathbf{h} + \mathbf{x}^T\mathbf{R}\mathbf{x}, \tag{10.12}$$

where

$$c = E[t^2], \mathbf{h} = E[t\mathbf{z}] \text{ and } \mathbf{R} = E[\mathbf{z}\mathbf{z}^T]. \tag{10.13}$$

Correlation Matrix

Here the vector $\mathbf{h}$ gives the cross-correlation between the input vector and its associated target, while $\mathbf{R}$ is the input *correlation matrix*. The diagonal elements of this matrix are equal to the mean square values of the elements of the input vectors.

Take a close look at Eq. (10.12), and compare it with the general form of the quadratic function given in Eq. (8.35) and repeated here:

$$F(\mathbf{x}) = c + \mathbf{d}^T\mathbf{x} + \frac{1}{2}\mathbf{x}^T\mathbf{A}\mathbf{x}. \tag{10.14}$$

We can see that the mean square error performance index for the ADALINE network is a quadratic function, where

$$\mathbf{d} = -2\mathbf{h} \text{ and } \mathbf{A} = 2\mathbf{R}. \tag{10.15}$$

This is a very important result, because we know from Chapter 8 that the characteristics of the quadratic function depend primarily on the Hessian matrix $\mathbf{A}$. For example, if the eigenvalues of the Hessian are all positive, then the function will have one unique global minimum.

In this case the Hessian matrix is twice the correlation matrix $\mathbf{R}$, and it can be shown that all correlation matrices are either positive definite or positive semidefinite, which means that they can never have negative eigenvalues. We are left with two possibilities. If the correlation matrix has only positive eigenvalues, the performance index will have one unique global minimum (see Figure 8.7). If the correlation matrix has some zero eigenvalues, the performance index will either have a weak minimum (see Figure 8.9) or no minimum (see Problem P8.7), depending on the vector $\mathbf{d} = -2\mathbf{h}$.

Now let's locate the stationary point of the performance index. From our previous discussion of quadratic functions we know that the gradient is

$$\nabla F(\mathbf{x}) = \nabla\left(c + \mathbf{d}^T\mathbf{x} + \frac{1}{2}\mathbf{x}^T\mathbf{A}\mathbf{x}\right) = \mathbf{d} + \mathbf{A}\mathbf{x} = -2\mathbf{h} + 2\mathbf{R}\mathbf{x}. \tag{10.16}$$

The stationary point of $F(\mathbf{x})$ can be found by setting the gradient equal to zero:

$$-2\mathbf{h} + 2\mathbf{R}\mathbf{x} = 0. \tag{10.17}$$

Therefore, if the correlation matrix is positive definite there will be a unique stationary point, which will be a strong minimum:

$$\mathbf{x}^* = \mathbf{R}^{-1}\mathbf{h}. \tag{10.18}$$

It is worth noting here that the existence of a unique solution depends only on the correlation matrix $\mathbf{R}$. Therefore the characteristics of the input vectors determine whether or not a unique solution exists.

LMS Algorithm

Now that we have analyzed our performance index, the next step is to design an algorithm to locate the minimum point. If we could calculate the statistical quantities $\mathbf{h}$ and $\mathbf{R}$, we could find the minimum point directly from Eq. (10.18). If we did not want to calculate the inverse of $\mathbf{R}$, we could use the steepest descent algorithm, with the gradient calculated from Eq. (10.16). In general, however, it is not desirable or convenient to calculate $\mathbf{h}$ and $\mathbf{R}$. For this reason we will use an approximate steepest descent algorithm, in which we use an estimated gradient.

The key insight of Widrow and Hoff was that they could estimate the mean square error $F(\mathbf{x})$ by

$$\hat{F}(\mathbf{x}) = (t(k) - a(k))^2 = e^2(k), \tag{10.19}$$

where the expectation of the squared error has been replaced by the squared error at iteration k. Then, at each iteration we have a gradient estimate of the form:

$$\hat{\nabla} F(\mathbf{x}) = \nabla e^2(k). \tag{10.20}$$

Stochastic Gradient

This is sometimes referred to as the *stochastic gradient*. When this is used in a gradient descent algorithm, it is referred to as "on-line" or incremental learning, since the weights are updated as each input is presented to the network.

The first R elements of $\nabla e^2(k)$ are derivatives with respect to the network weights, while the $(R+1)$st element is the derivative with respect to the bias. Thus we have

$$[\nabla e^2(k)]_j = \frac{\partial e^2(k)}{\partial w_{1,j}} = 2e(k)\frac{\partial e(k)}{\partial w_{1,j}} \text{ for } j = 1, 2, \ldots, R, \tag{10.21}$$

and

$$[\nabla e^2(k)]_{R+1} = \frac{\partial e^2(k)}{\partial b} = 2e(k)\frac{\partial e(k)}{\partial b}. \tag{10.22}$$

Now consider the partial derivative terms at the ends of these equations. First evaluate the partial derivative of $e(k)$ with respect to the weight $w_{1,j}$:

$$\frac{\partial e(k)}{\partial w_{1,j}} = \frac{\partial [t(k) - a(k)]}{\partial w_{1,j}} = \frac{\partial}{\partial w_{1,j}}[t(k) - ({}_1\mathbf{w}^T \mathbf{p}(k) + b)]$$

$$= \frac{\partial}{\partial w_{1,j}}\left[t(k) - \left(\sum_{i=1}^{R} w_{1,i} p_i(k) + b\right)\right] \tag{10.23}$$

where $p_i(k)$ is the ith element of the input vector at the kth iteration. This simplifies to

$$\frac{\partial e(k)}{\partial w_{1,j}} = -p_j(k). \tag{10.24}$$

In a similar way we can obtain the final element of the gradient:

$$\frac{\partial e(k)}{\partial b} = -1. \tag{10.25}$$

Note that $p_j(k)$ and 1 are the elements of the input vector $\mathbf{z}$, so the gradient of the squared error at iteration k can be written

$$\hat{\nabla} F(\mathbf{x}) = \nabla e^2(k) = -2e(k)\mathbf{z}(k). \tag{10.26}$$

Now we can see the beauty of approximating the mean square error by the single error at iteration k, as in Eq. (10.19). To calculate this approximate gradient we need only multiply the error times the input.

This approximation to $\nabla F(\mathbf{x})$ can now be used in the steepest descent algorithm. From Eq. (9.10) the steepest descent algorithm, with constant learning rate, is

$$\mathbf{x}_{k+1} = \mathbf{x}_k - \alpha \nabla F(\mathbf{x})\big|_{\mathbf{x} = \mathbf{x}_k}. \tag{10.27}$$

If we substitute $\hat{\nabla} F(\mathbf{x})$, from Eq. (10.26), for $\nabla F(\mathbf{x})$ we find

$$\mathbf{x}_{k+1} = \mathbf{x}_k + 2\alpha e(k)\mathbf{z}(k), \tag{10.28}$$

or

$${}_1\mathbf{w}(k+1) = {}_1\mathbf{w}(k) + 2\alpha e(k)\mathbf{p}(k), \tag{10.29}$$

and

$$b(k+1) = b(k) + 2\alpha e(k). \tag{10.30}$$

These last two equations make up the least mean square (LMS) algorithm. This is also referred to as the delta rule or the Widrow-Hoff learning algorithm.

The preceding results can be modified to handle the case where we have multiple outputs, and therefore multiple neurons, as in Figure 10.1. To update the ith row of the weight matrix use

$$_{i}\mathbf{w}(k+1) = {}_{i}\mathbf{w}(k) + 2\alpha e_i(k)\mathbf{p}(k), \tag{10.31}$$

where $e_i(k)$ is the ith element of the error at iteration k. To update the ith element of the bias we use

$$b_i(k+1) = b_i(k) + 2\alpha e_i(k). \tag{10.32}$$

LMS Algorithm

The *LMS algorithm* can be written conveniently in matrix notation:

$$\mathbf{W}(k+1) = \mathbf{W}(k) + 2\alpha\mathbf{e}(k)\mathbf{p}^T(k), \tag{10.33}$$

and

$$\mathbf{b}(k+1) = \mathbf{b}(k) + 2\alpha\mathbf{e}(k). \tag{10.34}$$

Note that the error $\mathbf{e}$ and the bias $\mathbf{b}$ are now vectors.

Analysis of Convergence

The stability of the steepest descent algorithm was investigated in Chapter 9. There we found that the maximum stable learning rate for quadratic functions is $\alpha < 2/\lambda_{max}$, where λ_{max} is the largest eigenvalue of the Hessian matrix. Now we want to investigate the convergence of the LMS algorithm, which is approximate steepest descent. We will find that the result is the same.

To begin, note that in the LMS algorithm, Eq. (10.28), $\mathbf{x}_k$ is a function only of $\mathbf{z}(k-1), \mathbf{z}(k-2), \ldots, \mathbf{z}(0)$. If we assume that successive input vectors are statistically independent, then $\mathbf{x}_k$ is independent of $\mathbf{z}(k)$. We will show in the following development that for stationary input processes meeting this condition, the expected value of the weight vector will converge to

$$\mathbf{x}^* = \mathbf{R}^{-1}\mathbf{h}. \tag{10.35}$$

This is the minimum mean square error $\{E[e_k^2]\}$ solution, as we saw in Eq. (10.18).

Recall the LMS algorithm (Eq. (10.28)):

$$\mathbf{x}_{k+1} = \mathbf{x}_k + 2\alpha e(k)\mathbf{z}(k). \tag{10.36}$$

Now take the expectation of both sides:

$$E[\mathbf{x}_{k+1}] = E[\mathbf{x}_k] + 2\alpha E[e(k)\mathbf{z}(k)]. \quad (10.37)$$

Substitute $t(k) - \mathbf{x}_k^T\mathbf{z}(k)$ for the error to give

$$E[\mathbf{x}_{k+1}] = E[\mathbf{x}_k] + 2\alpha\{E[t(k)\mathbf{z}(k)] - E[(\mathbf{x}_k^T\mathbf{z}(k))\mathbf{z}(k)]\}. \quad (10.38)$$

Finally, substitute $\mathbf{z}^T(k)\mathbf{x}_k$ for $\mathbf{x}_k^T\mathbf{z}(k)$ and rearrange terms to give

$$E[\mathbf{x}_{k+1}] = E[\mathbf{x}_k] + 2\alpha\{E[t_k\mathbf{z}(k)] - E[(\mathbf{z}(k)\mathbf{z}^T(k))\mathbf{x}_k]\}. \quad (10.39)$$

Since $\mathbf{x}_k$ is independent of $\mathbf{z}(k)$:

$$E[\mathbf{x}_{k+1}] = E[\mathbf{x}_k] + 2\alpha\{\mathbf{h} - \mathbf{R}E[\mathbf{x}_k]\}. \quad (10.40)$$

This can be written as

$$E[\mathbf{x}_{k+1}] = [\mathbf{I} - 2\alpha\mathbf{R}]E[\mathbf{x}_k] + 2\alpha\mathbf{h}. \quad (10.41)$$

This dynamic system will be stable if all of the eigenvalues of $[\mathbf{I} - 2\alpha\mathbf{R}]$ fall inside the unit circle (see [Brog91]). Recall from Chapter 9 that the eigenvalues of $[\mathbf{I} - 2\alpha\mathbf{R}]$ will be $1 - 2\alpha\lambda_i$, where the λ_i are the eigenvalues of $\mathbf{R}$. Therefore, the system will be stable if

$$1 - 2\alpha\lambda_i > -1. \quad (10.42)$$

Since $\lambda_i > 0$, $1 - 2\alpha\lambda_i$ is always less than 1. The condition on stability is therefore

$$\alpha < 1/\lambda_i \quad \text{for all } i, \quad (10.43)$$

or

$$0 < \alpha < 1/\lambda_{max}. \quad (10.44)$$

Note that this condition is equivalent to the condition we derived in Chapter 9 for the steepest descent algorithm, although in that case we were using the eigenvalues of the Hessian matrix $\mathbf{A}$. Now we are using the eigenvalues of the input correlation matrix $\mathbf{R}$. (Recall that $\mathbf{A} = 2\mathbf{R}$.)

If this condition on stability is satisfied, the steady state solution is

$$E[\mathbf{x}_{ss}] = [\mathbf{I} - 2\alpha\mathbf{R}]E[\mathbf{x}_{ss}] + 2\alpha\mathbf{h}, \quad (10.45)$$

or

$$E[\mathbf{x}_{ss}] = \mathbf{R}^{-1}\mathbf{h} = \mathbf{x}^*. \quad (10.46)$$

Thus the LMS solution, obtained by applying one input vector at a time, is the same as the minimum mean square error solution of Eq. (10.18).

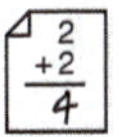

To test the ADALINE network and the LMS algorithm consider again the apple/orange recognition problem originally discussed in Chapter 3. For simplicity we will assume that the ADALINE network has a zero bias.

The LMS weight update algorithm of Eq. (10.29) will be used to calculate the new weights at each step in the network training:

$$\mathbf{W}(k+1) = \mathbf{W}(k) + 2\alpha e(k)\mathbf{p}^T(k). \tag{10.47}$$

First let's compute the maximum stable learning rate α. We can get such a value by finding the eigenvalues of the input correlation matrix. Recall that the orange and apple vectors and their associated targets are

$$\left\{\mathbf{p}_1 = \begin{bmatrix} 1 \\ -1 \\ -1 \end{bmatrix}, t_1 = \begin{bmatrix} -1 \end{bmatrix}\right\} \quad \left\{\mathbf{p}_2 = \begin{bmatrix} 1 \\ 1 \\ -1 \end{bmatrix}, t_2 = \begin{bmatrix} 1 \end{bmatrix}\right\}. \tag{10.48}$$

If we assume that the input vectors are generated randomly with equal probability, we can compute the input correlation matrix:

$$\mathbf{R} = E[\mathbf{p}\mathbf{p}^T] = \frac{1}{2}\mathbf{p}_1\mathbf{p}_1^T + \frac{1}{2}\mathbf{p}_2\mathbf{p}_2^T$$

$$= \frac{1}{2}\begin{bmatrix} 1 \\ -1 \\ -1 \end{bmatrix}\begin{bmatrix} 1 & -1 & -1 \end{bmatrix} + \frac{1}{2}\begin{bmatrix} 1 \\ 1 \\ -1 \end{bmatrix}\begin{bmatrix} 1 & 1 & -1 \end{bmatrix} = \begin{bmatrix} 1 & 0 & -1 \\ 0 & 1 & 0 \\ -1 & 0 & 1 \end{bmatrix}. \tag{10.49}$$

The eigenvalues of $\mathbf{R}$ are

$$\lambda_1 = 1.0, \qquad \lambda_2 = 0.0, \qquad \lambda_3 = 2.0. \tag{10.50}$$

Thus, the maximum stable learning rate is

$$\alpha < \frac{1}{\lambda_{max}} = \frac{1}{2.0} = 0.5. \tag{10.51}$$

To be conservative we will pick $\alpha = 0.2$. (Note that in practical applications it might not be practical to calculate $\mathbf{R}$, and α could be selected by trial and error. Other techniques for choosing α are given in [WiSt85].)

We will start, arbitrarily, with all the weights set to zero, and then will apply inputs $\mathbf{p}_1$, $\mathbf{p}_2$, $\mathbf{p}_1$, $\mathbf{p}_2$, etc., in that order, calculating the new weights after each input is presented. (The presentation of the weights in alternating order is not necessary. A random sequence would be fine.) Presenting $\mathbf{p}_1$, the orange, and using its target of -1 we get

$$a(0) = \mathbf{W}(0)\mathbf{p}(0) = \mathbf{W}(0)\mathbf{p}_1 = \begin{bmatrix} 0 & 0 & 0 \end{bmatrix} \begin{bmatrix} 1 \\ -1 \\ -1 \end{bmatrix} = 0, \tag{10.52}$$

and

$$e(0) = t(0) - a(0) = t_1 - a(0) = -1 - 0 = -1. \tag{10.53}$$

Now we can calculate the new weight matrix:

$$\mathbf{W}(1) = \mathbf{W}(0) + 2\alpha e(0)\mathbf{p}^T(0)$$

$$= \begin{bmatrix} 0 & 0 & 0 \end{bmatrix} + 2(0.2)(-1) \begin{bmatrix} 1 \\ -1 \\ -1 \end{bmatrix}^T = \begin{bmatrix} -0.4 & 0.4 & 0.4 \end{bmatrix}. \tag{10.54}$$

According to plan, we will next present the apple, $\mathbf{p}_2$, and its target of 1:

$$a(1) = \mathbf{W}(1)\mathbf{p}(1) = \mathbf{W}(1)\mathbf{p}_2 = \begin{bmatrix} -0.4 & 0.4 & 0.4 \end{bmatrix} \begin{bmatrix} 1 \\ 1 \\ -1 \end{bmatrix} = -0.4, \tag{10.55}$$

and so the error is

$$e(1) = t(1) - a(1) = t_2 - a(1) = 1 - (-0.4) = 1.4. \tag{10.56}$$

Now we calculate the new weights:

$$\mathbf{W}(2) = \mathbf{W}(1) + 2\alpha e(1)\mathbf{p}^T(1)$$

$$= \begin{bmatrix} -0.4 & 0.4 & 0.4 \end{bmatrix} + 2(0.2)(1.4) \begin{bmatrix} 1 \\ 1 \\ -1 \end{bmatrix}^T = \begin{bmatrix} 0.16 & 0.96 & -0.16 \end{bmatrix}. \tag{10.57}$$

Next we present the orange again:

$$a(2) = \mathbf{W}(2)\mathbf{p}(2) = \mathbf{W}(2)\mathbf{p}_1 = \begin{bmatrix} 0.16 & 0.96 & -0.16 \end{bmatrix} \begin{bmatrix} 1 \\ -1 \\ -1 \end{bmatrix} = -0.64. \tag{10.58}$$

The error is

$$e(2) = t(2) - a(2) = t_1 - a(2) = -1 - (-0.64) = -0.36 . \quad (10.59)$$

The new weights are

$$\mathbf{W}(3) = \mathbf{W}(2) + 2\alpha e(2)\mathbf{p}^T(2) = \begin{bmatrix} 0.016 & 1.1040 & -0.0160 \end{bmatrix} . \quad (10.60)$$

If we continue this procedure, the algorithm converges to

$$\mathbf{W}(\infty) = \begin{bmatrix} 0 & 1 & 0 \end{bmatrix} . \quad (10.61)$$

Compare this result with the result of the perceptron learning rule in Chapter 4. You will notice that the ADALINE has produced the same decision boundary that we designed in Chapter 3 for the apple/orange problem. This boundary falls halfway between the two reference patterns. The perceptron rule did not produce such a boundary. This is because the perceptron rule stops as soon as the patterns are correctly classified, even though some patterns may be close to the boundaries. The LMS algorithm minimizes the mean square error. Therefore it tries to move the decision boundaries as far from the reference patterns as possible.

Adaptive Filtering

As we mentioned at the beginning of this chapter, the ADALINE network has the same major limitation as the perceptron network; it can only solve linearly separable problems. In spite of this, the ADALINE has been much more widely used than the perceptron network. In fact, it is safe to say that it is one of the most widely used neural networks in practical applications. One of the major application areas of the ADALINE has been adaptive filtering, where it is still used extensively. In this section we will demonstrate an adaptive filtering example.

Tapped Delay Line

In order to use the ADALINE network as an adaptive filter, we need to introduce a new building block, the tapped delay line. A *tapped delay line* with R outputs is shown in Figure 10.4.

The input signal enters from the left. At the output of the tapped delay line we have an R-dimensional vector, consisting of the input signal at the current time and at delays of from 1 to $R-1$ time steps.

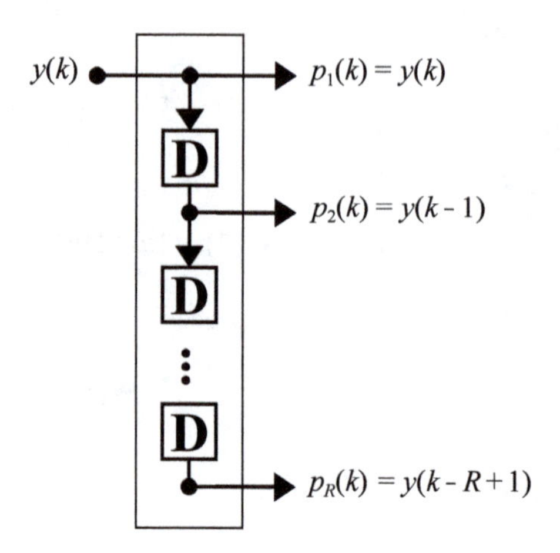

Figure 10.4 Tapped Delay Line

Adaptive Filter

If we combine a tapped delay line with an ADALINE network, we can create an *adaptive filter*, as is shown in Figure 10.5. The output of the filter is given by

$$a(k) = purelin(\mathbf{W}\mathbf{p} + b) = \sum_{i=1}^{R} w_{1,i} y(k - i + 1) + b . \qquad (10.62)$$

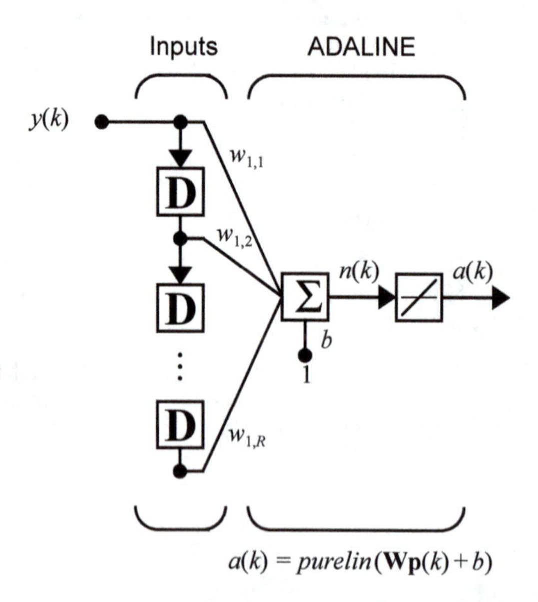

Figure 10.5 Adaptive Filter ADALINE

If you are familiar with digital signal processing, you will recognize the network of Figure 10.5 as a finite impulse response (FIR) filter [WiSt85]. It is beyond the scope of this text to review the field of digital signal processing, but we can demonstrate the usefulness of this adaptive filter through a simple, but practical, example.

Adaptive Noise Cancellation

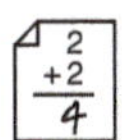

An adaptive filter can be used in a variety of novel ways. In the following example we will use it for noise cancellation. Take some time to look at this example, for it is a little different from what you might expect. For instance, the output "error" that the network tries to minimize is actually an approximation to the signal we are trying to recover!

Let's suppose that a doctor, in trying to review the electroencephalogram (EEG) of a distracted graduate student, finds that the signal he would like to see has been contaminated by a 60-Hz noise source. He is examining the patient on-line and wants to view the best signal that can be obtained. Figure 10.6 shows how an adaptive filter can be used to remove the contaminating signal.

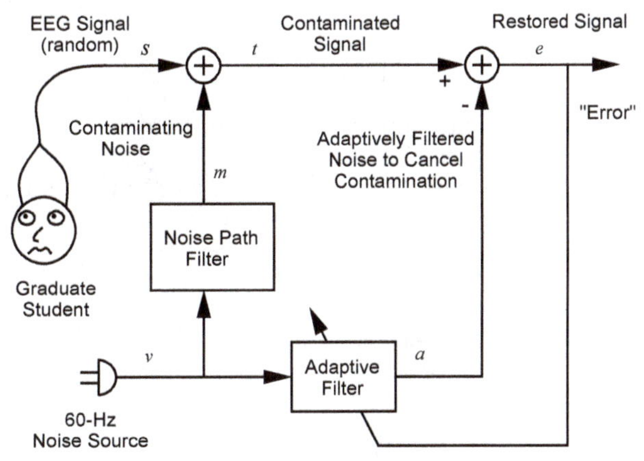

Figure 10.6 Noise Cancellation System

As shown, a sample of the original 60-Hz signal is fed to an adaptive filter, whose elements are adjusted so as to minimize the "error" e. The desired output of the filter is the contaminated EEG signal t. The adaptive filter will do its best to reproduce this contaminated signal, but it only knows about the original noise source, v. Thus, it can only reproduce the part of t that is linearly correlated with v, which is m. In effect, the adaptive filter will attempt to mimic the noise path filter, so that the output of the filter

a will be close to the contaminating noise m. In this way the error e will be close to the original uncontaminated EEG signal s.

In this simple case of a single sine wave noise source, a neuron with two weights and no bias is sufficient to implement the filter. The inputs to the filter are the current and previous values of the noise source. Such a two-input filter can attenuate and phase-shift the noise v in the desired way. The filter is shown in Figure 10.7.

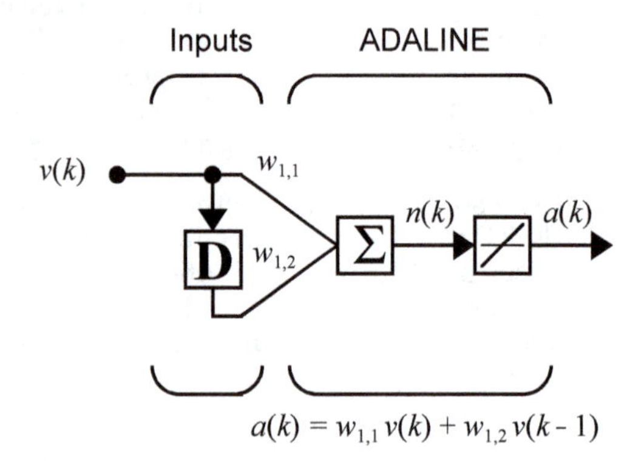

Figure 10.7 Adaptive Filter for Noise Cancellation

We can apply the mathematical relationships developed in the previous sections of this chapter to analyze this system. In order to do so, we will first need to find the input correlation matrix $\mathbf{R}$ and the input/target cross-correlation vector $\mathbf{h}$:

$$\mathbf{R} = [\mathbf{z}\mathbf{z}^T] \text{ and } \mathbf{h} = E[t\mathbf{z}]. \tag{10.63}$$

In our case the input vector is given by the current and previous values of the noise source:

$$\mathbf{z}(k) = \begin{bmatrix} v(k) \\ v(k-1) \end{bmatrix}, \tag{10.64}$$

while the target is the sum of the current signal and filtered noise:

$$t(k) = s(k) + m(k). \tag{10.65}$$

Now expand the expressions for $\mathbf{R}$ and $\mathbf{h}$ to give

$$\mathbf{R} = \begin{bmatrix} E[v^2(k)] & E[v(k)v(k-1)] \\ E[v(k-1)v(k)] & E[v^2(k-1)] \end{bmatrix}, \tag{10.66}$$

and

$$\mathbf{h} = \begin{bmatrix} E[(s(k)+m(k))v(k)] \\ E[(s(k)+m(k))v(k-1)] \end{bmatrix}. \quad (10.67)$$

To obtain specific values for these two quantities we must define the noise signal v, the EEG signal s and the filtered noise m. For this exercise we will assume: the EEG signal is a white (uncorrelated from one time step to the next) random signal uniformly distributed between the values -0.2 and +0.2, the noise source (60-Hz sine wave sampled at 180 Hz) is given by

$$v(k) = 1.2 \sin\left(\frac{2\pi k}{3}\right), \quad (10.68)$$

and the filtered noise that contaminates the EEG is the noise source attenuated by a factor of 10 and shifted in phase by $\pi/2$:

$$m(k) = 0.12 \sin\left(\frac{2\pi k}{3} + \frac{\pi}{2}\right). \quad (10.69)$$

Now calculate the elements of the input correlation matrix $\mathbf{R}$:

$$E[v^2(k)] = (1.2)^2 \frac{1}{3} \sum_{k=1}^{3} \left(\sin\left(\frac{2\pi k}{3}\right)\right)^2 = (1.2)^2 0.5 = 0.72, \quad (10.70)$$

$$E[v^2(k-1)] = E[v^2(k)] = 0.72, \quad (10.71)$$

$$E[v(k)v(k-1)] = \frac{1}{3} \sum_{k=1}^{3} \left(1.2 \sin\frac{2\pi k}{3}\right)\left(1.2 \sin\frac{2\pi(k-1)}{3}\right)$$

$$= (1.2)^2 0.5 \cos\left(\frac{2\pi}{3}\right) = -0.36 \quad (10.72)$$

(where we have used some trigonometric identities).

Thus $\mathbf{R}$ is

$$\mathbf{R} = \begin{bmatrix} 0.72 & -0.36 \\ -0.36 & 0.72 \end{bmatrix}. \quad (10.73)$$

The terms of $\mathbf{h}$ can be found in a similar manner. We will consider the top term in Eq. (10.67) first:

$$E[(s(k)+m(k))v(k)] = E[s(k)v(k)] + E[m(k)v(k)]. \quad (10.74)$$

Here the first term on the right is zero because $s(k)$ and $v(k)$ are independent and zero mean. The second term is also zero:

$$E[m(k)v(k)] = \frac{1}{3}\sum_{k=1}^{3}\left(0.12\sin\left(\frac{2\pi k}{3}+\frac{\pi}{2}\right)\right)\left(1.2\sin\frac{2\pi k}{3}\right) = 0 \qquad (10.75)$$

Thus, the first element of $\mathbf{h}$ is zero.

Next consider the second element of $\mathbf{h}$:

$$E[(s(k)+m(k))v(k-1)] = E[s(k)v(k-1)] + E[m(k)v(k-1)] . \qquad (10.76)$$

As with the first element of $\mathbf{h}$, the first term on the right is zero because $s(k)$ and $v(k-1)$ are independent and zero mean. The second term is evaluated as follows:

$$E[m(k)v(k-1)] = \frac{1}{3}\sum_{k=1}^{3}\left(0.12\sin\left(\frac{2\pi k}{3}+\frac{\pi}{2}\right)\right)\left(1.2\sin\frac{2\pi(k-1)}{3}\right)$$

$$= -0.0624 . \qquad (10.77)$$

Thus, $\mathbf{h}$ is

$$\mathbf{h} = \begin{bmatrix} 0 \\ -0.0624 \end{bmatrix} . \qquad (10.78)$$

The minimum mean square error solution for the weights is given by Eq. (10.18):

$$\mathbf{x}^* = \mathbf{R}^{-1}\mathbf{h} = \begin{bmatrix} 0.72 & -0.36 \\ -0.36 & 0.72 \end{bmatrix}^{-1} \begin{bmatrix} 0 \\ -0.0624 \end{bmatrix} = \begin{bmatrix} -0.0578 \\ -0.1156 \end{bmatrix} . \qquad (10.79)$$

Now, what kind of error will we have at the minimum solution? To find this error recall Eq. (10.12):

$$F(\mathbf{x}) = c - 2\mathbf{x}^T\mathbf{h} + \mathbf{x}^T\mathbf{R}\mathbf{x} . \qquad (10.80)$$

We have just found $\mathbf{x}^*$, $\mathbf{h}$ and $\mathbf{R}$, so we only need to find c:

$$c = E[t^2(k)] = E[(s(k)+m(k))^2]$$

$$= E[s^2(k)] + 2E[s(k)m(k)] + E[m^2(k)] . \qquad (10.81)$$

The middle term is zero because $s(k)$ and $m(k)$ are independent and zero mean. The first term, the mean squared value of the random signal, can be calculated as follows:

$$E[s^2(k)] = \frac{1}{0.4}\int_{-0.2}^{0.2} s^2 ds = \frac{1}{3(0.4)} s^3 \Big|_{-0.2}^{0.2} = 0.0133 . \qquad (10.82)$$

The mean square value of the filtered noise is

$$E[m^2(k)] = \frac{1}{3}\sum_{k=1}^{3} \left\{0.12 \sin\left(\frac{2\pi}{3} + \frac{\pi}{2}\right)\right\}^2 = 0.0072 , \qquad (10.83)$$

so that

$$c = 0.0133 + 0.0072 = 0.0205 . \qquad (10.84)$$

Substituting $\mathbf{x}^*$, $\mathbf{h}$ and $\mathbf{R}$ into Eq. (10.80), we find that the minimum mean square error is

$$F(\mathbf{x}^*) = 0.0205 - 2(0.0072) + 0.0072 = 0.0133 . \qquad (10.85)$$

The minimum mean square error is the same as the mean square value of the EEG signal. This is what we expected, since the "error" of this adaptive noise canceller is in fact the reconstructed EEG signal.

Figure 10.8 illustrates the trajectory of the LMS algorithm in the weight space with learning rate $\alpha = 0.1$. The system weights $w_{1,1}$ and $w_{1,2}$ in this simulation were initialized arbitrarily to 0 and -2, respectively. You can see from this figure that the LMS trajectory looks like a noisy version of steepest descent.

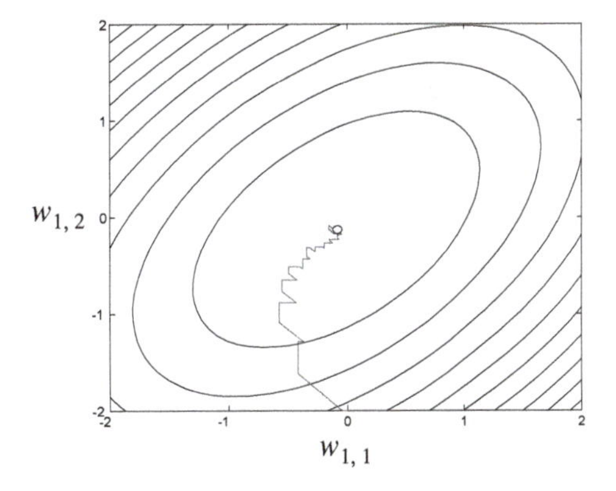

Figure 10.8 LMS Trajectory for $\alpha = 0.1$

Note that the contours in this figure reflect the fact that the eigenvalues and eigenvectors of the Hessian matrix ($\mathbf{A} = 2\mathbf{R}$) are

$$\lambda_1 = 2.16 \,,\ \mathbf{z}_1 = \begin{bmatrix} -0.7071 \\ 0.7071 \end{bmatrix} ,\ \lambda_2 = 0.72 \,,\ \mathbf{z}_2 = \begin{bmatrix} -0.7071 \\ -0.7071 \end{bmatrix} . \qquad (10.86)$$

(Refer back to our discussion in Chapter 8 on the eigensystem of the Hessian matrix.)

If the learning rate is decreased, the LMS trajectory is smoother than that shown in Figure 10.8, but the learning proceeds more slowly. If the learning rate is increased, the trajectory is more jagged and oscillatory. In fact, as noted earlier in this chapter, if the learning rate is increased too much the system does not converge at all. The maximum stable learning rate is $\alpha < 2/2.16 = 0.926$.

In order to judge the performance of our noise canceller, consider Figure 10.9. This figure illustrates how the filter adapts to cancel the noise. The top graph shows the restored and original EEG signals. At first the restored signal is a poor approximation of the original EEG signal. It takes about 0.2 second (with $\alpha = 0.1$) for the filter to adjust to give a reasonable restored signal. The mean square difference between the original and restored signal over the last half of the experiment was 0.002. This compares favorably with the signal mean square value of 0.0133. The difference between the original and restored signal is shown in the lower graph.

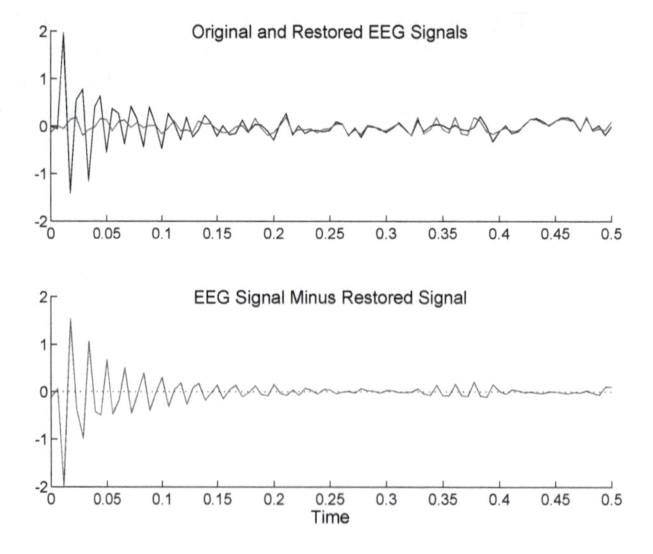

Figure 10.9 Adaptive Filter Cancellation of Contaminating Noise

You might wonder why the error does not go to zero. This is because the LMS algorithm is approximate steepest descent; it uses an estimate of the gradient, not the true gradient, to update the weights. The estimate of the gradient is a noisy version of the true gradient. This will cause the weights to continue to change slightly, even after the mean square error is at the minimum point. You can see this effect in Figure 10.8.

To experiment with the use of this adaptive noise cancellation filter, use the MATLAB® Neural Network Design Demonstration Adaptive Noise Cancellation (nnd10nc). *A more complex noise source and actual EEG data are used in the Demonstration* Electroencephalogram Noise Cancellation (nnd10eeg).

Echo Cancellation

Another very important practical application of adaptive noise cancellation is echo cancellation. Echoes are common in long distance telephone lines because of impedance mismatch at the "hybrid" device that forms the junction between the long distance line and the customer's local line. You may have experienced this effect on international telephone calls.

Figure 10.10 illustrates how an adaptive noise cancellation filter can be used to reduce these echoes [WiWi85]. At the end of the long distance line the incoming signal is sent to an adaptive filter, as well as to the hybrid device. The target output of the filter is the output of the hybrid. The filter thus tries to cancel the part of the hybrid output that is correlated with the input signal — the echo.

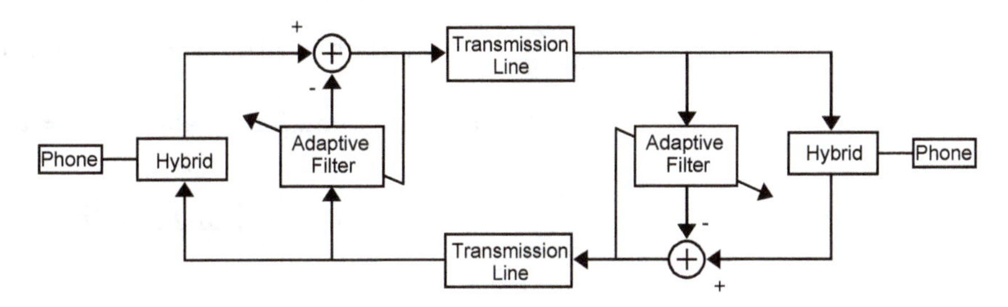

Figure 10.10 Echo Cancellation System

Summary of Results

ADALINE

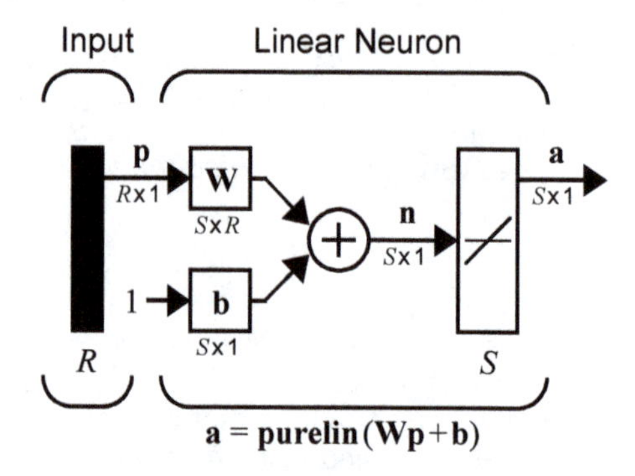

Mean Square Error

$$F(\mathbf{x}) = E[e^2] = E[(t-a)^2] = E[(t-\mathbf{x}^T\mathbf{z})^2]$$

$$F(\mathbf{x}) = c - 2\mathbf{x}^T\mathbf{h} + \mathbf{x}^T\mathbf{R}\mathbf{x},$$

$$c = E[t^2], \mathbf{h} = E[t\mathbf{z}] \text{ and } \mathbf{R} = E[\mathbf{z}\mathbf{z}^T]$$

Unique minimum, if it exists, is $\mathbf{x}^* = \mathbf{R}^{-1}\mathbf{h}$.

Where $\mathbf{x} = \begin{bmatrix} {}_1\mathbf{w} \\ b \end{bmatrix}$ and $\mathbf{z} = \begin{bmatrix} \mathbf{p} \\ 1 \end{bmatrix}$.

LMS Algorithm

$$\mathbf{W}(k+1) = \mathbf{W}(k) + 2\alpha\mathbf{e}(k)\mathbf{p}^T(k)$$

$$\mathbf{b}(k+1) = \mathbf{b}(k) + 2\alpha\mathbf{e}(k)$$

Convergence Point

$$\mathbf{x}^* = \mathbf{R}^{-1}\mathbf{h}$$

Stable Learning Rate

$$0 < \alpha < 1/\lambda_{max} \quad \text{where } \lambda_{max} \text{ is the maximum eigenvalue of } \mathbf{R}$$

Tapped Delay Line

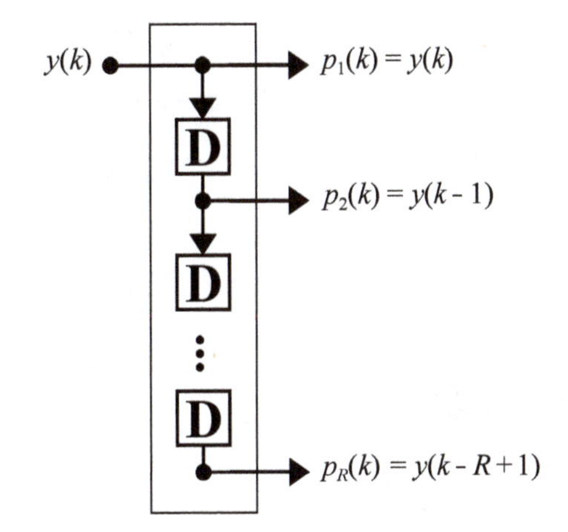

Adaptive Filter ADALINE

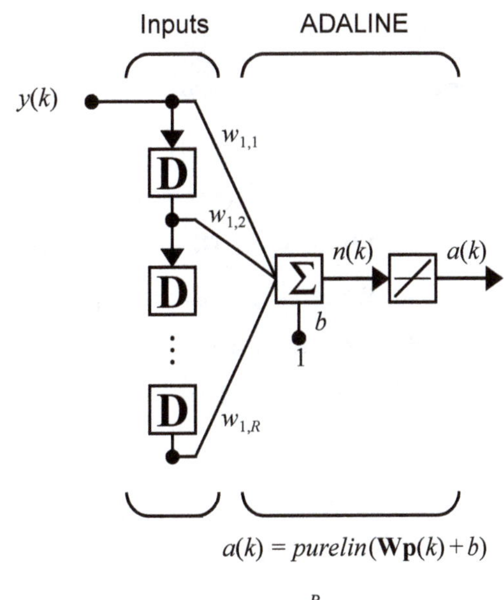

$$a(k) = purelin(\mathbf{W}\mathbf{p} + b) = \sum_{i=1}^{R} w_{1,i}\, y(k - i + 1) + b$$

Solved Problems

P10.1 Consider the ADALINE filter in Figure P10.1.

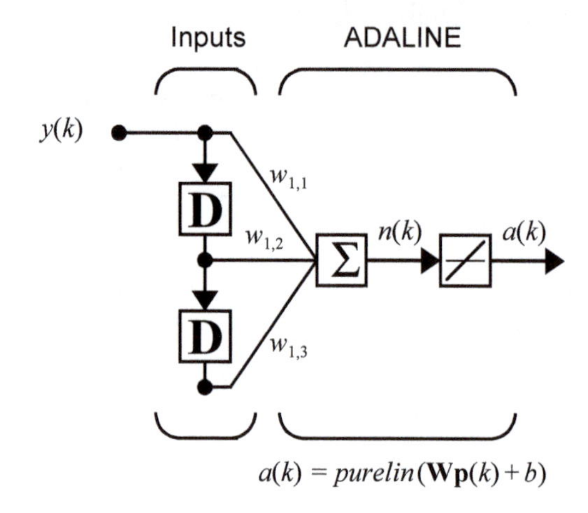

Figure P10.1 ADALINE Filter

Suppose that

$$w_{1,1} = 2, \quad w_{1,2} = -1, \quad w_{1,3} = 3,$$

and the input sequence is

$$\{y(k)\} = \{\ldots, 0, 0, 0, 5, -4, 0, 0, 0, \ldots\}$$

where $y(0) = 5$**,** $y(1) = -4$**, etc.**

i. **What is the filter output just prior to** $k = 0$**?**

ii. **What is the filter output from** $k = 0$ **to** $k = 5$**?**

iii. **How long does** $y(0)$ **contribute to the output?**

i. Just prior to $k = 0$ three zeros have entered the filter, and the output is zero.

ii. At $k = 0$ the digit "5" has entered the filter, and it will be multiplied by $w_{1,1}$, which has the value 2, so that $a(0) = 10$. This can be viewed as the matrix operation:

$$a(0) = \mathbf{W}\mathbf{p}(0) = \begin{bmatrix} w_{1,1} & w_{1,2} & w_{1,3} \end{bmatrix} \begin{bmatrix} y(0) \\ y(-1) \\ y(-2) \end{bmatrix} = \begin{bmatrix} 2 & -1 & 3 \end{bmatrix} \begin{bmatrix} 5 \\ 0 \\ 0 \end{bmatrix} = 10 .$$

Similarly, one can calculate the next outputs as

$$a(1) = \mathbf{W}\mathbf{p}(1) = \begin{bmatrix} 2 & -1 & 3 \end{bmatrix} \begin{bmatrix} -4 \\ 5 \\ 0 \end{bmatrix} = -13$$

$$a(2) = \mathbf{W}\mathbf{p}(2) = \begin{bmatrix} 2 & -1 & 3 \end{bmatrix} \begin{bmatrix} 0 \\ -4 \\ 5 \end{bmatrix} = 19$$

$$a(3) = \mathbf{W}\mathbf{p}(3) = \begin{bmatrix} 2 & -1 & 3 \end{bmatrix} \begin{bmatrix} 0 \\ 0 \\ -4 \end{bmatrix} = -12 , \; a(4) = \mathbf{W}\mathbf{p}(4) = \begin{bmatrix} 2 & -1 & 3 \end{bmatrix} \begin{bmatrix} 0 \\ 0 \\ 0 \end{bmatrix} = 0 .$$

All remaining outputs will be zero.

iii. The effects of $y(0)$ last from $k = 0$ through $k = 2$, so it will have an influence for three time intervals. This corresponds to the length of the impulse response of this filter.

P10.2 Suppose that we want to design an ADALINE network to distinguish between various categories of input vectors. Let us first try the categories listed below:

Category I: $\mathbf{p}_1 = \begin{bmatrix} 1 & 1 \end{bmatrix}^T$ **and** $\mathbf{p}_2 = \begin{bmatrix} -1 & -1 \end{bmatrix}^T$

Category II: $\mathbf{p}_3 = \begin{bmatrix} 2 & 2 \end{bmatrix}^T$.

i. **Can an ADALINE network be designed to make such a distinction?**

ii. **If the answer to part (i) is yes, what set of weights and bias might be used?**

Next consider a different set of categories.

Category III: $\mathbf{p}_1 = \begin{bmatrix} 1 & 1 \end{bmatrix}^T$ **and** $\mathbf{p}_2 = \begin{bmatrix} 1 & -1 \end{bmatrix}^T$

Category IV: $\mathbf{p}_3 = \begin{bmatrix} 1 & 0 \end{bmatrix}^T$.

iii. Can an ADALINE network be designed to make such a distinction?

iv. If the answer to part (iii) is yes, what set of weights and bias might be used?

i. The input vectors are plotted in Figure P10.2.

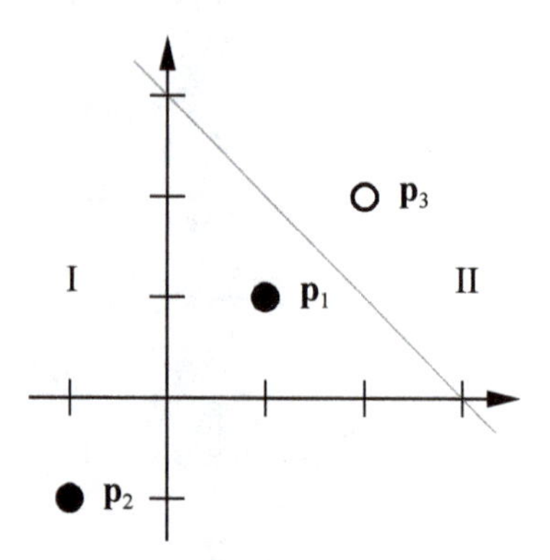

Figure P10.2 Input Vectors for Problem P10.1 (i)

The blue line in this figure is a decision boundary that separates the two categories successfully. Since they are linearly separable, an ADALINE network will do the job.

ii. The decision boundary passes through the points $(3, 0)$ and $(0, 3)$. We know these points to be the intercepts $-b/w_{1,1}$ and $-b/w_{1,2}$. Thus, a solution

$$b = 3, \; w_{1,1} = -1, \; w_{1,2} = -1,$$

is satisfactory. Note that if the output of the ADALINE is positive or zero the input vector is classified as Category I, and if the output is negative the input vector is classified as Category II. This solution also provides for error, since the decision boundary bisects the line between $\mathbf{p}_1$ and $\mathbf{p}_3$.

iii. The input vectors to be distinguished are shown in Figure P10.3. The vectors in the figure are not linearly separable, so an ADALINE network cannot distinguish between them.

iv. As noted in part (iii), an ADALINE cannot do the job, so there are no values for the weights and bias that are satisfactory.

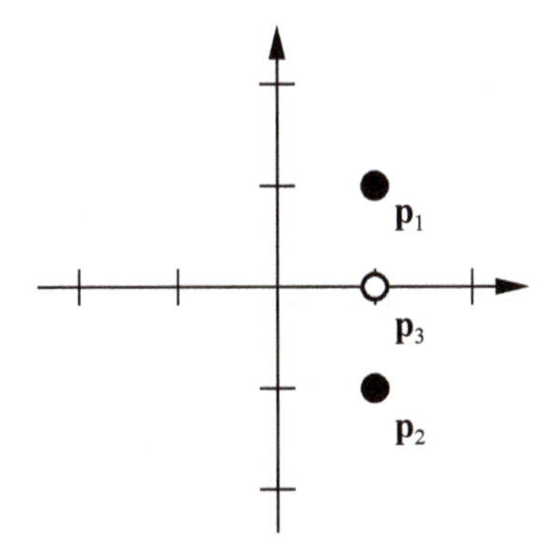

Figure P10.3 Input Vectors for Problem P10.1 (iii)

P10.3 Suppose that we have the following input/target pairs:

$$\left\{\mathbf{p}_1 = \begin{bmatrix} 1 \\ 1 \end{bmatrix}, t_1 = 1\right\}, \left\{\mathbf{p}_2 = \begin{bmatrix} 1 \\ -1 \end{bmatrix}, t_2 = -1\right\}.$$

These patterns occur with equal probability, and they are used to train an ADALINE network with no bias. What does the mean square error performance surface look like?

First we need to calculate the various terms of the quadratic function. Recall from Eq. (10.11) that the performance index can be written as

$$F(\mathbf{x}) = c - 2\mathbf{x}^T\mathbf{h} + \mathbf{x}^T\mathbf{R}\mathbf{x}.$$

Therefore we need to calculate c, $\mathbf{h}$ and $\mathbf{R}$.

The probability of each input occurring is 0.5, so the probability of each target is also 0.5. Thus, the expected value of the square of the targets is

$$c = E[t^2] = (1)^2(0.5) + (-1)^2(0.5) = 1.$$

In a similar way, the cross-correlation between the input and the target can be calculated:

$$\mathbf{h} = E[t\mathbf{z}] = (0.5)(1)\begin{bmatrix} 1 \\ 1 \end{bmatrix} + (0.5)(-1)\begin{bmatrix} 1 \\ -1 \end{bmatrix} = \begin{bmatrix} 0 \\ 1 \end{bmatrix}.$$

Finally, the input correlation matrix $\mathbf{R}$ is

$$\mathbf{R} = E[\mathbf{z}\mathbf{z}^T] = \mathbf{p}_1\mathbf{p}_1^T(0.5) + \mathbf{p}_2\mathbf{p}_2^T(0.5)$$

$$= (0.5)\left[\begin{bmatrix}1\\1\end{bmatrix}\begin{bmatrix}1 & 1\end{bmatrix} + \begin{bmatrix}1\\-1\end{bmatrix}\begin{bmatrix}1 & -1\end{bmatrix}\right] = \begin{bmatrix}1 & 0\\0 & 1\end{bmatrix}$$

Therefore the mean square error performance index is

$$F(\mathbf{x}) = c - 2\mathbf{x}^T\mathbf{h} + \mathbf{x}^T\mathbf{R}\mathbf{x}$$

$$= 1 - 2\begin{bmatrix}w_{1,1} & w_{1,2}\end{bmatrix}\begin{bmatrix}0\\1\end{bmatrix} + \begin{bmatrix}w_{1,1} & w_{1,2}\end{bmatrix}\begin{bmatrix}1 & 0\\0 & 1\end{bmatrix}\begin{bmatrix}w_{1,1}\\w_{1,2}\end{bmatrix}$$

$$= 1 - 2w_{1,2} + w_{1,1}^2 + w_{1,2}^2$$

The Hessian matrix of $F(\mathbf{x})$, which is equal to $2\mathbf{R}$, has both eigenvalues at 2. Therefore the contours of the performance surface will be circular. To find the center of the contours (the minimum point), we need to solve Eq. (10.18):

$$\mathbf{x}^* = \mathbf{R}^{-1}\mathbf{h} = \begin{bmatrix}1 & 0\\0 & 1\end{bmatrix}^{-1}\begin{bmatrix}0\\1\end{bmatrix} = \begin{bmatrix}0\\1\end{bmatrix}.$$

Thus we have a minimum at $w_{1,1} = 0$, $w_{1,2} = 1$. The resulting mean square error performance surface is shown in Figure P10.4.

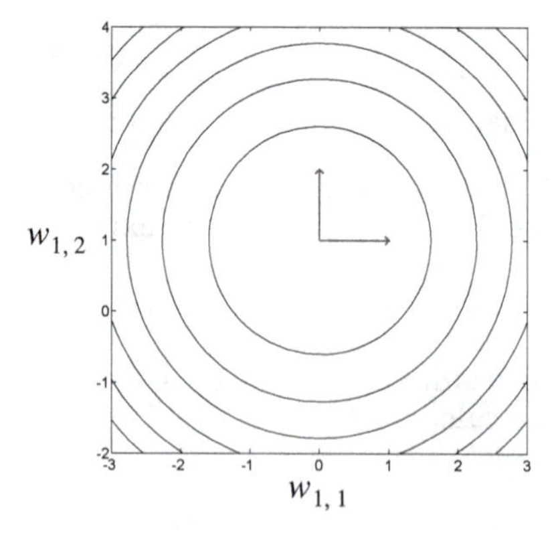

Figure P10.4 Contour Plot of $F(\mathbf{x})$ for Problem P10.3

P10.4 Consider the system of Problem P10.3 again. Train the network using the LMS algorithm, with the initial guess set to zero and a learning rate $\alpha = 0.25$. Apply each reference pattern only once during training. Draw the decision boundary at each stage.

Assume the input vector $\mathbf{p}_1$ is presented first. The output, error and new weights are calculated as follows:

$$a(0) = purelin\left(\begin{bmatrix} 0 & 0 \end{bmatrix}\begin{bmatrix} 1 \\ 1 \end{bmatrix}\right) = 0,$$

$$e(0) = t(0) - a(0) = 1 - 0 = 1,$$

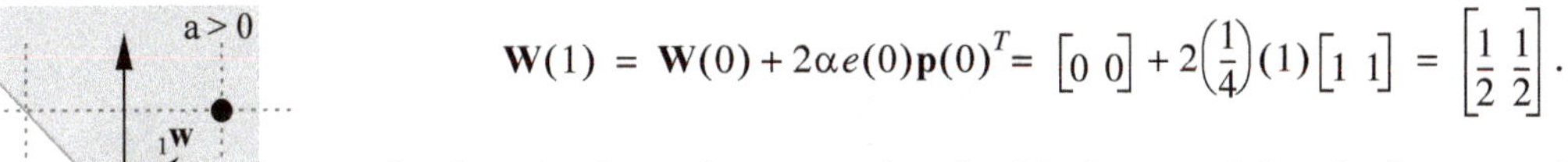

$$\mathbf{W}(1) = \mathbf{W}(0) + 2\alpha e(0)\mathbf{p}(0)^T = \begin{bmatrix} 0 & 0 \end{bmatrix} + 2\left(\frac{1}{4}\right)(1)\begin{bmatrix} 1 & 1 \end{bmatrix} = \begin{bmatrix} \frac{1}{2} & \frac{1}{2} \end{bmatrix}.$$

The decision boundary associated with these weights is shown to the left.

Now apply the second input vector:

$$a(1) = purelin\left\{\begin{bmatrix} \frac{1}{2} & \frac{1}{2} \end{bmatrix}\begin{bmatrix} 1 \\ -1 \end{bmatrix}\right\} = 0,$$

$$e(1) = t(1) - a(1) = -1 - 0 = -1,$$

$$\mathbf{W}(2) = \mathbf{W}(1) + 2\alpha e(1)\mathbf{p}(1)^T = \begin{bmatrix} \frac{1}{2} & \frac{1}{2} \end{bmatrix} + 2\left(\frac{1}{4}\right)(-1)\begin{bmatrix} 1 & -1 \end{bmatrix} = \begin{bmatrix} 0 & 1 \end{bmatrix}.$$

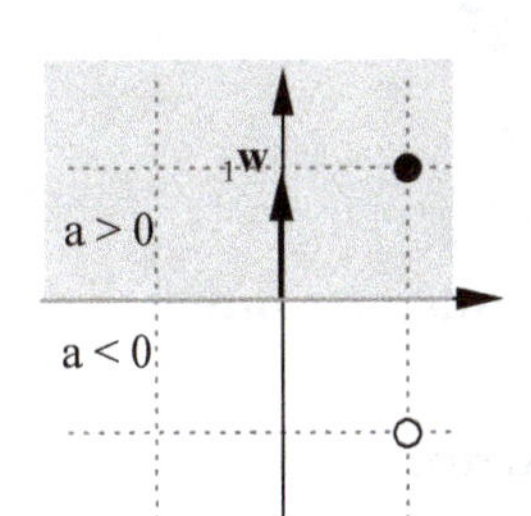

The decision boundary associated with these weights is shown to the left. This boundary shows real promise. It is exactly halfway between the input vectors. You might verify for yourself that each input vector, when applied, yields its correct associated target. (What set of weights would be optimal if the targets associated with the two input vectors were exchanged?)

P10.5 Now consider the convergence of the system of Problems P10.3 and P10.4. What is the maximum stable learning rate for the LMS algorithm?

```
» 2 + 2
ans =
   4
```

The LMS convergence is determined by the learning rate α, which should not exceed the reciprocal of the largest eigenvalue of $\mathbf{R}$. We can determine this limit by finding these eigenvalues using MATLAB.

```
[V,D] = eig (R)
V =
        1      0
        0      1

D=
        1      0
        0      1
```

The diagonal terms of matrix D give the eigenvalues, 1 and 1, while the columns of V show the eigenvectors. Note, incidentally, that the eigenvectors have the same direction as those shown in Figure P10.4.

The largest eigenvalue, $\lambda_{max} = 1$, sets the upper limit on the learning rate at

$$\alpha < 1/\lambda_{max} = 1/1 = 1 .$$

The suggested learning rate in the previous problem was 0.25, and you found (perhaps) that the LMS algorithm converged quickly. What do you suppose happens when the learning rate is 1.0 or larger?

P10.6 Consider the adaptive filter ADALINE shown in Figure P10.5. The purpose of this filter is to predict the next value of the input signal from the two previous values. Suppose that the input signal is a stationary random process, with autocorrelation function given by

$$C_y(n) = E[y(k)y(k+n)]$$

$$C_y(0) = 3 ,\ C_y(1) = -1 ,\ C_y(2) = -1 .$$

i. **Sketch the contour plot of the performance index (mean square error).**

ii. **What is the maximum stable value of the learning rate (α) for the LMS algorithm?**

iii. **Assume that a very small value is used for α. Sketch the path of the weights for the LMS algorithm, starting with initial guess $\mathbf{W}(0) = \begin{bmatrix} 0.75 & 0 \end{bmatrix}^T$. Explain your procedure for sketching the path.**

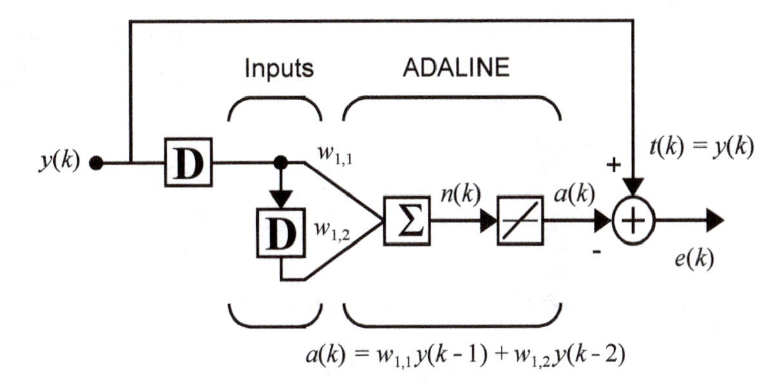

Figure P10.5 Adaptive Predictor

i. To sketch the contour plot we first need to find the performance index and the eigenvalues and eigenvectors of the Hessian matrix. First note that the input vector is given by

$$\mathbf{z}(k) = \mathbf{p}(k) = \begin{bmatrix} y(k-1) \\ y(k-2) \end{bmatrix}.$$

Now consider the performance index. Recall from Eq. (10.12) that

$$F(\mathbf{x}) = c - 2\mathbf{x}^T\mathbf{h} + \mathbf{x}^T\mathbf{R}\mathbf{x}.$$

We can calculate the constants in the performance index as shown below:

$$c = E[t^2(k)] = E[y^2(k)] = C_y(0) = 3,$$

$$\mathbf{R} = E[\mathbf{z}\mathbf{z}^T] = E\begin{bmatrix} y^2(k-1) & y(k-1)y(k-2) \\ y(k-1)y(k-2) & y^2(k-2) \end{bmatrix}$$

$$= \begin{bmatrix} C_y(0) & C_y(1) \\ C_y(1) & C_y(0) \end{bmatrix} = \begin{bmatrix} 3 & -1 \\ -1 & 3 \end{bmatrix}$$

$$\mathbf{h} = E\left[t\,\mathbf{z}\right] = E\begin{bmatrix} y(k)y(k-1) \\ y(k)y(k-2) \end{bmatrix} = \begin{bmatrix} C_y(1) \\ C_y(2) \end{bmatrix} = \begin{bmatrix} -1 \\ -1 \end{bmatrix}.$$

The optimal weights are

$$\mathbf{x}^* = \mathbf{R}^{-1}\mathbf{h} = \begin{bmatrix} 3 & -1 \\ -1 & 3 \end{bmatrix}^{-1} \begin{bmatrix} -1 \\ -1 \end{bmatrix} = \begin{bmatrix} 3/8 & 1/8 \\ 4/8 & 3/8 \end{bmatrix} \begin{bmatrix} -1 \\ -1 \end{bmatrix} = \begin{bmatrix} -1/2 \\ -1/2 \end{bmatrix}.$$

The Hessian matrix is

$$\nabla^2 F(\mathbf{x}) = \mathbf{A} = 2\mathbf{R} = \begin{bmatrix} 6 & -2 \\ -2 & 6 \end{bmatrix}.$$

Now we can get the eigenvalues:

$$\left| \mathbf{A} - \lambda\mathbf{I} \right| = \begin{vmatrix} 6-\lambda & -2 \\ -2 & 6-\lambda \end{vmatrix} = \lambda^2 - 12\lambda + 32 = (\lambda - 8)(\lambda - 4).$$

Thus,

$$\lambda_1 = 4, \qquad \lambda_2 = 8.$$

To find the eigenvectors we use

$$[\mathbf{A} - \lambda\mathbf{I}]\mathbf{v} = 0.$$

For $\lambda_1 = 4$,

$$\begin{bmatrix} 2 & -2 \\ -2 & 2 \end{bmatrix} \mathbf{v}_1 = 0 \qquad \mathbf{v}_1 = \begin{bmatrix} -1 \\ -1 \end{bmatrix},$$

and for $\lambda_2 = 8$,

$$\begin{bmatrix} -2 & -2 \\ -2 & -2 \end{bmatrix} \mathbf{v}_2 = 0 \qquad \mathbf{v}_2 = \begin{bmatrix} -1 \\ 1 \end{bmatrix}.$$

Therefore the contours of $F(\mathbf{x})$ will be elliptical, with the long axis of each ellipse along the first eigenvector, since the first eigenvalue has the smallest magnitude. The ellipses will be centered at $\mathbf{x}^*$. The contour plot is shown in Figure P10.6.

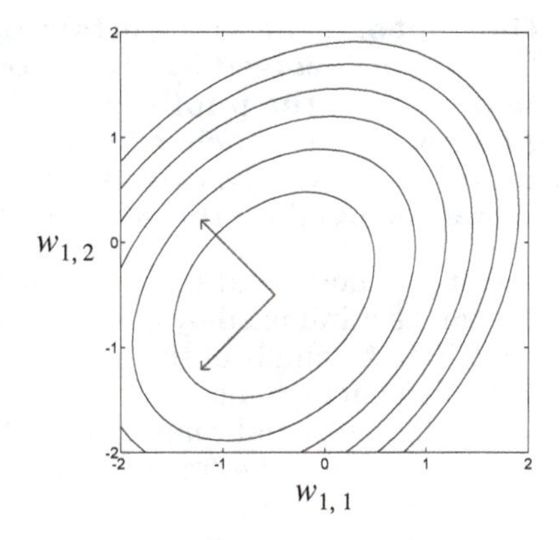

Figure P10.6 Error Contour for Problem P10.6

You might check your sketch by writing a MATLAB M-file to plot the contours.

ii. The maximum stable learning rate is the reciprocal of the maximum eigenvalue of $\mathbf{R}$, which is the same as twice the reciprocal of the largest eigenvalue of the Hessian matrix $\nabla^2 F(\mathbf{x}) = \mathbf{A}$:

$$\alpha < 2/\lambda_{max} = 2/8 = 0.25 .$$

iii. The LMS algorithm is approximate steepest descent, so the trajectory for small learning rates will move perpendicular to the contour lines, as shown in Figure P10.7.

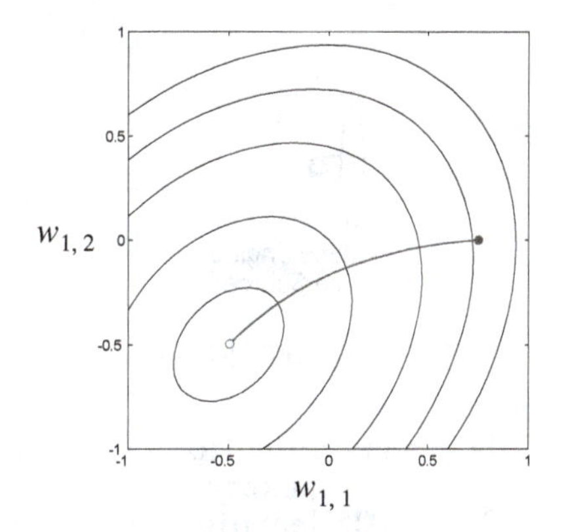

Figure P10.7 LMS Weight Trajectory

P10.7 The pilot of an airplane is talking into a microphone in his cockpit. The sound received by the air traffic controller in the tower is garbled because the pilot's voice signal has been contaminated by engine noise that reaches his microphone. Can you suggest an adaptive ADALINE filter that might help reduce the noise in the signal received by the control tower? Explain your system.

The engine noise that has been inadvertently added to the microphone input can be minimized by using the adaptive filtering system shown in Figure P10.8. A sample of the engine noise is supplied to an adaptive filter through a microphone in the cockpit. The desired output of the filter is the contaminated signal coming from the pilot's microphone. The filter attempts to reduce the "error" signal to a minimum. It can do this only by subtracting the component of the contaminated signal that is linearly correlated with the engine noise (and presumably uncorrelated with the pilot's voice). The result is that a clear voice signal is sent to the control tower, in spite of the fact that the engine noise got into the pilot's microphone along with his voice signal. (See [WiSt85] for discussion of similar noise cancellation systems.)

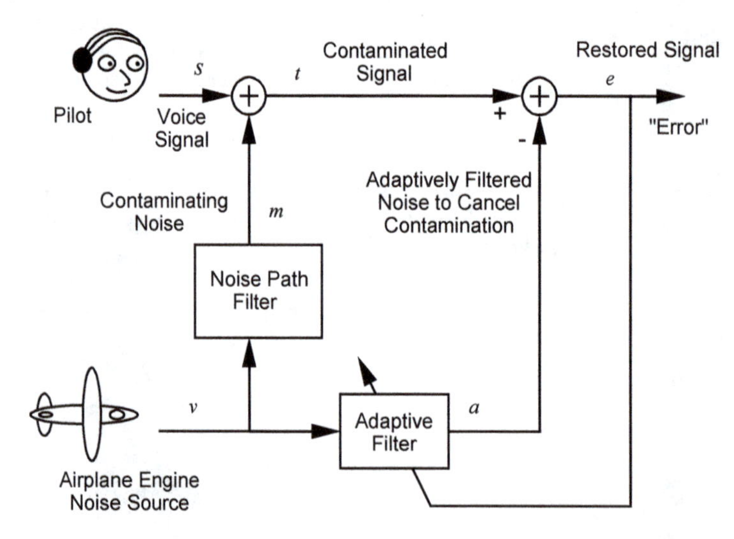

Figure P10.8 Filtering Engine Noise from Pilot's Voice Signal

P10.8 This is a classification problem like that described in Problems P4.3 and P4.5, except that here we will use an ADALINE network and the LMS learning rule rather than the perceptron learning rule. First we will describe the problem.

We have a classification problem with four classes of input vector. The four classes are

$$\text{class 1:}\left\{\mathbf{p}_1 = \begin{bmatrix}1\\1\end{bmatrix}, \mathbf{p}_2 = \begin{bmatrix}1\\2\end{bmatrix}\right\}, \text{class 2: }\left\{\mathbf{p}_3 = \begin{bmatrix}2\\-1\end{bmatrix}, \mathbf{p}_4 = \begin{bmatrix}2\\0\end{bmatrix}\right\},$$

$$\text{class 3:}\left\{\mathbf{p}_5 = \begin{bmatrix}-1\\2\end{bmatrix}, \mathbf{p}_6 = \begin{bmatrix}-2\\1\end{bmatrix}\right\}, \text{class 4: }\left\{\mathbf{p}_7 = \begin{bmatrix}-1\\-1\end{bmatrix}, \mathbf{p}_8 = \begin{bmatrix}-2\\-2\end{bmatrix}\right\}.$$

Train an ADALINE network to solve this problem using the LMS learning rule. Assume that each pattern occurs with probability $1/8$.

Let's begin by displaying the input vectors, as in Figure P10.9. The light circles ○ indicate class 1 vectors, the light squares □ indicate class 2 vectors, the dark circles ● indicate class 3 vectors, and the dark squares ■ indicate class 4 vectors. These input vectors can be plotted as shown in Figure P10.9.

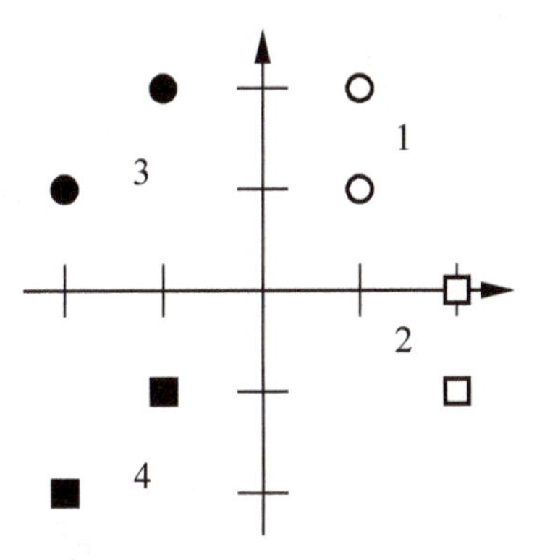

Figure P10.9 Input Vectors for Problem P10.8

We will use target vectors similar to the ones we introduced in Problem P4.3, except that we will replace any targets of 0 by targets of -1. (The perceptron could only output 0 or 1.) Thus, the training set will be:

$$\left\{\mathbf{p}_1 = \begin{bmatrix}1\\1\end{bmatrix}, \mathbf{t}_1 = \begin{bmatrix}-1\\-1\end{bmatrix}\right\}\left\{\mathbf{p}_2 = \begin{bmatrix}1\\2\end{bmatrix}, \mathbf{t}_2 = \begin{bmatrix}-1\\-1\end{bmatrix}\right\}\left\{\mathbf{p}_3 = \begin{bmatrix}2\\-1\end{bmatrix}, \mathbf{t}_3 = \begin{bmatrix}-1\\1\end{bmatrix}\right\}$$

$$\left\{\mathbf{p}_4 = \begin{bmatrix}2\\0\end{bmatrix}, \mathbf{t}_4 = \begin{bmatrix}-1\\1\end{bmatrix}\right\}\left\{\mathbf{p}_5 = \begin{bmatrix}-1\\2\end{bmatrix}, \mathbf{t}_5 = \begin{bmatrix}1\\-1\end{bmatrix}\right\}\left\{\mathbf{p}_6 = \begin{bmatrix}-2\\1\end{bmatrix}, \mathbf{t}_6 = \begin{bmatrix}1\\-1\end{bmatrix}\right\}$$

$$\left\{\mathbf{p}_7 = \begin{bmatrix}-1\\-1\end{bmatrix}, \mathbf{t}_7 = \begin{bmatrix}1\\1\end{bmatrix}\right\}\left\{\mathbf{p}_8 = \begin{bmatrix}-2\\-2\end{bmatrix}, \mathbf{t}_8 = \begin{bmatrix}1\\1\end{bmatrix}\right\}$$

Also, we will begin as in Problem P4.5 with the following initial weights and biases:

$$\mathbf{W}(0) = \begin{bmatrix} 1 & 0 \\ 0 & 1 \end{bmatrix}, \mathbf{b}(0) = \begin{bmatrix} 1 \\ 1 \end{bmatrix}.$$

Now we are almost ready to train an ADALINE network using the LMS rule. We will use a learning rate of $\alpha = 0.04$, and we will present the input vectors in order according to their subscripts. The first iteration is

$$\mathbf{a}(0) = purelin(\mathbf{W}(0)\mathbf{p}(0) + \mathbf{b}(0)) = purelin\left(\begin{bmatrix} 1 & 0 \\ 0 & 1 \end{bmatrix}\begin{bmatrix} 1 \\ 1 \end{bmatrix} + \begin{bmatrix} 1 \\ 1 \end{bmatrix}\right) = \begin{bmatrix} 2 \\ 2 \end{bmatrix}$$

$$\mathbf{e}(0) = \mathbf{t}(0) - \mathbf{a}(0) = \begin{bmatrix} -1 \\ -1 \end{bmatrix} - \begin{bmatrix} 2 \\ 2 \end{bmatrix} = \begin{bmatrix} -3 \\ -3 \end{bmatrix}$$

$$\mathbf{W}(1) = \mathbf{W}(0) + 2\alpha\mathbf{e}(0)\mathbf{p}^T(0)$$

$$= \begin{bmatrix} 1 & 0 \\ 0 & 1 \end{bmatrix} + 2(0.04)\begin{bmatrix} -3 \\ -3 \end{bmatrix}\begin{bmatrix} 1 & 1 \end{bmatrix} = \begin{bmatrix} 0.76 & -0.24 \\ -0.24 & 0.76 \end{bmatrix}$$

$$\mathbf{b}(1) = \mathbf{b}(0) + 2\alpha\mathbf{e}(0) = \begin{bmatrix} 1 \\ 1 \end{bmatrix} + 2(0.04)\begin{bmatrix} -3 \\ -3 \end{bmatrix} = \begin{bmatrix} 0.76 \\ 0.76 \end{bmatrix}.$$

The second iteration is

$$\mathbf{a}(1) = purelin(\mathbf{W}(1)\mathbf{p}(1) + \mathbf{b}(1))$$

$$= purelin\left(\begin{bmatrix} 0.76 & -0.24 \\ -0.24 & 0.76 \end{bmatrix}\begin{bmatrix} 1 \\ 2 \end{bmatrix} + \begin{bmatrix} 0.76 \\ 0.76 \end{bmatrix}\right) = \begin{bmatrix} 1.04 \\ 2.04 \end{bmatrix}$$

$$\mathbf{e}(1) = \mathbf{t}(1) - \mathbf{a}(1) = \begin{bmatrix} -1 \\ -1 \end{bmatrix} - \begin{bmatrix} 1.04 \\ 2.04 \end{bmatrix} = \begin{bmatrix} -2.04 \\ -3.04 \end{bmatrix}$$

$$\mathbf{W}(2) = \mathbf{W}(1) + 2\alpha\mathbf{e}(1)\mathbf{p}^T(1)$$

$$= \begin{bmatrix} 0.76 & -0.24 \\ -0.24 & 0.76 \end{bmatrix} + 2(0.04)\begin{bmatrix} -2.04 \\ -3.04 \end{bmatrix}\begin{bmatrix} 1 & 2 \end{bmatrix} = \begin{bmatrix} 0.5968 & -0.5664 \\ -0.4832 & 0.2736 \end{bmatrix}$$

$$\mathbf{b}(2) = \mathbf{b}(1) + 2\alpha\mathbf{e}(1) = \begin{bmatrix} 0.76 \\ 0.76 \end{bmatrix} + 2(0.04)\begin{bmatrix} -2.04 \\ -3.04 \end{bmatrix} = \begin{bmatrix} 0.5968 \\ 0.5168 \end{bmatrix}.$$

If we continue until the weights converge we find

$$\mathbf{W}(\infty) = \begin{bmatrix} -0.5948 & -0.0523 \\ 0.1667 & -0.6667 \end{bmatrix}, \mathbf{b}(\infty) = \begin{bmatrix} 0.0131 \\ 0.1667 \end{bmatrix}.$$

The resulting decision boundaries are shown in Figure P10.10. Compare this result with the final decision boundaries created by the perceptron learning rule in Problem P4.5 (Figure P4.7). The perceptron rule stops training when all the patterns are classified correctly. The LMS algorithm moves the boundaries as far from the patterns as possible.

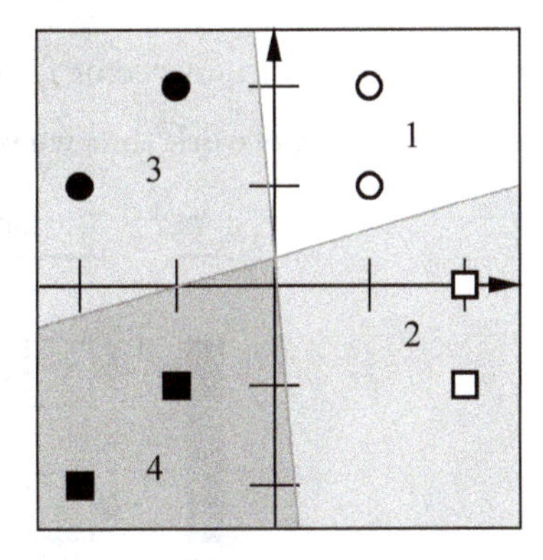

Figure P10.10 Final Decision Boundaries for Problem P10.8

P10.9 Repeat the work of Widrow and Hoff on a pattern recognition problem from their classic 1960 paper [WiHo60]. They wanted to design a recognition system that would classify the six patterns shown in Figure P10.11.

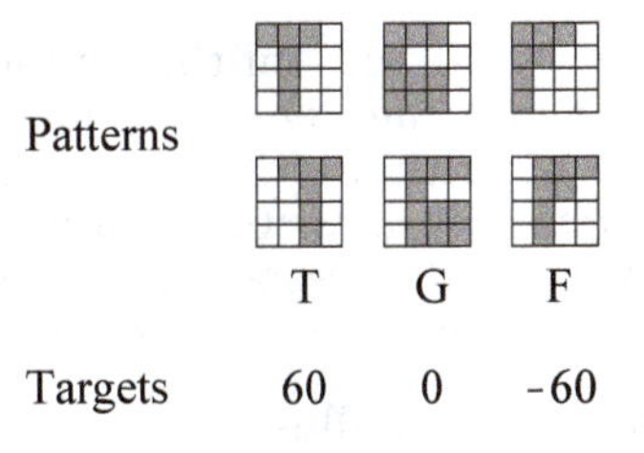

Figure P10.11 Patterns and Their Classification Targets

These patterns represent the letters T, G and F, in an original form on the top and in a shifted form on the bottom. The targets for these letters (in their original and shifted forms) are +60, 0 and -60, respectively. (The values of 60, 0 and -60 were nice for use on the face of a meter that Widrow and Hoff used to display their network output.) The objective is to train a network so that it will classify the six patterns into the appropriate T, G or F groups.

The blue squares in the letters will be assigned the value +1, and the white squares will be assigned the value -1. First we convert each of the letters into a single 16-element vector. We choose to do this by starting at the upper left corner, going down the left column, then going down the second column, etc. For example, the vector corresponding to the unshifted letter T is

$$\mathbf{p}_1 = \begin{bmatrix} 1 & -1 & -1 & -1 & 1 & 1 & 1 & 1 & -1 & -1 & -1 & -1 & -1 & -1 & -1 \end{bmatrix}^T$$

We have such an input vector for each of the six letters.

The ADALINE network that we will use is shown in Figure P10.12.

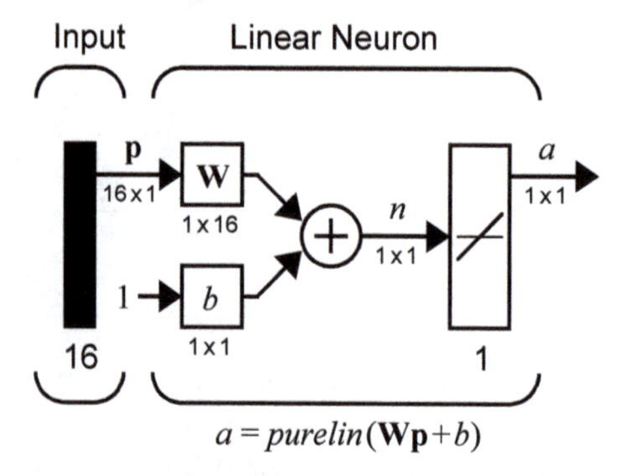

Figure P10.12 Adaptive Pattern Classifier

(Widrow and Hoff built their own machine to realize this ADALINE. According to them, it was "about the size of a lunch pail.")

Now we will present the six vectors to the network in a random sequence and adjust the weights of the network after each presentation using the LMS algorithm with a learning rate of $\alpha = 0.03$. After each adjustment of weights, all six vectors will be presented to the network to generate their outputs and corresponding errors. The sum of the squares of the errors will be examined as a measure of the quality of the network.

Figure P10.13 illustrates the convergence of the network. The network is trained to recognize these six characters in about 60 presentations, or roughly 10 for each of the possible input vectors.

The results shown in Figure P10.13 are quite like those obtained and published by Widrow and Hoff some 35 years ago. Widrow and Hoff did good science. One can indeed duplicate their work, even decades later (without a lunch pail).

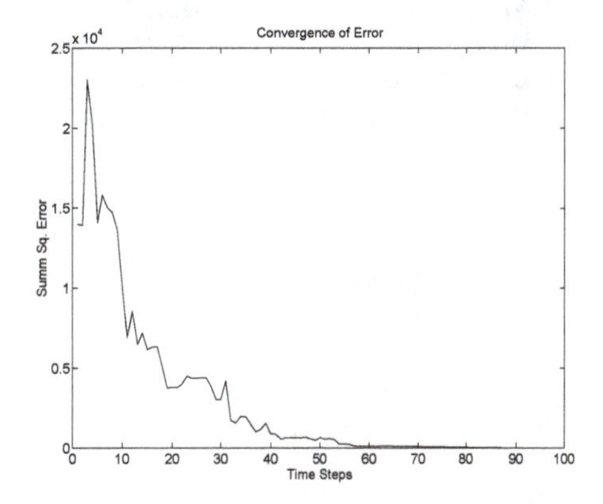

Figure P10.13 Error Convergence with Learning Rate of 0.03

To experiment with this character recognition problem, use the MATLAB® Neural Network Design Demonstration Linear Pattern Classification (`nnd10lc`). *Notice the sensitivity of the network to noise in the input pattern.*

Epilogue

In this chapter we have presented the ADALINE neural network and the LMS learning rule. The ADALINE network is very similar to the perceptron network of Chapter 4, and it has the same fundamental limitation: it can only classify linearly separable patterns. In spite of this limitation on the network, the LMS algorithm is in fact more powerful than the perceptron learning rule. Because it minimizes mean square error, the algorithm is able to create decision boundaries that are more robust to noise than those of the perceptron learning rule.

The ADALINE network and the LMS algorithm have found many practical applications. Even though they were first presented in the late 1950s, they are still very much in use in adaptive filtering applications. For example, echo cancellers using the LMS algorithm are currently employed on many long distance telephone lines. (Chapter 14 provides more extensive coverage of dynamic networks, which are widely used for filtering, prediction and control.)

In addition to its importance as a practical solution to many adaptive filtering problems, the LMS algorithm is also important because it is the forerunner of the backpropagation algorithm, which we will discuss in Chapter 11 through Chapter 14. Like the LMS algorithm, backpropagation is an approximate steepest descent algorithm that minimizes mean square error. The only difference between the two algorithms is in the manner in which the derivatives are calculated. Backpropagation is a generalization of the LMS algorithm that can be used for multilayer networks. These more complex networks are not limited to linearly separable problems. They can solve arbitrary classification problems.

Further Reading

[AnRo89] J. A. Anderson, E. Rosenfeld, *Neurocomputing: Foundations of Research*, Cambridge, MA: MIT Press, 1989.

Neurocomputing is a fundamental reference book. It contains over forty of the most important neurocomputing writings. Each paper is accompanied by an introduction that summarizes its results and gives a perspective on the position of the paper in the history of the field.

[StDo84] W. D. Stanley, G. R. Dougherty, R. Dougherty, *Digital Signal Processing*, Reston VA: Reston, 1984

[WiHo60] B. Widrow, M. E. Hoff, "Adaptive switching circuits," *1960 IRE WESCON Convention Record*, New York: IRE Part 4, pp. 96–104.

This seminal paper describes an adaptive perceptron-like network that can learn quickly and accurately. The authors assumed that the system had inputs, a desired output classification for each input, and that the system could calculate the error between the actual and desired output. The weights are adjusted, using a gradient descent method, so as to minimize the mean square error. (Least mean square error or LMS algorithm.)

This paper is reprinted in [AnRo88].

[WiSt 85] B. Widrow and S. D. Stearns, *Adaptive Signal Processing*, Englewood Cliffs, NJ: Prentice-Hall, 1985.

This informative book describes the theory and application of adaptive signal processing. The authors include a review of the mathematical background that is needed, give details on their adaptive algorithms, and then discuss practical information about many applications.

[WiWi 88] B. Widrow and R. Winter, "Neural nets for adaptive filtering and adaptive pattern recognition," *IEEE Computer Magazine*, March 1988, pp. 25–39.

This is a particularly readable paper that summarizes applications of adaptive multilayer neural networks. The networks are applied to system modeling, statistical prediction, echo cancellation, inverse modeling and pattern recognition.

Exercises

E10.1 An adaptive filter ADALINE is shown in Figure E10.1. Suppose that the weights of the network are given by

$$w_{1,1} = 1, \; w_{1,2} = -4, \; w_{1,3} = 2,$$

and the input to the filter is

$$\{y(k)\} = \{\ldots, 0, 0, 0, 1, 1, 2, 0, 0, \ldots\}.$$

Find the response $\{a(k)\}$ of the filter.

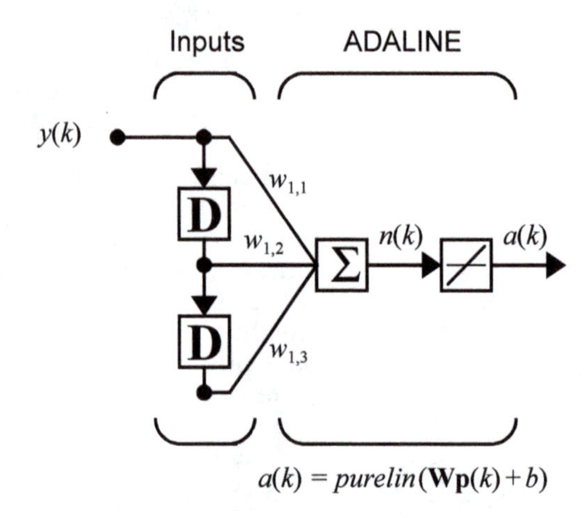

Figure E10.1 Adaptive Filter ADALINE for Exercise E10.1

E10.2 In Figure E10.2 two classes of patterns are given.

i. Use the LMS algorithm to train an ADALINE network to distinguish between class I and class II patterns (we want the network to identify horizontal and vertical lines).

ii. Can you explain why the ADALINE network might have difficulty with this problem?

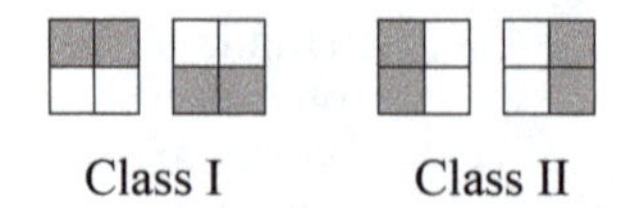

Class I Class II

Figure E10.2 Pattern Classification Problem for Exercise E10.2

E10.3 Suppose that we have the following two reference patterns and their targets:

$$\left\{\mathbf{p}_1 = \begin{bmatrix} 1 \\ 1 \end{bmatrix}, t_1 = 1\right\}, \left\{\mathbf{p}_2 = \begin{bmatrix} 1 \\ -1 \end{bmatrix}, t_2 = -1\right\}.$$

In Problem P10.3 these input vectors to an ADALINE were assumed to occur with equal probability. Now suppose that the probability of vector $\mathbf{p}_1$ is 0.75 and that the probability of vector $\mathbf{p}_2$ is 0.25. Does this change of probabilities change the mean square error surface? If yes, what does the surface look like now? What is the maximum stable learning rate?

E10.4 In this exercise we will modify the reference pattern $\mathbf{p}_2$ from Problem P10.3:

$$\left\{\mathbf{p}_1 = \begin{bmatrix} 1 \\ 1 \end{bmatrix}, t_1 = 1\right\}, \left\{\mathbf{p}_2 = \begin{bmatrix} -1 \\ -1 \end{bmatrix}, t_2 = -1\right\}.$$

i. Assume that the patterns occur with equal probability. Find the mean square error and sketch the contour plot.

ii. Find the maximum stable learning rate.

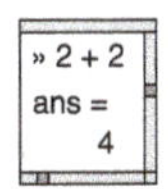

iii. Write a MATLAB M-file to implement the LMS algorithm for this problem. Take 40 steps of the algorithm for a stable learning rate. Use the zero vector as the initial guess. Sketch the trajectory on the contour plot.

iv. Take 40 steps of the algorithm after setting the initial values of both parameters to 1. Sketch the final decision boundary.

v. Compare the final parameters from parts (iii) and (iv). Explain your results.

E10.5 We again use the reference patterns and targets from Problem P10.3, and assume that they occur with equal probability. This time we want to train an ADALINE network with a bias. We now have three parameters to find: $w_{1,1}$, $w_{1,2}$ and b.

i. Find the mean square error and the maximum stable learning rate.

» 2 + 2
ans =
4

ii. Write a MATLAB M-file to implement the LMS algorithm for this problem. Take 40 steps of the algorithm for a stable learning rate. Use the zero vector as the initial guess. Sketch the final decision boundary.

iii. Take 40 steps of the algorithm after setting the initial values of all parameters to 1. Sketch the final decision boundary.

iv. Compare the final parameters and the decision boundaries from parts (iii) and (iv). Explain your results.

E10.6 We have two categories of vectors. Category I consists of

$$\left\{\begin{bmatrix}1\\1\end{bmatrix}, \begin{bmatrix}-1\\2\end{bmatrix}\right\}.$$

Category II consists of

$$\left\{\begin{bmatrix}0\\-1\end{bmatrix}, \begin{bmatrix}-4\\1\end{bmatrix}\right\}.$$

We want to train a single-neuron ADALINE network without a bias to recognize these categories ($t = 1$ for Category I and $t = -1$ for Category II). Assume that each pattern occurs with equal probability.

i. Draw the network diagram.

ii. Take four steps of the LMS algorithm, using the zero vector as the initial guess. (one pass through the four vectors above - present each vector once). Use a learning rate of 0.1.

iii. What are the optimal weights?

iv. Sketch the optimal decision boundary.

v. How do you think the boundary would change if the network were allowed to have a bias? If the boundary would change, indicate the approximate new position on your sketch of part iv. You do not need to perform any calculations here - just explain your reasoning.

E10.7 Suppose that we have the following three reference patterns and their targets:

$$\left\{\mathbf{p}_1 = \begin{bmatrix}3\\6\end{bmatrix}, t_1 = \begin{bmatrix}75\end{bmatrix}\right\}, \left\{\mathbf{p}_2 = \begin{bmatrix}6\\3\end{bmatrix}, t_2 = \begin{bmatrix}75\end{bmatrix}\right\}, \left\{\mathbf{p}_3 = \begin{bmatrix}-6\\3\end{bmatrix}, t_3 = \begin{bmatrix}-75\end{bmatrix}\right\}.$$

Each pattern is equally likely.

i. Draw the network diagram for an ADALINE network with no bias that could be trained on these patterns.

ii. We want to train the ADALINE network with no bias using these patterns. Sketch the contour plot of the mean square error performance index.

iii. Find the maximum stable learning rate for the LMS algorithm.

iv. Sketch the trajectory of the LMS algorithm on your contour plot. Assume a very small learning rate, and start with all weights equal to zero. This does not require any calculations.

E10.8 Suppose that we have the following two reference patterns and their targets:

$$\left\{\mathbf{p}_1 = \begin{bmatrix}1\\2\end{bmatrix}, t_1 = \begin{bmatrix}-1\end{bmatrix}\right\}, \left\{\mathbf{p}_2 = \begin{bmatrix}-2\\1\end{bmatrix}, t_2 = \begin{bmatrix}1\end{bmatrix}\right\}.$$

The probability of vector $\mathbf{p}_1$ is 0.5 and the probability of vector $\mathbf{p}_2$ is 0.5.We want to train an ADALINE network without a bias on this data set.

i. Sketch the contour plot of the mean square error performance index.

ii. Sketch the optimal decision boundary.

iii. Find the maximum stable learning rate.

iv. Sketch the trajectory of the LMS algorithm on your contour plot. Assume a very small learning rate, and start with initial weights $\mathbf{W}(0) = \begin{bmatrix}0 & 1\end{bmatrix}$.

E10.9 We have the following input/target pairs:

$$\left\{\mathbf{p}_1 = \begin{bmatrix}4\\2\end{bmatrix}, t_1 = 5\right\}, \left\{\mathbf{p}_2 = \begin{bmatrix}2\\-4\end{bmatrix}, t_2 = -2\right\}, \left\{\mathbf{p}_3 = \begin{bmatrix}-4\\4\end{bmatrix}, t_3 = 9\right\}.$$

The first two pair each occurs with probability of 0.25, and the third pair occurs with probability 0.5. We want to train a single-neuron ADALINE network without a bias to perform the desired mapping.

i. Draw the network diagram.

ii. What is the maximum stable learning rate?

iii. Perform one iteration of the LMS algorithm. Apply the input $\mathbf{p}_1$ and use a learning rate of $\alpha = 0.1$. Start from the initial weights $\mathbf{x}_0 = \begin{bmatrix}0 & 0\end{bmatrix}^T$.

E10.10 Repeat E10.9 for the following input/target pairs:

$$\left\{\mathbf{p}_1 = \begin{bmatrix} 2 \\ -4 \end{bmatrix}, t_1 = 1\right\}, \left\{\mathbf{p}_2 = \begin{bmatrix} -4 \\ 4 \end{bmatrix}, t_2 = -1\right\}, \left\{\mathbf{p}_3 = \begin{bmatrix} 4 \\ 2 \end{bmatrix}, t_3 = 1\right\}.$$

The first two pair each occurs with probability of 0.25, and the third pair occurs with probability 0.5. We want to train a single-neuron ADALINE network without a bias to perform the desired mapping.

E10.11 We want to train a single-neuron ADALINE network without a bias, using the following training set, which categorizes vectors into two classes. Each pattern occurs with equal probability.

$$\left\{\mathbf{p}_1 = \begin{bmatrix} -1 \\ 2 \end{bmatrix}, t_1 = -1\right\}\left\{\mathbf{p}_2 = \begin{bmatrix} 2 \\ -1 \end{bmatrix}, t_2 = -1\right\}\left\{\mathbf{p}_3 = \begin{bmatrix} 0 \\ -1 \end{bmatrix}, t_3 = 1\right\}\left\{\mathbf{p}_4 = \begin{bmatrix} -1 \\ 0 \end{bmatrix}, t_4 = 1\right\}$$

i. Draw the network diagram.

ii. Take one step of the LMS algorithm (present $\mathbf{p}_1$ only) starting from the initial weight $\mathbf{W}(0) = \begin{bmatrix} 0 & 0 \end{bmatrix}$. Use a learning rate of 0.1.

iii. What are the optimal weights? Show all calculations.

iv. Sketch the optimal decision boundary.

v. How do you think the boundary would change if the network were allowed to have a bias? Indicate the approximate new position on your sketch of part iv.

vi. What is the maximum stable learning rate for the LMS algorithm?

vii. Sketch the contour plot of the mean square error performance surface.

viii. On your contour plot of part vii, sketch the path of the LMS algorithm for a very small learning rate (e.g., 0.001) starting from the initial condition $\mathbf{W}(0) = \begin{bmatrix} 2 & 0 \end{bmatrix}$. This does not require any calculations, but explain how you obtained your answer.

E10.12 Suppose that we have the following three reference patterns and their targets:

$$\left\{\mathbf{p}_1 = \begin{bmatrix} 2 \\ 4 \end{bmatrix}, t_1 = \begin{bmatrix} 26 \end{bmatrix}\right\}, \left\{\mathbf{p}_2 = \begin{bmatrix} 4 \\ 2 \end{bmatrix}, t_2 = \begin{bmatrix} 26 \end{bmatrix}\right\}, \left\{\mathbf{p}_3 = \begin{bmatrix} -2 \\ -2 \end{bmatrix}, t_3 = \begin{bmatrix} -26 \end{bmatrix}\right\}.$$

The probability of vector $\mathbf{p}_1$ is 0.25, the probability of vector $\mathbf{p}_2$ is 0.25 and the probability of vector $\mathbf{p}_3$ is 0.5.

i. Draw the network diagram for an ADALINE network with no bias that could be trained on these patterns.

ii. Sketch the contour plot of the mean square error performance index.

iii. Show the optimal decision boundary (for the weights that minimize mean square error) and verify that it separates the patterns into the appropriate categories.

iv. Find the maximum stable learning rate for the LMS algorithm. If the target values are changed from 26 and -26 to 2 and -2, how would this change the maximum stable learning rate?

v. Perform one iteration of the LMS algorithm, starting with all weights equal to zero, and presenting input vector $\mathbf{p}_1$. Use a learning rate of $\alpha = 0.5$.

vi. Sketch the trajectory of the LMS algorithm on your contour plot. Assume a very small learning rate, and start with all weights equal to zero.

E10.13 Consider the adaptive predictor in Figure E10.3.

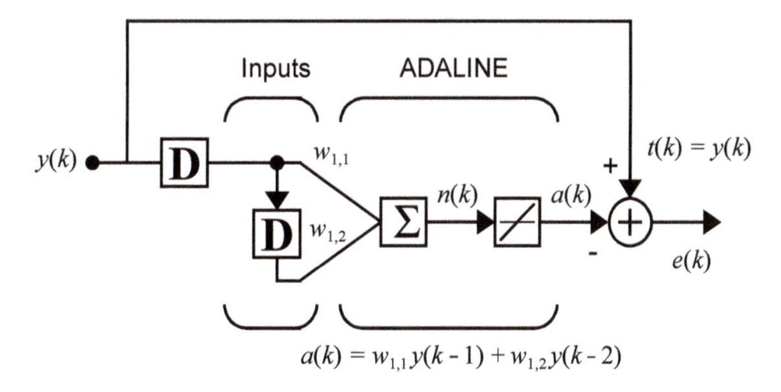

Figure E10.3 Adaptive Predictor for Exercise E10.13

Assume that $y(k)$ is a stationary process with autocorrelation function

$$C_y(n) = E[y(k)(y(k+n))] .$$

i. Write an expression for the mean square error in terms of $C_y(n)$.

ii. Give a specific expression for the mean square error when

$$y(k) = \sin\left(\frac{k\pi}{5}\right).$$

iii. Find the eigenvalues and eigenvectors of the Hessian matrix for the mean square error. Locate the minimum point and sketch a rough contour plot.

iv. Find the maximum stable learning rate for the LMS algorithm.

v. Take three steps of the LMS algorithm by hand, using a stable learning rate. Use the zero vector as the initial guess.

» 2 + 2
ans =
4

vi. Write a MATLAB M-file to implement the LMS algorithm for this problem. Take 40 steps of the algorithm for a stable learning rate and sketch the trajectory on the contour plot. Use the zero vector as the initial guess. Verify that the algorithm is converging to the optimal point.

vii. Verify experimentally that the algorithm is unstable for learning rates greater than that found in part (iv).

E10.14 Repeat Problem P10.9, but use the numerals "1", "2" and "4", instead of the letters "T", "G" and "F". Test the trained network on each reference pattern and on noisy patterns. Discuss the sensitivity of the network. (*Use the Neural Network Design Demonstration Linear Pattern Classification* (`nnd10lc`).)

11 Backpropagation

Objectives

In this chapter we continue our discussion of performance learning, which we began in Chapter 8, by presenting a generalization of the LMS algorithm of Chapter 10. This generalization, called backpropagation, can be used to train multilayer networks. As with the LMS learning law, backpropagation is an approximate steepest descent algorithm, in which the performance index is mean square error. The difference between the LMS algorithm and backpropagation is only in the way in which the derivatives are calculated. For a single-layer linear network the error is an explicit linear function of the network weights, and its derivatives with respect to the weights can be easily computed. In multilayer networks with nonlinear transfer functions, the relationship between the network weights and the error is more complex. In order to calculate the derivatives, we need to use the chain rule of calculus. In fact, this chapter is in large part a demonstration of how to use the chain rule.

Theory and Examples

The perceptron learning rule of Frank Rosenblatt and the LMS algorithm of Bernard Widrow and Marcian Hoff were designed to train single-layer perceptron-like networks. As we have discussed in previous chapters, these single-layer networks suffer from the disadvantage that they are only able to solve linearly separable classification problems. Both Rosenblatt and Widrow were aware of these limitations and proposed multilayer networks that could overcome them, but they were not able to generalize their algorithms to train these more powerful networks.

Apparently the first description of an algorithm to train multilayer networks was contained in the thesis of Paul Werbos in 1974 [Werbo74]. This thesis presented the algorithm in the context of general networks, with neural networks as a special case, and was not disseminated in the neural network community. It was not until the mid 1980s that the backpropagation algorithm was rediscovered and widely publicized. It was rediscovered independently by David Rumelhart, Geoffrey Hinton and Ronald Williams [RuHi86], David Parker [Park85], and Yann Le Cun [LeCu85]. The algorithm was popularized by its inclusion in the book *Parallel Distributed Processing* [RuMc86], which described the work of the Parallel Distributed Processing Group led by psychologists David Rumelhart and James McClelland. The publication of this book spurred a torrent of research in neural networks. The multilayer perceptron, trained by the backpropagation algorithm, is currently the most widely used neural network.

In this chapter we will first investigate the capabilities of multilayer networks and then present the backpropagation algorithm.

Multilayer Perceptrons

We first introduced the notation for multilayer networks in Chapter 2. For ease of reference we have reproduced the diagram of the three-layer perceptron in Figure 11.1. Note that we have simply cascaded three perceptron networks. The output of the first network is the input to the second network, and the output of the second network is the input to the third network. Each layer may have a different number of neurons, and even a different transfer function. Recall from Chapter 2 that we are using superscripts to identify the layer number. Thus, the weight matrix for the first layer is written as $\mathbf{W}^1$ and the weight matrix for the second layer is written $\mathbf{W}^2$.

To identify the structure of a multilayer network, we will sometimes use the following shorthand notation, where the number of inputs is followed by the number of neurons in each layer:

$$R - S^1 - S^2 - S^3 . \qquad (11.1)$$

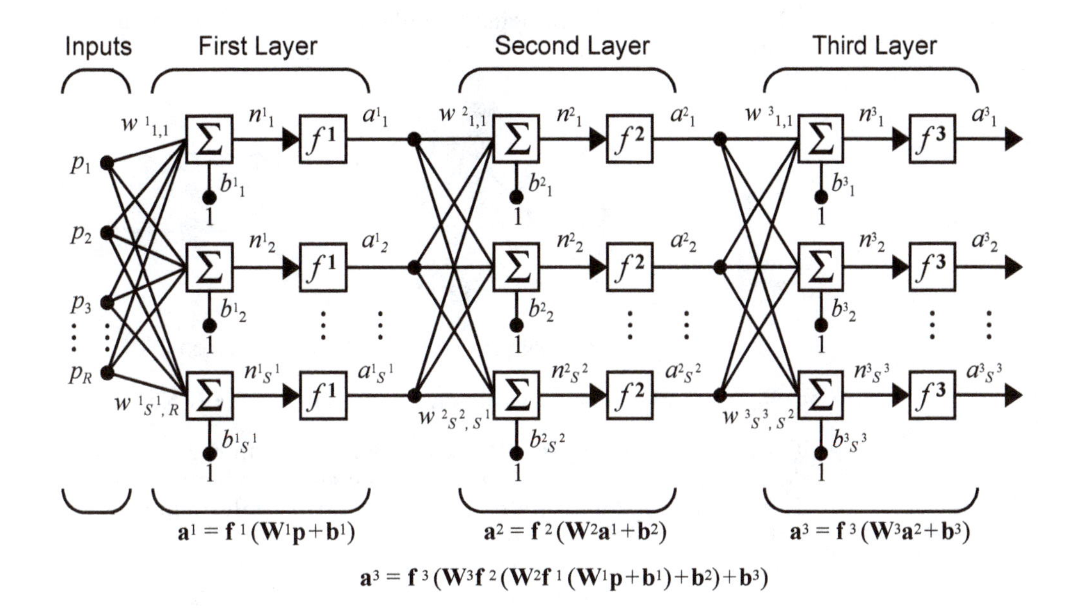

Figure 11.1 Three-Layer Network

Let's now investigate the capabilities of these multilayer perceptron networks. First we will look at the use of multilayer networks for pattern classification, and then we will discuss their application to function approximation.

Pattern Classification

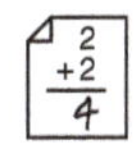

To illustrate the capabilities of the multilayer perceptron for pattern classification, consider the classic exclusive-or (XOR) problem. The input/target pairs for the XOR gate are

$$\left\{\mathbf{p}_1 = \begin{bmatrix}0\\0\end{bmatrix}, t_1 = 0\right\} \left\{\mathbf{p}_2 = \begin{bmatrix}0\\1\end{bmatrix}, t_2 = 1\right\} \left\{\mathbf{p}_3 = \begin{bmatrix}1\\0\end{bmatrix}, t_3 = 1\right\} \left\{\mathbf{p}_4 = \begin{bmatrix}1\\1\end{bmatrix}, t_4 = 0\right\}.$$

This problem, which is illustrated graphically in the figure to the left, was used by Minsky and Papert in 1969 to demonstrate the limitations of the single-layer perceptron. Because the two categories are not linearly separable, a single-layer perceptron cannot perform the classification.

A two-layer network can solve the XOR problem. In fact, there are many different multilayer solutions. One solution is to use two neurons in the first layer to create two decision boundaries. The first boundary separates $\mathbf{p}_1$ from the other patterns, and the second boundary separates $\mathbf{p}_4$. Then the second layer is used to combine the two boundaries together using an

AND operation. The decision boundaries for each first-layer neuron are shown in Figure 11.2.

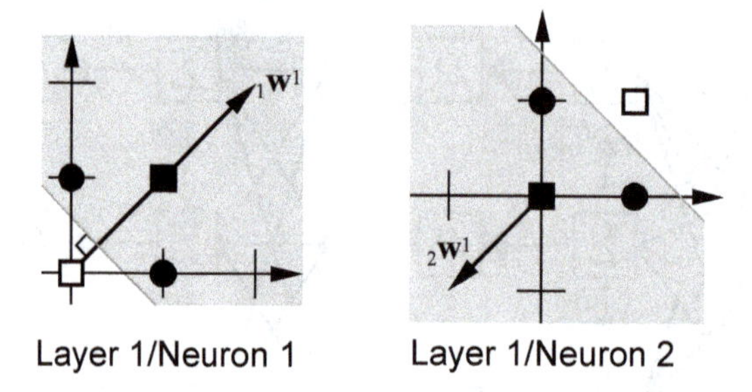

Figure 11.2 Decision Boundaries for XOR Network

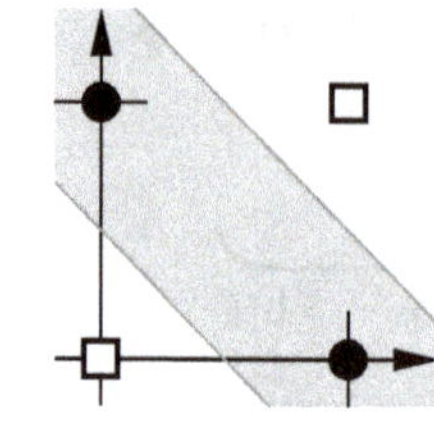

The resulting two-layer, 2-2-1 network is shown in Figure 11.3. The overall decision regions for this network are shown in the figure in the left margin. The shaded region indicates those inputs that will produce a network output of 1.

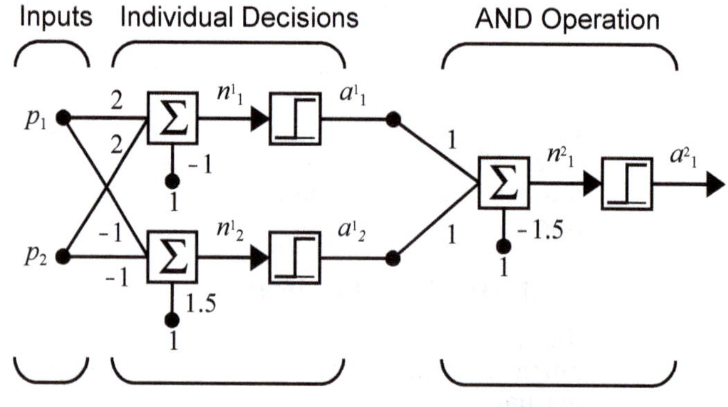

Figure 11.3 Two-Layer XOR Network

See Problems P11.1 and P11.2 for more on the use of multilayer networks for pattern classification.

Function Approximation

Up to this point in the text we have viewed neural networks mainly in the context of pattern classification. It is also instructive to view networks as function approximators. In control systems, for example, the objective is to find an appropriate feedback function that maps from measured outputs to control inputs. In adaptive filtering (Chapter 10) the objective is to find a function that maps from delayed values of an input signal to an appropriate output signal. The following example will illustrate the flexibility of the multilayer perceptron for implementing functions.

2
+2
4

Consider the two-layer, 1-2-1 network shown in Figure 11.4. For this example the transfer function for the first layer is log-sigmoid and the transfer function for the second layer is linear. In other words,

$$f^1(n) = \frac{1}{1+e^{-n}} \text{ and } f^2(n) = n. \tag{11.2}$$

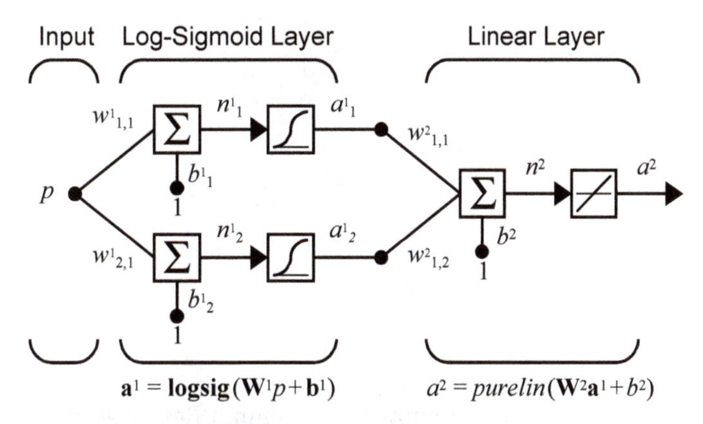

Figure 11.4 Example Function Approximation Network

Suppose that the nominal values of the weights and biases for this network are

$$w^1_{1,1} = 10, w^1_{2,1} = 10, b^1_1 = -10, b^1_2 = 10,$$

$$w^2_{1,1} = 1, w^2_{1,2} = 1, b^2 = 0.$$

The network response for these parameters is shown in Figure 11.5, which plots the network output a^2 as the input p is varied over the range $[-2, 2]$.

Notice that the response consists of two steps, one for each of the log-sigmoid neurons in the first layer. By adjusting the network parameters we can change the shape and location of each step, as we will see in the following discussion.

The centers of the steps occur where the net input to a neuron in the first layer is zero:

$$n^1_1 = w^1_{1,1}p + b^1_1 = 0 \Rightarrow p = -\frac{b^1_1}{w^1_{1,1}} = -\frac{-10}{10} = 1, \tag{11.3}$$

$$n_2^1 = w_{2,1}^1 p + b_2^1 = 0 \quad \Rightarrow \quad p = -\frac{b_2^1}{w_{2,1}^1} = -\frac{10}{10} = -1 . \tag{11.4}$$

The steepness of each step can be adjusted by changing the network weights.

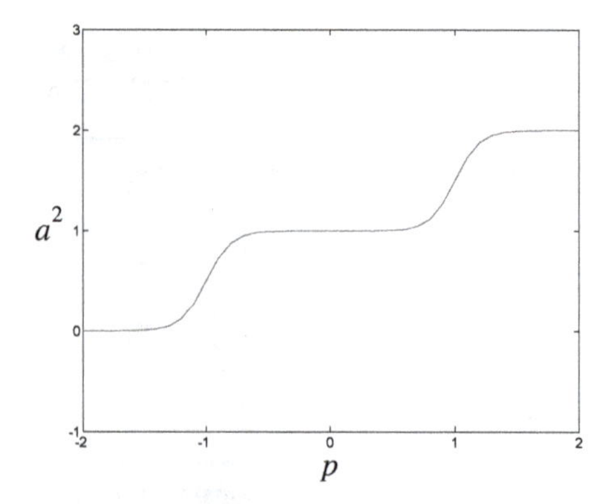

Figure 11.5 Nominal Response of Network of Figure 11.4

Figure 11.6 illustrates the effects of parameter changes on the network response. The blue curve is the nominal response. The other curves correspond to the network response when one parameter at a time is varied over the following ranges:

$$-1 \le w_{1,1}^2 \le 1 , \; -1 \le w_{1,2}^2 \le 1 , \; 0 \le b_2^1 \le 20 , \; -1 \le b^2 \le 1 . \tag{11.5}$$

Figure 11.6 (a) shows how the network biases in the first (hidden) layer can be used to locate the position of the steps. Figure 11.6 (b) illustrates how the weights determine the slope of the steps. The bias in the second (output) layer shifts the entire network response up or down, as can be seen in Figure 11.6 (d).

From this example we can see how flexible the multilayer network is. It would appear that we could use such networks to approximate almost any function, if we had a sufficient number of neurons in the hidden layer. In fact, it has been shown that two-layer networks, with sigmoid transfer functions in the hidden layer and linear transfer functions in the output layer, can approximate virtually any function of interest to any degree of accuracy, provided sufficiently many hidden units are available (see [HoSt89]).

To experiment with the response of this two-layer network, use the MATLAB® Neural Network Design Demonstration Network Function (`nnd11nf`).

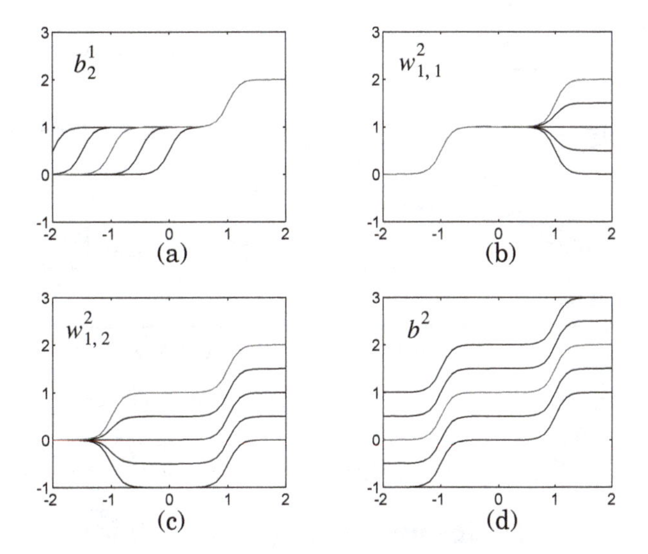

Figure 11.6 Effect of Parameter Changes on Network Response

Now that we have some idea of the power of multilayer perceptron networks for pattern recognition and function approximation, the next step is to develop an algorithm to train such networks.

The Backpropagation Algorithm

It will simplify our development of the backpropagation algorithm if we use the abbreviated notation for the multilayer network, which we introduced in Chapter 2. The three-layer network in abbreviated notation is shown in Figure 11.7.

As we discussed earlier, for multilayer networks the output of one layer becomes the input to the following layer. The equations that describe this operation are

$$\mathbf{a}^{m+1} = \mathbf{f}^{m+1}(\mathbf{W}^{m+1}\mathbf{a}^{m} + \mathbf{b}^{m+1}) \text{ for } m = 0, 1, \ldots, M-1, \tag{11.6}$$

where M is the number of layers in the network. The neurons in the first layer receive external inputs:

$$\mathbf{a}^{0} = \mathbf{p}, \tag{11.7}$$

which provides the starting point for Eq. (11.6). The outputs of the neurons in the last layer are considered the network outputs:

$$\mathbf{a} = \mathbf{a}^{M}. \tag{11.8}$$

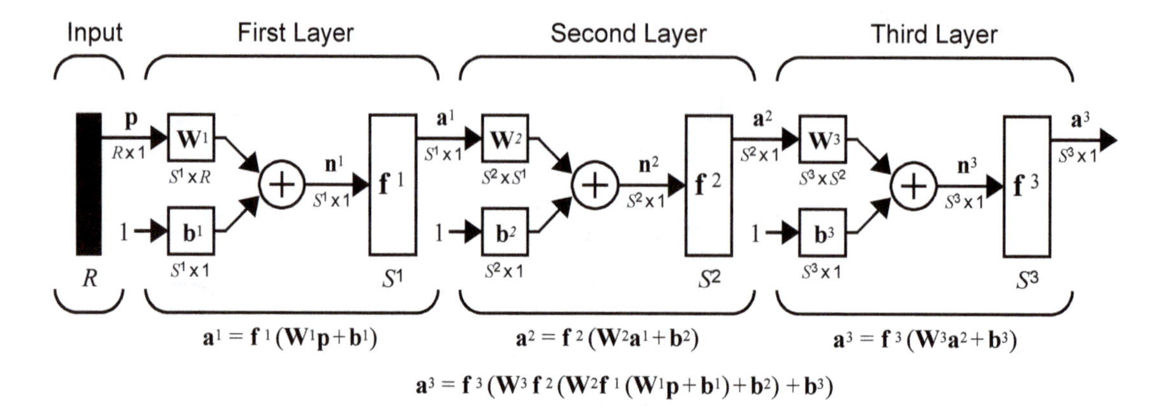

Figure 11.7 Three-Layer Network, Abbreviated Notation

Performance Index

The backpropagation algorithm for multilayer networks is a generalization of the LMS algorithm of Chapter 10, and both algorithms use the same performance index: *mean square error*. The algorithm is provided with a set of examples of proper network behavior:

$$\{\mathbf{p}_1, \mathbf{t}_1\}, \{\mathbf{p}_2, \mathbf{t}_2\}, \ldots, \{\mathbf{p}_Q, \mathbf{t}_Q\}, \tag{11.9}$$

where $\mathbf{p}_q$ is an input to the network, and $\mathbf{t}_q$ is the corresponding target output. As each input is applied to the network, the network output is compared to the target. The algorithm should adjust the network parameters in order to minimize the mean square error:

$$F(\mathbf{x}) = E[e^2] = E[(t-a)^2]. \tag{11.10}$$

where $\mathbf{x}$ is the vector of network weights and biases (as in Chapter 10). If the network has multiple outputs this generalizes to

$$F(\mathbf{x}) = E[\mathbf{e}^T\mathbf{e}] = E[(\mathbf{t}-\mathbf{a})^T(\mathbf{t}-\mathbf{a})]. \tag{11.11}$$

As with the LMS algorithm, we will approximate the mean square error by

$$\hat{F}(\mathbf{x}) = (\mathbf{t}(k)-\mathbf{a}(k))^T(\mathbf{t}(k)-\mathbf{a}(k)) = \mathbf{e}^T(k)\mathbf{e}(k), \tag{11.12}$$

where the expectation of the squared error has been replaced by the squared error at iteration k.

The steepest descent algorithm for the approximate mean square error (stochastic gradient descent) is

$$w_{i,j}^m(k+1) = w_{i,j}^m(k) - \alpha \frac{\partial \hat{F}}{\partial w_{i,j}^m}, \quad (11.13)$$

$$b_i^m(k+1) = b_i^m(k) - \alpha \frac{\partial \hat{F}}{\partial b_i^m}, \quad (11.14)$$

where α is the learning rate.

So far, this development is identical to that for the LMS algorithm. Now we come to the difficult part – the computation of the partial derivatives.

Chain Rule

For a single-layer linear network (the ADALINE) these partial derivatives are conveniently computed using Eq. (10.33) and Eq. (10.34). For the multilayer network the error is not an explicit function of the weights in the hidden layers, therefore these derivatives are not computed so easily.

Because the error is an indirect function of the weights in the hidden layers, we will use the chain rule of calculus to calculate the derivatives. To review the chain rule, suppose that we have a function f that is an explicit function only of the variable n. We want to take the derivative of f with respect to a third variable w. The chain rule is then:

$$\frac{df(n(w))}{dw} = \frac{df(n)}{dn} \times \frac{dn(w)}{dw}. \quad (11.15)$$

For example, if

$$f(n) = e^n \text{ and } n = 2w \text{, so that } f(n(w)) = e^{2w}, \quad (11.16)$$

then

$$\frac{df(n(w))}{dw} = \frac{df(n)}{dn} \times \frac{dn(w)}{dw} = (e^n)(2). \quad (11.17)$$

We will use this concept to find the derivatives in Eq. (11.13) and Eq. (11.14):

$$\frac{\partial \hat{F}}{\partial w_{i,j}^m} = \frac{\partial \hat{F}}{\partial n_i^m} \times \frac{\partial n_i^m}{\partial w_{i,j}^m}, \quad (11.18)$$

$$\frac{\partial \hat{F}}{\partial b_i^m} = \frac{\partial \hat{F}}{\partial n_i^m} \times \frac{\partial n_i^m}{\partial b_i^m}. \quad (11.19)$$

The second term in each of these equations can be easily computed, since the net input to layer m is an explicit function of the weights and bias in that layer:

$$n_i^m = \sum_{j=1}^{S^{m-1}} w_{i,j}^m a_j^{m-1} + b_i^m . \qquad (11.20)$$

Therefore

$$\frac{\partial n_i^m}{\partial w_{i,j}^m} = a_j^{m-1}, \quad \frac{\partial n_i^m}{\partial b_i^m} = 1 . \qquad (11.21)$$

If we now define

$$s_i^m \equiv \frac{\partial \hat{F}}{\partial n_i^m}, \qquad (11.22)$$

Sensitivity

(the *sensitivity* of $\hat{F}$ to changes in the ith element of the net input at layer m), then Eq. (11.18) and Eq. (11.19) can be simplified to

$$\frac{\partial \hat{F}}{\partial w_{i,j}^m} = s_i^m a_j^{m-1}, \qquad (11.23)$$

$$\frac{\partial \hat{F}}{\partial b_i^m} = s_i^m . \qquad (11.24)$$

We can now express the approximate steepest descent algorithm as

$$w_{i,j}^m(k+1) = w_{i,j}^m(k) - \alpha s_i^m a_j^{m-1}, \qquad (11.25)$$

$$b_i^m(k+1) = b_i^m(k) - \alpha s_i^m . \qquad (11.26)$$

In matrix form this becomes:

$$\mathbf{W}^m(k+1) = \mathbf{W}^m(k) - \alpha \mathbf{s}^m (\mathbf{a}^{m-1})^T, \qquad (11.27)$$

$$\mathbf{b}^m(k+1) = \mathbf{b}^m(k) - \alpha \mathbf{s}^m, \qquad (11.28)$$

where

$$\mathbf{s}^m \equiv \frac{\partial \hat{F}}{\partial \mathbf{n}^m} = \begin{bmatrix} \frac{\partial \hat{F}}{\partial n_1^m} \\ \frac{\partial \hat{F}}{\partial n_2^m} \\ \vdots \\ \frac{\partial \hat{F}}{\partial n_{S^m}^m} \end{bmatrix}. \tag{11.29}$$

(Note the close relationship between this algorithm and the LMS algorithm of Eq. (10.33) and Eq. (10.34)).

Backpropagating the Sensitivities

It now remains for us to compute the sensitivities $\mathbf{s}^m$, which requires another application of the chain rule. It is this process that gives us the term *backpropagation*, because it describes a recurrence relationship in which the sensitivity at layer m is computed from the sensitivity at layer $m+1$.

To derive the recurrence relationship for the sensitivities, we will use the following Jacobian matrix:

$$\frac{\partial \mathbf{n}^{m+1}}{\partial \mathbf{n}^m} \equiv \begin{bmatrix} \frac{\partial n_1^{m+1}}{\partial n_1^m} & \frac{\partial n_1^{m+1}}{\partial n_2^m} & \cdots & \frac{\partial n_1^{m+1}}{\partial n_{S^m}^m} \\ \frac{\partial n_2^{m+1}}{\partial n_1^m} & \frac{\partial n_2^{m+1}}{\partial n_2^m} & \cdots & \frac{\partial n_2^{m+1}}{\partial n_{S^m}^m} \\ \vdots & \vdots & & \vdots \\ \frac{\partial n_{S^{m+1}}^{m+1}}{\partial n_1^m} & \frac{\partial n_{S^{m+1}}^{m+1}}{\partial n_2^m} & \cdots & \frac{\partial n_{S^{m+1}}^{m+1}}{\partial n_{S^m}^m} \end{bmatrix}. \tag{11.30}$$

Next we want to find an expression for this matrix. Consider the i,j element of the matrix:

$$\frac{\partial n_i^{m+1}}{\partial n_j^m} = \frac{\partial\left(\sum_{l=1}^{S^m} w_{i,l}^{m+1} a_l^m + b_i^{m+1}\right)}{\partial n_j^m} = w_{i,j}^{m+1} \frac{\partial a_j^m}{\partial n_j^m}$$

$$= w_{i,j}^{m+1} \frac{\partial f^m(n_j^m)}{\partial n_j^m} = w_{i,j}^{m+1} \dot{f}^m(n_j^m) , \tag{11.31}$$

where

$$\dot{f}^m(n_j^m) = \frac{\partial f^m(n_j^m)}{\partial n_j^m} . \tag{11.32}$$

Therefore the Jacobian matrix can be written

$$\frac{\partial \mathbf{n}^{m+1}}{\partial \mathbf{n}^m} = \mathbf{W}^{m+1} \dot{\mathbf{F}}^m(\mathbf{n}^m) , \tag{11.33}$$

where

$$\dot{\mathbf{F}}^m(\mathbf{n}^m) = \begin{bmatrix} \dot{f}^m(n_1^m) & 0 & \dots & 0 \\ 0 & \dot{f}^m(n_2^m) & \dots & 0 \\ \vdots & \vdots & & \vdots \\ 0 & 0 & \dots & \dot{f}^m(n_{S^m}^m) \end{bmatrix} . \tag{11.34}$$

We can now write out the recurrence relation for the sensitivity by using the chain rule in matrix form:

$$\mathbf{s}^m = \frac{\partial \hat{F}}{\partial \mathbf{n}^m} = \left(\frac{\partial \mathbf{n}^{m+1}}{\partial \mathbf{n}^m}\right)^T \frac{\partial \hat{F}}{\partial \mathbf{n}^{m+1}} = \dot{\mathbf{F}}^m(\mathbf{n}^m)(\mathbf{W}^{m+1})^T \frac{\partial \hat{F}}{\partial \mathbf{n}^{m+1}}$$

$$= \dot{\mathbf{F}}^m(\mathbf{n}^m)(\mathbf{W}^{m+1})^T \mathbf{s}^{m+1} . \tag{11.35}$$

Now we can see where the backpropagation algorithm derives its name. The sensitivities are propagated backward through the network from the last layer to the first layer:

$$\mathbf{s}^M \to \mathbf{s}^{M-1} \to \dots \to \mathbf{s}^2 \to \mathbf{s}^1 . \tag{11.36}$$

At this point it is worth emphasizing that the backpropagation algorithm uses the same approximate steepest descent technique that we used in the LMS algorithm. The only complication is that in order to compute the gradient we need to first backpropagate the sensitivities. The beauty of backpropagation is that we have a very efficient implementation of the chain rule.

We still have one more step to make in order to complete the backpropagation algorithm. We need the starting point, $\mathbf{s}^M$, for the recurrence relation of Eq. (11.35). This is obtained at the final layer:

$$s_i^M = \frac{\partial \hat{F}}{\partial n_i^M} = \frac{\partial (\mathbf{t}-\mathbf{a})^T(\mathbf{t}-\mathbf{a})}{\partial n_i^M} = \frac{\partial \sum_{j=1}^{S^M} (t_j - a_j)^2}{\partial n_i^M} = -2(t_i - a_i)\frac{\partial a_i}{\partial n_i^M}. \tag{11.37}$$

Now, since

$$\frac{\partial a_i}{\partial n_i^M} = \frac{\partial a_i^M}{\partial n_i^M} = \frac{\partial f^M(n_i^M)}{\partial n_i^M} = \dot{f}^M(n_i^M), \tag{11.38}$$

we can write

$$s_i^M = -2(t_i - a_i)\dot{f}^M(n_i^M). \tag{11.39}$$

This can be expressed in matrix form as

$$\mathbf{s}^M = -2\dot{\mathbf{F}}^M(\mathbf{n}^M)(\mathbf{t}-\mathbf{a}). \tag{11.40}$$

Summary

Let's summarize the backpropagation algorithm. The first step is to propagate the input forward through the network:

$$\mathbf{a}^0 = \mathbf{p}, \tag{11.41}$$

$$\mathbf{a}^{m+1} = \mathbf{f}^{m+1}(\mathbf{W}^{m+1}\mathbf{a}^m + \mathbf{b}^{m+1}) \text{ for } m = 0, 1, \ldots, M-1, \tag{11.42}$$

$$\mathbf{a} = \mathbf{a}^M. \tag{11.43}$$

The next step is to propagate the sensitivities backward through the network:

$$\mathbf{s}^M = -2\dot{\mathbf{F}}^M(\mathbf{n}^M)(\mathbf{t}-\mathbf{a}), \tag{11.44}$$

$$\mathbf{s}^m = \dot{\mathbf{F}}^m(\mathbf{n}^m)(\mathbf{W}^{m+1})^T\mathbf{s}^{m+1}\text{, for } m = M-1, \ldots, 2, 1\,. \tag{11.45}$$

Finally, the weights and biases are updated using the approximate steepest descent rule:

$$\mathbf{W}^m(k+1) = \mathbf{W}^m(k) - \alpha\mathbf{s}^m(\mathbf{a}^{m-1})^T, \tag{11.46}$$

$$\mathbf{b}^m(k+1) = \mathbf{b}^m(k) - \alpha\mathbf{s}^m\,. \tag{11.47}$$

Example

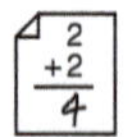

To illustrate the backpropagation algorithm, let's choose a network and apply it to a particular problem. To begin, we will use the 1-2-1 network that we discussed earlier in this chapter. For convenience we have reproduced the network in Figure 11.8.

Next we want to define a problem for the network to solve. Suppose that we want to use the network to approximate the function

$$g(p) = 1 + \sin\left(\frac{\pi}{4}p\right) \text{ for } -2 \le p \le 2\,. \tag{11.48}$$

To obtain our training set we will evaluate this function at several values of p.

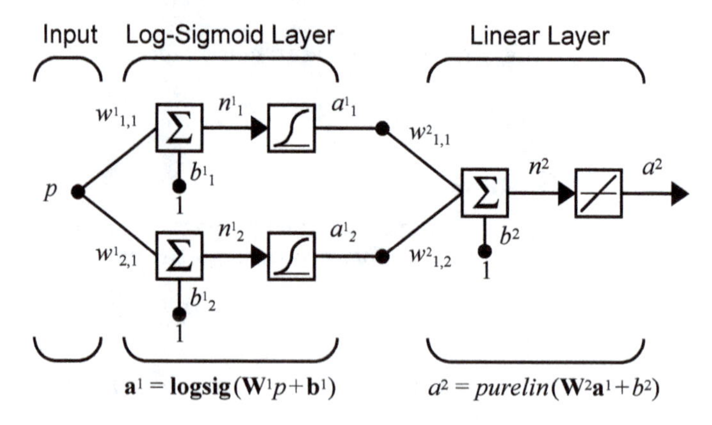

Figure 11.8 Example Function Approximation Network

Before we begin the backpropagation algorithm we need to choose some initial values for the network weights and biases. Generally these are chosen to be small random values. In the next chapter we will discuss some reasons for this. For now let's choose the values

$$\mathbf{W}^1(0) = \begin{bmatrix} -0.27 \\ -0.41 \end{bmatrix},\ \mathbf{b}^1(0) = \begin{bmatrix} -0.48 \\ -0.13 \end{bmatrix},\ \mathbf{W}^2(0) = \begin{bmatrix} 0.09 & -0.17 \end{bmatrix},\ \mathbf{b}^2(0) = \begin{bmatrix} 0.48 \end{bmatrix}.$$

The response of the network for these initial values is illustrated in Figure 11.9, along with the sine function we wish to approximate.

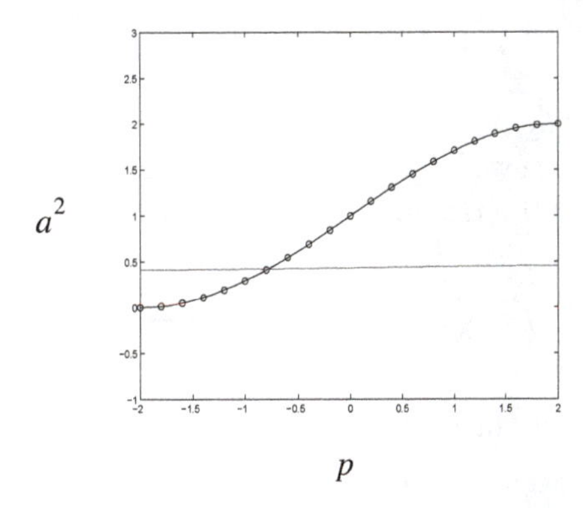

Figure 11.9 Initial Network Response

Next, we need to select a training set $\{p_1, t_1\}$, $\{p_2, t_2\}$, ... , $\{p_Q, t_Q\}$. In this case, we will sample the function at 21 points in the range [-2,2] at equally spaced intervals of 0.2. The training points are indicated by the circles in Figure 11.9.

Now we are ready to start the algorithm. The training points can be presented in any order, but they are often chosen randomly. For our initial input we will choose $p = 1$, which is the 16th training point:

$$a^0 = p = 1.$$

The output of the first layer is then

$$\mathbf{a}^1 = \mathbf{f}^1(\mathbf{W}^1\mathbf{a}^0 + \mathbf{b}^1) = \mathbf{logsig}\left(\begin{bmatrix} -0.27 \\ -0.41 \end{bmatrix}\begin{bmatrix} 1 \end{bmatrix} + \begin{bmatrix} -0.48 \\ -0.13 \end{bmatrix}\right) = \mathbf{logsig}\left(\begin{bmatrix} -0.75 \\ -0.54 \end{bmatrix}\right)$$

$$= \begin{bmatrix} \dfrac{1}{1 + e^{0.75}} \\ \dfrac{1}{1 + e^{0.54}} \end{bmatrix} = \begin{bmatrix} 0.321 \\ 0.368 \end{bmatrix}.$$

The second layer output is

$$a^2 = f^2(\mathbf{W}^2\mathbf{a}^1 + \mathbf{b}^2) = purelin\left(\begin{bmatrix}0.09 & -0.17\end{bmatrix}\begin{bmatrix}0.321\\0.368\end{bmatrix} + \begin{bmatrix}0.48\end{bmatrix}\right) = \begin{bmatrix}0.446\end{bmatrix}.$$

The error would then be

$$e = t - a = \left\{1 + \sin\left(\frac{\pi}{4}p\right)\right\} - a^2 = \left\{1 + \sin\left(\frac{\pi}{4}1\right)\right\} - 0.446 = 1.261.$$

The next stage of the algorithm is to backpropagate the sensitivities. Before we begin the backpropagation, recall that we will need the derivatives of the transfer functions, $\dot{f}^1(n)$ and $\dot{f}^2(n)$. For the first layer

$$\dot{f}^1(n) = \frac{d}{dn}\left(\frac{1}{1+e^{-n}}\right) = \frac{e^{-n}}{(1+e^{-n})^2} = \left(1 - \frac{1}{1+e^{-n}}\right)\left(\frac{1}{1+e^{-n}}\right) = (1 - a^1)(a^1).$$

For the second layer we have

$$\dot{f}^2(n) = \frac{d}{dn}(n) = 1.$$

We can now perform the backpropagation. The starting point is found at the second layer, using Eq. (11.44):

$$\mathbf{s}^2 = -2\dot{\mathbf{F}}^2(\mathbf{n}^2)(\mathbf{t} - \mathbf{a}) = -2\begin{bmatrix}\dot{f}^2(n^2)\end{bmatrix}(1.261) = -2\begin{bmatrix}1\end{bmatrix}(1.261) = -2.522.$$

The first layer sensitivity is then computed by backpropagating the sensitivity from the second layer, using Eq. (11.45):

$$\mathbf{s}^1 = \dot{\mathbf{F}}^1(\mathbf{n}^1)(\mathbf{W}^2)^T\mathbf{s}^2 = \begin{bmatrix}(1-a_1^1)(a_1^1) & 0\\ 0 & (1-a_2^1)(a_2^1)\end{bmatrix}\begin{bmatrix}0.09\\-0.17\end{bmatrix}\begin{bmatrix}-2.522\end{bmatrix}$$

$$= \begin{bmatrix}(1-0.321)(0.321) & 0\\ 0 & (1-0.368)(0.368)\end{bmatrix}\begin{bmatrix}0.09\\-0.17\end{bmatrix}\begin{bmatrix}-2.522\end{bmatrix}$$

$$= \begin{bmatrix}0.218 & 0\\ 0 & 0.233\end{bmatrix}\begin{bmatrix}-0.227\\0.429\end{bmatrix} = \begin{bmatrix}-0.0495\\0.0997\end{bmatrix}.$$

The final stage of the algorithm is to update the weights. For simplicity, we will use a learning rate $\alpha = 0.1$. (In Chapter 12 the choice of learning rate will be discussed in more detail.) From Eq. (11.46) and Eq. (11.47) we have

$$\mathbf{W}^2(1) = \mathbf{W}^2(0) - \alpha\mathbf{s}^2(\mathbf{a}^1)^T = \begin{bmatrix} 0.09 & -0.17 \end{bmatrix} - 0.1\begin{bmatrix} -2.522 \end{bmatrix}\begin{bmatrix} 0.321 & 0.368 \end{bmatrix}$$

$$= \begin{bmatrix} 0.171 & -0.0772 \end{bmatrix},$$

$$\mathbf{b}^2(1) = \mathbf{b}^2(0) - \alpha\mathbf{s}^2 = \begin{bmatrix} 0.48 \end{bmatrix} - 0.1\begin{bmatrix} -2.522 \end{bmatrix} = \begin{bmatrix} 0.732 \end{bmatrix},$$

$$\mathbf{W}^1(1) = \mathbf{W}^1(0) - \alpha\mathbf{s}^1(\mathbf{a}^0)^T = \begin{bmatrix} -0.27 \\ -0.41 \end{bmatrix} - 0.1\begin{bmatrix} -0.0495 \\ 0.0997 \end{bmatrix}\begin{bmatrix} 1 \end{bmatrix} = \begin{bmatrix} -0.265 \\ -0.420 \end{bmatrix},$$

$$\mathbf{b}^1(1) = \mathbf{b}^1(0) - \alpha\mathbf{s}^1 = \begin{bmatrix} -0.48 \\ -0.13 \end{bmatrix} - 0.1\begin{bmatrix} -0.0495 \\ 0.0997 \end{bmatrix} = \begin{bmatrix} -0.475 \\ -0.140 \end{bmatrix}.$$

This completes the first iteration of the backpropagation algorithm. We next proceed to randomly choose another input from the training set and perform another iteration of the algorithm. We continue to iterate until the difference between the network response and the target function reaches some acceptable level. (Note that this will generally take many passes through the entire training set.) We will discuss convergence criteria in more detail in Chapter 12.

To experiment with the backpropagation calculation for this two-layer network, use the MATLAB® Neural Network Design Demonstration Backpropagation Calculation (**nnd11bc**).

Batch vs. Incremental Training

Incremental Training

Batch Training

The algorithm described above is the stochastic gradient descent algorithm, which involves "on-line" or *incremental training*, in which the network weights and biases are updated after each input is presented (as with the LMS algorithm of Chapter 10). It is also possible to perform *batch training*, in which the complete gradient is computed (after all inputs are applied to the network) before the weights and biases are updated. For example, if each input occurs with equal probability, the mean square error performance index can be written

$$F(\mathbf{x}) = E[\mathbf{e}^T\mathbf{e}] = E[(\mathbf{t}-\mathbf{a})^T(\mathbf{t}-\mathbf{a})] = \frac{1}{Q}\sum_{q=1}^{Q}(\mathbf{t}_q - \mathbf{a}_q)^T(\mathbf{t}_q - \mathbf{a}_q). \quad (11.49)$$

The total gradient of this performance index is

$$\nabla F(\mathbf{x}) = \nabla\left\{\frac{1}{Q}\sum_{q=1}^{Q}(\mathbf{t}_q-\mathbf{a}_q)^T(\mathbf{t}_q-\mathbf{a}_q)\right\} = \frac{1}{Q}\sum_{q=1}^{Q}\nabla\{(\mathbf{t}_q-\mathbf{a}_q)^T(\mathbf{t}_q-\mathbf{a}_q)\}\,. \quad (11.50)$$

Therefore, the total gradient of the mean square error is the mean of the gradients of the individual squared errors. Therefore, to implement a batch version of the backpropagation algorithm, we would step through Eq. (11.41) through Eq. (11.45) for all of the inputs in the training set. Then, the individual gradients would be averaged to get the total gradient. The update equations for the batch steepest descent algorithm would then be

$$\mathbf{W}^m(k+1) = \mathbf{W}^m(k) - \frac{\alpha}{Q}\sum_{q=1}^{Q}\mathbf{s}_q^m(\mathbf{a}_q^{m-1})^T\,, \quad (11.51)$$

$$\mathbf{b}^m(k+1) = \mathbf{b}^m(k) - \frac{\alpha}{Q}\sum_{q=1}^{Q}\mathbf{s}_q^m\,. \quad (11.52)$$

Using Backpropagation

In this section we will present some issues relating to the practical implementation of backpropagation. We will discuss the choice of network architecture, and problems with network convergence and generalization. (We will discuss implementation issues again in Chapter 12, which investigates procedures for improving the algorithm.)

Choice of Network Architecture

As we discussed earlier in this chapter, multilayer networks can be used to approximate almost any function, if we have enough neurons in the hidden layers. However, we cannot say, in general, how many layers or how many neurons are necessary for adequate performance. In this section we want to use a few examples to provide some insight into this problem.

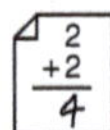

For our first example let's assume that we want to approximate the following functions:

$$g(p) = 1 + \sin\left(\frac{i\pi}{4}p\right) \text{ for } -2 \le p \le 2\,, \quad (11.53)$$

where i takes on the values 1, 2, 4 and 8. As i is increased, the function becomes more complex, because we will have more periods of the sine wave over the interval $-2 \le p \le 2$. It will be more difficult for a neural network with a fixed number of neurons in the hidden layers to approximate $g(p)$ as i is increased.

For this first example we will use a 1-3-1 network, where the transfer function for the first layer is log-sigmoid and the transfer function for the second layer is linear. Recall from our example on page 11-5 that this type of two-layer network can produce a response that is a sum of three log-sigmoid functions (or as many log-sigmoids as there are neurons in the hidden layer). Clearly there is a limit to how complex a function this network can implement. Figure 11.10 illustrates the response of the network after it has been trained to approximate $g(p)$ for $i = 1, 2, 4, 8$. The final network responses are shown by the blue lines.

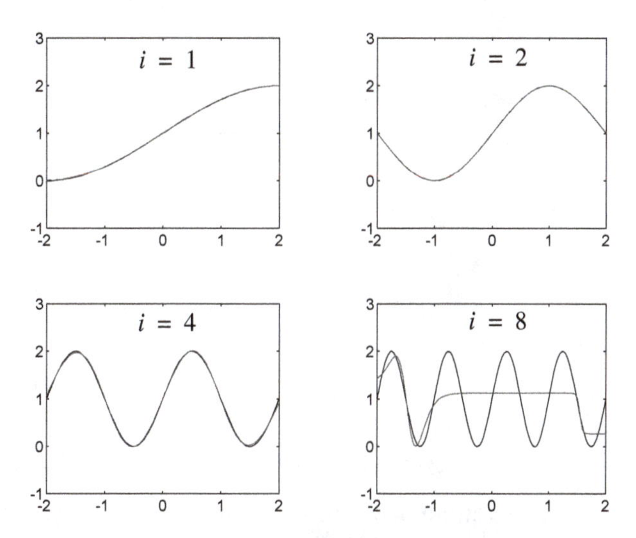

Figure 11.10 Function Approximation Using a 1-3-1 Network

We can see that for $i = 4$ the 1-3-1 network reaches its maximum capability. When $i > 4$ the network is not capable of producing an accurate approximation of $g(p)$. In the bottom right graph of Figure 11.10 we can see how the 1-3-1 network attempts to approximate $g(p)$ for $i = 8$. The mean square error between the network response and $g(p)$ is minimized, but the network response is only able to match a small part of the function.

In the next example we will approach the problem from a slightly different perspective. This time we will pick one function $g(p)$ and then use larger and larger networks until we are able to accurately represent the function. For $g(p)$ we will use

$$g(p) = 1 + \sin\left(\frac{6\pi}{4}p\right) \text{ for } -2 \le p \le 2 . \qquad (11.54)$$

To approximate this function we will use two-layer networks, where the transfer function for the first layer is log-sigmoid and the transfer function for the second layer is linear (1-S^1-1 networks). As we discussed earlier in this chapter, the response of this network is a superposition of S^1 sigmoid functions.

Figure 11.11 illustrates the network response as the number of neurons in the first layer (hidden layer) is increased. Unless there are at least five neurons in the hidden layer the network cannot accurately represent $g(p)$.

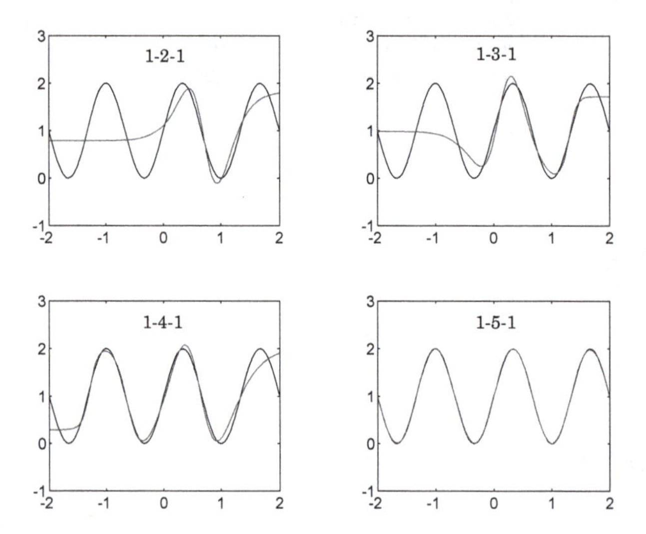

Figure 11.11 Effect of Increasing the Number of Hidden Neurons

To summarize these results, a 1-S^1-1 network, with sigmoid neurons in the hidden layer and linear neurons in the output layer, can produce a response that is a superposition of S^1 sigmoid functions. If we want to approximate a function that has a large number of inflection points, we will need to have a large number of neurons in the hidden layer.

Use the MATLAB® Neural Network Design Demonstration Function Approximation (`nnd11fa`) *to develop more insight into the capability of a two-layer network.*

Convergence

In the previous section we presented some examples in which the network response did not give an accurate approximation to the desired function, even though the backpropagation algorithm produced network parameters that minimized mean square error. This occurred because the capabilities of the network were inherently limited by the number of hidden neurons it contained. In this section we will provide an example in which the network is capable of approximating the function, but the learning algorithm does not produce network parameters that produce an accurate approximation. In the next chapter we will discuss this problem in more detail and explain why it occurs. For now we simply want to illustrate the problem.

The function that we want the network to approximate is

$$g(p) = 1 + \sin(\pi p) \text{ for } -2 \le p \le 2 . \qquad (11.55)$$

To approximate this function we will use a 1-3-1 network, where the transfer function for the first layer is log-sigmoid and the transfer function for the second layer is linear.

Figure 11.12 illustrates a case where the learning algorithm converges to a solution that minimizes mean square error. The thin blue lines represent intermediate iterations, and the thick blue line represents the final solution, when the algorithm has converged. (The numbers next to each curve indicate the sequence of iterations, where 0 represents the initial condition and 5 represents the final solution. The numbers do not correspond to the iteration number. There were many iterations for which no curve is represented. The numbers simply indicate an ordering.)

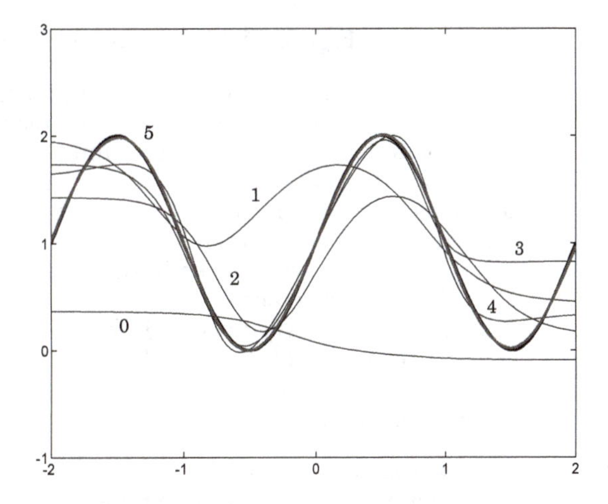

Figure 11.12 Convergence to a Global Minimum

Figure 11.13 illustrates a case where the learning algorithm converges to a solution that does not minimize mean square error. The thick blue line (marked with a 5) represents the network response at the final iteration. The gradient of the mean square error is zero at the final iteration, therefore we have a local minima, but we know that a better solution exists, as evidenced by Figure 11.12. The only difference between this result and the result shown in Figure 11.12 is the initial condition. From one initial condition the algorithm converged to a global minimum point, while from another initial condition the algorithm converged to a local minimum point.

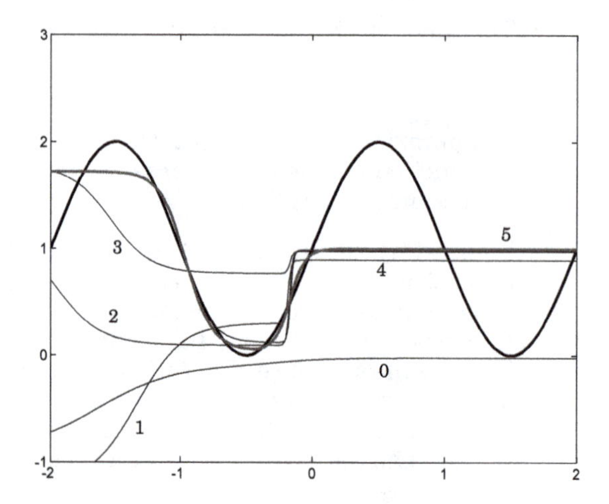

Figure 11.13 Convergence to a Local Minimum

Note that this result could not have occurred with the LMS algorithm. The mean square error performance index for the ADALINE network is a quadratic function with a single minimum point (under most conditions). Therefore the LMS algorithm is guaranteed to converge to the global minimum as long as the learning rate is small enough. The mean square error for the multilayer network is generally much more complex and has many local minima (as we will see in the next chapter). When the backpropagation algorithm converges we cannot be sure that we have an optimum solution. It is best to try several different initial conditions in order to ensure that an optimum solution has been obtained.

Generalization

In most cases the multilayer network is trained with a finite number of examples of proper network behavior:

$$\{\mathbf{p}_1, \mathbf{t}_1\}, \{\mathbf{p}_2, \mathbf{t}_2\}, \dots, \{\mathbf{p}_Q, \mathbf{t}_Q\}. \tag{11.56}$$

This training set is normally representative of a much larger class of possible input/output pairs. It is important that the network successfully *generalize* what it has learned to the total population.

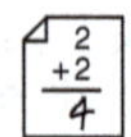

For example, suppose that the training set is obtained by sampling the following function:

$$g(p) = 1 + \sin\left(\frac{\pi}{4}p\right), \tag{11.57}$$

at the points $p = -2, -1.6, -1.2, \dots, 1.6, 2$. (There are a total of 11 input/target pairs.) In Figure 11.14 we see the response of a 1-2-1 network that has been trained on this data. The black line represents $g(p)$, the blue line represents the network response, and the '+' symbols indicate the training set.

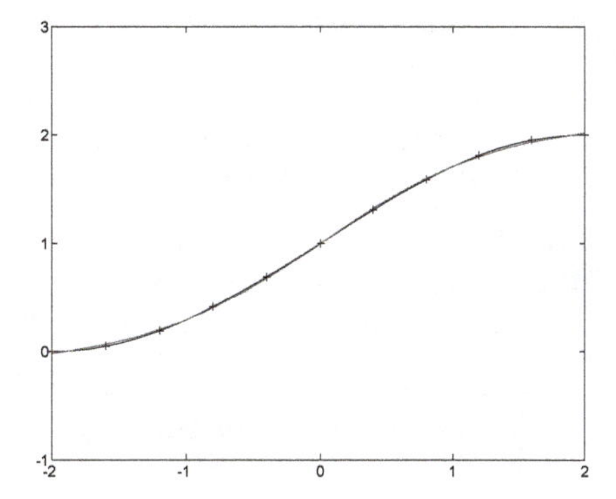

Figure 11.14 1-2-1 Network Approximation of $g(p)$

We can see that the network response is an accurate representation of $g(p)$. If we were to find the response of the network at a value of p that was not contained in the training set (e.g., $p = -0.2$), the network would still produce an output close to $g(p)$. This network generalizes well.

Now consider Figure 11.15, which shows the response of a 1-9-1 network that has been trained on the same data set. Note that the network response accurately models $g(p)$ at all of the training points. However, if we compute the network response at a value of p not contained in the training set (e.g., $p = -0.2$) the network might produce an output far from the true response $g(p)$. This network does not generalize well.

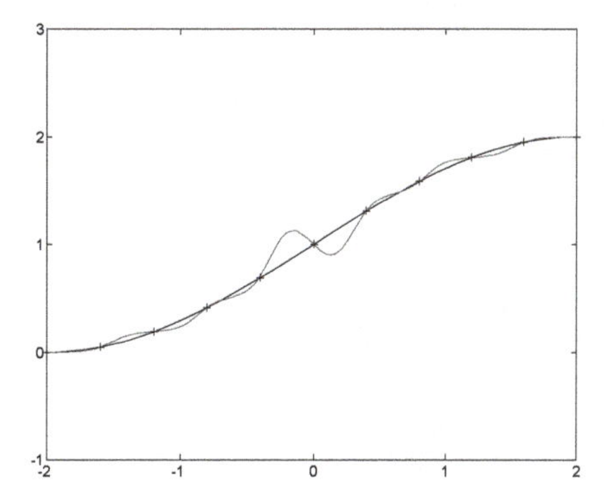

Figure 11.15 1-9-1 Network Approximation of $g(p)$

The 1-9-1 network has too much flexibility for this problem; it has a total of 28 adjustable parameters (18 weights and 10 biases), and yet there are only 11 data points in the training set. The 1-2-1 network has only 7 parameters and is therefore much more restricted in the types of functions that it can implement.

For a network to be able to generalize, it should have fewer parameters than there are data points in the training set. In neural networks, as in all modeling problems, we want to use the simplest network that can adequately represent the training set. Don't use a bigger network when a smaller network will work (a concept often referred to as Ockham's Razor).

An alternative to using the simplest network is to stop the training before the network overfits. A reference to this procedure and other techniques to improve generalization are given in Chapter 13.

To experiment with generalization in neural networks, use the MATLAB® Neural Network Design Demonstration Generalization (nnd11gn).

Summary of Results

Multilayer Network

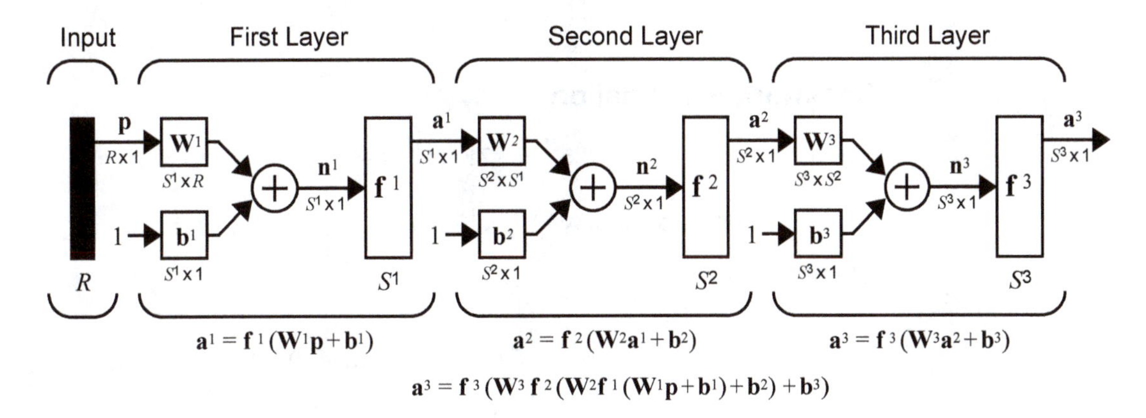

Backpropagation Algorithm

Performance Index

$$F(\mathbf{x}) = E[\mathbf{e}^T\mathbf{e}] = E[(\mathbf{t}-\mathbf{a})^T(\mathbf{t}-\mathbf{a})]$$

Approximate Performance Index

$$\hat{F}(\mathbf{x}) = \mathbf{e}^T(k)\mathbf{e}(k) = (\mathbf{t}(k)-\mathbf{a}(k))^T(\mathbf{t}(k)-\mathbf{a}(k))$$

Sensitivity

$$\mathbf{s}^m \equiv \frac{\partial \hat{F}}{\partial \mathbf{n}^m} = \begin{bmatrix} \dfrac{\partial \hat{F}}{\partial n_1^m} \\ \dfrac{\partial \hat{F}}{\partial n_2^m} \\ \vdots \\ \dfrac{\partial \hat{F}}{\partial n_{S^m}^m} \end{bmatrix}$$

Forward Propagation

$$\mathbf{a}^0 = \mathbf{p},$$

$$\mathbf{a}^{m+1} = \mathbf{f}^{m+1}(\mathbf{W}^{m+1}\mathbf{a}^m + \mathbf{b}^{m+1}) \text{ for } m = 0, 1, \dots, M-1,$$

$$\mathbf{a} = \mathbf{a}^M.$$

Backward Propagation

$$\mathbf{s}^M = -2\dot{\mathbf{F}}^M(\mathbf{n}^M)(\mathbf{t} - \mathbf{a}),$$

$$\mathbf{s}^m = \dot{\mathbf{F}}^m(\mathbf{n}^m)(\mathbf{W}^{m+1})^T\mathbf{s}^{m+1}, \text{ for } m = M-1, \dots, 2, 1,$$

where

$$\dot{\mathbf{F}}^m(\mathbf{n}^m) = \begin{bmatrix} \dot{f}^m(n_1^m) & 0 & \dots & 0 \\ 0 & \dot{f}^m(n_2^m) & \dots & 0 \\ \vdots & \vdots & & \vdots \\ 0 & 0 & & \dot{f}^m(n_{S^m}^m) \end{bmatrix},$$

$$\dot{f}^m(n_j^m) = \frac{\partial f^m(n_j^m)}{\partial n_j^m}.$$

Weight Update (Approximate Steepest Descent)

$$\mathbf{W}^m(k+1) = \mathbf{W}^m(k) - \alpha\mathbf{s}^m(\mathbf{a}^{m-1})^T,$$

$$\mathbf{b}^m(k+1) = \mathbf{b}^m(k) - \alpha\mathbf{s}^m.$$

Solved Problems

P11.1 Consider the two classes of patterns that are shown in Figure P11.1. Class I represents vertical lines and Class II represents horizontal lines.

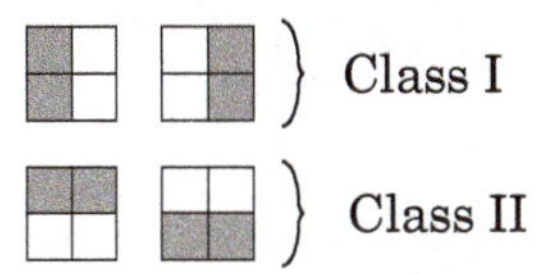

Figure P11.1 Pattern Classes for Problem P11.1

i. **Are these categories linearly separable?**

ii. **Design a multilayer network to distinguish these categories.**

i. Let's begin by converting the patterns to vectors by scanning each 2X2 grid one column at a time. Each white square will be represented by a "- 1" and each blue square by a "1". The vertical lines (Class I patterns) then become

$$\mathbf{p}_1 = \begin{bmatrix} 1 \\ 1 \\ -1 \\ -1 \end{bmatrix} \text{ and } \mathbf{p}_2 = \begin{bmatrix} -1 \\ -1 \\ 1 \\ 1 \end{bmatrix},$$

and the horizontal lines (Class II patterns) become

$$\mathbf{p}_3 = \begin{bmatrix} 1 \\ -1 \\ 1 \\ -1 \end{bmatrix} \text{ and } \mathbf{p}_4 = \begin{bmatrix} -1 \\ 1 \\ -1 \\ 1 \end{bmatrix}.$$

In order for these categories to be linearly separable we must be able to place a hyperplane between the two categories. This means there must be a weight matrix $\mathbf{W}$ and a bias b such that

$$\mathbf{W}\mathbf{p}_1 + b > 0 \,,\; \mathbf{W}\mathbf{p}_2 + b > 0 \,,\; \mathbf{W}\mathbf{p}_3 + b < 0 \,,\; \mathbf{W}\mathbf{p}_4 + b < 0 \,.$$

These conditions can be converted to

$$\begin{bmatrix} w_{1,1} & w_{1,2} & w_{1,3} & w_{1,4} \end{bmatrix} \begin{bmatrix} 1 \\ 1 \\ -1 \\ -1 \end{bmatrix} = \begin{bmatrix} w_{1,1} + w_{1,2} - w_{1,3} - w_{1,4} \end{bmatrix} > 0 ,$$

$$\begin{bmatrix} -w_{1,1} - w_{1,2} + w_{1,3} + w_{1,4} \end{bmatrix} > 0 ,$$

$$\begin{bmatrix} w_{1,1} - w_{1,2} + w_{1,3} - w_{1,4} \end{bmatrix} < 0 ,$$

$$\begin{bmatrix} -w_{1,1} + w_{1,2} - w_{1,3} + w_{1,4} \end{bmatrix} < 0 .$$

The first two conditions reduce to

$$w_{1,1} + w_{1,2} > w_{1,3} + w_{1,4} \text{ and } w_{1,3} + w_{1,4} > w_{1,1} + w_{1,2} ,$$

which are contradictory. The final two conditions reduce to

$$w_{1,1} + w_{1,3} > w_{1,2} + w_{1,4} \text{ and } w_{1,2} + w_{1,4} > w_{1,1} + w_{1,3} ,$$

which are also contradictory. Therefore there is no hyperplane that can separate these two categories.

ii. There are many different multilayer networks that could solve this problem. We will design a network by first noting that for the Class I vectors either the first two elements or the last two elements will be "1". The Class II vectors have alternating "1" and "-1" patterns. This leads to the network shown in Figure P11.2.

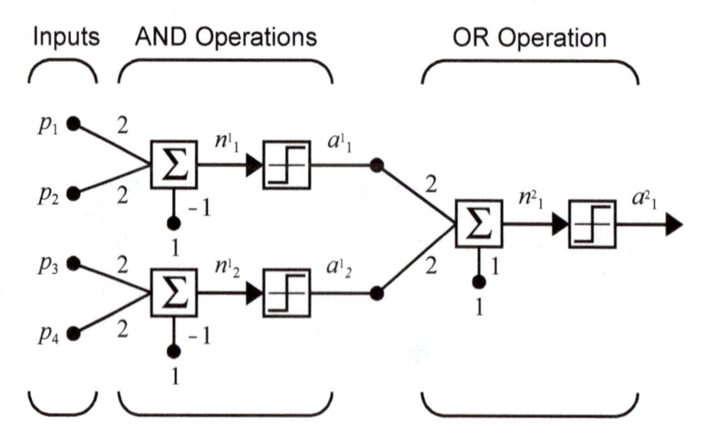

Figure P11.2 Network to Categorize Horizontal and Vertical Lines

The first neuron in the first layer tests the first two elements of the input vector. If they are both "1" it outputs a "1", otherwise it outputs a "-1". The second neuron in the first layer tests the last two elements of the input vector in the same way. Both of the neurons in the first layer perform AND operations. The second layer of the network tests whether either of the outputs of the first layer are "1". It performs an OR operation. In this way, the network will output a "1" if either the first two elements or the last two elements of the input vector are both "1".

P11.2 Figure P11.3 illustrates a classification problem, where Class I vectors are represented by light circles, and Class II vectors are represented by dark circles. These categories are not linearly separable. Design a multilayer network to correctly classify these categories.

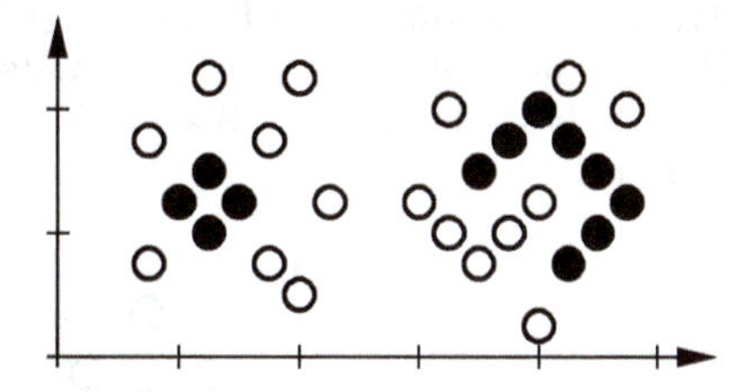

Figure P11.3 Classification Problem

We will solve this problem with a procedure that can be used for arbitrary classification problems. It requires a three-layer network, with hard-limiting neurons in each layer. In the first layer we create a set of linear decision boundaries that separate every Class I vector from every Class II vector. For this problem we used 11 such boundaries. They are shown in Figure P11.4.

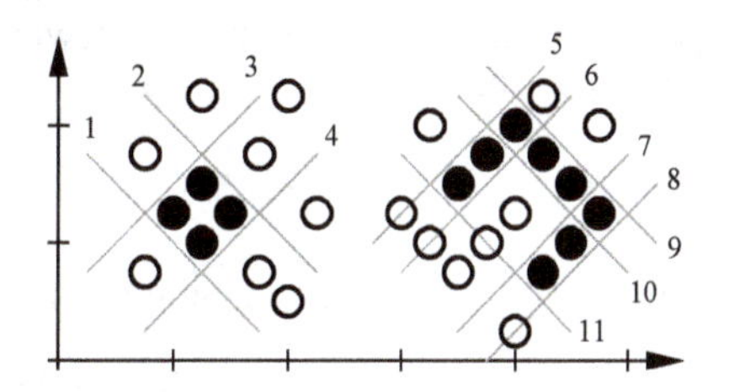

Figure P11.4 First Layer Decision Boundaries

Each row of the weight matrix in the first layer corresponds to one decision boundary. The weight matrix and bias vector for the first layer are

$$(\mathbf{W}^1)^T = \begin{bmatrix} 1 & -1 & 1 & -1 & 1 & -1 & 1 & -1 & -1 & 1 & 1 \\ 1 & -1 & -1 & 1 & -1 & 1 & -1 & 1 & -1 & 1 & 1 \end{bmatrix},$$

$$(\mathbf{b}^1)^T = \begin{bmatrix} -2 & 3 & 0.5 & 0.5 & -1.75 & 2.25 & -3.25 & 3.75 & 6.25 & -5.75 & -4.75 \end{bmatrix} .$$

(Review Chapter 3, 4 and 10 for procedures for calculating the appropriate weight matrix and bias for a given decision boundary.) Now we can combine the outputs of the 11 first layer neurons into groups with a second layer of AND neurons, such as those we used in the first layer of the network in Problem P11.1. The second layer weight matrix and bias are

$$\mathbf{W}^2 = \begin{bmatrix} 1 & 1 & 1 & 1 & 0 & 0 & 0 & 0 & 0 & 0 & 0 \\ 0 & 0 & 0 & 0 & 1 & 1 & 0 & 0 & 1 & 0 & 1 \\ 0 & 0 & 0 & 0 & 1 & 0 & 0 & 1 & 1 & 1 & 0 \\ 0 & 0 & 0 & 0 & 0 & 0 & 1 & 1 & 1 & 0 & 1 \end{bmatrix} , \mathbf{b}^T = \begin{bmatrix} -3 \\ -3 \\ -3 \\ -3 \end{bmatrix} .$$

The four decision boundaries for the second layer are shown in Figure P11.5. For example, the neuron 2 decision boundary is obtained by combining the boundaries 5, 6, 9 and 11 from layer 1. This can be seen by looking at row 2 of $\mathbf{W}^2$.

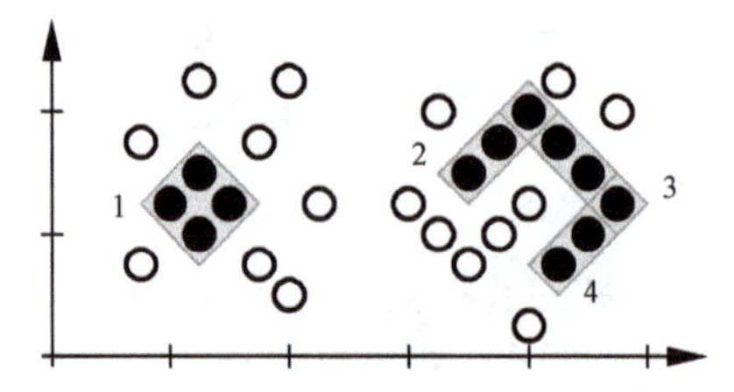

Figure P11.5 Second Layer Decision Regions

In the third layer of the network we will combine together the four decision regions of the second layer into one decision region using an OR operation, just as in the last layer of the network in Problem P11.1. The weight matrix and bias for the third layer are

$$\mathbf{W}^3 = \begin{bmatrix} 1 & 1 & 1 & 1 \end{bmatrix} , \mathbf{b}^3 = \begin{bmatrix} 3 \end{bmatrix} .$$

The complete network is shown in Figure P11.6.

The procedure that we used to develop this network can be used to solve classification problems with arbitrary decision boundaries as long as we have enough neurons in the hidden layers. The idea is to use the first layer to create a number of linear boundaries, which can be combined by using AND neurons in the second layer and OR neurons in the third layer. The decision regions of the second layer are convex, but the final decision boundaries created by the third layer can have arbitrary shapes.

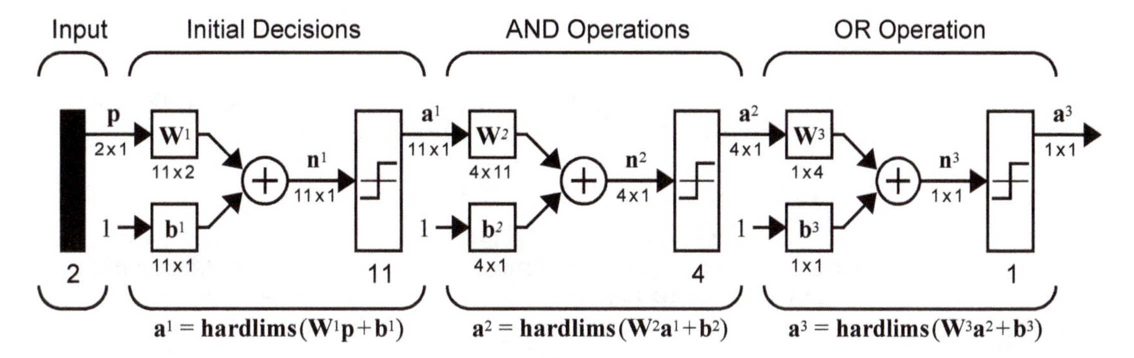

Figure P11.6 Network for Problem P11.2

The final network decision regions are given in Figure P11.7. Any vector in the shaded areas will produce a network output of 1, which corresponds to Class II. All other vectors will produce a network output of -1, which corresponds to Class I.

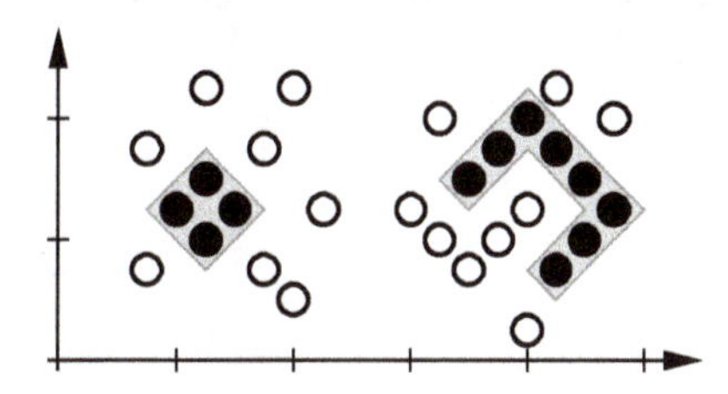

Figure P11.7 Final Decision Regions

P11.3 Show that a multilayer network with linear transfer functions is equivalent to a single-layer linear network.

For a multilayer linear network the forward equations would be

$$\mathbf{a}^1 = \mathbf{W}^1\mathbf{p} + \mathbf{b}^1$$

$$\mathbf{a}^2 = \mathbf{W}^2\mathbf{a}^1 + \mathbf{b}^2 = \mathbf{W}^2\mathbf{W}^1\mathbf{p} + [\mathbf{W}^2\mathbf{b}^1 + \mathbf{b}^2],$$

$$\mathbf{a}^3 = \mathbf{W}^3\mathbf{a}^2 + \mathbf{b}^3 = \mathbf{W}^3\mathbf{W}^2\mathbf{W}^1\mathbf{p} + [\mathbf{W}^3\mathbf{W}^2\mathbf{b}^1 + \mathbf{W}^3\mathbf{b}^2 + \mathbf{b}^3].$$

If we continue this process we can see that for an M-layer linear network, the equivalent single-layer linear network would have the following weight matrix and bias vector

$$\mathbf{W} = \mathbf{W}^M\mathbf{W}^{M-1}\ldots\mathbf{W}^2\mathbf{W}^1,$$

$$\mathbf{b} = [\mathbf{W}^M\mathbf{W}^{M-1}\ldots\mathbf{W}^2]\mathbf{b}^1 + [\mathbf{W}^M\mathbf{W}^{M-1}\ldots\mathbf{W}^3]\mathbf{b}^2 + \cdots + \mathbf{b}^M .$$

P11.4 The purpose of this problem is to illustrate the use of the chain rule. Consider the following dynamic system:

$$y(k+1) = f(y(k)) .$$

We want to choose the initial condition $y(0)$ so that at some final time $k = K$ the system output $y(K)$ will be as close as possible to some target output t. We will minimize the performance index

$$F(y(0)) = (t - y(K))^2$$

using steepest descent, so we need the gradient

$$\frac{\partial}{\partial y(0)}F(y(0)) .$$

Find a procedure for computing this using the chain rule.

The gradient is

$$\frac{\partial}{\partial y(0)}F(y(0)) = \frac{\partial(t-y(K))^2}{\partial y(0)} = 2(t-y(K))\left[-\frac{\partial}{\partial y(0)}y(K)\right] .$$

The key term is

$$\left[\frac{\partial}{\partial y(0)}y(K)\right] ,$$

which cannot be computed directly, since $y(K)$ is not an explicit function of $y(0)$. Let's define an intermediate term

$$r(k) \equiv \frac{\partial}{\partial y(0)}y(k) .$$

Then we can use the chain rule:

$$r(k+1) = \frac{\partial}{\partial y(0)}y(k+1) = \frac{\partial y(k+1)}{\partial y(k)} \times \frac{\partial y(k)}{\partial y(0)} = \frac{\partial y(k+1)}{\partial y(k)} \times r(k) .$$

From the system dynamics we know

$$\frac{\partial y(k+1)}{\partial y(k)} = \frac{\partial f(y(k))}{\partial y(k)} = \dot{f}(y(k)) .$$

Therefore the recursive equation for the computation of $r(k)$ is

$$r(k+1) = \dot{f}(y(k))r(k) .$$

This is initialized at $k = 0$:

$$r(0) = \frac{\partial y(0)}{\partial y(0)} = 1 .$$

The total procedure for computing the gradient is then

$$r(0) = 1 ,$$

$$r(k+1) = \dot{f}(y(k))r(k) \text{, for } k = 0, 1, \dots, K-1 ,$$

$$\frac{\partial}{\partial y(0)} F(y(0)) = 2(t - y(K))[-r(K)] .$$

P11.5 Consider the two-layer network shown in Figure P11.8. The initial weights and biases are set to

$$w^1 = 1 , b^1 = 1 , w^2 = -2 , b^2 = 1 .$$

An input/target pair is given to be

$$((p = 1),(t = 1)) .$$

i. **Find the squared error $(e)^2$ as an explicit function of all weights and biases.**

ii. **Using part (i) find $\partial(e)^2/\partial w^1$ at the initial weights and biases.**

iii. **Repeat part (ii) using backpropagation and compare results.**

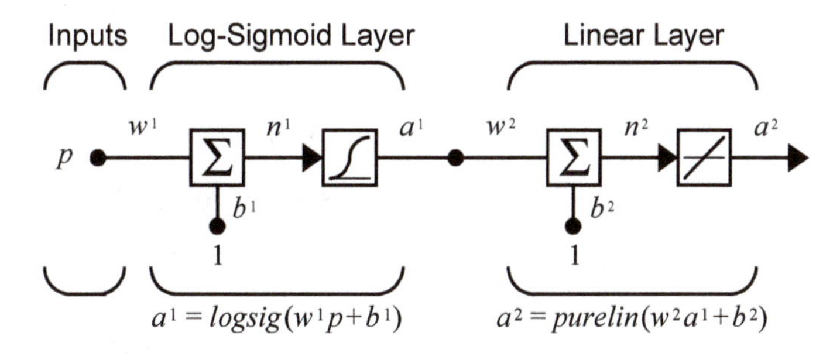

Figure P11.8 Two-Layer Network for Problem P11.5

i. The squared error is given by

$$(e)^2 = (t-a^2)^2 = \left(t - \left\{w^2 \frac{1}{(1+\exp(-(w^1 p + b^1)))} + b^2\right\}\right)^2 .$$

ii. The derivative is

$$\frac{\partial (e)^2}{\partial w^1} = 2e\frac{\partial e}{\partial w^1} = 2e\left\{w^2 \frac{1}{(1+\exp(-(w^1 p + b^1)))^2}\exp(-(w^1 p + b^1))(-p)\right\}$$

To evaluate this at the initial weights and biases we find

$$a^1 = \frac{1}{(1+\exp(-(w^1 p + b^1)))} = \frac{1}{(1+\exp(-(1(1)+1)))} = 0.8808$$

$$a^2 = w^2 a^1 + b^2 = (-2)0.8808 + 1 = -0.7616$$

$$e = (t - a^2) = (1 - (-0.7616)) = 1.7616$$

$$\frac{\partial (e)^2}{\partial w^1} = 2e\left\{w^2 \frac{1}{(1+\exp(-(w^1 p + b^1)))^2}\exp(-(w^1 p + b^1))(-p)\right\}$$

$$= 2(1.7616)\left\{(-2)\frac{1}{(1+\exp(-(1(1)+1)))^2}\exp(-(1(1)+1))(-1)\right\}$$

$$= 3.5232\left(0.2707\frac{1}{(1.289)^2}\right) = 0.7398 .$$

iii. To backpropagate the sensitivities we use Eq. (11.44) and Eq. (11.45):

$$\mathbf{s}^2 = -2\dot{\mathbf{F}}^2(\mathbf{n}^2)(\mathbf{t}-\mathbf{a}) = -2(1)(1-(-0.7616)) = -3.5232 ,$$

$$\mathbf{s}^1 = \dot{\mathbf{F}}^1(\mathbf{n}^1)(\mathbf{W}^2)^T\mathbf{s}^2 = [a^1(1-a^1)](-2)\mathbf{s}^2$$

$$= [0.8808(1-0.8808)](-2)(-3.5232) = 0.7398 .$$

From Eq. (11.23) we can compute $\partial(e)^2/\partial w^1$:

$$\frac{\partial (e)^2}{\partial w^1} = s^1 a^0 = s^1 p = (0.7398)(1) = 0.7398 .$$

This agrees with our result from part (ii).

P11.6 Earlier in this chapter we showed that if the neuron transfer function is log-sigmoid,

$$a = f(n) = \frac{1}{1+e^{-n}},$$

then the derivative can be conveniently computed by

$$\dot{f}(n) = a(1-a).$$

Find a convenient way to compute the derivative for the hyperbolic tangent sigmoid:

$$a = f(n) = tansig(n) = \frac{e^n - e^{-n}}{e^n + e^{-n}}.$$

Computing the derivative directly we find

$$\dot{f}(n) = \frac{df(n)}{dn} = \frac{d}{dn}\left(\frac{e^n - e^{-n}}{e^n + e^{-n}}\right) = -\frac{e^n - e^{-n}}{(e^n + e^{-n})^2}(e^n - e^{-n}) + \frac{e^n + e^{-n}}{e^n + e^{-n}}$$

$$= 1 - \frac{(e^n - e^{-n})^2}{(e^n + e^{-n})^2} = 1 - (a)^2.$$

P11.7 For the network shown in Figure P11.9 the initial weights and biases are chosen to be

$$w^1(0) = -1,\ b^1(0) = 1,\ w^2(0) = -2,\ b^2(0) = 1.$$

An input/target pair is given to be

$$((p = -1),(t = 1)).$$

Perform one iteration of backpropagation with $\alpha = 1$.

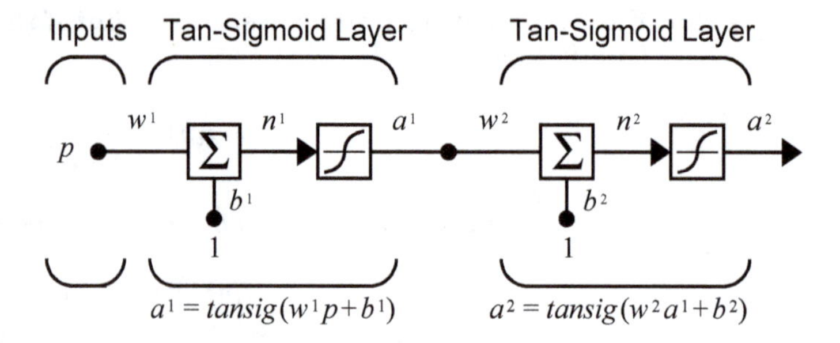

Figure P11.9 Two-Layer Tan-Sigmoid Network

The first step is to propagate the input through the network.

$$n^1 = w^1 p + b^1 = (-1)(-1) + 1 = 2$$

$$a^1 = tansig(n^1) = \frac{\exp(n^1) - \exp(-n^1)}{\exp(n^1) + \exp(-n^1)} = \frac{\exp(2) - \exp(-2)}{\exp(2) + \exp(-2)} = 0.964$$

$$n^2 = w^2 a^1 + b^2 = (-2)(0.964) + 1 = -0.928$$

$$a^2 = tansig(n^2) = \frac{\exp(n^2) - \exp(-n^2)}{\exp(n^2) + \exp(-n^2)} = \frac{\exp(-0.928) - \exp(0.928)}{\exp(-0.928) + \exp(0.928)}$$

$$= -0.7297$$

$$e = (t - a^2) = (1 - (-0.7297)) = 1.7297$$

Now we backpropagate the sensitivities using Eq. (11.44) and Eq. (11.45).

$$\mathbf{s}^2 = -2\dot{\mathbf{F}}^2(\mathbf{n}^2)(\mathbf{t} - \mathbf{a}) = -2[1 - (a^2)^2](e) = -2[1 - (-0.7297)^2]1.7297$$

$$= -1.6175$$

$$\mathbf{s}^1 = \dot{\mathbf{F}}^1(\mathbf{n}^1)(\mathbf{W}^2)^T \mathbf{s}^2 = [1 - (a^1)^2] w^2 \mathbf{s}^2 = [1 - (0.964)^2](-2)(-1.6175)$$

$$= 0.2285$$

Finally, the weights and biases are updated using Eq. (11.46) and Eq. (11.47):

$$w^2(1) = w^2(0) - \alpha s^2 (a^1)^T = (-2) - 1(-1.6175)(0.964) = -0.4407\,,$$

$$w^1(1) = w^1(0) - \alpha s^1 (a^0)^T = (-1) - 1(0.2285)(-1) = -0.7715,$$

$$b^2(1) = b^2(0) - \alpha s^2 = 1 - 1(-1.6175) = 2.6175,$$

$$b^1(1) = b^1(0) - \alpha s^1 = 1 - 1(0.2285) = 0.7715.$$

P11.8 In Figure P11.10 we have a network that is a slight modification to the standard two-layer feedforward network. It has a connection from the input directly to the second layer. Derive the backpropagation algorithm for this network.

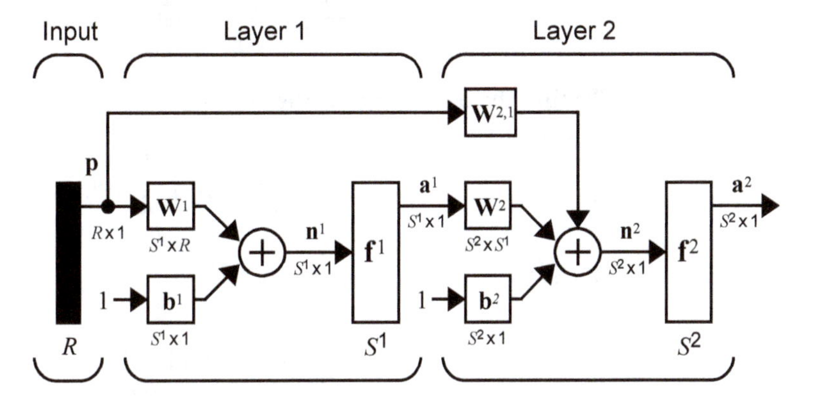

Figure P11.10 Network with Bypass Connection

We begin with the forward equations:

$$\mathbf{n}^1 = \mathbf{W}^1\mathbf{p} + \mathbf{b}^1,$$

$$\mathbf{a}^1 = \mathbf{f}^1(\mathbf{n}^1) = \mathbf{f}^1(\mathbf{W}^1\mathbf{p} + \mathbf{b}^1),$$

$$\mathbf{n}^2 = \mathbf{W}^2\mathbf{a}^1 + \mathbf{W}^{2,1}\mathbf{p} + \mathbf{b}^2,$$

$$\mathbf{a}^2 = \mathbf{f}^2(\mathbf{n}^2) = \mathbf{f}^2(\mathbf{W}^2\mathbf{a}^1 + \mathbf{W}^{2,1}\mathbf{p} + \mathbf{b}^2).$$

The backpropagation equations for the sensitivities will not change from those for a standard two-layer network. The sensitivities are the derivatives of the squared error with respect to the net inputs; these derivatives don't change, since we are simply adding a term to the net input.

Next we need the elements of the gradient for the weight update equations. For the standard weights and biases we have

$$\frac{\partial \hat{F}}{\partial w_{i,j}^m} = \frac{\partial \hat{F}}{\partial n_i^m} \times \frac{\partial n_i^m}{\partial w_{i,j}^m} = s_i^m a_j^{m-1},$$

$$\frac{\partial \hat{F}}{\partial b_i^m} = \frac{\partial \hat{F}}{\partial n_i^m} \times \frac{\partial n_i^m}{\partial b_i^m} = s_i^m.$$

Therefore the update equations for $\mathbf{W}^1$, $\mathbf{b}^1$, $\mathbf{W}^2$ and $\mathbf{b}^2$ do not change. We do need an additional equation for $\mathbf{W}^{2,1}$:

$$\frac{\partial \hat{F}}{\partial w_{i,j}^{2,1}} = \frac{\partial \hat{F}}{\partial n_i^2} \times \frac{\partial n_i^2}{\partial w_{i,j}^{2,1}} = s_i^2 \times \frac{\partial n_i^2}{\partial w_{i,j}^{2,1}}.$$

To find the derivative on the right-hand side of this equation note that

$$n_i^2 = \sum_{j=1}^{S^1} w_{i,j}^2 a_j^1 + \sum_{j=1}^{R} w_{i,j}^{2,1} p_j + b_i^2.$$

Therefore

$$\frac{\partial n_i^2}{\partial w_{i,j}^{2,1}} = p_j \text{ and } \frac{\partial \hat{F}}{\partial w_{i,j}^{2,1}} = s_i^2 p_j.$$

The update equations can thus be written in matrix form as:

$$\mathbf{W}^m(k+1) = \mathbf{W}^m(k) - \alpha \mathbf{s}^m (\mathbf{a}^{m-1})^T, \; m = 1, 2,$$

$$\mathbf{b}^m(k+1) = \mathbf{b}^m(k) - \alpha \mathbf{s}^m, \; m = 1, 2.$$

$$\mathbf{W}^{2,1}(k+1) = \mathbf{W}^{2,1}(k) - \alpha \mathbf{s}^2 (\mathbf{a}^0)^T = \mathbf{W}^{2,1}(k) - \alpha \mathbf{s}^2 (\mathbf{p})^T.$$

The main point of this problem is that the backpropagation concept can be used on networks more general than the standard multilayer feedforward network.

P11.9 Find an algorithm, based on the backpropagation concept, that can be used to update the weights w_1 and w_2 in the recurrent network shown in Figure P11.11.

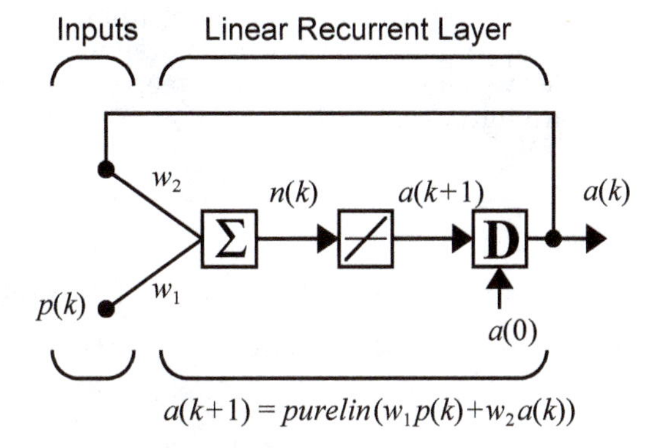

Figure P11.11 Linear Recurrent Network

The first step is to define our performance index. As with the multilayer networks, we will use squared error:

$$\hat{F}(\mathbf{x}) = (t(k) - a(k))^2 = (e(k))^2 .$$

For our weight updates we will use the steepest descent algorithm:

$$\Delta w_i = -\alpha \frac{\partial}{\partial w_i} \hat{F}(\mathbf{x}) .$$

These derivatives can be computed as follows:

$$\frac{\partial}{\partial w_i} \hat{F}(\mathbf{x}) = \frac{\partial}{\partial w_i} (t(k) - a(k))^2 = 2(t(k) - a(k)) \left\{ -\frac{\partial a(k)}{\partial w_i} \right\} .$$

Therefore, the key terms we need to compute are

$$\frac{\partial a(k)}{\partial w_i} .$$

To compute these terms we first need to write out the network equation:

$$a(k+1) = purelin(w_1 p(k) + w_2 a(k)) = w_1 p(k) + w_2 a(k) .$$

Next we take the derivative of both sides of this equation with respect to the network weights:

$$\frac{\partial a(k+1)}{\partial w_1} = p(k) + w_2 \frac{\partial a(k)}{\partial w_1} ,$$

$$\frac{\partial a(k+1)}{\partial w_2} = a(k) + w_2 \frac{\partial a(k)}{\partial w_2} .$$

(Note that we had to take account of the fact that $a(k)$ is itself a function of w_1 and w_2.) These two recursive equations are then used to compute the derivatives needed for the steepest descent weight update. The equations are initialized with

$$\frac{\partial a(0)}{\partial w_1} = 0, \frac{\partial a(0)}{\partial w_2} = 0,$$

since the initial condition is not a function of the weight.

To illustrate the process, let's say that $a(0) = 0$. The first network update would be

$$a(1) = w_1 p(0) + w_2 a(0) = w_1 p(0).$$

The first derivatives would be computed:

$$\frac{\partial a(1)}{\partial w_1} = p(0) + w_2 \frac{\partial a(0)}{\partial w_1} = p(0), \frac{\partial a(1)}{\partial w_2} = a(0) + w_2 \frac{\partial a(0)}{\partial w_2} = 0.$$

The first weight updates would be

$$\Delta w_i = -\alpha \frac{\partial}{\partial w_i} \hat{F}(\mathbf{x}) = -\alpha \left[2(t(1) - a(1)) \left\{ -\frac{\partial a(1)}{\partial w_i} \right\} \right]$$

$$\Delta w_1 = -2\alpha (t(1) - a(1)) \{-p(0)\}$$

$$\Delta w_2 = -2\alpha (t(1) - a(1)) \{0\} = 0.$$

This algorithm is a type of *dynamic backpropagation*, in which the gradient is computed by means of a difference equation.

P11.10 Show that backpropagation reduces to the LMS algorithm for a single-layer linear network (ADALINE).

The sensitivity calculation for a single-layer linear network would be:

$$\mathbf{s}^1 = -2\dot{\mathbf{F}}^1(\mathbf{n}^1)(\mathbf{t} - \mathbf{a}) = -2\mathbf{I}(\mathbf{t} - \mathbf{a}) = -2\mathbf{e},$$

The weight update (Eq. (11.46) and Eq. (11.47)) would be

$$\mathbf{W}^1(k+1) = \mathbf{W}^1(k) - \alpha \mathbf{s}^1 (\mathbf{a}^0)^T = \mathbf{W}^1(k) - \alpha(-2\mathbf{e})\mathbf{p}^T = \mathbf{W}^1(k) + 2\alpha \mathbf{e}\mathbf{p}^T$$

$$\mathbf{b}^1(k+1) = \mathbf{b}^1(k) - \alpha \mathbf{s}^1 = \mathbf{b}^1(k) - \alpha(-2\mathbf{e}) = \mathbf{b}^1(k) + 2\alpha \mathbf{e}.$$

This is identical to the LMS algorithm of Chapter 10.

Epilogue

In this chapter we have presented the multilayer perceptron network and the backpropagation learning rule. The multilayer network is a powerful extension of the single-layer perceptron network. Whereas the single-layer network is only able to classify linearly separable patterns, the multilayer network can be used for arbitrary classification problems. In addition, multilayer networks can be used as universal function approximators. It has been shown that a two-layer network, with sigmoid-type transfer functions in the hidden layer, can approximate any practical function, given enough neurons in the hidden layer.

The backpropagation algorithm is an extension of the LMS algorithm that can be used to train multilayer networks. Both LMS and backpropagation are approximate steepest descent algorithms that minimize squared error. The only difference between them is in the way in which the gradient is calculated. The backpropagation algorithm uses the chain rule in order to compute the derivatives of the squared error with respect to the weights and biases in the hidden layers. It is called backpropagation because the derivatives are computed first at the last layer of the network, and then propagated backward through the network, using the chain rule, to compute the derivatives in the hidden layers.

One of the major problems with backpropagation has been the long training times. It is not feasible to use the basic backpropagation algorithm on practical problems, because it can take weeks to train a network, even on a large computer. Since backpropagation was first popularized, there has been considerable work on methods to accelerate the convergence of the algorithm. In Chapter 12 we will discuss the reasons for the slow convergence of backpropagation and will present several techniques for improving the performance of the algorithm.

Another key problem in training multilayer networks is overfitting. The network may memorize the data in the training set, but fail to generalize to new situations. In Chapter 13 we will describe in detail training procedures that can be used to produce networks with excellent generalization.

This chapter has focused mainly on the theoretical development of the backpropagation learning rule for training multilayer networks. Practical aspects of training networks with this method are discussed in Chapter 17. Real-world case studies that demonstrate how to train and validate multilayer networks are provided in Chapter 18 (function approximation), Chapter 19 (probability estimation) and Chapter 20 (pattern recognition).

Further Reading

[HoSt89] K. M. Hornik, M. Stinchcombe and H. White, "Multilayer feedforward networks are universal approximators," *Neural Networks*, vol. 2, no. 5, pp. 359–366, 1989.

This paper proves that multilayer feedforward networks with arbitrary squashing functions can approximate any Borel integrable function from one finite dimensional space to another finite dimensional space.

[LeCu85] Y. Le Cun, "Une procedure d'apprentissage pour reseau a seuil assymetrique," *Cognitiva*, vol. 85, pp. 599–604, 1985.

Yann Le Cun discovered the backpropagation algorithm at about the same time as Parker and Rumelhart, Hinton and Williams. This paper describes his algorithm.

[Park85] D. B. Parker, "Learning-logic: Casting the cortex of the human brain in silicon," Technical Report TR-47, Center for Computational Research in Economics and Management Science, MIT, Cambridge, MA, 1985.

David Parker independently derived the backpropagation algorithm at about the same time as Le Cun and Rumelhart, Hinton and Williams. This report describes his algorithm.

[RuHi86] D. E. Rumelhart, G. E. Hinton and R. J. Williams, "Learning representations by back-propagating errors," *Nature*, vol. 323, pp. 533–536, 1986.

This paper contains the most widely publicized description of the backpropagation algorithm.

[RuMc86] D. E. Rumelhart and J. L. McClelland, eds., *Parallel Distributed Processing: Explorations in the Microstructure of Cognition*, vol. 1, Cambridge, MA: MIT Press, 1986.

This book was one of the two key influences in the resurgence of interest in the neural networks field during the 1980s. Among other topics, it presents the backpropagation algorithm for training multilayer neural networks.

[Werbo74] P. J. Werbos, "Beyond regression: New tools for prediction and analysis in the behavioral sciences," Ph.D. Thesis, Harvard University, Cambridge, MA, 1974.

This Ph.D. thesis contains what appears to be the first description of the backpropagation algorithm (although that name is not used). The algorithm is described here in the context of general networks, with neural networks as a special case. Backpropagation did not become widely known until it was rediscovered in the mid 1980s by Rumelhart, Hinton and Williams [RuHi86], David Parker [Park85] and Yann Le Cun [LeCu85].

Exercises

E11.1 Design multilayer networks to perform the classifications illustrated in Figure E11.1. The network should output a 1 whenever the input vector is in the shaded region (or on the boundary) and a -1 otherwise. Draw the network diagram in abbreviated notation and show the weight matrices and bias vectors.

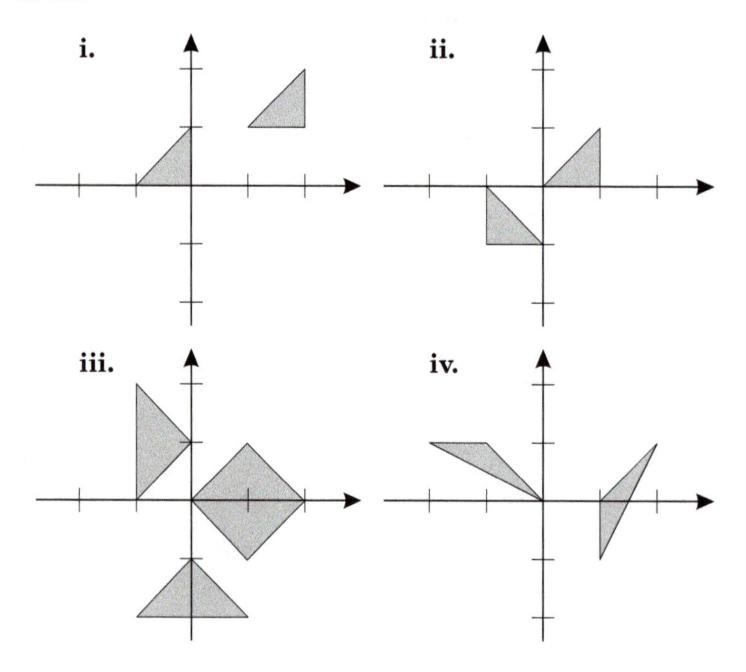

Figure E11.1 Pattern Classification Tasks

E11.2 Choose the weights and biases for the 1-2-1 network shown in Figure 11.4 so that the network response passes through the points indicated by the blue circles in Figure E11.2.

Use the MATLAB® Neural Network Design Demonstration Two-Layer Network Function (`nnd11nf`) *to check your result.*

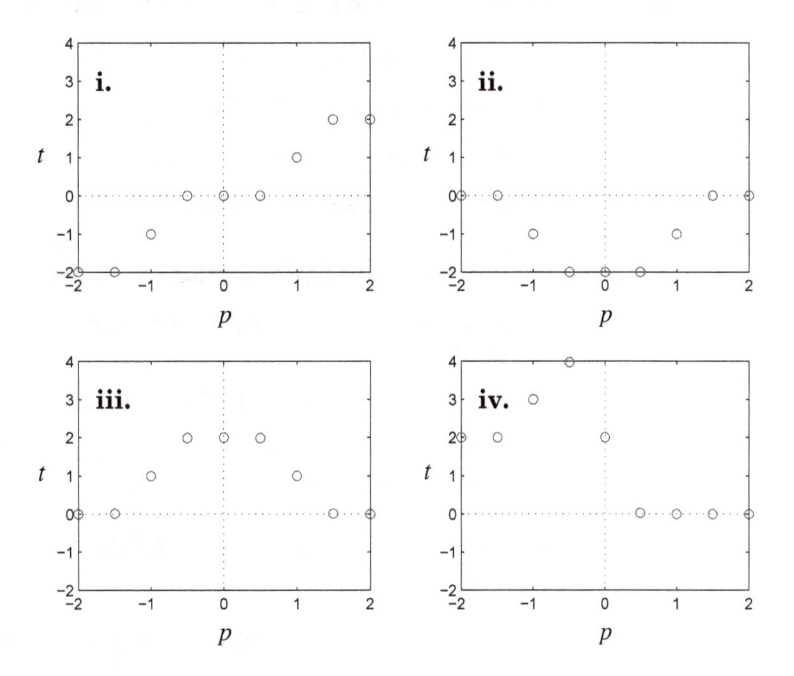

Figure E11.2 Function Approximation Tasks

E11.3 Find a single-layer network that has the same input/output characteristic as the network in Figure E11.3.

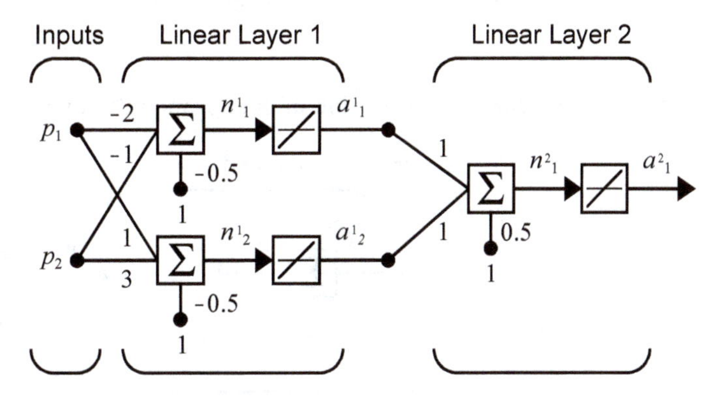

Figure E11.3 Two-Layer Linear Network

E11.4 Use the chain rule to find the derivative $\partial f / \partial w$ in the following cases:

i. $f(n) = \sin(n)$, $n(w) = w^2$.

ii. $f(n) = \tanh(n)$, $n(w) = 5w$.

iii. $f(n) = \exp(n)$, $n(w) = \cos(w)$.

iv. $f(n) = \mathrm{logsig}(n)$, $n(w) = \exp(w)$.

E11.5 Consider again the backpropagation example that begins on page 11-14.

i. Find the squared error $(e)^2$ as an explicit function of all weights and biases.

ii. Using part (i), find $\partial(e)^2 / \partial w^1_{1,1}$ at the initial weights and biases.

iii. Compare the results of part (ii) with the backpropagation results described in the text.

E11.6 For the network shown in Figure E11.4 the initial weights and biases are chosen to be

$$w^1(0) = 1,\ b^1(0) = -2,\ w^2(0) = 1,\ b^2(0) = 1.$$

The network transfer functions are

$$f^1(n) = (n)^2,\ f^2(n) = \frac{1}{n},$$

and an input/target pair is given to be

$$\{p = 1, t = 1\}.$$

Perform one iteration of backpropagation with $\alpha = 1$.

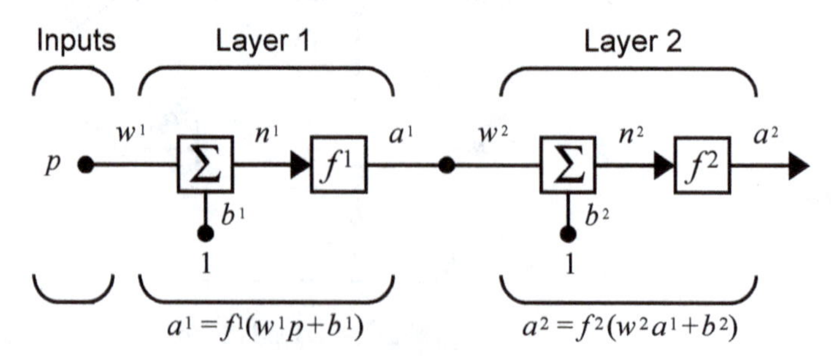

Figure E11.4 Two-Layer Network for Exercise E11.6

E11.7 Consider the two-layer network in Figure E11.5.

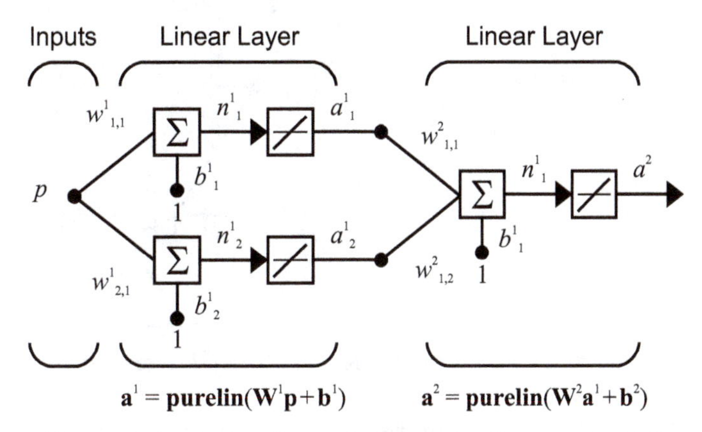

Figure E11.5 Two-Layer Network for Exercise E11.7

with the following input and target: $\{p_1 = 1, t_1 = 2\}$. The initial weights and biases are given by

$$\mathbf{W}^1(0) = \begin{bmatrix} 1 \\ -1 \end{bmatrix}, \mathbf{W}^2(0) = \begin{bmatrix} -1 & 1 \end{bmatrix}, \mathbf{b}^1(0) = \begin{bmatrix} 2 \\ 1 \end{bmatrix}, \mathbf{b}^2(0) = \begin{bmatrix} 3 \end{bmatrix}$$

i. Apply the input to the network and make one pass forward through the network to compute the output and the error.

ii. Compute the sensitivities by backpropagating through the network.

iii. Compute the derivative $\partial(e)^2/\partial w^1_{1,1}$ using the results of part ii. (Very little calculation is required here.)

E11.8 For the network shown in Figure E11.6 the neuron transfer function is

$$f^1(n) = (n)^2,$$

and an input/target pair is given to be

$$\left\{ \mathbf{p} = \begin{bmatrix} 1 \\ 1 \end{bmatrix}, \mathbf{t} = \begin{bmatrix} 8 \\ 2 \end{bmatrix} \right\}.$$

Perform one iteration of backpropagation with $\alpha = 1$.

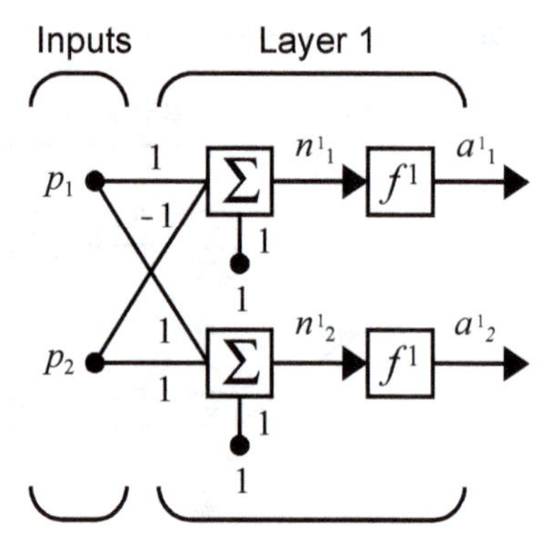

Figure E11.6 Single-Layer Network for Exercise E11.8

E11.9 We want to train the network in Figure E11.7 using the standard back-propagation algorithm (approximate steepest descent).

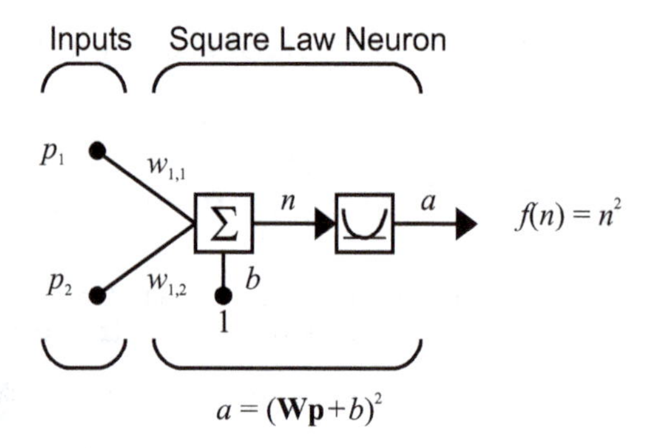

Figure E11.7 Square Law Neuron

The following input and target are given:

$$\left\{ \mathbf{p} = \begin{bmatrix} 1 \\ 1 \end{bmatrix}, t = \begin{bmatrix} 0 \end{bmatrix} \right\}$$

The initial weights and bias are

$$\mathbf{W}(0) = \begin{bmatrix} 1 & -1 \end{bmatrix},\ b(0) = 1 .$$

i. Propagate the input forward through the network.

ii. Compute the error.

iii. Propagate the sensitivities backward through the network.

iv. Compute the gradient of the squared error with respect to the weights and bias.

v. Update the weights and bias (assume a learning rate of $\alpha = 0.1$).

E11.10 Consider the following multilayer perceptron network. (The transfer function of the hidden layer is $f(n) = n^2$.)

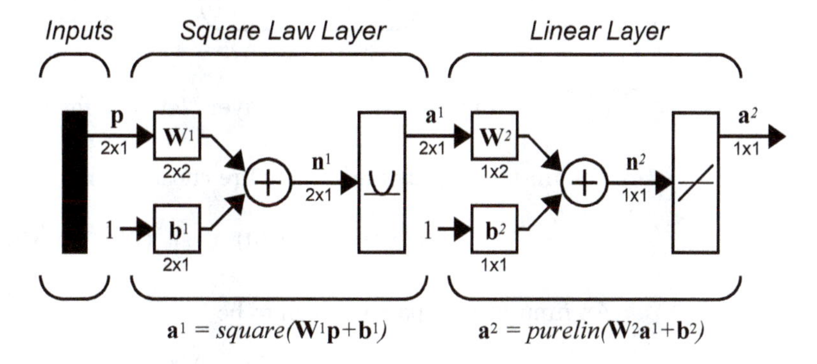

Figure E11.8 Two-Layer Square Law Network

The initial weights and biases are:

$$\mathbf{W}^1(0) = \begin{bmatrix} 1 & -1 \\ 1 & 0 \end{bmatrix},\ \mathbf{W}^2(0) = \begin{bmatrix} 2 & 1 \end{bmatrix},\ \mathbf{b}^1(0) = \begin{bmatrix} 1 \\ -1 \end{bmatrix},\ \mathbf{b}^2(0) = \begin{bmatrix} -1 \end{bmatrix}.$$

Perform one iteration of the standard steepest descent backpropagation (use matrix operations) with learning rate $\alpha = 0.5$ for the following input/target pair:

$$\left\{ \mathbf{p} = \begin{bmatrix} 1 \\ 1 \end{bmatrix}, t = \begin{bmatrix} 2 \end{bmatrix} \right\}$$

E11.11 Consider the network shown in Figure E11.9.

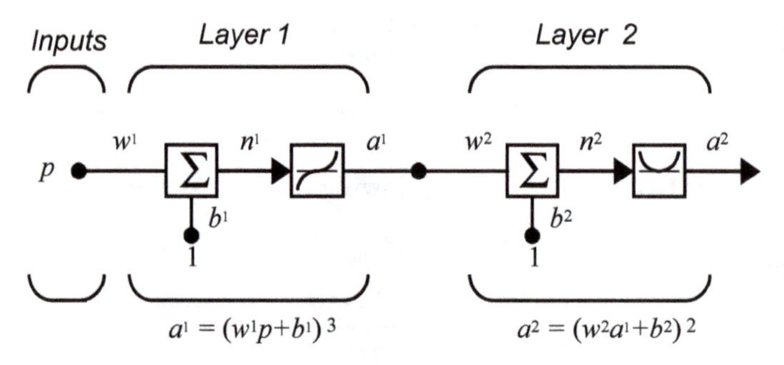

Figure E11.9 Two-Layer Network for Exercise E11.11

The initial weights and biases are chosen to be

$$w^1(0) = -2, b^1(0) = 1, w^2(0) = 1, b^2(0) = -2 .$$

An input/target pair is given to be

$$\{p_1 = 1, t_1 = 0\} ,$$

Perform one iteration of backpropagation (steepest descent) with $\alpha = 1$.

E11.12 Consider the multilayer perceptron network in Figure E11.10. (The transfer function of the hidden layer is $f(n) = n^3$.)

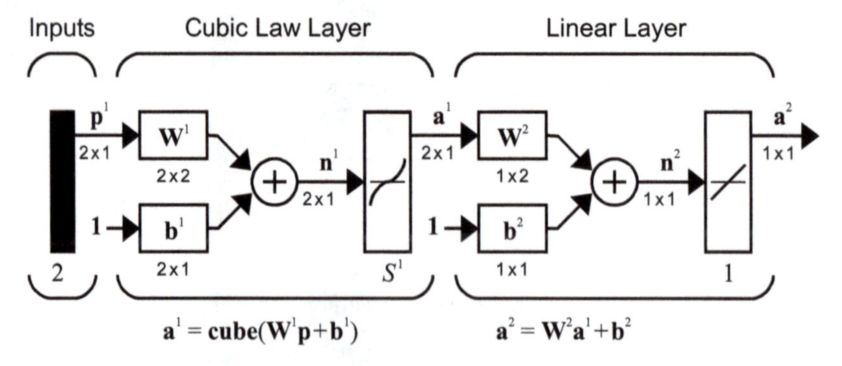

Figure E11.10 Cubic Law Neural Network

The initial weights and biases are:

$$\mathbf{W}^1(0) = \begin{bmatrix} 1 & -1 \\ 1 & 0 \end{bmatrix},\ \mathbf{b}^1(0) = \begin{bmatrix} 1 \\ 2 \end{bmatrix},\ \mathbf{W}^2(0) = \begin{bmatrix} 1 & 1 \end{bmatrix},\ \mathbf{b}^2(0) = \begin{bmatrix} 1 \end{bmatrix}.$$

Perform one iteration of the standard steepest descent backpropagation (use matrix operations) with learning rate $\alpha = 0.5$ for the following input/ target pair:

$$\left\{ \mathbf{p} = \begin{bmatrix} -1 \\ 1 \end{bmatrix}, t = \begin{bmatrix} -1 \end{bmatrix} \right\}.$$

E11.13 Someone has proposed that the standard multilayer network should be modified to include a scalar gain at each layer. This means that the net input at layer m would be computed as

$$\mathbf{n}^m = \beta^m[\mathbf{W}^m\mathbf{a}^{m-1} + \mathbf{b}^m],$$

where β^m is the scalar gain at layer m. This gain would be trained like the weights and biases of the network. Modify the backpropagation algorithm (Eq. (11.41) to Eq. (11.47)) for this new network. (There will need to be a new equation added to update β^m, but some of the other equations may have to be modified as well.)

E11.14 Consider the two-layer network shown in Figure E11.11.

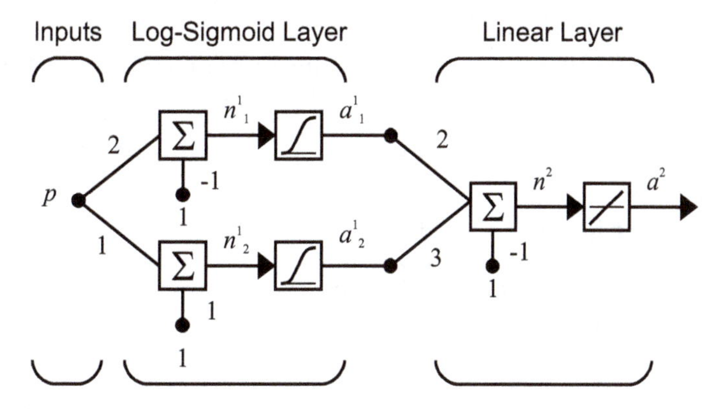

Figure E11.11 Two-Layer Network for Exercise E11.14

i. If $p = 1$, use a (slightly) modified form of backpropagation (as developed in Eq. (11.41) through Eq. (11.47)) to find

$$\frac{\partial a^2}{\partial n^2}, \frac{\partial a^2}{\partial n_1^1}, \frac{\partial a^2}{\partial n_2^1}.$$

ii. Use the results of i. and the chain rule to find $\frac{\partial a^2}{\partial p}$.

Your answers to both parts should be numerical.

E11.15 Consider the network shown in Figure E11.12, where the inputs to the neuron involve both the original inputs and their product. This is a type of higher-order network.

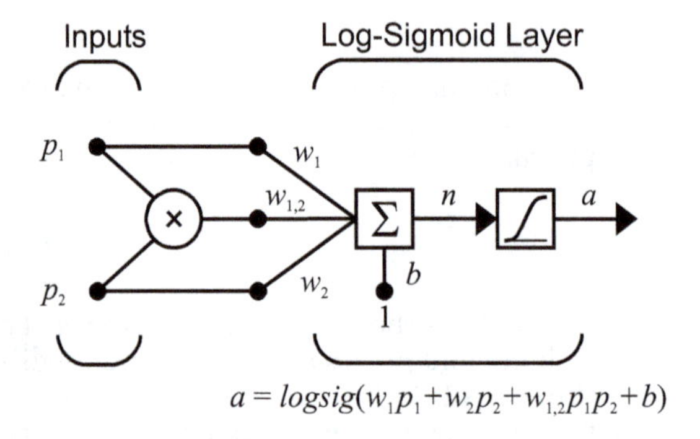

Figure E11.12 Higher-Order Network

i. Find a learning rule for the network parameters, using the approximate steepest descent algorithm (as was done for backpropagation).

ii. For the following initial parameter values, inputs and target, perform one iteration of your learning rule with learning rate $\alpha = 1$:

$$w_1 = 1, w_2 = -1, w_{1,2} = 0.5, b_1 = 1, p_1 = 0, p_2 = 1, t = 0.75$$

E11.16 In Figure E11.13 we have a two-layer network that has an additional connection from the input directly to the second layer. Derive the backpropagation algorithm for this network.

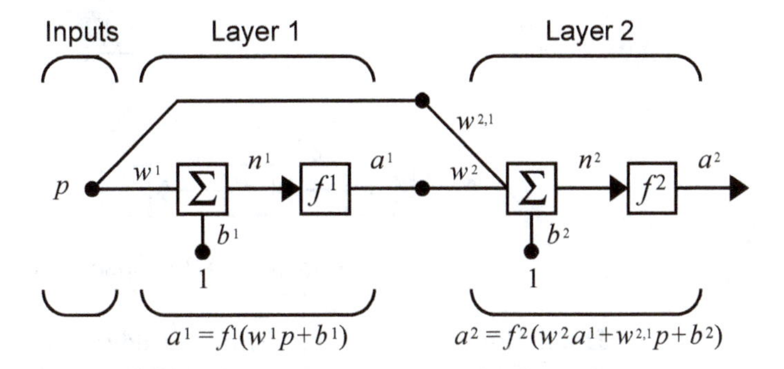

Figure E11.13 Two-Layer Network with Bypass Connection

E11.17 In the multilayer network, the net input is computed as follows

$$\mathbf{n}^{m+1} = \mathbf{W}^{m+1}\mathbf{a}^{m} + \mathbf{b}^{m+1} \text{ or } n_i^{m+1} = \sum_{j=1}^{S^m} w_{i,j}^{m+1} a_j^m + b_i^{m+1} .$$

If the net input calculation is changed to the following equation (squared distance calculation), how will the sensitivity backpropagation (Eq. (11.35)) change?

$$n_i^{m+1} = \sum_{j=1}^{S^m} (w_{i,j}^{m+1} - a_j^m)^2$$

E11.18 Consider again the net input calculation, as described in Exercise E11.17. If the net input calculation is changed to the following equation (multiply by the bias, instead of add), how will the sensitivity backpropagation (Eq. (11.35)) change?

$$n_i^{m+1} = \left(\sum_{j=1}^{S^m} w_{i,j}^{m+1} a_j^m \right) \times b_i^{m+1} .$$

E11.19 Consider the system shown in Figure E11.14. There are a series of stages, with different transfer functions in each stage. (There are no weights or biases.) We want to take the derivative of the output of this system (a^M) with respect to the input of the system (p). Derive a recursive algorithm that you can use to compute this derivative. Use the concepts that we used to derive the backpropagation algorithm, and use the following intermediate variable in your algorithm:

$$q^i = \frac{\partial a^M}{\partial a^i}.$$

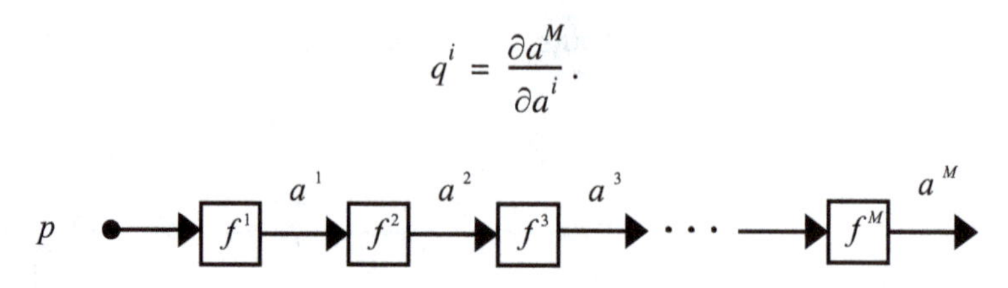

Figure E11.14 Cascade System

E11.20 The backpropagation algorithm is used to compute the gradient of the squared error with respect to the weights and biases of a multilayer network. How would the algorithm be changed if you wanted to compute the gradient with respect to the inputs of the network (i.e., with respect to the elements of the input vector $\mathbf{p}$)? Carefully explain all of your steps, and write out the final algorithm.

E11.21 With the standard backpropagation algorithm, we want to compute the derivative

$$\frac{\partial F}{\partial w}.$$

To calculate this derivative, we use the chain rule in the form

$$\frac{\partial F}{\partial w} = \frac{\partial F}{\partial n} \cdot \frac{\partial n}{\partial w}.$$

Suppose that we want to use Newton's method. We would need to find the second derivative

$$\frac{\partial^2 F}{\partial w^2}.$$

What form will the chain rule take in this case?

E11.22 The standard steepest descent backpropagation algorithm, which is summarized in Eq. (11.41) through Eq. (11.47), was designed to minimize the performance function that was the sum of squares of the network errors, as given in Eq. (11.12). Suppose that we want to change the performance function to the sum of the fourth powers of the errors (e^4) plus the sum of the squares of the weights and biases in the network. Show how Eq. (11.41) through Eq. (11.47) will change for this new performance function. (You don't need to rederive any steps which are already given in this chapter and do not change.)

E11.23 Repeat Problem P11.4 using the "backward" method described below.

In Problem P11.4. we had the dynamic system

$$y(k+1) = f(y(k)).$$

We had to choose the initial condition $y(0)$ so that at some final time $k = K$ the system output $y(K)$ would be as close as possible to some target output t. We minimized the performance index

$$F(y(0)) = (t - y(K))^2 = e^2(K)$$

using steepest descent, so we needed the gradient

$$\frac{\partial}{\partial y(0)} F(y(0)).$$

We developed a procedure for computing this gradient using the chain rule. The procedure involved a recursive equation for the term

$$r(k) \equiv \frac{\partial}{\partial y(0)} y(k),$$

which evolved forward in time. The gradient can also be computed in a different way by evolving the term

$$q(k) \equiv \frac{\partial}{\partial y(k)} e^2(K)$$

backward through time.

E11.24 Consider the recurrent neural network in Figure E11.15.

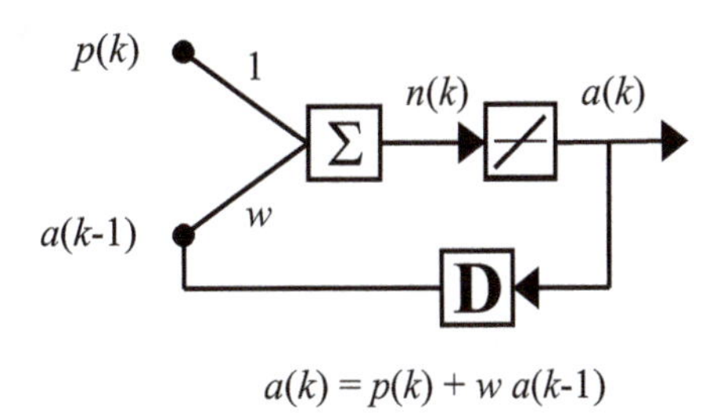

Figure E11.15 Recurrent Network

We want to find the weight value w so that at some final time $k = K$ the system output $a(K)$ will be as close as possible to some target output t. We will minimize the performance index $F(w) = (t - a(K))^2$ using steepest descent, so we need the gradient $\partial F(w)/\partial w$.

i. Find a general procedure to compute this gradient using the chain rule. Develop an equation to evolve the following term forward through time:

$$s(k) \equiv \frac{\partial}{\partial w} a(k) .$$

Show each step of your entire procedure carefully. This will involve updating $s(k)$ and also computing the gradient $\partial F(w)/\partial w$.

ii. Assume that $K = 3$. Write out the complete expression for $a(3)$ as a function of $p(1)$, $p(2)$, $p(3)$ and w (assuming $a(0) = 0$). Take the derivative of this expression with respect to w, and show that it equals $s(3)$.

» 2 + 2
ans =
4

E11.25 Write a MATLAB program to implement the backpropagation algorithm for a $1 - S^1 - 1$ network. Write the program using matrix operations, as in Eq. (11.41) to Eq. (11.47). Choose the initial weights and biases to be random numbers uniformly distributed between -0.5 and 0.5 (using the MATLAB function `rand`), and train the network to approximate the function

$$g(p) = 1 + \sin\left(\frac{\pi}{2}p\right) \text{ for } -2 \leq p \leq 2 .$$

Use $S^1 = 2$ and $S^1 = 10$. Experiment with several different values for the learning rate α, and use several different initial conditions. Discuss the convergence properties of the algorithm as the learning rate changes.

12 Variations on Backpropagation

Objectives

The backpropagation algorithm introduced in Chapter 11 was a major breakthrough in neural network research. However, the basic algorithm is too slow for most practical applications. In this chapter we present several variations of backpropagation that provide significant speedup and make the algorithm more practical.

We will begin by using a function approximation example to illustrate why the backpropagation algorithm is slow in converging. Then we will present several modifications to the algorithm. Recall that backpropagation is an approximate steepest descent algorithm. In Chapter 9 we saw that steepest descent is the simplest, and often the slowest, minimization method. The conjugate gradient algorithm and Newton's method generally provide faster convergence. In this chapter we will explain how these faster procedures can be used to speed up the convergence of backpropagation.

Theory and Examples

When the basic backpropagation algorithm is applied to a practical problem the training may take days or weeks of computer time. This has encouraged considerable research on methods to accelerate the convergence of the algorithm.

The research on faster algorithms falls roughly into two categories. The first category involves the development of heuristic techniques, which arise out of a study of the distinctive performance of the standard backpropagation algorithm. These heuristic techniques include such ideas as varying the learning rate, using momentum and rescaling variables (e.g., [VoMa88], [Jacob88], [Toll90] and [RiIr90]). In this chapter we will discuss the use of momentum and variable learning rates.

Another category of research has focused on standard numerical optimization techniques (e.g., [Shan90], [Barn92], [Batt92] and [Char92]). As we have discussed in Chapter 10 and 11, training feedforward neural networks to minimize squared error is simply a numerical optimization problem. Because numerical optimization has been an important research subject for 30 or 40 years (see Chapter 9), it seems reasonable to look for fast training algorithms in the large number of existing numerical optimization techniques. There is no need to "reinvent the wheel" unless absolutely necessary. In this chapter we will present two existing numerical optimization techniques that have been very successfully applied to the training of multilayer perceptrons: the conjugate gradient algorithm and the Levenberg-Marquardt algorithm (a variation of Newton's method).

We should emphasize that all of the algorithms that we will describe in this chapter use the backpropagation procedure, in which derivatives are processed from the last layer of the network to the first. For this reason they could all be called "backpropagation" algorithms. The differences between the algorithms occur in the way in which the resulting derivatives are used to update the weights. In some ways it is unfortunate that the algorithm we usually refer to as backpropagation is in fact a steepest descent algorithm. In order to clarify our discussion, for the remainder of this chapter we will refer to the basic backpropagation algorithm as steepest descent backpropagation (*SDBP*).

SDBP

In the next section we will use a simple example to explain why SDBP has problems with convergence. Then, in the following sections, we will present various procedures to improve the convergence of the algorithm.

Drawbacks of Backpropagation

Recall from Chapter 10 that the LMS algorithm is guaranteed to converge to a solution that minimizes the mean squared error, so long as the learning rate is not too large. This is true because the mean squared error for a single-layer linear network is a quadratic function. The quadratic function has only a single stationary point. In addition, the Hessian matrix of a quadratic function is constant, therefore the curvature of the function in a given direction does not change, and the function contours are elliptical.

SDBP is a generalization of the LMS algorithm. Like LMS, it is also an approximate steepest descent algorithm for minimizing the mean squared error. In fact, SDBP is equivalent to the LMS algorithm when used on a single-layer linear network. (See Problem P11.10.) When applied to multilayer networks, however, the characteristics of SDBP are quite different. This has to do with the differences between the mean squared error performance surfaces of single-layer linear networks and multilayer nonlinear networks. While the performance surface for a single-layer linear network has a single minimum point and constant curvature, the performance surface for a multilayer network may have many local minimum points, and the curvature can vary widely in different regions of the parameter space. This will become clear in the example that follows.

Performance Surface Example

To investigate the mean squared error performance surface for multilayer networks we will employ a simple function approximation example. We will use the 1-2-1 network shown in Figure 12.1, with log-sigmoid transfer functions in both layers.

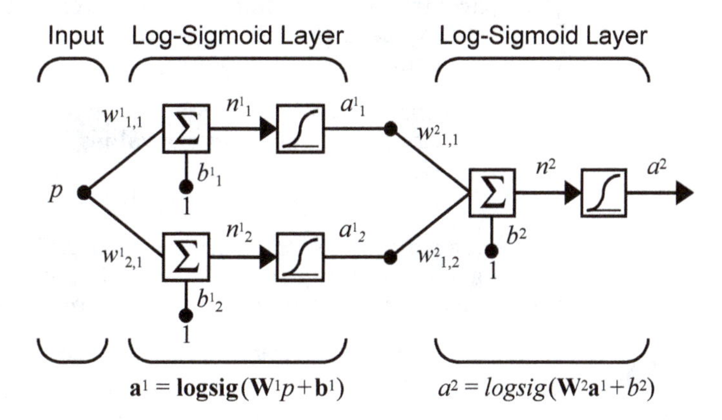

Figure 12.1 1-2-1 Function Approximation Network

In order to simplify our analysis, we will give the network a problem for which we know the optimal solution. The function we will approximate is the response of the same 1-2-1 network, with the following values for the weights and biases:

$$w^1_{1,1} = 10, \; w^1_{2,1} = 10, \; b^1_1 = -5, \; b^1_2 = 5, \tag{12.1}$$

$$w^2_{1,1} = 1, \; w^2_{1,2} = 1, \; b^2 = -1. \tag{12.2}$$

The network response for these parameters is shown in Figure 12.2, which plots the network output a^2 as the input p is varied over the range $[-2, 2]$.

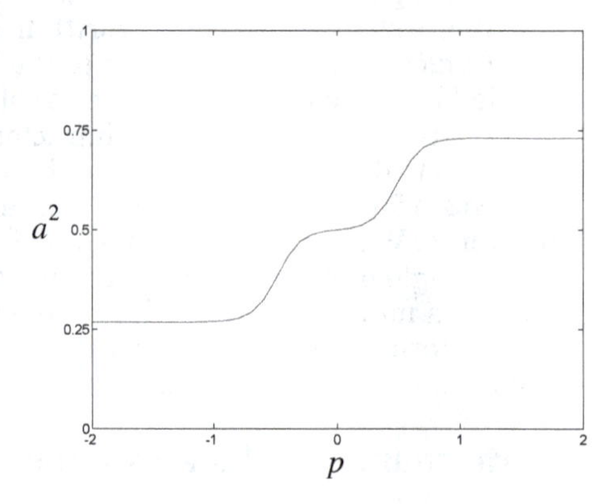

Figure 12.2 Nominal Function

We want to train the network of Figure 12.1 to approximate the function displayed in Figure 12.2. The approximation will be exact when the network parameters are set to the values given in Eq. (12.1) and Eq. (12.2). This is, of course, a very contrived problem, but it is simple and it illustrates some important concepts.

Let's now consider the performance index for our problem. We will assume that the function is sampled at the values

$$p = -2, -1.9, -1.8, \ldots, 1.9, 2, \tag{12.3}$$

and that each occurs with equal probability. The performance index will be the sum of the squared errors at these 41 points. (We won't bother to find the mean squared error, which just requires dividing by 41.)

In order to be able to graph the performance index, we will vary only two parameters at a time. Figure 12.3 illustrates the squared error when only $w^1_{1,1}$ and $w^2_{1,1}$ are being adjusted, while the other parameters are set to their optimal values given in Eq. (12.1) and Eq. (12.2). Note that the minimum error will be zero, and it will occur when $w^1_{1,1} = 10$ and $w^2_{1,1} = 1$, as indicated by the open blue circle in the figure.

There are several features to notice about this error surface. First, it is clearly not a quadratic function. The curvature varies drastically over the parameter space. For this reason it will be difficult to choose an appropriate learning rate for the steepest descent algorithm. In some regions the surface is very flat, which would allow a large learning rate, while in other regions the curvature is high, which would require a small learning rate. (Refer to discussions in Chapter 9 and 10 on the choice of learning rate for the steepest descent algorithm.)

It should be noted that the flat regions of the performance surface should not be unexpected, given the sigmoid transfer functions used by the network. The sigmoid is very flat for large inputs.

A second feature of this error surface is the existence of more than one local minimum point. The global minimum point is located at $w^1_{1,1} = 10$ and $w^2_{1,1} = 1$, along the valley that runs parallel to the $w^1_{1,1}$ axis. However, there is also a local minimum, which is located in the valley that runs parallel to the $w^2_{1,1}$ axis. (This local minimum is actually off the graph at $w^1_{1,1} = 0.88$, $w^2_{1,1} = 38.6$.) In the next section we will investigate the performance of backpropagation on this surface.

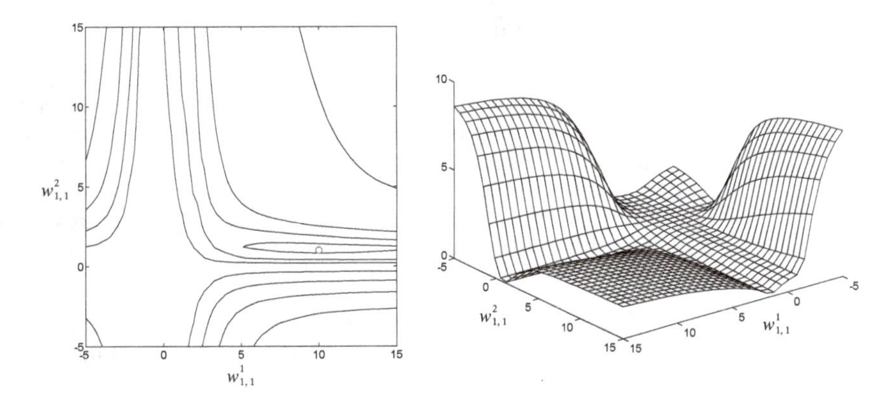

Figure 12.3 Squared Error Surface Versus $w^1_{1,1}$ and $w^2_{1,1}$

Figure 12.4 illustrates the squared error when $w^1_{1,1}$ and b^1_1 are being adjusted, while the other parameters are set to their optimal values. Note that the minimum error will be zero, and it will occur when $w^1_{1,1} = 10$ and $b^1_1 = -5$, as indicated by the open blue circle in the figure.

Again we find that the surface has a very contorted shape, steep in some regions and very flat in others. Surely the standard steepest descent algorithm will have some trouble with this surface. For example, if we have an initial guess of $w^1_{1,1} = 0$, $b^1_1 = -10$, the gradient will be very close to zero, and the steepest descent algorithm would effectively stop, even though it is not close to a local minimum point.

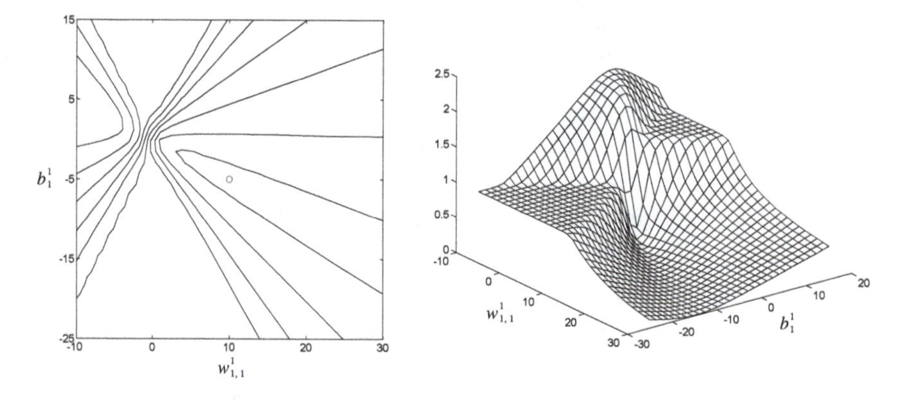

Figure 12.4 Squared Error Surface Versus $w_{1,1}^1$ and b_1^1

Figure 12.5 illustrates the squared error when b_1^1 and b_2^1 are being adjusted, while the other parameters are set to their optimal values. The minimum error is located at $b_1^1 = -5$ and $b_2^1 = 5$, as indicated by the open blue circle in the figure.

This surface illustrates an important property of multilayer networks: they have a symmetry to them. Here we see that there are two local minimum points and they both have the same value of squared error. The second solution corresponds to the same network being turned upside down (i.e., the top neuron in the first layer is exchanged with the bottom neuron). It is because of this characteristic of neural networks that we do not set the initial weights and biases to zero. The symmetry causes zero to be a saddle point of the performance surface.

This brief study of the performance surfaces for multilayer networks gives us some hints as to how to set the initial guess for the SDBP algorithm. First, we do not want to set the initial parameters to zero. This is because the origin of the parameter space tends to be a saddle point for the performance surface. Second, we do not want to set the initial parameters to large values. This is because the performance surface tends to have very flat regions as we move far away from the optimum point.

Typically we choose the initial weights and biases to be small random values. In this way we stay away from a possible saddle point at the origin without moving out to the very flat regions of the performance surface. (Another procedure for choosing the initial parameters is described in [NgWi90].) As we will see in the next section, it is also useful to try several different initial guesses, in order to be sure that the algorithm converges to a global minimum point.

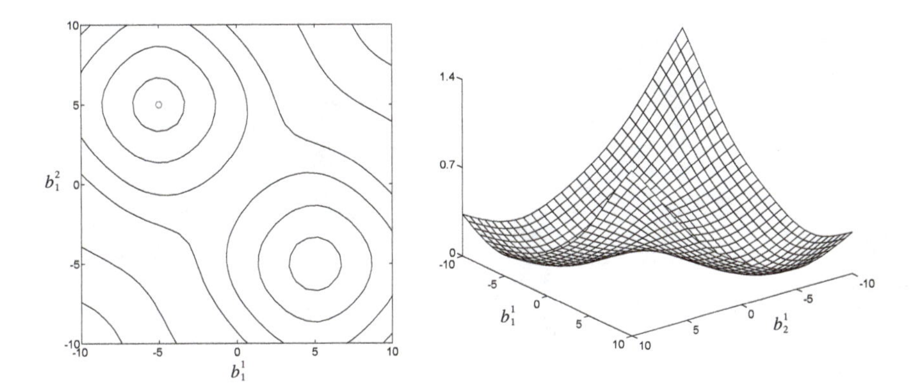

Figure 12.5 Squared Error Surface Versus b_1^1 and b_2^1

Convergence Example

Now that we have examined the performance surface, let's investigate the performance of SDBP. For this section we will use a variation of the standard algorithm, called *batching*, in which the parameters are updated only after the entire training set has been presented. The gradients calculated at each training example are averaged together to produce a more accurate estimate of the gradient. (If the training set is complete, i.e., covers all possible input/output pairs, then the gradient estimate will be exact.)

Batching

In Figure 12.6 we see two trajectories of SDBP (batch mode) when only two parameters, $w_{1,1}^1$ and $w_{1,1}^2$ are adjusted. For the initial condition labeled "a" the algorithm does eventually converge to the optimal solution, but the convergence is slow. The reason for the slow convergence is the change in curvature of the surface over the path of the trajectory. After an initial moderate slope, the trajectory passes over a very flat surface, until it falls into a very gently sloping valley. If we were to increase the learning rate, the algorithm would converge faster while passing over the initial flat surface, but would become unstable when falling into the valley, as we will see in a moment.

Trajectory "b" illustrates how the algorithm can converge to a local minimum point. The trajectory is trapped in a valley and diverges from the optimal solution. If allowed to continue the trajectory converges to $w_{1,1}^1 = 0.88$, $w_{1,1}^2 = 38.6$. The existence of multiple local minimum points is typical of the performance surface of multilayer networks. For this reason it is best to try several different initial guesses in order to ensure that a global minimum has been obtained. (Some of the local minimum points may have the same value of squared error, as we saw in Figure 12.5, so we would not expect the algorithm to converge to the same parameter values for each initial guess. We just want to be sure that the same minimum error is obtained.)

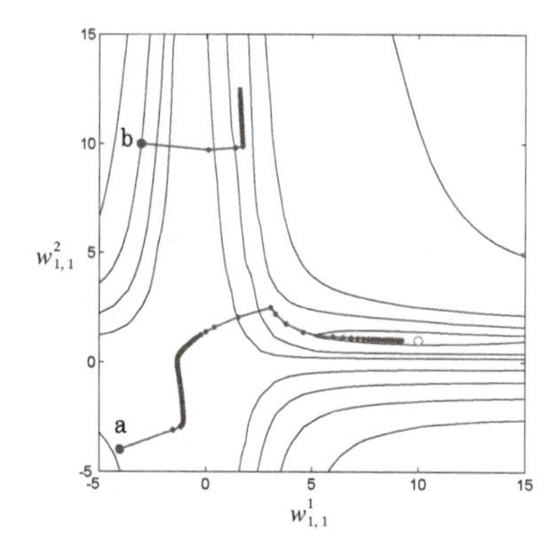

Figure 12.6 Two SDBP (Batch Mode) Trajectories

The progress of the algorithm can also be seen in Figure 12.7, which shows the squared error versus the iteration number. The curve on the left corresponds to trajectory "a" and the curve on the right corresponds to trajectory "b." These curves are typical of SDBP, with long periods of little progress and then short periods of rapid advance.

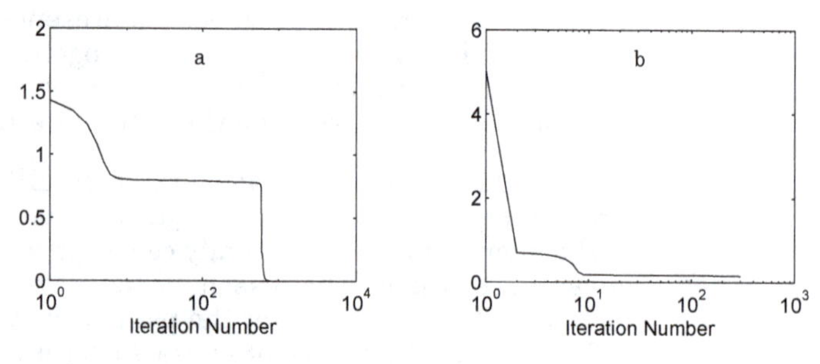

Figure 12.7 Squared Error Convergence Patterns

We can see that the flat sections in Figure 12.7 correspond to times when the algorithm is traversing a flat section of the performance surface, as shown in Figure 12.6. During these periods we would like to increase the learning rate, in order to speed up convergence. However, if we increase the learning rate the algorithm will become unstable when it reaches steeper portions of the performance surface.

This effect is illustrated in Figure 12.8. The trajectory shown here corresponds to trajectory "a" in Figure 12.6, except that a larger learning rate was used. The algorithm converges faster at first, but when the trajectory reaches the narrow valley that contains the minimum point the algorithm begins to diverge. This suggests that it would be useful to vary the learning rate. We could increase the learning rate on flat surfaces and then decrease the learning rate as the slope increased. The question is: "How will the algorithm know when it is on a flat surface?" We will discuss this in a later section.

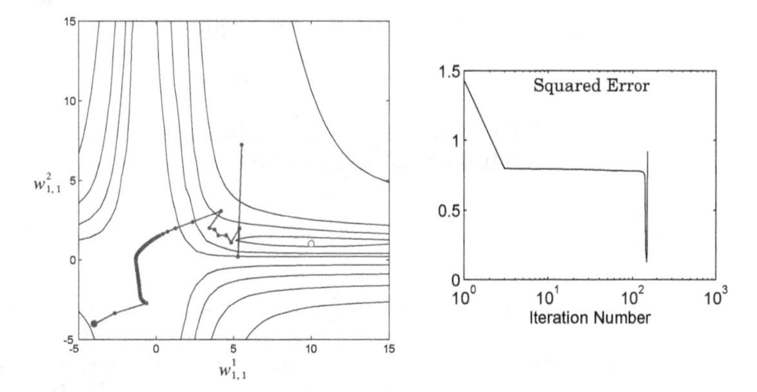

Figure 12.8 Trajectory with Learning Rate Too Large

Another way to improve convergence would be to smooth out the trajectory. Note in Figure 12.8 that when the algorithm begins to diverge it is oscillating back and forth across a narrow valley. If we could filter the trajectory, by averaging the updates to the parameters, this might smooth out the oscillations and produce a stable trajectory. We will discuss this procedure in the next section.

To experiment with this backpropagation example, use the MATLAB® Neural Network Design Demonstration Steepest Descent Backpropagation (**nnd12sd**).

Heuristic Modifications of Backpropagation

Now that we have investigated some of the drawbacks of backpropagation (steepest descent), let's consider some procedures for improving the algorithm. In this section we will discuss two heuristic methods. In a later section we will present two methods based on standard numerical optimization algorithms.

Momentum

The first method we will discuss is the use of momentum. This is a modification based on our observation in the last section that convergence might be improved if we could smooth out the oscillations in the trajectory. We can do this with a low-pass filter.

Before we apply momentum to a neural network application, let's investigate a simple example to illustrate the smoothing effect. Consider the following first-order filter:

$$y(k) = \gamma y(k-1) + (1-\gamma)w(k), \tag{12.4}$$

where $w(k)$ is the input to the filter, $y(k)$ is the output of the filter and γ is the momentum coefficient that must satisfy

$$0 \le \gamma < 1. \tag{12.5}$$

The effect of this filter is shown in Figure 12.9. For these examples the input to the filter was taken to be the sine wave:

$$w(k) = 1 + \sin\left(\frac{2\pi k}{16}\right), \tag{12.6}$$

and the momentum coefficient was set to $\gamma = 0.9$ (left graph) and $\gamma = 0.98$ (right graph). Here we can see that the oscillation of the filter output is less than the oscillation in the filter input (as we would expect for a low-pass filter). In addition, as γ is increased the oscillation in the filter output is reduced. Notice also that the average filter output is the same as the average filter input, although as γ is increased the filter output is slower to respond.

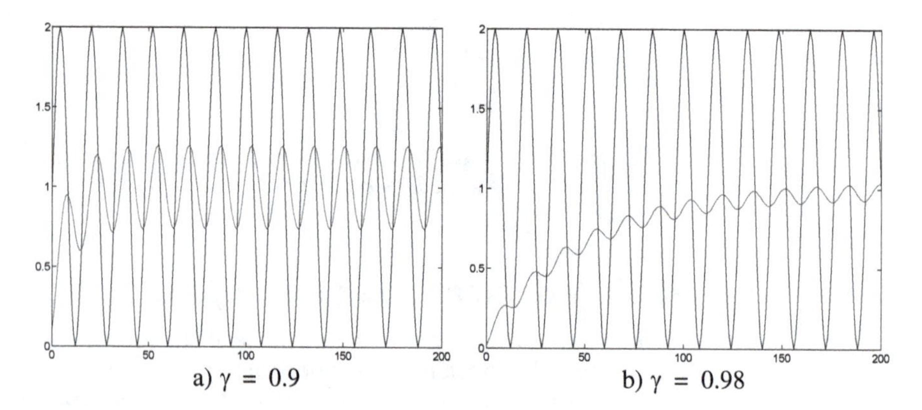

a) $\gamma = 0.9$ b) $\gamma = 0.98$

Figure 12.9 Smoothing Effect of Momentum

To summarize, the filter tends to reduce the amount of oscillation, while still tracking the average value. Now let's see how this works on the neural network problem. First, recall that the parameter updates for SDBP (Eq. (11.46) and Eq. (11.47)) are

$$\Delta \mathbf{W}^m(k) = -\alpha \mathbf{s}^m (\mathbf{a}^{m-1})^T, \tag{12.7}$$

$$\Delta \mathbf{b}^m(k) = -\alpha \mathbf{s}^m. \tag{12.8}$$

Momentum

MOBP

When the *momentum* filter is added to the parameter changes, we obtain the following equations for the momentum modification to backpropagation (*MOBP*):

$$\Delta\mathbf{W}^m(k) = \gamma\Delta\mathbf{W}^m(k-1) - (1-\gamma)\alpha\mathbf{s}^m(\mathbf{a}^{m-1})^T, \tag{12.9}$$

$$\Delta\mathbf{b}^m(k) = \gamma\Delta\mathbf{b}^m(k-1) - (1-\gamma)\alpha\mathbf{s}^m. \tag{12.10}$$

If we now apply these modified equations to the example in the preceding section, we obtain the results shown in Figure 12.10. (For this example we have used a batching form of MOBP, in which the parameters are updated only after the entire training set has been presented. The gradients calculated at each training example are averaged together to produce a more accurate estimate of the gradient.) This trajectory corresponds to the same initial condition and learning rate shown in Figure 12.8, but with a momentum coefficient of $\gamma = 0.8$. We can see that the algorithm is now stable. By the use of momentum we have been able to use a larger learning rate, while maintaining the stability of the algorithm. Another feature of momentum is that it tends to accelerate convergence when the trajectory is moving in a consistent direction.

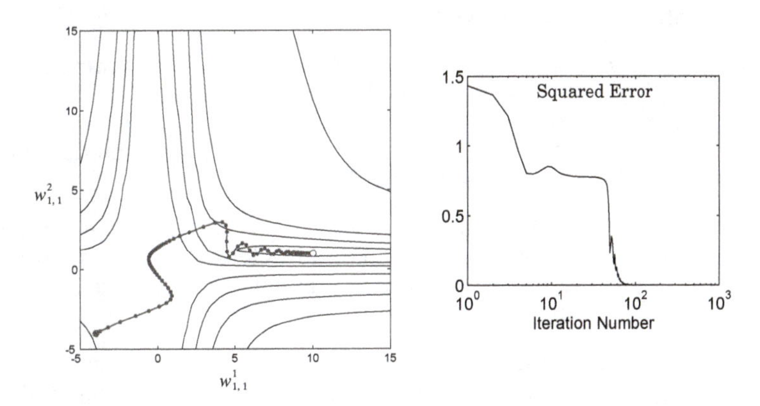

Figure 12.10 Trajectory with Momentum

If you look carefully at the trajectory in Figure 12.10, you can see why the procedure is given the name *momentum*. It tends to make the trajectory continue in the same direction. The larger the value of γ, the more "momentum" the trajectory has.

To experiment with momentum, use the MATLAB® Neural Network Design Demonstration Momentum Backpropagation (**nnd12mo**).

Variable Learning Rate

We suggested earlier in this chapter that we might be able to speed up convergence if we increase the learning rate on flat surfaces and then decrease the learning rate when the slope increases. In this section we want to explore this concept.

Recall that the mean squared error performance surface for single-layer linear networks is always a quadratic function, and the Hessian matrix is therefore constant. The maximum stable learning rate for the steepest descent algorithm is two divided by the maximum eigenvalue of the Hessian matrix. (See Eq. (9.25).)

As we have seen, the error surface for the multilayer network is not a quadratic function. The shape of the surface can be very different in different regions of the parameter space. Perhaps we can speed up convergence by adjusting the learning rate during the course of training. The trick will be to determine when to change the learning rate and by how much.

Variable Learning Rate

There are many different approaches for varying the learning rate. We will describe a very straightforward batching procedure [VoMa88], where the learning rate is varied according to the performance of the algorithm. The rules of the *variable learning rate* backpropagation algorithm (*VLBP*) are:

1. If the squared error (over the entire training set) increases by more than some set percentage ζ (typically one to five percent) after a weight update, then the weight update is discarded, the learning rate is multiplied by some factor $0 < \rho < 1$, and the momentum coefficient γ (if it is used) is set to zero.

2. If the squared error decreases after a weight update, then the weight update is accepted and the learning rate is multiplied by some factor $\eta > 1$. If γ has been previously set to zero, it is reset to its original value.

3. If the squared error increases by less than ζ, then the weight update is accepted but the learning rate is unchanged. If γ has been previously set to zero, it is reset to its original value.

(See Problem P12.3 for a numerical example of VLBP.)

To illustrate VLBP, let's apply it to the function approximation problem of the previous section. Figure 12.11 displays the trajectory for the algorithm using the same initial guess, initial learning rate and momentum coefficient as was used in Figure 12.10. The new parameters were assigned the values

$$\eta = 1.05, \rho = 0.7 \text{ and } \zeta = 4\%. \tag{12.11}$$

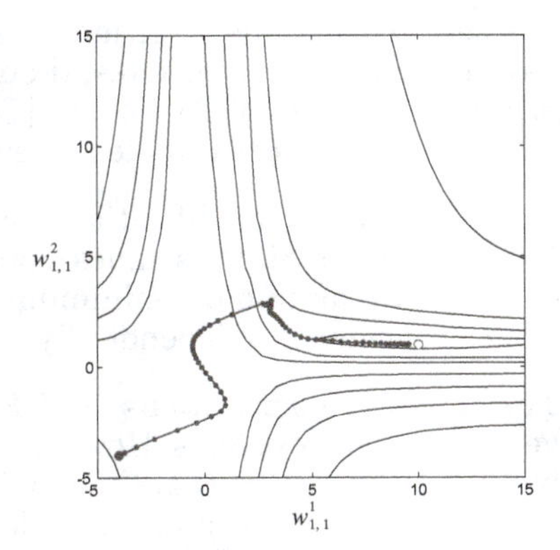

Figure 12.11 Variable Learning Rate Trajectory

Notice how the learning rate, and therefore the step size, tends to increase when the trajectory is traveling in a straight line with constantly decreasing error. This effect can also be seen in Figure 12.12, which shows the squared error and the learning rate versus iteration number.

When the trajectory reaches a narrow valley, the learning rate is rapidly decreased. Otherwise the trajectory would have become oscillatory, and the error would have increased dramatically. For each potential step where the error would have increased by more than 4% the learning rate is reduced and the momentum is eliminated, which allows the trajectory to make the quick turn to follow the valley toward the minimum point. The learning rate then increases again, which accelerates the convergence. The learning rate is reduced again when the trajectory overshoots the minimum point when the algorithm has almost converged. This process is typical of a VLBP trajectory.

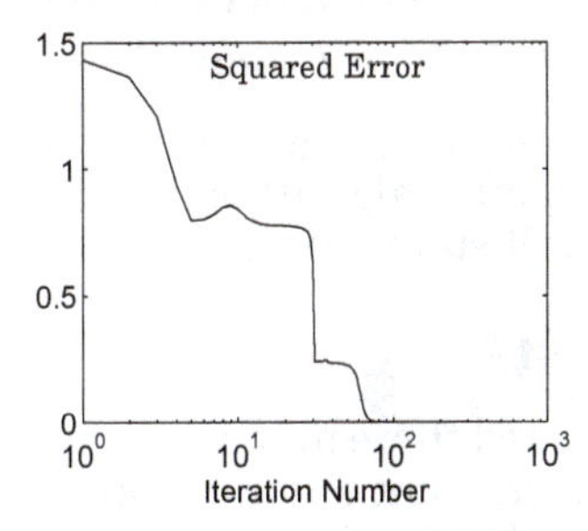

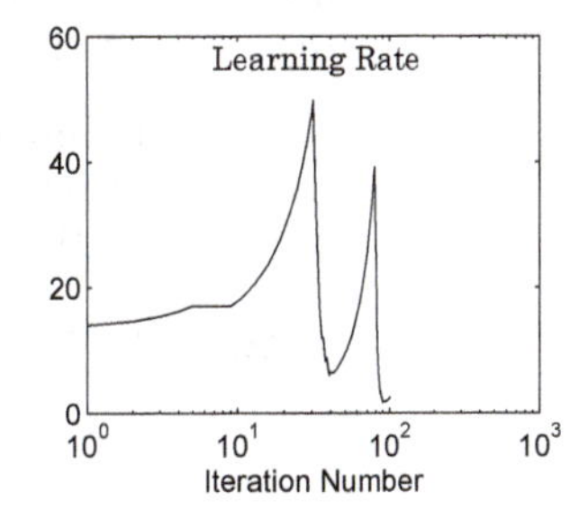

Figure 12.12 Convergence Characteristics of Variable Learning Rate

There are many variations on this variable learning rate algorithm. Jacobs [Jaco88] proposed the *delta-bar-delta* learning rule, in which each network parameter (weight or bias) has its own learning rate. The algorithm increases the learning rate for a network parameter if the parameter change has been in the same direction for several iterations. If the direction of the parameter change alternates, then the learning rate is reduced. The *SuperSAB* algorithm of Tollenaere [Toll90] is similar to the delta-bar-delta rule, but it has more complex rules for adjusting the learning rates.

Another heuristic modification to SDBP is the Quickprop algorithm of Fahlman [Fahl88]. It assumes that the error surface is parabolic and concave upward around the minimum point and that the effect of each weight can be considered independently.

The heuristic modifications to SDBP can often provide much faster convergence for some problems. However, there are two main drawbacks to these methods. The first is that the modifications require that several parameters be set (e.g., ζ, ρ and γ), while the only parameter required for SDBP is the learning rate. Some of the more complex heuristic modifications can have five or six parameters to be selected. Often the performance of the algorithm is sensitive to changes in these parameters. The choice of parameters is also problem dependent. The second drawback to these modifications to SDBP is that they can sometimes fail to converge on problems for which SDBP will eventually find a solution. Both of these drawbacks tend to occur more often when using the more complex algorithms.

To experiment with VLBP, use the MATLAB® Neural Network Design Demonstration Variable Learning Rate Backpropagation (`nnd12vl`).

Numerical Optimization Techniques

Now that we have investigated some of the heuristic modifications to SDBP, let's consider those methods that are based on standard numerical optimization techniques. We will investigate two techniques: conjugate gradient and Levenberg-Marquardt. The conjugate gradient algorithm for quadratic functions was presented in Chapter 9. We need to add two procedures to this algorithm in order to apply it to more general functions.

The second numerical optimization method we will discuss in this chapter is the Levenberg-Marquardt algorithm, which is a modification to Newton's method that is well-suited to neural network training.

Conjugate Gradient

In Chapter 9 we presented three numerical optimization techniques: steepest descent, conjugate gradient and Newton's method. Steepest descent is the simplest algorithm, but is often slow in converging. Newton's method is much faster, but requires that the Hessian matrix and its inverse be calculated. The conjugate gradient algorithm is something of a compromise; it does not require the calculation of second derivatives, and yet it still has the quadratic convergence property. (It converges to the minimum of a quadratic function in a finite number of iterations.) In this section we will describe how the conjugate gradient algorithm can be used to train

CGBP

multilayer networks. We will call this algorithm *conjugate gradient backpropagation (CGBP)*.

Let's begin by reviewing the conjugate gradient algorithm. For ease of reference, we will repeat the algorithm steps from Chapter 9 (page 9-18):

1. Select the first search direction $\mathbf{p}_0$ to be the negative of the gradient, as in Eq. (9.59):

$$\mathbf{p}_0 = -\mathbf{g}_0 , \tag{12.12}$$

where

$$\mathbf{g}_k \equiv \nabla F(\mathbf{x})\big|_{\mathbf{x} = \mathbf{x}_k} . \tag{12.13}$$

2. Take a step according to Eq. (9.57), selecting the learning rate α_k to minimize the function along the search direction:

$$\mathbf{x}_{k+1} = \mathbf{x}_k + \alpha_k \mathbf{p}_k . \tag{12.14}$$

3. Select the next search direction according to Eq. (9.60), using Eq. (9.61), Eq. (9.62), or Eq. (9.63) to calculate β_k:

$$\mathbf{p}_k = -\mathbf{g}_k + \beta_k \mathbf{p}_{k-1} , \tag{12.15}$$

with

$$\beta_k = \frac{\Delta\mathbf{g}_{k-1}^T \mathbf{g}_k}{\Delta\mathbf{g}_{k-1}^T \mathbf{p}_{k-1}} \text{ or } \beta_k = \frac{\mathbf{g}_k^T \mathbf{g}_k}{\mathbf{g}_{k-1}^T \mathbf{g}_{k-1}} \text{ or } \beta_k = \frac{\Delta\mathbf{g}_{k-1}^T \mathbf{g}_k}{\mathbf{g}_{k-1}^T \mathbf{g}_{k-1}} . \tag{12.16}$$

4. If the algorithm has not converged, continue from step 2.

This conjugate gradient algorithm cannot be applied directly to the neural network training task, because the performance index is not quadratic. This affects the algorithm in two ways. First, we will not be able to use Eq. (9.31) to minimize the function along a line, as required in step 2. Second, the exact minimum will not normally be reached in a finite number of steps, and therefore the algorithm will need to be reset after some set number of iterations.

Let's address the linear search first. We need to have a general procedure for locating the minimum of a function in a specified direction. This will involve two steps: interval location and interval reduction. The purpose of the interval location step is to find some initial interval that contains a local minimum. The interval reduction step then reduces the size of the initial interval until the minimum is located to the desired accuracy.

Interval Location

We will use a function comparison method [Scal85] to perform the *interval location* step. This procedure is illustrated in Figure 12.13. We begin by evaluating the performance index at an initial point, represented by a_1 in the figure. This point corresponds to the current values of the network weights and biases. In other words, we are evaluating

$$F(\mathbf{x}_0). \tag{12.17}$$

The next step is to evaluate the function at a second point, represented by b_1 in the figure, which is a distance ε from the initial point, along the first search direction $\mathbf{p}_0$. In other words, we are evaluating

$$F(\mathbf{x}_0 + \varepsilon\mathbf{p}_0). \tag{12.18}$$

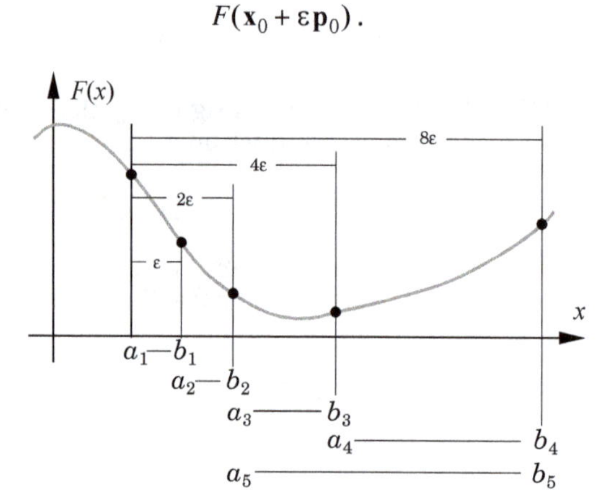

Figure 12.13 Interval Location

We then continue to evaluate the performance index at new points b_i, successively doubling the distance between points. This process stops when the function increases between two consecutive evaluations. In Figure 12.13 this is represented by b_3 to b_4. At this point we know that the minimum is bracketed by the two points a_5 and b_5. We cannot narrow the interval any further, because the minimum may occur either in the interval $[a_4, b_4]$ or in the interval $[a_3, b_3]$. These two possibilities are illustrated in Figure 12.14 (a).

Interval Reduction

Now that we have located an interval containing the minimum, the next step in the linear search is *interval reduction*. This will involve evaluating the function at points inside the interval $[a_5, b_5]$, which was selected in the interval location step. From Figure 12.14 we can see that we will need to evaluate the function at two internal points (at least) in order to reduce the size of the interval of uncertainty. Figure 12.14 (a) shows that one internal function evaluation does not provide us with any information on the location of the minimum. However, if we evaluate the function at two points c and d, as in Figure 12.14 (b), we can reduce the interval of uncertainty. If

$F(c) > F(d)$, as shown in Figure 12.14 (b), then the minimum must occur in the interval $[c, b]$. Conversely, if $F(c) < F(d)$, then the minimum must occur in the interval $[a, d]$. (Note that we are assuming that there is a single minimum located in the initial interval. More about that later.)

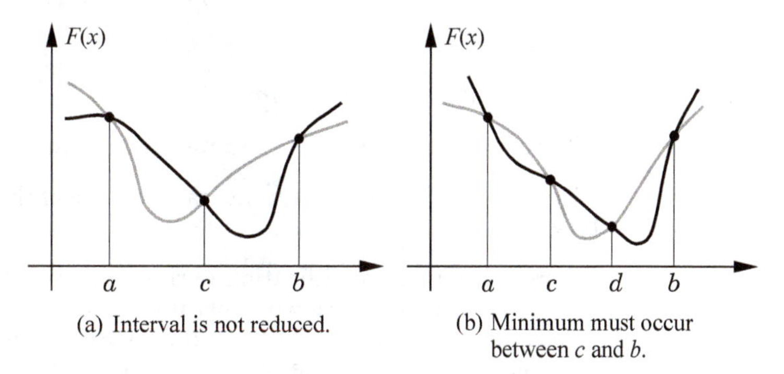

(a) Interval is not reduced.

(b) Minimum must occur between c and b.

Figure 12.14 Reducing the Size of the Interval of Uncertainty

Golden Section Search

The procedure described above suggests a method for reducing the size of the interval of uncertainty. We now need to decide how to determine the locations of the internal points c and d. There are several ways to do this (see [Scal85]). We will use a method called the *Golden Section search*, which is designed to reduce the number of function evaluations required. At each iteration one new function evaluation is required. For example, in the case illustrated in Figure 12.14 (b), point a would be discarded and point c would become the new a. Then point d would become the new point c, and a new d would be placed between the original points d and b. The trick is to place the new point so that the interval of uncertainty will be reduced as quickly as possible.

The algorithm for the Golden Section search is as follows [Scal85]:

$\tau = 0.618$

Set $c_1 = a_1 + (1 - \tau)(b_1 - a_1)$, $F_c = F(c_1)$.

$d_1 = b_1 - (1 - \tau)(b_1 - a_1)$, $F_d = F(d_1)$.

For $k = 1, 2, \ldots$ repeat

If $F_c < F_d$ then

Set $a_{k+1} = a_k$; $b_{k+1} = d_k$; $d_{k+1} = c_k$

$c_{k+1} = a_{k+1} + (1 - \tau)(b_{k+1} - a_{k+1})$

$F_d = F_c$; $F_c = F(c_{k+1})$

else

Set $a_{k+1} = c_k;\ b_{k+1} = b_k;\ c_{k+1} = d_k$

$d_{k+1} = b_{k+1} - (1-\tau)(b_{k+1} - a_{k+1})$

$F_c = F_d;\ F_d = F(d_{k+1})$

end

end until $b_{k+1} - a_{k+1} < tol$

Where tol is the accuracy tolerance set by the user.

(See Problem P12.4 for a numerical example of the interval location and interval reduction procedures.)

There is one more modification to the conjugate gradient algorithm that needs to be made before we apply it to neural network training. For quadratic functions the algorithm will converge to the minimum in at most n iterations, where n is the number of parameters being optimized. The mean squared error performance index for multilayer networks is not quadratic, therefore the algorithm would not normally converge in n iterations. The development of the conjugate gradient algorithm does not indicate what search direction to use once a cycle of n iterations has been completed. There have been many procedures suggested, but the simplest method is to reset the search direction to the steepest descent direction (negative of the gradient) after n iterations [Scal85]. We will use this method.

Let's now apply the conjugate gradient algorithm to the function approximation example that we have been using to demonstrate the other neural network training algorithms. We will use the backpropagation algorithm to compute the gradient (using Eq. (11.23) and Eq. (11.24)) and the conjugate gradient algorithm to determine the weight updates. This is a batch mode algorithm, as the gradient is computed after the entire training set has been presented to the network.

Figure 12.15 shows the intermediate steps of the CGBP algorithm for the first three iterations. The interval location process is illustrated by the open blue circles; each one represents one evaluation of the function. The final interval is indicated by the larger open black circles. The black dots in Figure 12.15 indicate the location of the new interior points during the Golden Section search, one for each iteration of the procedure. The final point is indicated by a blue dot.

Figure 12.16 shows the total trajectory to convergence. Notice that the CGBP algorithm converges in many fewer iterations than the other algorithms that we have tested. This is a little deceiving, since each iteration of CGBP requires more computations than the other methods; there are many function evaluations involved in each iteration of CGBP. Even so, CGBP has been shown to be one of the fastest batch training algorithms for multilayer networks [Char92].

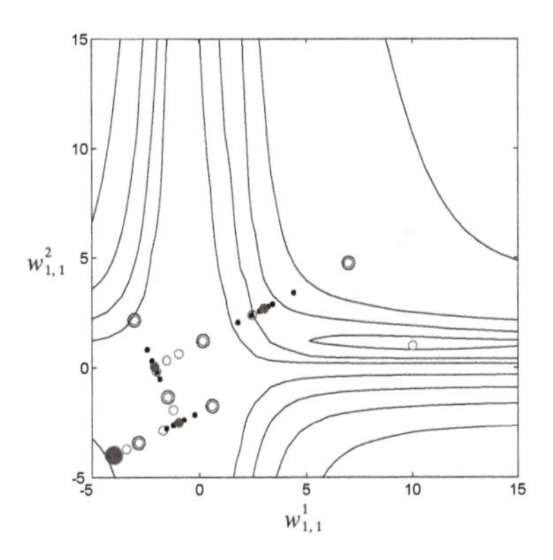

Figure 12.15 Intermediate Steps of CGBP

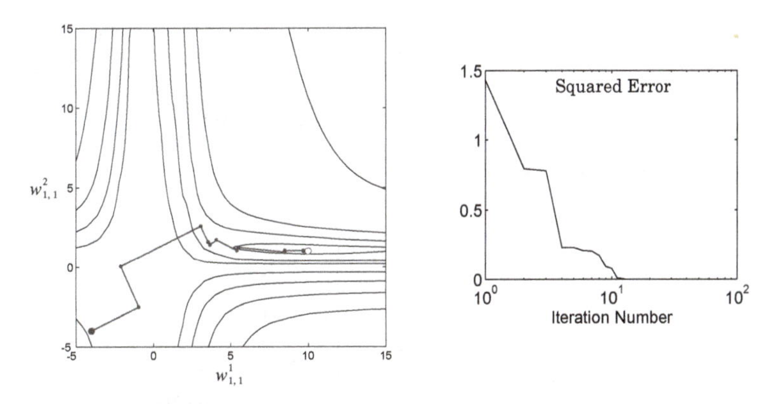

Figure 12.16 Conjugate Gradient Trajectory

To experiment with CGBP, use the MATLAB® Neural Network Design Demonstrations Conjugate Gradient Line Search (**nnd12ls**) *and Conjugate Gradient Backpropagation* (**nnd12cg**).

Levenberg-Marquardt Algorithm

The Levenberg-Marquardt algorithm is a variation of Newton's method that was designed for minimizing functions that are sums of squares of other nonlinear functions. This is very well suited to neural network training where the performance index is the mean squared error.

Basic Algorithm

Let's begin by considering the form of Newton's method where the performance index is a sum of squares. Recall from Chapter 9 that Newton's method for optimizing a performance index $F(\mathbf{x})$ is

$$\mathbf{x}_{k+1} = \mathbf{x}_k - \mathbf{A}_k^{-1}\mathbf{g}_k , \tag{12.19}$$

where $\mathbf{A}_k \equiv \nabla^2 F(\mathbf{x})\big|_{\mathbf{x} = \mathbf{x}_k}$ and $\mathbf{g}_k \equiv \nabla F(\mathbf{x})\big|_{\mathbf{x} = \mathbf{x}_k}$.

If we assume that $F(\mathbf{x})$ is a sum of squares function:

$$F(\mathbf{x}) = \sum_{i=1}^{N} v_i^2(\mathbf{x}) = \mathbf{v}^T(\mathbf{x})\mathbf{v}(\mathbf{x}) , \tag{12.20}$$

then the jth element of the gradient would be

$$[\nabla F(\mathbf{x})]_j = \frac{\partial F(\mathbf{x})}{\partial x_j} = 2\sum_{i=1}^{N} v_i(\mathbf{x})\frac{\partial v_i(\mathbf{x})}{\partial x_j} . \tag{12.21}$$

The gradient can therefore be written in matrix form:

$$\nabla F(\mathbf{x}) = 2\mathbf{J}^T(\mathbf{x})\mathbf{v}(\mathbf{x}) , \tag{12.22}$$

where

$$\mathbf{J}(\mathbf{x}) = \begin{bmatrix} \frac{\partial v_1(\mathbf{x})}{\partial x_1} & \frac{\partial v_1(\mathbf{x})}{\partial x_2} & \cdots & \frac{\partial v_1(\mathbf{x})}{\partial x_n} \\ \frac{\partial v_2(\mathbf{x})}{\partial x_1} & \frac{\partial v_2(\mathbf{x})}{\partial x_2} & \cdots & \frac{\partial v_2(\mathbf{x})}{\partial x_n} \\ \vdots & \vdots & & \vdots \\ \frac{\partial v_N(\mathbf{x})}{\partial x_1} & \frac{\partial v_N(\mathbf{x})}{\partial x_2} & \cdots & \frac{\partial v_N(\mathbf{x})}{\partial x_n} \end{bmatrix} . \tag{12.23}$$

Jacobian Matrix

is the *Jacobian matrix.*

Next we want to find the Hessian matrix. The k, j element of the Hessian matrix would be

$$[\nabla^2 F(\mathbf{x})]_{k,j} = \frac{\partial^2 F(\mathbf{x})}{\partial x_k \partial x_j} = 2\sum_{i=1}^{N} \left\{ \frac{\partial v_i(\mathbf{x})}{\partial x_k}\frac{\partial v_i(\mathbf{x})}{\partial x_j} + v_i(\mathbf{x})\frac{\partial^2 v_i(\mathbf{x})}{\partial x_k \partial x_j} \right\} . \tag{12.24}$$

The Hessian matrix can then be expressed in matrix form:

$$\nabla^2 F(\mathbf{x}) = 2\mathbf{J}^T(\mathbf{x})\mathbf{J}(\mathbf{x}) + 2\mathbf{S}(\mathbf{x}) , \tag{12.25}$$

where

$$\mathbf{S}(\mathbf{x}) = \sum_{i=1}^{N} v_i(\mathbf{x}) \nabla^2 v_i(\mathbf{x}) . \qquad (12.26)$$

If we assume that $\mathbf{S}(\mathbf{x})$ is small, we can approximate the Hessian matrix as

$$\nabla^2 F(\mathbf{x}) \cong 2\mathbf{J}^T(\mathbf{x})\mathbf{J}(\mathbf{x}) . \qquad (12.27)$$

Gauss-Newton

If we then substitute Eq. (12.27) and Eq. (12.22) into Eq. (12.19), we obtain the *Gauss-Newton* method:

$$\mathbf{x}_{k+1} = \mathbf{x}_k - [2\mathbf{J}^T(\mathbf{x}_k)\mathbf{J}(\mathbf{x}_k)]^{-1} 2\mathbf{J}^T(\mathbf{x}_k)\mathbf{v}(\mathbf{x}_k)$$

$$= \mathbf{x}_k - [\mathbf{J}^T(\mathbf{x}_k)\mathbf{J}(\mathbf{x}_k)]^{-1} \mathbf{J}^T(\mathbf{x}_k)\mathbf{v}(\mathbf{x}_k) . \qquad (12.28)$$

Note that the advantage of Gauss-Newton over the standard Newton's method is that it does not require calculation of second derivatives.

One problem with the Gauss-Newton method is that the matrix $\mathbf{H} = \mathbf{J}^T\mathbf{J}$ may not be invertible. This can be overcome by using the following modification to the approximate Hessian matrix:

$$\mathbf{G} = \mathbf{H} + \mu\mathbf{I} . \qquad (12.29)$$

To see how this matrix can be made invertible, suppose that the eigenvalues and eigenvectors of $\mathbf{H}$ are $\{\lambda_1, \lambda_2, \ldots, \lambda_n\}$ and $\{\mathbf{z}_1, \mathbf{z}_2, \ldots, \mathbf{z}_n\}$. Then

$$\mathbf{G}\mathbf{z}_i = [\mathbf{H} + \mu\mathbf{I}]\mathbf{z}_i = \mathbf{H}\mathbf{z}_i + \mu\mathbf{z}_i = \lambda_i\mathbf{z}_i + \mu\mathbf{z}_i = (\lambda_i + \mu)\mathbf{z}_i . \qquad (12.30)$$

Therefore the eigenvectors of $\mathbf{G}$ are the same as the eigenvectors of $\mathbf{H}$, and the eigenvalues of $\mathbf{G}$ are $(\lambda_i + \mu)$. $\mathbf{G}$ can be made positive definite by increasing μ until $(\lambda_i + \mu) > 0$ for all i, and therefore the matrix will be invertible.

Levenberg-Marquardt

This leads to the *Levenberg-Marquardt* algorithm [Scal85]:

$$\mathbf{x}_{k+1} = \mathbf{x}_k - [\mathbf{J}^T(\mathbf{x}_k)\mathbf{J}(\mathbf{x}_k) + \mu_k\mathbf{I}]^{-1} \mathbf{J}^T(\mathbf{x}_k)\mathbf{v}(\mathbf{x}_k) . \qquad (12.31)$$

or

$$\Delta\mathbf{x}_k = -[\mathbf{J}^T(\mathbf{x}_k)\mathbf{J}(\mathbf{x}_k) + \mu_k\mathbf{I}]^{-1} \mathbf{J}^T(\mathbf{x}_k)\mathbf{v}(\mathbf{x}_k) . \qquad (12.32)$$

This algorithm has the very useful feature that as μ_k is increased it approaches the steepest descent algorithm with small learning rate:

$$\mathbf{x}_{k+1} \cong \mathbf{x}_k - \frac{1}{\mu_k}\mathbf{J}^T(\mathbf{x}_k)\mathbf{v}(\mathbf{x}_k) = \mathbf{x}_k - \frac{1}{2\mu_k}\nabla F(\mathbf{x}) \text{, for large } \mu_k, \tag{12.33}$$

while as μ_k is decreased to zero the algorithm becomes Gauss-Newton.

The algorithm begins with μ_k set to some small value (e.g., $\mu_k = 0.01$). If a step does not yield a smaller value for $F(\mathbf{x})$, then the step is repeated with μ_k multiplied by some factor $\vartheta > 1$ (e.g., $\vartheta = 10$). Eventually $F(\mathbf{x})$ should decrease, since we would be taking a small step in the direction of steepest descent. If a step does produce a smaller value for $F(\mathbf{x})$, then μ_k is divided by ϑ for the next step, so that the algorithm will approach Gauss-Newton, which should provide faster convergence. The algorithm provides a nice compromise between the speed of Newton's method and the guaranteed convergence of steepest descent.

Now let's see how we can apply the Levenberg-Marquardt algorithm to the multilayer network training problem. The performance index for multilayer network training is the mean squared error (see Eq. (11.11)). If each target occurs with equal probability, the mean squared error is proportional to the sum of squared errors over the Q targets in the training set:

$$F(\mathbf{x}) = \sum_{q=1}^{Q} (\mathbf{t}_q - \mathbf{a}_q)^T(\mathbf{t}_q - \mathbf{a}_q)$$

$$= \sum_{q=1}^{Q} \mathbf{e}_q^T\mathbf{e}_q = \sum_{q=1}^{Q}\sum_{j=1}^{S^M} (e_{j,q})^2 = \sum_{i=1}^{N} (v_i)^2, \tag{12.34}$$

where $e_{j,q}$ is the j th element of the error for the q th input/target pair.

Eq. (12.34) is equivalent to the performance index, Eq. (12.20), for which Levenberg-Marquardt was designed. Therefore it should be a straightforward matter to adapt the algorithm for network training. It turns out that this is true in concept, but it does require some care in working out the details.

Jacobian Calculation

The key step in the Levenberg-Marquardt algorithm is the computation of the Jacobian matrix. To perform this computation we will use a variation of the backpropagation algorithm. Recall that in the standard backpropagation procedure we compute the derivatives of the squared errors, with respect to the weights and biases of the network. To create the Jacobian matrix we need to compute the derivatives of the errors, instead of the derivatives of the squared errors.

It is a simple matter conceptually to modify the backpropagation algorithm to compute the elements of the Jacobian matrix. Unfortunately, although the basic concept is simple, the details of the implementation can be a little tricky. For that reason you may want to skim through the rest of this section on your first reading, in order to obtain an overview of the general flow of the presentation, and return later to pick up the details. It may also be helpful to review the development of the backpropagation algorithm in Chapter 11 before proceeding.

Before we present the procedure for computing the Jacobian, let's take a closer look at its form (Eq. (12.23)). Note that the error vector is

$$\mathbf{v}^T = \begin{bmatrix} v_1 & v_2 & \dots & v_N \end{bmatrix} = \begin{bmatrix} e_{1,1} & e_{2,1} & \dots & e_{S^M,1} & e_{1,2} & \dots & e_{S^M,Q} \end{bmatrix}, \tag{12.35}$$

the parameter vector is

$$\mathbf{x}^T = \begin{bmatrix} x_1 & x_2 & \dots & x_n \end{bmatrix} = \begin{bmatrix} w^1_{1,1} & w^1_{1,2} & \dots & w^1_{S^1,R} & b^1_1 & \dots & b^1_{S^1} & w^2_{1,1} & \dots & b^M_{S^M} \end{bmatrix}, \tag{12.36}$$

$N = Q \times S^M$ and $n = S^1(R+1) + S^2(S^1+1) + \dots + S^M(S^{M-1}+1)$.

Therefore, if we make these substitutions into Eq. (12.23), the Jacobian matrix for multilayer network training can be written

$$\mathbf{J}(\mathbf{x}) = \begin{bmatrix} \frac{\partial e_{1,1}}{\partial w^1_{1,1}} & \frac{\partial e_{1,1}}{\partial w^1_{1,2}} & \dots & \frac{\partial e_{1,1}}{\partial w^1_{S^1,R}} & \frac{\partial e_{1,1}}{\partial b^1_1} & \dots \\ \frac{\partial e_{2,1}}{\partial w^1_{1,1}} & \frac{\partial e_{2,1}}{\partial w^1_{1,2}} & \dots & \frac{\partial e_{2,1}}{\partial w^1_{S^1,R}} & \frac{\partial e_{2,1}}{\partial b^1_1} & \dots \\ \vdots & \vdots & & \vdots & \vdots & \\ \frac{\partial e_{S^M,1}}{\partial w^1_{1,1}} & \frac{\partial e_{S^M,1}}{\partial w^1_{1,2}} & \dots & \frac{\partial e_{S^M,1}}{\partial w^1_{S^1,R}} & \frac{\partial e_{S^M,1}}{\partial b^1_1} & \dots \\ \frac{\partial e_{1,2}}{\partial w^1_{1,1}} & \frac{\partial e_{1,2}}{\partial w^1_{1,2}} & \dots & \frac{\partial e_{1,2}}{\partial w^1_{S^1,R}} & \frac{\partial e_{1,2}}{\partial b^1_1} & \dots \\ \vdots & \vdots & & \vdots & \vdots & \end{bmatrix}. \tag{12.37}$$

The terms in this Jacobian matrix can be computed by a simple modification to the backpropagation algorithm.

Standard backpropagation calculates terms like

$$\frac{\partial \hat{F}(\mathbf{x})}{\partial x_l} = \frac{\partial \mathbf{e}_q^T \mathbf{e}_q}{\partial x_l}. \tag{12.38}$$

For the elements of the Jacobian matrix that are needed for the Levenberg-Marquardt algorithm we need to calculate terms like

$$[\mathbf{J}]_{h,l} = \frac{\partial v_h}{\partial x_l} = \frac{\partial e_{k,q}}{\partial x_l}. \tag{12.39}$$

Recall from Eq. (11.18) in our derivation of backpropagation that

$$\frac{\partial \hat{F}}{\partial w_{i,j}^m} = \frac{\partial \hat{F}}{\partial n_i^m} \times \frac{\partial n_i^m}{\partial w_{i,j}^m}, \tag{12.40}$$

where the first term on the right-hand side was defined as the sensitivity:

$$s_i^m \equiv \frac{\partial \hat{F}}{\partial n_i^m}. \tag{12.41}$$

The backpropagation process computed the sensitivities through a recurrence relationship from the last layer backward to the first layer. We can use the same concept to compute the terms needed for the Jacobian matrix (Eq. (12.37)) if we define a new *Marquardt sensitivity*:

Marquardt Sensitivity

$$\tilde{s}_{i,h}^m \equiv \frac{\partial v_h}{\partial n_{i,q}^m} = \frac{\partial e_{k,q}}{\partial n_{i,q}^m}, \tag{12.42}$$

where, from Eq. (12.35), $h = (q-1)S^M + k$.

Now we can compute elements of the Jacobian by

$$[\mathbf{J}]_{h,l} = \frac{\partial v_h}{\partial x_l} = \frac{\partial e_{k,q}}{\partial w_{i,j}^m} = \frac{\partial e_{k,q}}{\partial n_{i,q}^m} \times \frac{\partial n_{i,q}^m}{\partial w_{i,j}^m} = \tilde{s}_{i,h}^m \times \frac{\partial n_{i,q}^m}{\partial w_{i,j}^m} = \tilde{s}_{i,h}^m \times a_{j,q}^{m-1}, \tag{12.43}$$

or if x_l is a bias,

$$[\mathbf{J}]_{h,l} = \frac{\partial v_h}{\partial x_l} = \frac{\partial e_{k,q}}{\partial b_i^m} = \frac{\partial e_{k,q}}{\partial n_{i,q}^m} \times \frac{\partial n_{i,q}^m}{\partial b_i^m} = \tilde{s}_{i,h}^m \times \frac{\partial n_{i,q}^m}{\partial b_i^m} = \tilde{s}_{i,h}^m. \tag{12.44}$$

The Marquardt sensitivities can be computed through the same recurrence relations as the standard sensitivities (Eq. (11.35)) with one modification at the final layer, which for standard backpropagation is computed with Eq. (11.40). For the Marquardt sensitivities at the final layer we have

$$\tilde{s}^{M}_{i,h} = \frac{\partial v_h}{\partial n^{M}_{i,q}} = \frac{\partial e_{k,q}}{\partial n^{M}_{i,q}} = \frac{\partial(t_{k,q} - a^{M}_{k,q})}{\partial n^{M}_{i,q}} = -\frac{\partial a^{M}_{k,q}}{\partial n^{M}_{i,q}}$$

$$= \begin{cases} -\dot{f}^{M}(n^{M}_{i,q}) & \text{for } i = k \\ 0 & \text{for } i \neq k \end{cases} . \tag{12.45}$$

Therefore when the input $\mathbf{p}_q$ has been applied to the network and the corresponding network output $\mathbf{a}^{M}_{q}$ has been computed, the Levenberg-Marquardt backpropagation is initialized with

$$\tilde{\mathbf{S}}^{M}_{q} = -\dot{\mathbf{F}}^{M}(\mathbf{n}^{M}_{q}), \tag{12.46}$$

where $\dot{\mathbf{F}}^{M}(\mathbf{n}^{M})$ is defined in Eq. (11.34). Each column of the matrix $\tilde{\mathbf{S}}^{M}_{q}$ must be backpropagated through the network using Eq. (11.35) to produce one row of the Jacobian matrix. The columns can also be backpropagated together using

$$\tilde{\mathbf{S}}^{m}_{q} = \dot{\mathbf{F}}^{m}(\mathbf{n}^{m}_{q})(\mathbf{W}^{m+1})^{T}\tilde{\mathbf{S}}^{m+1}_{q} . \tag{12.47}$$

The total Marquardt sensitivity matrices for each layer are then created by augmenting the matrices computed for each input:

$$\tilde{\mathbf{S}}^{m} = \left[\tilde{\mathbf{S}}^{m}_{1} \middle| \tilde{\mathbf{S}}^{m}_{2} \middle| \ldots \middle| \tilde{\mathbf{S}}^{m}_{Q} \right] . \tag{12.48}$$

Note that for each input that is presented to the network we will backpropagate S^{M} sensitivity vectors. This is because we are computing the derivatives of each individual error, rather than the derivative of the sum of squares of the errors. For every input applied to the network there will be S^{M} errors (one for each element of the network output). For each error there will be one row of the Jacobian matrix.

After the sensitivities have been backpropagated, the Jacobian matrix is computed using Eq. (12.43) and Eq. (12.44). See Problem P12.5 for a numerical illustration of the Jacobian computation.

LMBP

The iterations of the Levenberg-Marquardt backpropagation algorithm (*LMBP*) can be summarized as follows:

1. Present all inputs to the network and compute the corresponding network outputs (using Eq. (11.41) and Eq. (11.42)) and the errors $\mathbf{e}_q = \mathbf{t}_q - \mathbf{a}^{M}_{q}$. Compute the sum of squared errors over all inputs, $F(\mathbf{x})$,

using Eq. (12.34).

2. Compute the Jacobian matrix, Eq. (12.37). Calculate the sensitivities with the recurrence relations Eq. (12.47), after initializing with Eq. (12.46). Augment the individual matrices into the Marquardt sensitivities using Eq. (12.48). Compute the elements of the Jacobian matrix with Eq. (12.43) and Eq. (12.44).

3. Solve Eq. (12.32) to obtain $\Delta\mathbf{x}_k$.

4. Recompute the sum of squared errors using $\mathbf{x}_k + \Delta\mathbf{x}_k$. If this new sum of squares is smaller than that computed in step 1, then divide μ by ϑ, let $\mathbf{x}_{k+1} = \mathbf{x}_k + \Delta\mathbf{x}_k$ and go back to step 1. If the sum of squares is not reduced, then multiply μ by ϑ and go back to step 3.

The algorithm is assumed to have converged when the norm of the gradient, Eq. (12.22), is less than some predetermined value, or when the sum of squares has been reduced to some error goal.

To illustrate LMBP, let's apply it to the function approximation problem introduced at the beginning of this chapter. We will begin by looking at the basic Levenberg-Marquardt step. Figure 12.17 illustrates the possible steps the LMBP algorithm could take on the first iteration.

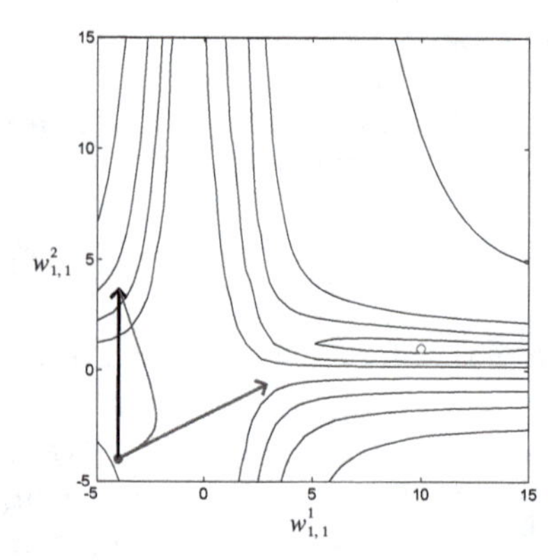

Figure 12.17 Levenberg-Marquardt Step

The black arrow represents the direction taken for small μ_k, which corresponds to the Gauss-Newton direction. The blue arrow represents the direction taken for large μ_k, which corresponds to the steepest descent direction. (This was the initial direction taken by all of the previous algorithms discussed.) The blue curve represents the Levenberg-Marquardt step for all intermediate values of μ_k. Note that as μ_k is increased the algorithm moves toward a small step in the direction of steepest descent. This guarantees that the algorithm will always be able to reduce the sum of squares at each iteration.

Figure 12.18 shows the path of the LMBP trajectory to convergence, with $\mu_0 = 0.01$ and $\vartheta = 5$. Note that the algorithm converges in fewer iterations than any of the methods we have discussed so far. Of course this algorithm also requires more computation per iteration than any of the other algorithms, since it involves a matrix inversion. Even given the large number of computations, however, the LMBP algorithm appears to be the fastest neural network training algorithm for moderate numbers of network parameters [HaMe94].

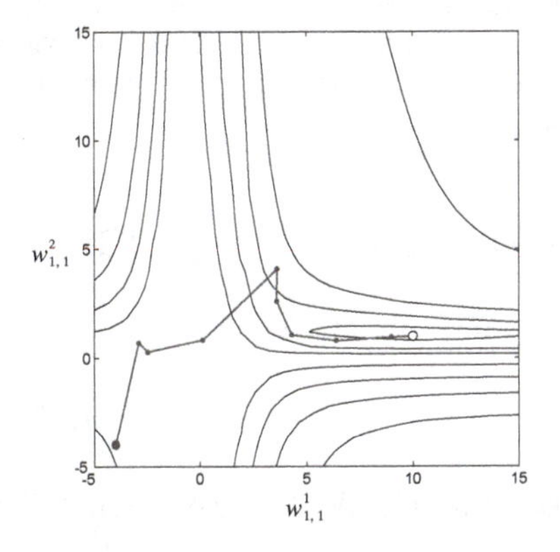

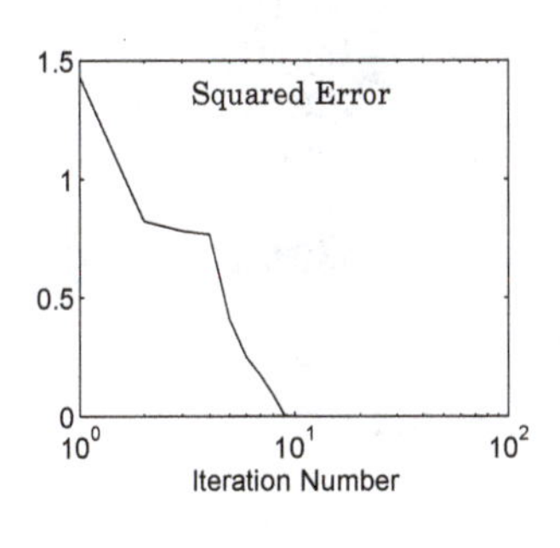

Figure 12.18 LMBP Trajectory

To experiment with the LMBP algorithm, use the MATLAB® Neural Network Design Demonstrations Marquardt Step (`nnd12ms`) *and Marquardt Backpropagation* (`nnd12m`).

The key drawback of the LMBP algorithm is the storage requirement. The algorithm must store the approximate Hessian matrix $\mathbf{J}^T\mathbf{J}$. This is an $n \times n$ matrix, where n is the number of parameters (weights and biases) in the network. Recall that the other methods discussed need only store the gradient, which is an n-dimensional vector. When the number of parameters is very large, it may be impractical to use the Levenberg-Marquardt algorithm. (What constitutes "very large" depends on the available memory on your computer, but typically a few thousand parameters is an upper limit.)

Summary of Results

Heuristic Variations of Backpropagation

Batching

The parameters are updated only after the entire training set has been presented. The gradients calculated for each training example are averaged together to produce a more accurate estimate of the gradient. (If the training set is complete, i.e., covers all possible input/output pairs, then the gradient estimate will be exact.)

Backpropagation with Momentum (MOBP)

$$\Delta \mathbf{W}^m(k) = \gamma \Delta \mathbf{W}^m(k-1) - (1-\gamma)\alpha \mathbf{s}^m (\mathbf{a}^{m-1})^T$$

$$\Delta \mathbf{b}^m(k) = \gamma \Delta \mathbf{b}^m(k-1) - (1-\gamma)\alpha \mathbf{s}^m$$

Variable Learning Rate Backpropagation (VLBP)

1. If the squared error (over the entire training set) increases by more than some set percentage ζ (typically one to five percent) after a weight update, then the weight update is discarded, the learning rate is multiplied by some factor $\rho < 1$, and the momentum coefficient γ (if it is used) is set to zero.

2. If the squared error decreases after a weight update, then the weight update is accepted and the learning rate is multiplied by some factor $\eta > 1$. If γ has been previously set to zero, it is reset to its original value.

3. If the squared error increases by less than ζ, then the weight update is accepted but the learning rate and the momentum coefficient are unchanged.

Numerical Optimization Techniques

Conjugate Gradient

Interval Location

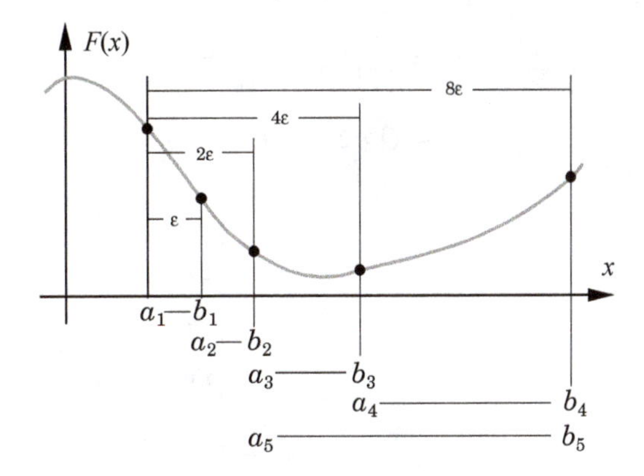

Interval Reduction (Golden Section Search)

$\tau = 0.618$

Set $c_1 = a_1 + (1 - \tau)(b_1 - a_1)$, $F_c = F(c_1)$.

$d_1 = b_1 - (1 - \tau)(b_1 - a_1)$, $F_d = F(d_1)$.

For $k = 1, 2, \ldots$ repeat

If $F_c < F_d$ then

Set $a_{k+1} = a_k$; $b_{k+1} = d_k$; $d_{k+1} = c_k$

$c_{k+1} = a_{k+1} + (1 - \tau)(b_{k+1} - a_{k+1})$

$F_d = F_c$; $F_c = F(c_{k+1})$

else

Set $a_{k+1} = c_k$; $b_{k+1} = b_k$; $c_{k+1} = d_k$

$d_{k+1} = b_{k+1} - (1 - \tau)(b_{k+1} - a_{k+1})$

$F_c = F_d$; $F_d = F(d_{k+1})$

end

end until $b_{k+1} - a_{k+1} < tol$

Levenberg-Marquardt Backpropagation (LMBP)

$$\Delta \mathbf{x}_k = -[\mathbf{J}^T(\mathbf{x}_k)\mathbf{J}(\mathbf{x}_k) + \mu_k \mathbf{I}]^{-1} \mathbf{J}^T(\mathbf{x}_k)\mathbf{v}(\mathbf{x}_k)$$

$$\mathbf{v}^T = \begin{bmatrix} v_1 & v_2 & \dots & v_N \end{bmatrix} = \begin{bmatrix} e_{1,1} & e_{2,1} & \dots & e_{S^M,1} & e_{1,2} & \dots & e_{S^M,Q} \end{bmatrix}$$

$$\mathbf{x}^T = \begin{bmatrix} x_1 & x_2 & \dots & x_n \end{bmatrix} = \begin{bmatrix} w^1_{1,1} & w^1_{1,2} & \dots & w^1_{S^1,R} & b^1_1 & \dots & b^1_{S^1} & w^2_{1,1} & \dots & b^M_{S^M} \end{bmatrix}$$

$$N = Q \times S^M \text{ and } n = S^1(R+1) + S^2(S^1+1) + \dots + S^M(S^{M-1}+1)$$

$$\mathbf{J}(\mathbf{x}) = \begin{bmatrix} \frac{\partial e_{1,1}}{\partial w^1_{1,1}} & \frac{\partial e_{1,1}}{\partial w^1_{1,2}} & \dots & \frac{\partial e_{1,1}}{\partial w^1_{S^1,R}} & \frac{\partial e_{1,1}}{\partial b^1_1} & \dots \\ \frac{\partial e_{2,1}}{\partial w^1_{1,1}} & \frac{\partial e_{2,1}}{\partial w^1_{1,2}} & \dots & \frac{\partial e_{2,1}}{\partial w^1_{S^1,R}} & \frac{\partial e_{2,1}}{\partial b^1_1} & \dots \\ \vdots & \vdots & & \vdots & \vdots & \\ \frac{\partial e_{S^M,1}}{\partial w^1_{1,1}} & \frac{\partial e_{S^M,1}}{\partial w^1_{1,2}} & \dots & \frac{\partial e_{S^M,1}}{\partial w^1_{S^1,R}} & \frac{\partial e_{S^M,1}}{\partial b^1_1} & \dots \\ \frac{\partial e_{1,2}}{\partial w^1_{1,1}} & \frac{\partial e_{1,2}}{\partial w^1_{1,2}} & \dots & \frac{\partial e_{1,2}}{\partial w^1_{S^1,R}} & \frac{\partial e_{1,2}}{\partial b^1_1} & \dots \\ \vdots & \vdots & & \vdots & \vdots & \end{bmatrix}$$

$$[\mathbf{J}]_{h,l} = \frac{\partial v_h}{\partial x_l} = \frac{\partial e_{k,q}}{\partial w^m_{i,j}} = \frac{\partial e_{k,q}}{\partial n^m_{i,q}} \times \frac{\partial n^m_{i,q}}{\partial w^m_{i,j}} = \tilde{s}^m_{i,h} \times \frac{\partial n^m_{i,q}}{\partial w^m_{i,j}} = \tilde{s}^m_{i,h} \times a^{m-1}_{j,q} \text{ for weight } x_l$$

$$[\mathbf{J}]_{h,l} = \frac{\partial v_h}{\partial x_l} = \frac{\partial e_{k,q}}{\partial b^m_i} = \frac{\partial e_{k,q}}{\partial n^m_{i,q}} \times \frac{\partial n^m_{i,q}}{\partial b^m_i} = \tilde{s}^m_{i,h} \times \frac{\partial n^m_{i,q}}{\partial b^m_i} = \tilde{s}^m_{i,h} \text{ for bias } x_l$$

$$\tilde{s}^m_{i,h} \equiv \frac{\partial v_h}{\partial n^m_{i,q}} = \frac{\partial e_{k,q}}{\partial n^m_{i,q}} \text{ (Marquardt Sensitivity) where } h = (q-1)S^M + k$$

$$\tilde{\mathbf{S}}^M_q = -\dot{\mathbf{F}}^M(\mathbf{n}^M_q)$$

$$\tilde{\mathbf{S}}^m_q = \dot{\mathbf{F}}^m(\mathbf{n}^m_q)(\mathbf{W}^{m+1})^T \tilde{\mathbf{S}}^{m+1}_q$$

$$\tilde{\mathbf{S}}^m = \left[\tilde{\mathbf{S}}^m_1 \middle| \tilde{\mathbf{S}}^m_2 \middle| \dots \middle| \tilde{\mathbf{S}}^m_Q \right]$$

Levenberg-Marquardt Iterations

1. Present all inputs to the network and compute the corresponding network outputs (using Eq. (11.41) and Eq. (11.42)) and the errors $\mathbf{e}_q = \mathbf{t}_q - \mathbf{a}_q^M$. Compute the sum of squared errors over all inputs, $F(\mathbf{x})$, using Eq. (12.34).

2. Compute the Jacobian matrix, Eq. (12.37). Calculate the sensitivities with the recurrence relations Eq. (12.47), after initializing with Eq. (12.46). Augment the individual matrices into the Marquardt sensitivities using Eq. (12.48). Compute the elements of the Jacobian matrix with Eq. (12.43) and Eq. (12.44).

3. Solve Eq. (12.32) to obtain $\Delta\mathbf{x}_k$.

4. Recompute the sum of squared errors using $\mathbf{x}_k + \Delta\mathbf{x}_k$. If this new sum of squares is smaller than that computed in step 1, then divide μ by ϑ, let $\mathbf{x}_{k+1} = \mathbf{x}_k + \Delta\mathbf{x}_k$ and go back to step 1. If the sum of squares is not reduced, then multiply μ by ϑ and go back to step 3.

Solved Problems

P12.1 We want to train the network shown in Figure P12.1 on the training set

$$\left\{(\mathbf{p}_1 = \begin{bmatrix}-3\end{bmatrix}),(\mathbf{t}_1 = \begin{bmatrix}0.5\end{bmatrix})\right\},\ \left\{(\mathbf{p}_2 = \begin{bmatrix}2\end{bmatrix}),(\mathbf{t}_2 = \begin{bmatrix}1\end{bmatrix})\right\},$$

starting from the initial guess

$$w(0) = 0.4\,,\ b(0) = 0.15\,.$$

Demonstrate the effect of batching by computing the direction of the initial step for SDBP with and without batching.

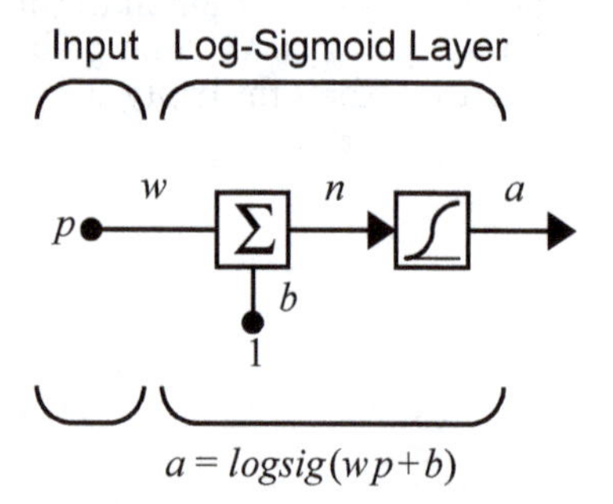

Figure P12.1 Network for Problem P12.1

Let's begin by computing the direction of the initial step if batching is not used. In this case the first step is computed from the first input/target pair. The forward and backpropagation steps are

$$a = logsig(wp+b) = \frac{1}{1+\exp(-(0.4(-3)+0.15))} = 0.2592$$

$$e = t - a = 0.5 - 0.2592 = 0.2408$$

$$s = -2\dot{f}(n)e = -2a(1-a)e = -2(0.2592)(1-0.2592)0.2408 = -0.0925\,.$$

The direction of the initial step is the negative of the gradient. For the weight this will be

$$-sp = -(-0.0925)(-3) = -0.2774\,.$$

For the bias we have

$$-s = -(-0.0925) = 0.0925\,.$$

Therefore the direction of the initial step in the (w, b) plane would be

$$\begin{bmatrix} -0.2774 \\ 0.0925 \end{bmatrix}.$$

Now let's consider the initial direction for the batch mode algorithm. In this case the gradient is found by adding together the individual gradients found from the two sets of input/target pairs. For this we need to apply the second input to the network and perform the forward and backpropagation steps:

$$a = logsig(wp + b) = \frac{1}{1 + \exp(-(0.4(2) + 0.15))} = 0.7211$$

$$e = t - a = 1 - 0.7211 = 0.2789$$

$$s = -2\dot{f}(n)e = -2a(1-a)e = -2(0.7211)(1 - 0.7211)0.2789 = -0.1122.$$

The direction of the step is the negative of the gradient. For the weight this will be

$$-sp = -(-0.1122)(2) = 0.2243.$$

For the bias we have

$$-s = -(-0.1122) = 0.1122.$$

The partial gradient for the second input/target pair is therefore

$$\begin{bmatrix} 0.2243 \\ 0.1122 \end{bmatrix}.$$

If we now add the results from the two input/target pairs we find the direction of the first step of the batch mode SDBP to be

$$\frac{1}{2}\left(\begin{bmatrix} -0.2774 \\ 0.0925 \end{bmatrix} + \begin{bmatrix} 0.2243 \\ 0.1122 \end{bmatrix}\right) = \frac{1}{2}\begin{bmatrix} -0.0531 \\ 0.2047 \end{bmatrix} = \begin{bmatrix} -0.0265 \\ 0.1023 \end{bmatrix}.$$

The results are illustrated in Figure P12.2. The blue circle indicates the initial guess. The two blue arrows represent the directions of the partial gradients for each of the two input/target pairs, and the black arrow represents the direction of the total gradient. The function that is plotted is the sum of squared errors for the entire training set. Note that the individual partial gradients can point in quite different directions than the true gradient. However, on the average, over several iterations, the path will generally follow the steepest descent trajectory.

The relative effectiveness of the batch mode over the incremental approach depends very much on the particular problem. The incremental approach requires less storage, and, if the inputs are presented randomly to the network, the trajectory is stochastic, which makes the algorithm somewhat less likely to be trapped in a local minimum. It may also take longer to converge than the batch mode algorithm.

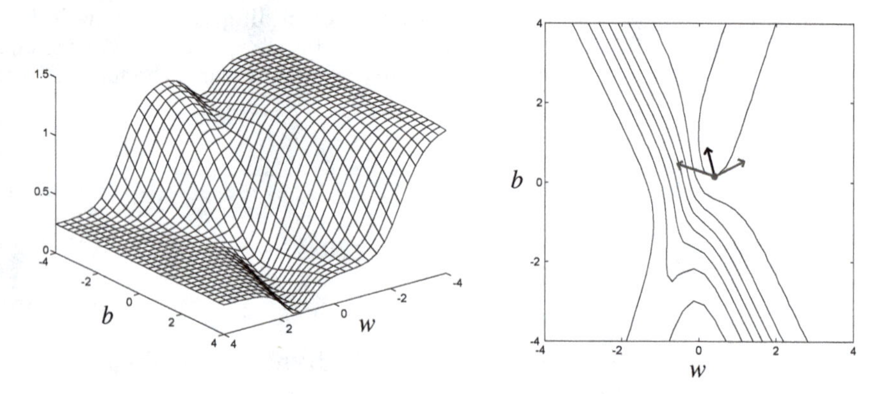

Figure P12.2 Effect of Batching in Problem P12.1

P12.2 In Chapter 9 we proved that the steepest descent algorithm, when applied to a quadratic function, would be stable if the learning rate was less than 2 divided by the maximum eigenvalue of the Hessian matrix. Show that if a momentum term is added to the steepest descent algorithm there will always be a momentum coefficient that will make the algorithm stable, regardless of the learning rate. Follow the format of the proof on page 9-6.

The standard steepest descent algorithm is

$$\Delta \mathbf{x}_k = -\alpha \nabla F(\mathbf{x}_k) = -\alpha \mathbf{g}_k ,$$

If we add momentum this becomes

$$\Delta \mathbf{x}_k = \gamma \Delta \mathbf{x}_{k-1} - (1-\gamma)\alpha \mathbf{g}_k .$$

Recall from Chapter 8 that the quadratic function has the form

$$F(\mathbf{x}) = \frac{1}{2}\mathbf{x}^T \mathbf{A} \mathbf{x} + \mathbf{d}^T \mathbf{x} + c ,$$

and the gradient of the quadratic function is

$$\nabla F(\mathbf{x}) = \mathbf{A}\mathbf{x} + \mathbf{d} .$$

If we now insert this expression into our expression for the steepest descent algorithm with momentum we obtain

$$\Delta \mathbf{x}_k = \gamma \Delta \mathbf{x}_{k-1} - (1-\gamma)\alpha(\mathbf{A}\mathbf{x}_k + \mathbf{d}).$$

Using the definition $\Delta \mathbf{x}_k = \mathbf{x}_{k+1} - \mathbf{x}_k$ this can be rewritten

$$\mathbf{x}_{k+1} - \mathbf{x}_k = \gamma(\mathbf{x}_k - \mathbf{x}_{k-1}) - (1-\gamma)\alpha(\mathbf{A}\mathbf{x}_k + \mathbf{d})$$

or

$$\mathbf{x}_{k+1} = [(1+\gamma)\mathbf{I} - (1-\gamma)\alpha\mathbf{A}]\mathbf{x}_k - \gamma\mathbf{x}_{k-1} - (1-\gamma)\alpha\mathbf{d}.$$

Now define a new vector

$$\tilde{\mathbf{x}}_k = \begin{bmatrix} \mathbf{x}_{k-1} \\ \mathbf{x}_k \end{bmatrix}.$$

The momentum variation of steepest descent can then be written

$$\tilde{\mathbf{x}}_{k+1} = \begin{bmatrix} \mathbf{0} & \mathbf{I} \\ -\gamma\mathbf{I} & [(1+\gamma)\mathbf{I} - (1-\gamma)\alpha\mathbf{A}] \end{bmatrix} \tilde{\mathbf{x}}_k + \begin{bmatrix} \mathbf{0} \\ -(1-\gamma)\alpha\mathbf{d} \end{bmatrix} = \mathbf{W}\tilde{\mathbf{x}}_k + \mathbf{v}.$$

This is a linear dynamic system that will be stable if the eigenvalues of $\mathbf{W}$ are less than one in magnitude. We will find the eigenvalues of $\mathbf{W}$ in stages. First, rewrite $\mathbf{W}$ as

$$\mathbf{W} = \begin{bmatrix} \mathbf{0} & \mathbf{I} \\ -\gamma\mathbf{I} & \mathbf{T} \end{bmatrix} \text{ where } \mathbf{T} = [(1+\gamma)\mathbf{I} - (1-\gamma)\alpha\mathbf{A}].$$

The eigenvalues and eigenvectors of $\mathbf{W}$ should satisfy

$$\mathbf{W}\mathbf{z}^w = \lambda^w \mathbf{z}^w, \text{ or } \begin{bmatrix} \mathbf{0} & \mathbf{I} \\ -\gamma\mathbf{I} & \mathbf{T} \end{bmatrix} \begin{bmatrix} \mathbf{z}_1^w \\ \mathbf{z}_2^w \end{bmatrix} = \lambda^w \begin{bmatrix} \mathbf{z}_1^w \\ \mathbf{z}_2^w \end{bmatrix}.$$

This means that

$$\mathbf{z}_2^w = \lambda^w \mathbf{z}_1^w \text{ and } -\gamma\mathbf{z}_1^w + \mathbf{T}\mathbf{z}_2^w = \lambda^w \mathbf{z}_2^w.$$

At this point we will choose $\mathbf{z}_2^w$ to be an eigenvector of the matrix $\mathbf{T}$, with corresponding eigenvalue λ^t. (If this choice is not appropriate it will lead to a contradiction.) Therefore the previous equations become

$$\mathbf{z}_2^w = \lambda^w \mathbf{z}_1^w \text{ and } -\gamma \mathbf{z}_1^w + \lambda^t \mathbf{z}_2^w = \lambda^w \mathbf{z}_2^w .$$

If we substitute the first equation into the second equation we find

$$-\frac{\gamma}{\lambda^w}\mathbf{z}_2^w + \lambda^t \mathbf{z}_2^w = \lambda^w \mathbf{z}_2^w \text{ or } [(\lambda^w)^2 - \lambda^t(\lambda^w) + \gamma]\mathbf{z}_2^w = 0 .$$

Therefore for each eigenvalue λ^t of $\mathbf{T}$ there will be two eigenvalues λ^w of $\mathbf{W}$ that are roots of the quadratic equation

$$(\lambda^w)^2 - \lambda^t(\lambda^w) + \gamma = 0 .$$

From the quadratic formula we have

$$\lambda^w = \frac{\lambda^t \pm \sqrt{(\lambda^t)^2 - 4\gamma}}{2} .$$

For the algorithm to be stable the magnitude of each eigenvalue must be less than 1. We will show that there always exists some range of γ for which this is true.

Note that if the eigenvalues λ^w are complex then their magnitude will be $\sqrt{\gamma}$:

$$|\lambda^w| = \sqrt{\frac{(\lambda^t)^2}{4} + \frac{4\gamma - (\lambda^t)^2}{4}} = \sqrt{\gamma} .$$

(This is true only for real λ^t. We will show later that λ^t is real.) Since γ is between 0 and 1, the magnitude of the eigenvalue must be less than 1. It remains to show that there exists some range of γ for which all of the eigenvalues are complex.

In order for λ^w to be complex we must have

$$(\lambda^t)^2 - 4\gamma < 0 \text{ or } |\lambda^t| < 2\sqrt{\gamma} .$$

Let's now consider the eigenvalues λ^t of $\mathbf{T}$. These eigenvalues can be expressed in terms of the eigenvalues of $\mathbf{A}$. Let $\{\lambda_1, \lambda_2, \dots, \lambda_n\}$ and $\{\mathbf{z}_1, \mathbf{z}_2, \dots, \mathbf{z}_n\}$ be the eigenvalues and eigenvectors of the Hessian matrix. Then

$$\mathbf{T}\mathbf{z}_i = [(1+\gamma)\mathbf{I} - (1-\gamma)\alpha\mathbf{A}]\mathbf{z}_i = (1+\gamma)\mathbf{z}_i - (1-\gamma)\alpha\mathbf{A}\mathbf{z}_i$$

$$= (1+\gamma)\mathbf{z}_i - (1-\gamma)\alpha\lambda_i\mathbf{z}_i = \{(1+\gamma) - (1-\gamma)\alpha\lambda_i\}\mathbf{z}_i = \lambda_i^t\mathbf{z}_i .$$

Therefore the eigenvectors of $\mathbf{T}$ are the same as the eigenvectors of $\mathbf{A}$, and the eigenvalues of $\mathbf{T}$ are

$$\lambda_i^t = \{(1+\gamma) - (1-\gamma)\alpha\lambda_i\}.$$

(Note that λ_i^t is real, since γ, α and λ_i for symmetric $\mathbf{A}$ are real.) Therefore, in order for λ^w to be complex we must have

$$|\lambda^t| < 2\sqrt{\gamma} \text{ or } |(1+\gamma) - (1-\gamma)\alpha\lambda_i| < 2\sqrt{\gamma}.$$

For $\gamma = 1$ both sides of the inequality will equal 2. The function on the right of the inequality, as a function of γ, has a slope of 1 at $\gamma = 1$. The function on the left of the inequality has a slope of $1 + \alpha\lambda_i$. Since the eigenvalues of the Hessian will be positive real numbers if the function has a strong minimum, and the learning rate is a positive number, this slope must be greater than 1. This shows that the inequality will always hold for γ close enough to 1.

To summarize the results, we have shown that if a momentum term is added to the steepest descent algorithm on a quadratic function, then there will always be a momentum coefficient that will make the algorithm stable, regardless of the learning rate. In addition we have shown that if γ is close enough to 1, then the magnitudes of the eigenvalues of $\mathbf{W}$ will be $\sqrt{\gamma}$. It can be shown (see [Brog91]) that the magnitudes of the eigenvalues determine how fast the algorithm will converge. The smaller the magnitude, the faster the convergence. As the magnitude approaches 1, the convergence time increases.

We can demonstrate these results using the example on page 9-7. There we showed that the steepest descent algorithm, when applied to the function $F(\mathbf{x}) = x_1^2 + 25x_2^2$, was unstable for a learning rate $\alpha \geq 0.4$. In Figure P12.3 we see the steepest descent trajectory (with momentum) with $\alpha = 0.041$ and $\gamma = 0.2$. Compare this trajectory with Figure 9.3, which uses the same learning rate but no momentum.

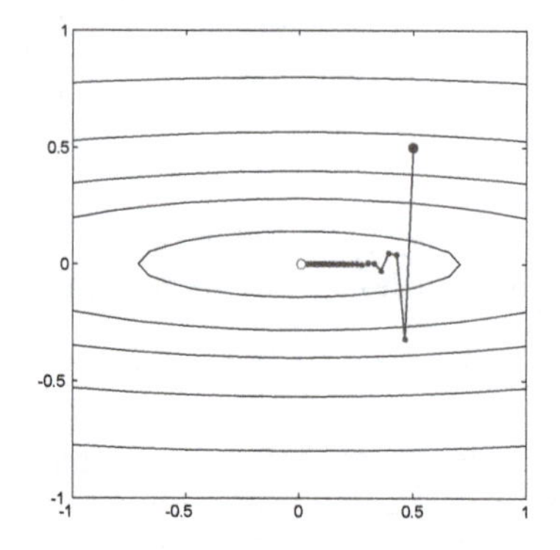

Figure P12.3 Trajectory for $\alpha = 0.041$ and $\gamma = 0.2$

P12.3 Execute three iterations of the variable learning rate algorithm on the following function (from the Chapter 9 example on page 9-7):

$$F(\mathbf{x}) = x_1^2 + 25x_2^2,$$

starting from the initial guess

$$\mathbf{x}_0 = \begin{bmatrix} 0.5 \\ 0.5 \end{bmatrix},$$

and use the following values for the algorithm parameters:

$$\alpha = 0.05, \gamma = 0.2, \eta = 1.5, \rho = 0.5, \zeta = 5\%.$$

The first step is to evaluate the function at the initial guess:

$$F(\mathbf{x}_0) = \frac{1}{2}\mathbf{x}_0^T \begin{bmatrix} 2 & 0 \\ 0 & 50 \end{bmatrix} \mathbf{x}_0 = \frac{1}{2}\begin{bmatrix} 0.5 & 0.5 \end{bmatrix} \begin{bmatrix} 2 & 0 \\ 0 & 50 \end{bmatrix} \begin{bmatrix} 0.5 \\ 0.5 \end{bmatrix} = 6.5.$$

The next step is to find the gradient:

$$\nabla F(\mathbf{x}) = \begin{bmatrix} \frac{\partial}{\partial x_1} F(\mathbf{x}) \\ \frac{\partial}{\partial x_2} F(\mathbf{x}) \end{bmatrix} = \begin{bmatrix} 2x_1 \\ 50x_2 \end{bmatrix}.$$

If we evaluate the gradient at the initial guess we find:

$$\mathbf{g}_0 = \nabla F(\mathbf{x})\Big|_{\mathbf{x}=\mathbf{x}_0} = \begin{bmatrix} 1 \\ 25 \end{bmatrix}.$$

With the initial learning rate of $\alpha = 0.05$, the tentative first step of the algorithm is

$$\Delta\mathbf{x}_0 = \gamma\Delta\mathbf{x}_{-1} - (1-\gamma)\alpha\mathbf{g}_0 = 0.2\begin{bmatrix} 0 \\ 0 \end{bmatrix} - 0.8(0.05)\begin{bmatrix} 1 \\ 25 \end{bmatrix} = \begin{bmatrix} -0.04 \\ -1 \end{bmatrix}$$

$$\mathbf{x}_1^t = \mathbf{x}_0 + \Delta\mathbf{x}_0 = \begin{bmatrix} 0.5 \\ 0.5 \end{bmatrix} + \begin{bmatrix} -0.04 \\ -1 \end{bmatrix} = \begin{bmatrix} 0.46 \\ -0.5 \end{bmatrix}.$$

To verify that this is a valid step we must test the value of the function at this new point:

$$F(\mathbf{x}_1^t) = \frac{1}{2}(\mathbf{x}_1^t)^T\begin{bmatrix} 2 & 0 \\ 0 & 50 \end{bmatrix}\mathbf{x}_1^t = \frac{1}{2}\begin{bmatrix} 0.46 & -0.5 \end{bmatrix}\begin{bmatrix} 2 & 0 \\ 0 & 50 \end{bmatrix}\begin{bmatrix} 0.46 \\ -0.5 \end{bmatrix} = 6.4616.$$

This is less than $F(\mathbf{x}_0)$. Therefore this tentative step is accepted and the learning rate is increased:

$$\mathbf{x}_1 = \mathbf{x}_1^t = \begin{bmatrix} 0.46 \\ -0.5 \end{bmatrix},\ F(\mathbf{x}_1) = 6.4616 \text{ and } \alpha = \eta\alpha = 1.5(0.05) = 0.075.$$

The tentative second step of the algorithm is

$$\Delta\mathbf{x}_1 = \gamma\Delta\mathbf{x}_0 - (1-\gamma)\alpha\mathbf{g}_1 = 0.2\begin{bmatrix} -0.04 \\ -1 \end{bmatrix} - 0.8(0.075)\begin{bmatrix} 0.92 \\ -25 \end{bmatrix} = \begin{bmatrix} -0.0632 \\ 1.3 \end{bmatrix}$$

$$\mathbf{x}_2^t = \mathbf{x}_1 + \Delta\mathbf{x}_1 = \begin{bmatrix} 0.46 \\ -0.5 \end{bmatrix} + \begin{bmatrix} -0.0632 \\ 1.3 \end{bmatrix} = \begin{bmatrix} 0.3968 \\ 0.8 \end{bmatrix}.$$

We evaluate the function at this point:

$$F(\mathbf{x}_2^t) = \frac{1}{2}(\mathbf{x}_2^t)^T\begin{bmatrix} 2 & 0 \\ 0 & 50 \end{bmatrix}\mathbf{x}_2^t = \frac{1}{2}\begin{bmatrix} 0.3968 & 0.8 \end{bmatrix}\begin{bmatrix} 2 & 0 \\ 0 & 50 \end{bmatrix}\begin{bmatrix} 0.3968 \\ 0.8 \end{bmatrix} = 16.157.$$

Since this is more than 5% larger than $F(\mathbf{x}_1)$, we reject this step, reduce the learning rate and set the momentum coefficient to zero.

$$\mathbf{x}_2 = \mathbf{x}_1,\ F(\mathbf{x}_2) = F(\mathbf{x}_1) = 6.4616,\ \alpha = \rho\alpha = 0.5(0.075) = 0.0375,\ \gamma = 0$$

Now a new tentative step is computed (momentum is zero).

$$\Delta \mathbf{x}_2 = -\alpha \mathbf{g}_2 = -(0.0375)\begin{bmatrix} 0.92 \\ -25 \end{bmatrix} = \begin{bmatrix} -0.0345 \\ 0.9375 \end{bmatrix}$$

$$\mathbf{x}_3^t = \mathbf{x}_2 + \Delta \mathbf{x}_2 = \begin{bmatrix} 0.46 \\ -0.5 \end{bmatrix} + \begin{bmatrix} -0.0345 \\ 0.9375 \end{bmatrix} = \begin{bmatrix} 0.4255 \\ 0.4375 \end{bmatrix}$$

$$F(\mathbf{x}_3^t) = \frac{1}{2}(\mathbf{x}_3^t)^T \begin{bmatrix} 2 & 0 \\ 0 & 50 \end{bmatrix} \mathbf{x}_3^t = \frac{1}{2}\begin{bmatrix} 0.4255 & 0.4375 \end{bmatrix} \begin{bmatrix} 2 & 0 \\ 0 & 50 \end{bmatrix} \begin{bmatrix} 0.4255 \\ 0.4375 \end{bmatrix} = 4.966$$

This is less than $F(\mathbf{x}_2)$. Therefore this step is accepted, the momentum is reset to its original value, and the learning rate is increased.

$$\mathbf{x}_3 = \mathbf{x}_3^t, \ \gamma = 0.2, \ \alpha = \eta\alpha = 1.5(0.0375) = 0.05625$$

This completes the third iteration.

P12.4 Recall the example from Chapter 9 that we used to demonstrate the conjugate gradient algorithm (page 9-18):

$$F(\mathbf{x}) = \frac{1}{2}\mathbf{x}^T \begin{bmatrix} 2 & 1 \\ 1 & 2 \end{bmatrix} \mathbf{x},$$

with initial guess

$$\mathbf{x}_0 = \begin{bmatrix} 0.8 \\ -0.25 \end{bmatrix}.$$

Perform one iteration of the conjugate gradient algorithm. For the linear minimization use interval location by function evaluation and interval reduction by the Golden Section search.

The gradient of this function is

$$\nabla F(\mathbf{x}) = \begin{bmatrix} 2x_1 + x_2 \\ x_1 + 2x_2 \end{bmatrix}.$$

As with steepest descent, the first search direction for the conjugate gradient algorithm is the negative of the gradient:

$$\mathbf{p}_0 = -\mathbf{g}_0 = -\nabla F(\mathbf{x})^T \Big|_{\mathbf{x} = \mathbf{x}_0} = \begin{bmatrix} -1.35 \\ -0.3 \end{bmatrix} .$$

For the first iteration we need to minimize $F(\mathbf{x})$ along the line

$$\mathbf{x}_1 = \mathbf{x}_0 + \alpha_0 \mathbf{p}_0 = \begin{bmatrix} 0.8 \\ -0.25 \end{bmatrix} + \alpha_0 \begin{bmatrix} -1.35 \\ -0.3 \end{bmatrix} .$$

The first step is interval location. Assume that the initial step size is $\varepsilon = 0.075$. Then the interval location would proceed as follows:

$$F(a_1) = F\left(\begin{bmatrix} 0.8 \\ -0.25 \end{bmatrix}\right) = 0.5025 ,$$

$$b_1 = \varepsilon = 0.075 ,\ F(b_1) = F\left(\begin{bmatrix} 0.8 \\ -0.25 \end{bmatrix} + 0.075 \begin{bmatrix} -1.35 \\ -0.3 \end{bmatrix}\right) = 0.3721$$

$$b_2 = 2\varepsilon = 0.15 ,\ F(b_2) = F\left(\begin{bmatrix} 0.8 \\ -0.25 \end{bmatrix} + 0.15 \begin{bmatrix} -1.35 \\ -0.3 \end{bmatrix}\right) = 0.2678$$

$$b_3 = 4\varepsilon = 0.3 ,\ F(b_3) = F\left(\begin{bmatrix} 0.8 \\ -0.25 \end{bmatrix} + 0.3 \begin{bmatrix} -1.35 \\ -0.3 \end{bmatrix}\right) = 0.1373$$

$$b_4 = 8\varepsilon = 0.6 ,\ F(b_4) = F\left(\begin{bmatrix} 0.8 \\ -0.25 \end{bmatrix} + 0.6 \begin{bmatrix} -1.35 \\ -0.3 \end{bmatrix}\right) = 0.1893 .$$

Since the function increases between two consecutive evaluations we know that the minimum must occur in the interval $[0.15, 0.6]$. This process is illustrated by the open blue circles in Figure P12.4, and the final interval is indicated by the large open black circles.

The next step in the linear minimization is interval reduction using the Golden Section search. This proceeds as follows:

$$c_1 = a_1 + (1 - \tau)(b_1 - a_1) = 0.15 + (0.382)(0.6 - 0.15) = 0.3219 ,$$

$$d_1 = b_1 - (1 - \tau)(b_1 - a_1) = 0.6 - (0.382)(0.6 - 0.15) = 0.4281 ,$$

$$F_a = 0.2678 ,\ F_b = 0.1893 ,\ F_c = 0.1270 ,\ F_d = 0.1085 .$$

Since $F_c > F_d$, we have

$$a_2 = c_1 = 0.3219, \; b_2 = b_1 = 0.6, \; c_2 = d_1 = 0.4281$$

$$d_2 = b_2 - (1-\tau)(b_2 - a_2) = 0.6 - (0.382)(0.6 - 0.3219) = 0.4938,$$

$$F_a = F_c = 0.1270, \; F_c = F_d = 0.1085, \; F_d = F(d_2) = 0.1232.$$

This time $F_c < F_d$, therefore

$$a_3 = a_2 = 0.3219, \; b_3 = d_2 = 0.4938, \; d_3 = c_2 = 0.4281,$$

$$c_3 = a_3 + (1-\tau)(b_3 - a_3) = 0.3219 + (0.382)(0.4938 - 0.3219) = 0.3876,$$

$$F_b = F_d = 0.1232, \; F_d = F_c = 0.1085, \; F_c = F(c_3) = 0.1094.$$

This routine continues until $b_{k+1} - a_{k+1} < tol$. The black dots in Figure P12.4 indicate the location of the new interior points, one for each iteration of the procedure. The final point is indicated by a blue dot. Compare this result with the first iteration shown in Figure 9.10.

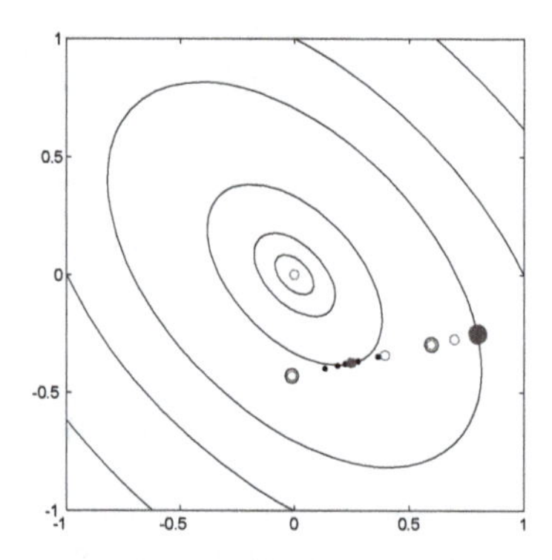

Figure P12.4 Linear Minimization Example

P12.5 To illustrate the computation of the Jacobian matrix for the Levenberg-Marquardt method, consider using the network of Figure P12.5 for function approximation. The network transfer functions are chosen to be

$$f^1(n) = (n)^2, \; f^2(n) = n.$$

Therefore their derivatives are

$$\dot{f}^1(n) = 2n, \; \dot{f}^2(n) = 1.$$

Assume that the training set consists of

$$\{(\mathbf{p}_1 = \begin{bmatrix}1\end{bmatrix}), (\mathbf{t}_1 = \begin{bmatrix}1\end{bmatrix})\} \ , \ \{(\mathbf{p}_2 = \begin{bmatrix}2\end{bmatrix}), (\mathbf{t}_2 = \begin{bmatrix}2\end{bmatrix})\} \ ,$$

and that the parameters are initialized to

$$\mathbf{W}^1 = \begin{bmatrix}1\end{bmatrix}, \ \mathbf{b}^1 = \begin{bmatrix}0\end{bmatrix}, \ \mathbf{W}^2 = \begin{bmatrix}2\end{bmatrix}, \ \mathbf{b}^1 = \begin{bmatrix}1\end{bmatrix} .$$

Find the Jacobian matrix for the first step of the Levenberg-Marquardt method.

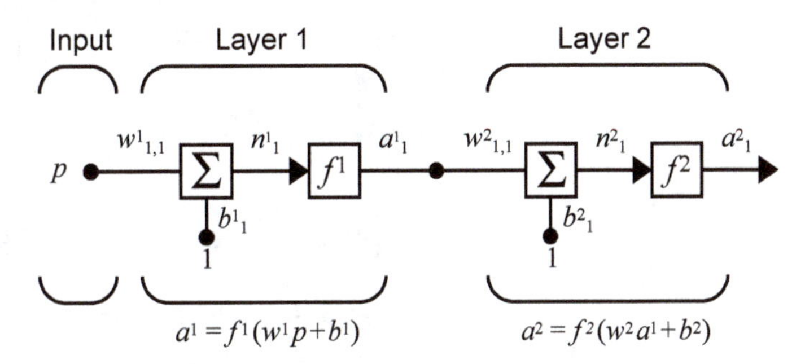

Figure P12.5 Two-Layer Network for LMBP Demonstration

The first step is to propagate the inputs through the network and compute the errors.

$$\mathbf{a}_1^0 = \mathbf{p}_1 = \begin{bmatrix}1\end{bmatrix}$$

$$\mathbf{n}_1^1 = \mathbf{W}^1\mathbf{a}_1^0 + \mathbf{b}^1 = \begin{bmatrix}1\end{bmatrix}\begin{bmatrix}1\end{bmatrix} + \begin{bmatrix}0\end{bmatrix} = \begin{bmatrix}1\end{bmatrix}, \ \mathbf{a}_1^1 = \mathbf{f}^1(\mathbf{n}_1^1) = (\begin{bmatrix}1\end{bmatrix})^2 = \begin{bmatrix}1\end{bmatrix}$$

$$\mathbf{n}_1^2 = \mathbf{W}^2\mathbf{a}_1^1 + \mathbf{b}^2 = (\begin{bmatrix}2\end{bmatrix}\begin{bmatrix}1\end{bmatrix} + \begin{bmatrix}1\end{bmatrix}) = \begin{bmatrix}3\end{bmatrix}, \ \mathbf{a}_1^2 = \mathbf{f}^2(\mathbf{n}_1^2) = (\begin{bmatrix}3\end{bmatrix}) = \begin{bmatrix}3\end{bmatrix}$$

$$\mathbf{e}_1 = (\mathbf{t}_1 - \mathbf{a}_1^2) = (\begin{bmatrix}1\end{bmatrix} - \begin{bmatrix}3\end{bmatrix}) = \begin{bmatrix}-2\end{bmatrix}$$

$$\mathbf{a}_2^0 = \mathbf{p}_2 = \begin{bmatrix}2\end{bmatrix}$$

$$\mathbf{n}_2^1 = \mathbf{W}^1\mathbf{a}_2^0 + \mathbf{b}^1 = \begin{bmatrix}1\end{bmatrix}\begin{bmatrix}2\end{bmatrix} + \begin{bmatrix}0\end{bmatrix} = \begin{bmatrix}2\end{bmatrix}, \ \mathbf{a}_2^1 = \mathbf{f}^1(\mathbf{n}_2^1) = (\begin{bmatrix}2\end{bmatrix})^2 = \begin{bmatrix}4\end{bmatrix}$$

$$\mathbf{n}_2^2 = \mathbf{W}^2\mathbf{a}_2^1 + \mathbf{b}^2 = (\begin{bmatrix}2\end{bmatrix}\begin{bmatrix}4\end{bmatrix} + \begin{bmatrix}1\end{bmatrix}) = \begin{bmatrix}9\end{bmatrix}, \ \mathbf{a}_2^2 = \mathbf{f}^2(\mathbf{n}_2^2) = (\begin{bmatrix}9\end{bmatrix}) = \begin{bmatrix}9\end{bmatrix}$$

$$\mathbf{e}_2 = (\mathbf{t}_2 - \mathbf{a}_2^2) = \left(\begin{bmatrix}2\end{bmatrix} - \begin{bmatrix}9\end{bmatrix}\right) = \begin{bmatrix}-7\end{bmatrix}$$

The next step is to initialize and backpropagate the Marquardt sensitivities using Eq. (12.46) and Eq. (12.47).

$$\tilde{\mathbf{S}}_1^2 = -\dot{\mathbf{F}}^2(\mathbf{n}_1^2) = -\begin{bmatrix}1\end{bmatrix}$$

$$\tilde{\mathbf{S}}_1^1 = \dot{\mathbf{F}}^1(\mathbf{n}_1^1)(\mathbf{W}^2)^T\tilde{\mathbf{S}}_1^2 = \begin{bmatrix}2n_{1,1}^1\end{bmatrix}\begin{bmatrix}2\end{bmatrix}\begin{bmatrix}-1\end{bmatrix} = \begin{bmatrix}2(1)\end{bmatrix}\begin{bmatrix}2\end{bmatrix}\begin{bmatrix}-1\end{bmatrix} = \begin{bmatrix}-4\end{bmatrix}$$

$$\tilde{\mathbf{S}}_2^2 = -\dot{\mathbf{F}}^2(\mathbf{n}_2^2) = -\begin{bmatrix}1\end{bmatrix}$$

$$\tilde{\mathbf{S}}_2^1 = \dot{\mathbf{F}}^1(\mathbf{n}_2^1)(\mathbf{W}^2)^T\tilde{\mathbf{S}}_2^2 = \begin{bmatrix}2n_{1,2}^2\end{bmatrix}\begin{bmatrix}2\end{bmatrix}\begin{bmatrix}-1\end{bmatrix} = \begin{bmatrix}2(2)\end{bmatrix}\begin{bmatrix}2\end{bmatrix}\begin{bmatrix}-1\end{bmatrix} = \begin{bmatrix}-8\end{bmatrix}$$

$$\tilde{\mathbf{S}}^1 = \begin{bmatrix}\tilde{\mathbf{S}}_1^1 \,|\, \tilde{\mathbf{S}}_2^1\end{bmatrix} = \begin{bmatrix}-4 & -8\end{bmatrix},\ \tilde{\mathbf{S}}^2 = \begin{bmatrix}\tilde{\mathbf{S}}_1^2 \,|\, \tilde{\mathbf{S}}_2^2\end{bmatrix} = \begin{bmatrix}-1 & -1\end{bmatrix}$$

We can now compute the Jacobian matrix using Eq. (12.43), Eq. (12.44) and Eq. (12.37).

$$\mathbf{J}(\mathbf{x}) = \begin{bmatrix}\frac{\partial v_1}{\partial x_1} & \frac{\partial v_1}{\partial x_2} & \frac{\partial v_1}{\partial x_3} & \frac{\partial v_1}{\partial x_4} \\ \frac{\partial v_2}{\partial x_1} & \frac{\partial v_2}{\partial x_2} & \frac{\partial v_2}{\partial x_3} & \frac{\partial v_2}{\partial x_4}\end{bmatrix} = \begin{bmatrix}\frac{\partial e_{1,1}}{\partial w_{1,1}^1} & \frac{\partial e_{1,1}}{\partial b_1^1} & \frac{\partial e_{1,1}}{\partial w_{1,1}^2} & \frac{\partial e_{1,1}}{\partial b_1^2} \\ \frac{\partial e_{1,2}}{\partial w_{1,1}^1} & \frac{\partial e_{1,2}}{\partial b_1^1} & \frac{\partial e_{1,2}}{\partial w_{1,1}^2} & \frac{\partial e_{1,2}}{\partial b_1^2}\end{bmatrix}$$

$$[\mathbf{J}]_{1,1} = \frac{\partial v_1}{\partial x_1} = \frac{\partial e_{1,1}}{\partial w_{1,1}^1} = \frac{\partial e_{1,1}}{\partial n_{1,1}^1} \times \frac{\partial n_{1,1}^1}{\partial w_{1,1}^1} = \tilde{s}_{1,1}^1 \times \frac{\partial n_{1,1}^1}{\partial w_{1,1}^1} = \tilde{s}_{1,1}^1 \times a_{1,1}^0$$

$$= (-4)(1) = -4$$

$$[\mathbf{J}]_{1,2} = \frac{\partial v_1}{\partial x_2} = \frac{\partial e_{1,1}}{\partial b_1^1} = \frac{\partial e_{1,1}}{\partial n_{1,1}^1} \times \frac{\partial n_{1,1}^1}{\partial b_1^1} = \tilde{s}_{1,1}^1 \times \frac{\partial n_{1,1}^1}{\partial b_1^1} = \tilde{s}_{1,1}^1 = -4$$

$$[\mathbf{J}]_{1,3} = \frac{\partial v_1}{\partial x_3} = \frac{\partial e_{1,1}}{\partial n_{1,1}^2} \times \frac{\partial n_{1,1}^2}{\partial w_{1,1}^2} = \tilde{s}_{1,1}^2 \times \frac{\partial n_{1,1}^2}{\partial w_{1,1}^2} = \tilde{s}_{1,1}^2 \times a_{1,1}^1 = (-1)(1) = -1$$

$$[\mathbf{J}]_{1,4} = \frac{\partial v_1}{\partial x_4} = \frac{\partial e_{1,1}}{\partial n^2_{1,1}} \times \frac{\partial n^1_{1,1}}{\partial b^2_1} = \tilde{s}^2_{1,1} \times \frac{\partial n^2_{1,1}}{\partial b^2_1} = \tilde{s}^2_{1,1} = -1$$

$$[\mathbf{J}]_{2,1} = \frac{\partial v_2}{\partial x_1} = \frac{\partial e_{1,2}}{\partial n^1_{1,2}} \times \frac{\partial n^1_{1,2}}{\partial w^1_{1,1}} = \tilde{s}^1_{1,2} \times \frac{\partial n^1_{1,2}}{\partial w^1_{1,1}} = \tilde{s}^1_{1,2} \times a^0_{1,2} = (-8)(2) = -16$$

$$[\mathbf{J}]_{2,2} = \frac{\partial v_2}{\partial x_2} = \frac{\partial e_{1,2}}{\partial b^1_1} = \frac{\partial e_{1,2}}{\partial n^1_{1,2}} \times \frac{\partial n^1_{1,2}}{\partial b^1_1} = \tilde{s}^1_{1,2} \times \frac{\partial n^1_{1,2}}{\partial b^1_1} = \tilde{s}^1_{1,2} = -8$$

$$[\mathbf{J}]_{2,3} = \frac{\partial v_2}{\partial x_3} = \frac{\partial e_{1,2}}{\partial n^2_{1,2}} \times \frac{\partial n^2_{1,2}}{\partial w^2_{1,1}} = \tilde{s}^2_{1,2} \times \frac{\partial n^2_{1,2}}{\partial w^2_{1,1}} = \tilde{s}^2_{1,2} \times a^1_{1,2} = (-1)(4) = -4$$

$$[\mathbf{J}]_{2,4} = \frac{\partial v_2}{\partial x_4} = \frac{\partial e_{1,2}}{\partial b^2_1} = \frac{\partial e_{1,2}}{\partial n^2_{1,2}} \times \frac{\partial n^2_{1,2}}{\partial b^2_1} = \tilde{s}^2_{1,2} \times \frac{\partial n^2_{1,2}}{\partial b^2_1} = \tilde{s}^2_{1,2} = -1$$

Therefore the Jacobian matrix is

$$\mathbf{J}(\mathbf{x}) = \begin{bmatrix} -4 & -4 & -1 & -1 \\ -16 & -8 & -4 & -1 \end{bmatrix}.$$

Epilogue

One of the major problems with the basic backpropagation algorithm (steepest descent backpropagation — SDBP) has been the long training times. It is not feasible to use SDBP on practical problems, because it can take weeks to train a network, even on a large computer. Since backpropagation was first popularized, there has been considerable work on methods to accelerate the convergence of the algorithm. In this chapter we have discussed the reasons for the slow convergence of SDBP and have presented several techniques for improving the performance of the algorithm.

The techniques for speeding up convergence have fallen into two main categories: heuristic methods and standard numerical optimization methods. We have discussed two heuristic methods: momentum (MOBP) and variable learning rate (VLBP). MOBP is simple to implement, can be used in batch mode or incremental mode and is significantly faster than SDBP. It does require the selection of the momentum coefficient, but γ is limited to the range $[0, 1]$ and the algorithm is not extremely sensitive to this choice.

The VLBP algorithm is faster than MOBP but must be used in batch mode. For this reason it requires more storage. VLBP also requires the selection of a total of five parameters. The algorithm is reasonably robust, but the choice of the parameters can affect the convergence speed and is problem dependent.

We also presented two standard numerical optimization techniques: conjugate gradient (CGBP) and Levenberg-Marquardt (LMBP). CGBP is generally faster than VLBP. It is a batch mode algorithm, which requires a linear search at each iteration, but its storage requirements are not significantly different than VLBP. There are many variations of the conjugate gradient algorithm proposed for neural network applications. We have presented only one.

The LMBP algorithm is the fastest algorithm that we have tested for training multilayer networks of moderate size, even though it requires a matrix inversion at each iteration. It requires that two parameters be selected, but the algorithm does not appear to be sensitive to this selection. The main drawback of LMBP is the storage requirement. The $\mathbf{J}^T\mathbf{J}$ matrix, which must be inverted, is $n \times n$, where n is the total number of weights and biases in the network. If the network has more than a few thousand parameters, the LMBP algorithm becomes impractical on current machines.

Further Reading

[Barn92] E. Barnard, "Optimization for training neural nets," *IEEE Trans. on Neural Networks*, vol. 3, no. 2, pp. 232–240, 1992.

A number of optimization algorithms that have promise for neural network training are discussed in this paper.

[Batt92] R. Battiti, "First- and second-order methods for learning: Between steepest descent and Newton's method," *Neural Computation*, vol. 4, no. 2, pp. 141–166, 1992.

This paper is an excellent survey of the current optimization algorithms that are suitable for neural network training.

[Char92] C. Charalambous, "Conjugate gradient algorithm for efficient training of artificial neural networks," *IEE Proceedings*, vol. 139, no. 3, pp. 301–310, 1992.

This paper explains how the conjugate gradient algorithm can be used to train multilayer networks. Comparisons are made to other training algorithms.

[Fahl88] S. E. Fahlman, "Faster-learning variations on back-propagation: An empirical study," In D. Touretsky, G. Hinton & T. Sejnowski, eds., *Proceedings of the 1988 Connectionist Models Summer School*, San Mateo, CA: Morgan Kaufmann, pp. 38–51, 1988.

The QuickProp algorithm, which is described in this paper, is one of the more popular heuristic modifications to backpropagation. It assumes that the error curve can be approximated by a parabola, and that the effect of each weight can be considered independently. QuickProp provides significant speedup over standard backpropagation on many problems.

[HaMe94] M. T. Hagan and M. Menhaj, "Training feedforward networks with the Marquardt algorithm," *IEEE Transactions on Neural Networks*, vol. 5, no. 6, 1994.

This paper describes the use of the Levenberg-Marquardt algorithm for training multilayer networks and compares the performance of the algorithm with variable learning rate backpropagation and conjugate gradient. The Levenberg-Marquardt algorithm is faster, but requires more storage.

[Jaco88] R. A. Jacobs, "Increased rates of convergence through learning rate adaptation," *Neural Networks*, vol. 1, no. 4, pp. 295–308, 1988.

This is another early paper discussing the use of variable learning rate backpropagation. The procedure described here is called the delta-bar-delta learning rule, in which each network parameter has its own learning rate that varies at each iteration.

[NgWi90] D. Nguyen and B. Widrow, "Improving the learning speed of 2-layer neural networks by choosing initial values of the adaptive weights," *Proceedings of the IJCNN*, vol. 3, pp. 21–26, July 1990.

This paper describes a procedure for setting the initial weights and biases for the backpropagation algorithm. It uses the shape of the sigmoid transfer function and the range of the input variables to determine how large the weights should be, and then uses the biases to center the sigmoids in the operating region. The convergence of backpropagation is improved significantly by this procedure.

[RiIr90] A. K. Rigler, J. M. Irvine and T. P. Vogl, "Rescaling of variables in back propagation learning," *Neural Networks*, vol. 4, no. 2, pp. 225–230, 1991.

This paper notes that the derivative of a sigmoid function is very small on the tails. This means that the elements of the gradient associated with the first few layers will generally be smaller that those associated with the last layer. The terms in the gradient are then scaled to equalize them.

[Scal85] L. E. Scales, *Introduction to Non-Linear Optimization*. New York: Springer-Verlag, 1985.

Scales has written a very readable text describing the major optimization algorithms. The book emphasizes methods of optimization rather than existence theorems and proofs of convergence. Algorithms are presented with intuitive explanations, along with illustrative figures and examples. Pseudocode is presented for most algorithms.

[Shan90] D. F. Shanno, "Recent advances in numerical techniques for large-scale optimization," *Neural Networks for Control*, Miller, Sutton and Werbos, eds., Cambridge MA: MIT Press, 1990.

This paper discusses some conjugate gradient and quasi-Newton optimization algorithms that could be used for neural network training.

[Toll90] T. Tollenaere, "SuperSAB: Fast adaptive back propagation with good scaling properties," *Neural Networks*, vol. 3, no. 5, pp. 561–573, 1990.

This paper presents a variable learning rate backpropagation algorithm in which different learning rates are used for each weight.

[VoMa88] T. P. Vogl, J. K. Mangis, A. K. Zigler, W. T. Zink and D. L. Alkon, "Accelerating the convergence of the backpropagation method," *Biological Cybernetics.*, vol. 59, pp. 256–264, Sept. 1988.

This was one of the first papers to introduce several heuristic techniques for accelerating the convergence of backpropagation. It included batching, momentum and variable learning rate.

Exercises

E12.1 We want to train the network shown in Figure E12.1 on the training set

$$\left\{(\mathbf{p}_1 = \begin{bmatrix}-2\end{bmatrix}), (\mathbf{t}_1 = \begin{bmatrix}0.8\end{bmatrix})\right\}, \left\{(\mathbf{p}_2 = \begin{bmatrix}2\end{bmatrix}), (\mathbf{t}_2 = \begin{bmatrix}1\end{bmatrix})\right\},$$

where each pair is equally likely to occur.

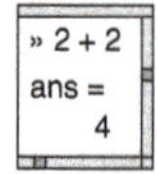

Write a MATLAB M-file to create a contour plot for the mean squared error performance index.

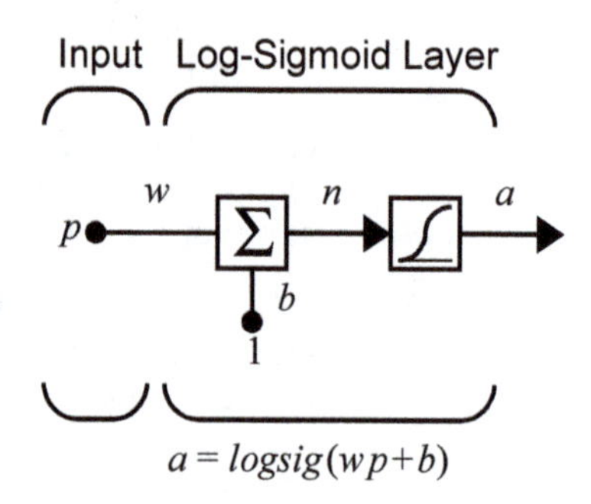

Figure E12.1 Network for Exercise E12.1

E12.2 Demonstrate the effect of batching by computing the direction of the initial step for SDBP with and without batching for the problem described in Exercise E12.1, starting from the initial guess

$$w(0) = 0, \; b(0) = 0.5 .$$

E12.3 Recall the quadratic function used in Problem P9.1:

$$F(\mathbf{x}) = \frac{1}{2}\mathbf{x}^T\begin{bmatrix}10 & -6\\ -6 & 10\end{bmatrix}\mathbf{x} + \begin{bmatrix}4 & 4\end{bmatrix}\mathbf{x} .$$

We want to use the steepest descent algorithm with momentum to minimize this function.

i. Suppose that the learning rate is $\alpha = 0.2$. Find a value for the momentum coefficient γ for which the algorithm will be stable. Use the ideas presented in Problem P12.2.

ii. Suppose that the learning rate is $\alpha = 20$. Find a value for the momentum coefficient γ for which the algorithm will be stable.

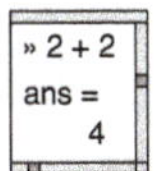

iii. Write a MATLAB program to plot the trajectories of the algorithm for the α and γ values of both part (i) and part (ii) on the contour plot of $F(\mathbf{x})$, starting from the initial guess

$$\mathbf{x}_0 = \begin{bmatrix} -1 \\ -2.5 \end{bmatrix}.$$

E12.4 Consider the following quadratic function.

$$F(\mathbf{x}) = \frac{1}{2}\mathbf{x}^T \begin{bmatrix} 3 & -1 \\ -1 & 3 \end{bmatrix} \mathbf{x} + \begin{bmatrix} 4 & -4 \end{bmatrix} \mathbf{x}.$$

We want to use the steepest descent algorithm with momentum to minimize this function.

i. Perform two iterations (finding $\mathbf{x}_1$ and $\mathbf{x}_2$) of steepest descent with momentum, starting from the initial condition $\mathbf{x}_0 = \begin{bmatrix} 0 & 0 \end{bmatrix}^T$. Use a learning rate of $\alpha = 1$ and a momentum coefficient of $\gamma = 0.75$.

ii. Is the algorithm stable with this learning rate and this momentum? Use the ideas presented in Problem P12.2.

iii. Would the algorithm be stable with this learning rate, if the momentum were zero?

E12.5 Consider the following quadratic function.

$$F(\mathbf{x}) = \frac{1}{2}\mathbf{x}^T \begin{bmatrix} 3 & 1 \\ 1 & 3 \end{bmatrix} \mathbf{x} + \begin{bmatrix} 1 & 2 \end{bmatrix} \mathbf{x} + 2.$$

We want to use the steepest descent algorithm with momentum to minimize this function.

i. Suppose the learning rate is $\alpha = 1$. Is the algorithm stable, if the momentum coefficient is $\gamma = 0$? Use the ideas presented in Problem P12.2.

ii. Suppose the learning rate is $\alpha = 1$. Is the algorithm stable, if the momentum coefficient is $\gamma = 0.6$?

E12.6 Consider the following quadratic function.

$$F(\mathbf{x}) = \frac{1}{2}\mathbf{x}^T\begin{bmatrix} 2 & 1 \\ 1 & 2 \end{bmatrix}\mathbf{x} + \begin{bmatrix} 1 & 2 \end{bmatrix}\mathbf{x} + 2 .$$

We want to use the steepest descent algorithm with momentum to minimize this function. Suppose the learning rate is $\alpha = 1$. Find a value for the momentum coefficient γ so that the algorithm will be stable. Use the ideas presented in Problem Eq. P12.2.

E12.7 For the function of Exercise E12.3, perform three iterations of the variable learning rate algorithm, with initial guess

$$\mathbf{x}_0 = \begin{bmatrix} -1 \\ -2.5 \end{bmatrix} .$$

Plot the algorithm trajectory on a contour plot of $F(\mathbf{x})$. Use the algorithm parameters

$$\alpha = 0.4 ,\ \gamma = 0.1 ,\ \eta = 1.5 ,\ \rho = 0.5 ,\ \zeta = 5\% .$$

E12.8 Consider the following quadratic function:

$$F(\mathbf{x}) = x_1^2 + 2x_2^2 .$$

Perform three iterations of the variable learning rate algorithm, with initial guess

$$\mathbf{x}_0 = \begin{bmatrix} 0 \\ -1 \end{bmatrix} .$$

Use the algorithm parameters

$$\alpha = 1 ,\ \gamma = 0.2 ,\ \eta = 1.5 ,\ \rho = 0.5 ,\ \zeta = 5\% .$$

(Count an iteration each time the function is evaluated after the initial guess.)

E12.9 For the function of Exercise E12.3, perform one iteration of the conjugate gradient algorithm, with initial guess

$$\mathbf{x}_0 = \begin{bmatrix} -1 \\ -2.5 \end{bmatrix} .$$

For the linear minimization use interval location by function evaluation and interval reduction by the Golden Section search. Plot the path of the search on a contour plot of $F(\mathbf{x})$.

E12.10 Consider the following quadratic function.

$$F(\mathbf{x}) = \frac{1}{2}\mathbf{x}^T\begin{bmatrix} 4 & 0 \\ 0 & 2 \end{bmatrix}\mathbf{x} + \begin{bmatrix} -2 & -1 \end{bmatrix}\mathbf{x}.$$

We want to minimize this function along the line

$$\mathbf{x} = \begin{bmatrix} 0 \\ 0 \end{bmatrix} + \alpha\begin{bmatrix} 1 \\ 1 \end{bmatrix}.$$

i. Sketch this line in the x_1, x_2 plane.

ii. The learning rate α must fall somewhere between 0 and 3. Perform one iteration of the golden section search. You should find a_2, b_2, c_2 and d_2, and indicate these points along the line that you drew in part i.

E12.11 Consider the following quadratic function.

$$F(\mathbf{x}) = \frac{1}{2}\mathbf{x}^T\begin{bmatrix} 1 & 1 \\ 1 & 1 \end{bmatrix}\mathbf{x} + \begin{bmatrix} 1 & 1 \end{bmatrix}\mathbf{x}.$$

We want to minimize this function along the line

$$\mathbf{x} = \begin{bmatrix} 0 \\ 0 \end{bmatrix} + \alpha\begin{bmatrix} -1 \\ 0 \end{bmatrix}.$$

i. Use the method described on page 12-16 to determine an initial interval containing the minimum. Use $\varepsilon = 0.5$.

ii. Take one iteration of the golden section search to reduce the interval you obtained in part i.

E12.12 Consider the following quadratic function.

$$F(\mathbf{x}) = \frac{1}{2}\mathbf{x}^T\begin{bmatrix} 1 & 0 \\ 0 & 2 \end{bmatrix}\mathbf{x}.$$

We want to minimize this function along the line

$$\mathbf{x} = \begin{bmatrix} 1 \\ 1 \end{bmatrix} + \alpha \begin{bmatrix} 1 \\ -1 \end{bmatrix}.$$

Perform two iterations of the Golden Section search ($k = 1, 2$) to find the interval $[a_3, b_3]$. Assume that the initial interval is defined by $a_1 = 0$ and $b_1 = 1$. Make a rough sketch of the contour plot of $F(\mathbf{x})$, draw the search line in the same figure and indicate your search points (points where you evaluated $F(\mathbf{x})$) on the line.

E12.13 Consider the following quadratic function.

$$F(\mathbf{x}) = \frac{1}{2}\mathbf{x}^T \begin{bmatrix} 2 & 0 \\ 0 & 1 \end{bmatrix} \mathbf{x}.$$

We want to minimize this function along the line

$$\mathbf{x} = \begin{bmatrix} 0 \\ 1 \end{bmatrix} + \alpha \begin{bmatrix} 1 \\ -1 \end{bmatrix}.$$

Perform two iterations of the Golden Section search ($k = 1, 2$) to find the interval $[a_3, b_3]$. Assume that the initial interval is defined by $a_1 = 0$ and $b_1 = 1$. Make a rough sketch of the contour plot of $F(\mathbf{x})$, draw the search line in the same figure and indicate your search points (points where you evaluated $F(\mathbf{x})$) on the line.

E12.14 We want to use the network of Figure E12.2 to approximate the function

$$g(p) = 1 + \sin\left(\frac{\pi}{4}p\right) \text{ for } -2 \le p \le 2.$$

The initial network parameters are chosen to be

$$\mathbf{w}^1(0) = \begin{bmatrix} -0.27 \\ -0.41 \end{bmatrix},\ \mathbf{b}^1(0) = \begin{bmatrix} -0.48 \\ -0.13 \end{bmatrix},\ \mathbf{w}^2(0) = \begin{bmatrix} 0.09 & -0.17 \end{bmatrix},\ \mathbf{b}^2(0) = \begin{bmatrix} 0.48 \end{bmatrix}.$$

To create the training set we sample the function $g(p)$ at the points $p = 1$ and $p = 0$. Find the Jacobian matrix for the first step of the LMBP algorithm. (Some of the information you will need has been computed in the example starting on page 11-14.)

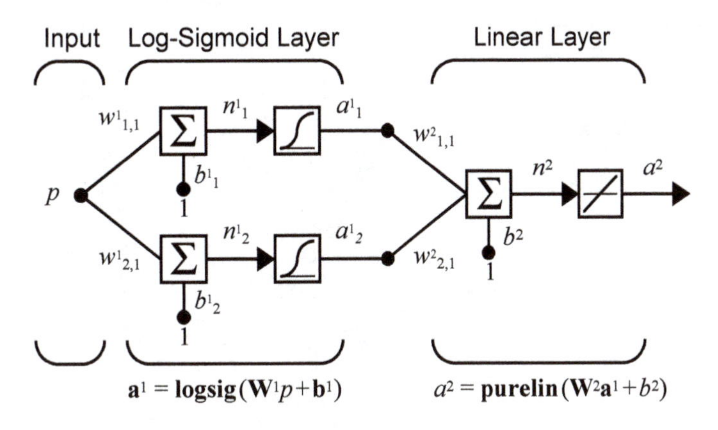

Figure E12.2 Network for Exercise E12.14

E12.15 Show that for a linear network the LMBP algorithm will converge to an optimum solution in one iteration if $\mu = 0$.

E12.16 In Exercise E11.25 you wrote a MATLAB program to implement the SDBP algorithm for a $1 - S^1 - 1$ network, and trained the network to approximate the function

$$g(p) = 1 + \sin\left(\frac{\pi}{4}p\right) \text{ for } -2 \le p \le 2 .$$

Repeat this exercise, modifying your program to use the training procedures discussed in this chapter: batch mode SDBP, MOBP, VLBP, CGBP and LMBP. Compare the convergence results of the various methods.

13 Generalization

Objectives

One of the key issues in designing a multilayer network is determining the number of neurons to use. In effect, that is the objective of this chapter.

In Chapter 11 we showed that if the number of neurons is too large, the network will overfit the training data. This means that the error on the training data will be very small, but the network will fail to perform as well when presented with new data. A network that generalizes well will perform as well on new data as it does on the training data.

The complexity of a neural network is determined by the number of free parameters that it has (weights and biases), which in turn is determined by the number of neurons. If a network is too complex for a given data set, then it is likely to overfit and to have poor generalization.

In this chapter we will see that we can adjust the complexity of a network to fit the complexity of the data. In addition, this can be done without changing the number of neurons. We can adjust the effective number of free parameters without changing the actual number of free parameters.

Theory and Examples

Mark Twain once said "We should be careful to get out of an experience only the wisdom that is in it-and stop there; lest we be like the cat that sits down on a hot stove-lid. She will never sit down on a hot stove-lid again-and that is well; but also she will never sit down on a cold one any more." (From *Following the Equator*, 1897.)

Generalization

That is the objective of this chapter. We want to train neural networks to get out of the data only the wisdom that is in it. This concept is called *generalization*. A network trained to generalize will perform as well in new situations as it does on the data on which it was trained.

Ockham's Razor

The key strategy we will use for obtaining good generalization is to find the simplest model that explains the data. This is a variation of a principle called *Ockham's razor*, which is named after the English logician William of Ockham, who worked in the 14th Century. The idea is that the more complexity you have in your model, the greater the possibility for errors.

In terms of neural networks, the simplest model is the one that contains the smallest number of free parameters (weights and biases), or, equivalently, the smallest number of neurons. To find a network that generalizes well, we need to find the simplest network that fits the data.

There are at least five different approaches that people have used to produce simple networks: growing, pruning, global searches, regularization, and early stopping. Growing methods start with no neurons in the network and then add neurons until the performance is adequate. Pruning methods start with large networks, which likely overfit, and then remove neurons (or weights) one at a time until the performance degrades significantly. Global searches, such as genetic algorithms, search the space of all possible network architectures to locate the simplest model that explains the data.

The final two approaches, regularization and early stopping, keep the network small by constraining the *magnitude* of the network weights, rather than by constraining the *number* of network weights. In this chapter we will concentrate on these two approaches. We will begin by defining the problem of generalization and by showing examples of both good and poor generalization. We will then describe the regularization and early stopping methods for training neural networks. Finally, we will demonstrate how these two methods are, in effect, performing the same operation.

Problem Statement

Let's begin our discussion of generalization by defining the problem. We start with a training set of example network inputs and corresponding target outputs:

$$\{\mathbf{p}_1, \mathbf{t}_1\}, \{\mathbf{p}_2, \mathbf{t}_2\}, \dots, \{\mathbf{p}_Q, \mathbf{t}_Q\}. \qquad (13.1)$$

For our development of the concept of generalization, we will assume that the target outputs are generated by

$$\mathbf{t}_q = \mathbf{g}(\mathbf{p}_q) + \varepsilon_q, \qquad (13.2)$$

where $\mathbf{g}(.)$ is some unknown function, and ε_q is a random, independent and zero mean noise source. Our training objective will be to produce a neural network that approximates $\mathbf{g}(.)$, while ignoring the noise.

The standard performance index for neural network training is the sum squared error on the training set:

$$F(\mathbf{x}) = E_D = \sum_{q=1}^{Q} (\mathbf{t}_q - \mathbf{a}_q)^T (\mathbf{t}_q - \mathbf{a}_q), \qquad (13.3)$$

where $\mathbf{a}_q$ is the network output for input $\mathbf{p}_q$. We are using the variable E_D to represent the sum squared error on the training data, because later we will modify the performance index to include an additional term.

Overfitting

The problem of *overfitting* is illustrated in Figure 13.1. The blue curve represents the function $\mathbf{g}(.)$. The large open circles represent the noisy target points. The black curve represents the trained network response, and the smaller circles filled with crosses represent the network response at the training points. In this figure we can see that the network response exactly matches the training points. However, it does a very poor job of matching the underlying function. It overfits.

There are actually two kinds of errors that occur in Figure 13.1. The first type of error, which is caused by overfitting, occurs for input values between -3 and 0. This is the region where all of the training data points occur. The network response in this region overfits the training data and will fail to perform well for input values that are not in the training set. The network does a poor job of *interpolation*; it fails to accurately approximate the function near the training points.

Interpolation

The second type of error occurs for inputs in the region between 0 and 3. The network fails to perform well in this region, not because it is overfitting, but because there is no training data there. The network is *extrapolating* beyond the range of the input data.

Extrapolation

In this chapter we will discuss methods for preventing errors of interpolation (overfitting). There is no way to prevent errors of extrapolation, unless the data that is used to train the network covers all regions of the input space where the network will be used. The network has no way of knowing what the true function looks like in regions where there is no data.

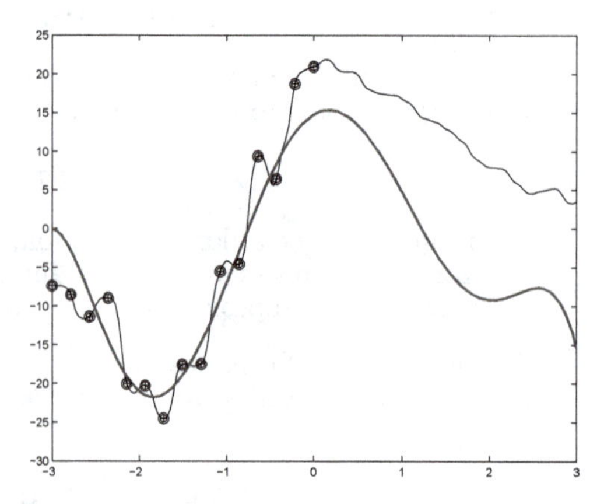

Figure 13.1 Example of Overfitting and Poor Extrapolation

In Figure 13.2 we have an example of a network that has been trained to generalize well. The network has the same number of weights as the network of Figure 13.1, and it was trained using the same data set, but it has been trained in such a way that it does not fully use all of the weights that are available. It only uses as many weights as necessary to fit the data. The network response does not fit the function perfectly, but it does the best job it can, based on limited and noisy data.

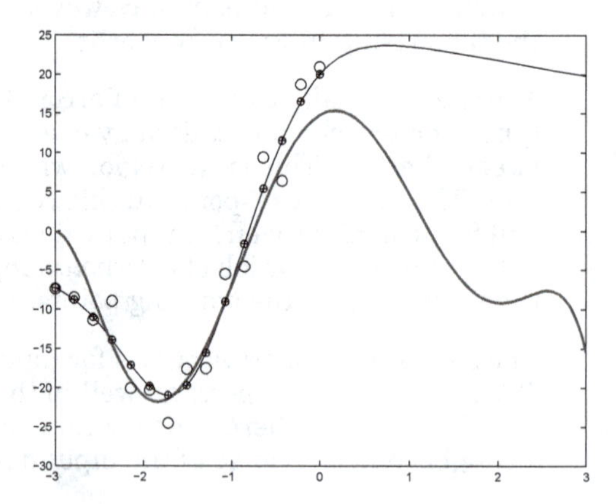

Figure 13.2 Example of Good Interpolation and Poor Extrapolation

In both Figure 13.1 and Figure 13.2 we can see that the network fails to extrapolate accurately. This is understandable, since the network has been provided with no information about the characteristics of the function outside of the range $-3 \le p \le 0$. The network response outside this range will be unpredictable. This is why it is important to have training data for all regions of the input space where the network will be used. It is usually not difficult to determine the required input range when the network has a single input, as in this example. However, when the network has many inputs, it becomes more difficult to determine when the network is interpolating and when it is extrapolating.

This problem is illustrated in a simple way in Figure 13.3. On the left side of this figure we see the function that is to be approximated. The range for the input variables is $-3 \le p_1 \le 3$ and $-3 \le p_2 \le 3$. The neural network was trained over these ranges of the two variables, but only for $p_1 \le p_2$. Therefore, both p_1 and p_2 cover their individual ranges, but only half of the total input space is covered. When $p_1 \ge p_2$, the network is extrapolating, and we can see on the right side of Figure 13.3 that the network performs poorly in this region. (See Problem P13.4 for another example of extrapolation.) If there are many input variables, it will be quite difficult to determine when the network is interpolating and when it is extrapolating. We will discuss some practical ways of dealing with this problem in Chapter 17.

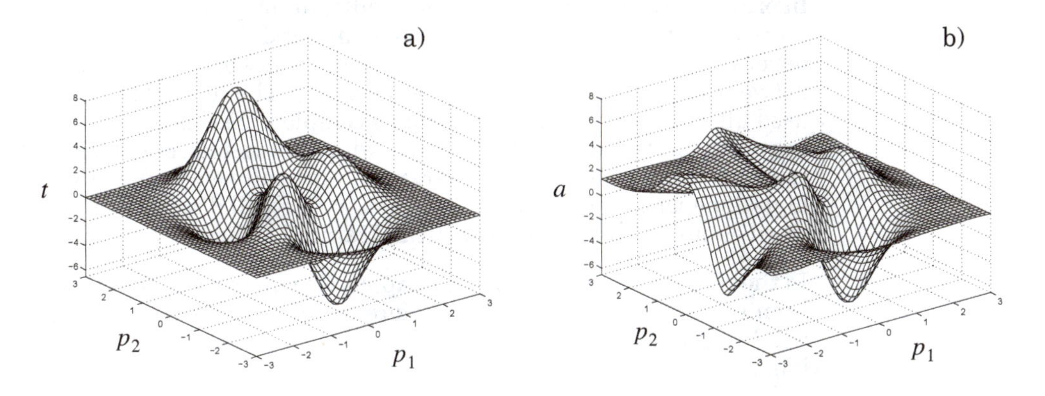

Figure 13.3 Function (a) and Neural Network Approximation (b)

Methods for Improving Generalization

The remainder of this chapter will discuss methods for improving the generalization capability of neural networks. As we discussed earlier, there are a number of approaches to this problem - all of which try to find the simplest network that will fit the data. These approaches fit into two general categories: restricting the number of weights (or, equivalently, the number of neurons) in the network, or restricting the magnitude of the weights. We will concentrate on two methods that we have found to be particularly useful: early stopping and regularization. Both of these approaches attempt to restrict the magnitude of the weights, although they do so in very different ways. At the end of this chapter, we will demonstrate the approximate equivalence of the two methods.

We should note that in this chapter we are assuming that there is a limited amount of data with which to train the network. If the amount of data is unlimited, which in practical terms means that the number of data points is significantly larger than the number of network parameters, then there will not be a problem of overfitting.

Estimating Generalization Error - The Test Set

Before we discuss methods for improving the generalization capability of neural networks, we should first discuss how we can estimate this error for a specific neural network. Given a limited amount of available data, it is important to hold aside a certain subset during the training process. After the network has been trained, we will compute the errors that the trained network makes on this *test set*. The test set errors will then give us an indication of how the network will perform in the future; they are a measure of the generalization capability of the network.

Test Set

In order for the test set to be a valid indicator of generalization capability, there are two important things to keep in mind. First, the test set must never be used in any way to train the neural network, or even to select one network from a group of candidate networks. The test set should only be used after all training and selection is complete. Second, the test set must be representative of all situations for which the network will be used. This can sometimes be difficult to guarantee, especially when the input space is high-dimensional or has a complex shape. We will discuss this problem in more detail in Chapter 17, Practical Training Issues.

In the remaining sections of this chapter, we will assume that a test set has been removed from the data set before training begins, and that this set will be used at the completion of training to measure generalization capability.

Early Stopping

The first method we will discuss for improving generalization is also the simplest method. It is called early stopping [WaVe94]. The idea behind this method is that as training progresses the network uses more and more of its weights, until all weights are fully used when training reaches a minimum of the error surface. By increasing the number of iterations of training, we are increasing the complexity of the resulting network. If training is stopped before the minimum is reached, then the network will effectively be using fewer parameters and will be less likely to overfit. In a later section of this chapter we will demonstrate how the number of parameters changes as the number of iterations increases.

In order to use early stopping effectively, we need to know when to stop the training. We will describe a method, called *cross-validation*, that uses a

Cross-Validation

Validation Set

validation set to decide when to stop [Sarl95]. The available data (after removing the test set, as described above) is divided into two parts: a training set and a validation set. The training set is used to compute gradients or Jacobians and to determine the weight update at each iteration. The validation set is an indicator of what is happening to the network function "in between" the training points, and its error is monitored during the training process. When the error on the validation set goes up for several iterations, the training is stopped, and the weights that produced the minimum error on the validation set are used as the final trained network weights.

This process is illustrated in Figure 13.4. The graph at the bottom of this figure shows the progress of the training and validation performance indices, F (the sum squared errors), during training. Although the training error continues to go down throughout the training process, a minimum of the validation error occurs at the point labeled "a," which corresponds to training iteration 14. The graph at the upper left shows the network response at this *early stopping* point. The resulting network provides a good fit to the true function. The graph at the upper right demonstrates the network response if we continue to train to point "b," where the validation error has increased and the network is overfitting.

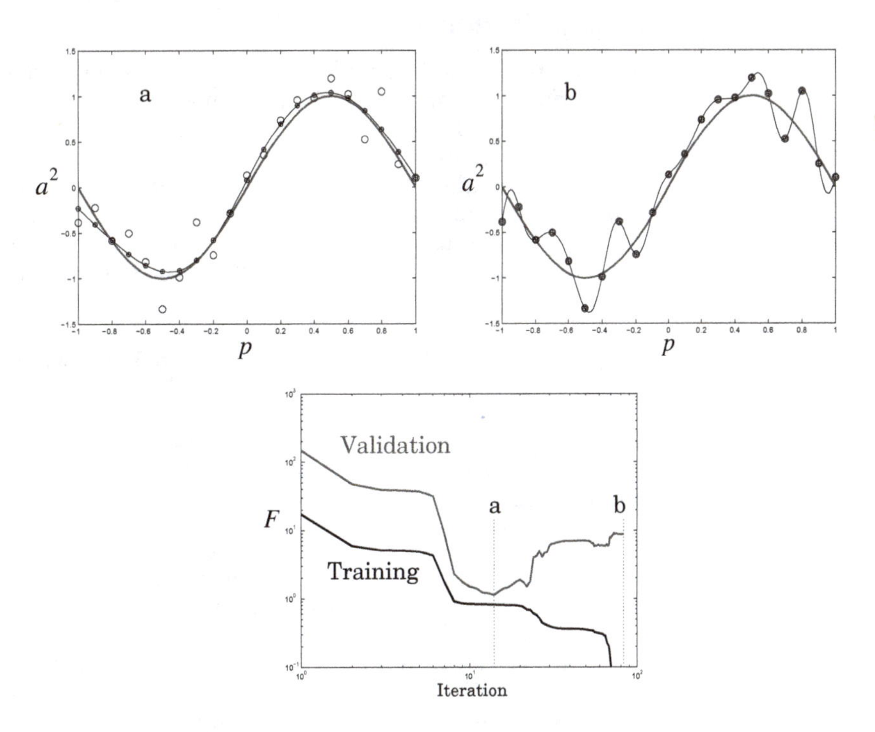

Figure 13.4 Illustration of Early Stopping

The basic concept for early stopping is simple, but there are several practical issues to be addressed. First, the validation set must be chosen so that it is representative of all situations for which the network will be used. This is also true for the test and training sets, as we mentioned earlier. Each set must be roughly equivalent in its coverage of the input space, although the size of each set may be different.

When we divide the data, approximately 70% is typically used for training, with 15% for validation and 15% for testing. These are only approximate numbers. A complete discussion of how to select the amount of data for the validation set is given in [AmMu97].

Another practical point to be made about early stopping is that we should use a relatively slow training method. During training, the network will use more and more of the available network parameters (as we will explain in the last section of this chapter). If the training method is too fast, it will likely jump past the point at which the validation error is minimized.

To experiment with the effect of early stopping, use the MATLAB® Neural Network Design Demonstration Early Stopping (**nnd13es**).

Regularization

The second method we will discuss for improving generalization is called regularization. For this method, we modify the sum squared error performance index of Eq. (13.3) to include a term that penalizes network complexity. This concept was introduced by Tikhonov [Tikh63]. He added a penalty, or regularization, term that involved the derivatives of the approximating function (neural network in our case), which forced the resulting function to be smooth. Under certain conditions, this regularization term can be written as the sum of squares of the network weights, as in

$$F(\mathbf{x}) = \beta E_D + \alpha E_W = \beta \sum_{q=1}^{Q} (\mathbf{t}_q - \mathbf{a}_q)^T (\mathbf{t}_q - \mathbf{a}_q) + \alpha \sum_{i=1}^{n} x_i^2 , \qquad (13.4)$$

where the ratio α / β controls the effective complexity of the network solution. The larger this ratio is, the smoother the network response. (Note that we could have used a single parameter here, but developments in later sections will require two parameters.)

Why do we want to penalize the sum squared weights, and how is this similar to reducing the number of neurons? Consider again the example multilayer network shown in Figure 11.4. Recall how increasing a weight increased the slope of the network function. You can see this effect again in Figure 13.5, where we have changed the weight $w_{1,1}^2$ from 0 to 2. When the weights are large, the function created by the network can have large slopes, and is therefore more likely to overfit the training data. If we restrict the weights to be small, then the network function will create a smooth interpolation through the training data - just as if the network had a small number of neurons.

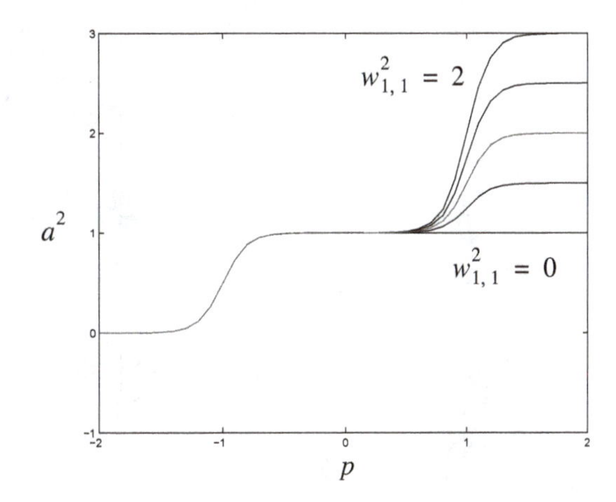

Figure 13.5 Effect of Weight on Network Response

To experiment with the effect of weight changes on the network function, use the MATLAB® Neural Network Design Demonstration Network Function (`nnd11nf`).

The key to the success of the regularization method in producing a network that generalizes well is the correct choice of the regularization ratio α/β. Figure 13.6 illustrates the effect of changing this ratio. Here we have trained a 1-20-1 network on 21 noisy samples of a sine wave.

In the figure, the blue line represents the true function, and the large open circles represent the noisy data. The black curve represents the trained network response, and the smaller circles filled with crosses represent the network response at the training points. From the figure, we can see that the ratio $\alpha/\beta = 0.01$ produces the best fit to the true function. For ratios larger than this, the network response is too smooth, and for ratios smaller than this, the network overfits.

There are several techniques for setting the regularization parameter. One approach is to use a validation set, such as we described in the section on early stopping; the regularization parameter is set to minimize the squared error on the validation set [GoLa98]. In the next two sections we will describe a different technique for automatically setting the regularization parameter. It is called Bayesian regularization.

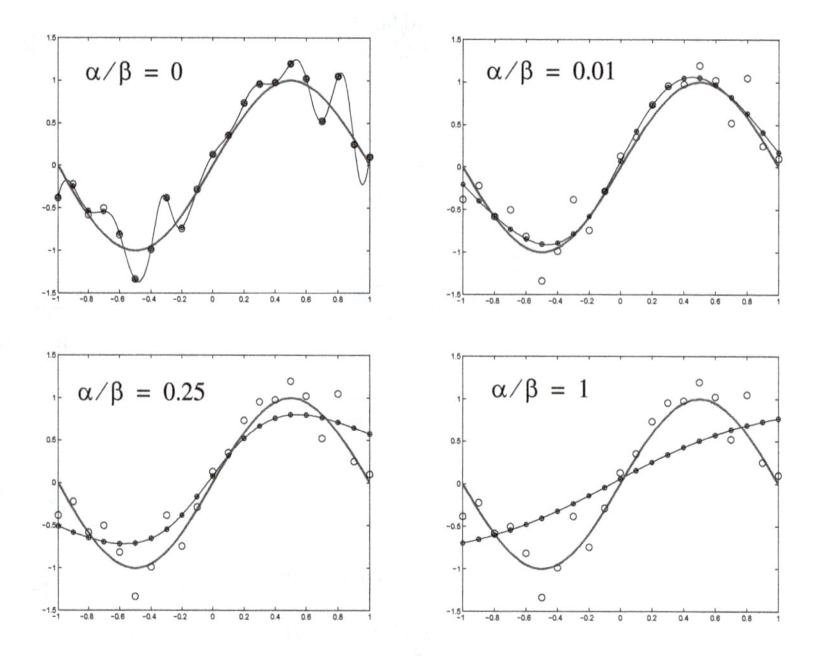

Figure 13.6 Effect of Regularization Ratio

To experiment with the effect of regularization, use the MATLAB® Neural Network Design Demonstration Regularization (nnd13reg).

Bayesian Analysis

Thomas Bayes was a Presbyterian minister who lived in England during the 1700's. He was also an amateur mathematician. His most important work was published after his death. In it, he presented what is now known as Bayes' Theorem. The theorem states that if you have two random events, A and B, then the conditional probability of the occurrence of A, given the occurrence of B can be computed as

$$P(A|B) = \frac{P(B|A)P(A)}{P(B)} . \tag{13.5}$$

Eq. (13.5) is called Bayes' rule. Each of the terms in this expression has a name by which it is commonly referred. $P(A)$ is called the *prior* probability. It tells us what we know about A before we know the outcome of B. $P(A|B)$ is called the *posterior* probability. This tells us what we know about A after we learn about B. $P(B|A)$ is the conditional probability of B given A. Normally this term is given by our knowledge of the system that describes the relationship between B and A. $P(B)$ is the marginal probability of the event B, and it acts as a normalization factor in Bayes' rule.

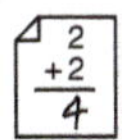

To illustrate how Bayes' rule can be used, consider the following medical situation. Assume that 1% of the population have a certain disease. There is a test that can be performed to detect the presence of this disease. The test is 80% accurate in detecting the disease in people who have it. However, 10% of the time, someone without the disease will register a positive test. If you take the test and register positive, your question would be: What is the probability that I actually have the disease? Most of us (including most physicians, as has been shown in many studies), would guess that the probability is very high, considering that the test is 80% accurate in detecting the disease in a sick person. However, this turns out not to be the case, and Bayes' rule can help us overcome this lack of intuition, when it comes to probability.

Let A represent the event that you have the disease. Let B represent the event that you have a positive test result. We can then use Bayes' rule to find $P(A|B)$, which is the probability that you have the disease, given that you have a positive test. We know that the prior probability $P(A)$ would be 0.01, because 1% of the population have the disease. $P(B|A)$ is 0.8, because the test is 80% accurate in detecting the disease in people who have it. (Notice that this conditional probability is based on our knowledge of the test procedure and its accuracy.) In order to use Bayes' rule, we need one more term, which is $P(B)$. This is the probability of getting a positive test, whether or not you have the disease. This can be obtained by adding the probability of having a positive test when you have the disease to the probability of having a positive test when you don't have the disease:

$$P(B) = P(A \cap B) + P(\bar{A} \cap B) = P(B|A)P(A) + P(B|\bar{A})P(\bar{A}), \quad (13.6)$$

where we have used the definition of conditional probability:

$$P(B|A) = \frac{P(A \cap B)}{P(A)}, \text{ or } P(A \cap B) = P(B|A)P(A). \quad (13.7)$$

If we plug in our known probabilities into Eq. (13.6), we find

$$P(B) = 0.8 \times 0.01 + 0.1 \times 0.99 = 0.107, \quad (13.8)$$

where $P(B|\bar{A})$ is 0.1, because 10% of healthy people register a positive test. We can now use Bayes' rule to find the posterior probability $P(A|B)$:

$$P(A|B) = \frac{P(B|A)P(A)}{P(B)} = \frac{0.8 \times 0.01}{0.107} = 0.0748. \quad (13.9)$$

This tells us that even if you get a positive test, you only have a 7.5% chance of having the disease. For most of us, this result is not intuitive.

The key to Bayes' rule is the prior probability $P(A)$. In this case, the prior odds of having the disease were only 1 in 100. If this number had been much higher, then our posterior probability $P(A|B)$ would have also increased significantly. It is important when using Bayes' rule to have the prior probability $P(A)$ accurately reflect our prior knowledge.

For another example of using Bayes' rule and the effect of the prior density, see Solved Problem P13.2 and its associated demonstration.

In the next section, we will apply Bayesian analysis to the training of multilayer networks. The advantage of Bayesian methods is that we can insert prior knowledge through the selection of the prior probability. For neural network training, we will make the prior assumption that the function we are approximating is smooth. This means that the weights cannot be too large, as was demonstrated in Figure 13.5. The trick will be to incorporate this prior knowledge into an appropriate choice for the prior probability.

Bayesian Regularization

Although there have been many approaches to the automatic selection of the regularization parameter, we will concentrate on one developed by David MacKay [MacK92]. This approach puts the training of neural networks into a Bayesian statistical framework. This framework is useful for many aspects of training, in addition to the selection of the regularization parameter, so it is an important concept to become familiar with. There are two levels to this Bayesian analysis. We will begin with Level I.

Level I Bayesian Framework

The Bayesian framework begins with the assumption that the network weights are random variables. We then choose the weights that maximize the conditional probability of the weights given the data. Bayes' rule is used to find this probability function:

$$P(\mathbf{x} \mid D, \alpha, \beta, M) = \frac{P(D \mid \mathbf{x}, \beta, M) P(\mathbf{x} \mid \alpha, M)}{P(D \mid \alpha, \beta, M)}, \qquad (13.10)$$

where $\mathbf{x}$ is the vector containing all of the weights and biases in the network, D represents the training data set, α and β are parameters associated with the density functions $P(D|\mathbf{x}, \beta, M)$ and $P(\mathbf{x}|\alpha, M)$, and M is the selected model - the architecture of the network we have chosen (i.e., how many layers and how may neurons in each layer).

It is worth taking some time to investigate each of the terms in Eq. (13.10). First, $P(D|\mathbf{x}, \beta, M)$ is the probability density for the data, given a certain set of weights $\mathbf{x}$, the parameter β (which we will explain shortly), and the choice of model M. If we assume that the noise terms in Eq. (13.2) are independent and have a Gaussian distribution, then

$$P(D|\mathbf{x}, \beta, M) = \frac{1}{Z_D(\beta)} \exp(-\beta E_D), \qquad (13.11)$$

where $\beta = 1/(2\sigma_\varepsilon^2)$, σ_ε^2 is the variance of each element of ε_q, E_D is the squared error (as defined in Eq. (13.3)), and

$$Z_D(\beta) = (2\pi\sigma_\varepsilon^2)^{N/2} = (\pi/\beta)^{N/2}, \tag{13.12}$$

where N is $Q \times S^M$, as in Eq. (12.34).

Likelihood Function

Maximum Likelihood

Eq. (13.11) is called the *likelihood function*. It is a function of the network weights $\mathbf{x}$, and it describes how likely a given data set is to occur, given a specific set of weights. The *maximum likelihood* method selects the weights so as to maximize the likelihood function, which in this Gaussian case is the same as minimizing the squared error E_D. Therefore, our standard sum squared error performance index can be derived statistically with the assumption of Gaussian noise in the training set, and our standard choice for the weights is the maximum likelihood estimate.

Prior Density

Now consider the second term on the right side of Eq. (13.10): $P(\mathbf{x}|\alpha, M)$. This is called the *prior density*. It embodies our knowledge about the network weights before we collect any data. Bayesian statistics allows us to incorporate prior knowledge through the prior density. For example, if we assume that the weights are small values centered around zero, we might select a zero-mean Gaussian prior density:

$$P(\mathbf{x}|\alpha, M) = \frac{1}{Z_W(\alpha)}\exp(-\alpha E_W) \tag{13.13}$$

where $\alpha = 1/(2\sigma_w^2)$, σ_w^2 is the variance of each of the weights, E_W is the sum squared weights (as defined in Eq. (13.4)), and

$$Z_W(\alpha) = (2\pi\sigma_w^2)^{n/2} = (\pi/\alpha)^{n/2}, \tag{13.14}$$

where n is the number of weights and biases in the network, as in Eq. (12.35).

Evidence

Posterior Density

The final term on the right side of Eq. (13.10) is $P(D|\alpha, \beta, M)$. This is called the *evidence*, and it is a normalizing term that is not a function of $\mathbf{x}$. If our objective is to find the weights $\mathbf{x}$ that maximize the *posterior density* $P(\mathbf{x}|D, \alpha, \beta, M)$, then we do not need to be concerned with $P(D|\alpha, \beta, M)$. (However, it will be important later for estimating α and β.)

With the Gaussian assumptions that we made earlier, we can rewrite the posterior density, using Eq. (13.10), in the following form:

$$P(\mathbf{x}|D, \alpha, \beta, M) = \frac{\frac{1}{Z_W(\alpha)}\frac{1}{Z_D(\beta)}\exp(-(\beta E_D + \alpha E_W))}{\text{Normalization Factor}} \qquad (13.15)$$

$$= \frac{1}{Z_F(\alpha, \beta)}\exp(-F(\mathbf{x}))$$

where $Z_F(\alpha, \beta)$ is a function of α and β (but not a function of $\mathbf{x}$), and $F(\mathbf{x})$ is our regularized performance index, which we defined in Eq. (13.4). To find the most probable value for the weights, we should maximize the posterior density $P(\mathbf{x}|D, \alpha, \beta, M)$. This is equivalent to minimizing the regularized performance index $F(\mathbf{x}) = \beta E_D + \alpha E_W$.

Therefore, our regularized performance index can be derived using Bayesian statistics, with the assumption of Gaussian noise in the training set and a Gaussian prior density for the network weights. We will identify the weights that maximize the posterior density as $\mathbf{x}^{MP}$, or *most probable*. This is to be contrasted with the weights that maximize the likelihood function: $\mathbf{x}^{ML}$.

Most Probable

Note how this statistical framework provides a physical meaning for the parameters α and β. The parameter β is inversely proportional to the variance in the measurement noise ε_q. Therefore, if the noise variance is large, β will be small, and the regularization ratio α/β will be large. This will force the resulting weights to be small and the network function to be smooth (as seen in Figure 13.6). The larger the measurement noise, the more we will smooth the network function, in order to average out the affects of the noise.

The parameter α is inversely proportional to the variance in the prior distribution for the network weights. If this variance is large, it means that we have very little certainty about the values of the network weights, and, therefore, they might be very large. The parameter α will then be small, and the regularization ratio α/β will also be small. This will allow the network weights to be large, and the network function will be allowed to have more variation (as seen in Figure 13.6). The larger the variance in the prior density for the network weights, the more variation the network function will be allowed to have.

Level II Bayesian Framework

So far we have an interesting statistical derivation of the regularized performance index and some new insight into the meanings of the parameters α and β, but what we really want to find is a way to estimate these parameters from the data. In order to do this, we need to take the Bayesian analysis to another level. If we want to estimate α and β using Bayesian analysis, we need the probability density $P(\alpha, \beta|D, M)$. Using Bayes' rule this can written

$$P(\alpha, \beta | D, M) = \frac{P(D | \alpha, \beta, M) P(\alpha, \beta | M)}{P(D | M)} . \quad (13.16)$$

This has the same format as Eq. (13.10), with the likelihood function and the prior density in the numerator of the right hand side. If we assume a uniform (constant) prior density $P(\alpha, \beta | M)$ for the regularization parameters α and β, then maximizing the posterior is achieved by maximizing the likelihood function $P(D | \alpha, \beta, M)$. However, note that this likelihood function is the normalization factor (evidence) from Eq. (13.10). Since we have assumed that all probabilities have a Gaussian form, we know the form for the posterior density of Eq. (13.10). It is shown in Eq. (13.15). Now we can solve Eq. (13.10) for the normalization factor (evidence).

$$\begin{aligned} P(D | \alpha, \beta, M) &= \frac{P(D | \mathbf{x}, \beta, M) P(\mathbf{x} | \alpha, M)}{P(\mathbf{x} | D, \alpha, \beta, M)} \\ &= \frac{\left[\frac{1}{Z_D(\beta)} \exp(-\beta E_D)\right]\left[\frac{1}{Z_W(\alpha)} \exp(-\alpha E_W)\right]}{\frac{1}{Z_F(\alpha, \beta)} \exp(-F(\mathbf{x}))} \\ &= \frac{Z_F(\alpha, \beta)}{Z_D(\beta) Z_W(\alpha)} \cdot \frac{\exp(-\beta E_D - \alpha E_W)}{\exp(-F(\mathbf{x}))} = \frac{Z_F(\alpha, \beta)}{Z_D(\beta) Z_W(\alpha)} \end{aligned} \quad (13.17)$$

Note that we know the constants $Z_D(\beta)$ and $Z_W(\alpha)$ from Eq. (13.12) and Eq. (13.14). The only part we do not know is $Z_F(\alpha, \beta)$. However, we can estimate it by using a Taylor series expansion.

Since the objective function has the shape of a quadratic in a small area surrounding a minimum point, we can expand $F(\mathbf{x})$ in a second order Taylor series (see Eq. (8.9)) around its minimum point, $\mathbf{x}^{MP}$, where the gradient is zero:

$$F(\mathbf{x}) \approx F(\mathbf{x}^{MP}) + \frac{1}{2} (\mathbf{x} - \mathbf{x}^{MP})^T \mathbf{H}^{MP} (\mathbf{x} - \mathbf{x}^{MP}) , \quad (13.18)$$

where $\mathbf{H} = \beta \nabla^2 E_D + \alpha \nabla^2 E_W$ is the Hessian matrix of $F(\mathbf{x})$, and $\mathbf{H}^{MP}$ is the Hessian evaluated at $\mathbf{x}^{MP}$. We can now substitute this approximation into the expression for the posterior density, Eq. (13.15):

$$P(\mathbf{x} | D, \alpha, \beta, M) \approx \frac{1}{Z_F} \exp\left[- F(\mathbf{x}^{MP}) - \frac{1}{2} (\mathbf{x} - \mathbf{x}^{MP})^T \mathbf{H}^{MP} (\mathbf{x} - \mathbf{x}^{MP})\right], \quad (13.19)$$

which can be rewritten as

$$P(\mathbf{x}|D, \alpha, \beta, M) \approx \left\{ \frac{1}{Z_F} \exp(-F(\mathbf{x}^{MP})) \right\} \exp\left[-\frac{1}{2} (\mathbf{x} - \mathbf{x}^{MP})^T \mathbf{H}^{MP} (\mathbf{x} - \mathbf{x}^{MP}) \right]. \quad (13.20)$$

The standard form of the Gaussian density is

$$P(\mathbf{x}) = \frac{1}{\sqrt{(2\pi)^n \left| (\mathbf{H}^{MP})^{-1} \right|}} \exp\left(-\frac{1}{2} (\mathbf{x} - \mathbf{x}^{MP})^T \mathbf{H}^{MP} (\mathbf{x} - \mathbf{x}^{MP}) \right). \quad (13.21)$$

Therefore, equating Eq. (13.21) with Eq. (13.20), we can solve for $Z_F(\alpha, \beta)$:

$$Z_F(\alpha, \beta) \approx (2\pi)^{n/2} (\det((\mathbf{H}^{MP})^{-1}))^{1/2} \exp(-F(\mathbf{x}^{MP})). \quad (13.22)$$

Placing this result into Eq. (13.17), we can solve for the optimal values for α and β at the minimum point. We do this by taking the derivative with respect to each of the log of Eq. (13.17) and set them equal to zero. This yields (see Solved Problem P13.3):

$$\alpha^{MP} = \frac{\gamma}{2E_W(\mathbf{x}^{MP})} \text{ and } \beta^{MP} = \frac{N - \gamma}{2E_D(\mathbf{x}^{MP})}, \quad (13.23)$$

Effective # of Parameters

where $\gamma = n - 2\alpha^{MP} \mathrm{tr}(\mathbf{H}^{MP})^{-1}$ is called the *effective number of parameters*, and n is the total number of parameters in the network. The term γ is a measure of how many parameters (weights and biases) in the neural network are effectively used in reducing the error function. It can range from zero to n. (See the example on page 13-23 for more analysis of γ.)

Bayesian Regularization Algorithm

The Bayesian optimization of the regularization parameters requires the computation of the Hessian matrix of $F(\mathbf{x})$ at the minimum point $\mathbf{x}^{MP}$. We propose using the Gauss-Newton approximation to the Hessian matrix [FoHa97], which is readily available if the Levenberg-Marquardt optimization algorithm is used to locate the minimum point (see Eq. (12.31)). The additional computation required for optimization of the regularization is minimal.

Here are the steps required for Bayesian optimization of the regularization parameters, with the Gauss-Newton approximation to the Hessian matrix:

0. Initialize α, β and the weights. The weights are initialized randomly, and then E_D and E_W are computed. Set $\gamma = n$, and compute α and β using Eq. (13.23).
1. Take one step of the Levenberg-Marquardt algorithm toward minimizing the objective function $F(\mathbf{x}) = \beta E_D + \alpha E_W$.
2. Compute the effective number of parameters $\gamma = n - 2\alpha \mathrm{tr}(\mathbf{H})^{-1}$, mak-

ing use of the Gauss-Newton approximation to the Hessian available in the Levenberg-Marquardt training algorithm: $\mathbf{H} = \nabla^2 F(\mathbf{x}) \approx 2\beta \mathbf{J}^T \mathbf{J} + 2\alpha \mathbf{I}_n$, where $\mathbf{J}$ is the Jacobian matrix of the training set errors (see Eq. (12.37)).

3. Compute new estimates for the regularization parameters $\alpha = \frac{\gamma}{2E_W(\mathbf{x})}$ and $\beta = \frac{N-\gamma}{2E_D(\mathbf{x})}$.
4. Now iterate steps 1 through 3 until convergence.

Bear in mind that with each reestimate of the regularization parameters α and β the objective function $F(\mathbf{x})$ changes; therefore, the minimum point is moving. If traversing the performance surface generally moves toward the next minimum point, then the new estimates for the regularization parameters will be more precise. Eventually, the precision will be good enough that the objective function will not significantly change in subsequent iterations. Thus, we will obtain convergence.

GNBR

When this Gauss-Newton approximation to Bayesian regularization (*GNBR*) algorithm is used, the best results are obtained if the training data is first mapped into the range [-1,1] (or some similar region). We will discuss this preprocessing of the training data in Chapter 17.

In Figure 13.7 you can see the results of training a 1-20-1 network with GNBR on the same data set represented in Figure 13.4 and Figure 13.6. The network has fit the underlying function, without overfitting to the noise. The fit looks similar to that obtained in Figure 13.6, with the regularization ratio set to $\alpha/\beta = 0.01$. In fact, at the completion of training with GNBR, the final regularization ratio for this example was $\alpha/\beta = 0.0137$.

The training process for this example is illustrated in Figure 13.8. In the upper left of this figure, you see the squared error on the training set. Notice that it does not necessarily go down at each iteration. In the upper right of the figure, you see the squared testing error. This was obtained by comparing the network function to the true function at a number of points between -1 and 1. It is a measure of the generalization capability of the network. (This would not be possible in a practical case, where the true function was unknown.) Note that the testing error is at its minimum at the completion of training.

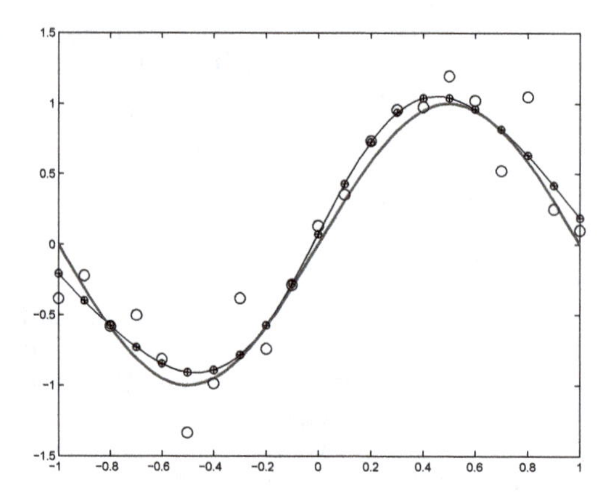

Figure 13.7 Bayesian Regularization Fit

Figure 13.8 also shows the regularization ratio α/β and the effective number of parameters γ during training. These parameters have no particular meaning during the training process, but at the completion of training they are significant. As we mentioned earlier, the final regularization ratio was $\alpha/\beta = 0.0137$, which is consistent with our earlier investigation of regularization—illustrated in Figure 13.6. The final effective number of parameters was $\gamma = 5.2$. This is out of a total of 61 total weights and biases in the network.

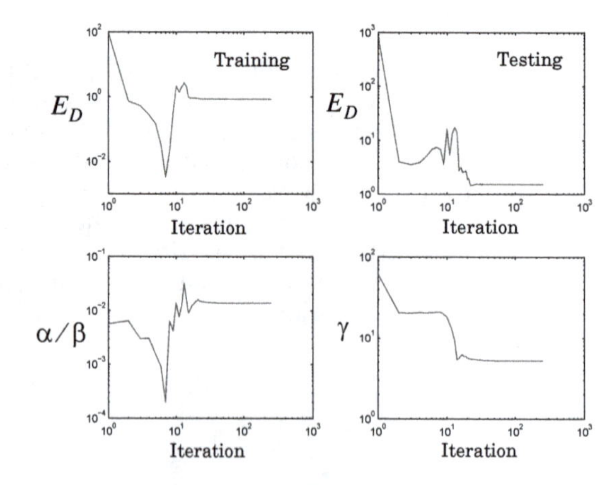

Figure 13.8 Bayesian Regularization Training Process

The fact that in this example the effective number of parameters is much less than the total number of parameters (6 versus 61) means that we might well have been able to use a smaller network to fit this data. There are two disadvantages of a large network: 1) it may overfit the data, and 2) it requires more computation to calculate the network output. We have overcome the first disadvantage by training with GNBR; although the network has 61 parameters, it is equivalent to a network with only 6 parameters. The second disadvantage is only important if the calculation time for the network response is critical to the application. This is not usually the case, since the time to calculate a network response to a particular input is measured in milliseconds. In those cases where the calculation time is significant, you can train a smaller network on the data.

On the other hand, when the effective number of parameters is close to the total number of parameters, this can mean that the network is not large enough to fit the data. In this case, you should increase the size of the network and retrain on the data set.

To experiment with Bayesian Regularization, use the MATLAB® Neural Network Design Demonstration Bayesian Regularization (`nnd17breg`).

Relationship Between Early Stopping and Regularization

We have discussed two techniques for improving network generalization: early stopping and regularization. These two methods were developed in very different ways, but they both improve generalization by restricting the network weights and, therefore, producing a network with fewer effective parameters. Early stopping restricts the network weights by stopping the training before the weights have converged to the minimum of the squared error. Regularization restricts the weights by adding a term to the squared error that penalizes large weights. In this section we want to demonstrate, using a linear example, an approximate equivalence of these two methods. During the process, we will also shed some light on the meaning of the effective number of parameters, γ. This development is based on the more general procedures described in [SjLj94].

Early Stopping Analysis

Consider the single layer linear network shown in Figure 10.1. We have shown in Eq. (10.12) and Eq. (10.14) that the mean square error performance function for this linear network is quadratic, of the form

$$F(\mathbf{x}) = c + \mathbf{d}^T\mathbf{x} + \frac{1}{2}\mathbf{x}^T\mathbf{A}\mathbf{x}, \qquad (13.24)$$

where $\mathbf{A}$ is the Hessian matrix. In order to study the performance of early stopping, we will analyze the evolution of the steepest descent algorithm on this linear network. From Eq. (10.16), we know that the gradient of the performance index is

$$\nabla F(\mathbf{x}) = \mathbf{A}\mathbf{x} + \mathbf{d}. \qquad (13.25)$$

Therefore, the steepest descent algorithm (see Eq. (9.10)) will be

$$\mathbf{x}_{k+1} = \mathbf{x}_k - \alpha \mathbf{g}_k = \mathbf{x}_k - \alpha(\mathbf{A}\mathbf{x}_k + \mathbf{d}). \tag{13.26}$$

We want to know how close we come to the minimum of the squared error at each iteration. For quadratic performance indices, we know that the minimum will occur at the following point (see Eq. (8.62)):

$$\mathbf{x}^{ML} = -\mathbf{A}^{-1}\mathbf{d}, \tag{13.27}$$

where the superscript ML indicates that this result maximizes the likelihood function, in addition to minimizing the squared error, as we saw in Eq. (13.11).

We can now rewrite Eq. (13.26) as

$$\mathbf{x}_{k+1} = \mathbf{x}_k - \alpha\mathbf{A}(\mathbf{x}_k + \mathbf{A}^{-1}\mathbf{d}) = \mathbf{x}_k - \alpha\mathbf{A}(\mathbf{x}_k - \mathbf{x}^{ML}). \tag{13.28}$$

With some additional algebra we can find

$$\mathbf{x}_{k+1} = [\mathbf{I} - \alpha\mathbf{A}]\mathbf{x}_k + \alpha\mathbf{A}\mathbf{x}^{ML} = \mathbf{M}\mathbf{x}_k + [\mathbf{I} - \mathbf{M}]\mathbf{x}^{ML}, \tag{13.29}$$

where $\mathbf{M} = [\mathbf{I} - \alpha\mathbf{A}]$. The next step is to relate $\mathbf{x}_{k+1}$ to the initial guess $\mathbf{x}_0$. Starting at the first iteration, using Eq. (13.29), we have

$$\mathbf{x}_1 = \mathbf{M}\mathbf{x}_0 + [\mathbf{I} - \mathbf{M}]\mathbf{x}^{ML}, \tag{13.30}$$

where the initial guess $\mathbf{x}_0$ usually consists of random values near zero. Continuing to the second iteration:

$$\begin{aligned}\mathbf{x}_2 &= \mathbf{M}\mathbf{x}_1 + [\mathbf{I} - \mathbf{M}]\mathbf{x}^{ML} \\ &= \mathbf{M}^2\mathbf{x}_0 + \mathbf{M}[\mathbf{I} - \mathbf{M}]\mathbf{x}^{ML} + [\mathbf{I} - \mathbf{M}]\mathbf{x}^{ML} \\ &= \mathbf{M}^2\mathbf{x}_0 + \mathbf{M}\mathbf{x}^{ML} - \mathbf{M}^2\mathbf{x}^{ML} + \mathbf{x}^{ML} - \mathbf{M}\mathbf{x}^{ML} \\ &= \mathbf{M}^2\mathbf{x}_0 + \mathbf{x}^{ML} - \mathbf{M}^2\mathbf{x}^{ML} = \mathbf{M}^2\mathbf{x}_0 + [\mathbf{I} - \mathbf{M}^2]\mathbf{x}^{ML}\end{aligned} \tag{13.31}$$

Following similar steps, at the kth iteration we have

$$\mathbf{x}_k = \mathbf{M}^k\mathbf{x}_0 + [\mathbf{I} - \mathbf{M}^k]\mathbf{x}^{ML}, \tag{13.32}$$

This key result shows how far we progress from the initial guess to the maximum likelihood weights in k iterations. We will use this result later to compare with regularization.

Regularization Analysis

Recall from Eq. (13.4) that the regularized performance index adds a penalty term to the sum squared error, as in

$$F(\mathbf{x}) = \beta E_D + \alpha E_W . \tag{13.33}$$

For the following analysis, it will more convenient to consider the following equivalent (because the minimum occurs at the same place) performance index

$$F^*(\mathbf{x}) = \frac{F(\mathbf{x})}{\beta} = E_D + \frac{\alpha}{\beta} E_W = E_D + \rho E_W , \tag{13.34}$$

which has only one regularization parameter.

The sum squared weight penalty term E_W can be written

$$E_W = (\mathbf{x} - \mathbf{x}_0)^T (\mathbf{x} - \mathbf{x}_0) , \tag{13.35}$$

where the nominal value $\mathbf{x}_0$ is normally taken to be the zero vector.

In order to locate the minimum of the regularized performance index, which is also the most probable value $\mathbf{x}^{MP}$, we will set the gradient equal to zero:

$$\nabla F^*(\mathbf{x}) = \nabla E_D + \rho \nabla E_W = \mathbf{0} . \tag{13.36}$$

The gradient of the penalty term, Eq. (13.35), is

$$\nabla E_W = 2(\mathbf{x} - \mathbf{x}_0) . \tag{13.37}$$

From Eq. (13.25) and Eq. (13.28), the gradient of the sum squared error is

$$\nabla E_D = \mathbf{A}\mathbf{x} + \mathbf{d} = \mathbf{A}(\mathbf{x} + \mathbf{A}^{-1}\mathbf{d}) = \mathbf{A}(\mathbf{x} - \mathbf{x}^{ML}) . \tag{13.38}$$

We can now set the total gradient to zero:

$$\nabla F^*(\mathbf{x}) = \mathbf{A}(\mathbf{x} - \mathbf{x}^{ML}) + 2\rho(\mathbf{x} - \mathbf{x}_0) = \mathbf{0} . \tag{13.39}$$

The solution of Eq. (13.39) is the most probable value for the weights, $\mathbf{x}^{MP}$. We can make that substitution and perform some algebra to obtain

$$\begin{aligned} \mathbf{A}(\mathbf{x}^{MP} - \mathbf{x}^{ML}) &= -2\rho(\mathbf{x}^{MP} - \mathbf{x}_0) = -2\rho(\mathbf{x}^{MP} - \mathbf{x}^{ML} + \mathbf{x}^{ML} - \mathbf{x}_0) \\ &= -2\rho(\mathbf{x}^{MP} - \mathbf{x}^{ML}) - 2\rho(\mathbf{x}^{ML} - \mathbf{x}_0) \end{aligned} \tag{13.40}$$

Now combine the terms multiplying $(\mathbf{x}^{MP} - \mathbf{x}^{ML})$:

$$(\mathbf{A} + 2\rho\mathbf{I})(\mathbf{x}^{MP} - \mathbf{x}^{ML}) = 2\rho(\mathbf{x}_0 - \mathbf{x}^{ML}). \quad (13.41)$$

Solving for $(\mathbf{x}^{MP} - \mathbf{x}^{ML})$, we find

$$(\mathbf{x}^{MP} - \mathbf{x}^{ML}) = 2\rho(\mathbf{A} + 2\rho\mathbf{I})^{-1}(\mathbf{x}_0 - \mathbf{x}^{ML}) = \mathbf{M}_\rho(\mathbf{x}_0 - \mathbf{x}^{ML}), \quad (13.42)$$

where $\mathbf{M}_\rho = 2\rho(\mathbf{A} + 2\rho\mathbf{I})^{-1}$.

We want to know the relationship between the regularized solution $\mathbf{x}^{MP}$ and the minimum of the squared error $\mathbf{x}^{ML}$, so we can solve Eq. (13.42) for $\mathbf{x}^{MP}$:

$$\mathbf{x}^{MP} = \mathbf{M}_\rho\mathbf{x}_0 + [\mathbf{I} - \mathbf{M}_\rho]\mathbf{x}^{ML}. \quad (13.43)$$

This is the key result that describes the relationship between the regularized solution and the minimum of the squared error. By comparing Eq. (13.43) with Eq. (13.32), we can investigate the relationship between early stopping and regularization. We will do that in the next section.

Connection Between Early Stopping and Regularization

To compare early stopping and regularization, we need to compare Eq. (13.43) and Eq. (13.32). They are summarized in Figure 13.9. We would like to find out when these two solutions are equal. In other words, when do early stopping and regularization produce the same weights?

Early Stopping	*Regularization*
$\mathbf{x}_k = \mathbf{M}^k\mathbf{x}_0 + [\mathbf{I} - \mathbf{M}^k]\mathbf{x}^{ML}$	$\mathbf{x}^{MP} = \mathbf{M}_\rho\mathbf{x}_0 + [\mathbf{I} - \mathbf{M}_\rho]\mathbf{x}^{ML}$
$\mathbf{M} = [\mathbf{I} - \alpha\mathbf{A}]$	$\mathbf{M}_\rho = 2\rho(\mathbf{A} + 2\rho\mathbf{I})^{-1}$

Figure 13.9 Early Stopping and Regularization Solutions

The key matrix for early stopping is $\mathbf{M}^k = [\mathbf{I} - \alpha\mathbf{A}]^k$. The key matrix for regularization is $\mathbf{M}_\rho = 2\rho(\mathbf{A} + 2\rho\mathbf{I})^{-1}$. If these two matrices are equal, then the weights for early stopping will be the same as the weights for regularization. In Eq. (9.22) we showed that the eigenvectors of $\mathbf{M}$ are the same as the eigenvectors of $\mathbf{A}$ and that the eigenvalues of $\mathbf{M}$ are $(1 - \alpha\lambda_i)$, where the eigenvalues of $\mathbf{A}$ are λ_i. The eigenvalues of $\mathbf{M}^k$ are then

$$eig(\mathbf{M}^k) = (1 - \alpha\lambda_i)^k. \quad (13.44)$$

Now let's consider the matrix $\mathbf{M}_\rho$. First, using the same procedures that led to Eq. (9.22), we can show that the eigenvectors of $(\mathbf{A} + 2\rho\mathbf{I})$ are the same as the eigenvectors of $\mathbf{A}$, and the eigenvalues of $(\mathbf{A} + 2\rho\mathbf{I})$ are

$(2\rho + \lambda_i)$. Also, the eigenvectors of the inverse of a matrix are the same as the eigenvectors of the original matrix, and the eigenvalues of the inverse are the reciprocals of the original eigenvalues. Therefore, the eigenvectors of $\mathbf{M}_\rho$ are the same as the eigenvectors of $\mathbf{A}$, and the eigenvalues of $\mathbf{M}_\rho$ are

$$eig(\mathbf{M}_\rho) = \frac{2\rho}{(\lambda_i + 2\rho)}. \quad (13.45)$$

Therefore, in order for $\mathbf{M}^k$ to equal $\mathbf{M}_\rho$, they just need to have equal eigenvalues:

$$\frac{2\rho}{(\lambda_i + 2\rho)} = (1 - \alpha\lambda_i)^k. \quad (13.46)$$

Take the logarithm of both sides:

$$-\log\left(1 + \frac{\lambda_i}{2\rho}\right) = k\log(1 - \alpha\lambda_i). \quad (13.47)$$

These expressions are equal at $\lambda_i = 0$, so they will always be equal if their derivatives are equal. Taking derivatives of both sides, we have

$$-\frac{1}{\left(1 + \frac{\lambda_i}{2\rho}\right)}\frac{1}{\rho} = \frac{k}{1 - \alpha\lambda_i}(-\alpha), \quad (13.48)$$

or

$$\alpha k = \frac{1}{2\rho}\frac{(1 - \alpha\lambda_i)}{(1 + \lambda_i/(2\rho))}. \quad (13.49)$$

If $\alpha\lambda_i$ is small (slow, stable learning) and $\lambda_i/(2\rho)$ is small, then we have the approximate result

$$\alpha k \cong \frac{1}{2\rho}. \quad (13.50)$$

Therefore, early stopping is approximately equivalent to regularization. Increasing the number of iterations k is approximately the same as decreasing the regularization parameter ρ. This makes intuitive sense, because increasing the number of iterations, or decreasing the regularization parameter, can lead to overfitting.

Example, Interpretation of Effective Number of Parameters

2 +2 4

We will illustrate this result with a simple example. Suppose that we have a single layer, linear network with no bias. The input/target pairs are given by

$$\left\{\mathbf{p}_1 = \begin{bmatrix}1\\1\end{bmatrix}, t_1 = 1\right\}, \left\{\mathbf{p}_2 = \begin{bmatrix}-1\\1\end{bmatrix}, t_2 = -1\right\},$$

where the probability of the first pair is 0.75, and the probability of the second pair is 0.25. Following Eq. (10.13) and Eq. (10.15), we can find the quadratic mean square error performance index as

$$c = E[t^2] = (1)^2(0.75) + (-1)^2(0.25) = 1,$$

$$\mathbf{h} = E[t\mathbf{z}] = (0.75)(1)\begin{bmatrix}1\\1\end{bmatrix} + (0.25)(-1)\begin{bmatrix}-1\\1\end{bmatrix} = \begin{bmatrix}1\\0.5\end{bmatrix},$$

$$\mathbf{d} = -2\mathbf{h} = (-2)\begin{bmatrix}1\\0.5\end{bmatrix} = \begin{bmatrix}-2\\-1\end{bmatrix},$$

$$\mathbf{A} = 2\mathbf{R} = 2(E[\mathbf{z}\mathbf{z}^T]) = 2\left((0.75)\begin{bmatrix}1\\1\end{bmatrix}\begin{bmatrix}1 & 1\end{bmatrix} + 0.25\begin{bmatrix}-1\\1\end{bmatrix}\begin{bmatrix}-1 & 1\end{bmatrix}\right) = \begin{bmatrix}2 & 1\\1 & 2\end{bmatrix},$$

$$E_D = c + \mathbf{x}^T\mathbf{d} + \frac{1}{2}\mathbf{x}^T\mathbf{A}\mathbf{x}.$$

The minimum of the mean squared error occurs at

$$\mathbf{x}^{ML} = -\mathbf{A}^{-1}\mathbf{d} = \mathbf{R}^{-1}\mathbf{h} = \begin{bmatrix}1 & 0.5\\0.5 & 1\end{bmatrix}^{-1}\begin{bmatrix}1\\0.5\end{bmatrix} = \begin{bmatrix}1\\0\end{bmatrix}.$$

Now let's investigate the eigensystem of the Hessian matrix of E_D:

$$\nabla^2 E_D(\mathbf{x}) = \mathbf{A} = 2\mathbf{R} = \begin{bmatrix}2 & 1\\1 & 2\end{bmatrix}.$$

To find the eigenvalues:

$$\left|\mathbf{A} - \lambda\mathbf{I}\right| = \begin{vmatrix}2-\lambda & 1\\1 & 2-\lambda\end{vmatrix} = \lambda^2 - 4\lambda + 3 = (\lambda - 1)(\lambda - 3),$$

$$\lambda_1 = 1, \qquad \lambda_2 = 3.$$

To find the eigenvectors:

$$[\mathbf{A} - \lambda \mathbf{I}]\mathbf{v} = 0 .$$

For $\lambda_1 = 1$,

$$\begin{bmatrix} 1 & 1 \\ 1 & 1 \end{bmatrix} \mathbf{v}_1 = 0 \qquad \mathbf{v}_1 = \begin{bmatrix} 1 \\ -1 \end{bmatrix},$$

and for $\lambda_2 = 3$,

$$\begin{bmatrix} -1 & 1 \\ 1 & -1 \end{bmatrix} \mathbf{v}_2 = 0 \qquad \mathbf{v}_2 = \begin{bmatrix} 1 \\ 1 \end{bmatrix}.$$

The contour plot for E_D is shown in Figure 13.10

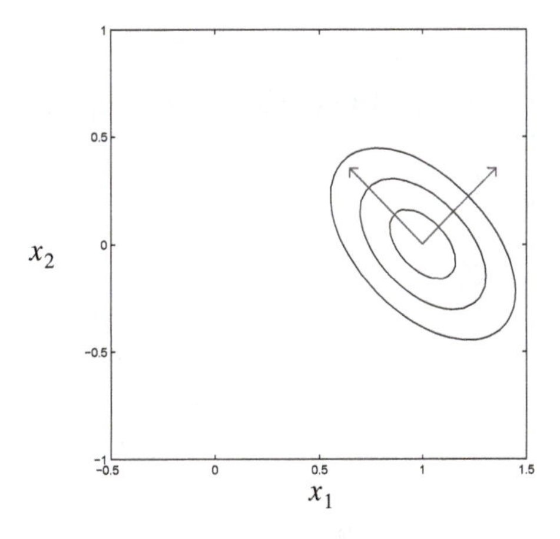

Figure 13.10 Contour Plot for E_D

Now consider the regularized performance index of Eq. (13.34). Its Hessian matrix will be

$$\nabla^2 F^*(\mathbf{x}) = \nabla^2 E_D + \rho \nabla^2 E_W = \nabla^2 E_D + 2\rho \mathbf{I} = \begin{bmatrix} 2 & 1 \\ 1 & 2 \end{bmatrix} + \rho \begin{bmatrix} 2 & 0 \\ 0 & 2 \end{bmatrix} = \begin{bmatrix} 2+2\rho & 1 \\ 1 & 2+2\rho \end{bmatrix}.$$

In Figure 13.11 we have contour plots for F as ρ is equal to 0, 1 and ∞.

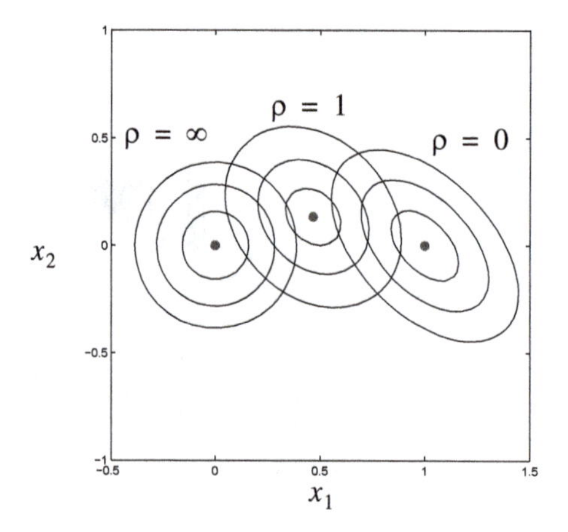

Figure 13.11 Contour Plot for F

In Figure 13.12 the blue curve represents the movement of $\mathbf{x}^{MP}$ as ρ is varied.

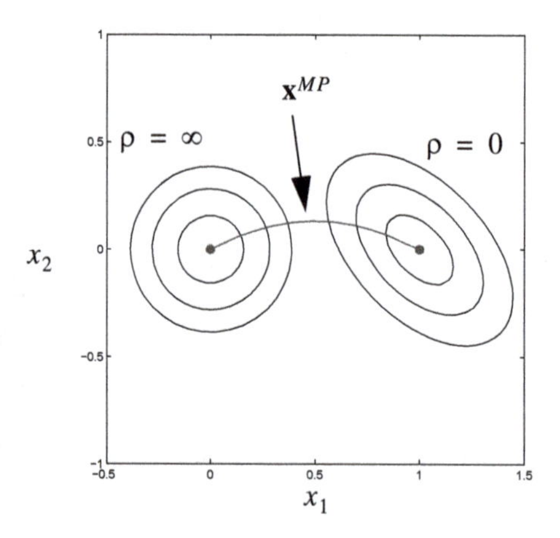

Figure 13.12 $\mathbf{x}^{MP}$ as ρ is Varied

Now let's compare this regularization result with early stopping. Figure 13.13 shows the steepest descent trajectory for minimizing E_D, starting from very small values for the weights. If we stop early, the result will fall along the blue curve. Notice that this curve is very close to the regularization curve in Figure 13.12. If the number of iterations is very small, this is equivalent to a very large value for ρ. As the number of iterations increases, it is equivalent to reducing ρ.

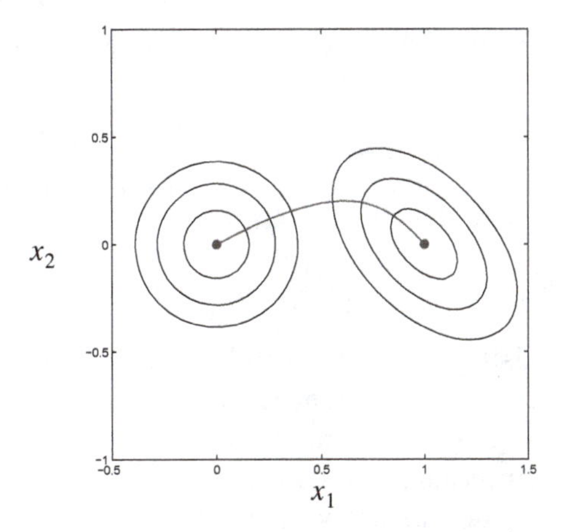

Figure 13.13 Steepest Descent Trajectory

To experiment with the relationship between Early Stopping and Regularization, use the MATLAB® Neural Network Design Demonstration Early Stopping/Regularization (**nnd17esr**).

It is useful to consider the relationship between the eigenvalues and eigenvectors of the Hessian matrix $\nabla^2 E_D(\mathbf{x})$ and the results of regularization and early stopping. In this example, λ_2 is larger than λ_1, so E_D has higher curvature in the $\mathbf{v}_2$ direction. This means that we will get a quicker reduction in the squared error if we move in that direction first. This is shown in Figure 13.13, as the initial steepest descent movement is almost in the direction of $\mathbf{v}_2$. Note also that in regularization, as shown in Figure 13.12, as ρ decreases from a large value, the weights move first in the $\mathbf{v}_2$ direction. For a given change in the weights, this direction provides the largest reduction in the squared error.

Since the eigenvalue λ_1 is smaller than λ_2, we only move in the $\mathbf{v}_1$ direction after achieving significant reduction in E_D in the $\mathbf{v}_2$ direction. This would be even more pronounced if the difference between λ_1 and λ_2 were greater. In the limiting case, where $\lambda_1 = 0$, we would not have to move in the $\mathbf{v}_1$ direction at all. We would only need to move in the $\mathbf{v}_2$ direction to get the complete reduction in the squared error. (This would be the case of the stationary valley, as in Figure 8.9.) Note that in this case we would only be effectively using one parameter, even though the network has two weights. (Of course, this one effective parameter is some combination of the two weights.) Therefore, the effective number of parameters is related to the number of eigenvalues of $\nabla^2 E_D(\mathbf{x})$ that are significantly different than zero. We will analyze this in detail in the next section.

Effective Number of Parameters

Recall the previous definition for the effective number of parameters:

$$\gamma = n - 2\alpha^{MP}\mathrm{tr}\{(\mathbf{H}^{MP})^{-1}\} \tag{13.51}$$

We can express this in terms of the eigenvalues of $\nabla^2 E_D(\mathbf{x})$. First, we can write the Hessian matrix as

$$\mathbf{H}(\mathbf{x}) = \nabla^2 F(\mathbf{x}) = \beta\nabla^2 E_D + \alpha\nabla^2 E_W = \beta\nabla^2 E_D + 2\alpha\mathbf{I}. \tag{13.52}$$

Using arguments similar to those leading to Eq. (13.44), we can show that the eigenvalues of $\mathbf{H}(\mathbf{x})$ are $(\beta\lambda_i + 2\alpha)$. We can then use two properties of eigenvalues to compute $\mathrm{tr}\{\mathbf{H}^{-1}\}$. First, the eigenvalues of $\mathbf{H}^{-1}$ are the reciprocals of the eigenvalues of $\mathbf{H}$, and, second, the trace of a matrix is equal to the sum of its eigenvalues. Using these two properties, we can write

$$\mathrm{tr}\{\mathbf{H}^{-1}\} = \sum_{i=1}^{n} \frac{1}{\beta\lambda_i + 2\alpha}. \tag{13.53}$$

We can now write the effective number of parameters as

$$\gamma = n - 2\alpha^{MP}\mathrm{tr}\{(\mathbf{H}^{MP})^{-1}\} = n - \sum_{i=1}^{n} \frac{2\alpha}{\beta\lambda_i + 2\alpha} = \sum_{i=1}^{n} \frac{\beta\lambda_i}{\beta\lambda_i + 2\alpha}, \tag{13.54}$$

or

$$\gamma = \sum_{i=1}^{n} \frac{\beta\lambda_i}{\beta\lambda_i + 2\alpha} = \sum_{i=1}^{n} \gamma_i, \tag{13.55}$$

where

$$\gamma_i = \frac{\beta\lambda_i}{\beta\lambda_i + 2\alpha}. \tag{13.56}$$

Note that $0 \le \gamma_i \le 1$, so the effective number of parameters γ must fall between zero and n. If all of the eigenvalues of $\nabla^2 E_D(\mathbf{x})$ are large, then the effective number of parameters will equal the total number of parameters. If some of the eigenvalues are very small, then the effective number of parameters will equal the number of large eigenvalues, as was also demonstrated by our example in the previous section. Large eigenvalues mean large curvature, which means that the performance index changes rapidly along those eigenvectors. Every large eigenvector represents a productive direction for optimizing performance.

Summary of Results

Problem Statement

A network trained to generalize will perform as well in new situations as it does on the data on which it was trained.

$$E_D = \sum_{q=1}^{Q} (\mathbf{t}_q - \mathbf{a}_q)^T (\mathbf{t}_q - \mathbf{a}_q)$$

Methods for Improving Generalization

Estimating Generalization Error - The Test Set

Given a limited amount of available data, it is important to hold aside a certain subset during the training process. After the network has been trained, we will compute the errors that the trained network makes on this *test set*. The test set errors will then give us an indication of how the network will perform in the future; they are a measure of the generalization capability of the network.

Early Stopping

The available data (after removing the test set) is divided into two parts: a training set and a validation set. The training set is used to compute gradients or Jacobians and to determine the weight update at each iteration. When the error on the validation set goes up for several iterations, the training is stopped, and the weights that produced the minimum error on the validation set are used as the final trained network weights.

Regularization

$$F(\mathbf{x}) = \beta E_D + \alpha E_W = \beta \sum_{q=1}^{Q} (\mathbf{t}_q - \mathbf{a}_q)^T (\mathbf{t}_q - \mathbf{a}_q) + \alpha \sum_{i=1}^{n} x_i^2$$

Bayesian Regularization

Level I Bayesian Framework

$$P(\mathbf{x} | D, \alpha, \beta, M) = \frac{P(D|\mathbf{x}, \beta, M) P(\mathbf{x}|\alpha, M)}{P(D|\alpha, \beta, M)}$$

$$P(D|\mathbf{x}, \beta, M) = \frac{1}{Z_D(\beta)}\exp(-\beta E_D)\,,\ \beta = 1/(2\sigma_\varepsilon^2)$$

$$Z_D(\beta) = (2\pi\sigma_\varepsilon^2)^{N/2} = (\pi/\beta)^{N/2}$$

$$P(\mathbf{x}\,|\,\alpha, M) = \frac{1}{Z_W(\alpha)}\exp(-\alpha E_W)\,,\ \alpha = 1/(2\sigma_w^2)$$

$$Z_W(\alpha) = (2\pi\sigma_w^2)^{n/2} = (\pi/\alpha)^{n/2}$$

$$P(\mathbf{x}\,|\,D, \alpha, \beta, M) = \frac{1}{Z_F(\alpha, \beta)}\exp(-F(\mathbf{x}))$$

Level II Bayesian Framework

$$P(\alpha, \beta|D, M) = \frac{P(D|\alpha, \beta, M)P(\alpha, \beta|M)}{P(D|M)}$$

$$\alpha^{MP} = \frac{\gamma}{2E_W(\mathbf{x}^{MP})} \text{ and } \beta^{MP} = \frac{N-\gamma}{2E_D(\mathbf{x}^{MP})}$$

$$\gamma = n - 2\alpha^{MP}\operatorname{tr}(\mathbf{H}^{MP})^{-1}$$

Bayesian Regularization Algorithm

0. Initialize α, β and the weights. The weights are initialized randomly, and then E_D and E_W are computed. Set $\gamma = n$, and compute α and β using Eq. (13.23).
1. Take one step of the Levenberg-Marquardt algorithm toward minimizing the objective function $F(\mathbf{x}) = \beta E_D + \alpha E_W$.
2. Compute the effective number of parameters $\gamma = N - 2\alpha\operatorname{tr}(\mathbf{H})^{-1}$, making use of the Gauss-Newton approximation to the Hessian available in the Levenberg-Marquardt training algorithm:
   $\mathbf{H} = \nabla^2 F(\mathbf{x}) \approx 2\beta\mathbf{J}^T\mathbf{J} + 2\alpha\mathbf{I}_n$, where $\mathbf{J}$ is the Jacobian matrix of the training set errors (see Eq. (12.37)).
3. Compute new estimates for the regularization parameters $\alpha = \dfrac{\gamma}{2E_W(\mathbf{x})}$ and $\beta = \dfrac{N-\gamma}{2E_D(\mathbf{x})}$.
4. Now iterate steps 1 through 3 until convergence.

Relationship Between Early Stopping and Regularization

Early Stopping	*Regularization*
$\mathbf{x}_k = \mathbf{M}^k\mathbf{x}_0 + [\mathbf{I} - \mathbf{M}^k]\mathbf{x}^{ML}$	$\mathbf{x}^{MP} = \mathbf{M}_\rho\mathbf{x}_0 + [\mathbf{I} - \mathbf{M}_\rho]\mathbf{x}^{ML}$
$\mathbf{M} = [\mathbf{I} - \alpha\mathbf{A}]$	$\mathbf{M}_\rho = 2\rho(\mathbf{A} + 2\rho\mathbf{I})^{-1}$

$$eig(\mathbf{M}^k) = (1 - \alpha\lambda_i)^k$$

$$eig(\mathbf{M}_\rho) = \frac{2\rho}{(\lambda_i + 2\rho)}$$

$$\alpha k \cong \frac{1}{2\rho}$$

Effective Number of Parameters

$$\gamma = \sum_{i=1}^{n} \frac{\beta\lambda_i}{\beta\lambda_i + 2\alpha}$$

$$0 \le \gamma \le n$$

Solved Problems

P13.1 In this problem and in the following one we want to investigate the relationship between maximum likelihood methods and Bayesian methods. Suppose that we have a random variable that is uniformly distributed between 0 and *x*. We take a series of *Q* independent samples of the random variable. Find the maximum likelihood estimate of *x*.

Before we begin this problem, let's review the Level I Bayesian formulation of Eq. (13.10). We will not need the Level II formulation for this simple problem, so we do not need the regularization parameters. Also, we only have a single parameter to estimate, so x is a scalar. Eq. (13.10) can then be simplified to

$$P(x|D) = \frac{P(D|x)P(x)}{P(D)}.$$

We are interested in the maximum likelihood estimate for this problem, so we need to find the value of x that maximizes the likelihood term $P(D|x)$. The data is the Q independent samples from the uniformly distributed random variable. A graph of the uniform density function is given in Figure P13.1.

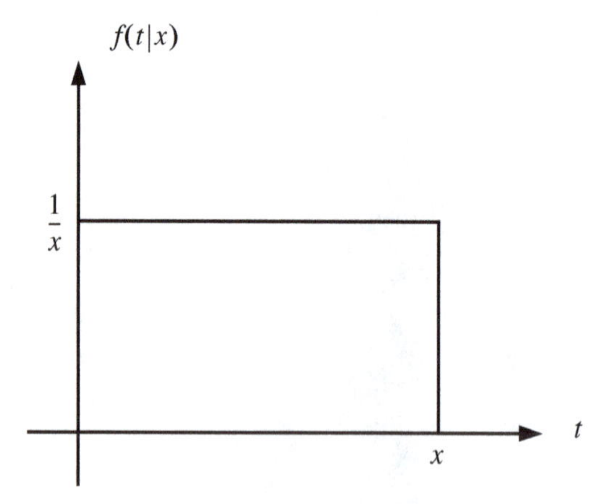

Figure P13.1 Uniform Density Function

The definition can be written

$$f(t|x) = \begin{cases} \frac{1}{x}, & 0 \le t \le x \\ 0, & \text{elsewhere} \end{cases}.$$

If we have Q independent samples of the random variable, then we can multiply each of the individual probabilities to get the joint probability of all samples:

$$P(D|x) = \prod_{i=1}^{Q} f(t_i|x) = \begin{cases} \frac{1}{x^Q}, 0 \le t_i \le x, \text{ for all } i \\ 0, \text{ elsewhere} \end{cases} = \begin{cases} \frac{1}{x^Q}, x \ge max(t_i) \\ 0, x < max(t_i) \end{cases}$$

The plot of the resulting likelihood function is shown in Figure P13.1.

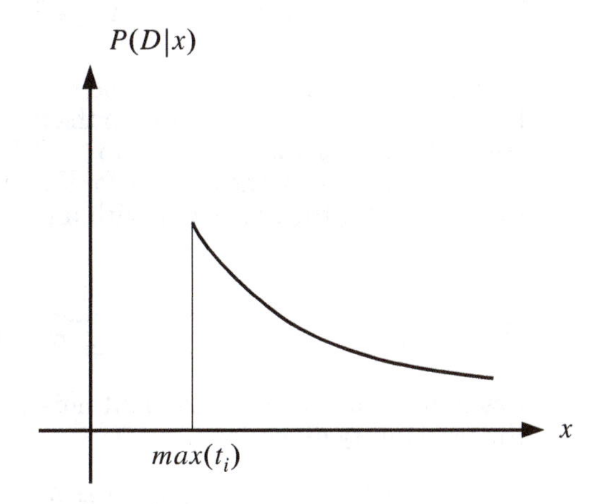

Figure P13.2 Likelihood Function for Solved Problem P13.1

From this plot, we can see that the value of x that maximizes the likelihood function is

$$x^{ML} = max(t_i).$$

Therefore, the maximum likelihood estimate of x is the maximum value obtained from the Q independent samples of the random variable. This seems like a reasonable estimate of x, which is the upper limit of the random variable.

P13.2 In this problem we will compare the maximum likelihood and Bayesian estimators. Assume that we have a series of measurements of a random signal in noise:

$$t_i = x + \varepsilon_i.$$

Assume that the noise has a Gaussian density, with zero mean:

$$f(\varepsilon_i) = \frac{1}{\sqrt{2\pi}\sigma}\exp\left(-\frac{\varepsilon_i^2}{2\sigma^2}\right)$$

i. **Find the maximum likelihood estimate of x.**

ii. **Find the most probable estimate of x. Assume that x is a zero-mean random variable, with Gaussian prior density:**

$$f(x) = \frac{1}{\sqrt{2\pi}\sigma_x}\exp\left(-\frac{x^2}{2\sigma_x^2}\right) = \frac{1}{Z_W(\alpha)}\exp(-\alpha E_W)$$

i. To find the maximum likelihood estimate, we need to find the likelihood function $P(D|x)$. This represents the density of the data, given x. The first step is to use the noise density to find the density of the measurement. Since, with x given, the density for the measurement would be the same as the density for the noise, but with a mean of x, we have

$$f(t_i|x) = \frac{1}{\sqrt{2\pi}\sigma}\exp\left(-\frac{(t_i - x)^2}{2\sigma^2}\right).$$

Assuming that the measurement noises are independent, we can multiply the probability densities:

$$P(D|x) = f(t_1, t_2, \ldots, t_Q|x) = f(t_1|x)f(t_2|x)\ldots f(t_Q|x) = P(D|x)$$

$$= \frac{1}{(2\pi)^{Q/2}\sigma^Q}\exp\left(-\frac{\sum_{i=1}^{Q}(t_i - x)^2}{2\sigma^2}\right) = \frac{1}{Z(\beta)}\exp(-\beta E_D)$$

where

$$\beta = \frac{1}{2\sigma^2},\ E_D = \sum_{i=1}^{Q}(t_i - x)^2 = \sum_{i=1}^{Q} e_i^2,\ Z(\beta) = (\pi/\beta)^{Q/2}.$$

To maximize the likelihood, we should minimize E_D. Setting the derivative to zero, we find

$$\frac{dE_D}{dx} = \frac{d}{dx}\sum_{i=1}^{Q}(t_i - x)^2 = -2\sum_{i=1}^{Q}(t_i - x) = -2\left[\left(\sum_{i=1}^{Q} t_i\right) - Qx\right] = 0.$$

Solving for x, we find the maximum likelihood estimate:

$$x^{ML} = \frac{1}{Q}\sum_{i=1}^{Q} t_i$$

ii. To find the most probable estimate, we need to use Bayes' rule (Eq. (13.10)) to find the posterior density:

$$P(x|D) = \frac{P(D|x)P(x)}{P(D)}.$$

The likelihood function $P(D|x)$ was found above to be

$$P(D|x) = \frac{1}{Z(\beta)}\exp(-\beta E_D)$$

The prior density is

$$P(x) = f(x) = \frac{1}{\sqrt{2\pi}\sigma_x}\exp\left(-\frac{x^2}{2\sigma_x^2}\right) = \frac{1}{Z_W(\alpha)}\exp(-\alpha E_W),$$

where

$$\alpha = \frac{1}{2\sigma_x^2},\ Z_W(\alpha) = (\pi/\alpha)^{1/2},\ E_W = x^2.$$

The posterior density can then be computed as

$$P(x|D) = f(x|t_1, t_2, \ldots, t_Q)$$

$$= \frac{f(t_1, t_2, \ldots, t_Q|x)f(x)}{f(t_1, t_2, \ldots, t_Q)}$$

$$= \frac{\frac{1}{Z_D(\beta)}\frac{1}{Z_W(\alpha)}\exp(-(\beta E_D + \alpha E_W))}{\text{Normalization Factor}}$$

To find the most probable value for x, we maximize the posterior density. This is equivalent to minimizing

$$\beta E_D + \alpha E_W = \beta\sum_{i=1}^{Q}(t_i - x)^2 + \alpha x^2.$$

To find the minimum, we take the derivative with respect to x and set it equal to zero:

$$\frac{d}{dx}(\beta E_D + \alpha E_W) = \frac{d}{dx}\left(\beta \sum_{i=1}^{Q} (t_i - x)^2 + \alpha x^2\right) = -2\beta \sum_{i=1}^{Q} (t_i - x) + 2\alpha x$$

$$= -2\beta\left[\left(\sum_{i=1}^{Q} t_i\right) - Qx\right] + 2\alpha x$$

$$= -2\left[\beta\left(\sum_{i=1}^{Q} t_i\right) - (\alpha + Q\beta)x\right] = 0$$

Solving for μ, we obtain

$$x^{MP} = \frac{\beta\left(\sum_{i=1}^{Q} t_i\right)}{\alpha + Q\beta}$$

Notice that as α goes to zero (variance σ_x^2 goes to infinity), x^{MP} approaches x^{ML}. Increasing the variance of the prior density represents increased uncertainty in our prior knowledge about x. With large prior uncertainty, we rely on the data for our estimate of x, which leads to the maximum likelihood estimate.

Figure P13.3 illustrates $P(D|x)$, $P(x)$ and $P(x|D)$ for the case where $\sigma_x^2 = 2$, $\sigma^2 = 1$, $Q = 1$ and $t_1 = 1$. Here the variance associated with the measurement is smaller than the variance associated with our prior density for x, so x^{MP} is closer to $x^{ML} = t_1 = 1$ than it is to the maximum of the prior density, which occurs at 0.

To experiment with this signal in noise example, use the MATLAB® Neural Network Design Demonstration Signal Plus Noise (**nnd13spn**).

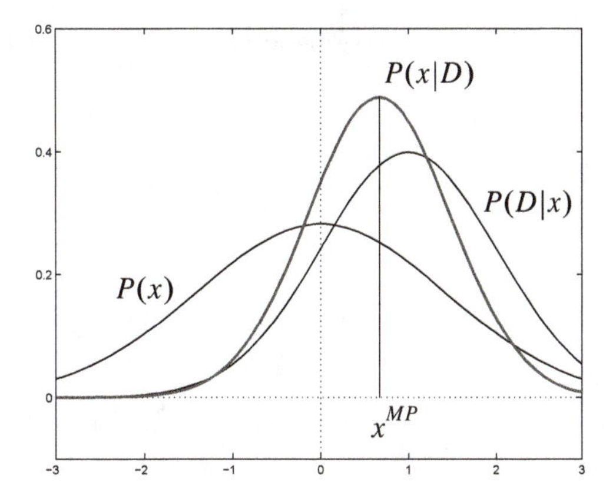

Figure P13.3 Prior and Posterior Density Functions

P13.3 Derive Eq. (13.23).

To solve for α^{MP} and β^{MP}, we will take the derivatives of the log of $P(D|\alpha, \beta, M)$, given in Eq. (13.17), with respect to α and β, and set the derivatives to zero. Taking the log of Eq. (13.17), and substituting Eq. (13.12), Eq. (13.14) and Eq. (13.22), we obtain

$$\log P(D|\alpha, \beta, M) = \log(Z_F) - \log(Z_D(\beta)) - \log(Z_W(\alpha))$$

$$= \frac{n}{2}\log(2\pi) - \frac{1}{2}\log\det(\mathbf{H}^{MP}) - F(\mathbf{x}^{MP}) - \frac{N}{2}\log\left(\frac{\pi}{\beta}\right) - \frac{n}{2}\log\left(\frac{\pi}{\alpha}\right)$$

$$= -F(\mathbf{x}^{MP}) - \frac{1}{2}\log\det(\mathbf{H}^{MP}) + \frac{N}{2}\log(\beta) + \frac{n}{2}\log(\alpha) + \frac{n}{2}\log 2 - \frac{N}{2}\log\pi$$

We will consider first the second term in this expression. Since $\mathbf{H}$ is the Hessian of F in Eq. (13.4), we can write it as $\mathbf{H} = \nabla^2 F = \nabla^2(\beta E_D) + \nabla^2(\alpha E_W) = \beta\mathbf{B} + 2\alpha\mathbf{I}$, where $\mathbf{B} = \nabla^2 E_D$. If we let λ^h be an eigenvalue of $\mathbf{H}$ and λ^b be an eigenvalue of $\beta\mathbf{B}$, then $\lambda^h = \lambda^b + 2\alpha$ for all corresponding eigenvalues. Now we take the derivative of the second term in the above equation with respect to α. Since the determinant of a matrix can be expressed as the product of its eigenvalues, we can reduce it as shown below, where $\mathrm{tr}(\mathbf{H}^{-1})$ is the trace of the inverse of the Hessian $\mathbf{H}$.

$$\frac{\partial}{\partial\alpha}\frac{1}{2}\log\det\mathbf{H} = \frac{1}{2\det\mathbf{H}}\frac{\partial}{\partial\alpha}\left(\prod_{k=1}^{n}\lambda^{h}\right)$$

$$= \frac{1}{2\det\mathbf{H}}\frac{\partial}{\partial\alpha}\left(\prod_{i=1}^{n}(\lambda_i^b + 2\alpha)\right)$$

$$= \frac{1}{2\det\mathbf{H}}\left[\sum_{i=1}^{n}\left(\prod_{j\neq i}(\lambda_j^b + 2\alpha)\right)\frac{\partial}{\partial\alpha}(\lambda_i^b + 2\alpha)\right]$$

$$= \frac{\sum_{i=1}^{n}\left(\prod_{j\neq i}(\lambda_j^b + 2\alpha)\right)}{\prod_{i=1}^{n}(\lambda_i^b + 2\alpha)}$$

$$= \sum_{i=1}^{n}\frac{1}{\lambda_i^b + 2\alpha} = \mathrm{tr}(\mathbf{H}^{-1})$$

Next, we will take the derivative of the same term with respect to β. First, define the parameter γ, as shown below, and expand it for use in our next step. The parameter γ is referred to as the effective number of parameters.

$$\gamma \equiv n - 2\alpha\mathrm{tr}(\mathbf{H}^{-1})$$

$$= n - 2\alpha\sum_{i=1}^{N}\frac{1}{\lambda_i^b + 2\alpha} = \sum_{i=1}^{n}\left(1 - \frac{2\alpha}{\lambda_i^b + 2\alpha}\right) = \sum_{i=1}^{n}\left(\frac{\lambda_i^b}{\lambda_i^b + 2\alpha}\right) = \sum_{i=1}^{n}\frac{\lambda_i^b}{\lambda_i^h}$$

Now take the derivative of $\frac{1}{2}\log\det(\mathbf{H}^{MP})$ with respect to β.

$$\frac{\partial}{\partial\beta}\frac{1}{2}\log\det\mathbf{H} = \frac{1}{2\det\mathbf{H}}\frac{\partial}{\partial\beta}\left(\prod_{k=1}^{n}\lambda_k^h\right)$$

$$= \frac{1}{2\det\mathbf{H}}\frac{\partial}{\partial\beta}\left(\prod_{i=1}^{n}(\lambda_i^b+2\alpha)\right)$$

$$= \frac{1}{2\det\mathbf{H}}\left[\sum_{i=1}^{n}\left(\prod_{j\neq i}(\lambda_j^b+2\alpha)\right)\frac{\partial}{\partial\beta}(\lambda_i^b+2\alpha)\right]$$

$$= \frac{1}{2}\frac{\sum_{i=1}^{n}\left(\prod_{j\neq i}(\lambda_j^b+2\alpha)\left(\frac{\lambda_i^b}{\beta}\right)\right)}{\prod_{i=1}^{n}(\lambda_i^b+2\alpha)}$$

$$= \frac{1}{2\beta}\sum_{i=1}^{n}\frac{\lambda_i^b}{\lambda_i^b+2\alpha} = \frac{\gamma}{2\beta}$$

where the fourth step is derived from the fact that λ_i^b is an eigenvalue of $\beta\mathbf{B}$, and therefore the derivative of λ_i^b with respect to β is just the eigenvalue of $\mathbf{B}$ which is λ_i^b/β.

Now we are finally ready to take the derivatives of all terms in $\log P(D|\alpha, \beta, M)$ and set them equal to zero. The derivative with respect to α will be

$$\frac{\partial}{\partial\alpha}\log P(D|\alpha,\beta,M) = -\frac{\partial}{\partial\alpha}F(\mathbf{w}^{MP}) - \frac{\partial}{\partial\alpha}\frac{1}{2}\log\det(\mathbf{H}^{MP}) + \frac{\partial}{\partial\alpha}\frac{n}{2}\log\alpha$$

$$= -\frac{\partial}{\partial\alpha}(\alpha E_W(\mathbf{w}^{MP})) - \mathrm{tr}(\mathbf{H}^{MP})^{-1} + \frac{n}{2\alpha^{MP}}$$

$$= -E_W(\mathbf{w}^{MP}) - \mathrm{tr}(\mathbf{H}^{MP})^{-1} + \frac{n}{2\alpha^{MP}} = 0$$

Rearranging terms, and using our definition of γ, we have

$$E_W(\mathbf{w}^{MP}) = \frac{n}{2\alpha^{MP}} - \mathrm{tr}(\mathbf{H}^{MP})^{-1}$$

$$2\alpha^{MP}E_W(\mathbf{w}^{MP}) = n - 2\alpha^{MP}\mathrm{tr}(\mathbf{H}^{MP})^{-1} = \gamma$$

$$\alpha^{MP} = \frac{\gamma}{2E_W(\mathbf{w}^{MP})}$$

We now repeat the process for β.

$$\begin{aligned}\frac{\partial}{\partial\beta}\log P(D|\alpha,\beta,M) &= \frac{\partial}{\partial\beta}F(\mathbf{w}^{MP}) - \frac{\partial}{\partial\beta}\frac{1}{2}\log\det(\mathbf{H}^{MP}) + \frac{\partial}{\partial\beta}\frac{N}{2}\log\beta \\ &= \frac{\partial}{\partial\beta}(\beta E_D(\mathbf{w}^{MP})) - \frac{\gamma}{2\beta^{MP}} + \frac{N}{2\beta^{MP}} \\ &= -E_D(\mathbf{w}^{MP}) - \frac{\gamma}{2\beta^{MP}} + \frac{N}{2\beta^{MP}} = 0\end{aligned}$$

Rearranging terms,

$$\begin{aligned}E_D(\mathbf{w}^{MP}) &= \frac{N}{2\beta^{MP}} - \frac{\gamma}{2\beta^{MP}} \\ \beta^{MP} &= \frac{N-\gamma}{2E_D(\mathbf{w}^{MP})}\end{aligned}$$

P13.4 Demonstrate that extrapolation can occur in a region that is surrounded by training data.

Consider the function displayed in Figure 13.3. In that example, extrapolation occurred in the upper left region of the input space, because all of the training data was in the lower right. Let's provide training data around the outside of the input space, but without data in the region

$$-1.5 < p_1 < 1.5 \qquad -1.5 < p_2 < 1.5 .$$

The training data is distributed as shown in Figure P13.4.

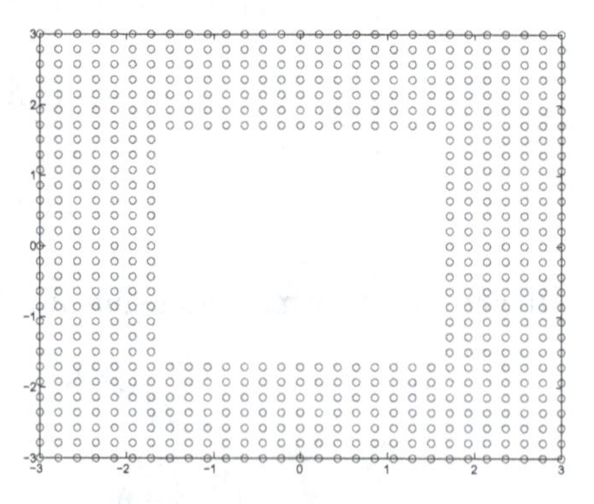

Figure P13.4 Training Data Locations

The result of the training is shown in Figure P13.5. The neural network approximation significantly overestimates the true function in the region without training data, even though surrounded by regions with training data. In addition, this result is random. With a different set of initial random weights, the network might underestimate the true function in this region. Extrapolation occurs because there is a significantly large region without training data. When the input space is of high dimension, it can be very difficult to tell when a network is extrapolating. It cannot be done by simply checking the individual ranges of each input variable.

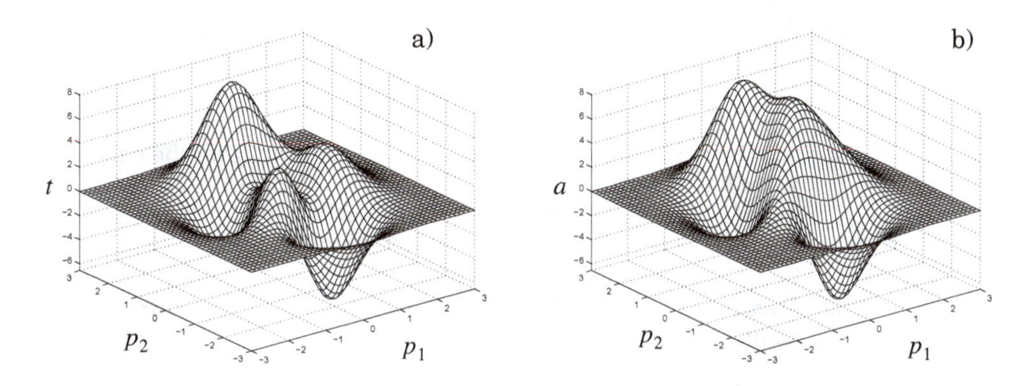

Figure P13.5 Function (a) and Neural Network Approximation (b)

P13.5 Consider the example starting on page 13-23. Find the effective number of parameters if $\rho = 1$.

To find the effective number of parameters, we can use Eq. (13.55):

$$\gamma = \sum_{i=1}^{n} \frac{\beta\lambda_i}{\beta\lambda_i + 2\alpha} .$$

We earlier found the eigenvalues to be $\lambda_1 = 1$, $\lambda_2 = 3$. The regularization parameter is

$$\rho = \frac{\alpha}{\beta} = 1 .$$

We can rewrite γ in terms of ρ as follows

$$\gamma = \sum_{i=1}^{n} \frac{\beta\lambda_i}{\beta\lambda_i + 2\alpha} = \sum_{i=1}^{n} \frac{\lambda_i}{\lambda_i + 2\frac{\alpha}{\beta}} = \sum_{i=1}^{n} \frac{\lambda_i}{\lambda_i + 2\rho} .$$

Substituting our numbers, we find

$$\gamma = \sum_{i=1}^{n} \frac{\lambda_i}{\lambda_i + 2\rho} = \frac{1}{1+2} + \frac{3}{3+2} = \frac{1}{3} + \frac{3}{5} = \frac{14}{15}.$$

Therefore, we are using approximately one of the two available parameters. The network has two parameters: $w_{1,1}$ and $w_{1,2}$. The parameter we are using is not one of these two, but rather a combination. As we can see from Figure 13.11, we move in the direction of the second eigenvector:

$$\mathbf{v}_2 = \begin{bmatrix} 1 \\ 1 \end{bmatrix},$$

which means that we are changing $w_{1,1}$ and $w_{1,2}$ by the same amount. Although there are two parameters, we are effectively using only one. Since $\mathbf{v}_2$ is the eigenvector with the largest eigenvalue, we move in that direction to obtain the greatest reduction in the squared error.

P13.6 Demonstrate overfitting with polynomials. Consider fitting a polynomial

$$g_k(p) = x_0 + x_1 p + x_2 p^2 + \ldots + x_k p^k$$

to a set of data $\{p_1, t_1\}, \{p_2, t_2\}, \ldots, \{p_Q, t_Q\}$ so as to minimize the following squared error performance function.

$$F(\mathbf{x}) = \sum_{q=1}^{Q} (t_q - g_k(p_q))^2$$

First, we want to express the problem in matrix form. Define the following vectors.

$$\mathbf{t} = \begin{bmatrix} t_1 \\ t_2 \\ \vdots \\ t_Q \end{bmatrix} \quad \mathbf{G} = \begin{bmatrix} 1 & p_1 & \cdots & p_1^k \\ 1 & p_2 & \cdots & p_2^k \\ \vdots & \vdots & & \vdots \\ 1 & p_Q & \cdots & p_Q^k \end{bmatrix} \quad \mathbf{x} = \begin{bmatrix} x_0 \\ x_1 \\ \vdots \\ x_k \end{bmatrix}$$

We can then write the performance index as follows.

$$F(\mathbf{x}) = [\mathbf{t} - \mathbf{G}\mathbf{x}]^T [\mathbf{t} - \mathbf{G}\mathbf{x}] = \mathbf{t}^T \mathbf{t} - 2\mathbf{x}^T \mathbf{G}^T t + \mathbf{x}^T \mathbf{G}^T \mathbf{G}\mathbf{x}$$

To locate the minimum, we take the gradient and set it equal to zero.

$$\nabla F(\mathbf{x}) = -2\mathbf{G}^T t + 2\mathbf{G}^T\mathbf{G}\mathbf{x} = 0$$

Solving for the weights, we obtain the least squares solution (maximum likelihood for the Gaussian noise case).

$$[\mathbf{G}^T\mathbf{G}]\mathbf{x}^{ML} = \mathbf{G}^T t \qquad \Rightarrow \qquad \mathbf{x}^{ML} = [\mathbf{G}^T\mathbf{G}]^{-1}\mathbf{G}^T t$$

To demonstrate the operation of the polynomial fitting, we will use the simple linear function $t = p$. To create the data set, we will sample the function at five different points and will add noise as follows

$$t_i = p_i + \varepsilon_i,\ p = \{-1, -0.5, 0, 0.5, 1\},$$

where ε_i has a uniform density with range $[-0.25, 0.25]$. The code below shows how to generate the data and fit a 4th order polynomial. The results of fitting 2nd and 4th order polynomials are shown in Figure P13.6. The 4th order polynomial has five parameters, which allow it to exactly fit the five noisy data points, but it doesn't produce an accurate approximation of the true function.

» 2 + 2
ans =
4

```
p = -1:.5:1;
t = p + 0.5*(rand(size(p))-0.5);
Q = length(p);
ord = 4;
G = ones(Q,1);
for i=1:ord,
      G = [G (p').^i];
end
x = (G'*G)\G'*t'; % Could also use x = G\t';
```

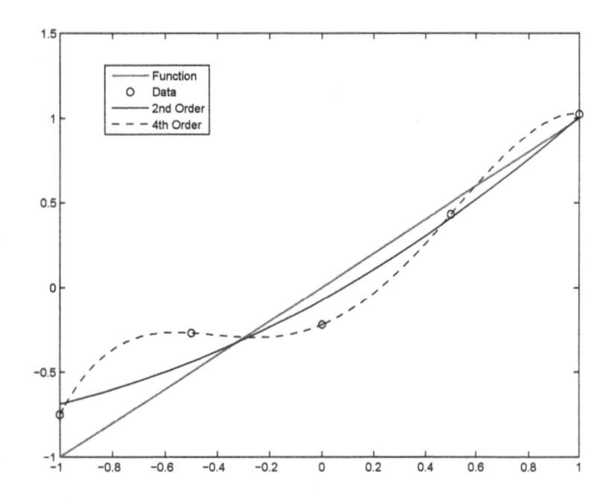

Figure P13.6 Polynomial Approximations to a Straight Line

Epilogue

The focus of this chapter has been the development of algorithms for training multilayer neural networks so that they generalize well. A network that generalizes well will perform as well in new situations as it performs on the data for which it was trained.

The basic approach to producing networks that generalize well is to find the simplest network that can represent the data. A simple network is one that has a small number of weights and biases.

The two methods that we presented in this chapter, early stopping and regularization, produce simple networks by constraining the weights, rather than by reducing the number of weights. We showed in this chapter that constraining the weights is equivalent to reducing the number of weights.

Chapter 18 presents a case study that uses Bayesian regularization to prevent overfitting in a practical function approximation problem. Chapter 20 presents a case study that uses early stopping to prevent overfitting in a practical pattern recognition problem.

Further Reading

[AmMu97] S. Amari, N. Murata, K.-R. Muller, M. Finke, and H. H. Yang, "Asymptotic Statistical Theory of Overtraining and Cross-Validation," *IEEE Transactions on Neural Networks*, vol. 8, no. 5, 1997.

When using early stopping, it is important to decide on the number of data points to place in the validation set. This paper provides a theoretical basis for the choice of validation set size.

[FoHa97] D. Foresee and M. Hagan, "Gauss-Newton Approximation to Bayesian Learning," *Proceedings of the 1997 International Joint Conference on Neural Networks*, vol. 3, pp. 1930 - 1935, 1997.

This paper describes a method for implementing Bayesian regularization by using the Gauss-Newton approximation to the Hessian matrix.

[GoLa98] C. Goutte and J. Larsen, "Adaptive Regularization of Neural Networks Using Conjugate Gradient," *Proceedings of the IEEE International Conference on Acoustics, Speech and Signal Processing*, vol. 2, pp. 1201-1204, 1998.

When using regularization, the important step is setting the regularization parameter. This paper describes a procedure for setting the regularization parameter to minimize the validation set error.

[MacK92] D. J. C. MacKay, "Bayesian Interpolation," *Neural Computation*, vol. 4, pp. 415-447, 1992.

Bayesian approaches have been used for many years in statistics. This paper presents one of the first developments of a Bayesian framework for training neural networks. MacKay followed this paper with many others describing refinements of the approach.

[Sarle95] W. S. Sarle, "Stopped training and other remedies for overfitting," In *Proceedings of the 27th Symposium on Interface*, 1995.

This is one of the early papers on the use of early stopping with a validation set to prevent overfitting. The paper describes simulation results comparing early stopping with other methods for improving generalization.

[SjLj94] J. Sjoberg and L. Ljung, "Overtraining, regularization and searching for minimum with application to neural networks," Linkoping University, Sweden, Tech. Rep. LiTH-ISY-R-1567, 1994.

This report explains how early stopping and regularization are approximately equivalent processes. It demonstrates that the number of iterations of training is inversely proportional to the regularization parameter.

[Tikh63] A. N. Tikhonov, "The solution of ill-posed problems and the regularization method," *Dokl. Acad. Nauk USSR*, vol. 151, no. 3, pp. 501-504, 1963.

Regularization is a method by which a squared error performance index is augmented by a penalty on the complexity of the approximating function. This is the original paper that introduced the concept of regularization. The penalty involved the derivatives of the approximating function.

[WaVe94] C. Wang, S. S. Venkatesh, and J. S. Judd, "Optimal Stopping and Effective Machine Complexity in Learning," *Advances in Neural Information Processing Systems*, J. D. Cowan, G. Tesauro, and J. Alspector, Eds., vol. 6, pp. 303-310, 1994.

This paper describes how the effective number of network parameters changes during the training process and how the generalization capability of the network can be improved by stopping the training early.

Exercises

E13.1 Consider fitting a polynomial (kth order)

$$g_k(p) = x_0 + x_1 p + x_2 p^2 + \ldots + x_k p^k$$

to a set of data $\{p_1, t_1\}$, $\{p_2, t_2\}$, ..., $\{p_Q, t_Q\}$. It has been proposed that minimizing a performance index that penalizes the derivatives of the polynomial will provide improved generalization. Investigate the relationship between this technique and regularization using squared weights.

i. Derive the least squares solution for the weights x_i, which minimizes the following squared error performance index. (See Solved Problem P13.6.)

$$F(\mathbf{x}) = \sum_{q=1}^{Q} (t_q - g_k(p_q))^2$$

ii. Derive the regularized least squares solution, with a squared weight penalty.

$$F(\mathbf{x}) = \sum_{q=1}^{Q} (t_q - g_k(p_q))^2 + \rho \sum_{i=0}^{k} x_i^2$$

iii. Derive a solution for the weights that minimizes a sum of the squared error plus a sum of squared derivatives.

$$F(\mathbf{x}) = \sum_{q=1}^{Q} (t_q - g_k(p_q))^2 + \rho \sum_{i=1}^{Q} \left[\frac{d}{dp} g_k(p_q) \right]^2$$

iv. Derive a solution for the weights that minimizes a sum of the squared error plus a sum of squared second derivatives.

$$F(\mathbf{x}) = \sum_{q=1}^{Q} (t_q - g_k(p_q))^2 + \rho \sum_{i=1}^{Q} \left[\frac{d^2}{dp^2} g_k(p_q) \right]^2$$

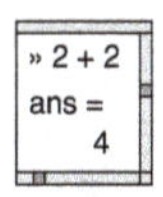

E13.2 Write a MATLAB program to implement the solutions you found in E13.1 i. through iv. Using the following data points, adjust the ρ values to obtain the best results. Use $k = 8$ for all cases. Plot the data points, the noise-free function ($t = p$) and the polynomial approximation in each case. Compare the four approximations. Which do you think produces the best results? Which cases produce similar results?

$$t_i = p_i + \varepsilon_i,\ p = \{-1, -0.5, 0, 0.5, 1\},$$

where ε_i has a uniform density with range $[-0.1, 0.1]$ (use the rand command in MATLAB).

E13.3 Consider fitting a polynomial (1st order), $g_1(p) = x_0 + x_1 p$, to the following data set:

$$\left\{p_1 = \begin{bmatrix}1\end{bmatrix}, t_1 = \begin{bmatrix}4\end{bmatrix}\right\}, \left\{p_2 = \begin{bmatrix}2\end{bmatrix}, t_2 = \begin{bmatrix}6\end{bmatrix}\right\}.$$

i. Find the least squares solutions for the weights x_0 and x_1 that minimize the following sum squared error performance index:

$$F(\mathbf{x}) = \sum_{q=1}^{2} (t_q - g_1(p_q))^2.$$

ii. Find the regularized least squares solution for the weights x_0 and x_1 when the following squared weight penalty is used:

$$F(\mathbf{x}) = \sum_{q=1}^{2} (t_q - g_1(p_q))^2 + \sum_{i=0}^{1} x_i^2.$$

E13.4 Investigate the extrapolation characteristics of neural networks and polynomials. Consider the problem described in E11.25, where a sine wave is fit over the range $-2 \le p \le 2$. Select 11 training points evenly spaced over this interval.

» 2 + 2
ans =
4

i. After fitting the 1-2-1 neural network over this range, plot the actual sine function and the neural network approximation over the range $-4 \le p \le 4$.

ii. Fit a fifth-order polynomial (which has the same number of free parameters as the 1-2-1 network) to the sine wave over the range $-2 \le p \le 2$ (using your results from E13.1 i.). Plot the actual function and the polynomial approximation over the range $-4 \le p \le 4$.

iii. Discuss the extrapolation characteristics of the neural network and the polynomial.

E13.5 Suppose that we have a random variable t that is distributed according to the following density function. We take a series of Q independent samples of the random variable. Find the maximum likelihood estimate of x (x^{ML}).

$$f(t|x) = \frac{t}{x^2}\exp\left(-\frac{t}{x}\right) \qquad t \geq 0$$

E13.6 For the random variable t given in E13.5, suppose that x is a random variable with the following prior density function. Find the most probable estimate of x (x^{MP}).

$$f(x) = \exp(-x) \qquad x \geq 0$$

E13.7 Repeat E13.6 for the following prior density function. Under what conditions will $x^{MP} = x^{ML}$?

$$f(x) = \frac{1}{\sqrt{2\pi}\sigma_x}\exp\left(-\frac{(x-\mu_x)^2}{2\sigma_x^2}\right)$$

E13.8 In the signal plus noise example given in Solved Problem P13.2, find x^{MP} for the following prior density functions.

i. $f(x) = \exp(-x) \qquad x \geq 0$

ii. $f(x) = \frac{1}{2}\exp(-|x-\mu|)$

E13.9 Suppose that we have a random variable t that is distributed according to the following density function. We take a series of $Q = 2$ independent samples of the random variable.

$$f(t|x) = \begin{cases} \exp(-(t-x)) & t \geq x \\ 0 & t < x \end{cases}$$

i. Find the likelihood function $f(t_1, t_2|x)$, and sketch versus x.

ii. Suppose that the two measurements are $t_1 = 1$ and $t_2 = 1$. Find the maximum likelihood estimate of x (x^{ML}).

For the random variable t above, suppose that x is a random variable with the following prior density function.

$$f(x) = \frac{1}{\sqrt{\pi/2}}\exp(-2x^2)$$

iii. Sketch the posterior density $f(x|t_1, t_2)$. (You do not need to compute the denominator, just find the general shape. Assume the same measurements from part ii.)

iv. Find the most probable estimate of x (x^{MP}).

E13.10 We have a coin that is not fair. (The probability of heads is not the same as the probability of tails.) We want to estimate the probability of heads (x).

i. If we flip the coin 10 times, the probability of getting exactly t heads, given that the probability of heads is x, is given below. Find the maximum likelihood estimate of x (x^{ML}). (Hint: Take the natural log of $p(t|x)$ before finding the maximum.) Is it reasonable? Explain.

$$p(t|x) = \binom{10}{t} x^t (1-x)^{(10-t)}, \text{ where } \binom{10}{t} = \frac{10!}{t!(10-t)}$$

ii. Assume that the probability of heads, x, is a random variable with the following prior density function. Find the most probable estimate of x (x^{MP}). (Hint: Take the natural log of $p(t|x)p(x)$ before finding the maximum.) Explain why x^{ML} is different than x^{MP}.

$$p(x) = 12x^2(1-x),\ 0 \le x \le 1 .$$

E13.11 Suppose that the prior density in the level I Bayesian analysis (see page 13-12) has nonzero mean, μ_x. Find the new performance index.

E13.12 Suppose that we have the following inputs and targets:

$$\left\{\mathbf{p}_1 = \begin{bmatrix} 2 \\ 1 \end{bmatrix}, t_1 = 1\right\}, \left\{\mathbf{p}_2 = \begin{bmatrix} -2 \\ -1 \end{bmatrix}, t_2 = 3\right\}$$

We want to train a single-layer linear network without a bias on this training set. Assume that each input vector occurs with equal probability. We will train the network using the regularized performance index of Eq. (13.34).

i. Find $\mathbf{x}^{MP}$ and the effective number of parameters γ, if $\rho = 1$.

ii. Find $\mathbf{x}^{MP}$ and γ as $\rho \to \infty$. Explain the difference between $\mathbf{x}^{MP}$ and $\mathbf{x}^{ML}$. (Your answer should be specific to this problem—not a general discussion of the difference between $\mathbf{x}^{MP}$ and $\mathbf{x}^{ML}$.)

E13.13 Suppose we have the following input pattern and target:

$$\left\{\mathbf{p}_1 = \begin{bmatrix} 2 \\ 1 \end{bmatrix}, t_1 = 1\right\}.$$

This pattern is used to train a single layer, linear network with no bias.

i. The network is to be trained with a regularized performance index, with the regularization parameter set to $\rho = \alpha/\beta = 1/2$. Find an expression for the regularized performance index.

ii. Find x^{MP} and the effective number of parameters γ.

iii. Sketch the contour plot of the regularized performance index.

iv. Find the maximum stable learning rate if steepest descent is used to train the network.

v. Find the initial direction for the steepest descent algorithm trajectory, if both initial weights are set to zero.

vi. Sketch an approximate complete trajectory (on your contour plot from part iii.) for the steepest algorithm, with very small learning rate, from the initial conditions where both weights are set to zero. Explain your procedure for sketching the trajectory.

E13.14 Suppose that we have a single layer, linear network with no bias. The input/target pairs of the training set are given by

$$\left\{\mathbf{p}_1 = \begin{bmatrix} -1 \\ 1 \end{bmatrix}, t_1 = -2\right\}, \left\{\mathbf{p}_2 = \begin{bmatrix} 1 \\ 2 \end{bmatrix}, t_2 = 2\right\}, \left\{\mathbf{p}_3 = \begin{bmatrix} 2 \\ 1 \end{bmatrix}, t_3 = 4\right\},$$

where each pair occurs with equal probability. We want to minimize the regularized performance index of Eq. (13.34).

i. Find the effective number of parameters γ, for $\rho = 1$.

ii. Starting with zero initial weights, approximately how many iterations of the steepest descent algorithm would need to be made on the mean square performance index E_D to produce results that would be equivalent to minimizing the regularized performance index with $\rho = 1$? Assume a learning rate of $\alpha = 0.01$.

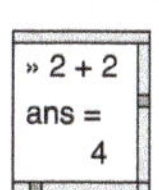

E13.15 Repeat E11.25, but modify your program to use early stopping and to use 30 neurons. Select 10 training points and 5 validation points. Add noise to the validation and testing points that is uniformly distributed between −0.1 and 0.1 (using the MATLAB function `rand`). Measure the mean square error of the trained network on a testing set consisting of 20 equally-spaced points of the noise-free function. Try 10 different random sets of training and validation data. Compare the results with early-stopping with the results without early stopping.

E13.16 Repeat E13.15, but use regularization instead of early stopping. This will require modifying your program to compute the gradient of the regularized performance index. Add the standard gradient of the squared error, which is computed by the standard backpropagation algorithm, to the gradient of ρ times the squared weights. Try three different values of ρ. Compare these results with the early stopping results.

E13.17 Consider again the problem described in E10.4

i. Find the regularized performance index for $\rho = 0, 1, \infty$. Sketch the contour plot in each case. Indicate the location of the optimal weights in each case.

ii. Find the effective number of parameters for $\rho = 0, 1, \infty$.

iii. Starting with zero initial weights, approximately how many iterations of the steepest descent algorithm would need to be made on the mean square performance index to produce results that would be equivalent to minimizing the regularized performance index with $\rho = 1$? Assume a learning rate of $\alpha = 0.01$.

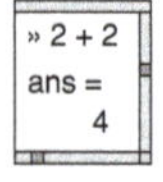

iv. Write a MATLAB M-file to implement the steepest descent algorithm to minimize the mean square error performance index that you found in part i. (This is a quadratic function.) Start the algorithm with zero initial conditions, and use a learning rate of $\alpha = 0.01$. Sketch the trajectory on a contour plot of the mean square error (the contour plot was found in E10.4). Verify that at the iteration you computed in part iii., the weights are close to the same values you found to minimize the regularized performance index with $\rho = 1$ in part i.

14 Dynamic Networks

Objectives

Neural networks can be classified into static and dynamic categories. The multilayer network that we have discussed in the last three chapters is a static network. This means that the output can be calculated directly from the input through feedforward connections. In dynamic networks, the output depends not only on the current input to the network, but also on the current or previous inputs, outputs or states of the network. For example, the adaptive filter networks we discussed in Chapter 10 are dynamic networks, since the output is computed from a tapped delay line of previous inputs. The Hopfield network we discussed in Chapter 3 is also a dynamic network. It has recurrent (feedback) connections, which means that the current output is a function of outputs at previous times.

We will begin this chapter with a brief introduction to the operation of dynamic networks, and then we will describe how these types of networks can be trained. The training will be based on optimization algorithms that use gradients (as in steepest descent and conjugate gradient algorithms) or Jacobians (as in Gauss-Newton and Levenberg-Marquardt algorithms) These algorithms were described in Chapter 10, 11 and 12 for static networks. The difference between the training of static and dynamic networks is in the manner in which the gradient or Jacobian is computed. In this chapter, we will present methods for computing gradients for dynamic networks.

Theory and Examples

Dynamic Networks

Dynamic networks are networks that contain delays (or integrators, for continuous-time networks) and that operate on a sequence of inputs. (In other words, the ordering of the inputs is important to the operation of the network.) These dynamic networks can have purely feedforward connections, like the adaptive filters of Chapter 10, or they can also have some feedback (*recurrent*) connections, like the Hopfield network of Chapter 3. Dynamic networks have memory. Their response at any given time will depend not only on the current input, but on the history of the input sequence.

Recurrent

Because dynamic networks have memory, they can be trained to learn sequential or time-varying patterns. Instead of approximating functions, like the static multilayer perceptron network of Chapter 11, a dynmic network can approximate a dynamic system. This has applications in such diverse areas as control of dynamic systems, prediction in financial markets, channel equalization in communication systems, phase detection in power systems, sorting, fault detection, speech recognition, learning of grammars in natural languages, and even the prediction of protein structure in genetics.

Dynamic networks can be trained using the standard optimization methods that we have discussed in Chapter 9 through 12. However, the gradients and Jacobians that are required for these methods cannot be computed using the standard backpropagation algorithm. In this chapter we will present the dynamic backpropagation algorithms that are required for computing the gradients for dynamic networks.

There are two general approaches (with many variations) to gradient and Jacobian calculations in dynamic networks: backpropagation-through-time (BPTT) [Werb90] and real-time recurrent learning (RTRL) [WiZi89]. In the BPTT algorithm, the network response is computed for all time points, and then the gradient is computed by starting at the last time point and working backward in time. This algorithm is efficient for the gradient calculation, but it is difficult to implement on-line, because the algorithm works backward in time from the last time step.

In the RTRL algorithm, the gradient can be computed at the same time as the network response, since it is computed by starting at the first time point, and then working forward through time. RTRL requires more calculations than BPTT for calculating the gradient, but RTRL allows a convenient framework for on-line implementation. For Jacobian calculations, the RTRL algorithm is generally more efficient than the BPTT algorithm.

In order to more easily present general BPTT and RTRL algorithms, it will be helpful to introduce modified notation for networks that can have recurrent connections. In the next section we will introduce this notation, and then the remainder of the chapter will present general BPTT and RTRL algorithms for dynamic networks.

Layered Digital Dynamic Networks

In this section we want to introduce the neural network framework that we will use to represent general dynamic networks. We call this framework Layered Digital Dynamic Networks (LDDN). It is an extension of the notation that we have used to represent static multilayer networks. With this new notation, we can conveniently represent networks with multiple recurrent (feedback) connections and tapped delay lines.

LDDN

To help us introduce the LDDN notation, consider the example dynamic network given in Figure 14.1.

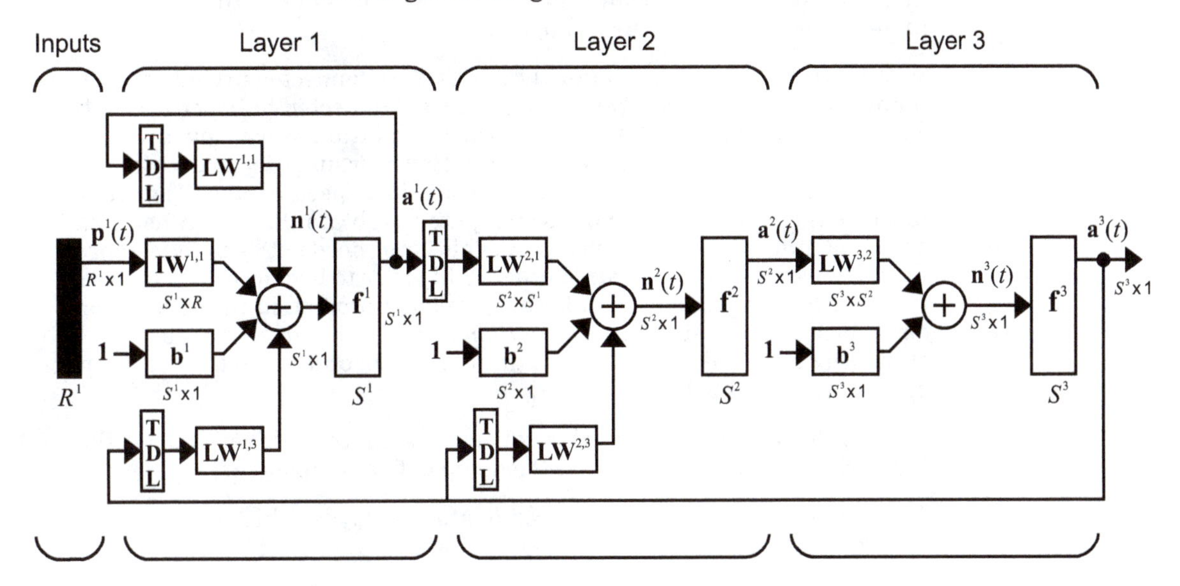

Figure 14.1 Example Dynamic Network

The general equations for the computation of the net input $\mathbf{n}^m(t)$ for layer m of an LDDN are

$$\mathbf{n}^m(t) = \sum_{l \in L_m^f} \sum_{d \in DL_{m,l}} \mathbf{LW}^{m,l}(d)\mathbf{a}^l(t-d) + \sum_{l \in I_m} \sum_{d \in DI_{m,l}} \mathbf{IW}^{m,l}(d)\mathbf{p}^l(t-d) + \mathbf{b}^m \qquad (14.1)$$

where $\mathbf{p}^l(t)$ is the lth input vector to the network at time t, $\mathbf{IW}^{m,l}$ is the *input weight* between input l and layer m, $\mathbf{LW}^{m,l}$ is the *layer weight* between layer l and layer m, $\mathbf{b}^m$ is the bias vector for layer m, $DL_{m,l}$ is the set of all delays in the tapped delay line between Layer l and Layer m, $DI_{m,l}$ is the set of all delays in the tapped delay line between Input l and Layer m,

Input Weight

Layer Weight

I_m is the set of indices of input vectors that connect to layer m, and L_m^f is the set of indices of layers that directly connect *forward* to layer m. The output of layer m is then computed as

$$\mathbf{a}^m(t) = \mathbf{f}^m(\mathbf{n}^m(t)) . \qquad (14.2)$$

Compare this with the static multilayer network of Eq. (11.6). LDDN networks can have several layers connecting to layer m. Some of the connections can be recurrent through tapped delay lines. An LDDN can also have multiple input vectors, and the input vectors can be connected to any layer in the network; for static multilayer networks, we assumed that the single input vector connected only to Layer 1.

With static multilayer networks, the layers were connected to each other in numerical order. In other words, Layer 1 was connected to Layer 2, which was connected to Layer 3, etc. Within the LDDN framework, any layer can connect to any other layer, even to itself. However, in order to use Eq. (14.1), we need to compute the layer outputs in a specific order. The order in which the layer outputs must be computed to obtain the correct network output is called the *simulation order*. (This order need not be unique; there may be several valid simulation orders.) In order to backpropagate the derivatives for the gradient calculations, we must proceed in the opposite order, which is called the *backpropagation order*. In Figure 14.1, the standard numerical order, 1-2-3, is the simulation order, and the backpropagation order is 3-2-1.

Simulation Order

Backpropagation Order

As with the multilayer network, the fundamental unit of the LDDN is the layer. Each layer in the LDDN is made up of five components:

1. a set of weight matrices that come into that layer (which may connect from other layers or from external inputs),
2. any tapped delay lines (represented by $DL_{m,l}$ or $DI_{m,l}$) that appear at the input of a set of weight matrices (Any set of weight matrices can be preceded by a TDL. For example, Layer 1 of Figure 14.1 contains the weights $\mathbf{LW}^{1,3}(d)$ and the corresponding TDL.),
3. a bias vector,
4. a summing junction, and
5. a transfer function.

The output of the LDDN is a function not only of the weights, biases, and current network inputs, but also of some layer outputs at previous points in time. For this reason, it is not a simple matter to calculate the gradient of the network output with respect to the weights and biases. The weights and biases have two different effects on the network output. The first is the direct effect, which can be calculated using the standard backpropagation algorithm from Chapter 11. The second is an indirect effect, since some of the inputs to the network are previous outputs, which are also functions of the weights and biases. The main development of the next two sections is a general gradient calculation for arbitrary LDDNs.

Example Dynamic Networks

Before we introduce dynamic training, let's get a feeling for the types of responses we can expect to see from dynamic networks. Consider first the feedforward dynamic network shown in Figure 14.2.

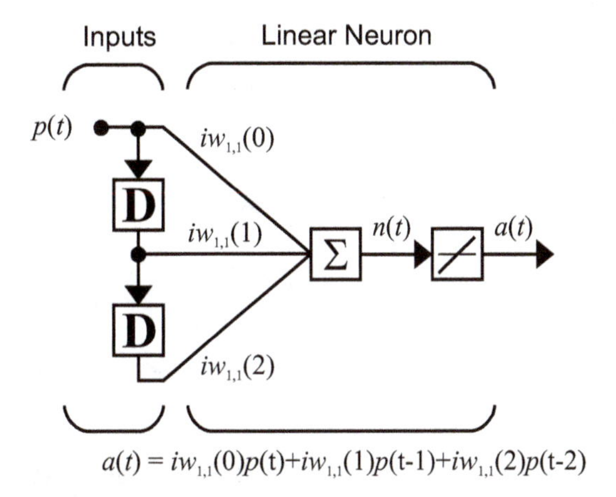

Figure 14.2 Example Feedforward Dynamic Network

This is an ADALINE filter, which we discussed in Chapter 10 (see Figure 10.5). Here we are representing it in the LDDN framework. The network has a TDL on the input, with $DI_{1,1} = \{0, 1, 2\}$. To demonstrate the operation of this network, we will apply a square wave as input, and we will set all of the weight values equal to 1/3:

$$iw_{1,1}(0) = \frac{1}{3},\ iw_{1,1}(1) = \frac{1}{3},\ iw_{1,1}(2) = \frac{1}{3}. \quad (14.3)$$

The network response is calculated from:

$$\begin{aligned}\mathbf{a}(t) = \mathbf{n}(t) &= \sum_{d=0}^{2} \mathbf{IW}(d)\mathbf{p}(t-d) \\ &= n_1(t) = iw_{1,1}(0)p(t) + iw_{1,1}(1)p(t-1) + iw_{1,1}(2)p(t-2)\end{aligned} \quad (14.4)$$

where we have left off the superscripts on the weight and the input, since there is only one input and only one layer.

The response of the network is shown in Figure 14.3. The open circles represent the square-wave input signal $p(t)$. The dots represent the network response $a(t)$. For this dynamic network, the response at any time point depends on the previous three input values. If the input is constant, the output will become constant after three time steps. This type of linear network is called a Finite Impulse Response (FIR) filter.

FIR

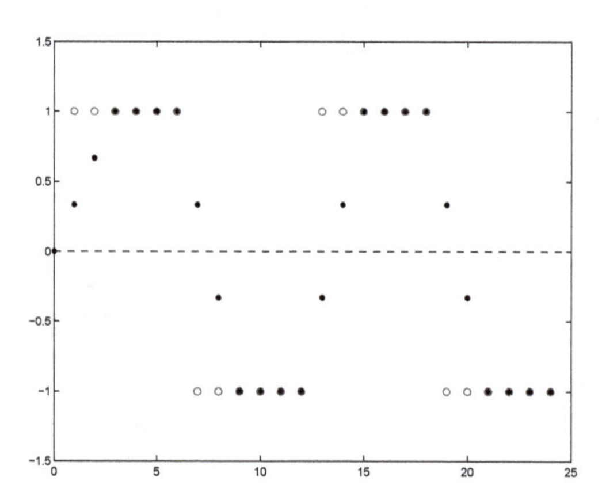

Figure 14.3 Response of ADALINE Filter Network

This dynamic network has memory. Its response at any given time will depend not only on the current input, but on the history of the input sequence. If the network does not have any feedback connections, then only a finite amount of history will affect the response. In the next example, we will consider a network that has an infinite memory.

To experiment with the finite impulse response example, use the Neural Network Design Demonstration Finite Impulse Response Network (`nnd14fir`).

Now consider another simple linear dynamic network, but one that has a recurrent connection. The network in Figure 14.4 is a recurrent dynamic network. The equation of operation of the network is

$$\begin{aligned} \mathbf{a}^1(t) &= \mathbf{n}^1(t) = \mathbf{LW}^{1,1}(1)\mathbf{a}^1(t-1) + \mathbf{IW}^{1,1}(0)\mathbf{p}^1(t) \\ &= lw_{1,1}(1)a(t-1) + iw_{1,1}p(t) \end{aligned} \quad (14.5)$$

where, in the last line, we have left off the superscripts, since there is only one neuron and one layer in the network. To demonstrate the operation of this network, we will set the weight values to

$$lw_{1,1}(1) = \frac{1}{2} \text{ and } iw_{1,1} = \frac{1}{2}. \quad (14.6)$$

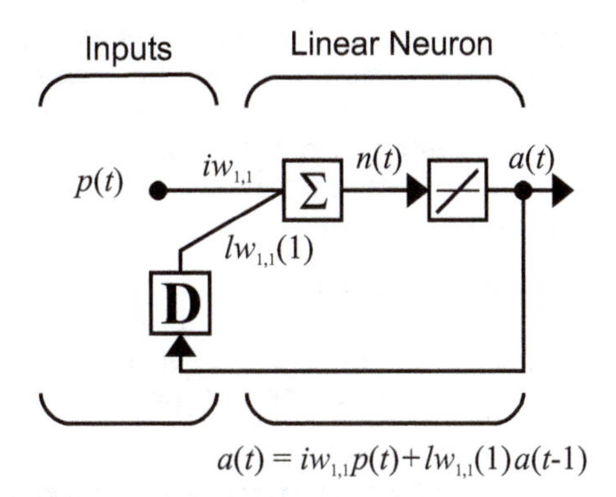

Figure 14.4 Recurrent Linear Neuron

The response of this network to the square wave input is shown in Figure 14.5. The network responds exponentially to a change in the input sequence. Unlike the FIR filter network of Figure 14.2, the exact response of the network at any given time is a function of the infinite history of inputs to the network.

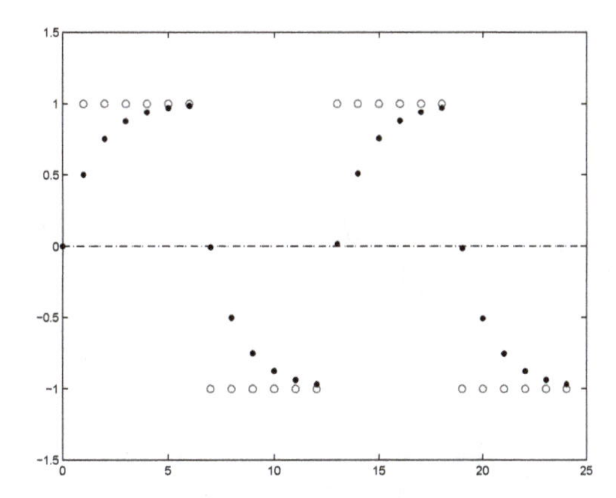

Figure 14.5 Recurrent Neuron Response

To experiment with this infinite impulse response example, use the Neural Network Design Demonstration Infinite Impulse Response Network (`nnd14iir`).

Compare the dynamic networks of the previous two examples with the static, two-layer perceptron of Figure 11.4. Static networks can be trained to approximate static functions, like $\sin(p)$, where the output can be computed directly from the current input. Dynamic networks, on the other hand, can be trained to approximate dynamic systems, such as robot arms, aircraft, biological processes and economic systems, where the current system output depends on a history of previous inputs and outputs. Because dynamic systems are more complex than static functions, we expect that the training process for dynamic networks will be more challenging than static network training.

In the following section, we will discuss the computation of gradients for the training of dynamic networks. For static networks, these gradients were computed using the standard backpropagation algorithm. For dynamic networks, the backpropagation algorithm must be modified.

Principles of Dynamic Learning

Before we get into the details of training dynamic networks, let's first investigate a simple example. Consider again the recurrent network of Figure 14.4. Suppose that we want to train the network using steepest descent. The first step is to compute the gradient of the performance function. For this example we will use sum squared error:

$$F(\mathbf{x}) = \sum_{t=1}^{Q} e^2(t) = \sum_{t=1}^{Q} (t(t) - a(t))^2 . \tag{14.7}$$

The two elements of the gradient will be

$$\frac{\partial F(\mathbf{x})}{\partial lw_{1,1}(1)} = \sum_{t=1}^{Q} \frac{\partial e^2(t)}{\partial lw_{1,1}(1)} = -2\sum_{t=1}^{Q} e(t)\frac{\partial a(t)}{\partial lw_{1,1}(1)} , \tag{14.8}$$

$$\frac{\partial F(\mathbf{x})}{\partial iw_{1,1}} = \sum_{t=1}^{Q} \frac{\partial e^2(t)}{\partial iw_{1,1}} = -2\sum_{t=1}^{Q} e(t)\frac{\partial a(t)}{\partial iw_{1,1}} \tag{14.9}$$

The key terms in these equations are the derivatives of the network output with respect to the weights:

$$\frac{\partial a(t)}{\partial lw_{1,1}(1)} \text{ and } \frac{\partial a(t)}{\partial iw_{1,1}} . \tag{14.10}$$

If we had a static network, then these terms would be very easy to compute. They would correspond to $a(t-1)$ and $p(t)$, respectively. However, for recurrent networks, the weights have two effects on the network output. The first is the direct effect, which is also seen in the corresponding static network. The second is an indirect effect, caused by the fact that one of the network inputs is a previous network output. Let's compute the derivatives of the network output, in order to demonstrate these two effects.

The equation of operation of the network is

$$a(t) = lw_{1,1}(1)a(t-1) + iw_{1,1}p(t). \tag{14.11}$$

We can compute the terms in Eq. (14.10) by taking the derivatives of Eq. (14.11):

$$\frac{\partial a(t)}{\partial lw_{1,1}(1)} = a(t-1) + lw_{1,1}(1)\frac{\partial a(t-1)}{\partial lw_{1,1}(1)}, \tag{14.12}$$

$$\frac{\partial a(t)}{\partial iw_{1,1}} = p(t) + lw_{1,1}(1)\frac{\partial a(t-1)}{\partial iw_{1,1}}. \tag{14.13}$$

The first term in each of these equations represents the direct effect that each weight has on the network output. The second term represents the indirect effect. Note that unlike the gradient computation for static networks, the derivative at each time point depends on the derivative at previous time points (or at future time points, as we will see later).

The following figures illustrate the dynamic derivatives. In Figure 14.6 a) we see the total derivatives $\partial a(t)/\partial iw_{1,1}$ and also the static portions of the derivatives. Note that if we consider only the static portion, we will underestimate the effect of a change in the weight. In Figure 14.6 b) we see the original response of the network (which was also shown in Figure 14.5) and a new response, in which $iw_{1,1}$ is increased from 0.5 to 0.6. By comparing the two parts of Figure 14.6, we can see how the derivative indicates the effect on the network response of a change in the weight $iw_{1,1}$.

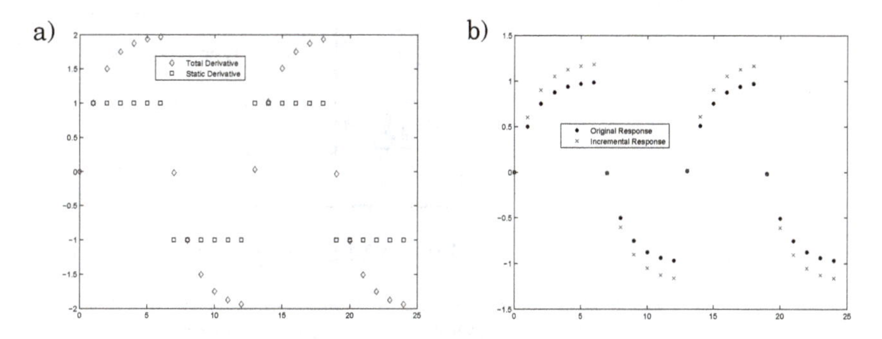

Figure 14.6 Derivatives for $iw_{1,1}$ and Response of Network in Figure 14.4

In Figure 14.7 we see similar results for the weight $lw_{1,1}(1)$. The key ideas to get from this example are: 1) the derivatives have static and dynamic components, and 2) the dynamic component depends on other time points.

To experiment with dynamic derivatives, use the Neural Network Design Demonstration Dynamic Derivatives (**nnd14dynd**).

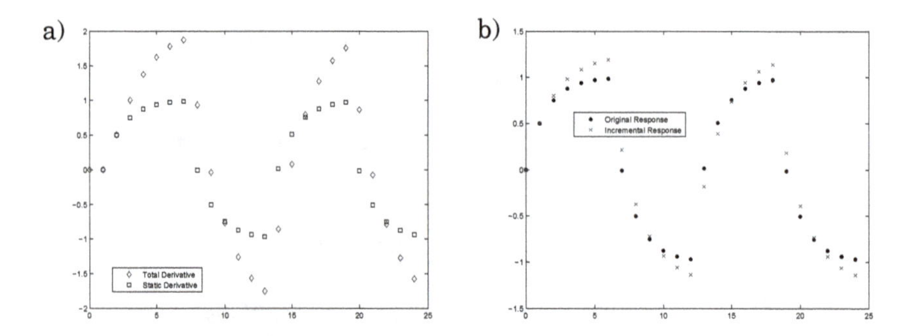

Figure 14.7 Derivatives for $lw_{1,1}(1)$ and Response of Network in Figure 14.4

Having made this initial investigation of a single-neuron network, let's consider the slightly more complex dynamic network that is shown in Figure 14.8. It consists of a static multilayer network with a single feedback loop added from the output of the network to the input of the network through a single delay. In this figure, the vector $\mathbf{x}$ represents all of the network parameters (weights and biases), and the vector $\mathbf{a}(t)$ represents the output of the multilayer network at time step t. This network will help us demonstrate the key steps of dynamic training.

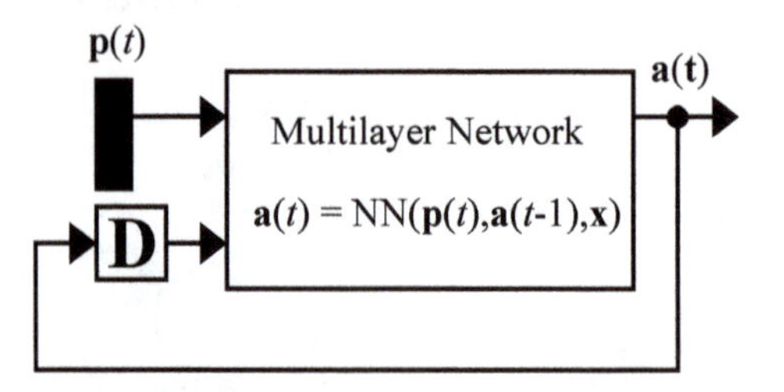

Figure 14.8 Simple Dynamic Network

As with a standard multilayer network, we want to adjust the weights and biases of the network to minimize the performance index, $F(\mathbf{x})$, which is normally chosen to be the mean squared error. In Chapter 11, we derived the backpropagation algorithm for computing the gradient of $F(\mathbf{x})$, which we could then use with any of the optimization methods from Chapter 12 to minimize $F(\mathbf{x})$. With dynamic networks, we need to modify the standard backpropagation algorithm. There are two different approaches to this problem. They both use the chain rule, but are implemented in different ways:

$$\frac{\partial F}{\partial \mathbf{x}} = \sum_{t=1}^{Q} \left[\frac{\partial \mathbf{a}(t)}{\partial \mathbf{x}^T}\right]^T \times \frac{\partial^e F}{\partial \mathbf{a}(t)}, \tag{14.14}$$

or

$$\frac{\partial F}{\partial \mathbf{x}} = \sum_{t=1}^{Q} \left[\frac{\partial^e \mathbf{a}(t)}{\partial \mathbf{x}^T}\right]^T \times \frac{\partial F}{\partial \mathbf{a}(t)}. \tag{14.15}$$

where the superscript e indicates an explicit derivative, not accounting for indirect effects through time. The explicit derivatives can be obtained with the standard backpropagation algorithm of Chapter 11. To find the complete derivatives that are required in Eq. (14.14) and Eq. (14.15), we need the additional equations:

$$\frac{\partial \mathbf{a}(t)}{\partial \mathbf{x}^T} = \frac{\partial^e \mathbf{a}(t)}{\partial \mathbf{x}^T} + \frac{\partial^e \mathbf{a}(t)}{\partial \mathbf{a}^T(t-1)} \times \frac{\partial \mathbf{a}(t-1)}{\partial \mathbf{x}^T} \tag{14.16}$$

and

$$\frac{\partial F}{\partial \mathbf{a}(t)} = \frac{\partial^e F}{\partial \mathbf{a}(t)} + \frac{\partial^e \mathbf{a}(t+1)}{\partial \mathbf{a}^T(t)} \times \frac{\partial F}{\partial \mathbf{a}(t+1)}. \tag{14.17}$$

RTRL

Eq. (14.14) and Eq. (14.16) make up the real-time recurrent learning (RTRL) algorithm. Note that the key term is

$$\frac{\partial \mathbf{a}(t)}{\partial \mathbf{x}^T}, \tag{14.18}$$

BPTT

which must be propagated forward through time. Eq. (14.15) and Eq. (14.17) make up the backpropagation-through-time (BPTT) algorithm. Here the key term is

$$\frac{\partial F}{\partial \mathbf{a}(t)} \tag{14.19}$$

which must be propagated backward through time.

In general, the RTRL algorithm requires somewhat more computation than the BPTT algorithm to compute the gradient. However, the BPTT algorithm cannot be conveniently implemented in real time, since the outputs must be computed for all time steps, and then the derivatives must be backpropagated back to the initial time point. The RTRL algorithm is well suited for real time implementation, since the derivatives can be calculated at each time step. (For Jacobian calculations, which are needed for Levenberg-Marquardt algorithms, the RTRL algorithm is often more efficient than the BPTT algorithm. See [DeHa07].)

Dynamic Backpropagation

In this section, we will develop general RTRL and BPTT algorithms for dynamic networks represented in the LDDN framework. This development will involve generalizing Eq. (14.14) through Eq. (14.17).

Preliminary Definitions

Input Layer

Output Layer

In order to simplify the description of the training algorithm, some layers of the LDDN will be assigned as network outputs, and some will be assigned as network inputs. A layer is an *input layer* if it has an input weight, or if it contains any delays with any of its weight matrices. A layer is an *output layer* if its output will be compared to a target during training, or if it is connected to an input layer through a matrix that has any delays associated with it.

For example, the LDDN shown in Figure 14.1 has two output layers (1 and 3) and two input layers (1 and 2). For this network the simulation order is 1-2-3, and the backpropagation order is 3-2-1. As an aid in later derivations, we will define U as the set of all output layer numbers and X as the set of all input layer numbers. For the LDDN in Figure 14.1, U={1,3} and X={1,2}.

The general equations for simulating an arbitrary LDDN network are given in Eq. (14.1) and Eq. (14.2). At each time point, these equations are iterated forward through the layers, as m is incremented through the simulation order. Time is then incremented from $t = 1$ to $t = Q$.

Real Time Recurrent Learning

In this subsection we will generalize the RTRL algorithm, given in Eq. (14.14) and Eq. (14.16), for LDDN networks. This development will follow in many respects the development of the backpropagation algorithm for static multilayer networks in Chapter 11. You may want to quickly review that material before proceeding.

Eq. (14.14)

The first step in developing the RTRL algorithm is to generalize Eq. (14.14). For the general LDDN network, we can calculate the terms of the gradient by using the chain rule, as in

$$\frac{\partial F}{\partial \mathbf{x}} = \sum_{t=1}^{Q} \sum_{u \in U} \left[\left[\frac{\partial \mathbf{a}^u(t)}{\partial \mathbf{x}^T} \right]^T \times \frac{\partial^e F}{\partial \mathbf{a}^u(t)} \right]. \tag{14.20}$$

If we compare this equation with Eq. (14.14), we notice that in addition to each time step, we also have a term in the sum for each output layer. However, if the performance index $F(\mathbf{x})$ is not explicitly a function of a specific output $\mathbf{a}^u(t)$, then that explicit derivative will be zero.

Eq. (14.16)

The next step of the development of the RTRL algorithm is the generalization of Eq. (14.16). Again, we use the chain rule:

$$\frac{\partial \mathbf{a}^{u}(t)}{\partial \mathbf{x}^{T}} = \frac{\partial^{e} \mathbf{a}^{u}(t)}{\partial \mathbf{x}^{T}} + \sum_{u' \in U} \sum_{x \in X} \sum_{d \in DL_{x,u'}} \frac{\partial^{e} \mathbf{a}^{u}(t)}{\partial \mathbf{n}^{x}(t)^{T}} \times \frac{\partial^{e} \mathbf{n}^{x}(t)}{\partial \mathbf{a}^{u'}(t-d)^{T}} \times \frac{\partial \mathbf{a}^{u'}(t-d)}{\partial \mathbf{x}^{T}} . \quad (14.21)$$

In Eq. (14.16) we only had one delay in the system. Now we need to account for each output and also for the number of times each output is delayed before it is input to another layer. That is the reason for the first two summations in Eq. (14.21). These equations must be updated forward in time, as t is varied from 1 to Q. The terms

$$\frac{\partial \mathbf{a}^{u}(t)}{\partial \mathbf{x}^{T}} \quad (14.22)$$

are generally set to zero for $t \le 0$.

To implement Eq. (14.21), we need to compute the terms

$$\frac{\partial^{e} \mathbf{a}^{u}(t)}{\partial \mathbf{n}^{x}(t)^{T}} \times \frac{\partial^{e} \mathbf{n}^{x}(t)}{\partial \mathbf{a}^{u'}(t-d)^{T}} . \quad (14.23)$$

To find the second term on the right, we can use

$$n_k^x(t) = \sum_{l \in L_x^f} \sum_{d' \in DL_{x,l}} \left[\sum_{i=1}^{S^l} lw_{k,i}^{x,l}(d') a_i^l(t-d') \right] + \sum_{l \in I_x} \sum_{d' \in DI_{x,l}} \left[\sum_{i=1}^{R^l} iw_{k,i}^{x,l}(d') p_i^l(t-d') \right] + b_k^x \quad (14.24)$$

(Compare with Eq. (11.20).) We can now write

$$\frac{\partial^{e} n_k^x(t)}{\partial a_j^{u'}(t-d)} = lw_{k,j}^{x,u'}(d) . \quad (14.25)$$

If we define the following sensitivity term

$$s_{k,i}^{u,m}(t) \equiv \frac{\partial^{e} a_k^u(t)}{\partial n_i^m(t)} , \quad (14.26)$$

which can be used to make up the following matrix

$$\mathbf{S}^{u,m}(t) = \frac{\partial^e \mathbf{a}^u(t)}{\partial \mathbf{n}^m(t)^T} = \begin{bmatrix} s_{1,1}^{u,m}(t) & s_{1,2}^{u,m}(t) & \dots & s_{1,S_m}^{u,m}(t) \\ s_{2,1}^{u,m}(t) & s_{2,2}^{u,m}(t) & \dots & s_{1,S_m}^{u,m}(t) \\ \vdots & \vdots & & \vdots \\ s_{S_u,1}^{u,m}(t) & s_{S_u,2}^{u,m}(t) & \dots & s_{S_u,S_m}^{u,m}(t) \end{bmatrix}, \tag{14.27}$$

then we can write Eq. (14.23) as

$$\left[\frac{\partial^e \mathbf{a}^u(t)}{\partial \mathbf{n}^x(t)^T} \times \frac{\partial^e \mathbf{n}^x(t)}{\partial \mathbf{a}^{u'}(t-d)^T}\right]_{i,j} = \sum_{k=1}^{S^x} s_{i,k}^{u,x}(t+d) \times lw_{k,j}^{x,u'}(d), \tag{14.28}$$

or in matrix form

$$\frac{\partial^e \mathbf{a}^u(t)}{\partial \mathbf{n}^x(t)^T} \times \frac{\partial^e \mathbf{n}^x(t)}{\partial \mathbf{a}^{u'}(t-d)^T} = \mathbf{S}^{u,x}(t) \times \mathbf{LW}^{x,u'}(d). \tag{14.29}$$

Therefore Eq. (14.21) can be written

$$\frac{\partial \mathbf{a}^u(t)}{\partial \mathbf{x}^T} = \frac{\partial^e \mathbf{a}^u(t)}{\partial \mathbf{x}^T} + \sum_{u' \in U} \sum_{x \in X} \sum_{d \in DL_{x,u'}} \mathbf{S}^{u,x}(t) \times \mathbf{LW}^{x,u'}(d) \times \frac{\partial \mathbf{a}^{u'}(t-d)}{\partial \mathbf{x}^T} \tag{14.30}$$

Many of the terms in the summation on the right hand side of Eq. (14.30) will be zero and will not have to be computed. To take advantage of these efficiencies, we introduce some indicator sets. They are sets that tell us for which layers the weights and the sensitivities are nonzero.

The first type of indicator set contains all of the output layers that connect to a specified layer x (which will always be an input layer) with at least some nonzero delay:

$$E_{LW}^U(x) = \{u \in U \ni \exists(\mathbf{LW}^{x,u}(d) \neq 0, d \neq 0)\}, \tag{14.31}$$

where $\ni$ means "such that," and $\exists$ means "there exists."

The second type of indicator set contains the input layers that have a nonzero sensitivity with a specified layer u:

$$E_S^X(u) = \{x \in X \ni \exists(\mathbf{S}^{u,x} \neq 0)\}. \tag{14.32}$$

(When $\mathbf{S}^{u,x}$ is nonzero, there is a static connection from layer x to ouput layer u.) The third type of indicator set contains the layers that have a nonzero sensitivity with a specified layer u:

$$E_S(u) = \{x \ni \exists(\mathbf{S}^{u,x} \neq 0)\}. \tag{14.33}$$

The difference between $E_S^X(u)$ and $E_S(u)$ is that $E_S^X(u)$ contains only input layers. $E_S(u)$ will not be needed in the simplification of Eq. (14.30), but it will be used for the calculation of sensitivities in Eq. (14.38).

Using Eq. (14.31) and Eq. (14.32), we can rearrange the order of the summations in Eq. (14.30) and sum only over nonzero terms:

$$\frac{\partial \mathbf{a}^u(t)}{\partial \mathbf{x}^T} = \frac{\partial^e \mathbf{a}^u(t)}{\partial \mathbf{x}^T} + \sum_{x \in E_S^X(u)} \mathbf{S}^{u,x}(t) \sum_{u' \in E_{LW}^U(x)} \sum_{d \in DL_{x,u'}} \mathbf{LW}^{x,u'}(d) \times \frac{\partial \mathbf{a}^{u'}(t-d)}{\partial \mathbf{x}^T}. \quad (14.34)$$

Eq. (14.34) makes up the generalization of Eq. (14.16) for the LDDN network. It remains to compute the sensitivity matrices $\mathbf{S}^{u,m}(t)$ and the explicit derivatives $\partial^e \mathbf{a}^u(t)/\partial w$, which are described in the next two subsections.

Sensitivities

In order to compute the elements of the sensitivity matrix, we use a form of standard static backpropagation. The sensitivities at the outputs of the network can be computed as

$$s_{k,i}^{u,u}(t) = \frac{\partial^e a_k^u(t)}{\partial n_i^u(t)} = \begin{cases} \dot{f}^u(n_i^u(t)) & \text{for } i = k \\ 0 & \text{for } i \neq k \end{cases}, \quad u \in U, \quad (14.35)$$

or, in matrix form,

$$\mathbf{S}^{u,u}(t) = \dot{\mathbf{F}}^u(\mathbf{n}^u(t)), \quad (14.36)$$

where $\dot{\mathbf{F}}^u(\mathbf{n}^u(t))$ is defined as

$$\dot{\mathbf{F}}^u(\mathbf{n}^u(t)) = \begin{bmatrix} \dot{f}^u(n_1^u(t)) & 0 & \dots & 0 \\ 0 & \dot{f}^u(n_2^u(t)) & \dots & 0 \\ \vdots & \vdots & & \vdots \\ 0 & 0 & \dots & \dot{f}^u(n_{S^u}^u(t)) \end{bmatrix} \quad (14.37)$$

(see also Eq. (11.34)). The matrices $\mathbf{S}^{u,m}(t)$ can be computed by backpropagating through the network, from each network output, using

$$\mathbf{S}^{u,m}(t) = \left[\sum_{l \in E_S(u) \cap L_m^b} \mathbf{S}^{u,l}(t) \mathbf{LW}^{l,m}(0) \right] \dot{\mathbf{F}}^m(\mathbf{n}^m(t)), \quad u \in U, \quad (14.38)$$

where m is decremented from u through the backpropagation order, and L_m^b is the set of indices of layers that are directly connected backwards to layer m (or to which layer m connects forward) and that contain no delays in the connection. The backpropagation step given in Eq. (14.38) is essentially the same as that given in Eq. (11.45), but it is generalized to allow for arbitrary connections between layers.

Explicit Derivatives

We also need to compute the explicit derivatives

$$\frac{\partial^e \mathbf{a}^u(t)}{\partial \mathbf{x}^T} . \tag{14.39}$$

Using the chain rule of calculus, we can derive the following expansion of Eq. (14.39) for input weights:

$$\frac{\partial^e a_k^u(t)}{\partial iw_{i,j}^{m,l}(d)} = \frac{\partial^e a_k^u(t)}{\partial n_i^m(t)} \times \frac{\partial^e n_i^m(t)}{\partial iw_{i,j}^{m,l}(d)} = s_{k,i}^{u,m}(t) \times p_j^l(t-d) . \tag{14.40}$$

In vector form we can write

$$\frac{\partial^e \mathbf{a}^u(t)}{\partial iw_{i,j}^{m,l}(d)} = \mathbf{s}_i^{u,m}(t) \times p_j^l(t-d) . \tag{14.41}$$

In matrix form we have

$$\frac{\partial^e \mathbf{a}^u(t)}{\partial vec(\mathbf{IW}^{m,l}(d))^T} = [\mathbf{p}^l(t-d)]^T \otimes \mathbf{S}^{u,m}(t) , \tag{14.42}$$

and in a similar way we can derive the derivatives for layer weights and biases:

$$\frac{\partial^e \mathbf{a}^u(t)}{\partial vec(\mathbf{LW}^{m,l}(d))^T} = [\mathbf{a}^l(t-d)]^T \otimes \mathbf{S}^{u,m}(t) , \tag{14.43}$$

$$\frac{\partial^e \mathbf{a}^u(t)}{\partial (\mathbf{b}^m)^T} = \mathbf{S}^{u,m}(t) , \tag{14.44}$$

where the *vec* operator transforms a matrix into a vector by stacking the columns of the matrix one underneath the other, and $\mathbf{A} \otimes \mathbf{B}$ is the Kronecker product of $\mathbf{A}$ and $\mathbf{B}$ [MaNe99].

The total RTRL algorithm for the LDDN network is summarized in the following pseudo code.

Real-Time Recurrent Learning Gradient

Initialize:

$$\frac{\partial \mathbf{a}^u(t)}{\partial \mathbf{x}^T} = \mathbf{0},\ t \le 0,\ \text{for all } u \in U,$$

For $t = 1$ to Q,

$U' = \varnothing$, $E_S(u) = \varnothing$ and $E_S^X(u) = \varnothing$ for all $u \in U$.

For m decremented through the BP order

For all $u \in U'$, if $E_S(u) \cap L_m^b \ne \varnothing$

$$\mathbf{S}^{u,m}(t) = \left[\sum_{l \in E_S(u) \cap L_m^b} \mathbf{S}^{u,l}(t)\mathbf{LW}^{l,m}(0) \right] \dot{\mathbf{F}}^m(\mathbf{n}^m(t))$$

add m to the set $E_S(u)$

if $m \in X$, add m to the set $E_S^X(u)$

EndFor u

If $m \in U$

$$\mathbf{S}^{m,m}(t) = \dot{\mathbf{F}}^m(\mathbf{n}^m(t))$$

add m to the sets U' and $E_S(m)$

if $m \in X$, add m to the set $E_S^X(m)$

EndIf m

EndFor m

For $u \in U$ incremented through the simulation order

For all weights and biases ($\mathbf{x}$ is a vector containing all weights and biases)

$$\frac{\partial^e \mathbf{a}^u(t)}{\partial vec(\mathbf{IW}^{m,l}(d))^T} = [\mathbf{p}^l(t-d)]^T \otimes \mathbf{S}^{u,m}(t)$$

$$\frac{\partial^e \mathbf{a}^u(t)}{\partial vec(\mathbf{LW}^{m,l}(d))^T} = [\mathbf{a}^l(t-d)]^T \otimes \mathbf{S}^{u,m}(t)$$

$$\frac{\partial^e \mathbf{a}^u(t)}{\partial (\mathbf{b}^m)^T} = \mathbf{S}^{u,m}(t)$$

EndFor weights and biases

$$\frac{\partial \mathbf{a}^u(t)}{\partial \mathbf{x}^T} = \frac{\partial^e \mathbf{a}^u(t)}{\partial \mathbf{x}^T} + \sum_{x \in E_S^X(u)} \mathbf{S}^{u,x}(t) \sum_{u' \in E_{LW}^U(x)} \sum_{d \in DL_{x,u'}} \mathbf{LW}^{x,u'}(d) \times \frac{\partial \mathbf{a}^{u'}(t-d)}{\partial \mathbf{x}^T}$$

EndFor u

EndFor t

Compute Gradients

$$\frac{\partial F}{\partial \mathbf{x}} = \sum_{t=1}^{Q} \sum_{u \in U} \left[\left[\frac{\partial \mathbf{a}^u(t)}{\partial \mathbf{x}^T} \right]^T \times \frac{\partial^e F}{\partial \mathbf{a}^u(t)} \right]$$

Example RTRL Implementations (FIR and IIR)

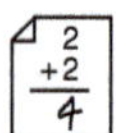

To demonstrate the RTRL algorithm, consider again the feedforward dynamic network of Figure 14.2. The equation of operation of this network is

$$a(t) = n(t) = iw_{1,1}(0)p(t) + iw_{1,1}(1)p(t-1) + iw_{1,1}(2)p(t-2).$$

The architecture of the network is defined by

$$U = \{1\}, X = \{1\}, I_1 = \{1\}, DI_{1,1} = \{0, 1, 2\}, L_1^f = \varnothing, E_{LW}^U(1) = \varnothing.$$

We will choose the following standard performance function with three time points:

$$F = \sum_{t=1}^{Q} (t(t) - a(t))^2 = \sum_{t=1}^{3} e^2(t) = e^2(1) + e^2(2) + e^2(3),$$

with the following inputs and targets:

$$\{p(1), t(1)\}, \{p(2), t(2)\}, \{p(3), t(3)\}.$$

The RTRL algorithm begins with some initialization:

$$U' = \varnothing, E_S(1) = \varnothing, E_S^X(1) = \varnothing.$$

In addition, the initial conditions for the delays, $p(0), p(-1)$, must be provided.

The network response is then computed for the first time step:

$$a(1) = n(1) = iw_{1,1}(0)p(1) + iw_{1,1}(1)p(0) + iw_{1,1}(2)p(-1)$$

Because the RTRL algorithm proceeds forward through time, we can immediately compute the derivatives for the first time step. We will see in the next section that the BPTT algorithm, which proceeds backward through time, will require that we proceed through all of the time points before we can compute the derivatives.

From the preceding pseudo-code, the first step in the derivative calculation will be

$$\mathbf{S}^{1,1}(1) = \dot{\mathbf{F}}^1(\mathbf{n}^1(1)) = 1,$$

since the transfer function is linear. We also update the following sets:

$$E_S^X(1) = \{1\}, E_S(1) = \{1\}.$$

The next step is the computation of the explicit derivatives from Eq. (14.42):

$$\frac{\partial^e \mathbf{a}^1(1)}{\partial vec(\mathbf{IW}^{1,1}(0))^T} = \frac{\partial^e a(1)}{\partial iw_{1,1}(0)} = [\mathbf{p}^1(1)]^T \otimes \mathbf{S}^{1,1}(t) = p(1),$$

$$\frac{\partial^e \mathbf{a}^1(1)}{\partial vec(\mathbf{IW}^{1,1}(1))^T} = \frac{\partial^e a(1)}{\partial iw_{1,1}(1)} = [\mathbf{p}^1(0)]^T \otimes \mathbf{S}^{1,1}(t) = p(0),$$

$$\frac{\partial^e \mathbf{a}^1(1)}{\partial vec(\mathbf{IW}^{1,1}(2))^T} = \frac{\partial^e a(1)}{\partial iw_{1,1}(2)} = [\mathbf{p}^1(-1)]^T \otimes \mathbf{S}^{1,1}(t) = p(-1).$$

The next step would be to compute the total derivative, using Eq. (14.34). However, since $E^U_{LW}(1) = \varnothing$, the total derivatives are equal to the explicit derivatives.

All of the above steps would be repeated for each time point, and then the final step is to compute the derivatives of the performance index with respect to the weights, using Eq. (14.20):

$$\frac{\partial F}{\partial \mathbf{x}} = \sum_{t=1}^{Q} \sum_{u \in U} \left[\left[\frac{\partial \mathbf{a}^u(t)}{\partial \mathbf{x}^T}\right]^T \times \frac{\partial^e F}{\partial \mathbf{a}^u(t)}\right] = \sum_{t=1}^{3} \left[\left[\frac{\partial \mathbf{a}^1(t)}{\partial \mathbf{x}^T}\right]^T \times \frac{\partial^e F}{\partial \mathbf{a}^1(t)}\right].$$

If we break this down for each weight, we have

$$\frac{\partial F}{\partial iw_{1,1}(0)} = p(1)(-2e(1)) + p(2)(-2e(2)) + p(3)(-2e(3)),$$

$$\frac{\partial F}{\partial iw_{1,1}(1)} = p(0)(-2e(1)) + p(1)(-2e(2)) + p(2)(-2e(3)),$$

$$\frac{\partial F}{\partial iw_{1,1}(2)} = p(-1)(-2e(1)) + p(0)(-2e(2)) + p(1)(-2e(3)).$$

We can then use this gradient in any of our standard optimization algorithms from Chapter 9 and Chapter 12. Note that if we use steepest descent, this result is a batch form of the LMS algorithm (see Eq. (10.33)).

2 +2 4

Let's now do an example using a recurrent neural network. Consider again the simple recurrent network in Figure 14.4. From Eq. (14.5), the equation of operation of this network is

$$a(t) = lw_{1,1}(1)a(t-1) + iw_{1,1}p(t)$$

The architecture of the network is defined by

$$U = \{1\}, X = \{1\}, I_1 = \{1\}, DI_{1,1} = \{0\},$$

$$DL_{1,1} = \{1\}, L_1^f = \{1\}, E_{LW}^U(1) = \{1\}.$$

We will choose the same performance function as the previous example:

$$F = \sum_{t=1}^{Q} (t(t) - a(t))^2 = \sum_{t=1}^{3} e^2(t) = e^2(1) + e^2(2) + e^2(3),$$

with the following inputs and targets:

$$\{p(1), t(1)\}, \{p(2), t(2)\}, \{p(3), t(3)\}.$$

We initialize with

$$U' = \varnothing, E_S(1) = \varnothing, E_S^X(1) = \varnothing.$$

In addition, the initial condition for the delay, $a(0)$, and the initial derivatives

$$\frac{\partial a(0)}{\partial iw_{1,1}} \text{ and } \frac{\partial a(0)}{\partial lw_{1,1}(1)}$$

must be provided. (The initial derivatives are usually set to zero.)

The network response is then computed for the first time step:

$$a(1) = lw_{1,1}(1)a(0) + iw_{1,1}p(1)$$

The derivative calculation begins with

$$\mathbf{S}^{1,1}(1) = \dot{\mathbf{F}}^1(\mathbf{n}^1(1)) = 1,$$

since the transfer function is linear. We also update the following sets:

$$E_S^X(1) = \{1\}, E_S(1) = \{1\}.$$

The next step is the computation of the explicit derivatives:

$$\frac{\partial^e \mathbf{a}^1(1)}{\partial vec(\mathbf{IW}^{1,1}(0))^T} = \frac{\partial^e a(1)}{\partial iw_{1,1}} = [\mathbf{p}^1(1)]^T \otimes \mathbf{S}^{1,1}(1) = p(1),$$

$$\frac{\partial^e \mathbf{a}^1(1)}{\partial vec(\mathbf{LW}^{1,1}(1))^T} = \frac{\partial^e a(1)}{\partial lw_{1,1}(1)} = [\mathbf{a}^1(0)]^T \otimes \mathbf{S}^{1,1}(1) = a(0).$$

The next step is to compute the total derivative, using Eq. (14.34):

$$\frac{\partial \mathbf{a}^1(t)}{\partial \mathbf{x}^T} = \frac{\partial^e \mathbf{a}^1(t)}{\partial \mathbf{x}^T} + \mathbf{S}^{1,1}(t)\mathbf{LW}^{1,1}(1)\frac{\partial \mathbf{a}^1(t-1)}{\partial \mathbf{x}^T}. \tag{14.45}$$

Replicating this formula for each of our weights for this network, for $t = 1$, we have

$$\frac{\partial a(1)}{\partial iw_{1,1}} = p(1) + lw_{1,1}(1)\frac{\partial a(0)}{\partial iw_{1,1}} = p(1),$$

$$\frac{\partial a(1)}{\partial lw_{1,1}(1)} = a(0) + lw_{1,1}(1)\frac{\partial a(0)}{\partial lw_{1,1}(1)} = a(0).$$

Note that unlike the corresponding equation in the previous example, these equations are recursive. The derivative at the current time depends on the derivative at the previous time. (Note that the two initial derivatives on the right side of this equation would normally be set to zero, but at the next time step they would be nonzero.) As we mentioned earlier, the weights in a recurrent network have two different effects on the network output. The first is the direct effect, which is represented by the explicit derivative in Eq. (14.45). The second is an indirect effect, since one of the inputs to the network is a previous output, which is also a function of the weights. This effect causes the second term in Eq. (14.45).

All of the above steps would be repeated for each time point:

$$\frac{\partial^e a(2)}{\partial iw_{1,1}} = p(2), \quad \frac{\partial^e a(2)}{\partial lw_{1,1}(1)} = a(1),$$

$$\frac{\partial a(2)}{\partial iw_{1,1}} = p(2) + lw_{1,1}(1)\frac{\partial a(1)}{\partial iw_{1,1}} = p(2) + lw_{1,1}(1)p(1),$$

$$\frac{\partial a(2)}{\partial lw_{1,1}(1)} = a(1) + lw_{1,1}(1)\frac{\partial a(1)}{\partial lw_{1,1}(1)} = a(1) + lw_{1,1}(1)a(0),$$

$$\frac{\partial^e a(3)}{\partial iw_{1,1}} = p(3), \quad \frac{\partial^e a(3)}{\partial lw_{1,1}(1)} = a(2),$$

$$\frac{\partial a(3)}{\partial iw_{1,1}} = p(3) + lw_{1,1}(1)\frac{\partial a(2)}{\partial iw_{1,1}} = p(3) + lw_{1,1}(1)p(2) + (lw_{1,1}(1))^2 p(1),$$

$$\frac{\partial a(3)}{\partial lw_{1,1}(1)} = a(2) + lw_{1,1}(1)\frac{\partial a(2)}{\partial lw_{1,1}(1)} = a(2) + lw_{1,1}(1)a(1) + (lw_{1,1}(1))^2 a(0).$$

The final step is to compute the derivatives of the performance index with respect to the weights, using Eq. (14.20):

$$\frac{\partial F}{\partial \mathbf{x}} = \sum_{t=1}^{Q} \sum_{u \in U} \left[\left[\frac{\partial \mathbf{a}^u(t)}{\partial \mathbf{x}^T} \right]^T \times \frac{\partial^e F}{\partial \mathbf{a}^u(t)} \right] = \sum_{t=1}^{3} \left[\left[\frac{\partial \mathbf{a}^1(t)}{\partial \mathbf{x}^T} \right]^T \times \frac{\partial^e F}{\partial \mathbf{a}^1(t)} \right].$$

If we break this down for each weight, we have

$$\frac{\partial F}{\partial iw_{1,1}} = \frac{\partial a(1)}{\partial iw_{1,1}}(-2e(1)) + \frac{\partial a(2)}{\partial iw_{1,1}}(-2e(2)) + \frac{\partial a(3)}{\partial iw_{1,1}}(-2e(3))$$
$$= -2e(1)[p(1)] - 2e(2)[p(2) + lw_{1,1}(1)p(1)]$$
$$-2e(3)[p(3) + lw_{1,1}(1)p(2) + (lw_{1,1}(1))^2 p(1)]$$

$$\frac{\partial F}{\partial lw_{1,1}(1)} = \frac{\partial a(1)}{\partial lw_{1,1}(1)}(-2e(1)) + \frac{\partial a(2)}{\partial lw_{1,1}(1)}(-2e(2)) + \frac{\partial a(3)}{\partial lw_{1,1}(1)}(-2e(3))$$
$$= -2e(1)[a(0)] - 2e(2)[a(1) + lw_{1,1}(1)a(0)]$$
$$-2e(3)[a(2) + lw_{1,1}(1)a(1) + (lw_{1,1}(1))^2 a(0)]$$

The expansions that we show in the final two lines of the above equations (and also in some of the previous equations) would not be necessary in practice, since the results would be numerical. We have included them here so that we can compare this result with the BPTT algorithm, which we present next.

Backpropagation-Through-Time

In this section we will generalize the Backpropagation-Through-Time (BPTT) algorithm, given in Eq. (14.15) and Eq. (14.17), for LDDN networks.

Eq. (14.15)

The first step is to generalize Eq. (14.15). For the general LDDN network, we can calculate the terms of the gradient by using the chain rule, as in

$$\frac{\partial F}{\partial lw_{i,j}^{m,l}(d)} = \sum_{t=1}^{Q} \left[\sum_{u \in U} \sum_{k=1}^{S^u} \frac{\partial F}{\partial a_k^u(t)} \times \frac{\partial^e a_k^u(t)}{\partial n_i^m(t)} \right] \frac{\partial^e n_i^m(t)}{\partial lw_{i,j}^{m,l}(d)} \tag{14.46}$$

(for the layer weights), where u is an output layer, U is the set of all output layers, and S^u is the number of neurons in layer u.

From Eq. (14.24) we can write

$$\frac{\partial^e n_i^m(t)}{\partial lw_{i,j}^{m,l}(d)} = a_j^l(t-d). \tag{14.47}$$

We will also define

$$d_i^m(t) = \sum_{u \in U} \sum_{k=1}^{S^u} \frac{\partial F}{\partial a_k^u(t)} \times \frac{\partial^e a_k^u(t)}{\partial n_i^m(t)}. \tag{14.48}$$

The terms of the gradient for the layer weights can then be written

$$\frac{\partial F}{\partial lw_{i,j}^{m,l}(d)} = \sum_{t=1}^{Q} d_i^m(t) a_j^l(t-d), \tag{14.49}$$

If we use the sensitivity term defined in Eq. (14.26),

$$s_{k,i}^{u,m}(t) \equiv \frac{\partial^e a_k^u(t)}{\partial n_i^m(t)}, \tag{14.50}$$

then the elements $d_i^m(t)$ can be written

$$d_i^m(t) = \sum_{u \in U} \sum_{k=1}^{S^u} \frac{\partial F}{\partial a_k^u(t)} \times s_{k,i}^{u,m}(t). \tag{14.51}$$

In matrix form this becomes

$$\mathbf{d}^m(t) = \sum_{u \in U} [\mathbf{S}^{u,m}(t)]^T \times \frac{\partial F}{\partial \mathbf{a}^u(t)} \tag{14.52}$$

where

$$\frac{\partial F}{\partial \mathbf{a}^u(t)} = \left[\frac{\partial F}{\partial a_1^u(t)} \; \frac{\partial F}{\partial a_2^u(t)} \; \cdots \; \frac{\partial F}{\partial a_{S_u}^u(t)} \right]^T \tag{14.53}$$

Now the gradient can be written in matrix form.

$$\frac{\partial F}{\partial \mathbf{LW}^{m,l}(d)} = \sum_{t=1}^{Q} \mathbf{d}^m(t) \times [\mathbf{a}^l(t-d)]^T, \tag{14.54}$$

and by similar steps we can find the derivatives for the biases and input weights:

$$\frac{\partial F}{\partial \mathbf{IW}^{m,l}(d)} = \sum_{t=1}^{Q} \mathbf{d}^{m}(t) \times [\mathbf{p}^{l}(t-d)]^{T}, \qquad (14.55)$$

$$\frac{\partial F}{\partial \mathbf{b}^{m}} = \sum_{t=1}^{Q} \mathbf{d}^{m}(t). \qquad (14.56)$$

Eq. (14.54) through Eq. (14.56) make up the generalization of Eq. (14.15) for the LDDN network.

Eq. (14.17)

The next step in the development of the BPTT algorithm is the generalization of Eq. (14.17). Again, we use the chain rule:

$$\frac{\partial F}{\partial \mathbf{a}^{u}(t)} = \frac{\partial^{e} F}{\partial \mathbf{a}^{u}(t)} + \sum_{u' \in U} \sum_{x \in X} \sum_{d \in DL_{x,u}} \left[\frac{\partial^{e} \mathbf{a}^{u'}(t+d)}{\partial \mathbf{n}^{x}(t+d)^{T}} \times \frac{\partial^{e} \mathbf{n}^{x}(t+d)}{\partial \mathbf{a}^{u}(t)^{T}} \right]^{T} \times \frac{\partial F}{\partial \mathbf{a}^{u'}(t+d)} \qquad (14.57)$$

(Many of the terms in these summations will be zero. We will provide a more efficient representation later in this section.) In Eq. (14.17) we only had one delay in the system. Now we need to account for each network output, how that network output is connected back through a network input, and also for the number of times each network output is delayed before it is applied to a network input. That is the reason for the three summations in Eq. (14.57). This equation must be updated backward in time, as t is varied from Q to 1. The terms

$$\frac{\partial F}{\partial \mathbf{a}^{u'}(t)} \qquad (14.58)$$

are generally set to zero for $t > Q$.

If we consider the matrix in the brackets on the right side of Eq. (14.57), from Eq. (14.29) we can write

$$\frac{\partial^{e} \mathbf{a}^{u'}(t+d)}{\partial \mathbf{n}^{x}(t+d)^{T}} \times \frac{\partial^{e} \mathbf{n}^{x}(t+d)}{\partial \mathbf{a}^{u}(t)^{T}} = \mathbf{S}^{u',x}(t+d) \times \mathbf{LW}^{x,u}(d). \qquad (14.59)$$

This allows us to write Eq. (14.57) as

$$\frac{\partial F}{\partial \mathbf{a}^u(t)} = \frac{\partial^e F}{\partial \mathbf{a}^u(t)} + \sum_{u' \in U} \sum_{x \in X} \sum_{d \in DL_{x,u}} [\mathbf{S}^{u',x}(t+d) \times \mathbf{LW}^{x,u}(d)]^T \times \frac{\partial F}{\partial \mathbf{a}^{u'}(t+d)} \tag{14.60}$$

Many of the terms in the summation on the right hand side of Eq. (14.60) will be zero and will not have to be computed. In order to provide a more efficient implementation of Eq. (14.60), we define the following indicator sets:

$$E_{LW}^X(u) = \{x \in X \ni \exists(\mathbf{LW}^{x,u}(d) \neq 0, d \neq 0)\}, \tag{14.61}$$

$$E_S^U(x) = \{u \in U \ni \exists(\mathbf{S}^{u,x} \neq 0)\}. \tag{14.62}$$

The first set contains all of the input layers that have a connection from output layer u with at least some nonzero delay. The second set contains output layers that have a nonzero sensitivity with input layer x. When the sensitivity $S^{u,x}$ is nonzero, there is a static connection from input layer x to output layer u.

We can now rearrange the order of the summation in Eq. (14.60) and sum only over the existing terms:

$$\frac{\partial F}{\partial \mathbf{a}^u(t)} = \frac{\partial^e F}{\partial \mathbf{a}^u(t)} + \sum_{x \in E_{LW}^X(u)} \sum_{d \in DL_{x,u}} \mathbf{LW}^{x,u}(d)^T \sum_{u' \in E_S^U(x)} \mathbf{S}^{u',x}(t+d)^T \times \frac{\partial F}{\partial \mathbf{a}^{u'}(t+d)} \tag{14.63}$$

Summary

The total BPTT algorithm is summarized in the following pseudo code.

Backpropagation-Through-Time Gradient

Initialize:

$$\frac{\partial F}{\partial \mathbf{a}^u(t)} = \mathbf{0},\ t > Q \text{, for all } u \in U,$$

For $t = Q$ to 1,

$U' = \varnothing$, $E_S(u) = \varnothing$, and $E_S^U(u) = \varnothing$ for all $u \in U$.

For m decremented through the BP order

For all $u \in U'$, if $E_S(u) \cap L_m^b \neq \varnothing$

$$\mathbf{S}^{u,m}(t) = \left[\sum_{l \in E_S(u) \cap L_m^b} \mathbf{S}^{u,l}(t)\mathbf{LW}^{l,m}(0) \right] \dot{\mathbf{F}}^m(\mathbf{n}^m(t))$$

add m to the set $E_S(u)$

add u to the set $E_S^U(m)$

EndFor u

If $m \in U$

$$\mathbf{S}^{m,m}(t) = \dot{\mathbf{F}}^m(\mathbf{n}^m(t))$$

add m to the sets U', $E_S(m)$ and $E_S^U(m)$

EndIf m

EndFor m

For $u \in U$ decremented through the BP order

$$\frac{\partial F}{\partial \mathbf{a}^u(t)} = \frac{\partial^e F}{\partial \mathbf{a}^u(t)} + \sum_{x \in E_{LW}^X(u)} \sum_{d \in DL_{x,u}} \mathbf{LW}^{x,u}(d)^T \sum_{u' \in E_S^U(x)} \mathbf{S}^{u',x}(t+d)^T \times \frac{\partial F}{\partial \mathbf{a}^{u'}(t+d)}$$

EndFor u

For all layers m

$$\mathbf{d}^m(t) = \sum_{u \in E_S^U(m)} [\mathbf{S}^{u,m}(t)]^T \times \frac{\partial F}{\partial \mathbf{a}^u(t)}$$

EndFor m

EndFor t

Compute Gradients

$$\frac{\partial F}{\partial \mathbf{LW}^{m,l}(d)} = \sum_{t=1}^{Q} \mathbf{d}^m(t) \times [\mathbf{a}^l(t-d)]^T$$

$$\frac{\partial F}{\partial \mathbf{IW}^{m,l}(d)} = \sum_{t=1}^{Q} \mathbf{d}^m(t) \times [\mathbf{p}^l(t-d)]^T$$

$$\frac{\partial F}{\partial \mathbf{b}^m} = \sum_{t=1}^{Q} \mathbf{d}^m(t)$$

Example BPTT Implementations (FIR and IIR)

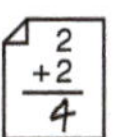

To demonstrate the BPTT algorithm, we will use the same example networks that we used for the RTRL algorithm. First, we use the feedforward dynamic network of Figure 14.2. We defined the network architecture on page 14-18.

Before the gradient can be computed using BPTT, the network response must be computed for all time steps:

$$a(1) = n(1) = iw_{1,1}(0)p(1) + iw_{1,1}(1)p(0) + iw_{1,1}(2)p(-1),$$

$$a(2) = n(2) = iw_{1,1}(0)p(2) + iw_{1,1}(1)p(1) + iw_{1,1}(2)p(0),$$

$$a(3) = n(3) = iw_{1,1}(0)p(3) + iw_{1,1}(2)p(0) + iw_{1,1}(2)p(1).$$

The BPTT algorithm begins with some initialization:

$$U' = \varnothing,\ E_S(1) = \varnothing,\ E_S^U(1) = \varnothing.$$

The first step in the derivative calculation will be the sensitivity calculation. For BPTT, we start at the last time point ($t = 3$):

$$\mathbf{S}^{1,1}(3) = \dot{\mathbf{F}}^1(\mathbf{n}^1(3)) = 1,$$

since the transfer function is linear. We also update the following sets:

$$E_S^U(1) = \{1\},\ E_S(1) = \{1\}.$$

The next step is the calculation of the following derivative using Eq. (14.63):

$$\frac{\partial F}{\partial \mathbf{a}^1(3)} = \frac{\partial^e F}{\partial \mathbf{a}^1(3)} = -2e(3).$$

The final step for $t = 3$ is Eq. (14.52):

$$\mathbf{d}^1(3) = [\mathbf{S}^{1,1}(3)]^T \times \frac{\partial F}{\partial \mathbf{a}^1(3)} = -2e(3).$$

We repeat the previous steps for $t = 2$ and $t = 1$, to obtain

$$\mathbf{d}^1(2) = [\mathbf{S}^{1,1}(2)]^T \times \frac{\partial F}{\partial \mathbf{a}^1(2)} = -2e(2),$$

$$\mathbf{d}^1(1) = [\mathbf{S}^{1,1}(1)]^T \times \frac{\partial F}{\partial \mathbf{a}^1(1)} = -2e(1).$$

Now, all time steps are combined in Eq. (14.55):

$$\frac{\partial F}{\partial \mathbf{IW}^{1,1}(0)} = \frac{\partial F}{\partial iw_{1,1}(0)} = \sum_{t=1}^{3} \mathbf{d}^1(t) \times [\mathbf{p}^1(t)]^T = \sum_{t=1}^{3} -2e(t) \times p(t),$$

$$\frac{\partial F}{\partial \mathbf{IW}^{1,1}(1)} = \frac{\partial F}{\partial iw_{1,1}(1)} = \sum_{t=1}^{3} \mathbf{d}^1(t) \times [\mathbf{p}^1(t-1)]^T = \sum_{t=1}^{3} -2e(t) \times p(t-1),$$

$$\frac{\partial F}{\partial \mathbf{IW}^{1,1}(2)} = \frac{\partial F}{\partial iw_{1,1}(2)} = \sum_{t=1}^{3} \mathbf{d}^1(t) \times [\mathbf{p}^1(t-2)]^T = \sum_{t=1}^{3} -2e(t) \times p(t-2).$$

Note that this is the same result we obtained for the RTRL algorithm example on page 14-19. RTRL and BPTT should always produce the same gradient. The only difference is in the implementation.

Let's now use our previous recurrent neural network example of Figure 14.4. We defined the architecture of this network on page 14-19.

Unlike the RTRL algorithm, where initial conditions for the derivatives must be provided, the BPTT algorithm requires final conditions for the derivatives:

$$\frac{\partial a(4)}{\partial iw_{1,1}} \text{ and } \frac{\partial a(4)}{\partial lw_{1,1}(1)},$$

which are normally set to zero.

The network response is then computed for all time steps:

$$a(1) = lw_{1,1}(1)a(0) + iw_{1,1}p(1)$$

$$a(2) = lw_{1,1}(1)a(1) + iw_{1,1}p(2)$$

$$a(3) = lw_{1,1}(1)a(2) + iw_{1,1}p(3)$$

The derivative calculation begins with

$$\mathbf{S}^{1,1}(3) = \dot{\mathbf{F}}^1(\mathbf{n}^1(3)) = 1,$$

since the transfer function is linear. We also update the following sets:

$$E_S^X(1) = \{1\},\ E_S(1) = \{1\}.$$

Next we compute the following derivative using Eq. (14.63):

$$\frac{\partial F}{\partial \mathbf{a}^1(t)} = \frac{\partial^e F}{\partial \mathbf{a}^1(t)} + \mathbf{LW}^{1,1}(1)^T \mathbf{S}^{1,1}(t+1)^T \times \frac{\partial F}{\partial \mathbf{a}^1(t+1)}$$

For $t = 3$, we find

$$\frac{\partial F}{\partial \mathbf{a}^1(3)} = \frac{\partial^e F}{\partial \mathbf{a}^1(3)} + lw_{1,1}(1)\mathbf{S}^{1,1}(4)^T \times \frac{\partial F}{\partial \mathbf{a}^1(4)} = \frac{\partial^e F}{\partial \mathbf{a}^1(3)} = -2e(3)$$

(where $\frac{\partial F}{\partial \mathbf{a}^1(4)}$ is struck through and marked 0)

and

$$\mathbf{d}^1(3) = [\mathbf{S}^{1,1}(3)]^T \times \frac{\partial F}{\partial \mathbf{a}^1(3)} = -2e(3)$$

Continuing to $t = 2$,

$$\mathbf{S}^{1,1}(2) = \dot{\mathbf{F}}^1(\mathbf{n}^1(2)) = 1,$$

$$\frac{\partial F}{\partial \mathbf{a}^1(2)} = \frac{\partial^e F}{\partial \mathbf{a}^1(2)} + lw_{1,1}(1)\mathbf{S}^{1,1}(3)^T \times \frac{\partial F}{\partial \mathbf{a}^1(3)}$$
$$= -2e(2) + lw_{1,1}(1)(-2e(3))$$

and

$$\mathbf{d}^1(2) = [\mathbf{S}^{1,1}(2)]^T \times \frac{\partial F}{\partial \mathbf{a}^1(2)} = -2e(2) + lw_{1,1}(1)(-2e(3))$$

Finally, for $t = 1$,

$$\mathbf{S}^{1,1}(1) = \dot{\mathbf{F}}^1(\mathbf{n}^1(1)) = 1,$$

$$\frac{\partial F}{\partial \mathbf{a}^1(1)} = \frac{\partial^e F}{\partial \mathbf{a}^1(1)} + lw_{1,1}(1)\mathbf{S}^{1,1}(2)^T \times \frac{\partial F}{\partial \mathbf{a}^1(2)}$$
$$= -2e(1) + lw_{1,1}(1)(-2e(2)) + (lw_{1,1}(1))^2(-2e(3))$$

and

$$\mathbf{d}^1(1) = [\mathbf{S}^{1,1}(1)]^T \times \frac{\partial F}{\partial \mathbf{a}^1(1)} = -2e(1) + lw_{1,1}(1)(-2e(2)) + (lw_{1,1}(1))^2(-2e(3))$$

Now we can compute the total gradient, using Eq. (14.54) and Eq. (14.55):

$$\frac{\partial F}{\partial \mathbf{LW}^{1,1}(1)} = \frac{\partial F}{\partial lw_{1,1}(1)} = \sum_{t=1}^{3} \mathbf{d}^{1}(t) \times [\mathbf{a}^{1}(t-1)]^{T}$$

$$= a(0)[-2e(1) + lw_{1,1}(1)(-2e(2)) + (lw_{1,1}(1))^{2}(-2e(3))]$$
$$+ a(1)[-2e(2) + lw_{1,1}(1)(-2e(3))] + a(0)[-2e(3)]$$

$$\frac{\partial F}{\partial \mathbf{IW}^{1,1}(0)} = \frac{\partial F}{\partial iw_{1,1}} = \sum_{t=1}^{3} \mathbf{d}^{1}(t) \times [\mathbf{p}^{1}(t)]^{T}$$

$$= p(1)[-2e(1) + lw_{1,1}(1)(-2e(2)) + (lw_{1,1}(1))^{2}(-2e(3))]$$
$$+ p(2)[-2e(2) + lw_{1,1}(1)(-2e(3))] + p(3)[-2e(3)]$$

This is the same result that we obtained with the RTRL algorithm on page 14-22.

Summary and Comments on Dynamic Training

The RTRL and BPTT algorithms represent two methods for computing the gradients for dynamic networks. Both algorithms compute the exact gradient, and therefore they produce the same final results. The RTRL algorithm performs the calculations from the first time point forward, which is suitable for on-line (real-time) implementation. The BPTT algorithm starts from the last time point and works backward in time. The BPTT algorithm generally requires fewer computations for the gradient calculation than RTRL, but BPTT usually requires more memory storage.

In addition to the gradient, versions of BPTT and RTRL can be used to compute Jacobian matrices, as are needed in the Levenberg-Marquardt described in Chapter 12. For Jacobian calculations, the RTRL algorithm is generally more efficient that the BPTT algorithm. See [DeHa07] for details.

Once the gradients or Jacobians are computed, many standard optimization algorithms can be used to train the networks. However, training dynamic networks is generally more difficult than training feedforward networks—for a number of reasons. First, a recurrent net can be thought of as a feedforward network, in which the recurrent network is unfolded in time. For example, consider the simple single-layer recurrent network of Figure 14.4. If this network were to be trained over five time steps, we could unfold the network to create 5 layers - one for each time step. If a sigmoid transfer function is used, then if the output of the network is near the saturation point for any time point, the resulting gradient could be quite small.

Another problem in training dynamic networks is the shape of the error surface. It has been shown (see [PhHa13]) that the error surfaces of recurrent networks can have spurious valleys that are not related to the dynamic system that is being approximated. The underlying cause of these valleys is the fact that recurrent networks have the potential for instabilities. For example, the network of Figure 14.4 will be unstable if $lw_{1,1}(1)$ is greater than one in magnitude. However, it is possible, for a particular input sequence, that the network output can be small for a particular value of $lw_{1,1}(1)$ greater than one in magnitude, or for certain combinations of values for $lw_{1,1}(1)$ and $iw_{1,1}$.

Finally, it is sometimes difficult to get adequate training data for dynamic networks. This is because the inputs to some layers will come from tapped delay lines. This means that the elements of the input vector cannot be selected independently, since the time sequence from which they are sampled is generally correlated in time. Unlike static networks, in which the network response depends only on the input to the network at the current time, dynamic network responses depend on the history of the input sequence. The data used to train the network must be representative of all situations for which the network will be used, both in terms of the ranges for each input, but also in terms of the variation of the inputs over time.

2 +2 4

To illustrate the training of dynamic networks, consider again the simple recurrent network of Figure 14.4, but let's use a nonlinear sigmoid transfer function, as shown in Figure 14.9.

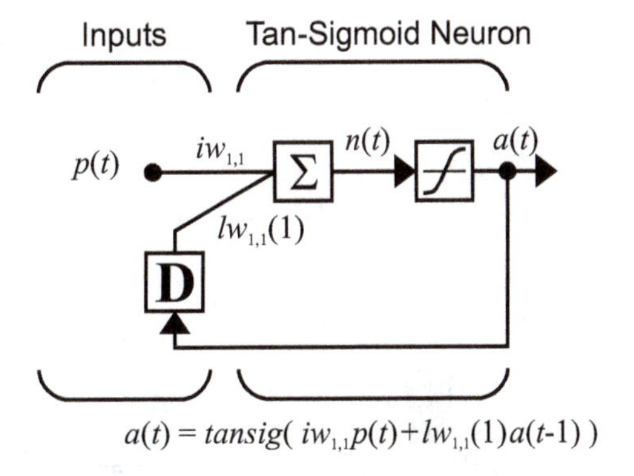

Figure 14.9 Nonlinear Recurrent Network

Recall from Chapter 11 that static multilayer networks can be used to approximate functions. Dynamic networks can be used to approximate dynamic systems. A function maps from one vector space (the domain) to another vector space (the range). A dynamic system maps from one set of time sequences (the input sequences $p(t)$) to another set of time sequences (the output sequences $a(t)$). For example, the network of Figure 14.9 is a dynamic system. It maps from input sequences to output sequences.

In order to simplify our analysis, we will give the network a problem for which we know the optimal solution. The dynamic system we will approximate is the same network, with the following values for the weights:

$$lw_{1,1}(1) = 0.5\,,\ iw_{1,1} = 0.5\,, \tag{14.64}$$

The input sequence that we use to train a dynamic network must be representative of all possible input sequences. Because this network is so simple, it is not difficult to find an appropriate input sequence, but for many practical networks it can be difficult. We will use a standard form of input sequence (called the skyline function), which consists of a series of pulses of varying height and width. The input and target sequences are shown in Figure 14.10. The circles represent the input sequence and the dots represent the target sequence. The targets were created by applying the given input sequence to the network of Figure 14.9, with the weights given by Eq. (14.64).

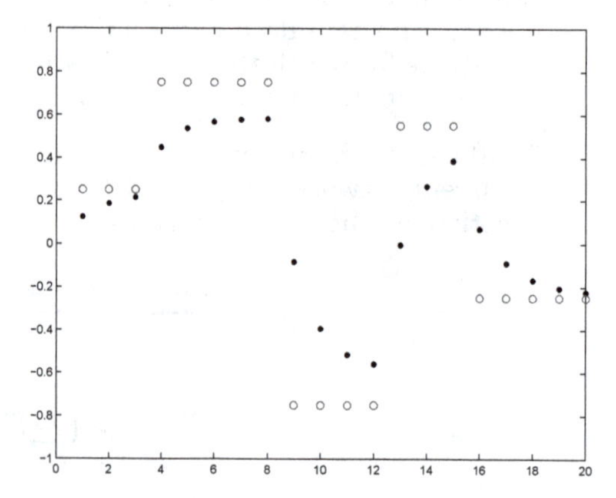

Figure 14.10 Input and Target Sequences

Figure 14.11 shows the squared error performance surface for this problem. Note that as the weight $lw_{1,1}(1)$ becomes greater than one in magnitude, the squared error grows steeply. This effect would be even more prominent, if the length of the training sequence were longer. However, we can also see some narrow valleys in the surface in the regions where $lw_{1,1}(1)$ is greater than one. (This is a very common result, as discussed in [PhHa13]. See Exercise E14.18 to investigate the cause of these valleys.)

The narrow valleys can have an effect on training, since the trajectory can be trapped or misdirected by the spurious valleys. On the left of Figure 14.11 we see a steepest descent path. The path is misdirected at the beginning of the trajectory, because of the narrow valley seen near the bottom of the contour plot.

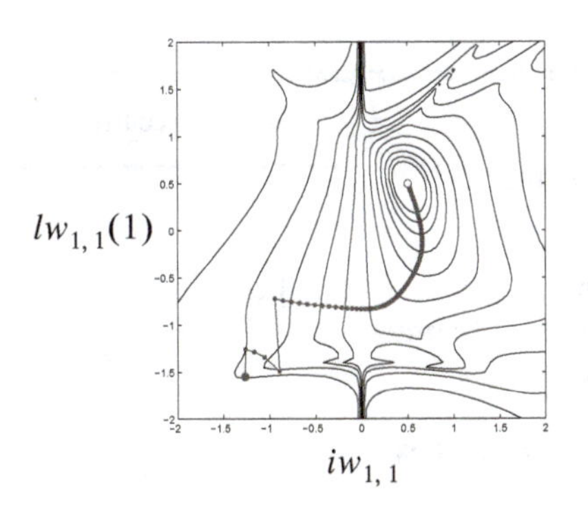

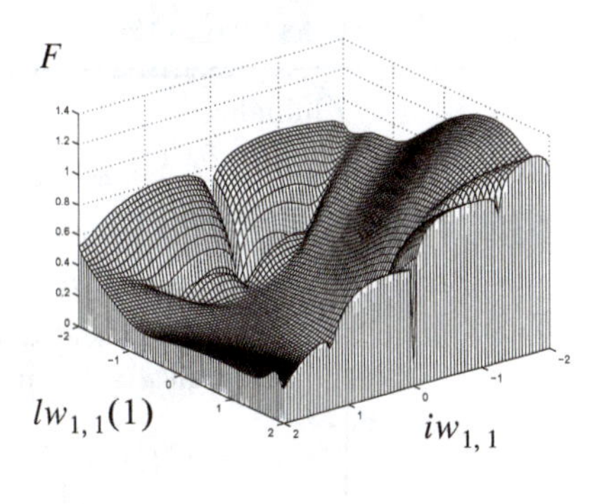

Figure 14.11 Performance Surface and Steepest Descent Trajectory

To experiment with the training of this recurrent network, use the Neural Network Design Demonstration Recurrent Network Training (`nnd14rnt`).

Summary of Results

Real-Time Recurrent Learning Gradient

Initialize:

$$\frac{\partial \mathbf{a}^u(t)}{\partial \mathbf{x}^T} = \mathbf{0},\ t \le 0,\ \text{for all } u \in U,$$

For $t = 1$ to Q,

$U' = \varnothing$, $E_S(u) = \varnothing$ and $E_S^X(u) = \varnothing$ for all $u \in U$.

For m decremented through the BP order

For all $u \in U'$, if $E_S(u) \cap L_m^b \neq \varnothing$

$$\mathbf{S}^{u,m}(t) = \left[\sum_{l \in E_S(u) \cap L_m^b} \mathbf{S}^{u,l}(t)\mathbf{LW}^{l,m}(0)\right]\dot{\mathbf{F}}^m(\mathbf{n}^m(t))$$

add m to the set $E_S(u)$

if $m \in X$, add m to the set $E_S^X(u)$

EndFor u

If $m \in U$

$$\mathbf{S}^{m,m}(t) = \dot{\mathbf{F}}^m(\mathbf{n}^m(t))$$

add m to the sets U' and $E_S(m)$

if $m \in X$, add m to the set $E_S^X(m)$

EndIf m

EndFor m

For $u \in U$ incremented through the simulation order

For all weights and biases ($\mathbf{x}$ is a vector containing all weights and biases)

$$\frac{\partial^e \mathbf{a}^u(t)}{\partial vec(\mathbf{IW}^{m,l}(d))^T} = [\mathbf{p}^l(t-d)]^T \otimes \mathbf{S}^{u,m}(t)$$

$$\frac{\partial^e \mathbf{a}^u(t)}{\partial vec(\mathbf{LW}^{m,l}(d))^T} = [\mathbf{a}^l(t-d)]^T \otimes \mathbf{S}^{u,m}(t)$$

$$\frac{\partial^e \mathbf{a}^u(t)}{\partial (\mathbf{b}^m)^T} = \mathbf{S}^{u,m}(t)$$

EndFor weights and biases

$$\frac{\partial \mathbf{a}^u(t)}{\partial \mathbf{x}^T} = \frac{\partial^e \mathbf{a}^u(t)}{\partial \mathbf{x}^T} + \sum_{x \in E_S^X(u)} \mathbf{S}^{u,x}(t) \sum_{u' \in E_{LW}^U(x)} \sum_{d \in DL_{x,u'}} \mathbf{LW}^{x,u'}(d) \times \frac{\partial \mathbf{a}^{u'}(t-d)}{\partial \mathbf{x}^T}$$

EndFor u

EndFor t

Compute Gradients

$$\frac{\partial F}{\partial \mathbf{x}} = \sum_{t=1}^{Q} \sum_{u \in U} \left[\left[\frac{\partial \mathbf{a}^u(t)}{\partial \mathbf{x}^T}\right]^T \times \frac{\partial^e F}{\partial \mathbf{a}^u(t)}\right]$$

Backpropagation-Through-Time Gradient

Initialize:

$$\frac{\partial F}{\partial \mathbf{a}^u(t)} = \mathbf{0},\, t > Q \text{, for all } u \in U,$$

For $t = Q$ to 1,

$U' = \varnothing$, $E_S(u) = \varnothing$, and $E_S^U(u) = \varnothing$ for all $u \in U$.

For m decremented through the BP order

For all $u \in U'$, if $E_S(u) \cap L_m^b \neq \varnothing$

$$\mathbf{S}^{u,m}(t) = \left[\sum_{l \in E_S(u) \cap L_m^b} \mathbf{S}^{u,l}(t)\mathbf{LW}^{l,m}(0) \right] \dot{\mathbf{F}}^m(\mathbf{n}^m(t))$$

add m to the set $E_S(u)$

add u to the set $E_S^U(m)$

EndFor u

If $m \in U$

$$\mathbf{S}^{m,m}(t) = \dot{\mathbf{F}}^m(\mathbf{n}^m(t))$$

add m to the sets U', $E_S(m)$ and $E_S^U(m)$

EndIf m

EndFor m

For $u \in U$ decremented through the BP order

$$\frac{\partial F}{\partial \mathbf{a}^u(t)} = \frac{\partial^e F}{\partial \mathbf{a}^u(t)} + \sum_{x \in E_{LW}^X(u)} \sum_{d \in DL_{x,u}} \mathbf{LW}^{x,u}(d)^T \sum_{u' \in E_S^U(x)} \mathbf{S}^{u',x}(t+d)^T \times \frac{\partial F}{\partial \mathbf{a}^{u'}(t+d)}$$

EndFor u

For all layers m

$$\mathbf{d}^m(t) = \sum_{u \in E_S^U(m)} [\mathbf{S}^{u,m}(t)]^T \times \frac{\partial F}{\partial \mathbf{a}^u(t)}$$

EndFor m

EndFor t

Compute Gradients

$$\frac{\partial F}{\partial \mathbf{LW}^{m,l}(d)} = \sum_{t=1}^{Q} \mathbf{d}^m(t) \times [\mathbf{a}^l(t-d)]^T$$

$$\frac{\partial F}{\partial \mathbf{IW}^{m,l}(d)} = \sum_{t=1}^{Q} \mathbf{d}^m(t) \times [\mathbf{p}^l(t-d)]^T$$

$$\frac{\partial F}{\partial \mathbf{b}^m} = \sum_{t=1}^{Q} \mathbf{d}^m(t)$$

Definitions/Notation

$\mathbf{p}^{l}(t)$ is the lth *input vector* to the network at time t.

$\mathbf{n}^{m}(t)$ is the *net input* for layer m.

$\mathbf{f}^{m}(\ )$ is the *transfer function* for layer m.

$\mathbf{a}^{m}(t)$ is the *output* for layer m.

$\mathbf{IW}^{m,l}$ is the *input weight* between input l and layer m.

$\mathbf{LW}^{m,l}$ is the *layer weight* between layer l and layer m.

$\mathbf{b}^{m}$ is the *bias vector* for layer m.

$DL_{m,l}$ is the set of all delays in the tapped delay line between Layer l and Layer m.

$DI_{m,l}$ is the set of all delays in the tapped delay line between Input l and Layer m.

I_m is the set of indices of input vectors that connect to layer m.

L_m^f is the set of indices of layers that directly connect *forward* to layer m.

L_m^b is the set of indices of layers that are directly connected backwards to layer m (or to which layer m connects forward) and that contain no delays in the connection.

A layer is an *input layer* if it has an input weight, or if it contains any delays with any of its weight matrices. The set of input layers is X.

A layer is an *output layer* if its output will be compared to a target during training, or if it is connected to an input layer through a matrix that has any delays associated with it. The set of output layers is U.

The *sensitivity* is defined $s_{k,i}^{u,m}(t) \equiv \dfrac{\partial^e a_k^u(t)}{\partial n_i^m(t)}$.

$$E_{LW}^{U}(x) = \{u \in U \ni \exists(\mathbf{LW}^{x,u}(d) \neq 0, d \neq 0)\}$$

$$E_S^X(u) = \{x \in X \ni \exists(\mathbf{S}^{u,x} \neq 0)\}$$

$$E_S(u) = \{x \ni \exists(\mathbf{S}^{u,x} \neq 0)\}$$

$$E_{LW}^{X}(u) = \{x \in X \ni \exists(\mathbf{LW}^{x,u}(d) \neq 0, d \neq 0)\}$$

$$E_S^U(x) = \{u \in U \ni \exists(\mathbf{S}^{u,x} \neq 0)\}$$

Solved Problems

P14.1 Before stating the problem, let's first introduce some notation that will allow us to efficiently represent dynamic networks:

Layer #1 with 2 tansig neurons.

Input vector with 3 elements.

Tapped delay from 1 to 9.

Tapped delay from 0 to 9.

Figure P14.1 Blocks for Dynamic Network Schematics

Using this notation, consider the following network

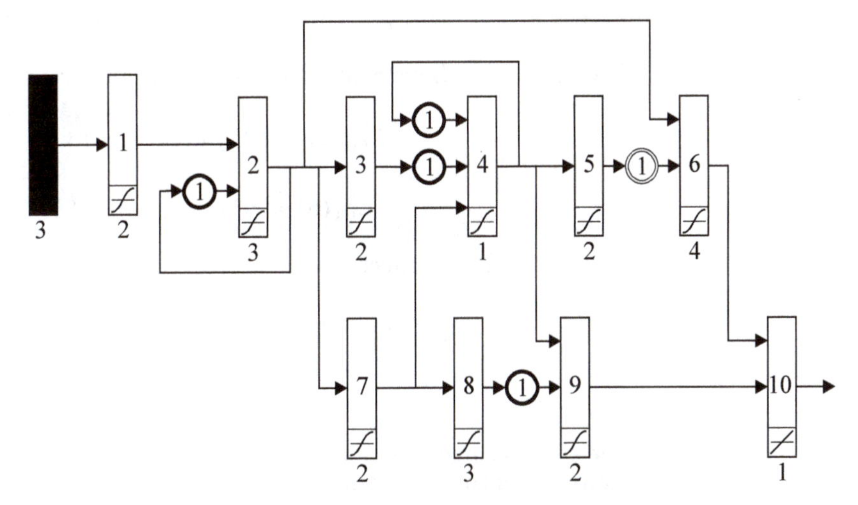

Figure P14.2 Example Dynamic Network for Problem P14.1

Define the network architecture, by showing U**,** X**,** I_m**,** $DI_{m,1}$**,** $DL_{m,l}$**,** L_m^f**,** L_m^b**,** $E_{LW}^U(x)$**,** $E_{LW}^X(u)$**. Also, select a simulation order and indicate the dimension of each weight matrix.**

The input layers have input weights, or have delays with their weight matrices. Therefore $X = \{1, 2, 4, 6, 9\}$. The output layers are compared to targets, or connect to an input layer through a matrix that has any delays. If we assume that only Layer 10 is compared to a target, then $U = \{2, 3, 4, 5, 8, 10\}$. Since the single input vector connects only to Layer 1, the only nonempty set of inputs will be $I_1 = \{1\}$. For the same reason, there will only be one nonempty set of input delays: $DI_{1,1} = \{0\}$. The connections between layers are defined by

$$L_1^f = \varnothing,\ L_2^f = \{1, 2\},\ L_3^f = \{2\},\ L_4^f = \{3, 4, 7\},\ L_5^f = \{4\},$$

$$L_6^f = \{2, 5\},\ L_7^f = \{2\},\ L_8^f = \{7\},\ L_9^f = \{4, 8\},\ L_{10}^f = \{6, 9\}.$$

$$L_1^b = \{2\},\ L_2^b = \{3, 6, 7\},\ L_3^b = \varnothing,\ L_4^b = \{5, 9\},\ L_5^b = \varnothing,$$

$$L_6^b = \{10\},\ L_7^b = \{4, 8\},\ L_8^b = \varnothing,\ L_9^b = \{10\},\ L_{10}^b = \varnothing.$$

Associated with these connections are the following layer delays

$$DL_{2,1} = \{0\},\ DL_{2,2} = \{1\},\ DL_{3,2} = \{0\},\ DL_{4,3} = \{1\},\ DL_{4,4} = \{1\},$$

$$DL_{4,7} = \{0\},\ DL_{5,4} = \{0\},\ DL_{6,2} = \{0\},\ DL_{6,5} = \{0, 1\},\ DL_{7,2} = \{0\},$$

$$DL_{8,7} = \{0\},\ DL_{9,4} = \{0\},\ DL_{9,8} = \{1\},\ DL_{10,6} = \{0\},\ DL_{10,9} = \{0\}.$$

The layers that have connections from output layers are

$$E_{LW}^U(2) = \{2\},\ E_{LW}^U(4) = \{3, 4\},$$

$$E_{LW}^U(6) = \{5\},\ E_{LW}^U(9) = \{8\}.$$

The layers that connect to input layers are

$$E_{LW}^X(2) = \{2\},\ E_{LW}^X(3) = \{4\},\ E_{LW}^X(4) = \{4\},$$

$$E_{LW}^X(5) = \{6\},\ E_{LW}^X(8) = \{9\}.$$

The simulation order can be chosen to be $\{1, 2, 3, 7, 4, 5, 6, 8, 9, 10\}$. The dimensions of the weight matrices are

$$\mathbf{IW}^{1,1}(0) \Rightarrow 2 \times 3,\ \mathbf{LW}^{2,1}(0) \Rightarrow 3 \times 2,\ \mathbf{LW}^{2,2}(1) \Rightarrow 3 \times 3,\ \mathbf{LW}^{3,2}(0) \Rightarrow 2 \times 3,$$

$$\mathbf{LW}^{4,3}(1) \Rightarrow 1 \times 2,\ \mathbf{LW}^{4,4}(1) \Rightarrow 1 \times 1,\ \mathbf{LW}^{4,7}(0) \Rightarrow 1 \times 2,\ \mathbf{LW}^{5,4}(0) \Rightarrow 2 \times 1,$$

$$\mathbf{LW}^{6,2}(0) \Rightarrow 4 \times 3,\ \mathbf{LW}^{6,5}(1) \Rightarrow 4 \times 2,\ \mathbf{LW}^{7,2}(0) \Rightarrow 2 \times 3,\ \mathbf{LW}^{8,7}(0) \Rightarrow 3 \times 2,$$

$$\mathbf{LW}^{9,4}(0) \Rightarrow 2 \times 1,\ \mathbf{LW}^{9,8}(1) \Rightarrow 2 \times 3,\ \mathbf{LW}^{10,6}(0) \Rightarrow 1 \times 4,\ \mathbf{LW}^{10,9}(d) \Rightarrow 1 \times 2.$$

P14.2 Write out the BPTT equations for the network presented in Problem P14.1.

We will assume that the network response has been computed for all time points, and we will demonstrate the process for one time step, since they all will be similar. We proceed through the layers according to the backpropagation order, which is the reverse of the simulation order: $\{10, 9, 8, 6, 5, 4, 7, 3, 2, 1\}$.

$$\mathbf{S}^{10,10}(t) = \dot{\mathbf{F}}^{10}(\mathbf{n}^{10}(t))$$

$$\frac{\partial F}{\partial \mathbf{a}^{10}(t)} = \frac{\partial^e F}{\partial \mathbf{a}^{10}(t)}$$

$$\mathbf{d}^{10}(t) = [\mathbf{S}^{10,10}(t)]^T \times \frac{\partial F}{\partial \mathbf{a}^{10}(t)}$$

$$\mathbf{S}^{10,9}(t) = \mathbf{S}^{10,10}(t)\mathbf{LW}^{10,9}(0)\dot{\mathbf{F}}^{9}(\mathbf{n}^{9}(t))$$

$$\mathbf{d}^{9}(t) = [\mathbf{S}^{10,9}(t)]^T \times \frac{\partial F}{\partial \mathbf{a}^{10}(t)}$$

$$\mathbf{S}^{8,8}(t) = \dot{\mathbf{F}}^{8}(\mathbf{n}^{8}(t))$$

$$\frac{\partial F}{\partial \mathbf{a}^{8}(t)} = \frac{\partial^e F}{\partial \mathbf{a}^{8}(t)} + \mathbf{LW}^{9,8}(1)^T \mathbf{S}^{10,9}(t+1)^T \times \frac{\partial F}{\partial \mathbf{a}^{10}(t+1)}$$

$$\mathbf{d}^{8}(t) = [\mathbf{S}^{8,8}(t)]^T \times \frac{\partial F}{\partial \mathbf{a}^{8}(t)}$$

$$\mathbf{S}^{10,6}(t) = \mathbf{S}^{10,10}(t)\mathbf{LW}^{10,6}(0)\dot{\mathbf{F}}^{6}(\mathbf{n}^{6}(t))$$

$$\mathbf{d}^{6}(t) = [\mathbf{S}^{10,6}(t)]^T \times \frac{\partial F}{\partial \mathbf{a}^{10}(t)}$$

$$\mathbf{S}^{5,5}(t) = \dot{\mathbf{F}}^{5}(\mathbf{n}^{5}(t))$$

$$\frac{\partial F}{\partial \mathbf{a}^{5}(t)} = \frac{\partial^e F}{\partial \mathbf{a}^{5}(t)} + \mathbf{LW}^{6,5}(1)^T \mathbf{S}^{10,6}(t+1)^T \times \frac{\partial F}{\partial \mathbf{a}^{10}(t+1)} + \mathbf{LW}^{6,5}(0)^T \mathbf{S}^{10,6}(t)^T \times \frac{\partial F}{\partial \mathbf{a}^{10}(t)}$$

$$\mathbf{d}^{5}(t) = [\mathbf{S}^{5,5}(t)]^T \times \frac{\partial F}{\partial \mathbf{a}^{5}(t)}$$

$$\mathbf{S}^{10,4}(t) = \mathbf{S}^{10,9}(t)\mathbf{LW}^{9,4}(0)\dot{\mathbf{F}}^4(\mathbf{n}^4(t))$$

$$\mathbf{S}^{5,4}(t) = \mathbf{S}^{5,5}(t)\mathbf{LW}^{5,4}(0)\dot{\mathbf{F}}^4(\mathbf{n}^4(t))$$

$$\mathbf{S}^{4,4}(t) = \dot{\mathbf{F}}^4(\mathbf{n}^4(t))$$

$$\frac{\partial F}{\partial \mathbf{a}^4(t)} = \frac{\partial^e F}{\partial \mathbf{a}^4(t)} + \mathbf{LW}^{4,4}(1)^T\left[\mathbf{S}^{4,4}(t+1)^T \times \frac{\partial F}{\partial \mathbf{a}^4(t+1)} + \mathbf{S}^{5,4}(t+1)^T \times \frac{\partial F}{\partial \mathbf{a}^5(t+1)} + \mathbf{S}^{10,4}(t+1)^T \times \frac{\partial F}{\partial \mathbf{a}^{10}(t+1)}\right]$$

$$\mathbf{d}^4(t) = [\mathbf{S}^{4,4}(t)]^T \times \frac{\partial F}{\partial \mathbf{a}^4(t)} + [\mathbf{S}^{5,4}(t)]^T \times \frac{\partial F}{\partial \mathbf{a}^5(t)} + [\mathbf{S}^{10,4}(t)]^T \times \frac{\partial F}{\partial \mathbf{a}^{10}(t)}$$

$$\mathbf{S}^{10,7}(t) = \mathbf{S}^{10,4}(t)\mathbf{LW}^{4,7}(0)\dot{\mathbf{F}}^7(\mathbf{n}^7(t))$$

$$\mathbf{S}^{8,7}(t) = \mathbf{S}^{8,8}(t)\mathbf{LW}^{8,7}(0)\dot{\mathbf{F}}^7(\mathbf{n}^7(t))$$

$$\mathbf{S}^{5,7}(t) = \mathbf{S}^{5,4}(t)\mathbf{LW}^{4,7}(0)\dot{\mathbf{F}}^7(\mathbf{n}^7(t))$$

$$\mathbf{S}^{4,7}(t) = \mathbf{S}^{4,4}(t)\mathbf{LW}^{4,7}(0)\dot{\mathbf{F}}^7(\mathbf{n}^7(t))$$

$$\mathbf{d}^7(t) = [\mathbf{S}^{10,7}(t)]^T \times \frac{\partial F}{\partial \mathbf{a}^{10}(t)} + [\mathbf{S}^{8,7}(t)]^T \times \frac{\partial F}{\partial \mathbf{a}^8(t)} + [\mathbf{S}^{5,7}(t)]^T \times \frac{\partial F}{\partial \mathbf{a}^5(t)} + [\mathbf{S}^{4,7}(t)]^T \times \frac{\partial F}{\partial \mathbf{a}^4(t)}$$

$$\mathbf{S}^{3,3}(t) = \dot{\mathbf{F}}^3(\mathbf{n}^3(t))$$

$$\frac{\partial F}{\partial \mathbf{a}^3(t)} = \frac{\partial^e F}{\partial \mathbf{a}^3(t)} + \mathbf{LW}^{4,3}(1)^T\left[\mathbf{S}^{4,4}(t+1)^T \times \frac{\partial F}{\partial \mathbf{a}^4(t+1)} + \mathbf{S}^{5,4}(t+1)^T \times \frac{\partial F}{\partial \mathbf{a}^5(t+1)} + \mathbf{S}^{10,4}(t+1)^T \times \frac{\partial F}{\partial \mathbf{a}^{10}(t+1)}\right]$$

$$\mathbf{d}^3(t) = [\mathbf{S}^{3,3}(t)]^T \times \frac{\partial F}{\partial \mathbf{a}^3(t)}$$

$$\mathbf{S}^{10,2}(t) = \mathbf{S}^{10,6}(t)\mathbf{LW}^{6,2}(0)\dot{\mathbf{F}}^2(\mathbf{n}^2(t)) + \mathbf{S}^{10,7}(t)\mathbf{LW}^{7,2}(0)\dot{\mathbf{F}}^2(\mathbf{n}^2(t))$$

$$\mathbf{S}^{8,2}(t) = \mathbf{S}^{8,7}(t)\mathbf{LW}^{7,2}(0)\dot{\mathbf{F}}^2(\mathbf{n}^2(t))$$

$$\mathbf{S}^{5,2}(t) = \mathbf{S}^{5,7}(t)\mathbf{LW}^{7,2}(0)\dot{\mathbf{F}}^2(\mathbf{n}^2(t))$$

$$\mathbf{S}^{4,2}(t) = \mathbf{S}^{4,7}(t)\mathbf{LW}^{7,2}(0)\dot{\mathbf{F}}^2(\mathbf{n}^2(t))$$

$$\mathbf{S}^{3,2}(t) = \mathbf{S}^{3,3}(t)\mathbf{LW}^{3,2}(0)\dot{\mathbf{F}}^2(\mathbf{n}^2(t))$$

$$\mathbf{S}^{2,2}(t) = \dot{\mathbf{F}}^2(\mathbf{n}^2(t))$$

$$\frac{\partial F}{\partial \mathbf{a}^2(t)} = \frac{\partial^e F}{\partial \mathbf{a}^2(t)} + \mathbf{LW}^{2,2}(1)^T \Big[\mathbf{S}^{2,2}(t+1)^T \times \frac{\partial F}{\partial \mathbf{a}^2(t+1)} + \mathbf{S}^{3,2}(t+1)^T \times \frac{\partial F}{\partial \mathbf{a}^3(t+1)}$$
$$+ \mathbf{S}^{4,2}(t+1)^T \times \frac{\partial F}{\partial \mathbf{a}^4(t+1)} + \mathbf{S}^{5,2}(t+1)^T \times \frac{\partial F}{\partial \mathbf{a}^5(t+1)}$$
$$+ \mathbf{S}^{8,2}(t+1)^T \times \frac{\partial F}{\partial \mathbf{a}^8(t+1)} + \mathbf{S}^{10,2}(t+1)^T \times \frac{\partial F}{\partial \mathbf{a}^{10}(t+1)} \Big]$$

$$\mathbf{d}^2(t) = [\mathbf{S}^{10,2}(t)]^T \times \frac{\partial F}{\partial \mathbf{a}^{10}(t)} + [\mathbf{S}^{8,2}(t)]^T \times \frac{\partial F}{\partial \mathbf{a}^8(t)} + [\mathbf{S}^{5,2}(t)]^T \times \frac{\partial F}{\partial \mathbf{a}^5(t)}$$
$$+ [\mathbf{S}^{4,2}(t)]^T \times \frac{\partial F}{\partial \mathbf{a}^4(t)} + [\mathbf{S}^{3,2}(t)]^T \times \frac{\partial F}{\partial \mathbf{a}^3(t)} + [\mathbf{S}^{2,2}(t)]^T \times \frac{\partial F}{\partial \mathbf{a}^2(t)}$$

$$\mathbf{S}^{10,1}(t) = \mathbf{S}^{10,2}(t)\mathbf{LW}^{2,1}(0)\dot{\mathbf{F}}^1(\mathbf{n}^1(t))$$

$$\mathbf{S}^{8,1}(t) = \mathbf{S}^{8,2}(t)\mathbf{LW}^{2,1}(0)\dot{\mathbf{F}}^1(\mathbf{n}^1(t))$$

$$\mathbf{S}^{5,1}(t) = \mathbf{S}^{5,2}(t)\mathbf{LW}^{2,1}(0)\dot{\mathbf{F}}^1(\mathbf{n}^1(t))$$

$$\mathbf{S}^{4,1}(t) = \mathbf{S}^{4,2}(t)\mathbf{LW}^{2,1}(0)\dot{\mathbf{F}}^1(\mathbf{n}^1(t))$$

$$\mathbf{S}^{3,1}(t) = \mathbf{S}^{3,2}(t)\mathbf{LW}^{2,1}(0)\dot{\mathbf{F}}^1(\mathbf{n}^1(t))$$

$$\mathbf{S}^{2,1}(t) = \mathbf{S}^{2,2}(t)\mathbf{LW}^{2,1}(0)\dot{\mathbf{F}}^1(\mathbf{n}^1(t))$$

$$\mathbf{d}^1(t) = [\mathbf{S}^{10,1}(t)]^T \times \frac{\partial F}{\partial \mathbf{a}^{10}(t)} + [\mathbf{S}^{8,1}(t)]^T \times \frac{\partial F}{\partial \mathbf{a}^8(t)} + [\mathbf{S}^{5,1}(t)]^T \times \frac{\partial F}{\partial \mathbf{a}^5(t)}$$
$$+ [\mathbf{S}^{4,1}(t)]^T \times \frac{\partial F}{\partial \mathbf{a}^4(t)} + [\mathbf{S}^{3,1}(t)]^T \times \frac{\partial F}{\partial \mathbf{a}^3(t)} + [\mathbf{S}^{2,1}(t)]^T \times \frac{\partial F}{\partial \mathbf{a}^2(t)}$$

After the preceding steps have been repeated for all time points, from the last time point back to the first, then the gradient can be computed as follows:

$$\frac{\partial F}{\partial \mathbf{LW}^{m,l}(d)} = \sum_{t=1}^{Q} \mathbf{d}^m(t) \times [\mathbf{a}^l(t-d)]^T$$

$$\frac{\partial F}{\partial \mathbf{IW}^{m,l}(d)} = \sum_{t=1}^{Q} \mathbf{d}^m(t) \times [\mathbf{p}^l(t-d)]^T$$

$$\frac{\partial F}{\partial \mathbf{b}^m} = \sum_{t=1}^{Q} \mathbf{d}^m(t)$$

P14.3 Write out the RTRL equations for the network presented in Problem P14.1.

As in the previous problem, we will demonstrate the process for one time step, since each step is similar. We will proceed through the layers according to the backpropagation order. The sensitivity matrices $\mathbf{S}^{u,m}(t)$ are computed in the same way for the RTRL algorithm as for the BPTT algorithm, so we won't repeat those steps from Problem P14.2.

The explicit derivative calculations for the input weight will be

$$\frac{\partial^e \mathbf{a}^u(t)}{\partial vec(\mathbf{IW}^{1,1}(0))^T} = [\mathbf{p}^1(t)]^T \otimes \mathbf{S}^{u,1}(t)$$

For the layer weights and the biases, the explicit derivatives are calculated by

$$\frac{\partial^e \mathbf{a}^u(t)}{\partial vec(\mathbf{LW}^{m,l}(d))^T} = [\mathbf{a}^l(t-d)]^T \otimes \mathbf{S}^{u,m}(t)$$

$$\frac{\partial^e \mathbf{a}^u(t)}{\partial (\mathbf{b}^m)^T} = \mathbf{S}^{u,m}(t)$$

For the total derivatives, we have

$$\frac{\partial \mathbf{a}^2(t)}{\partial \mathbf{x}^T} = \frac{\partial^e \mathbf{a}^2(t)}{\partial \mathbf{x}^T} + \mathbf{S}^{2,2}(t)\left[\mathbf{LW}^{2,2}(1) \times \frac{\partial \mathbf{a}^2(t-1)}{\partial \mathbf{x}^T}\right]$$

$$\frac{\partial \mathbf{a}^3(t)}{\partial \mathbf{x}^T} = \frac{\partial^e \mathbf{a}^3(t)}{\partial \mathbf{x}^T} + \mathbf{S}^{3,2}(t)\left[\mathbf{LW}^{2,2}(1) \times \frac{\partial \mathbf{a}^2(t-1)}{\partial \mathbf{x}^T}\right]$$

$$\frac{\partial \mathbf{a}^4(t)}{\partial \mathbf{x}^T} = \frac{\partial^e \mathbf{a}^4(t)}{\partial \mathbf{x}^T} + \mathbf{S}^{4,4}(t)\left[\mathbf{LW}^{4,4}(1) \times \frac{\partial \mathbf{a}^4(t-1)}{\partial \mathbf{x}^T} + \mathbf{LW}^{4,3}(1) \times \frac{\partial \mathbf{a}^3(t-1)}{\partial \mathbf{x}^T}\right]$$

$$+ \mathbf{S}^{4,2}(t)\left[\mathbf{LW}^{2,2}(1) \times \frac{\partial \mathbf{a}^2(t-1)}{\partial \mathbf{x}^T}\right]$$

$$\frac{\partial \mathbf{a}^5(t)}{\partial \mathbf{x}^T} = \frac{\partial^e \mathbf{a}^5(t)}{\partial \mathbf{x}^T} + \mathbf{S}^{5,4}(t)\left[\mathbf{LW}^{4,4}(1) \times \frac{\partial \mathbf{a}^4(t-1)}{\partial \mathbf{x}^T} + \mathbf{LW}^{4,3}(1) \times \frac{\partial \mathbf{a}^3(t-1)}{\partial \mathbf{x}^T}\right]$$

$$+ \mathbf{S}^{5,2}(t)\left[\mathbf{LW}^{2,2}(1) \times \frac{\partial \mathbf{a}^2(t-1)}{\partial \mathbf{x}^T}\right]$$

$$\frac{\partial \mathbf{a}^8(t)}{\partial \mathbf{x}^T} = \frac{\partial^e \mathbf{a}^8(t)}{\partial \mathbf{x}^T} + \mathbf{S}^{8,2}(t)\left[\mathbf{LW}^{2,2}(1) \times \frac{\partial \mathbf{a}^2(t-1)}{\partial \mathbf{x}^T}\right]$$

$$\frac{\partial \mathbf{a}^{10}(t)}{\partial \mathbf{x}^T} = \frac{\partial^e \mathbf{a}^{10}(t)}{\partial \mathbf{x}^T} + \mathbf{S}^{10,9}(t)\left[\mathbf{LW}^{9,8}(1) \times \frac{\partial \mathbf{a}^8(t-1)}{\partial \mathbf{x}^T}\right]$$

$$+ \mathbf{S}^{10,6}(t)\left[\mathbf{LW}^{6,5}(0) \times \frac{\partial \mathbf{a}^5(t)}{\partial \mathbf{x}^T} + \mathbf{LW}^{6,5}(1) \times \frac{\partial \mathbf{a}^5(t-1)}{\partial \mathbf{x}^T}\right]$$

$$+ \mathbf{S}^{10,4}(t)\left[\mathbf{LW}^{4,4}(1) \times \frac{\partial \mathbf{a}^4(t-1)}{\partial \mathbf{x}^T} + \mathbf{LW}^{4,3}(1) \times \frac{\partial \mathbf{a}^3(t-1)}{\partial \mathbf{x}^T}\right] + \mathbf{S}^{10,2}(t)\left[\mathbf{LW}^{2,2}(1) \times \frac{\partial \mathbf{a}^2(t-1)}{\partial \mathbf{x}^T}\right]$$

After the above steps are iterated through all time points, we can compute the gradient with

$$\frac{\partial F}{\partial \mathbf{x}^T} = \sum_{t=1}^{Q} \left[\left[\frac{\partial^e F}{\partial \mathbf{a}^2(t)}\right]^T \times \frac{\partial \mathbf{a}^2(t)}{\partial \mathbf{x}^T} + \left[\frac{\partial^e F}{\partial \mathbf{a}^3(t)}\right]^T \times \frac{\partial \mathbf{a}^3(t)}{\partial \mathbf{x}^T} + \left[\frac{\partial^e F}{\partial \mathbf{a}^4(t)}\right]^T \times \frac{\partial \mathbf{a}^4(t)}{\partial \mathbf{x}^T}\right.$$

$$\left. + \left[\frac{\partial^e F}{\partial \mathbf{a}^5(t)}\right]^T \times \frac{\partial \mathbf{a}^5(t)}{\partial \mathbf{x}^T} + \left[\frac{\partial^e F}{\partial \mathbf{a}^8(t)}\right]^T \times \frac{\partial \mathbf{a}^8(t)}{\partial \mathbf{x}^T} + \left[\frac{\partial^e F}{\partial \mathbf{a}^{10}(t)}\right]^T \times \frac{\partial \mathbf{a}^{10}(t)}{\partial \mathbf{x}^T} \right]$$

P14.4 From the previous problem, show the detail of the calculations involved in the explicit derivative term

$$\frac{\partial^e \mathbf{a}^2(t)}{\partial vec(\mathbf{IW}^{1,1}(0))^T} = [\mathbf{p}^1(t)]^T \otimes \mathbf{S}^{2,1}(t)$$

First, let's display the details of the individual vectors and matrices in this expression:

$$\mathbf{IW}^{1,1}(0) = \begin{bmatrix} iw^{1,1}_{1,1} & iw^{1,1}_{1,2} & iw^{1,1}_{1,3} \\ iw^{1,1}_{2,1} & iw^{1,1}_{2,2} & iw^{1,1}_{2,3} \end{bmatrix}$$

$$vec(\mathbf{IW}^{1,1}(0))^T = \begin{bmatrix} iw^{1,1}_{1,1} & iw^{1,1}_{2,1} & iw^{1,1}_{1,2} & iw^{1,1}_{2,2} & iw^{1,1}_{1,3} & iw^{1,1}_{2,3} \end{bmatrix}$$

$$\mathbf{p}^1(t) = \begin{bmatrix} p_1 \\ p_2 \\ p_3 \end{bmatrix} \quad \mathbf{S}^{2,1}(t) = \frac{\partial^e \mathbf{a}^2(t)}{\partial \mathbf{n}^1(t)^T} = \begin{bmatrix} s^{2,1}_{1,1} & s^{2,1}_{1,2} \\ s^{2,1}_{2,1} & s^{2,1}_{2,2} \\ s^{2,1}_{3,1} & s^{2,1}_{3,2} \end{bmatrix}$$

$$\frac{\partial^e \mathbf{a}^2(t)}{\partial vec(\mathbf{IW}^{1,1}(0))^T} = \begin{bmatrix} \frac{\partial a^2_1}{\partial iw^{1,1}_{1,1}} & \frac{\partial a^2_1}{\partial iw^{1,1}_{2,1}} & \frac{\partial a^2_1}{\partial iw^{1,1}_{1,2}} & \frac{\partial a^2_1}{\partial iw^{1,1}_{2,2}} & \frac{\partial a^2_1}{\partial iw^{1,1}_{1,3}} & \frac{\partial a^2_1}{\partial iw^{1,1}_{2,3}} \\ \frac{\partial a^2_2}{\partial iw^{1,1}_{1,1}} & \frac{\partial a^2_2}{\partial iw^{1,1}_{2,1}} & \frac{\partial a^2_2}{\partial iw^{1,1}_{1,2}} & \frac{\partial a^2_2}{\partial iw^{1,1}_{2,2}} & \frac{\partial a^2_2}{\partial iw^{1,1}_{1,3}} & \frac{\partial a^2_2}{\partial iw^{1,1}_{2,3}} \\ \frac{\partial a^2_3}{\partial iw^{1,1}_{1,1}} & \frac{\partial a^2_3}{\partial iw^{1,1}_{2,1}} & \frac{\partial a^2_3}{\partial iw^{1,1}_{1,2}} & \frac{\partial a^2_3}{\partial iw^{1,1}_{2,2}} & \frac{\partial a^2_3}{\partial iw^{1,1}_{1,3}} & \frac{\partial a^2_3}{\partial iw^{1,1}_{2,3}} \end{bmatrix}$$

The Kronecker product is defined as

$$\mathbf{A} \otimes \mathbf{B} = \begin{bmatrix} a_{1,1}\mathbf{B} & \dots & a_{1,m}\mathbf{B} \\ \vdots & & \vdots \\ a_{n,1}\mathbf{B} & \dots & a_{n,m}\mathbf{B} \end{bmatrix},$$

therefore

$$[\mathbf{p}^1(t)]^T \otimes \mathbf{S}^{2,1}(t) = \begin{bmatrix} p_1 s^{2,1}_{1,1} & p_1 s^{2,1}_{1,2} & p_2 s^{2,1}_{1,1} & p_2 s^{2,1}_{1,2} & p_3 s^{2,1}_{1,1} & p_3 s^{2,1}_{1,2} \\ p_1 s^{2,1}_{2,1} & p_1 s^{2,1}_{2,2} & p_2 s^{2,1}_{2,1} & p_2 s^{2,1}_{2,2} & p_3 s^{2,1}_{2,1} & p_3 s^{2,1}_{2,2} \\ p_1 s^{2,1}_{3,1} & p_1 s^{2,1}_{3,2} & p_2 s^{2,1}_{3,1} & p_2 s^{2,1}_{3,2} & p_3 s^{2,1}_{3,1} & p_3 s^{2,1}_{3,2} \end{bmatrix}.$$

P14.5 Find the computational complexity for the BPTT and RTRL algorithms applied to the sample network in Figure P14.3 as a function of the number of neurons in Layer 1 (S^1), the number of delays in the tapped delay line (D) and the length of the training sequence (Q).

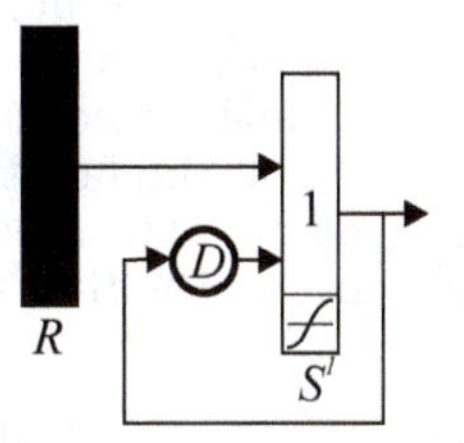

Figure P14.3 Sample Network for Exercise E14.1

The complexity of the BPTT gradient calculation is generally determined by Eq. (14.54). For this network, the most important weights will be $\mathbf{LW}^{1,1}(d)$:

$$\frac{\partial F}{\partial \mathbf{LW}^{1,1}(d)} = \sum_{t=1}^{Q} \mathbf{d}^1(t) \times [\mathbf{a}^1(t-d)]^T,$$

The outer product calculation involves $(S^1)^2$ operations, which must be done for Q time steps and for D delays, so the BPTT gradient calculation is $O[(S^1)^2 DQ]$.

The complexity of the RTRL gradient is based generally on Eq. (14.34). For this sample network, we can consider the equation for $u = 2$:

$$\frac{\partial \mathbf{a}^2(t)}{\partial \mathbf{x}^T} = \frac{\partial^e \mathbf{a}^2(t)}{\partial \mathbf{x}^T} + \mathbf{S}^{2,1}(t)\left[\sum_{d=1}^{D} \mathbf{LW}^{1,1}(d) \times \frac{\partial \mathbf{a}^1(t-d)}{\partial \mathbf{x}^T}\right]$$

Inside the summation we have a matrix multiplication involving an $S^1 \times S^1$ matrix times an $S^1 \times \{(DS^1+3)S^1+1\}$ matrix. This multiplication will be $O[(S^1)^4 D]$. It must be done for every d and every t, therefore the RTRL gradient calculations are $O[(S^1)^4 D^2 Q]$. The multiplication by the sensitivity matrix does not change the order of the complexity.

Epilogue

Dynamic networks can be trained using the same optimization procedures that we described in Chapter 12 for static multilayer networks. However, the calculation of the gradient for dynamic networks is more complex than for static networks. There are two basic approaches to the gradient calculation in dynamic networks. One approach, backpropagation through time (BPTT), starts at the last time point, and works backward in time to compute the gradient. The second approach, real-time recurrent learning (RTRL), starts at the first time point, and then works forward through time.

RTRL requires more calculations than BPTT for calculating the gradient, but RTRL allows a convenient framework for on-line implementation. Also, RTRL generally requires less storage than BPTT. For Jacobian calculations, RTRL is often more efficient than the BPTT algorithm.

Chapter 22 presents a real-world case study for using dynamic networks to solve a prediction problem.

Further Reading

[DeHa07] O. De Jesús and M. Hagan, "Backpropagation Algorithms for a Broad Class of Dynamic Networks," *IEEE Transactions on Neural Networks*, vol. 18, no. 1, pp., 2007.

This paper presents a general development of BPTT and RTRL algorithms for gradient and Jacobian calculations. Experimental results are presented that compare the computational complexities of the two algorithms for a variety of network architectures.

[MaNe99] J.R. Magnus and H. Neudecker, *Matrix Differential Calculus*, John Wiley & Sons, Ltd., Chichester, 1999.

This textbook provides a very clear and complete description of matrix theory and matrix differential calculus.

[PhHa13] M. Phan and M. Hagan, "Error Surface of Recurrent Networks," *IEEE Transactions on Neural Networks and Learning Systems*, Vol. 24, No. 11, pp. 1709 - 1721, October, 2013.

This paper describes spurious valleys that appear in the error surfaces of recurrent networks. It also describes some procedures that can be used to improve training of recurrent networks.

[Werb90] P. J. Werbos, "Backpropagation through time: What it is and how to do it," *Proceedings of the IEEE*, vol. 78, pp. 1550–1560, 1990.

The backpropagation through time algorithm is one of the two principal approaches to computing the gradients in recurrent neural network. This paper describes the general framework for backpropagation through time.

[WiZi89] R. J. Williams and D. Zipser, "A learning algorithm for continually running fully recurrent neural networks," *Neural Computation*, vol. 1, pp. 270–280, 1989.

This paper introduced the real-time, recurrent learning algorithm for computing the gradients of dynamic networks. Using this method, the gradients are computed by starting at the first time point, and then working forward through time. The algorithm is suitable for on-line, or real-time, implementation.

Exercises

E14.1 Put the network of Figure 14.1 into the schematic form, which we introduced in Problem P14.1.

E14.2 Consider the network in Figure 14.4, with weight values $iw_{1,1} = 2$ and $lw_{1,1}(1) = 0.5$. If $a(0) = 4$, and $p(1) = 2$, $p(2) = 3$, $p(3) = 2$, find $a(1)$, $a(2)$, $a(3)$.

E14.3 Consider the network in Figure P14.3, with $D = 2$, $S^1 = 1$, $R = 1$, $\mathbf{IW}^{1,1} = [1]$, $\mathbf{LW}^{1,1}(1) = [0.5]$, $\mathbf{LW}^{1,1}(2) = [0.2]$, $\mathbf{a}^1(0) = [2]$ and $\mathbf{a}^1(-1) = [1]$. If $\mathbf{p}^1(1) = [1]$, $\mathbf{p}^1(2) = [2]$ and $\mathbf{p}^1(3) = [-1]$, find $\mathbf{a}^1(1)$, $\mathbf{a}^1(2)$ and $\mathbf{a}^1(3)$.

E14.4 Consider the network in Figure E14.1. Define the network architecture, by showing U, X, I_m, $DI_{m,1}$, $DL_{m,l}$, L_m^f, L_m^b, $E_{LW}^U(x)$, $E_{LW}^X(u)$. Also, select a simulation order and indicate the dimension of each weight matrix.

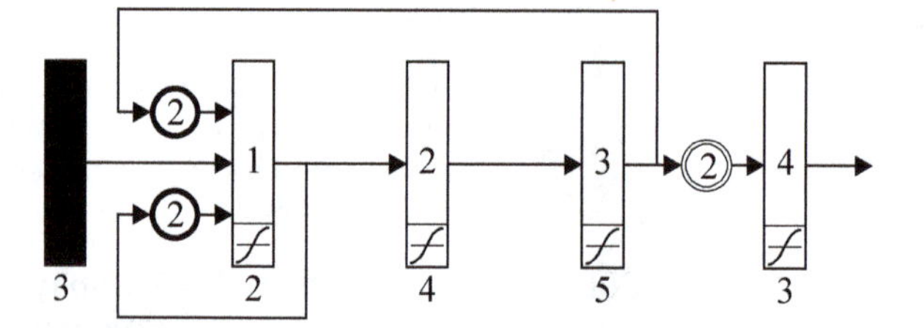

Figure E14.1 Dynamic Network for Exercise E14.4

E14.5 Write out the RTRL equations for the network in Figure E14.2

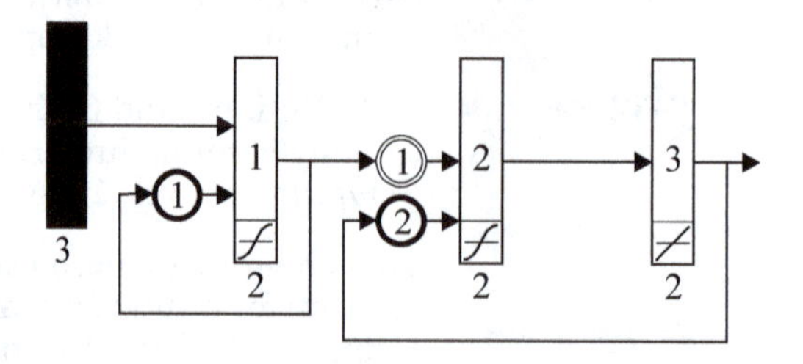

Figure E14.2 Dynamic Network for Exercise E14.5

E14.6 Write out the BPTT equations for the network in Figure E14.2.

E14.7 Write out the equations of operation for the network in Figure E14.3. Assume that all weights have a value of 0.5, and that all biases have a value of 0.

i. Assume that the initial network output is $a(0) = 0.5$, and that the initial network input is $p(1) = 1$. Solve for $a(1)$.

ii. Describe any problems that you would have in simulating this network. Will you be able to apply the BPTT and RTRL algorithms to compute the gradients for this network? What test should you apply to recurrent networks to determine if they can be simulated and trained?

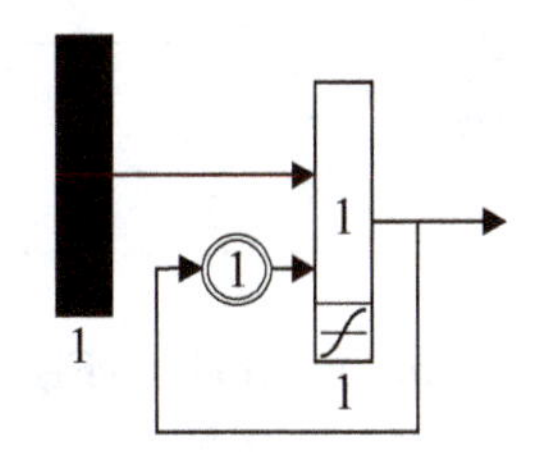

Figure E14.3 Dynamic Network for Exercise E14.7

E14.8 Consider the network in Figure E14.4.

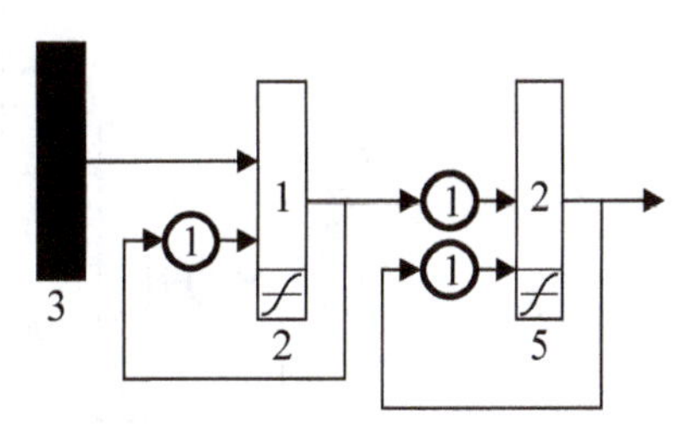

Figure E14.4 Dynamic Network for Exercise E14.8

i. Write out the equations for computing the network response.

ii. Write out the BPTT equations for the network.

iii. Write out the RTRL equations for the network.

E14.9 Repeat E14.8 for the following networks.

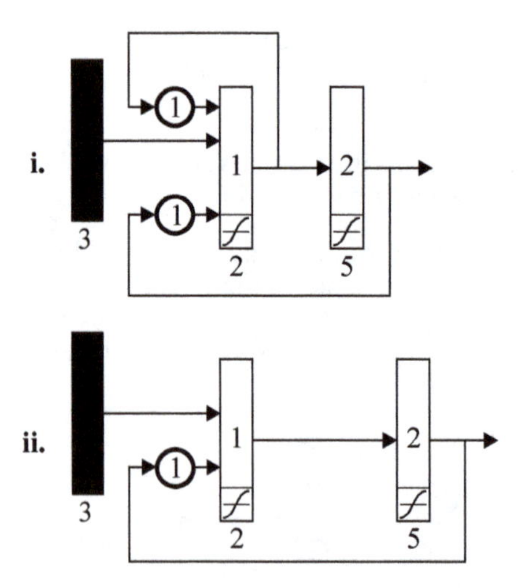

E14.10 Consider the network in Figure E14.5.

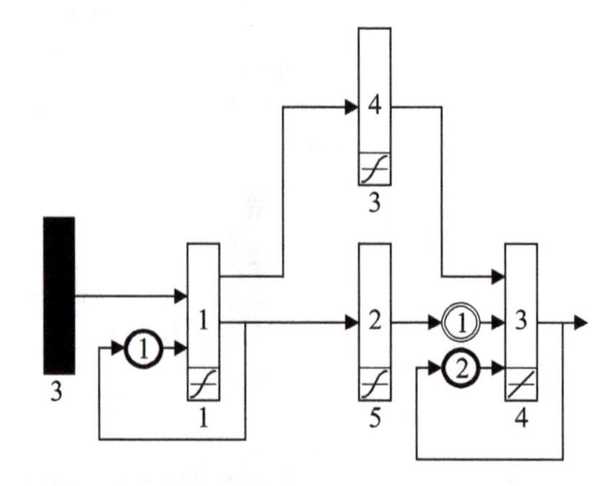

Figure E14.5 Recurrent Network for Exercise E14.10

i. Define the network architecture, by showing U, X, I_m, $DI_{m,1}$, $DL_{m,l}$, L_m^f, L_m^b, $E_{LW}^U(x)$, $E_{LW}^X(u)$.

ii. Select a simulation order and write out the equations needed to determine the network response.

iii. Which $S^{u,x}(t)$ will need to be calculated (i.e., for which u and which x)?

iv. Write out Eq. (14.63) specifically for the term $\partial F / \partial \mathbf{a}^3(t)$ and Eq. (14.34) specifically for the term $\partial \mathbf{a}^3(t) / \partial \mathbf{x}^T$. (Expand the summation to show exactly which terms are included.)

E14.11 Repeat E14.10 for the following networks, except, in part iv., change $\mathbf{a}^3$ to the indicated layer.

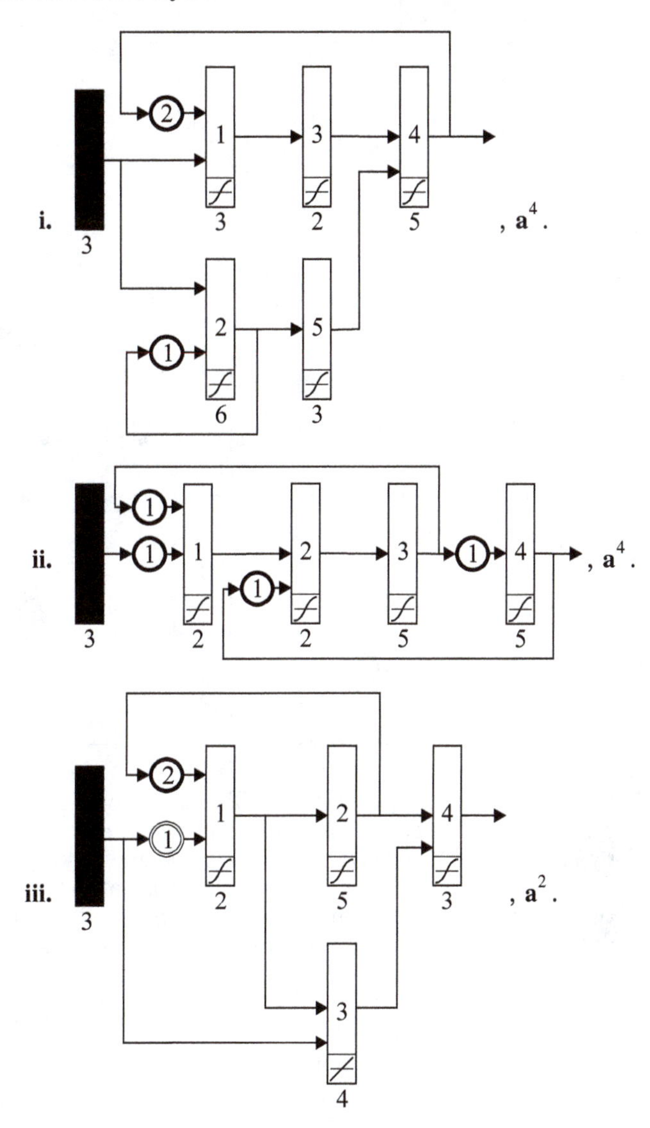

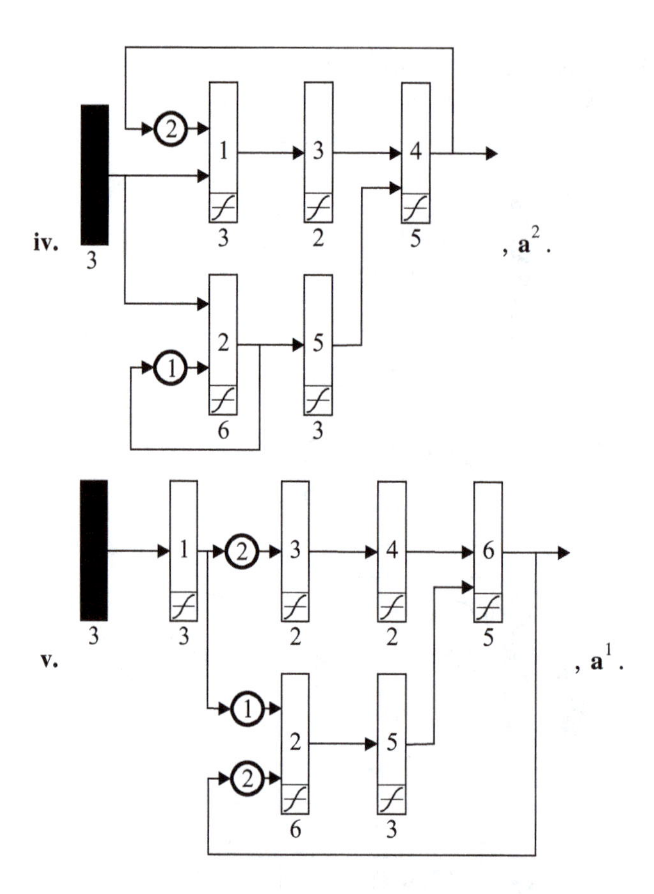

E14.12 One of the advantages of the RTRL algorithm is that the gradient can be computed at the same time as the network response, since it is computed by starting at the first time point, and then working forward through time. Therefore, RTRL allows a convenient framework for on-line implementation. Suppose that you are implementing the RTRL algorithm, and you update the weights of the network at each time step.

i. Discuss the accuracy of the gradient calculation if the weights are updated at each time step of a steepest descent algorithm.

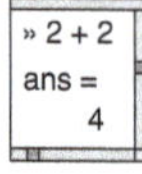

ii. Implement the RTRL algorithm in MATLAB for the network in Figure 14.4. Generate training data by using the network with the two weights set equal to 0.5. Use the same input sequence shown in Figure 14.10, and use the network responses $a(t)$ as the targets. Using this data set, train the network using the steepest descent algorithm, with a learning rate of $\alpha = 0.1$. Set the initial weights to zero. Update the weights at each time step. Save the gradient at the end of the eighth time step.

iii. Repeat part ii., but do not update the weights. Compute the gradient at the end of the eighth time step. Compare this gradient with the one you obtained in part ii. Explain any differences

E14.13 Consider again the recurrent network of Figure 14.4. Assume that $F(\mathbf{x})$ is the sum squared error for the first two time points.

i. Let $a(0) = 0$. Find $a(1)$ and $a(2)$ as a functions of $p(1)$, $p(2)$, and the weights in the network.

ii. Find the sum squared error for the first two time steps as an explicit function of $p(1)$, $p(2)$, $t(1)$, $t(2)$, and the network weights.

iii. Using part (ii), find $\dfrac{\partial F}{\partial lw_{1,1}(1)}$.

iv. Compare the results of part (iii) with the results determined by RTRL on page 14-22 and by BPTT on page 14-30.

E14.14 In the process of deriving the RTRL algorithm in Exercise E14.5, you should have produced the following expression

$$\frac{\partial^e \mathbf{a}^3(t)}{\partial vec(\mathbf{LW}^{2,1}(1))^T} = [\mathbf{a}^1(t-1)]^T \otimes \mathbf{S}^{3,2}(t)$$

If

$$\mathbf{a}^1(1) = \begin{bmatrix} 1 \\ -1 \end{bmatrix} \text{ and } \mathbf{S}^{3,2}(2) = \begin{bmatrix} 2 & 3 \\ 4 & -5 \end{bmatrix},$$

Find $\dfrac{\partial^e \mathbf{a}^3(2)}{\partial vec(\mathbf{LW}^{2,1}(1))^T}$ and indicate $\dfrac{\partial^e a_1^3(2)}{\partial vec(lw_{1,2}^{2,1}(1))^T}$.

E14.15 Each layer of a standard LDDN network has a summing junction, which combines contributions from inputs, other layers, and the bias, as in Eq. (14.1), which is repeated here:

$$\mathbf{n}^m(t) = \sum_{l \in L_m^f} \sum_{d \in DL_{m,l}} \mathbf{LW}^{m,l}(d)\mathbf{a}^l(t-d) + \sum_{l \in I_m} \sum_{d \in DI_{m,l}} \mathbf{IW}^{m,l}(d)\mathbf{p}^l(t-d) + \mathbf{b}^m .$$

If, instead of *summing*, the net input was computed as a *product* of the contributions, how would the RTRL and BPTT algorithms change?

E14.16 As discussed in Exercise E14.15, the contribution to the net input from other layers is computed as product of a layer matrix with a layer output, as in

$$\mathbf{LW}^{m,l}(d)\mathbf{a}^l(t-d).$$

If, instead of multiplying the layer matrix times the layer output, we were to compute the distance between each row of the weight matrix and the layer output, as in

$$n_i = -\|{}_i\mathbf{w} - \mathbf{a}\| .$$

How would the RTRL and BPTT algorithms change?

E14.17 Find and compare the computational complexity for the BPTT and RTRL algorithms applied to the sample network in Figure E14.6 as a function of the number of neurons in Layer 2 (S^2), the number of delays in the tapped delay line (D) and the length of the training sequence (Q).

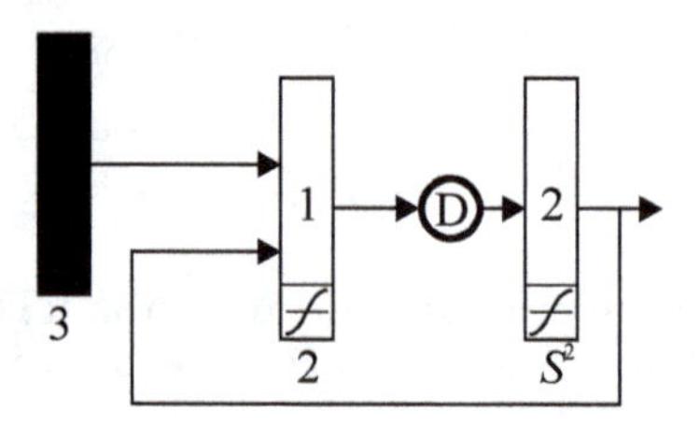

Figure E14.6 Recurrent Network for Exercise E14.17

E14.18 Consider again the network of Figure 14.4. Let the input weight of the network $iw_{1,1} = 1$. Assume that the initial network output is $a(0) = 0$.

i. Write the network output at time t as a function only of the layer weight $lw_{1,1}(1)$, and the input sequence. (The result should be a polynomial in $lw_{1,1}(1)$.)

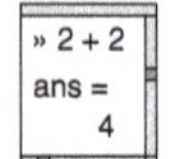

ii. Find the network output at time $t = 8$, using $lw_{1,1}(1) = -1.4$ and the following input sequence:

$$p(t) = \{3, 1, 1, 6, 3, 5, 1, 6\} .$$

iii. With $lw_{1,1}(1) = -1.4$, the network should be unstable, since this weight in the feedback loop is greater than one in magnitude. The output would generally grow with time for an unstable network. (This applies to linear networks.) In ii., you should have found a small value for $a(8)$. Can you explain this result? (Hint: Investigate the roots of the polynomial you found in part i. You can use the MATLAB command `roots`.) How might this result be related to the spurious valleys in the error surface discussed on page 14-32?

15 Competitive Networks

Objectives

The Hamming network, introduced in Chapter 3, demonstrated one technique for using a neural network for pattern recognition. It required that the prototype patterns be known beforehand and incorporated into the network as rows of a weight matrix.

In this chapter we will discuss networks that are very similar in structure and operation to the Hamming network. Unlike the Hamming network, however, they use associative learning rules to adaptively learn to classify patterns. Three such networks are introduced in this chapter: the competitive network, the feature map and the learning vector quantization (LVQ) network.

Theory and Examples

The Hamming network is one of the simplest examples of a competitive network. The neurons in the output layer of the Hamming network compete with each other to determine a winner. The winner indicates which prototype pattern is most representative of the input pattern. The competition is implemented by lateral inhibition — a set of negative connections between the neurons in the output layer. In this chapter we will illustrate how this competition can be combined with associative learning rules to produce powerful self-organizing (unsupervised) networks.

As early as 1959, Frank Rosenblatt created a simple "spontaneous" classifier, an unsupervised network based on the perceptron, which learned to classify input vectors into two classes with roughly equal members.

In the late 1960s and early 1970s, Stephen Grossberg introduced many competitive networks that used lateral inhibition to good effect. Some of the useful behaviors he obtained were noise suppression, contrast-enhancement and vector normalization.

In 1973, Christoph von der Malsburg introduced a self-organizing learning rule that allowed a network to classify inputs in such a way that neighboring neurons responded to similar inputs. The topology of his network mimicked, in some ways, the structures previously found in the visual cortex of cats by David Hubel and Torten Wiesel. His learning rule generated a great deal of interest, but it used a nonlocal calculation to ensure that weights were normalized. This made it less biologically plausible.

Grossberg extended von der Malsburg's work by rediscovering the instar rule. (The instar rule had previously been introduced by Nils Nilsson in his 1965 book *Learning Machines*.) Grossberg showed that the instar rule removed the necessity of re-normalizing weights, since weight vectors that learn to recognize normalized input vectors will automatically be normalized themselves.

The work of Grossberg and von der Malsburg emphasizes the biological plausibility of their networks. Another influential researcher, Teuvo Kohonen, has also been a strong proponent of competitive networks. However, his emphasis has been on engineering applications and efficient mathematical descriptions of the networks. During the 1970s he developed a simplified version of the instar rule and also, inspired by the work of von der Malsburg and Grossberg, found an efficient way to incorporate topology into a competitive network.

In this chapter we will concentrate on the Kohonen framework for competitive networks. His models illustrate the major features of competitive networks, and yet they are mathematically more tractable than the Grossberg networks. They provide a good introduction to competitive learning.

We will begin with the simple competitive network. Next we will present the *self-organizing feature map*, which incorporates a network topology. Finally, we will discuss *learning vector quantization*, which incorporates competition within a supervised learning framework.

Hamming Network

Since the competitive networks discussed in this chapter are closely related to the Hamming network (shown in Figure 15.1), it is worth reviewing the key concepts of that network first.

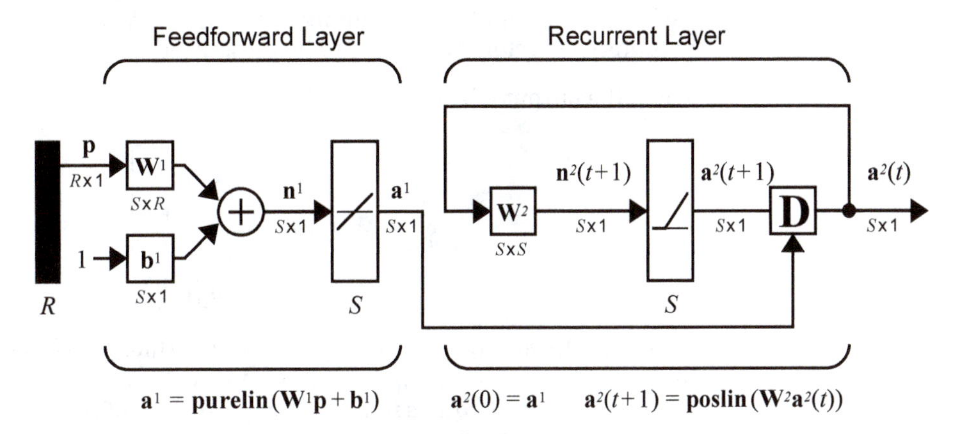

Figure 15.1 Hamming Network

The Hamming network consists of two layers. The first layer (which is a layer of instars) performs a correlation between the input vector and the prototype vectors. The second layer performs a competition to determine which of the prototype vectors is closest to the input vector.

Layer 1

A single instar is able to recognize only one pattern. In order to allow multiple patterns to be classified, we need to have multiple instars. This is accomplished in the Hamming network.

Suppose that we want the network to recognize the following prototype vectors:

$$\{\mathbf{p}_1, \mathbf{p}_2, \ldots, \mathbf{p}_Q\}. \tag{15.1}$$

Then the weight matrix, $\mathbf{W}^1$, and the bias vector, $\mathbf{b}^1$, for Layer 1 will be:

$$\mathbf{W}^1 = \begin{bmatrix} {}_1\mathbf{w}^T \\ {}_2\mathbf{w}^T \\ \vdots \\ {}_S\mathbf{w}^T \end{bmatrix} = \begin{bmatrix} \mathbf{p}_1^T \\ \mathbf{p}_2^T \\ \vdots \\ \mathbf{p}_Q^T \end{bmatrix}, \quad \mathbf{b}^1 = \begin{bmatrix} R \\ R \\ \vdots \\ R \end{bmatrix}, \tag{15.2}$$

where each row of $\mathbf{W}^1$ represents a prototype vector which we want to recognize, and each element of $\mathbf{b}^1$ is set equal to the number of elements in each input vector (R). (The number of neurons, S, is equal to the number of prototype vectors which are to be recognized, Q.)

Thus, the output of the first layer is

$$\mathbf{a}^1 = \mathbf{W}^1\mathbf{p} + \mathbf{b}^1 = \begin{bmatrix} \mathbf{p}_1^T\mathbf{p} + R \\ \mathbf{p}_2^T\mathbf{p} + R \\ \vdots \\ \mathbf{p}_Q^T\mathbf{p} + R \end{bmatrix}. \tag{15.3}$$

Note that the outputs of Layer 1 are equal to the inner products of the prototype vectors with the input, plus R. As we discussed in Chapter 3 (page 3-9), these inner products indicate how close each of the prototype patterns is to the input vector.

Layer 2

In the instar, a *hardlim* transfer function is used to decide if the input vector is close enough to the prototype vector. In Layer 2 of the Hamming network we have multiple instars, therefore we want to decide which prototype vector is closest to the input. Instead of the *hardlim* transfer function, we will use a competitive layer to choose the closest prototype.

Layer 2 is a competitive layer. The neurons in this layer are initialized with the outputs of the feedforward layer, which indicate the correlation between the prototype patterns and the input vector. Then the neurons compete with each other to determine a winner. After the competition, only one neuron will have a nonzero output. The winning neuron indicates which category of input was presented to the network (each prototype vector represents a category).

The first-layer output $\mathbf{a}^1$ is used to initialize the second layer.

$$\mathbf{a}^2(0) = \mathbf{a}^1 \tag{15.4}$$

Then the second-layer output is updated according to the following recurrence relation:

$$\mathbf{a}^2(t+1) = \mathbf{poslin}(\mathbf{W}^2\mathbf{a}^2(t)) . \tag{15.5}$$

The second-layer weights $\mathbf{W}^2$ are set so that the diagonal elements are 1, and the off-diagonal elements have a small negative value.

$$w_{ij}^2 = \begin{cases} 1, & \text{if } i = j \\ -\varepsilon, & \text{otherwise} \end{cases}, \text{ where } 0 < \varepsilon < \frac{1}{S-1} \tag{15.6}$$

Lateral Inhibition

This matrix produces *lateral inhibition*, in which the output of each neuron has an inhibitory effect on all of the other neurons. To illustrate this effect, substitute weight values of 1 and $-\varepsilon$ for the appropriate elements of $\mathbf{W}^2$, and rewrite Eq. (15.5) for a single neuron.

$$a_i^2(t+1) = poslin\left(a_i^2(t) - \varepsilon \sum_{j \neq i} a_j^2(t)\right) \tag{15.7}$$

At each iteration, each neuron's output will decrease in proportion to the sum of the other neurons' outputs (with a minimum output of 0). The output of the neuron with the largest initial condition will decrease more slowly than the outputs of the other neurons. Eventually that neuron will be the only one with a positive output. At this point the network has reached steady state. The index of the second-layer neuron with a stable positive output is the index of the prototype vector that best matched the input.

Winner-Take-All

This is called a *winner-take-all competition*, since only one neuron will have a nonzero output.

You may wish to experiment with the Hamming network and the apple/orange classification problem. The Neural Network Design Demonstration Hamming Classification (`nnd3hamc`) *was previously introduced in Chapter 3.*

Competitive Layer

Competition

The second-layer neurons in the Hamming network are said to be in *competition* because each neuron excites itself and inhibits all the other neurons. To simplify our discussions in the remainder of this chapter, we will define a transfer function that does the job of a recurrent competitive layer:

$$\mathbf{a} = \mathbf{compet}(\mathbf{n}) . \tag{15.8}$$

It works by finding the index i^* of the neuron with the largest net input, and setting its output to 1 (with ties going to the neuron with the lowest index). All other outputs are set to 0.

$$a_i = \begin{cases} 1, i = i^* \\ 0, i \neq i^* \end{cases}, \text{ where } n_{i^*} \geq n_i, \forall i, \text{ and } i^* \leq i, \forall n_i = n_{i^*} \qquad (15.9)$$

Replacing the recurrent layer of the Hamming network with a competitive transfer function on the first layer will simplify our presentations in this chapter. A competitive layer is displayed in Figure 15.2.

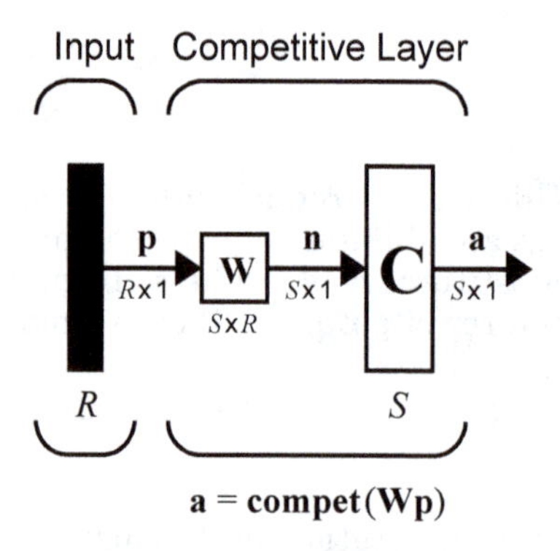

Figure 15.2 Competitive Layer

As with the Hamming network, the prototype vectors are stored in the rows of $\mathbf{W}$. The net input $\mathbf{n}$ calculates the distance between the input vector $\mathbf{p}$ and each prototype ${}_i\mathbf{w}$ (assuming vectors have normalized lengths of L). The net input n_i of each neuron i is proportional to the angle θ_i between $\mathbf{p}$ and the prototype vector ${}_i\mathbf{w}$:

$$\mathbf{n} = \mathbf{W}\mathbf{p} = \begin{bmatrix} {}_1\mathbf{w}^T \\ {}_2\mathbf{w}^T \\ \vdots \\ {}_S\mathbf{w}^T \end{bmatrix} \mathbf{p} = \begin{bmatrix} {}_1\mathbf{w}^T\mathbf{p} \\ {}_2\mathbf{w}^T\mathbf{p} \\ \vdots \\ {}_S\mathbf{w}^T\mathbf{p} \end{bmatrix} = \begin{bmatrix} L^2\cos\theta_1 \\ L^2\cos\theta_2 \\ \vdots \\ L^2\cos\theta_S \end{bmatrix}. \qquad (15.10)$$

The competitive transfer function assigns an output of 1 to the neuron whose weight vector points in the direction closest to the input vector:

$$\mathbf{a} = \mathbf{compet}(\mathbf{W}\mathbf{p}). \qquad (15.11)$$

To experiment with the competitive network and the apple/orange classification problem, use the Neural Network Design Demonstration Competitive Classification (nnd14cc).

Competitive Learning

We can now design a competitive network classifier by setting the rows of $\mathbf{W}$ to the desired prototype vectors. However, we would like to have a learning rule that could be used to train the weights in a competitive network, without knowing the prototype vectors. One such learning rule is the instar rule:

$$ {}_i\mathbf{w}(q) = {}_i\mathbf{w}(q-1) + \alpha a_i(q)(\mathbf{p}(q) - {}_i\mathbf{w}(q-1)). \tag{15.12} $$

For the competitive network, $\mathbf{a}$ is only nonzero for the winning neuron ($i = i^*$). Therefore, we can get the same results using the Kohonen rule.

$$ \begin{aligned} {}_i\mathbf{w}(q) &= {}_i\mathbf{w}(q-1) + \alpha(\mathbf{p}(q) - {}_i\mathbf{w}(q-1)) \\ &= (1-\alpha){}_i\mathbf{w}(q-1) + \alpha\mathbf{p}(q) \end{aligned} \tag{15.13} $$

and

$$ {}_i\mathbf{w}(q) = {}_i\mathbf{w}(q-1) \qquad i \neq i^* \tag{15.14} $$

Thus, the row of the weight matrix that is closest to the input vector (or has the largest inner product with the input vector) moves toward the input vector. It moves along a line between the old row of the weight matrix and the input vector, as shown in Figure 15.3.

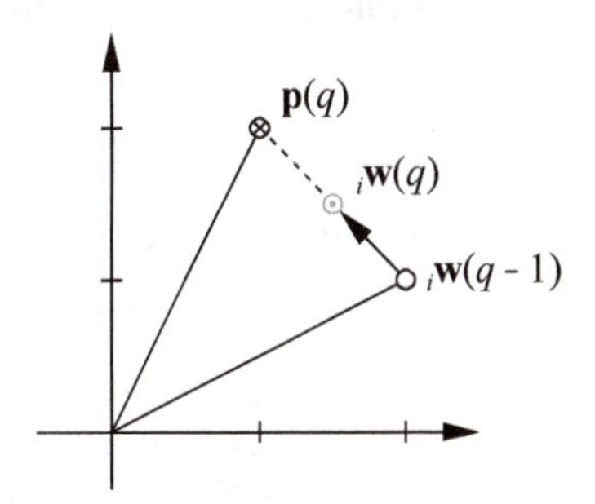

Figure 15.3 Graphical Representation of the Kohonen Rule

Let's use the six vectors in Figure 15.4 to demonstrate how a competitive layer learns to classify vectors. Here are the six vectors:

$$ \mathbf{p}_1 = \begin{bmatrix} -0.1961 \\ 0.9806 \end{bmatrix}, \mathbf{p}_2 = \begin{bmatrix} 0.1961 \\ 0.9806 \end{bmatrix}, \mathbf{p}_3 = \begin{bmatrix} 0.9806 \\ 0.1961 \end{bmatrix} \tag{15.15} $$

$$ \mathbf{p}_4 = \begin{bmatrix} 0.9806 \\ -0.1961 \end{bmatrix}, \mathbf{p}_5 = \begin{bmatrix} -0.5812 \\ -0.8137 \end{bmatrix}, \mathbf{p}_6 = \begin{bmatrix} -0.8137 \\ -0.5812 \end{bmatrix}. $$

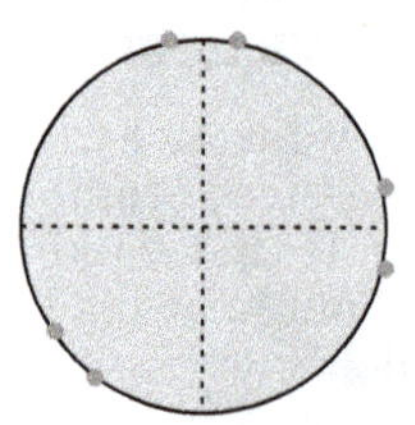

Figure 15.4 Sample Input Vectors

Our competitive network will have three neurons, and therefore it can classify vectors into three classes. Here are the "randomly" chosen normalized initial weights:

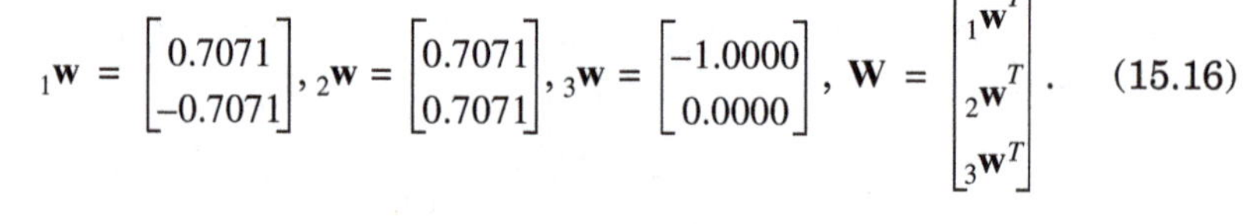

$$ {}_1\mathbf{w} = \begin{bmatrix} 0.7071 \\ -0.7071 \end{bmatrix}, {}_2\mathbf{w} = \begin{bmatrix} 0.7071 \\ 0.7071 \end{bmatrix}, {}_3\mathbf{w} = \begin{bmatrix} -1.0000 \\ 0.0000 \end{bmatrix}, \mathbf{W} = \begin{bmatrix} {}_1\mathbf{w}^T \\ {}_2\mathbf{w}^T \\ {}_3\mathbf{w}^T \end{bmatrix}. \tag{15.16} $$

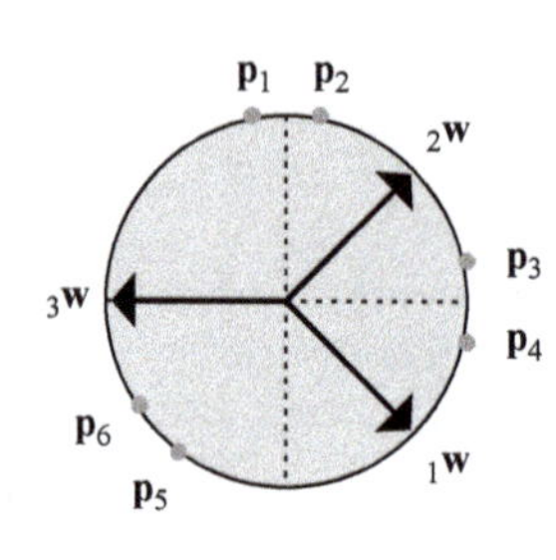

The data vectors are shown at left, with the weight vectors displayed as arrows. Let's present the vector $\mathbf{p}_2$ to the network:

$$ \mathbf{a} = \mathbf{compet}(\mathbf{W}\mathbf{p}_2) = \mathbf{compet}\left(\begin{bmatrix} 0.7071 & -0.7071 \\ 0.7071 & 0.7071 \\ -1.0000 & 0.0000 \end{bmatrix} \begin{bmatrix} 0.1961 \\ 0.9806 \end{bmatrix} \right) \tag{15.17} $$

$$ = \mathbf{compet}\left(\begin{bmatrix} -0.5547 \\ 0.8321 \\ -0.1961 \end{bmatrix} \right) = \begin{bmatrix} 0 \\ 1 \\ 0 \end{bmatrix}. $$

The second neuron's weight vector was closest to $\mathbf{p}_2$, so it won the competition ($i^* = 2$) and output a 1. We now apply the Kohonen learning rule to the winning neuron with a learning rate of $\alpha = 0.5$.

$$ {}_2\mathbf{w}^{new} = {}_2\mathbf{w}^{old} + \alpha(\mathbf{p}_2 - {}_2\mathbf{w}^{old}) \tag{15.18} $$

$$ = \begin{bmatrix} 0.7071 \\ 0.7071 \end{bmatrix} + 0.5\left(\begin{bmatrix} 0.1961 \\ 0.9806 \end{bmatrix} - \begin{bmatrix} 0.7071 \\ 0.7071 \end{bmatrix} \right) = \begin{bmatrix} 0.4516 \\ 0.8438 \end{bmatrix} $$

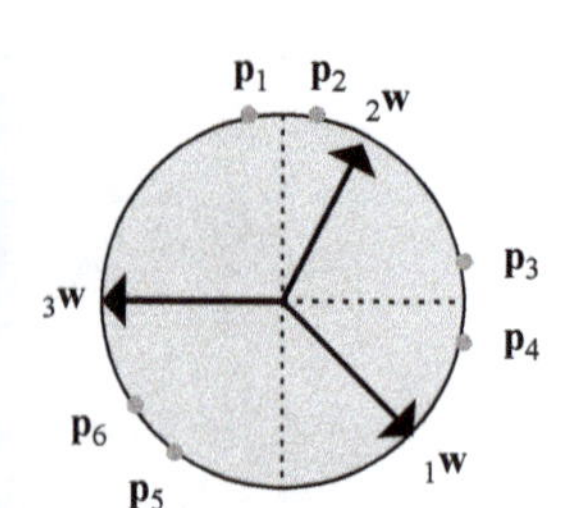

The Kohonen rule moves ${}_2\mathbf{w}$ closer to $\mathbf{p}_2$, as can be seen in the diagram at left. If we continue choosing input vectors at random and presenting them to the network, then at each iteration the weight vector closest to the input vector will move toward that vector. Eventually, each weight vector will point at a different cluster of input vectors. Each weight vector becomes a prototype for a different cluster.

This problem is simple enough that we can predict which weight vector will point at which cluster. The final weights will look something like those shown in Figure 15.5.

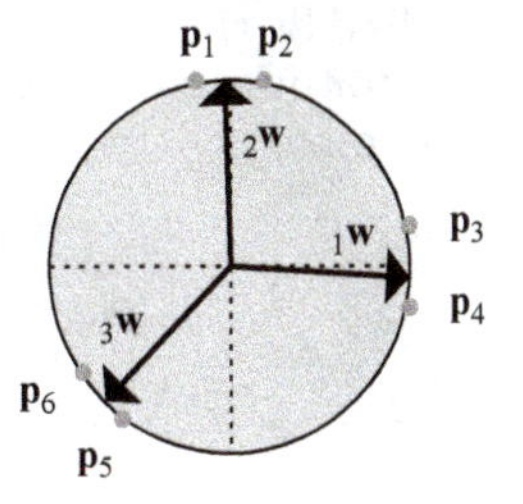

Figure 15.5 Final Weights

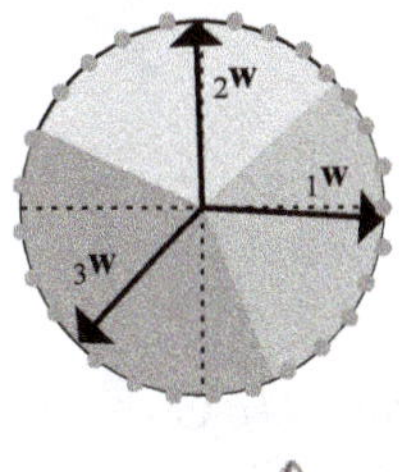

Once the network has learned to cluster the input vectors, it will classify new vectors accordingly.The diagram in the left margin uses shading to show which region each neuron will respond to. The competitive layer assigns each input vector $\mathbf{p}$ to one of these classes by producing an output of 1 for the neuron whose weight vector is closest to $\mathbf{p}$.

To experiment with the competitive learning use the Neural Network Design Demonstration Competitive Learning (`nnd14cl`).

Problems with Competitive Layers

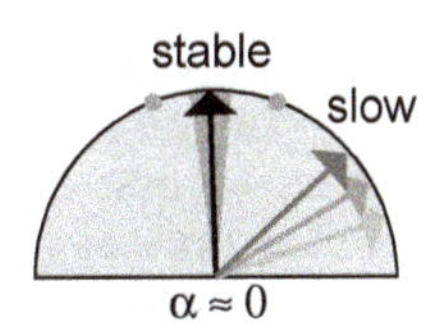

Competitive layers make efficient adaptive classifiers, but they do suffer from a few problems. The first problem is that the choice of learning rate forces a trade-off between the speed of learning and the stability of the final weight vectors. A learning rate near zero results in slow learning. However, once a weight vector reaches the center of a cluster it will tend to stay close to the center.

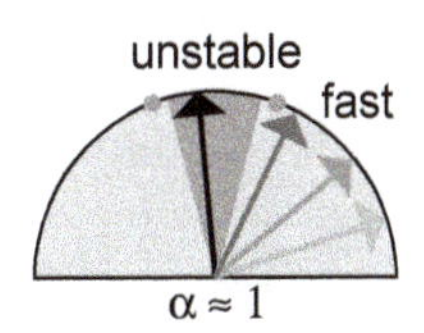

In contrast, a learning rate near 1.0 results in fast learning. However, once the weight vector has reached a cluster, it will continue to oscillate as different vectors in the cluster are presented.

Sometimes this trade-off between fast learning and stability can be used to advantage. Initial training can be done with a large learning rate for fast learning. Then the learning rate can be decreased as training progresses, to achieve stable prototype vectors. Unfortunately, this technique will not work if the network needs to continuously adapt to new arrangements of input vectors.

A more serious stability problem occurs when clusters are close together. In certain cases, a weight vector forming a prototype of one cluster may "invade" the territory of another weight vector, and therefore upset the current classification scheme.

The series of four diagrams in Figure 15.6 illustrate this problem. Two input vectors (shown with blue circles in diagram (a)) are presented several times. The result is that the weight vectors representing the middle and right clusters shift to the right. Eventually one of the right cluster vectors is reclassified by the center weight vector. Further presentations move the middle vector over to the right until it "loses" some of its vectors, which then become part of the class associated with the left weight vector.

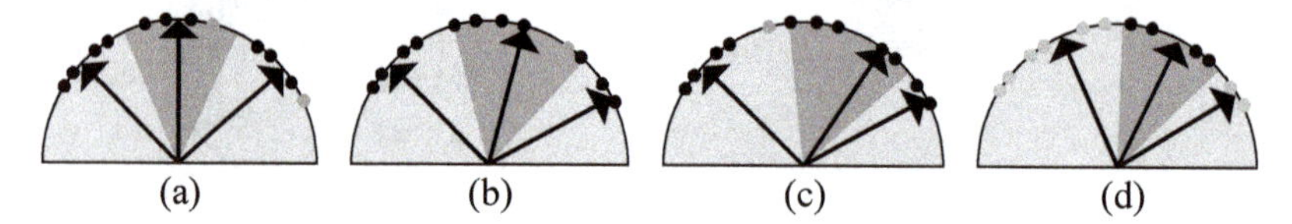

Figure 15.6 Example of Unstable Learning

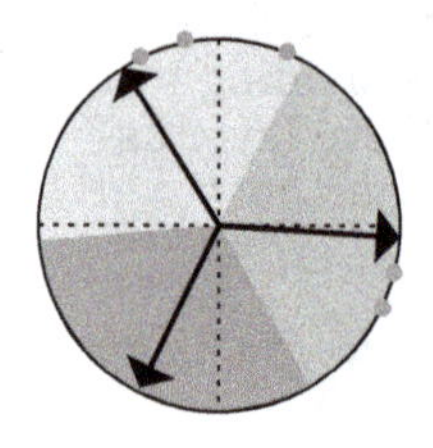

A third problem with competitive learning is that occasionally a neuron's initial weight vector is located so far from any input vectors that it never wins the competition, and therefore never learns. The result is a "dead" neuron, which does nothing useful. For example, the downward-pointing weight vector in the diagram to the left will never learn, regardless of the order in which vectors are presented. One solution to this problem consists of adding a negative bias to the net input of each neuron and then decreasing the bias each time the neuron wins. This will make it harder for a neuron to win the competition if it has won often. This mechanism is sometimes called a "conscience." (See Exercise E15.4.)

Finally, a competitive layer always has as many classes as it has neurons. This may not be acceptable for some applications, especially when the number of clusters is not known in advance. In addition, for competitive layers, each class consists of a convex region of the input space. Competitive layers cannot form classes with nonconvex regions or classes that are the union of unconnected regions.

Some of the problems discussed in this section are solved by the feature map and LVQ networks, which are introduced in later sections of this chapter.

Competitive Layers in Biology

In previous chapters we have made no mention of how neurons are physically organized within a layer (the topology of the network). In biological neural networks, neurons are typically arranged in two-dimensional layers, in which they are densely interconnected through lateral feedback. The diagram to the left shows a layer of twenty-five neurons arranged in a two-dimensional grid.

Often weights vary as a function of the distance between the neurons they connect. For example, the weights for Layer 2 of the Hamming network are assigned as follows:

$$w_{ij} = \begin{cases} 1, & \text{if } i = j \\ -\varepsilon, & \text{if } i \neq j \end{cases} . \tag{15.19}$$

Eq. (15.20) assigns the same values as Eq. (15.19), but in terms of the distances d_{ij} between neurons:

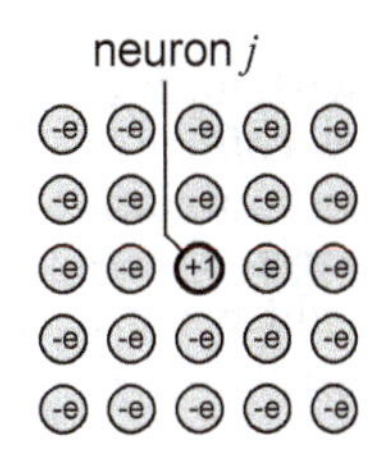

$$w_{ij} = \begin{cases} 1, & \text{if } d_{ij} = 0 \\ -\varepsilon, & \text{if } d_{ij} > 0 \end{cases} . \tag{15.20}$$

Either Eq. (15.19) or Eq. (15.20) will assign the weight values shown in the diagram at left. Each neuron i is labeled with the value of the weight w_{ij}, which comes from it to the neuron marked j.

On-center/ off-surround

The term *on-center / off-surround* is often used to describe such a connection pattern between neurons. Each neuron reinforces itself (center), while inhibiting all other neurons (surround).

It turns out that this is a crude approximation of biological competitive layers. In biology, a neuron reinforces not only itself, but also those neurons close to it. Typically, the transition from reinforcement to inhibition occurs smoothly as the distance between neurons increases.

Mexican-Hat Function

This transition is illustrated on the left side of Figure 15.7. This is a function that relates the distance between neurons to the weight connecting them. Those neurons that are close provide excitatory (reinforcing) connections, and the magnitude of the excitation decreases as the distance increases. Beyond a certain distance, the neurons begin to have inhibitory connections, and the inhibition increases as the distance increases. Because of its shape, the function is referred to as the *Mexican-hat function*. On the right side of Figure 15.7 is a two-dimensional illustration of the Mexican-hat (on-center/off-surround) function. Each neuron i is marked to show the sign and relative strength of its weight w_{ij} going to neuron j.

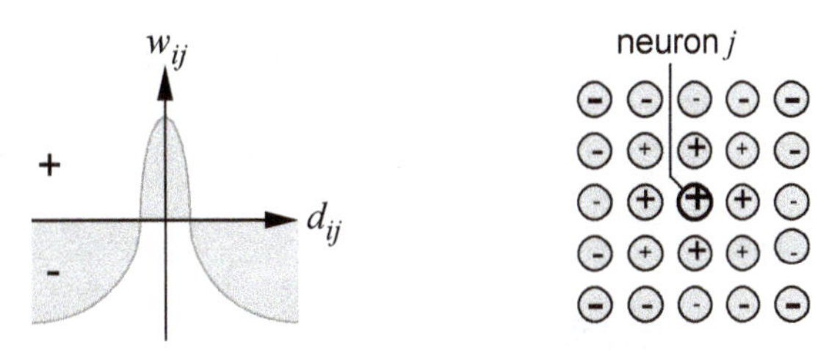

Figure 15.7 On-Center/Off-Surround Layer in Biology

Biological competitive systems, in addition to having a gradual transition between excitatory and inhibitory regions of the on-center/off-surround connection pattern, also have a weaker form of competition than the winner-take-all competition of the Hamming network. Instead of a single active neuron (winner), biological networks generally have "bubbles" of activity that are centered around the most active neuron. This is caused in part by the form of the on-center/off-surround connectivity pattern and also by nonlinear feedback connections.

Self-Organizing Feature Maps

In order to emulate the activity bubbles of biological systems, without having to implement the nonlinear on-center/off-surround feedback connections, Kohonen designed the following simplification. His self-organizing feature map (SOFM) network first determines the winning neuron i^* using the same procedure as the competitive layer. Next, the weight vectors for all neurons within a certain neighborhood of the winning neuron are updated using the Kohonen rule,

SOFM

$$\begin{aligned} {}_i\mathbf{w}(q) &= {}_i\mathbf{w}(q-1) + \alpha(\mathbf{p}(q) - {}_i\mathbf{w}(q-1)) \\ &= (1-\alpha)\,{}_i\mathbf{w}(q-1) + \alpha\mathbf{p}(q) \end{aligned} \qquad i \in N_{i^*}(d)\,, \tag{15.21}$$

where the *neighborhood* $N_{i^*}(d)$ contains the indices for all of the neurons that lie within a radius d of the winning neuron i^*:

Neighborhood

$$N_i(d) = \{j, d_{ij} \le d\}\,. \tag{15.22}$$

When a vector $\mathbf{p}$ is presented, the weights of the winning neuron *and* its neighbors will move toward $\mathbf{p}$. The result is that, after many presentations, neighboring neurons will have learned vectors similar to each other.

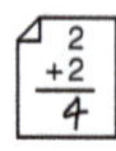

To demonstrate the concept of a neighborhood, consider the two diagrams shown in Figure 15.8. The left diagram illustrates a two-dimensional neighborhood of radius $d = 1$ around neuron 13. The right diagram shows a neighborhood of radius $d = 2$.

The definition of these neighborhoods would be

$$N_{13}(1) = \{8, 12, 13, 14, 18\}\,, \tag{15.23}$$

$$N_{13}(2) = \{3, 7, 8, 9, 11, 12, 13, 14, 15, 17, 18, 19, 23\}\,. \tag{15.24}$$

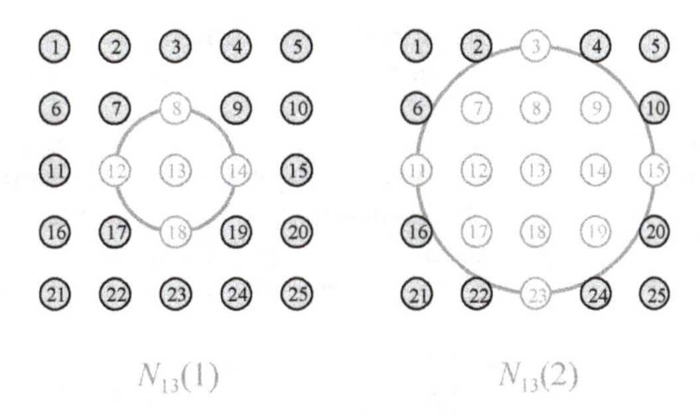

Figure 15.8 Neighborhoods

We should mention that the neurons in an SOFM do not have to be arranged in a two-dimensional pattern. It is possible to use a one-dimensional arrangement, or even three or more dimensions. For a one-dimensional SOFM, a neuron will only have two neighbors within a radius of 1 (or a single neighbor if the neuron is at the end of the line). It is also possible to define distance in different ways. For instance, Kohonen has suggested rectangular and hexagonal neighborhoods for efficient implementation. The performance of the network is not sensitive to the exact shape of the neighborhoods.

Now let's demonstrate the performance of an SOFM network. Figure 15.9 shows a feature map and the two-dimensional topology of its neurons.

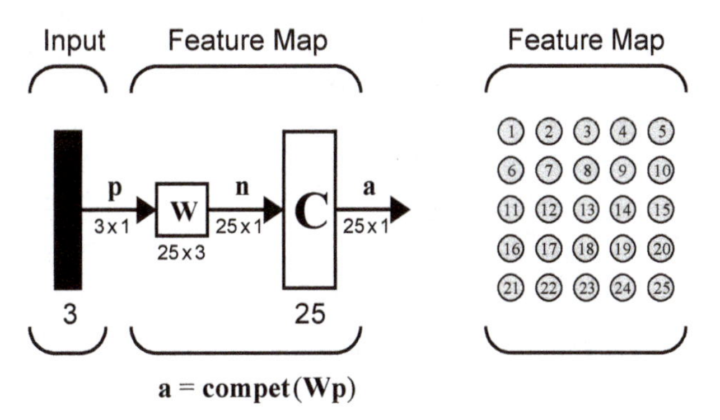

$\mathbf{a} = \mathbf{compet}(\mathbf{Wp})$

Figure 15.9 Self-Organizing Feature Map

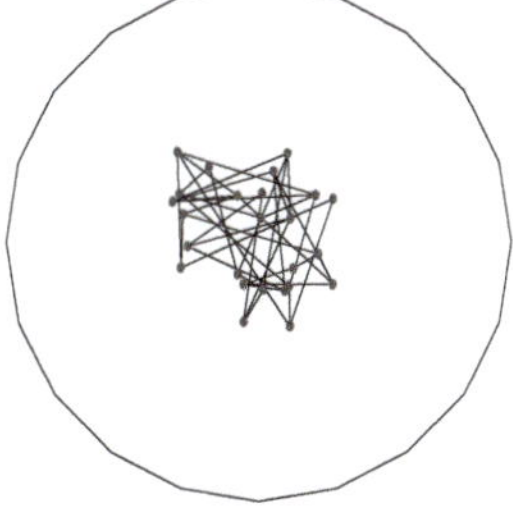

The diagram in the left margin shows the initial weight vectors for the feature map. Each three-element weight vector is represented by a dot on the sphere. (The weights are normalized, therefore they will fall on the surface of a sphere.) Dots of neighboring neurons are connected by lines so you can see how the physical topology of the network is arranged in the input space.

The diagram to the left shows a square region on the surface of the sphere. We will randomly pick vectors in this region and present them to the feature map.

Each time a vector is presented, the neuron with the closest weight vector will win the competition. The winning neuron and its neighbors move their weight vectors closer to the input vector (and therefore to each other). For this example we are using a neighborhood with a radius of 1.

The weight vectors have two tendencies: first, they spread out over the input space as more vectors are presented; second, they move toward the weight vectors of neighboring neurons. These two tendencies work together to rearrange the neurons in the layer so that they evenly classify the input space.

The series of diagrams in Figure 15.10 shows how the weights of the twenty-five neurons spread out over the active input space and organize themselves to match its topology.

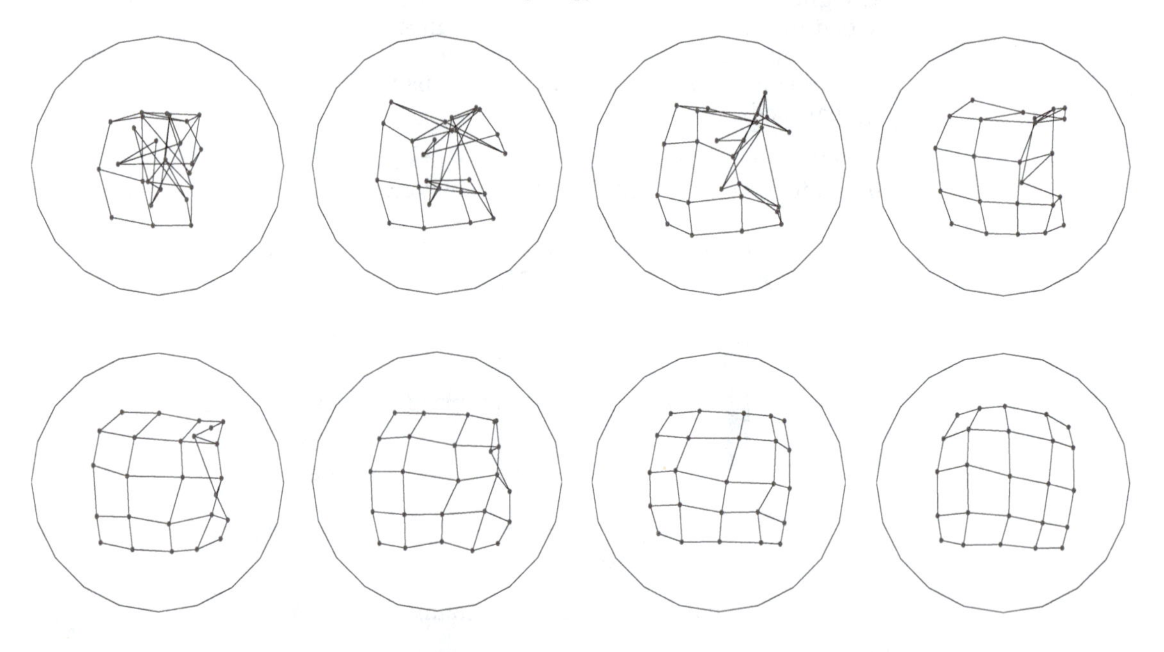

Figure 15.10 Self-Organization, 250 Iterations per Diagram

In this example, the input vectors were generated with equal probability from any point in the input space. Therefore, the neurons classify roughly equal areas of the input space.

Figure 15.11 provides more examples of input regions and the resulting feature maps after self-organization.

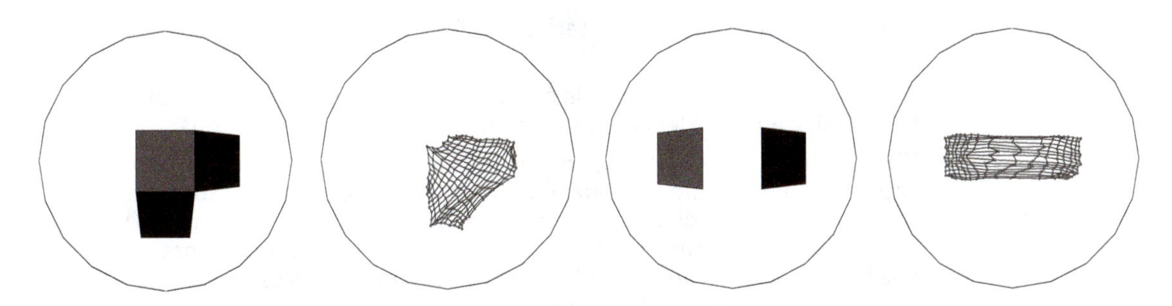

Figure 15.11 Other Examples of Feature Map Training

Occasionally feature maps can fail to properly fit the topology of their input space. This usually occurs when two parts of the net fit the topology of separate parts of the input space, but the net forms a twist between them. An example is given in Figure 15.12.

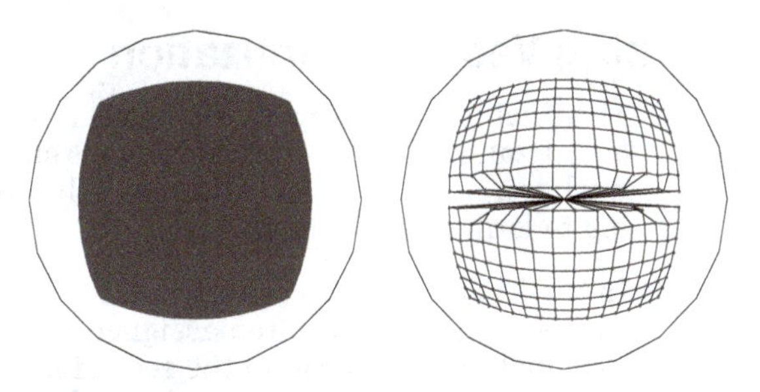

Figure 15.12 Feature Map with a Twist

It is unlikely that this twist will ever be removed, because the two ends of the net have formed stable classifications of different regions.

Improving Feature Maps

So far, we have described only the most basic algorithm for training feature maps. Now let's consider several techniques that can be used to speed up the self-organizing process and to make it more reliable.

One method to improve the performance of the feature map is to vary the size of the neighborhoods during training. Initially, the neighborhood size, d, is set large. As training progresses, d is gradually reduced, until it only includes the winning neuron. This speeds up self-organizing and makes twists in the map very unlikely.

The learning rate can also be varied over time. An initial rate of 1 allows neurons to quickly learn presented vectors. During training, the learning rate is decreased asymptotically toward 0, so that learning becomes stable.

(We discussed the use of this technique for competitive layers earlier in the chapter.)

Another alteration that speeds self-organization is to have the winning neuron use a larger learning rate than the neighboring neurons.

Finally, both competitive layers and feature maps often use an alternative expression for net input. Instead of using the inner product, they can directly compute the distance between the input vector and the prototype vectors. The advantage of using the distance is that input vectors do not need to be normalized. This alternative net input expression is introduced in the next section on LVQ networks.

Other enhancements to the SOFM are described in Chapter 21, including a batch version of the SOFM learning rule. That chapter is a case study of using the SOFM for clustering.

To experiment with feature maps use the Neural Network Design Demonstrations 1-D Feature Maps (`nnd14fm1`) *and* 2-D Feature Maps (`nnd14fm2`).

Learning Vector Quantization

The final network we will introduce in this chapter is the learning vector quantization (LVQ) network, which is shown in Figure 15.13. The LVQ network is a hybrid network. It uses both unsupervised and supervised learning to form classifications.

In the LVQ network, each neuron in the first layer is assigned to a class, with several neurons often assigned to the same class. Each class is then assigned to one neuron in the second layer. The number of neurons in the first layer, S^1, will therefore always be at least as large as the number of neurons in the second layer, S^2, and will usually be larger.

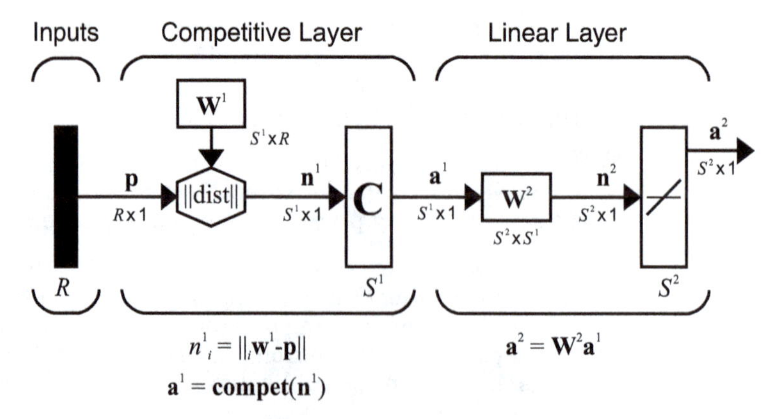

Figure 15.13 LVQ Network

As with the competitive network, each neuron in the first layer of the LVQ network learns a prototype vector, which allows it to classify a region of the input space. However, instead of computing the proximity of the input and weight vectors by using the inner product, we will simulate the LVQ networks by calculating the distance directly. One advantage of calculating the distance directly is that vectors need not be normalized. When the vectors are normalized, the response of the network will be the same, whether the inner product is used or the distance is directly calculated.

The net input of the first layer of the LVQ will be

$$n_i^1 = -\left\| {}_i\mathbf{w}^1 - \mathbf{p} \right\| , \qquad (15.25)$$

or, in vector form,

$$\mathbf{n}^1 = -\begin{bmatrix} \left\| {}_1\mathbf{w}^1 - \mathbf{p} \right\| \\ \left\| {}_2\mathbf{w}^1 - \mathbf{p} \right\| \\ \vdots \\ \left\| {}_{S^1}\mathbf{w}^1 - \mathbf{p} \right\| \end{bmatrix} . \qquad (15.26)$$

The output of the first layer of the LVQ is

$$\mathbf{a}^1 = \mathbf{compet}(\mathbf{n}^1) . \qquad (15.27)$$

Therefore the neuron whose weight vector is closest to the input vector will output a 1, and the other neurons will output 0.

Thus far, the LVQ network behaves exactly like the competitive network (at least for normalized vectors). There is a difference in interpretation, however. In the competitive network, the neuron with the nonzero output indicates which class the input vector belongs to. For the LVQ network, the winning neuron indicates a *subclass*, rather than a class. There may be several different neurons (subclasses) that make up each class.

Subclass

The second layer of the LVQ network is used to combine subclasses into a single class. This is done with the $\mathbf{W}^2$ matrix. The columns of $\mathbf{W}^2$ represent subclasses, and the rows represent classes. $\mathbf{W}^2$ has a single 1 in each column, with the other elements set to zero. The row in which the 1 occurs indicates which class the appropriate subclass belongs to.

$$(w_{ki}^2 = 1) \Rightarrow \text{subclass } i \text{ is a part of class } k \qquad (15.28)$$

The process of combining subclasses to form a class allows the LVQ network to create complex class boundaries. A standard competitive layer has the limitation that it can only create decision regions that are convex. The LVQ network overcomes this limitation.

LVQ Learning

The learning in the LVQ network combines competitive learning with supervision. As with all supervised learning algorithms, it requires a set of examples of proper network behavior:

$$\{\mathbf{p}_1, \mathbf{t}_1\}, \{\mathbf{p}_2, \mathbf{t}_2\}, \dots, \{\mathbf{p}_Q, \mathbf{t}_Q\}.$$

Each target vector must contain only zeros, except for a single 1. The row in which the 1 appears indicates the class to which the input vector belongs. For example, if we have a problem where we would like to classify a particular three-element vector into the second of four classes, we can express this as

$$\left\{ \mathbf{p}_1 = \begin{bmatrix} \sqrt{1/2} \\ 0 \\ \sqrt{1/2} \end{bmatrix}, \mathbf{t}_1 = \begin{bmatrix} 0 \\ 1 \\ 0 \\ 0 \end{bmatrix} \right\}. \qquad (15.29)$$

Before learning can occur, each neuron in the first layer is assigned to an output neuron. This generates the matrix $\mathbf{W}^2$. Typically, equal numbers of hidden neurons are connected to each output neuron, so that each class can be made up of the same number of convex regions. All elements of $\mathbf{W}^2$ are set to zero, except for the following:

$$\text{If hidden neuron } i \text{ is to be assigned to class } k, \text{ then set } w_{ki}^2 = 1. \qquad (15.30)$$

Once $\mathbf{W}^2$ is defined, it will never be altered. The hidden weights $\mathbf{W}^1$ are trained with a variation of the Kohonen rule.

The LVQ learning rule proceeds as follows. At each iteration, an input vector $\mathbf{p}$ is presented to the network, and the distance from $\mathbf{p}$ to each prototype vector is computed. The hidden neurons compete, neuron i^* wins the competition, and the i^* th element of $\mathbf{a}^1$ is set to 1. Next, $\mathbf{a}^1$ is multiplied by $\mathbf{W}^2$ to get the final output $\mathbf{a}^2$, which also has only one nonzero element, k^*, indicating that $\mathbf{p}$ is being assigned to class k^*.

The Kohonen rule is used to improve the hidden layer of the LVQ network in two ways. First, if $\mathbf{p}$ is classified correctly, then we move the weights ${}_{i^*}\mathbf{w}^1$ of the winning hidden neuron toward $\mathbf{p}$.

$${}_{i^*}\mathbf{w}^1(q) = {}_{i^*}\mathbf{w}^1(q-1) + \alpha(\mathbf{p}(q) - {}_{i^*}\mathbf{w}^1(q-1)), \text{ if } a_{k^*}^2 = t_{k^*} = 1 \qquad (15.31)$$

Second, if $\mathbf{p}$ was classified incorrectly, then we know that the wrong hidden neuron won the competition, and therefore we move its weights ${}_{i^*}\mathbf{w}^1$ *away* from $\mathbf{p}$.

$${}_{i^*}\mathbf{w}^1(q) = {}_{i^*}\mathbf{w}^1(q-1) - \alpha(\mathbf{p}(q) - {}_{i^*}\mathbf{w}^1(q-1)),\ \text{if } a^2_{k^*} = 1 \neq t_{k^*} = 0 \qquad (15.32)$$

The result will be that each hidden neuron moves toward vectors that fall into the class for which it forms a subclass and away from vectors that fall into other classes.

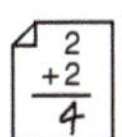

Let's take a look at an example of LVQ training. We would like to train an LVQ network to solve the following classification problem:

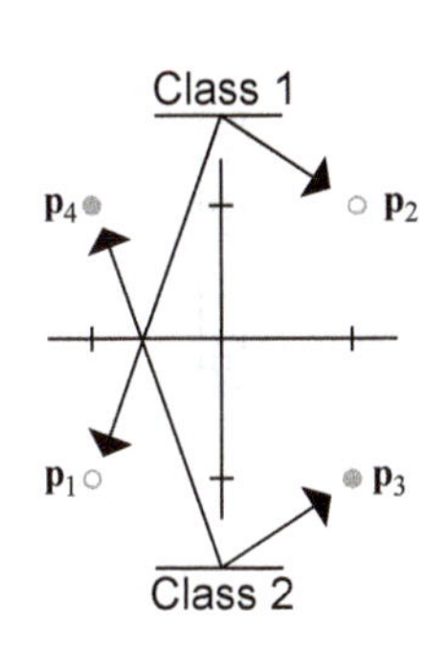

$$\text{class 1:}\left\{\mathbf{p}_1 = \begin{bmatrix}-1\\-1\end{bmatrix}, \mathbf{p}_2 = \begin{bmatrix}1\\1\end{bmatrix}\right\}, \text{class 2:}\left\{\mathbf{p}_3 = \begin{bmatrix}1\\-1\end{bmatrix}, \mathbf{p}_4 = \begin{bmatrix}-1\\1\end{bmatrix}\right\}, \qquad (15.33)$$

as illustrated by the figure in the left margin. We begin by assigning target vectors to each input:

$$\left\{\mathbf{p}_1 = \begin{bmatrix}-1\\-1\end{bmatrix}, \mathbf{t}_1 = \begin{bmatrix}1\\0\end{bmatrix}\right\}, \left\{\mathbf{p}_2 = \begin{bmatrix}1\\1\end{bmatrix}, \mathbf{t}_2 = \begin{bmatrix}1\\0\end{bmatrix}\right\}, \qquad (15.34)$$

$$\left\{\mathbf{p}_3 = \begin{bmatrix}1\\-1\end{bmatrix}, \mathbf{t}_3 = \begin{bmatrix}0\\1\end{bmatrix}\right\}, \left\{\mathbf{p}_4 = \begin{bmatrix}-1\\1\end{bmatrix}, \mathbf{t}_4 = \begin{bmatrix}0\\1\end{bmatrix}\right\}. \qquad (15.35)$$

We now must choose how many subclasses will make up each of the two classes. If we let each class be the union of two subclasses, we will end up with four neurons in the hidden layer. The output layer weight matrix will be

$$\mathbf{W}^2 = \begin{bmatrix}1 & 1 & 0 & 0\\0 & 0 & 1 & 1\end{bmatrix}. \qquad (15.36)$$

$\mathbf{W}^2$ connects hidden neurons 1 and 2 to output neuron 1. It connects hidden neurons 3 and 4 to output neuron 2. Each class will be made up of two convex regions.

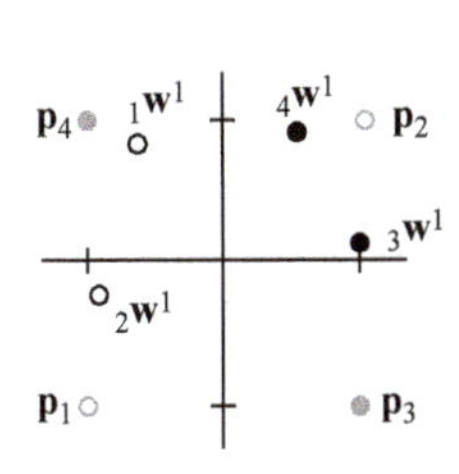

The row vectors in $\mathbf{W}^1$ are initially set to random values. They can be seen in the diagram at left. The weights belonging to the two hidden neurons that define class 1 are marked with hollow circles. The weights defining class 2 are marked with solid circles. The values for these weights are

$${}_1\mathbf{w}^1 = \begin{bmatrix}-0.543\\0.840\end{bmatrix}, {}_2\mathbf{w}^1 = \begin{bmatrix}-0.969\\-0.249\end{bmatrix}, {}_3\mathbf{w}^1 = \begin{bmatrix}0.997\\0.094\end{bmatrix}, {}_4\mathbf{w}^1 = \begin{bmatrix}0.456\\0.954\end{bmatrix}. \qquad (15.37)$$

At each iteration of the training process, we present an input vector, find its response, and then adjust the weights. In this case we will begin by presenting $\mathbf{p}_3$.

$$\mathbf{a}^1 = \mathbf{compet}(\mathbf{n}^1) = \mathbf{compet}\left(\begin{bmatrix} -\|{}_1\mathbf{w}^1 - \mathbf{p}_3\| \\ -\|{}_2\mathbf{w}^1 - \mathbf{p}_3\| \\ -\|{}_3\mathbf{w}^1 - \mathbf{p}_3\| \\ -\|{}_4\mathbf{w}^1 - \mathbf{p}_3\| \end{bmatrix}\right) \qquad (15.38)$$

$$= \mathbf{compet}\left(\begin{bmatrix} -\left\|\begin{bmatrix}-0.543 & 0.840\end{bmatrix}^T - \begin{bmatrix}1 & -1\end{bmatrix}^T\right\| \\ -\left\|\begin{bmatrix}-0.969 & -0.249\end{bmatrix}^T - \begin{bmatrix}1 & -1\end{bmatrix}^T\right\| \\ -\left\|\begin{bmatrix}0.997 & 0.094\end{bmatrix}^T - \begin{bmatrix}1 & -1\end{bmatrix}^T\right\| \\ -\left\|\begin{bmatrix}0.456 & 0.954\end{bmatrix}^T - \begin{bmatrix}1 & -1\end{bmatrix}^T\right\| \end{bmatrix}\right) = \mathbf{compet}\left(\begin{bmatrix}-2.40 \\ -2.11 \\ -1.09 \\ -2.03\end{bmatrix}\right) = \begin{bmatrix}0 \\ 0 \\ 1 \\ 0\end{bmatrix}$$

The third hidden neuron has the closest weight vector to $\mathbf{p}_3$. In order to determine which class this neuron belongs to, we multiply $\mathbf{a}^1$ by $\mathbf{W}^2$.

$$\mathbf{a}^2 = \mathbf{W}^2\mathbf{a}^1 = \begin{bmatrix}1 & 1 & 0 & 0 \\ 0 & 0 & 1 & 1\end{bmatrix}\begin{bmatrix}0 \\ 0 \\ 1 \\ 0\end{bmatrix} = \begin{bmatrix}0 \\ 1\end{bmatrix} \qquad (15.39)$$

This output indicates that $\mathbf{p}_3$ is a member of class 2. This is correct, so ${}_3\mathbf{w}^1$ is updated by moving it toward $\mathbf{p}_3$.

$${}_3\mathbf{w}^1(1) = {}_3\mathbf{w}^1(0) + \alpha(\mathbf{p}_3 - {}_3\mathbf{w}^1(0)) \qquad (15.40)$$

$$= \begin{bmatrix}0.997 \\ 0.094\end{bmatrix} + 0.5\left(\begin{bmatrix}1 \\ -1\end{bmatrix} - \begin{bmatrix}0.997 \\ 0.094\end{bmatrix}\right) = \begin{bmatrix}0.998 \\ -0.453\end{bmatrix}$$

The diagram on the left side of Figure 15.14 shows the weights after ${}_3\mathbf{w}^1$ was updated on the first iteration. The diagram on the right side of Figure 15.14 shows the weights after the algorithm has converged.

The diagram on the right side of Figure 15.14 also indicates how the regions of the input space will be classified. The regions that will be classified as class 1 are shown in gray, and the regions that will be classified as class 2 are shown in blue.

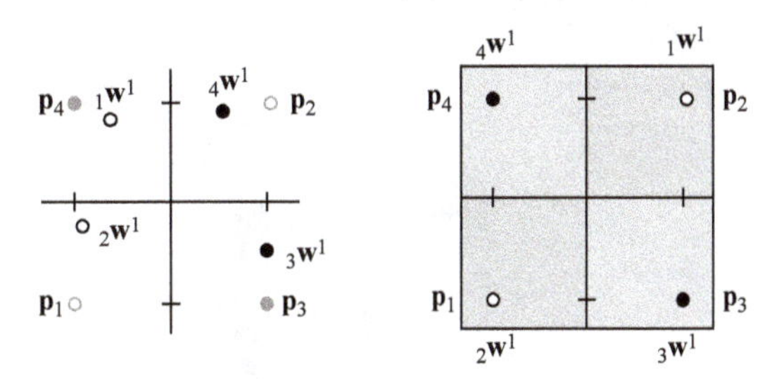

Figure 15.14 After First and Many Iterations

Improving LVQ Networks (LVQ2)

The LVQ network described above works well for many problems, but it does suffer from a couple of limitations. First, as with competitive layers, occasionally a hidden neuron in an LVQ network can have initial weight values that stop it from ever winning the competition. The result is a dead neuron that never does anything useful. This problem is solved with the use of a "conscience" mechanism, a technique discussed earlier for competitive layers, and also presented in Exercise E15.4.

Secondly, depending on how the initial weight vectors are arranged, a neuron's weight vector may have to travel through a region of a class that it doesn't represent, to get to a region that it does represent. Because the weights of such a neuron will be repulsed by vectors in the region it must cross, it may not be able to cross, and so it may never properly classify the region it is being attracted to. This is usually solved by applying the following modification to the Kohonen rule.

If the winning neuron in the hidden layer incorrectly classifies the current input, we move its weight vector away from the input vector, as before. However, we also adjust the weights of the closest neuron to the input vector that does classify it properly. The weights for this second neuron should be moved toward the input vector.

When the network correctly classifies an input vector, the weights of only one neuron are moved toward the input vector. However, if the input vector is incorrectly classified, the weights of two neurons are updated, one weight vector is moved away from the input vector, and the other one is moved toward the input vector. The resulting algorithm is called *LVQ2*.

LVQ2

To experiment with LVQ networks use the Neural Network Design Demonstrations LVQ1 Networks (`nnd14lv1`) *and* LVQ2 Networks (`nnd14lv2`).

Summary of Results

Competitive Layer

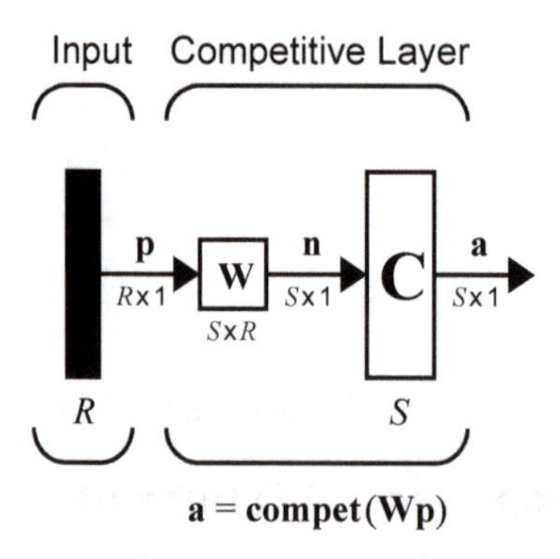

Competitive Learning with the Kohonen Rule

$$_{i*}\mathbf{w}(q) = {}_{i*}\mathbf{w}(q-1) + \alpha(\mathbf{p}(q) - {}_{i*}\mathbf{w}(q-1)) = (1-\alpha)\,{}_{i*}\mathbf{w}(q-1) + \alpha\mathbf{p}(q)$$

$$_{i*}\mathbf{w}(q) = {}_{i*}\mathbf{w}(q-1) \qquad i \neq i^*,$$

where i^* is the winning neuron.

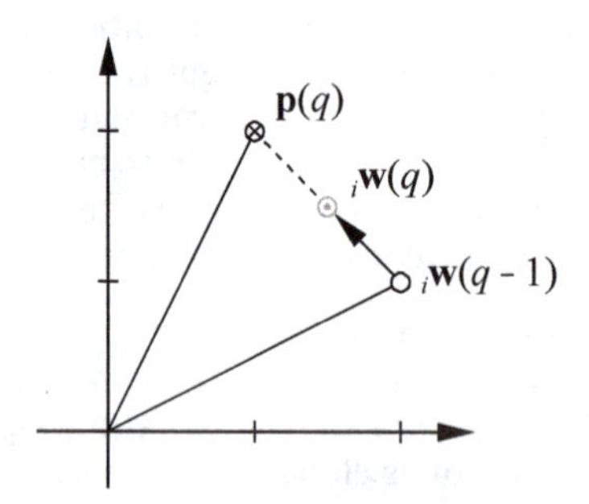

Self-Organizing Feature Map

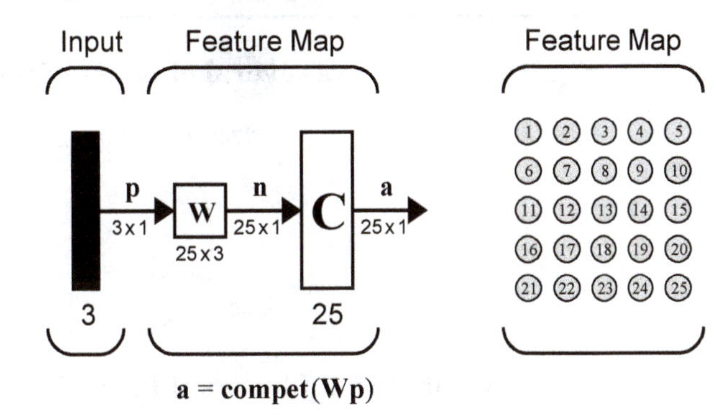

$$\mathbf{a} = \mathbf{compet}(\mathbf{Wp})$$

Self-Organizing with the Kohonen Rule

$$ {}_i\mathbf{w}(q) = {}_i\mathbf{w}(q-1) + \alpha(\mathbf{p}(q) - {}_i\mathbf{w}(q-1)) \qquad i \in N_{i*}(d)$$

$$ = (1-\alpha)\,{}_i\mathbf{w}(q-1) + \alpha\mathbf{p}(q)$$

$$N_i(d) = \{j, d_{ij} \le d\}$$

LVQ Network

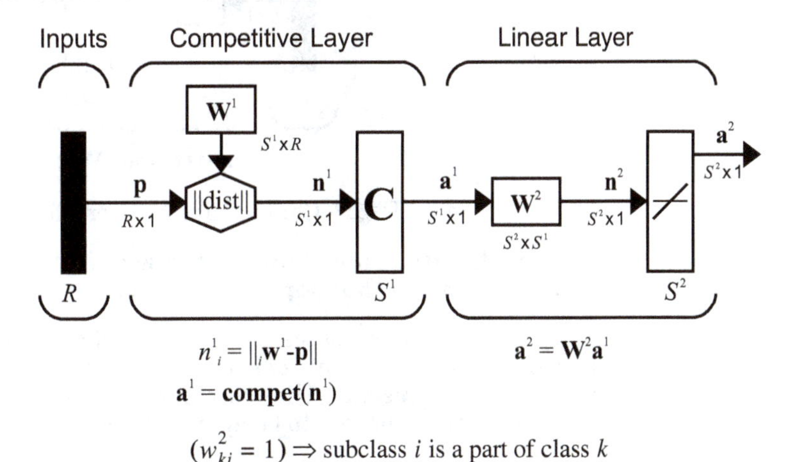

$$n^1_i = \|{}_i\mathbf{w}^1 - \mathbf{p}\| \qquad \mathbf{a}^2 = \mathbf{W}^2\mathbf{a}^1$$

$$\mathbf{a}^1 = \mathbf{compet}(\mathbf{n}^1)$$

$$(w^2_{ki} = 1) \Rightarrow \text{subclass } i \text{ is a part of class } k$$

LVQ Network Learning with the Kohonen Rule

$${}_{i*}\mathbf{w}^1(q) = {}_{i*}\mathbf{w}^1(q-1) + \alpha(\mathbf{p}(q) - {}_{i*}\mathbf{w}^1(q-1)),\ \text{if } a^2_{k*} = t_{k*} = 1$$

$${}_{i*}\mathbf{w}^1(q) = {}_{i*}\mathbf{w}^1(q-1) - \alpha(\mathbf{p}(q) - {}_{i*}\mathbf{w}^1(q-1)),\ \text{if } a^2_{k*} = 1 \ne t_{k*} = 0$$

Solved Problems

P15.1 Figure P15.1 shows several clusters of normalized vectors.

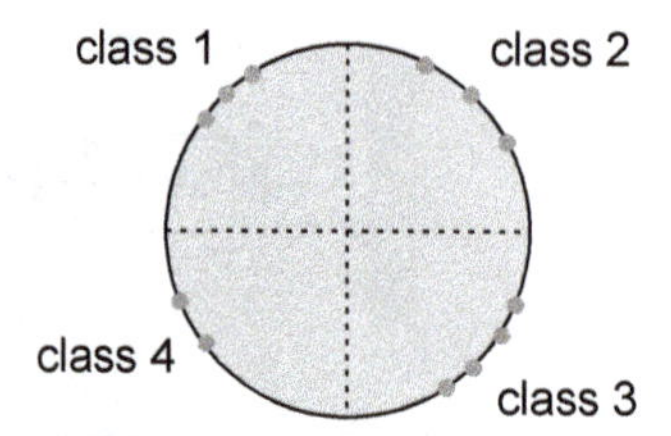

Figure P15.1 Clusters of Input Vectors for Problem P15.1

Design the weights of the competitive network shown in Figure P15.2, so that it classifies the vectors according to the classes indicated in the diagram and with the minimum number of neurons.

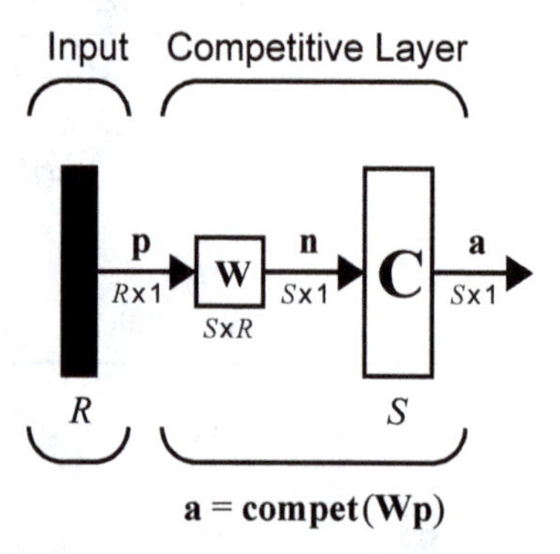

Figure P15.2 Competitive Network for Problem P15.1

Redraw the diagram showing the weights you chose and the decision boundaries that separate the region of each class.

Since there are four classes to be defined, the competitive layer will need four neurons. The weights of each neuron act as prototypes for the class that neuron represents. Therefore, for each neuron we will choose a prototype vector that appears to be approximately at the center of a cluster.

Classes 1, 2 and 3 each appear to be roughly centered at a multiple of $45°$. Given this, the following three vectors are normalized (as is required for the competitive layer) and point in the proper directions.

$$_1\mathbf{w} = \begin{bmatrix} -1/\sqrt{2} \\ 1/\sqrt{2} \end{bmatrix},\ _2\mathbf{w} = \begin{bmatrix} 1/\sqrt{2} \\ 1/\sqrt{2} \end{bmatrix},\ _3\mathbf{w} = \begin{bmatrix} 1/\sqrt{2} \\ -1/\sqrt{2} \end{bmatrix}$$

The center of the fourth cluster appears to be about twice as far from the vertical axis as it is from the horizontal axis. The resulting normalized weight vector is

$$_4\mathbf{w} = \begin{bmatrix} -2/\sqrt{5} \\ -1/\sqrt{5} \end{bmatrix}.$$

The weight matrix $\mathbf{W}$ for the competitive layer is simply the matrix of the transposed prototype vectors:

$$W = \begin{bmatrix} {}_1\mathbf{w}^T \\ {}_2\mathbf{w}^T \\ {}_3\mathbf{w}^T \\ {}_4\mathbf{w}^T \end{bmatrix} = \begin{bmatrix} -1/\sqrt{2} & 1/\sqrt{2} \\ 1/\sqrt{2} & 1/\sqrt{2} \\ 1/\sqrt{2} & -1/\sqrt{2} \\ -2/\sqrt{5} & -1/\sqrt{5} \end{bmatrix}.$$

We get Figure P15.3 by drawing these weight vectors with arrows and bisecting the circle between each adjacent weight vector to get the class regions.

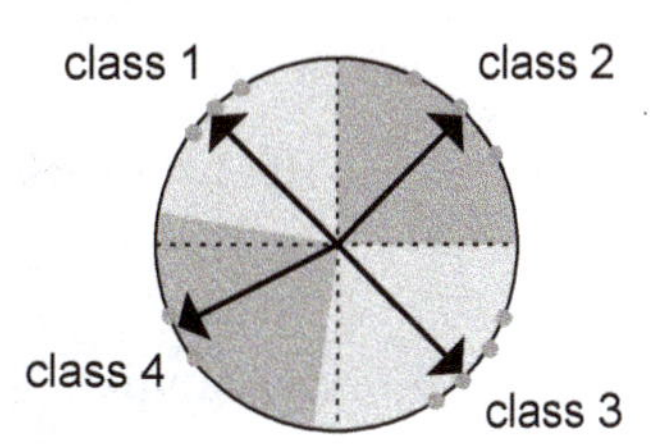

Figure P15.3 Final Classifications for Problem P15.1

P15.2 Figure P15.4 shows three input vectors and three initial weight vectors for a three-neuron competitive layer. Here are the values of the input vectors:

$$\mathbf{p}_1 = \begin{bmatrix} -1 \\ 0 \end{bmatrix},\ \mathbf{p}_2 = \begin{bmatrix} 0 \\ 1 \end{bmatrix},\ \mathbf{p}_3 = \begin{bmatrix} 1/\sqrt{2} \\ 1/\sqrt{2} \end{bmatrix}.$$

The initial values of the three weight vectors are

$$_1\mathbf{w} = \begin{bmatrix} 0 \\ -1 \end{bmatrix},\ _2\mathbf{w} = \begin{bmatrix} -2/\sqrt{5} \\ 1/\sqrt{5} \end{bmatrix},\ _3\mathbf{w} = \begin{bmatrix} -1/\sqrt{5} \\ 2/\sqrt{5} \end{bmatrix}.$$

Calculate the resulting weights found after training the competitive layer with the Kohonen rule and a learning rate α of 0.5, on the following series of inputs:

$$\mathbf{p}_1, \mathbf{p}_2, \mathbf{p}_3, \mathbf{p}_1, \mathbf{p}_2, \mathbf{p}_3 .$$

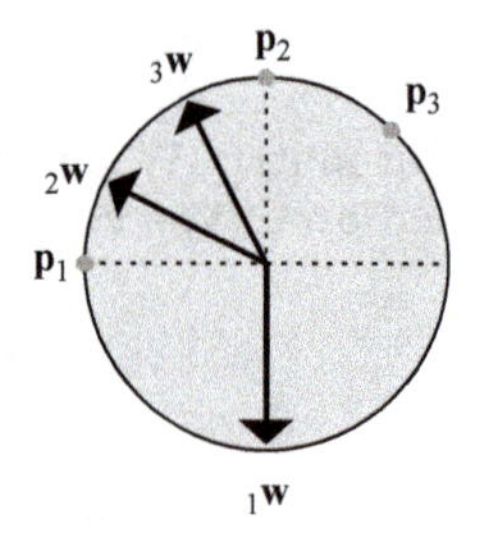

Figure P15.4 Input Vectors and Initial Weights for Problem P15.2

First we combine the weight vectors into the weight matrix $\mathbf{W}$.

$$\mathbf{W} = \begin{bmatrix} 0 & -1 \\ -2/\sqrt{5} & 1/\sqrt{5} \\ -1/\sqrt{5} & 2/\sqrt{5} \end{bmatrix}$$

Then we present the first vector $\mathbf{p}_1$.

$$\mathbf{a} = \mathbf{compet}(\mathbf{W}\mathbf{p}_1) = \mathbf{compet}\left(\begin{bmatrix} 0 & -1 \\ -2/\sqrt{5} & 1/\sqrt{5} \\ -1/\sqrt{5} & 2/\sqrt{5} \end{bmatrix}\begin{bmatrix} -1 \\ 0 \end{bmatrix}\right) = \mathbf{compet}\left(\begin{bmatrix} 0 \\ 0.894 \\ 0.447 \end{bmatrix}\right) = \begin{bmatrix} 0 \\ 1 \\ 0 \end{bmatrix}$$

The second neuron responded, since ${}_2\mathbf{w}$ was closest to $\mathbf{p}_1$. Therefore, we will update ${}_2\mathbf{w}$ with the Kohonen rule.

$${}_2\mathbf{w}^{new} = {}_2\mathbf{w}^{old} + \alpha(\mathbf{p}_1 - {}_2\mathbf{w}^{old}) = \begin{bmatrix} -2/\sqrt{5} \\ 1/\sqrt{5} \end{bmatrix} + 0.5\left(\begin{bmatrix} -1 \\ 0 \end{bmatrix} - \begin{bmatrix} -2/\sqrt{5} \\ 1/\sqrt{5} \end{bmatrix}\right) = \begin{bmatrix} -0.947 \\ 0.224 \end{bmatrix}$$

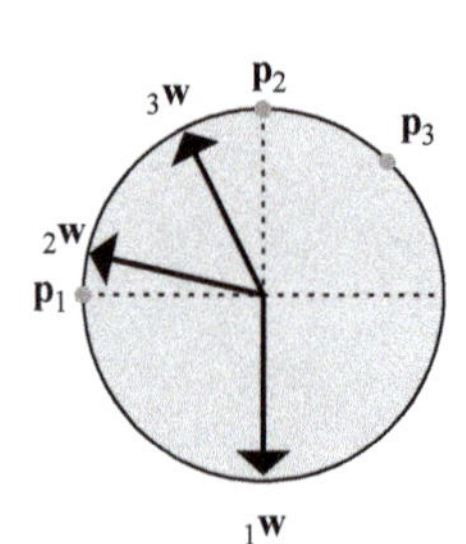

The diagram at left shows that the new ${}_2\mathbf{w}$ moved closer to $\mathbf{p}_1$.

We will now repeat this process for $\mathbf{p}_2$.

$$\mathbf{a} = \mathbf{compet}(\mathbf{W}\mathbf{p}_2) = \mathbf{compet}\left(\begin{bmatrix} 0 & -1 \\ -0.947 & 0.224 \\ -1/\sqrt{5} & 2/\sqrt{5} \end{bmatrix}\begin{bmatrix} 0 \\ 1 \end{bmatrix}\right) = \mathbf{compet}\left(\begin{bmatrix} -1 \\ 0.224 \\ 0.894 \end{bmatrix}\right) = \begin{bmatrix} 0 \\ 0 \\ 1 \end{bmatrix}$$

The third neuron won, so its weights move closer to $\mathbf{p}_2$.

$${}_3\mathbf{w}^{new} = {}_3\mathbf{w}^{old} + \alpha(\mathbf{p}_2 - {}_3\mathbf{w}^{old}) = \begin{bmatrix} -1/\sqrt{5} \\ 2/\sqrt{5} \end{bmatrix} + 0.5\left(\begin{bmatrix} 0 \\ 1 \end{bmatrix} - \begin{bmatrix} -1/\sqrt{5} \\ 2/\sqrt{5} \end{bmatrix}\right) = \begin{bmatrix} -0.224 \\ 0.947 \end{bmatrix}$$

We now present $\mathbf{p}_3$.

$$\mathbf{a} = \mathbf{compet}(\mathbf{W}\mathbf{p}_3) = \mathbf{compet}\left(\begin{bmatrix} 0 & -1 \\ -0.947 & 0.224 \\ -0.224 & 0.947 \end{bmatrix}\begin{bmatrix} 1/\sqrt{2} \\ 1/\sqrt{2} \end{bmatrix}\right)$$

$$= \mathbf{compet}\left(\begin{bmatrix} -0.707 \\ -0.512 \\ 0.512 \end{bmatrix}\right) = \begin{bmatrix} 0 \\ 0 \\ 1 \end{bmatrix}$$

The third neuron wins again.

$${}_3\mathbf{w}^{new} = {}_3\mathbf{w}^{old} + \alpha(\mathbf{p}_2 - {}_3\mathbf{w}^{old}) = \begin{bmatrix} -0.224 \\ 0.947 \end{bmatrix} + 0.5\left(\begin{bmatrix} 1/\sqrt{2} \\ 1/\sqrt{2} \end{bmatrix} - \begin{bmatrix} -0.224 \\ 0.947 \end{bmatrix}\right) = \begin{bmatrix} 0.2417 \\ 0.8272 \end{bmatrix}$$

After presenting $\mathbf{p}_1$ through $\mathbf{p}_3$ again, neuron 2 will again win once and neuron 3 twice. The final weights are

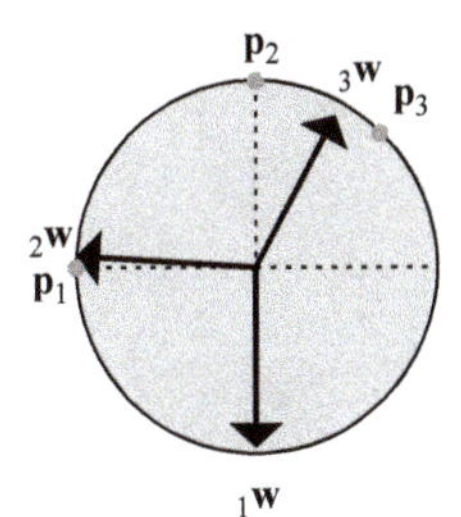

$$\mathbf{W} = \begin{bmatrix} 0 & -1 \\ -0.974 & 0.118 \\ 0.414 & 0.8103 \end{bmatrix}.$$

The final weights are also shown in the diagram at left.

Note that ${}_2\mathbf{w}$ has almost learned $\mathbf{p}_1$, and ${}_3\mathbf{w}$ is directly between $\mathbf{p}_2$ and $\mathbf{p}_3$. The other weight vector, ${}_1\mathbf{w}$, was never updated. The first neuron, which never won the competition, is a dead neuron.

P15.3 Consider the configuration of input vectors and initial weights shown in Figure P15.5. Train a competitive network to cluster

these vectors using the Kohonen rule with learning rate $\alpha = 0.5$. Find graphically the position of the weights after all of the input vectors (in the order shown) have been presented once.

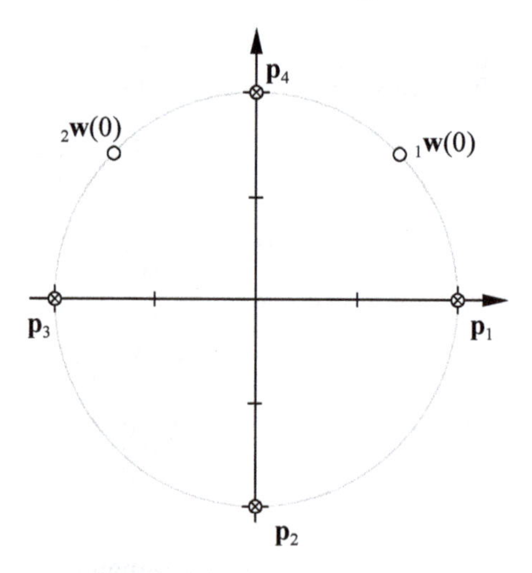

Figure P15.5 Input Vectors and Initial Weights for Problem P15.3

This problem can be solved graphically, without any computations. The results are displayed in Figure P15.6.

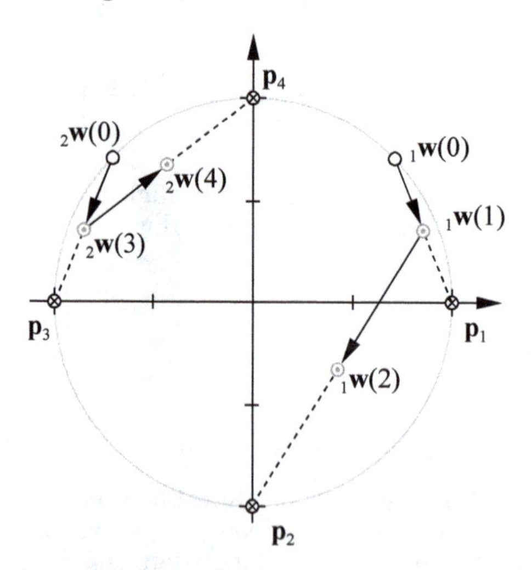

Figure P15.6 Solution for Problem P15.3

The input vector $\mathbf{p}_1$ is presented first. The weight vector ${}_1\mathbf{w}$ is closest to $\mathbf{p}_1$, therefore neuron 1 wins the competition and ${}_1\mathbf{w}$ is moved halfway to

$\mathbf{p}_1$, since $\alpha = 0.5$. Next, $\mathbf{p}_2$ is presented, and again neuron 1 wins the competition and ${}_1\mathbf{w}$ is moved halfway to $\mathbf{p}_2$. During these first two iterations, ${}_2\mathbf{w}$ is not changed.

On the third iteration, $\mathbf{p}_3$ is presented. This time ${}_2\mathbf{w}$ wins the competition and is moved halfway to $\mathbf{p}_3$. On the fourth iteration, $\mathbf{p}_4$ is presented, and neuron 2 again wins. The weight vector ${}_2\mathbf{w}$ is moved halfway to $\mathbf{p}_4$.

If we continue to train the network, neuron 1 will classify the input vectors $\mathbf{p}_1$ and $\mathbf{p}_2$, and neuron 2 will classify the input vectors $\mathbf{p}_3$ and $\mathbf{p}_4$. If the input vectors were presented in a different order, would the final classification be different?

P15.4 So far in this chapter we have only talked about feature maps whose neurons are arranged in two dimensions. The feature map shown in Figure P15.7 contains nine neurons arranged in one dimension.

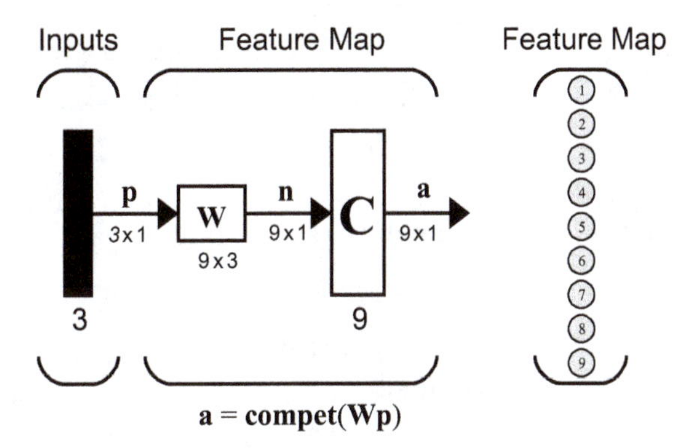

Figure P15.7 Nine-Neuron Feature Map

Given the following initial weights, draw a diagram of the weight vectors, with lines connecting weight vectors of neighboring neurons.

$$\mathbf{W} = \begin{bmatrix} 0.41 & 0.45 & 0.41 & 0 & 0 & 0 & -0.41 & -0.45 & -0.41 \\ 0.41 & 0 & -0.41 & 0.45 & 0 & -0.45 & 0.41 & 0 & -0.41 \\ 0.82 & 0.89 & 0.82 & 0.89 & 1 & 0.89 & 0.82 & 0.89 & 0.82 \end{bmatrix}^T$$

Train the feature map for one iteration, on the vector below, using a learning rate of 0.1 and a neighborhood of radius 1. Redraw the diagram for the new weight matrix.

$$\mathbf{p} = \begin{bmatrix} 0.67 \\ 0.07 \\ 0.74 \end{bmatrix}$$

The feature map diagram for the initial weights is given in Figure P15.8.

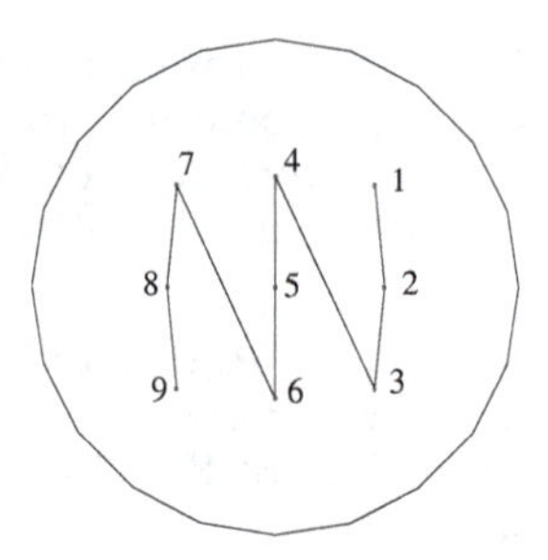

Figure P15.8 Original Feature Map

We start updating the network by presenting $\mathbf{p}$ to the network.

$$\mathbf{a} = \mathbf{compet}(\mathbf{Wp})$$

$$= \mathbf{compet}\left(\begin{bmatrix} 0.41 & 0.45 & 0.41 & 0 & 0 & 0 & -0.41 & -0.45 & -0.41 \\ 0.41 & 0 & -0.41 & 0.45 & 0 & -0.45 & 0.41 & 0 & -0.41 \\ 0.82 & 0.89 & 0.82 & 0.89 & 1 & 0.89 & 0.82 & 0.89 & 0.82 \end{bmatrix}^T \begin{bmatrix} 0.67 \\ 0.07 \\ 0.74 \end{bmatrix} \right)$$

$$= \mathbf{compet}(\begin{bmatrix} 0.91 & 0.96 & 0.85 & 0.70 & 0.74 & 0.63 & 0.36 & 0.36 & 0.3 \end{bmatrix}^T)$$

$$= \begin{bmatrix} 0 & 1 & 0 & 0 & 0 & 0 & 0 & 0 & 0 \end{bmatrix}^T$$

The second neuron won the competition. Looking at the network diagram, we see that the second neuron's neighbors, at a radius of 1, include neurons 1 and 3. We must update each of these neurons' weights with the Kohonen rule.

$$ {}_1\mathbf{w}(1) = {}_1\mathbf{w}(0) + \alpha(\mathbf{p} - {}_1\mathbf{w}(0)) = \begin{bmatrix} 0.41 \\ 0.41 \\ 0.82 \end{bmatrix} + 0.1\left(\begin{bmatrix} 0.67 \\ 0.07 \\ 0.74 \end{bmatrix} - \begin{bmatrix} 0.41 \\ 0.41 \\ 0.82 \end{bmatrix}\right) = \begin{bmatrix} 0.43 \\ 0.37 \\ 0.81 \end{bmatrix} $$

$$ {}_2\mathbf{w}(1) = {}_2\mathbf{w}(0) + \alpha(\mathbf{p} - {}_2\mathbf{w}(0)) = \begin{bmatrix} 0.45 \\ 0 \\ 0.89 \end{bmatrix} + 0.1\left(\begin{bmatrix} 0.67 \\ 0.07 \\ 0.74 \end{bmatrix} - \begin{bmatrix} 0.45 \\ 0 \\ 0.89 \end{bmatrix}\right) = \begin{bmatrix} 0.47 \\ 0.01 \\ 0.88 \end{bmatrix} $$

$$ {}_3\mathbf{w}(1) = {}_3\mathbf{w}(0) + \alpha(\mathbf{p} - {}_3\mathbf{w}(0)) = \begin{bmatrix} 0.41 \\ -0.41 \\ 0.82 \end{bmatrix} + 0.1\left(\begin{bmatrix} 0.67 \\ 0.07 \\ 0.74 \end{bmatrix} - \begin{bmatrix} 0.41 \\ -0.41 \\ 0.82 \end{bmatrix}\right) = \begin{bmatrix} 0.43 \\ -0.36 \\ 0.81 \end{bmatrix} $$

Figure P15.9 shows the feature map after the weights were updated.

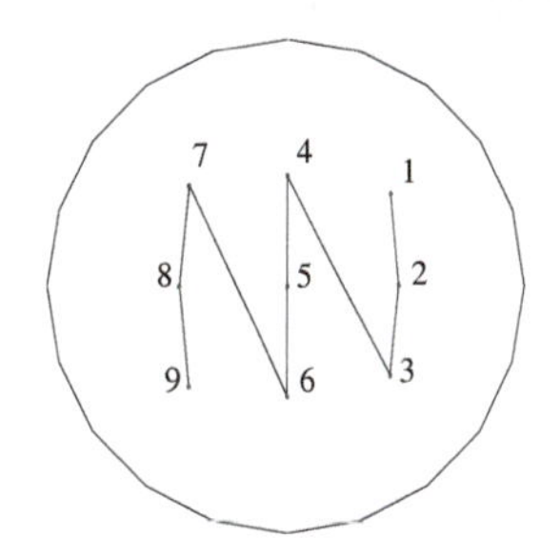

Figure P15.9 Feature Map after Update

P15.5 Given the LVQ network shown in Figure P15.10 and the weight values shown below, draw the regions of the input space that make up each class.

$$ \mathbf{W}^1 = \begin{bmatrix} 0 & 0 \\ 1 & -1 \\ 1 & 1 \\ -1 & 1 \\ -1 & -1 \end{bmatrix},\ \mathbf{W}^2 = \begin{bmatrix} 1 & 0 & 0 & 0 & 0 \\ 0 & 1 & 0 & 0 & 0 \\ 0 & 0 & 1 & 1 & 1 \end{bmatrix} $$

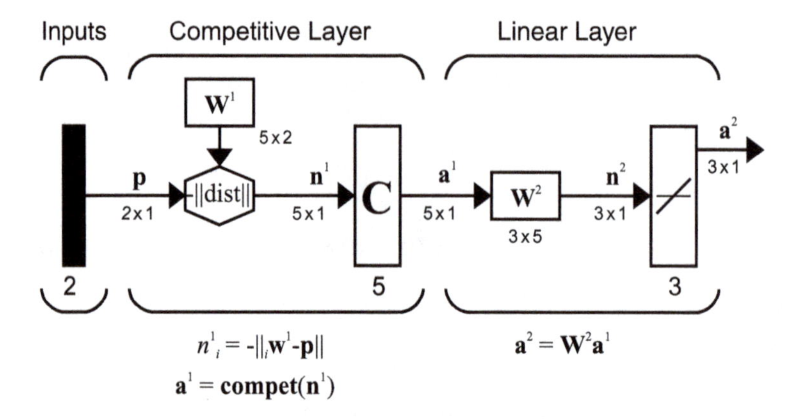

Figure P15.10 LVQ Network for Problem P15.5

We create the diagram shown in Figure P15.11 by marking each vector ${}_i\mathbf{w}$ in $\mathbf{W}^1$ according to the index k of the corresponding nonzero element in the ith column of $\mathbf{W}^2$, which indicates the class.

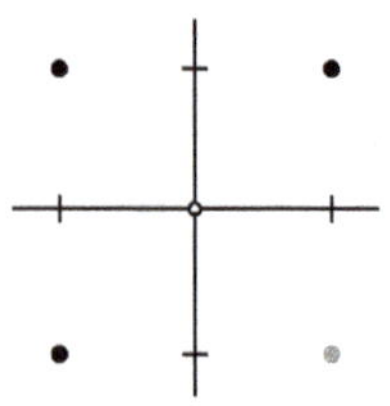

Figure P15.11 Prototype Vectors Marked by Class

The decision boundaries separating each class are found by drawing lines between each pair of prototype vectors, perpendicular to an imaginary line connecting them and equidistant from each vector.

In Figure P15.12, each convex region is colored according to the weight vector it is closest to.

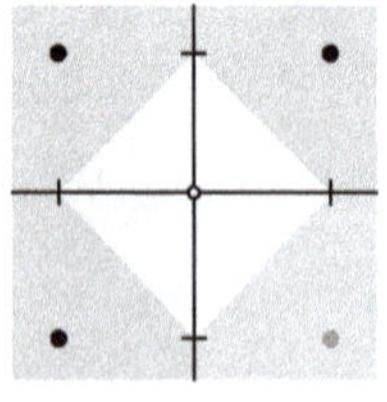

Figure P15.12 Class Regions and Decision Boundaries

P15.6 Design an LVQ network to solve the classification problem shown in Figure P15.13. The vectors in the diagram are to be classified into one of three classes, according to their color.

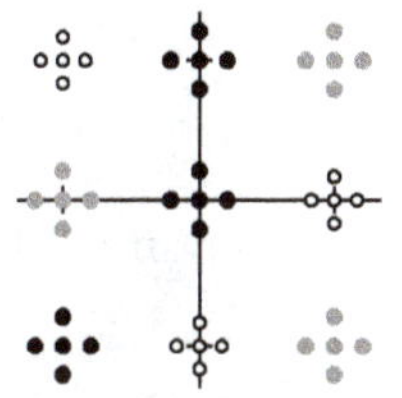

Figure P15.13 Classification Problem

When the design is complete, draw a diagram showing the region for each class.

We will begin by noting that since LVQ networks calculate the distance between vectors directly, instead of using the inner product, they can classify vectors that are not normalized, such as those above.

Next we will identify each color with a class:

- Class 1 will include all white dots.
- Class 2 will include all black dots.
- Class 3 will include all blue dots.

Now we can choose the dimensions of the LVQ network. Since there are three classes, the network must have three neurons in its output layer. There are nine subclasses (i.e., clusters). Therefore the hidden layer must have nine neurons. This gives us the network shown in Figure P15.14.

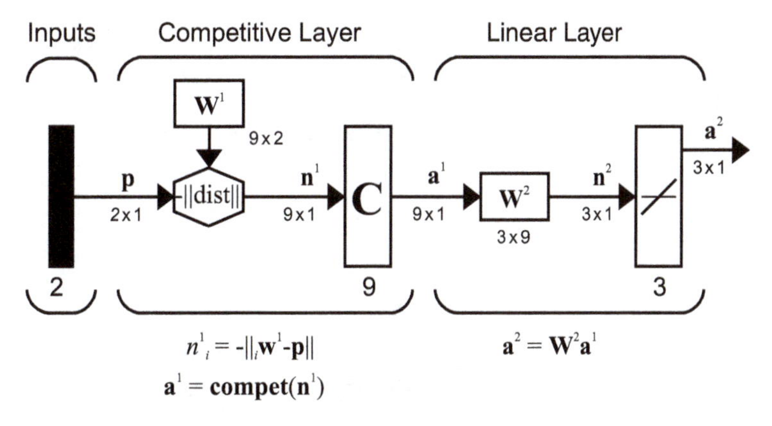

Figure P15.14 LVQ Network for Problem P15.6

We can now design the weight matrix $\mathbf{W}^1$ of the first layer by setting each row equal to a transposed prototype vector for one cluster. Picking prototype vectors at the center of each cluster gives us the following values:

$$\mathbf{W}^1 = \begin{bmatrix} -1 & 0 & 1 & -1 & 0 & 1 & -1 & 0 & 1 \\ 1 & 1 & 1 & 0 & 0 & 0 & -1 & -1 & -1 \end{bmatrix}^T .$$

Now each neuron in the first layer will respond to a different cluster.

Next we choose $\mathbf{W}^2$ so that each subclass is connected to the appropriate class. To do this we use the following rule:

$$\text{If subclass } i \text{ is to be assigned to class } k, \text{ then set } w_{ki}^2 = 1 .$$

For example, the first subclass is the top-left cluster in the vector diagram. The vectors in this cluster are white, so they belong in the first class. Therefore we should set $w_{1,1}^2$ to one.

Once we have done this for all nine subclasses we end up with these values:

$$\mathbf{W}^2 = \begin{bmatrix} 1 & 0 & 0 & 0 & 0 & 1 & 0 & 1 & 0 \\ 0 & 1 & 0 & 0 & 1 & 0 & 1 & 0 & 0 \\ 0 & 0 & 1 & 1 & 0 & 0 & 0 & 0 & 1 \end{bmatrix} .$$

We can test the network by presenting a vector to it. Here we calculate the output of the first layer for $\mathbf{p} = \begin{bmatrix} 1 & 0 \end{bmatrix}^T$:

$$\mathbf{a}^1 = \mathbf{compet}(\mathbf{n}^1) = \mathbf{compet}\left(\begin{bmatrix} -\sqrt{5} \\ -\sqrt{2} \\ -1 \\ -2 \\ -1 \\ 0 \\ -\sqrt{5} \\ -\sqrt{2} \\ -1 \end{bmatrix}\right) = \begin{bmatrix} 0 \\ 0 \\ 0 \\ 0 \\ 0 \\ 1 \\ 0 \\ 0 \\ 0 \end{bmatrix} .$$

The network says that the vector we presented is in the sixth subclass. Let's see what the second layer says.

$$\mathbf{a}^2 = \mathbf{W}^2\mathbf{a}^1 = \begin{bmatrix} 1 & 0 & 0 & 0 & 0 & 1 & 0 & 1 & 0 \\ 0 & 1 & 0 & 0 & 1 & 0 & 1 & 0 & 0 \\ 0 & 0 & 1 & 1 & 0 & 0 & 0 & 0 & 1 \end{bmatrix} \begin{bmatrix} 0 \\ 0 \\ 0 \\ 0 \\ 0 \\ 1 \\ 0 \\ 0 \\ 0 \end{bmatrix} = \begin{bmatrix} 1 \\ 0 \\ 0 \end{bmatrix}$$

The second layer indicates that the vector is in class 1, as indeed it is. The diagram of class regions and decision boundaries is shown in Figure P15.15.

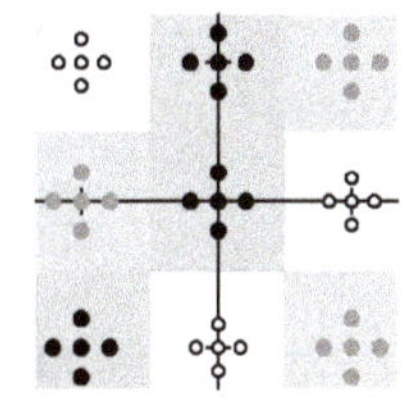

Figure P15.15 Class Regions and Decision Boundaries

P15.7 Competitive layers and feature maps require that input vectors be normalized. But what if the available data is not normalized?

One way to handle such data is simply to normalize it before giving it to the network. This has the disadvantage that the vector magnitude information, which may be important, is lost.

Another solution is to replace the inner product expression usually used to calculate net input,

$$\mathbf{a} = \mathbf{compet}(\mathbf{Wp}),$$

with a direct calculation of distance,

$$n_i = -\|{}_i\mathbf{w} - \mathbf{p}\| \text{ and } \mathbf{a} = \mathbf{compet}(\mathbf{n}),$$

as is done with the LVQ network. This works and saves the magnitude information.

However, a third solution is to append a constant of 1 to each input vector before normalizing it. Now the change in the added element will preserve the vector magnitude information.

Normalize the following vectors using this last method:

$$\mathbf{p}_1 = \begin{bmatrix} 1 \\ 1 \end{bmatrix}, \mathbf{p}_2 = \begin{bmatrix} 0 \\ 1 \end{bmatrix}, \mathbf{p}_3 = \begin{bmatrix} 0 \\ 0 \end{bmatrix}.$$

First we add an extra element with value 1 to each vector.

$$\mathbf{p}'_1 = \begin{bmatrix} 1 \\ 1 \\ 1 \end{bmatrix}, \mathbf{p}'_2 = \begin{bmatrix} 0 \\ 1 \\ 1 \end{bmatrix}, \mathbf{p}'_3 = \begin{bmatrix} 0 \\ 0 \\ 1 \end{bmatrix}$$

Then we normalize each vector.

$$\mathbf{p}''_1 = \begin{bmatrix} 1 \\ 1 \\ 1 \end{bmatrix} / \left\| \begin{bmatrix} 1 \\ 1 \\ 1 \end{bmatrix} \right\| = \begin{bmatrix} 1/\sqrt{3} \\ 1/\sqrt{3} \\ 1/\sqrt{3} \end{bmatrix}$$

$$\mathbf{p}''_2 = \begin{bmatrix} 0 \\ 1 \\ 1 \end{bmatrix} / \left\| \begin{bmatrix} 0 \\ 1 \\ 1 \end{bmatrix} \right\| = \begin{bmatrix} 0 \\ 1/\sqrt{2} \\ 1/\sqrt{2} \end{bmatrix}$$

$$\mathbf{p}''_3 = \begin{bmatrix} 0 \\ 0 \\ 1 \end{bmatrix} / \left\| \begin{bmatrix} 0 \\ 0 \\ 1 \end{bmatrix} \right\| = \begin{bmatrix} 0 \\ 0 \\ 1 \end{bmatrix}$$

Now the third element of each vector contains magnitude information, since it is equal to the inverse of the magnitude of the extended vectors.

Epilogue

In this chapter we have demonstrated how the associative instar learning rule can be combined with competitive networks, similar to the Hamming network of Chapter 3, to produce powerful self-organizing networks. By combining competition with the instar rule, each of the prototype vectors that are learned by the network become representative of a particular class of input vector. Thus the competitive networks learn to divide their input space into distinct classes. Each class is represented by one of the prototype vectors (rows of the weight matrix).

Three types of networks, all developed by Tuevo Kohonen, were discussed in this chapter. The first is the standard competitive layer. Its simple operation makes it a practical network for many problems.

The self-organizing feature map is very similar to the competitive layer, but more closely models biological on-center/off-surround networks. The result is a network that not only learns to classify input vectors, but also learns the topology of the input space.

The third network, the LVQ network, uses both unsupervised and supervised learning to recognize clusters. It uses a second layer to combine multiple convex regions into classes that can have any shape. LVQ networks can even be trained to recognize classes made up of multiple unconnected regions.

Chapter 17 presents practical tips for training competitive networks, and Chapter 21 is a case study of using self organizing feature maps on a real-world clustering problem.

Further Reading

[FrSk91] J. Freeman and D. Skapura, *Neural Networks: Algorithms, Applications, and Programming Techniques*, Reading, MA: Addison-Wesley, 1991.

This text contains code fragments for network algorithms, making the details of the networks clear.

[Koho87] T. Kohonen, *Self-Organization and Associative Memory*, 2nd Ed., Berlin: Springer-Verlag, 1987.

Kohonen introduces the Kohonen rule and several networks that use it. It provides a complete analysis of linear associative models and gives many extensions and examples.

[Hech90] R. Hecht-Nielsen, *Neurocomputing*, Reading, MA: Addison-Wesley, 1990.

This book contains a section on the history and mathematics of competitive learning.

[RuMc86] D. Rumelhart, J. McClelland et al., *Parallel Distributed Processing*, vol. 1, Cambridge, MA: MIT Press, 1986.

Both volumes of this set are classics in neural network literature. The first volume contains a chapter describing competitive layers and how they learn to detect features.

Exercises

E15.1 Suppose that the weight matrix for layer 2 of the Hamming network is given by

$$\mathbf{W}^2 = \begin{bmatrix} 1 & -\frac{3}{4} & -\frac{3}{4} \\ -\frac{3}{4} & 1 & -\frac{3}{4} \\ -\frac{3}{4} & -\frac{3}{4} & 1 \end{bmatrix}.$$

This matrix violates Eq. (15.6), since

$$\varepsilon = \frac{3}{4} > \frac{1}{S-1} = \frac{1}{2}.$$

Give an example of an output from Layer 1 for which Layer 2 will fail to operate correctly.

E15.2 Consider the input vectors and initial weights shown in Figure E15.1.

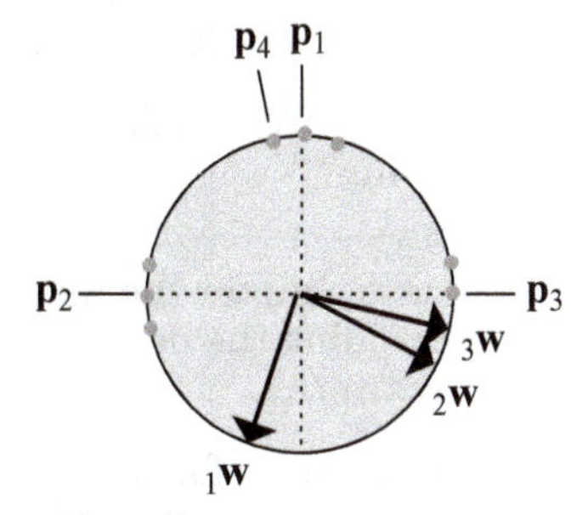

Figure E15.1 Cluster Data Vectors

i. Draw the diagram of a competitive network that could classify the data above so that each of the three clusters of vectors would have its own class.

ii. Train the network graphically (using the initial weights shown) by presenting the labeled vectors in the following order:

$$\mathbf{p}_1, \mathbf{p}_2, \mathbf{p}_3, \mathbf{p}_4.$$

Recall that the competitive transfer function chooses the neuron with the lowest index to win if more than one neuron has the same net input. The Kohonen rule is introduced graphically in Figure 15.3.

iii. Redraw the diagram in Figure E15.1, showing your final weight vectors and the decision boundaries between each region that represents a class.

E15.3 Train a competitive network using the following input patterns:

$$\mathbf{p}_1 = \begin{bmatrix} 1 \\ -1 \end{bmatrix}, \mathbf{p}_2 = \begin{bmatrix} 1 \\ 1 \end{bmatrix}, \mathbf{p}_3 = \begin{bmatrix} -1 \\ -1 \end{bmatrix}.$$

i. Use the Kohonen learning law with $\alpha = 0.5$, and train for one pass through the input patterns. (Present each input once, in the order given.) Display the results graphically. Assume the initial weight matrix is

$$\mathbf{W} = \begin{bmatrix} \sqrt{2} & 0 \\ 0 & \sqrt{2} \end{bmatrix}.$$

ii. After one pass through the input patterns, how are the patterns clustered? (In other words, which patterns are grouped together in the same class?) Would this change if the input patterns were presented in a different order? Explain.

iii. Repeat part (i) using $\alpha = 0.25$. How does this change affect the training?

E15.4 Earlier in this chapter the term "conscience" was used to refer to a technique for avoiding the dead neuron problem plaguing competitive layers and LVQ networks.

Neurons that are too far from input vectors to ever win the competition can be given a chance by using adaptive biases that get more negative each time a neuron wins the competition. The result is that neurons that win very often start to feel "guilty" until other neurons have a chance to win.

Figure E15.2 shows a competitive network with biases. A typical learning rule for the bias b_i of neuron i is

$$b_i^{new} = \begin{cases} 0.9b_i^{old}, \text{ if } i \neq i^* \\ b_i^{old} - 0.2, \text{ if } i = i^* \end{cases}.$$

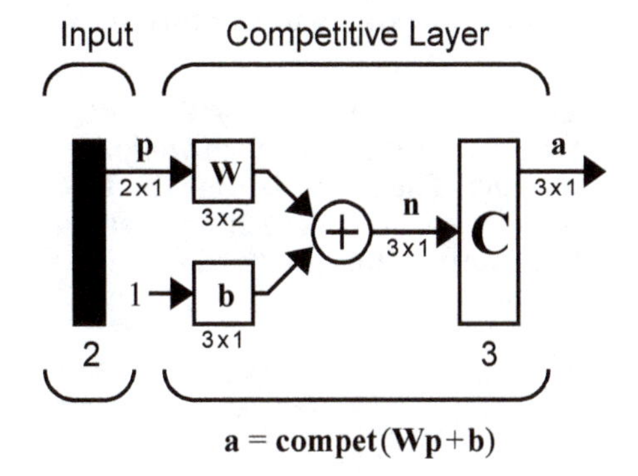

Figure E15.2 Competitive Layer with Biases

i. Examine the vectors in Figure E15.3. Is there any order in which the vectors can be presented that will cause $_1w$ to win the competition and move closer to one of the vectors? (Note: assume that adaptive biases are *not* being used.)

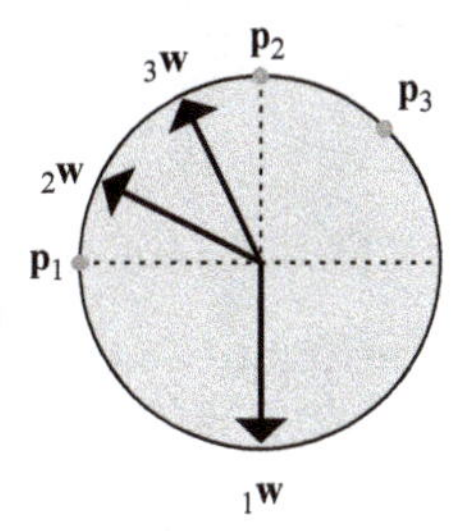

Figure E15.3 Input Vectors and Dead Neuron

ii. Given the input vectors and the initial weights and biases defined below, calculate the weights (using the Kohonen rule) and the biases (using the above bias rule). Repeat the sequence shown below until neuron 1 wins the competition.

$$\mathbf{p}_1 = \begin{bmatrix} -1 \\ 0 \end{bmatrix}, \mathbf{p}_2 = \begin{bmatrix} 0 \\ 1 \end{bmatrix}, \mathbf{p}_3 = \begin{bmatrix} 1/\sqrt{2} \\ 1/\sqrt{2} \end{bmatrix}$$

$${}_1\mathbf{w} = \begin{bmatrix} 0 \\ -1 \end{bmatrix}, {}_2\mathbf{w} = \begin{bmatrix} -2/\sqrt{5} \\ -1/\sqrt{5} \end{bmatrix}, {}_3\mathbf{w} = \begin{bmatrix} -1/\sqrt{5} \\ -2/\sqrt{5} \end{bmatrix}, b_1(0) = b_2(0) = b_3(0) = 0$$

Sequence of input vectors: $\mathbf{p}_1, \mathbf{p}_2, \mathbf{p}_3, \mathbf{p}_1, \mathbf{p}_2, \mathbf{p}_3, \ldots$

iii. How many presentations occur before ${}_1\mathbf{w}$ wins the competition?

E15.5 The net input expression for LVQ networks calculates the distance between the input and each weight vector directly, instead of using the inner product. The result is that the LVQ network does not require normalized input vectors. This technique can also be used to allow a competitive layer to classify nonnormalized vectors. Such a network is shown in Figure E15.4.

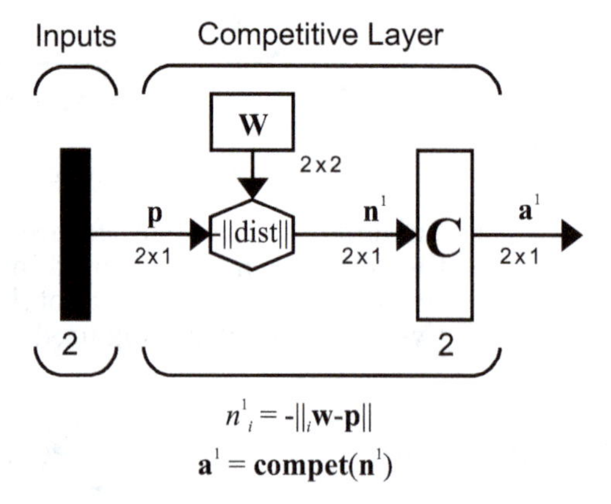

Figure E15.4 Competitive Layer with Alternate Net Input Expression

Use this technique to train a two-neuron competitive layer on the (nonnormalized) vectors below, using a learning rate, α, of 0.5.

$$\mathbf{p}_1 = \begin{bmatrix} 1 \\ 1 \end{bmatrix}, \mathbf{p}_2 = \begin{bmatrix} -1 \\ 2 \end{bmatrix}, \mathbf{p}_3 = \begin{bmatrix} -2 \\ -2 \end{bmatrix}$$

Present the vectors in the following order:

$$\mathbf{p}_1, \mathbf{p}_2, \mathbf{p}_3, \mathbf{p}_2, \mathbf{p}_3, \mathbf{p}_1.$$

Here are the initial weights of the network:

$${}_1\mathbf{w} = \begin{bmatrix} 0 \\ 1 \end{bmatrix}, {}_2\mathbf{w} = \begin{bmatrix} 1 \\ 0 \end{bmatrix}.$$

E15.6 Repeat E15.5 for the following inputs and initial weights. Show the movements of the weights graphically for each step. If the network is trained for a large number of iterations, how will the three vectors be clustered in the final configuration?

$$\mathbf{p}_1 = \begin{bmatrix} 2 \\ 0 \end{bmatrix}, \mathbf{p}_2 = \begin{bmatrix} 0 \\ 1 \end{bmatrix}, \mathbf{p}_3 = \begin{bmatrix} 2 \\ 2 \end{bmatrix}$$

$${}_1\mathbf{w} = \begin{bmatrix} 1 \\ 0 \end{bmatrix}, {}_2\mathbf{w} = \begin{bmatrix} -1 \\ 0 \end{bmatrix}.$$

E15.7 We have a competitive learning problem, where the input vectors are

$$\mathbf{p}_1 = \begin{bmatrix} 0 \\ 1 \end{bmatrix}, \mathbf{p}_2 = \begin{bmatrix} 0 \\ 2 \end{bmatrix}, \mathbf{p}_3 = \begin{bmatrix} 1 \\ 1 \end{bmatrix}, \mathbf{p}_4 = \begin{bmatrix} 2 \\ 2 \end{bmatrix},$$

and the initial weight matrix is

$$\mathbf{W} = \begin{bmatrix} 1 & -1 \\ -1 & 1 \end{bmatrix}.$$

i. Use the Kohonen learning law to train a competitive network using a learning rate of $\alpha = 0.5$. (Present each vector once, in the order shown.) Use the modified competitive network of Figure E15.4, which uses negative distance, instead of inner product.

ii. Display the results of part i graphically, as in Figure 15.3. (Show all four iterations.)

iii. Where will the weights eventually converge (approximately)? Explain. Sketch the approximate final decision boundaries.

E15.8 Show that the modified competitive network of Figure E15.4, which computes distance directly, will produce the same results as the standard competitive network, which uses the inner product, when the input vectors are normalized.

E15.9 We would like a classifier that divides the interval of the input space defined below into five classes.

$$0 \le p_1 \le 1$$

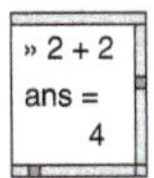

i. Use MATLAB to randomly generate 100 values in the interval shown above with a uniform distribution.

ii. Square each number so that the distribution is no longer uniform.

iii. Write a MATLAB M-file to implement a competitive layer. Use the M-file to train a five-neuron competitive layer on the squared values until its weights are fairly stable.

iv. How are the weight values of the competitive layer distributed? Is there some relationship between how the weights are distributed and how the squared input values are distributed?

E15.10 We would like a classifier that divides the square region defined below into sixteen classes of roughly equal size.

$$0 \le p_1 \le 1 \,,\; 2 \le p_2 \le 3$$

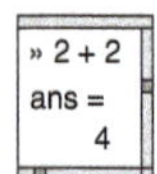

i. Use MATLAB to randomly generate 200 vectors in the region shown above.

ii. Write a MATLAB M-file to implement a competitive layer with Kohonen learning. Calculate the net input by finding the distance between the input and weight vectors directly, as is done by the LVQ network, so the vectors do not need to be normalized. Use the M-file to train a competitive layer to classify the 200 vectors. Try different learning rates and compare performance.

iii. Write a MATLAB M-file to implement a four-neuron by four-neuron (two-dimensional) feature map. Use the feature map to classify the same vectors. Use different learning rates and neighborhood sizes, then compare performance.

E15.11 We want to train the following 1-D feature map (which uses distance instead of inner product to compute the net input):

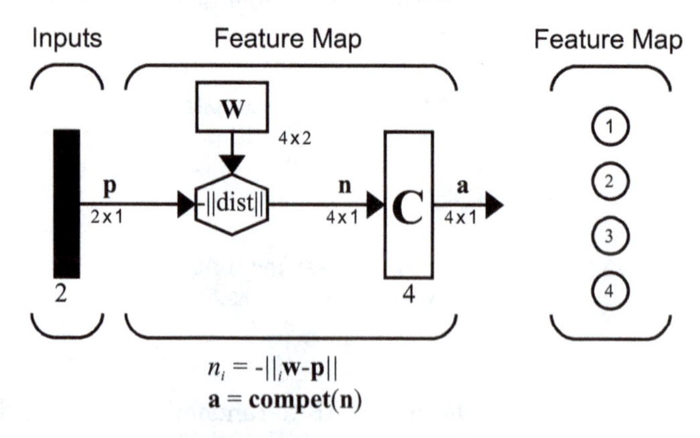

Figure E15.5 1-D Feature Map for Exercise E15.11

The initial weight matrix is $\mathbf{W}(0) = \begin{bmatrix} 2 & -1 & -1 & 1 \\ 2 & 1 & -2 & 0 \end{bmatrix}^T$.

i. Plot the initial weight vectors as dots, and connect the neighboring weight vectors as lines (as in Figure 15.10, except that this is a 1-D feature map).

ii. The following input vector is applied to the network. Perform one iteration of the feature map learning rule. (You can do this graphically.) Use a neighborhood size of 1 and a learning rate of $\alpha = 0.5$.

$$\mathbf{p}_1 = \begin{bmatrix} -2 & 0 \end{bmatrix}^T$$

iii. Plot the new weight vectors as dots, and connect the neighboring weight vectors as lines.

E15.12 Consider the following feature map, where distance is used instead of inner product to compute the net input.

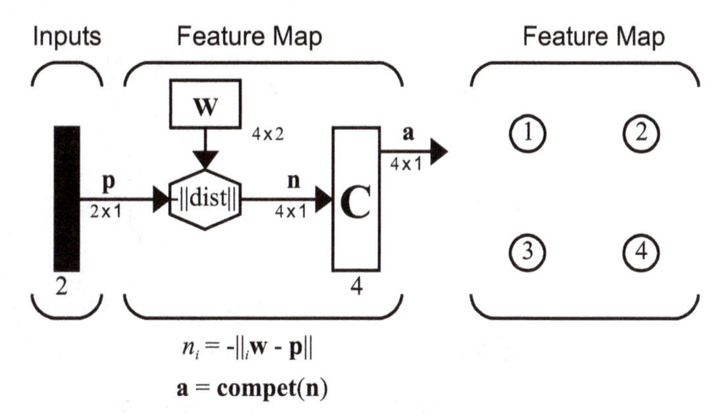

Figure E15.6 2-D Feature Map for Exercise E15.12

The initial weight matrix is

$$\mathbf{W} = \begin{bmatrix} 0 & 1 & 1 & 0 \\ 0 & 0 & 1 & -1 \end{bmatrix}^T$$

i. Plot the initial weights, and show their topological connections, as in Figure 15.10.

ii. Apply the input $\mathbf{p} = \begin{bmatrix} -1 & 1 \end{bmatrix}^T$, and perform one iteration of the feature map learning rule, with learning rate of $\alpha = 0.5$, and neighborhood radius of 1.

iii. Plot the weights after the first iteration, and show their topological connections.

E15.13 An LVQ network has the following weights:

$$\mathbf{W}^1 = \begin{bmatrix} 0 & 0 \\ 1 & 0 \\ -1 & 0 \\ 0 & 1 \\ 0 & -1 \end{bmatrix}, \quad \mathbf{W}^2 = \begin{bmatrix} 1 & 0 & 0 & 0 & 0 \\ 0 & 1 & 1 & 0 & 0 \\ 0 & 0 & 0 & 1 & 1 \end{bmatrix}.$$

i. How many classes does this LVQ network have? How many subclasses?

ii. Draw a diagram showing the first-layer weight vectors and the decision boundaries that separate the input space into subclasses.

iii. Label each subclass region to indicate which class it belongs to.

E15.14 We would like an LVQ network that classifies the following vectors according to the classes indicated:

$$\text{class 1: } \left\{ \begin{bmatrix} -1 \\ 1 \\ -1 \end{bmatrix}, \begin{bmatrix} 1 \\ -1 \\ -1 \end{bmatrix} \right\}, \text{ class 2: } \left\{ \begin{bmatrix} -1 \\ -1 \\ 1 \end{bmatrix}, \begin{bmatrix} 1 \\ -1 \\ 1 \end{bmatrix}, \begin{bmatrix} 1 \\ 1 \\ -1 \end{bmatrix} \right\}, \text{ class 3: } \left\{ \begin{bmatrix} -1 \\ -1 \\ -1 \end{bmatrix}, \begin{bmatrix} -1 \\ 1 \\ 1 \end{bmatrix} \right\}.$$

i. How many neurons are required in each layer of the LVQ network?

ii. Define the weights for the first layer.

iii. Define the weights for the second layer.

iv. Test your network for at least one vector from each class.

E15.15 We would like an LVQ network that classifies the following vectors according to the classes indicated:

$$\text{class 1: } \left\{ \mathbf{p}_1 = \begin{bmatrix} 1 \\ 1 \end{bmatrix}, \mathbf{p}_2 = \begin{bmatrix} 0 \\ 2 \end{bmatrix} \right\}, \text{ class 2: } \left\{ \mathbf{p}_3 = \begin{bmatrix} -1 \\ 1 \end{bmatrix}, \mathbf{p}_4 = \begin{bmatrix} 1 \\ 2 \end{bmatrix} \right\}$$

i. Could this classification problem be solved by a perceptron? Explain your answer.

ii. How many neurons must be in each layer of an LVQ network that can classify the above data, given that each class is made up of two convex-shaped subclasses?

iii. Define the second-layer weights for such a network.

iv. Initialize the first-layer weights of the network to all zeros and calculate the changes made to the weights by the Kohonen rule (with a learning rate α of 0.5) for the following series of vectors:

$$\mathbf{p}_4,\ \mathbf{p}_2,\ \mathbf{p}_3,\ \mathbf{p}_1,\ \mathbf{p}_2.$$

v. Draw a diagram showing the input vectors, the final weight vectors and the decision boundaries between the two classes.

E15.16 An LVQ network has the following weights and training data.

$$\mathbf{W}^1 = \begin{bmatrix} 1 & 0 \\ 0 & 1 \\ 0 & 0 \end{bmatrix},\ \mathbf{W}^2 = \begin{bmatrix} 1 & 1 & 0 \\ 0 & 0 & 1 \end{bmatrix},$$

$$\left\{\mathbf{p}_1 = \begin{bmatrix} -2 \\ 2 \end{bmatrix}, \mathbf{t}_1 = \begin{bmatrix} 1 \\ 0 \end{bmatrix}\right\},\ \left\{\mathbf{p}_2 = \begin{bmatrix} 2 \\ 0 \end{bmatrix}, \mathbf{t}_2 = \begin{bmatrix} 0 \\ 1 \end{bmatrix}\right\},\ \left\{\mathbf{p}_3 = \begin{bmatrix} 2 \\ -2 \end{bmatrix}, \mathbf{t}_3 = \begin{bmatrix} 1 \\ 0 \end{bmatrix}\right\},$$

$$\left\{\mathbf{p}_4 = \begin{bmatrix} -2 \\ 0 \end{bmatrix}, \mathbf{t}_4 = \begin{bmatrix} 0 \\ 1 \end{bmatrix}\right\}$$

i. Plot the training data input vectors and weight vectors (as in Figure 15.14).

ii. Perform four iterations of the LVQ learning rule, with learning rate $\alpha = 0.5$, as you present the following sequence of input vectors: $\mathbf{p}_1$, $\mathbf{p}_2$, $\mathbf{p}_3$, $\mathbf{p}_4$ (one iteration for each input). Do this graphically, on a separate diagram from part i.

iii. After completing the iterations in part ii, on a new diagram, sketch the regions of the input space that make up each subclass and each class. Label each region to indicate which class it belongs to.

E15.17 An LVQ network has the following weights:

$$\mathbf{W}^1 = \begin{bmatrix} 0 & 1 & -1 & 0 & 0 & -1 & -1 \\ 0 & 0 & 0 & 1 & -1 & -1 & 1 \end{bmatrix}^T,\ \mathbf{W}^2 = \begin{bmatrix} 1 & 0 & 1 & 0 & 1 & 1 & 0 \\ 0 & 1 & 0 & 1 & 0 & 0 & 1 \end{bmatrix}.$$

i. How many classes does this LVQ network have? How many subclasses?

ii. Draw a diagram showing the first-layer weight vectors and the decision boundaries that separate the input space into subclasses.

iii. Label each subclass region to indicate which class it belongs to.

iv. Suppose that an input $\mathbf{p} = \begin{bmatrix} 1 & 0.5 \end{bmatrix}^T$ from Class 1 is presented to the network. Perform one iteration of the LVQ algorithm, with $\alpha = 0.5$.

E15.18 An LVQ network has the following weights:

$$\mathbf{W}^1 = \begin{bmatrix} 0 & 0 & 2 & 1 & 1 & -1 \\ 0 & 2 & 2 & 1 & -1 & -1 \end{bmatrix}^T, \mathbf{W}^2 = \begin{bmatrix} 1 & 1 & 1 & 0 & 0 & 0 \\ 0 & 0 & 0 & 1 & 1 & 1 \end{bmatrix}.$$

i. How many classes does this LVQ network have? How many subclasses?

ii. Draw a diagram showing the first-layer weight vectors and the decision boundaries that separate the input space into subclasses.

iii. Label each subclass region to indicate which class it belongs to.

iv. Perform one iteration of the LVQ algorithm, with the following input/target pair: $\mathbf{p} = \begin{bmatrix} -1 & -2 \end{bmatrix}^T$, $\mathbf{t} = \begin{bmatrix} 1 & 0 \end{bmatrix}^T$. Use learning rate $\alpha = 0.5$.

16 Radial Basis Networks

Objectives

The multilayer networks discussed in Chapter 11 and Chapter 12 represent one type of neural network structure for function approximation and pattern recognition. As we saw in Chapter 11, multilayer networks with sigmoid transfer functions in the hidden layers and linear transfer functions in the output layer are universal function approximators. In this chapter we will discuss another type of universal approximation network, the radial basis function network. This network can be used for many of the same applications as multilayer networks.

This chapter will follow the structure of Chapter 11. We will begin by demonstrating, in an intuitive way, the universal approximation capabilities of the radial basis function network. Then we will describe three different techniques for training these networks. They can be trained by the same gradient-based algorithms discussed in Chapter 11 and Chapter 12, with derivatives computed using a form of backpropagation. However, they can also be trained using a two-stage process, in which the first layer weights are computed independently from the weights in the second layer. Finally, these networks can be built in an incremental way - one neuron at a time.

Theory and Examples

The radial basis function network is related to the multilayer perceptron network of Chapter 11. It is also a universal approximator and can be used for function approximation or pattern recognition. We will begin this chapter with a description of the network and a demonstration of its abilities for function approximation and pattern recognition.

The original work in radial basis functions was performed by Powell and others during the 1980's [Powe87]. In this original work, radial basis functions were used for exact interpolation in a multidimensional space. In other words, the function created by the radial basis interpolation was required to pass exactly through all targets in the training set. The use of radial basis functions for exact interpolation continues to be an important application area, and it is also an active area of research.

For our purposes, however, we will not be considering exact interpolation. Neural networks are often used on noisy data, and exact interpolation often results in overfitting when the training data is noisy, as we discussed in Chapter 13. Our interest is in the use of radial basis functions to provide robust approximations to unknown functions based on generally limited and noisy measurements. Broomhead and Lowe [BrLo88] were the first to develop the radial basis function neural network model, which produces a smooth interpolating function. No attempt is made to force the network response to exactly match target outputs. The emphasis is on producing networks that will generalize well to new situations.

In the next section we will demonstrate the capabilities of the radial basis function neural network. In the following sections we will describe procedures for training these networks.

Radial Basis Network

RBF

The radial basis network is a two-layer network. There are two major distinctions between the radial basis function (RBF) network and a two layer perceptron network. First, in layer 1 of the RBF network, instead of performing an inner product operation between the weights and the input (matrix multiplication), we calculate the distance between the input vector and the rows of the weight matrix. (This is similar to the LVQ network shown in Figure 15.13.) Second, instead of adding the bias, we multiply by the bias. Therefore, the net input for neuron i in the first layer is calculated as follows:

$$n_i^1 = \| \mathbf{p} - {}_i\mathbf{w}^1 \| b_i^1 . \quad (16.1)$$

Each row of the weight matrix acts as a center point - a point where the net input value will be zero. The bias performs a scaling operation on the transfer (basis) function, causing it to stretch or compress.

We should note that most papers and texts on RBF networks use the terms standard deviation, variance or spread constant, rather than bias. We have used the bias in order to maintain a consistency with other networks in this text. This is simply a matter of notation and pedagogy. The operation of the network is not affected. When a Gaussian transfer function is used, the bias is related to the standard deviation as follows: $b = 1/(\sigma\sqrt{2})$.

The transfer functions used in the first layer of the RBF network are different than the sigmoid functions generally used in the hidden layers of multilayer perceptrons (MLP). There are several different types of transfer function that can be used (see [BrLo88]), but for clarity of presentation we will consider only the Gaussian function, which is the one most commonly used in the neural network community. It is defined as follows

$$a = e^{-n^2}, \tag{16.2}$$

and it is plotted in Figure 16.1.

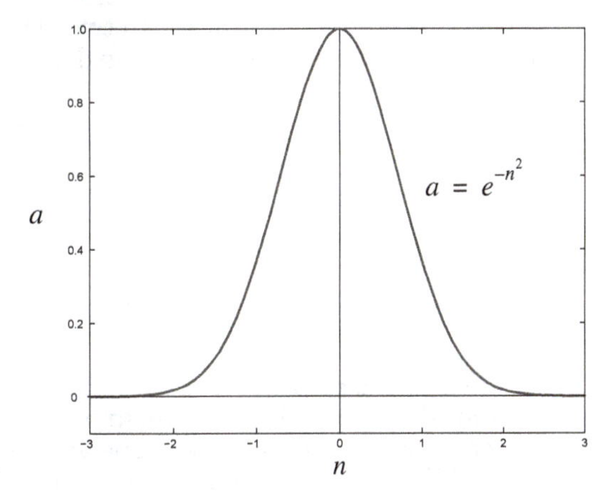

Figure 16.1 Gaussian Basis Function

Local Function

Global Function

A key property of this function is that it is *local*. This means that the output is close to zero if you move very far in either direction from the center point. This is in contrast to the *global* sigmoid functions, whose output remains close to 1 as the net input goes to infinity.

The second layer of the RBF network is a standard linear layer:

$$\mathbf{a}^2 = \mathbf{W}^2\mathbf{a}^1 + \mathbf{b}^2 \tag{16.3}$$

Figure 16.2 shows the complete RBF network.

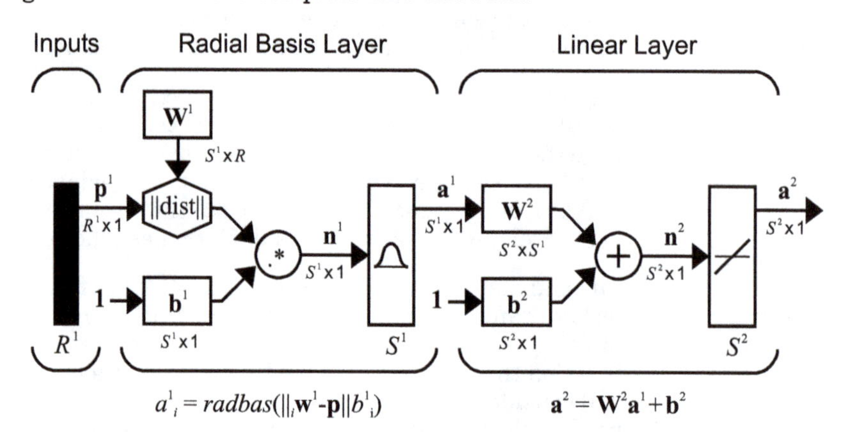

Figure 16.2 Radial Basis Network

Function Approximation

This RBF network has been shown to be a universal approximator [PaSa93], just like the MLP network. To illustrate the capability of this network, consider a network with two neurons in the hidden layer, one output neuron, and with the following default parameters:

$$w^1_{1,1} = -1 \, , \; w^1_{2,1} = 1 \, , \; b^1_1 = 2 \, , \; b^1_2 = 2 \, ,$$

$$w^2_{1,1} = 1 \, , \; w^2_{1,2} = 1 \, , \; b^2 = 0 \, .$$

The response of the network with the default parameters is shown in Figure 16.3, which plots the network output a^2 as the input p is varied over the range $[-2, 2]$.

Notice that the response consists of two hills, one for each of the Gaussian neurons (basis functions) in the first layer. By adjusting the network parameters, we can change the shape and location of each hill, as we will see in the following discussion. (As you proceed through this example, it may be helpful to compare the response of this sample RBF network with the response of the sample MLP network in Figure 11.5.)

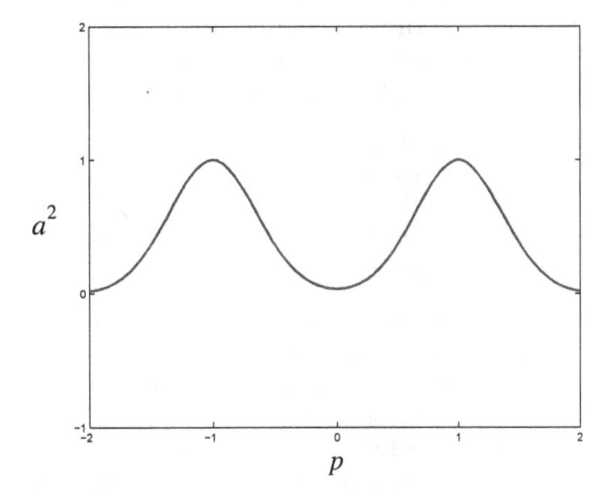

Figure 16.3 Default Network Response

Figure 16.4 illustrates the effects of parameter changes on the network response. The blue curve is the nominal response. The other curves correspond to the network response when one parameter at a time is varied over the following ranges:

$$0 \le w^1_{2,1} \le 2 \, , \; -1 \le w^2_{1,1} \le 1 \, , \; 0.5 \le b^1_2 \le 8 \, , \; -1 \le b^2 \le 1 \, . \qquad (16.4)$$

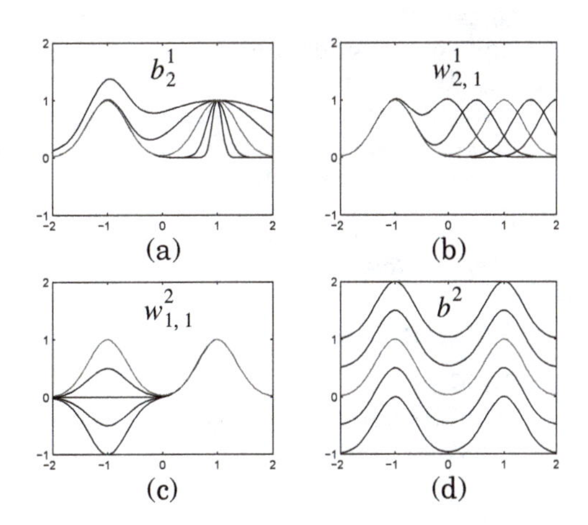

Figure 16.4 Effect of Parameter Changes on Network Response

Figure 16.4 (a) shows how the network biases in the first layer can be used to change the width of the hills - the larger the bias, the narrower the hill. Figure 16.4 (b) illustrates how the weights in the first layer determine the location of the hills; there will be a hill centered at each first layer weight. For multidimensional inputs there will be a hill centered at each row of the weight matrix. For this reason, each row of the first layer weight matrix is often called the *center* for the corresponding neuron (basis function).

Center

Notice that the effects of the weight and the bias in first layer of the RBF network are much different than for the MLP network, which was shown in Figure 11.6. In the MLP network, the sigmoid functions create steps. The weights change the slopes of the steps, and the biases change the locations of the steps.

Figure 16.4 (c) illustrates how the weights in the second layer scale the height of the hills. The bias in the second layer shifts the entire network response up or down, as can be seen in Figure 16.4 (d). The second layer of the RBF network is the same type of linear layer used in the MLP network of Figure 11.6, and it performs a similar function, which is to create a weighted sum of the outputs of the layer 1 neurons.

This example demonstrates the flexibility of the RBF network for function approximation. As with the MLP, it seems clear that if we have enough neurons in the first layer of the RBF network, we can approximate virtually any function of interest, and [PaSa93] provides a mathematical proof that this is the case. However, although both MLP and RBF networks are universal approximators, they perform their approximations in different ways. For the RBF network, each transfer function is only active over a small region of the input space - the response is *local*. If the input moves far from a given center, the output of the corresponding neuron will be close to zero. This has consequences for the design of RBF networks. We must have centers adequately distributed throughout the range of the network inputs, and we must select biases in such a way that all of the basis functions overlap in a significant way. (Recall that the biases change the width of each basis function.) We will discuss these design considerations in more detail in later sections.

To experiment with the response of this RBF network, use the MATLAB® Neural Network Design Demonstration RBF Network Function (**nnd17nf**).

Pattern Classification

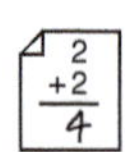

To illustrate the capabilities of the RBF network for pattern classification, consider again the classic exclusive-or (XOR) problem. The categories for the XOR gate are

$$\text{Category 1: } \left\{ \mathbf{p}_2 = \begin{bmatrix} -1 \\ 1 \end{bmatrix}, \mathbf{p}_3 = \begin{bmatrix} 1 \\ -1 \end{bmatrix} \right\}, \text{ Category 2: } \left\{ \mathbf{p}_1 = \begin{bmatrix} -1 \\ -1 \end{bmatrix}, \mathbf{p}_4 = \begin{bmatrix} 1 \\ 1 \end{bmatrix} \right\}.$$

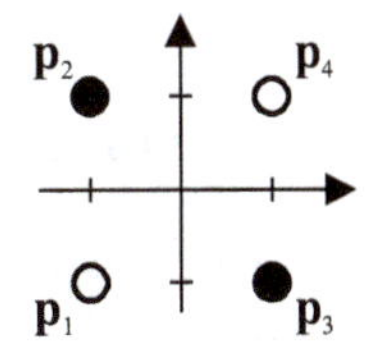

The problem is illustrated graphically in the figure to the left. Because the two categories are not linearly separable, a single-layer network cannot perform the classification.

RBF networks can classify these patterns. In fact, there are many different RBF solutions. We will consider one solution that demonstrates in a simple way how to use RBF networks for pattern classification. The idea will be to have the network produce outputs greater than zero when the input is near patterns $\mathbf{p}_2$ or $\mathbf{p}_3$, and outputs less than zero for all other inputs. (Note that the procedures we will use to design this example network are not suitable for complex problems, but they will help us illustrate the capabilities of the RBF network.)

From the problem statement, we know that the network will need to have two inputs and one output. For simplicity, we will use only two neurons in the first layer (two basis functions), since this will be sufficient to solve the XOR problem. As we discussed earlier, the rows of the first-layer weight matrix will create centers for the two basis functions. We will choose the centers to be equal to the patterns $\mathbf{p}_2$ and $\mathbf{p}_3$. By centering a basis function at each pattern, we can produce maximum network outputs there. The first layer weight matrix is then

$$\mathbf{W}^1 = \begin{bmatrix} \mathbf{p}_2^T \\ \mathbf{p}_3^T \end{bmatrix} = \begin{bmatrix} -1 & 1 \\ 1 & -1 \end{bmatrix}. \qquad (16.5)$$

The choice of the bias in the first layer depends on the width that we want for each basis function. For this problem, we would like the network function to have two distinct peaks at $\mathbf{p}_2$ and $\mathbf{p}_3$. Therefore, we don't want the basis functions to overlap too much. The centers of the basis functions are each a distance of $\sqrt{2}$ from the origin. We want the basis function to drop significantly from its peak in this distance. If we use a bias of 1, we would get the following reduction in that distance:

$$a = e^{-n^2} = e^{-(1 \cdot \sqrt{2})^2} = e^{-2} = 0.1353. \qquad (16.6)$$

Therefore, each basis function will have a peak of 1 at the centers, and will drop to 0.1353 at the origin. This will work for our problem, so we select the first layer bias vector to be

$$\mathbf{b}^1 = \begin{bmatrix} 1 \\ 1 \end{bmatrix}. \qquad (16.7)$$

The original basis function response ranges from 0 to 1 (see Figure 16.1). We want the output to be negative for inputs much different than $\mathbf{p}_2$ and $\mathbf{p}_3$, so we will use a bias of -1 for the second layer, and we will use a value of 2 for the second layer weights, in order to bring the peaks back up to 1. The second layer weights and biases then become

$$\mathbf{W}^2 = \begin{bmatrix} 2 & 2 \end{bmatrix}, \; b^2 = \begin{bmatrix} -1 \end{bmatrix}. \tag{16.8}$$

For the network parameter values given in (16.5), (16.7) and (16.8), the network response is shown in Figure 16.5. This figure also shows where the surface intersects the plane at $a^2 = 0$, which is where the decision takes place. This is also indicated by the contours shown underneath the surface. These are the function contours where $a^2 = 0$. They are almost circles that surround the $\mathbf{p}_2$ and $\mathbf{p}_3$ vectors. This means that the network output will be greater than 0 only when the input vector is near the $\mathbf{p}_2$ and $\mathbf{p}_3$ vectors.

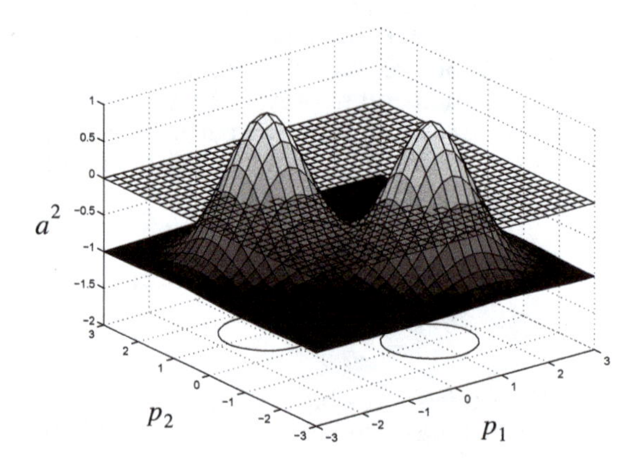

Figure 16.5 Example 2-Input RBF Function Surface

Figure 16.6 illustrates more clearly the decision boundaries. Whenever the input falls in the blue regions, the output of the network will be greater than zero. Whenever the network input falls outside the blue regions, the network output will be less than zero.

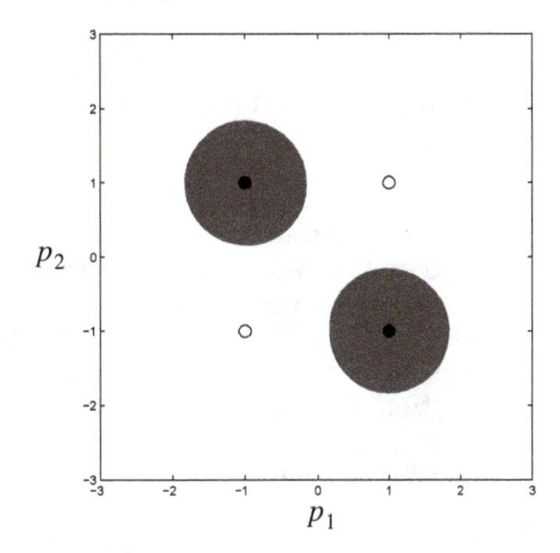

Figure 16.6 RBF Example Decision Regions

This network, therefore, classifies the patterns correctly. It is not the best solution, in the sense that it does not always assign input patterns to the closest prototype vector, unlike the MLP solution shown in Figure 11.2. You will notice that the decision regions for this RBF network are circles, unlike the linear boundaries that we see in single layer perceptrons. The MLP can put linear boundaries together to create arbitrary decision regions. The RBF network can put circular boundaries together to create arbitrary decision regions. In this problem, the linear boundaries are more efficient. Of course, when many neurons are used, and the centers are close together, the elementary RBF boundaries are no longer purely circular, and the elementary MLP boundaries are no longer purely linear. However, associating circular boundaries with RBF networks and linear boundaries with MLP networks can be helpful in understanding their operation as pattern classifiers.

To experiment with the RBF network for pattern classification, use the MATLAB® Neural Network Design Demonstration RBF Pattern Classification (**nnd17pc**).

Now that we see the power of RBF networks for function approximation and pattern recognition, the next step is to develop general training algorithms for these networks.

Global vs. Local

Before we discuss the training algorithms, we should say a final word about the advantages and disadvantages of the global transfer functions used by the MLP networks and the local transfer functions used by the RBF networks. The MLP creates a distributed representation, because all of the transfer functions overlap in their activity. At any given input value, many sigmoid functions in the first layer will have significant outputs. They must sum or cancel in the second layer in order to produce the appropriate response at each point. In the RBF network, each basis function is only active over a small range of the input. For any given input, only a few basis functions will be active.

There are advantages and disadvantages to each approach. The global approach tends to require fewer neurons in the hidden layer, since each neuron contributes to the response over a large part of the input space. For the RBF network, however, basis centers must be spread throughout the range of the input space in order to provide an accurate approximation. This leads to the problem of the "curse of dimensionality," which we will discuss in the next section. Also, if more neurons, and therefore more parameters, are used, then there is a greater likelihood that the network will overfit the training data and fail to generalize well to new situations.

On the other hand, the local approach generally leads to faster training, especially when the two-stage algorithms, which will be discussed in the next section, are used. Also, the local approach can be very useful for adaptive training, in which the network continues to be incrementally trained while it is being used, as in adaptive filters (nonlinear versions of the filters in Chapter 10) or controllers. If, for a period of time, training data only appears in a certain region of the input space, then a global representation will tend to improve its accuracy in those regions at the expense of its representation in other regions. Local representations will not have this problem to the same extent. Because each neuron is only active in a small region of the input space, its weights will not be adjusted if the input falls outside that region.

Training RBF Networks

Unlike the MLP network, which is almost always trained by some gradient-based algorithm (steepest descent, conjugate gradient, Levenberg-Marquardt, etc.), the RBF network can be trained by a variety of approaches.

RBF networks can be trained using gradient-based algorithms. However, because of the local nature of the transfer function and the way in which the first layer weights and biases operate, there tend to be many more unsatisfactory local minima in the error surfaces of RBF networks than in those of MLP networks. For this reason, gradient-based algorithms are often unsatisfactory for the complete training of RBF networks. They are used on occasion, however, for fine-tuning of the network after it has been initially trained using some other method. Later in this chapter we will discuss the modifications to the backpropagation equations in Chapter 11 that are needed to compute the gradients for RBF networks.

The most commonly used RBF training algorithms have two stages, which treat the two layers of the RBF network separately. The algorithms differ mainly in how the first layer weights and biases are selected. Once the first layer weights and biases have been selected, the second layer weights can be computed in one step, using a linear least-squares algorithm. We will discuss linear least squares in the next section.

The simplest of the two-stage algorithms arranges the centers (first layer weights) in a grid pattern throughout the input range and then chooses a constant bias so that the basis functions have some degree of overlap. This procedure is not optimal, because the most efficient approximation would place more basis functions in regions of the input space where the function to be approximated is most complex. Also, for many practical cases the full range of the input space is not used, and therefore many basis functions could be wasted. One of the drawbacks of the RBF network, especially when the centers are selected on a grid, is that they suffer from the *curse of dimensionality*. This means that as the dimension of the input space increases, the number of basis functions required increases geometrically. For example, suppose that we have one input variable, and we specify a grid of 10 basis functions evenly spaced across the range of that input variable. Now, increase the number of input variables to 2. To maintain the same grid coverage for both input variables, we would need 10^2, or 100 basis functions.

Curse of Dimensionality

Another method for selecting the centers is to select some random subset of the input vectors in the training set. This ensures that basis centers will be placed in areas where they will be useful to the network. However, due to the randomness of the selection, this procedure is not optimal. A more efficient approach is to use a method such as the Kohonen competitive layer or the feature map, described in Chapter 15, to cluster the input space. The cluster centers then become basis function centers. This ensures that the basis functions are placed in regions with significant activity. We will discuss this method in a later section.

A final procedure that we will discuss for RBF training is called orthogonal least squares. It is based on a general method for building linear models called subset selection. This method starts with a large number of possible centers—typically all of the input vectors from the training data. At each stage of the procedure, it selects one center to add to the first layer weight. The selection is based on how much the new neuron will reduce the sum squared error. Neurons are added until some criteria is met. The criteria is typically chosen to maximize the generalization capability of the network.

Linear Least Squares

In this section we will assume that the first layer weights and biases of the RBF network are fixed. This can be done by fixing the centers on a grid, or by randomly selecting the centers from the input vectors in the training data set (or by using the clustering method which is described in a later section). When the centers are randomly selected, all of the biases can be computed using the following formula [Lowe89]:

$$b_i^1 = \frac{\sqrt{S^1}}{d_{max}}, \tag{16.9}$$

where d_{max} is the maximum distance between neighboring centers. This is designed to ensure an appropriate degree of overlap between the basis functions. Using this method, all of the biases have the same value. There are other methods which use different values for each bias. We will discuss one such method later, in the section on clustering.

Once the first layer parameters have been set, the training of the second layer weights and biases is equivalent to training a linear network, as in Chapter 10. For example, consider that we have the following training points

$$\{\mathbf{p}_1, \mathbf{t}_1\}, \{\mathbf{p}_2, \mathbf{t}_2\}, \ldots, \{\mathbf{p}_Q, \mathbf{t}_Q\}, \tag{16.10}$$

where $\mathbf{p}_q$ is an input to the network, and $\mathbf{t}_q$ is the corresponding target output. The output of the first layer for each input $\mathbf{p}_q$ in the training set can be computed as

$$n_{i,q}^1 = \|\mathbf{p}_q - {}_i\mathbf{w}^1\| b_i^1, \tag{16.11}$$

$$\mathbf{a}_q^1 = \mathbf{radbas}(\mathbf{n}_q^1). \tag{16.12}$$

Since the first layer weights and biases will not be adjusted, the training data set for the second layer then becomes

$$\{\mathbf{a}_1^1, \mathbf{t}_1\}, \{\mathbf{a}_2^1, \mathbf{t}_2\}, \ldots, \{\mathbf{a}_Q^1, \mathbf{t}_Q\}. \tag{16.13}$$

The second layer response is linear:

$$\mathbf{a}^2 = \mathbf{W}^2\mathbf{a}^1 + \mathbf{b}^2. \tag{16.14}$$

We want to select the weights and biases in this layer to minimize the sum square error performance index over the training set:

$$F(\mathbf{x}) = \sum_{q=1}^{Q} (\mathbf{t}_q - \mathbf{a}_q^2)^T(\mathbf{t}_q - \mathbf{a}_q^2) \tag{16.15}$$

Our derivation of the solution to this linear least squares problem will follow the linear network derivation starting with Eq. (10.6). To simplify the discussion, we will assume a scalar target, and we will lump all of the parameters we are adjusting, including the bias, into one vector:

$$\mathbf{x} = \begin{bmatrix} {}_1\mathbf{w}^2 \\ b^2 \end{bmatrix}. \quad (16.16)$$

Similarly, we include the bias input "1" as a component of the input vector

$$\mathbf{z}_q = \begin{bmatrix} \mathbf{a}_q^1 \\ 1 \end{bmatrix}. \quad (16.17)$$

Now the network output, which we usually write in the form

$$a_q^2 = ({}_1\mathbf{w}^2)^T \mathbf{a}_q^1 + b^2, \quad (16.18)$$

can be written as

$$a_q = \mathbf{x}^T \mathbf{z}_q. \quad (16.19)$$

This allows us to conveniently write out an expression for the sum square error:

$$F(\mathbf{x}) = \sum_{q=1}^{Q} (e_q)^2 = \sum_{q=1}^{Q} (t_q - a_q)^2 = \sum_{q=1}^{Q} (t_q - \mathbf{x}^T \mathbf{z}_q)^2. \quad (16.20)$$

To express this in matrix form, we define the following matrices:

$$\mathbf{t} = \begin{bmatrix} t_1 \\ t_2 \\ \vdots \\ t_Q \end{bmatrix}, \mathbf{U} = \begin{bmatrix} {}_1\mathbf{u}^T \\ {}_2\mathbf{u}^T \\ \vdots \\ {}_Q\mathbf{u}^T \end{bmatrix} = \begin{bmatrix} \mathbf{z}_1^T \\ \mathbf{z}_2^T \\ \vdots \\ \mathbf{z}_Q^T \end{bmatrix}, \mathbf{e} = \begin{bmatrix} e_1 \\ e_2 \\ \vdots \\ e_Q \end{bmatrix}. \quad (16.21)$$

The error can now be written

$$\mathbf{e} = \mathbf{t} - \mathbf{U}\mathbf{x}, \quad (16.22)$$

and the performance index become

$$F(\mathbf{x}) = (\mathbf{t} - \mathbf{U}\mathbf{x})^T (\mathbf{t} - \mathbf{U}\mathbf{x}). \quad (16.23)$$

If we use regularization, as we discussed in Chapter 13, to help in preventing overfitting, we obtain the following form for the performance index:

$$F(\mathbf{x}) = (\mathbf{t}-\mathbf{U}\mathbf{x})^T(\mathbf{t}-\mathbf{U}\mathbf{x}) + \rho\sum_{i=1}^{n} x_i^2 = (\mathbf{t}-\mathbf{U}\mathbf{x})^T(\mathbf{t}-\mathbf{U}\mathbf{x}) + \rho\mathbf{x}^T\mathbf{x}, \quad (16.24)$$

where $\rho = \alpha/\beta$ from Eq. (13.4). Let's expand this expression to obtain

$$\begin{aligned} F(\mathbf{x}) &= (\mathbf{t}-\mathbf{U}\mathbf{x})^T(\mathbf{t}-\mathbf{U}\mathbf{x}) + \rho\mathbf{x}^T\mathbf{x} = \mathbf{t}^T\mathbf{t} - 2\mathbf{t}^T\mathbf{U}\mathbf{x} + \mathbf{x}^T\mathbf{U}^T\mathbf{U}\mathbf{x} + \rho\mathbf{x}^T\mathbf{x} \\ &= \mathbf{t}^T\mathbf{t} - 2\mathbf{t}^T\mathbf{U}\mathbf{x} + \mathbf{x}^T[\mathbf{U}^T\mathbf{U} + \rho\mathbf{I}]\mathbf{x} \end{aligned} \quad (16.25)$$

Take a close look at Eq. (16.25), and compare it with the general form of the quadratic function, given in Eq. (8.35) and repeated here:

$$F(\mathbf{x}) = c + \mathbf{d}^T\mathbf{x} + \frac{1}{2}\mathbf{x}^T\mathbf{A}\mathbf{x}. \quad (16.26)$$

Our performance function is a quadratic function, where

$$c = \mathbf{t}^T\mathbf{t},\ \mathbf{d} = -2\mathbf{U}^T\mathbf{t} \text{ and } \mathbf{A} = 2[\mathbf{U}^T\mathbf{U} + \rho\mathbf{I}]. \quad (16.27)$$

From Chapter 8 we know that the characteristics of the quadratic function depend primarily on the Hessian matrix $\mathbf{A}$. For example, if the eigenvalues of the Hessian are all positive, then the function will have one unique global minimum.

In this case the Hessian matrix is $2[\mathbf{U}^T\mathbf{U} + \rho\mathbf{I}]$, and it can be shown that this matrix is either positive definite or positive semidefinite (see Exercise E16.4), which means that it can never have negative eigenvalues. We are left with two possibilities. If the Hessian matrix has only positive eigenvalues, the performance index will have one unique global minimum (see Figure 8.7). If the Hessian matrix has some zero eigenvalues, the performance index will either have a weak minimum (see Figure 8.9) or no minimum (see Problem P8.7), depending on the vector $\mathbf{d}$. In this case, it must have a minimum, since $F(\mathbf{x})$ is a sum square function, which cannot be negative.

Now let's locate the stationary point of the performance index. From our previous discussion of quadratic functions in Chapter 8, we know that the gradient is

$$\nabla F(\mathbf{x}) = \nabla\left(c + \mathbf{d}^T\mathbf{x} + \frac{1}{2}\mathbf{x}^T\mathbf{A}\mathbf{x}\right) = \mathbf{d} + \mathbf{A}\mathbf{x} = -2\mathbf{U}^T\mathbf{t} + 2[\mathbf{U}^T\mathbf{U} + \rho\mathbf{I}]\mathbf{x}. \quad (16.28)$$

The stationary point of $F(\mathbf{x})$ can be found by setting the gradient equal to zero:

$$-2\mathbf{Z}^T\mathbf{t} + 2[\mathbf{U}^T\mathbf{U} + \rho\mathbf{I}]\mathbf{x} = 0 \quad \Rightarrow \quad [\mathbf{U}^T\mathbf{U} + \rho\mathbf{I}]\mathbf{x} = \mathbf{U}^T\mathbf{t}. \quad (16.29)$$

Therefore, the optimum weights $\mathbf{x}^*$ can be computed from

$$[\mathbf{U}^T\mathbf{U} + \rho\mathbf{I}]\mathbf{x}^* = \mathbf{U}^T\mathbf{t}. \tag{16.30}$$

If the Hessian matrix is positive definite, there will be a unique stationary point, which will be a strong minimum:

$$\mathbf{x}^* = [\mathbf{U}^T\mathbf{U} + \rho\mathbf{I}]^{-1}\mathbf{U}^T\mathbf{t} \tag{16.31}$$

Let's demonstrate this procedure with a simple problem.

Example

To illustrate the least squares algorithm, let's choose a network and apply it to a particular problem. We will use an RBF network with three neurons in the first layer to approximate the following function

$$g(p) = 1 + \sin\left(\frac{\pi}{4}p\right) \text{ for } -2 \le p \le 2. \tag{16.32}$$

To obtain our training set we will evaluate this function at six values of p:

$$p = \{-2, -1.2, -0.4, 0.4, 1.2, 2\}. \tag{16.33}$$

This produces the targets

$$t = \{0, 0.19, 0.69, 1.3, 1.8, 2\}. \tag{16.34}$$

We will choose the basis function centers to be spaced equally throughout the input range: -2, 0 and 2. For simplicity, we will choose the bias to be the reciprocal of the spacing between points. This produces the following first layer weight and bias.

$$\mathbf{W}^1 = \begin{bmatrix} -2 \\ 0 \\ 2 \end{bmatrix}, \mathbf{b}^1 = \begin{bmatrix} 0.5 \\ 0.5 \\ 0.5 \end{bmatrix}. \tag{16.35}$$

The next step is to compute the output of the first layer, using the following equations.

$$n^1_{i,q} = \left\| \mathbf{p}_q - {}_i\mathbf{w}^1 \right\| b^1_i, \tag{16.36}$$

$$\mathbf{a}^1_q = \mathbf{radbas}(\mathbf{n}^1_q). \tag{16.37}$$

This produces the following $\mathbf{a}^1$ vectors

$$\mathbf{a}^1 = \left\{ \begin{bmatrix} 1 \\ 0.368 \\ 0.018 \end{bmatrix}, \begin{bmatrix} 0.852 \\ 0.698 \\ 0.077 \end{bmatrix}, \begin{bmatrix} 0.527 \\ 0.961 \\ 0.237 \end{bmatrix}, \begin{bmatrix} 0.237 \\ 0.961 \\ 0.527 \end{bmatrix}, \begin{bmatrix} 0.077 \\ 0.698 \\ 0.852 \end{bmatrix}, \begin{bmatrix} 0.018 \\ 0.368 \\ 1 \end{bmatrix} \right\} \tag{16.38}$$

We can use Eq. (16.17) and Eq. (16.21) to create the **U** and **t** matrices

$$\mathbf{U}^T = \begin{bmatrix} 1 & 0.852 & 0.527 & 0.237 & 0.077 & 0.018 \\ 0.368 & 0.698 & 0.961 & 0.961 & 0.698 & 0.368 \\ 0.018 & 0.077 & 0.237 & 0.527 & 0.852 & 1 \\ 1 & 1 & 1 & 1 & 1 & 1 \end{bmatrix}, \tag{16.39}$$

$$\mathbf{t}^T = \begin{bmatrix} 0 & 0.19 & 0.69 & 1.3 & 1.8 & 2 \end{bmatrix}. \tag{16.40}$$

The next step is to solve for the weights and biases in the second layer using Eq. (16.30). We will begin with the regularization parameter set to zero.

$$\mathbf{x}^* = [\mathbf{U}^T\mathbf{U} + \rho\mathbf{I}]^{-1}\mathbf{U}^T\mathbf{t}$$
$$= \begin{bmatrix} 2.07 & 1.76 & 0.42 & 2.71 \\ 1.76 & 3.09 & 1.76 & 4.05 \\ 0.42 & 1.76 & 2.07 & 2.71 \\ 2.71 & 4.05 & 2.71 & 6 \end{bmatrix}^{-1} \begin{bmatrix} 1.01 \\ 4.05 \\ 4.41 \\ 6 \end{bmatrix} = \begin{bmatrix} -1.03 \\ 0 \\ 1.03 \\ 1 \end{bmatrix} \tag{16.41}$$

The second layer weight and bias are therefore

$$\mathbf{W}^2 = \begin{bmatrix} -1.03 & 0 & 1.03 \end{bmatrix}, \mathbf{b}^2 = \begin{bmatrix} 1 \end{bmatrix}. \tag{16.42}$$

Figure 16.7 illustrates the operation of this RBF network. The blue line represents the RBF approximation, and the circles represent the six data points. The dotted lines in the upper axis represent the individual basis functions scaled by the corresponding weights in the second layer (including the constant bias term). The sum of the dotted lines will produce the blue line. In the small axis at the bottom, you can see the unscaled basis functions, which are the outputs of the first layer.

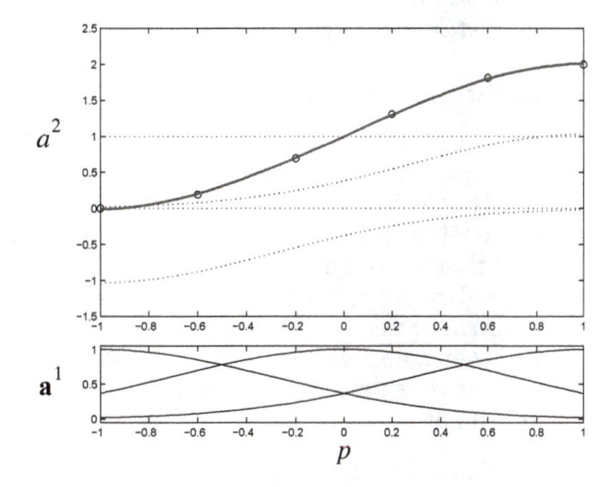

Figure 16.7 RBF Sine Approximation

The RBF network design process can be sensitive to the choice of the center locations and the bias. For example, if we select six basis functions and six data points, and if we choose the first layer biases to be 8, instead of 0.5, then the network response will be as shown in Figure 16.8. The spread of the basis function decreases as the inverse of the bias. When the bias is this large, there is not sufficient overlap in the basis functions to provide a smooth approximation. We match each data point exactly. However, because of the local nature of the basis function, the approximation to the true function is not accurate between the training data points.

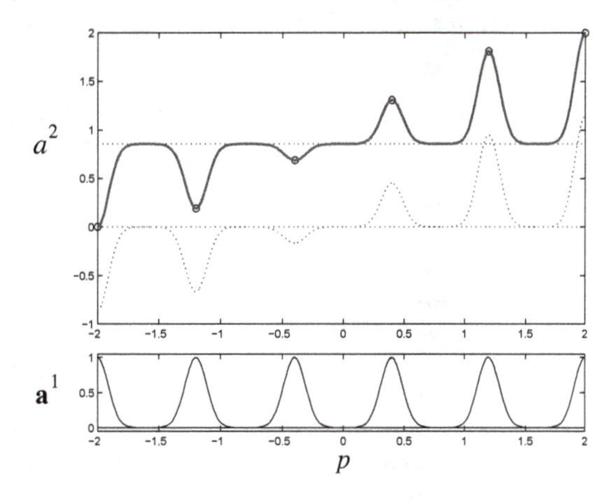

Figure 16.8 RBF Response with Bias Too Large

To experiment with the linear least squares fitting, use the MATLAB® Neural Network Design Demonstration RBF Linear Least Squares (`nnd17lls`).

Orthogonal Least Squares

In the previous section we assumed that the weights and biases in the first layer were fixed. (For example, the centers could be fixed on a grid, or randomly selected from the input vectors in the training set.) In this section we consider a different approach for selecting the centers. We will assume that there exists a number of potential centers. These centers could include the entire set of input vectors in the training set, vectors chosen in a grid pattern, or vectors chosen by any other procedure one might think of. We will then select vectors one at a time from this set of potential centers, until the network performance is satisfactory. We will build up the network one neuron at a time.

Subset Selection

The basic idea behind this method comes from statistics, and it is called *subset selection* [Mill90]. The general objective of subset selection is to choose an appropriate subset of independent variables to provide the most efficient prediction of a target dependent variable. For example, suppose that we have 10 independent variables, and we want to use them to predict our target dependent variable. We want to create the simplest predictor possible, so we want to use the minimum number of independent variables for the prediction. Which subset of the 10 independent variables should we use? The optimal approach, called an exhaustive search, tries all combinations of subsets and finds the smallest one that provides satisfactory performance. (We will define later what we mean by satisfactory performance.)

Unfortunately, this strategy is not practical. If we have Q variables in our original set, the following expression gives the number of distinct subsets:

$$\sum_{q=1}^{Q} \frac{Q!}{q!(Q-q)!} . \quad (16.43)$$

If $Q = 10$, this number is 1023. If $Q = 20$, the number is more than 1 million. We need to have a less expensive strategy than the exhaustive search. There are several suboptimal procedures. They are not guaranteed to find the optimal subset, but they require significantly less computation. One procedure is called *forward selection*. This method begins with an empty model and then adds variables one at a time. At each stage, we add the independent variable that provides the largest reduction in squared error. We stop adding variables when the performance is adequate. Another approach, called *backward elimination*, starts with all independent variables selected for the model. At each stage we eliminate the variable that would cause the least increase in the squared error. The process continues until the performance is inadequate. There are other approaches which combine

Forward Selection

Backward Elimination

forward selection and backward elimination, so that variables can be added and deleted at each iteration.

Any of the standard subset selection techniques can be used for selecting RBF centers. For purposes of illustration, we will consider one specific form of forward selection, called orthogonal least squares [ChCo91]. Its main feature is that it efficiently calculates the error reduction provided by the addition of each potential center to the RBF network.

To develop the orthogonal least squares algorithm, we begin with Eq. (16.22), repeated here in slightly different form:

$$\mathbf{t} = \mathbf{U}\mathbf{x} + \mathbf{e}. \tag{16.44}$$

We will use our standard notation for matrix rows and columns to individually identify both the rows and the columns of the matrix $\mathbf{U}$:

$$\mathbf{U} = \begin{bmatrix} {}_1\mathbf{u}^T \\ {}_2\mathbf{u}^T \\ \vdots \\ {}_Q\mathbf{u}^T \end{bmatrix} = \begin{bmatrix} \mathbf{z}_1^T \\ \mathbf{z}_2^T \\ \vdots \\ \mathbf{z}_Q^T \end{bmatrix} = \begin{bmatrix} \mathbf{u}_1 & \mathbf{u}_2 & \dots & \mathbf{u}_n \end{bmatrix} \tag{16.45}$$

Here each row of the matrix $\mathbf{U}$ represents the output of layer 1 of the RBF network for one input vector from the training set. There will be a column of the matrix $\mathbf{U}$ for each neuron (basis function) in layer 1 plus the bias term ($n = S^1 + 1$). Note that for the OLS algorithm, the potential centers for the basis functions are often chosen to be all of the input vectors in the training set. In this case, n will equal $Q + 1$, since the constant "1" for the bias term is included in $\mathbf{z}$, as shown in Eq. (16.17).

Regression Matrix

Eq. (16.44) is in the form of a standard linear regression model. The matrix $\mathbf{U}$ is called the *regression matrix*, and the columns of $\mathbf{U}$ are called the regressor vectors.

The objective of OLS is to determine how many columns of $\mathbf{U}$ (numbers of neurons or basis functions) should be used. The first step is to calculate how much each potential column would reduce the squared error. The problem is that the columns are generally correlated with each other, and so it is difficult to determine how much each individual column would reduce the error. For this reason, we need to first orthogonalize the columns. Orthogonalizing the columns means that we can decompose $\mathbf{U}$ as follows:

$$\mathbf{U} = \mathbf{M}\mathbf{R}, \tag{16.46}$$

where $\mathbf{R}$ is an upper triangular matrix, with ones on the diagonal:

$$\mathbf{R} = \begin{bmatrix} 1 & r_{1,2} & r_{1,3} & \dots & r_{1,n} \\ 0 & 1 & r_{2,3} & \dots & r_{2,n} \\ \vdots & \vdots & \vdots & \dots & \vdots \\ 0 & 0 & 0 & \dots & r_{n-1,n} \\ 0 & 0 & 0 & \dots & 1 \end{bmatrix}, \tag{16.47}$$

and $\mathbf{M}$ is a matrix with orthogonal columns $\mathbf{m}_i$. This means that $\mathbf{M}$ has the following properties:

$$\mathbf{M}^T\mathbf{M} = \mathbf{V} = \begin{bmatrix} v_{1,1} & 0 & \dots & 0 \\ 0 & v_{2,2} & \dots & 0 \\ \vdots & \vdots & & \vdots \\ 0 & 0 & \dots & v_{n,n} \end{bmatrix} = \begin{bmatrix} \mathbf{m}_1^T\mathbf{m}_1 & 0 & \dots & 0 \\ 0 & \mathbf{m}_2^T\mathbf{m}_2 & \dots & 0 \\ \vdots & \vdots & & \vdots \\ 0 & 0 & \dots & \mathbf{m}_n^T\mathbf{m}_n \end{bmatrix}. \tag{16.48}$$

Now Eq. (16.44) can be written

$$\mathbf{t} = \mathbf{MRx} + \mathbf{e} = \mathbf{Mh} + \mathbf{e}, \tag{16.49}$$

where

$$\mathbf{h} = \mathbf{Rx}. \tag{16.50}$$

The least squares solution for Eq. (16.49) is

$$\mathbf{h}^* = [\mathbf{M}^T\mathbf{M}]^{-1}\mathbf{M}^T\mathbf{t} = [\mathbf{V}]^{-1}\mathbf{M}^T\mathbf{t}, \tag{16.51}$$

and because $\mathbf{V}$ is diagonal, the elements of $\mathbf{h}^*$ can be computed

$$h_i^* = \frac{\mathbf{m}_i^T\mathbf{t}}{v_{i,i}} = \frac{\mathbf{m}_i^T\mathbf{t}}{\mathbf{m}_i^T\mathbf{m}_i}. \tag{16.52}$$

From $\mathbf{h}^*$ we can compute $\mathbf{x}^*$ using Eq. (16.50). Since $\mathbf{R}$ is upper-triangular, Eq. (16.50) can be solved by back-substitution and does not require a matrix inversion.

There are a number of ways to obtain the orthogonal vectors $\mathbf{m}_i$, but we will use the Gram-Schmidt orthogonalization process of Eq. (5.20), starting with the original columns of $\mathbf{U}$.

$$\mathbf{m}_1 = \mathbf{u}_1, \tag{16.53}$$

$$\mathbf{m}_k = \mathbf{u}_k - \sum_{i=1}^{k-1} r_{i,k}\mathbf{m}_i , \qquad (16.54)$$

where

$$r_{i,k} = \frac{\mathbf{m}_i^T \mathbf{u}_k}{\mathbf{m}_i^T \mathbf{m}_i} , \; i = 1, \ldots, k-1 . \qquad (16.55)$$

Now let's see how orthogonalizing the columns of $\mathbf{U}$ enables us to efficiently calculate the squared error contribution of each basis vector. Using Eq. (16.49), the total sum square value of the targets is given by

$$\mathbf{t}^T\mathbf{t} = [\mathbf{Mh} + \mathbf{e}]^T[\mathbf{Mh} + \mathbf{e}] = \mathbf{h}^T\mathbf{M}^T\mathbf{Mh} + \mathbf{e}^T\mathbf{Mh} + \mathbf{h}^T\mathbf{M}^T\mathbf{e} + \mathbf{e}^T\mathbf{e} . \qquad (16.56)$$

Consider the second term in the sum:

$$\mathbf{e}^T\mathbf{Mh} = [\mathbf{t} - \mathbf{Mh}]^T\mathbf{Mh} = \mathbf{t}^T\mathbf{Mh} - \mathbf{h}^T\mathbf{M}^T\mathbf{Mh} . \qquad (16.57)$$

If we use the optimal $\mathbf{h}^*$ from Eq. (16.51), we find

$$\mathbf{e}^T\mathbf{Mh}^* = \mathbf{t}^T\mathbf{Mh}^* - \mathbf{t}^T\mathbf{MV}^{-1}\mathbf{M}^T\mathbf{Mh}^* = \mathbf{t}^T\mathbf{Mh}^* - \mathbf{t}^T\mathbf{Mh}^* = 0 . \qquad (16.58)$$

Therefore the total sum square value from Eq. (16.56) becomes

$$\mathbf{t}^T\mathbf{t} = \mathbf{h}^T\mathbf{M}^T\mathbf{Mh} + \mathbf{e}^T\mathbf{e} = \mathbf{h}^T\mathbf{Vh} + \mathbf{e}^T\mathbf{e} = \sum_{i=1}^{n} h_i^2\mathbf{m}_i^T\mathbf{m}_i + \mathbf{e}^T\mathbf{e} . \qquad (16.59)$$

The first term on the right of Eq. (16.59) is the contribution to the sum squared value explained by the regressors, and the second term is the remaining sum squared value that is not explained by the regressors. Therefore, regressor (basis function) i contributes

$$h_i^2\mathbf{m}_i^T\mathbf{m}_i \qquad (16.60)$$

to the squared value. This also represents how much the squared error can be reduced by including the corresponding basis function in the network. We will use this number, after normalizing by the total squared value, to determine the next basis function to include at each iteration:

$$o_i = \frac{h_i^2\mathbf{m}_i^T\mathbf{m}_i}{\mathbf{t}^T\mathbf{t}} . \qquad (16.61)$$

This number always falls between zero and one.

Now let's put all these ideas together into an algorithm for selecting centers.

The OLS Algorithm

To begin the algorithm, we start with all potential basis functions included in the regression matrix $\mathbf{U}$. (As we explained below Eq. (16.45), if all input vectors in the training set are to be considered potential basis function centers, then the $\mathbf{U}$ matrix will be Q by $Q+1$.) This matrix represents only potential basis functions, since we start with no basis functions included in the network.

The first stage of the OLS algorithm consists of the following three steps, for $i = 1, \ldots, Q$:

$$\mathbf{m}_1^{(i)} = \mathbf{u}_i, \tag{16.62}$$

$$h_1^{(i)} = \frac{\mathbf{m}_1^{(i)T}\mathbf{t}}{\mathbf{m}_1^{(i)T}\mathbf{m}_1^{(i)}}, \tag{16.63}$$

$$o_1^{(i)} = \frac{(h_1^{(i)})^2\mathbf{m}_1^{(i)T}\mathbf{m}_1^{(i)}}{\mathbf{t}^T\mathbf{t}}. \tag{16.64}$$

We then select the basis function that creates the largest reduction in error:

$$o_1 = o_1^{(i_1)} = max\{o_1^{(i)}\}, \tag{16.65}$$

$$\mathbf{m}_1 = \mathbf{m}_1^{(i_1)} = \mathbf{u}_{i_1}. \tag{16.66}$$

The remaining iterations of the algorithm continue as follows (for iteration k):

For $i = 1, \ldots, Q$, $i \neq i_1$, ..., $i \neq i_{k-1}$

$$r_{j,k}^{(i)} = \frac{\mathbf{m}_j^T\mathbf{u}_i}{\mathbf{m}_j^T\mathbf{m}_j}, \; j = 1, \ldots, k-1, \tag{16.67}$$

$$\mathbf{m}_k^{(i)} = \mathbf{u}_i - \sum_{j=1}^{k-1} r_{j,k}^{(i)}\mathbf{m}_j, \tag{16.68}$$

$$h_k^{(i)} = \frac{\mathbf{m}_k^{(i)T}\mathbf{t}}{\mathbf{m}_k^{(i)T}\mathbf{m}_k^{(i)}}, \tag{16.69}$$

$$o_k^{(i)} = \frac{(h_k^{(i)})^2 \mathbf{m}_k^{(i)T}\mathbf{m}_k^{(i)}}{\mathbf{t}^T\mathbf{t}}, \tag{16.70}$$

$$o_k = o_k^{(i_k)} = max\{o_k^{(i)}\}, \tag{16.71}$$

$$r_{j,k} = r_{j,k}^{(i_k)}, j = 1, \ldots, k-1. \tag{16.72}$$

$$\mathbf{m}_k = \mathbf{m}_k^{(i_k)}. \tag{16.73}$$

The iterations continue until some stopping criterion is met. One choice of stopping criterion is

$$1 - \sum_{j=1}^{k} o_j < \delta, \tag{16.74}$$

where δ is some small number. However, if δ is chosen too small, we can have overfitting, since the network will become too complex. An alternative is to use a validation set, as we discussed in the chapter on generalization. We would stop when the error on the validation set increased.

After the algorithm has converged, the original weights $\mathbf{x}$ can be computed from the transformed weights $\mathbf{h}$ by using Eq. (16.50). This produces, by back substitution,

$$x_n = h_n, x_k = h_k - \sum_{j=k+1}^{n} r_{j,k}x_j, \tag{16.75}$$

where n is the final number of weights and biases in the second layer (adjustable parameters).

To experiment with orthogonal least squares learning, use the MATLAB® Neural Network Design Demonstration RBF Orthogonal Least Squares (`nnd17ols`).

Clustering

There is another approach [MoDa89] for selecting the weights and biases in the first layer of the RBF network. This method uses the competitive networks described in Chapter 15. Recall that the competitive layer of Kohonen (see Figure 15.2) and the Self Organizing Feature Map (see Figure

15.9) perform a clustering operation on the input vectors of the training set. After training, the rows of the competitive networks contain prototypes, or cluster centers. This provides an approach for locating centers and selecting biases for the first layer of the RBF network. If we take the input vectors from the training set and perform a clustering operation on them, the resulting prototypes (cluster centers) could be used as centers for the RBF network. In addition, we could compute the variance of each individual cluster and use that number to calculate an appropriate bias to be used for the corresponding neuron.

Consider again the following training set:

$$\{\mathbf{p}_1, \mathbf{t}_1\}, \{\mathbf{p}_2, \mathbf{t}_2\}, \ldots, \{\mathbf{p}_Q, \mathbf{t}_Q\} . \tag{16.76}$$

We want to perform a clustering of the input vectors from this training set:

$$\{\mathbf{p}_1, \mathbf{p}_2, \ldots, \mathbf{p}_Q\} . \tag{16.77}$$

We will train the first layer weights of the RBF network to perform a clustering of these vectors, using the Kohonen learning rule of Eq. (15.13), and repeated here:

$${}_{i*}\mathbf{w}^1(q) = {}_{i*}\mathbf{w}^1(q-1) + \alpha(\mathbf{p}(q) - {}_{i*}\mathbf{w}^1(q-1)) , \tag{16.78}$$

where $\mathbf{p}(q)$ is one of the input vectors in the training set, and ${}_{i*}\mathbf{w}^1(q-1)$ is the weight vector that was closest to $\mathbf{p}(q)$. (We could also use other clustering algorithms, such as the Self Organizing Feature Map, or the k-means clustering algorithm, which was suggested in [MoDa89].) As described in Chapter 15, Eq. (16.78) is repeated until the weights have converged. The resulting converged weights will represent cluster centers of the training set input vectors. This will insure that we will have basis functions located in areas where input vectors are most likely to occur.

In addition to selecting the first layer weights, the clustering process can provide us with a method for determining the first layer biases. For each neuron (basis function), locate the n_c input vectors from the training set that are closest to the corresponding weight vector (center). Then compute the average distance between the center and its neighbors.

$$dist_i = \frac{1}{n_c}\left(\sum_{j=1}^{n_c} \left\|\mathbf{p}_j^i - {}_i\mathbf{w}^1\right\|^2\right)^{\frac{1}{2}} \tag{16.79}$$

where $\mathbf{p}_1^i$ is the input vector that closest to ${}_i\mathbf{w}^1$, and is $\mathbf{p}_2^i$ the next closest input vector. From these distances, [MoDa89] recommends setting the first layer biases as follows:

$$b_i^1 = \frac{1}{\sqrt{2}dist_i}. \tag{16.80}$$

Therefore, when a cluster is wide, the corresponding basis function will be wide as well. Notice that in this case each bias in the first layer will be different. This should provide a network that is more efficient in its use of basis functions than a network with equal biases.

After the weights and biases of the first layer are determined, linear least squares is used to find the second layer weights and biases.

There is a potential drawback to the clustering method for designing the first layer of the RBF network. The method only takes into account the distribution of the input vectors; it does not consider the targets. It is possible that the function we are trying to approximate is more complex in regions for which there are fewer inputs. For this case, the clustering method will not distribute the centers appropriately. On the other hand, one would hope that the training data is located in regions where the network will be most used, and therefore the function approximation will be most accurate in those areas.

Nonlinear Optimization

It is also possible to train RBF networks in the same manner as MLP networks—using nonlinear optimization techniques, in which all weights and biases in the network are adjusted at the same time. These methods are not generally used for the full training of RBF networks, because these networks tend to have many more unsatisfactory local minima in their error surfaces. However, nonlinear optimization can be used for the fine-tuning of the network parameters, after initial training by one of the two-stage methods we presented in earlier sections.

We will not present the nonlinear optimization methods in their entirety here, since they were treated extensively in Chapter 11 and Chapter 12. Instead, we will simply indicate how the basic backpropagation algorithm for computing the gradient in MLP networks can be modified for RBF networks.

The derivation of the gradient for RBF networks follows the same pattern as the gradient development for MLP networks, starting with Eq. (11.9), which you may wish to review at this time. Here we will only discuss the one step where the two derivations differ. The difference occurs with Eq. (11.20). The net input for the second layer of the RBF network has the same form as its counterpart in the MLP network, but the first layer net input has a different form (as given in Eq. (16.1) and repeated here):

$$n_i^1 = \|\mathbf{p} - {}_i\mathbf{w}^1\| b_i^1 = b_i^1 \sqrt{\sum_{j=1}^{S^1} (p_j - w_{i,j}^1)^2} . \quad (16.81)$$

If we take the derivative of this function with respect to the weights and biases, we obtain

$$\frac{\partial n_i^1}{\partial w_{i,j}^1} = b_i^1 \frac{1/2}{\sqrt{\sum_{j=1}^{S^1} (p_j - w_{i,j}^1)^2}} 2(p_j - w_{i,j}^1)(-1) = \frac{b_i^1 (w_{i,j}^1 - p_j)}{\|\mathbf{p} - {}_i\mathbf{w}^1\|} , \quad (16.82)$$

$$\frac{\partial n_i^1}{\partial b_i^1} = \|\mathbf{p} - {}_i\mathbf{w}^1\| . \quad (16.83)$$

This produces the modified gradient equations (compare with Eq. (11.23) and Eq. (11.24)) for Layer 1 of the RBF network

$$\frac{\partial \hat{F}}{\partial w_{i,j}^1} = s_i^1 \frac{b_i^1 (w_{i,j}^1 - p_j)}{\|\mathbf{p} - {}_i\mathbf{w}^1\|} , \quad (16.84)$$

$$\frac{\partial \hat{F}}{\partial b_i^1} = s_i^1 \|\mathbf{p} - {}_i\mathbf{w}^1\| . \quad (16.85)$$

Therefore, if we look at the summary of the gradient descent backpropagation algorithm for MLP networks, from Eq. (11.44) to Eq. (11.47), we find that the only difference for RBF networks is that we substitute Eq. (16.84) and Eq. (16.85) for Eq. (11.46) and Eq. (11.47) when $m = 1$. When $m = 2$ the original equations remain the same.

To experiment with nonlinear optimization learning, use the MATLAB® Neural Network Design Demonstration RBF Nonlinear Optimization (nnd17no).

Other Training Techniques

In this chapter we have only touched the surface of the variety of training techniques that have been proposed for RBF networks. We have attempted to present the principal concepts, but there are many variations. For example, the OLS algorithm has been extended to handle multiple outputs [ChCo92] and regularized performance indices [ChCh96]. It has also been used in combination with a genetic algorithm [ChCo99], which was used to select the first layer biases and the regularization parameter. The expectation maximization algorithm has also been suggested by several authors for optimizing the center locations, starting with [Bish91]. [OrHa00] used a regression tree approach for center selection. There have also been many variations on the use of clustering and on the combination of clustering for initialization and nonlinear optimization for fine-tuning. The architecture of the RBF network lends itself to many training approaches.

Summary of Results

Radial Basis Network

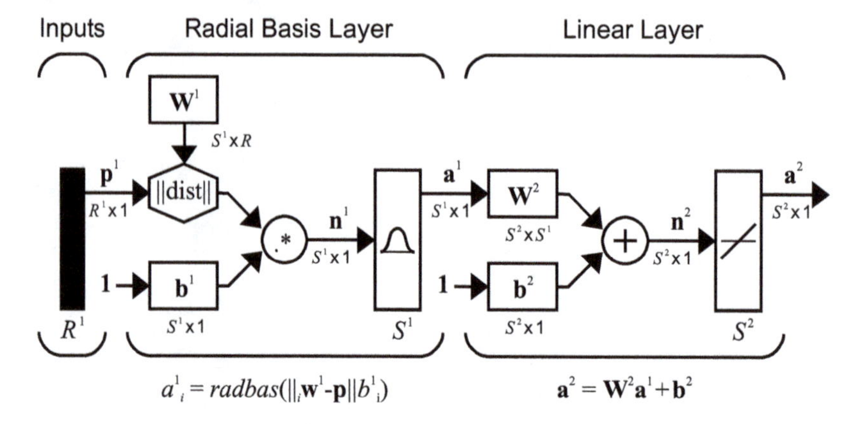

Training RBF Networks

Linear Least Squares

$$\mathbf{x} = \begin{bmatrix} {}_1\mathbf{w}^2 \\ b^2 \end{bmatrix}, \quad \mathbf{z}_q = \begin{bmatrix} \mathbf{a}_q^1 \\ 1 \end{bmatrix}$$

$$\mathbf{t} = \begin{bmatrix} t_1 \\ t_2 \\ \vdots \\ t_Q \end{bmatrix}, \quad \mathbf{U} = \begin{bmatrix} {}_1\mathbf{u}^T \\ {}_2\mathbf{u}^T \\ \vdots \\ {}_Q\mathbf{u}^T \end{bmatrix} = \begin{bmatrix} \mathbf{z}_1^T \\ \mathbf{z}_2^T \\ \vdots \\ \mathbf{z}_Q^T \end{bmatrix}, \quad \mathbf{e} = \begin{bmatrix} e_1 \\ e_2 \\ \vdots \\ e_Q \end{bmatrix}$$

$$F(\mathbf{x}) = (\mathbf{t} - \mathbf{U}\mathbf{x})^T(\mathbf{t} - \mathbf{U}\mathbf{x}) + \rho\mathbf{x}^T\mathbf{x}$$

$$[\mathbf{U}^T\mathbf{U} + \rho\mathbf{I}]\mathbf{x}^* = \mathbf{U}^T\mathbf{t}$$

Orthogonal Least Squares

Step 1

$$\mathbf{m}_1^{(i)} = \mathbf{u}_i ,$$

$$h_1^{(i)} = \frac{\mathbf{m}_1^{(i)T}\mathbf{t}}{\mathbf{m}_1^{(i)T}\mathbf{m}_1^{(i)}} ,$$

$$o_1^{(i)} = \frac{(h_1^{(i)})^2\mathbf{m}_1^{(i)T}\mathbf{m}_1^{(i)}}{\mathbf{t}^T\mathbf{t}} .$$

$$o_1 = o_1^{(i_1)} = max\{o_1^{(i)}\}$$

$$\mathbf{m}_1 = \mathbf{m}_1^{(i_1)} = \mathbf{u}_{i_1} .$$

Step *k*

For $i = 1, \ldots, Q,\ i \neq i_1,\ \ldots,\ i \neq i_{k-1}$

$$r_{j,k}^{(i)} = \frac{\mathbf{m}_j^T\mathbf{u}_k}{\mathbf{m}_j^T\mathbf{m}_j} ,\ j = 1, \ldots, k ,$$

$$\mathbf{m}_k^{(i)} = \mathbf{u}_i - \sum_{j=1}^{k-1} r_{j,k}^{(i)}\mathbf{m}_j ,$$

$$h_k^{(i)} = \frac{\mathbf{m}_k^{(i)T}\mathbf{t}}{\mathbf{m}_k^{(i)T}\mathbf{m}_k^{(i)}} ,$$

$$o_k^{(i)} = \frac{(h_k^{(i)})^2\mathbf{m}_k^{(i)T}\mathbf{m}_k^{(i)}}{\mathbf{t}^T\mathbf{t}} ,$$

$$o_k = o_k^{(i_k)} = max\{o_k^{(i)}\} ,$$

$$\mathbf{m}_k = \mathbf{m}_k^{(i_k)} .$$

Clustering

Training the weights

$$ {}_{i*}\mathbf{w}^1(q) = {}_{i*}\mathbf{w}^1(q-1) + \alpha(\mathbf{p}(q) - {}_{i*}\mathbf{w}^1(q-1)) $$

Selecting the bias

$$ dist_i = \frac{1}{n_c}\left(\sum_{j=1}^{n_c} \left\|\mathbf{p}_j^i - {}_i\mathbf{w}^1\right\|^2\right)^{\frac{1}{2}} $$

$$ b_i^1 = \frac{1}{\sqrt{2}dist_i} $$

Nonlinear Optimization

Replace Eq. (11.46) and Eq. (11.47) in standard backpropagation with

$$ \frac{\partial \hat{F}}{\partial w_{i,j}^1} = s_i^1 \frac{b_i^1(w_{i,j}^1 - p_j)}{\left\|\mathbf{p} - {}_i\mathbf{w}^1\right\|}, $$

$$ \frac{\partial \hat{F}}{\partial b_i^1} = s_i^1 \left\|\mathbf{p} - {}_i\mathbf{w}^1\right\|. $$

Solved Problems

P16.1 Use the OLS algorithm, to approximate the following function:

$$g(p) = \cos(\pi p) \text{ for } -1 \le p \le 1 .$$

To obtain our training set we will evaluate this function at five values of p:

$$p = \{-1, -0.5, 0, 0.5, 1\} .$$

This produces the targets

$$t = \{-1, 0, 1, 0, -1\} .$$

Perform one iteration of the OLS algorithm. Assume that the inputs in the training set are the potential centers and that the biases are all equal to 1.

First, we compute the outputs of the first layer:

$$n^1_{i,q} = \| \mathbf{p}_q - {}_i\mathbf{w}^1 \| b^1_i ,$$

$$\mathbf{a}^1_q = \mathbf{radbas}(\mathbf{n}^1_q) ,$$

$$\mathbf{a}^1 = \left\{ \begin{bmatrix} 1.000 \\ 0.779 \\ 0.368 \\ 0.105 \\ 0.018 \end{bmatrix}, \begin{bmatrix} 0.779 \\ 1.000 \\ 0.779 \\ 0.368 \\ 0.105 \end{bmatrix}, \begin{bmatrix} 0.368 \\ 0.779 \\ 1.000 \\ 0.779 \\ 0.368 \end{bmatrix}, \begin{bmatrix} 0.105 \\ 0.368 \\ 0.779 \\ 1.000 \\ 0.779 \end{bmatrix}, \begin{bmatrix} 0.018 \\ 0.105 \\ 0.368 \\ 0.779 \\ 1.000 \end{bmatrix} \right\} .$$

We can use Eq. (16.17) and Eq. (16.21) to create the $\mathbf{U}$ and $\mathbf{t}$ matrices:

$$\mathbf{U}^T = \begin{bmatrix} 1.000 & 0.779 & 0.368 & 0.105 & 0.018 \\ 0.779 & 1.000 & 0.779 & 0.368 & 0.105 \\ 0.368 & 0.779 & 1.000 & 0.779 & 0.368 \\ 0.105 & 0.368 & 0.779 & 1.000 & 0.779 \\ 0.018 & 0.105 & 0.368 & 0.779 & 1.000 \\ 1.000 & 1.000 & 1.000 & 1.000 & 1.000 \end{bmatrix} ,$$

$$\mathbf{t}^T = \begin{bmatrix} -1 & 0 & 1 & 0 & -1 \end{bmatrix} .$$

Now we perform step one of the algorithm:

$$\mathbf{m}_1^{(i)} = \mathbf{u}_i ,$$

$$\mathbf{m}_1^{(1)} = \begin{bmatrix} 1.000 \\ 0.779 \\ 0.368 \\ 0.105 \\ 0.018 \end{bmatrix}, \mathbf{m}_1^{(2)} = \begin{bmatrix} 0.779 \\ 1.000 \\ 0.779 \\ 0.368 \\ 0.105 \end{bmatrix}, \mathbf{m}_1^{(3)} = \begin{bmatrix} 0.368 \\ 0.779 \\ 1.000 \\ 0.779 \\ 0.368 \end{bmatrix}, \mathbf{m}_1^{(4)} = \begin{bmatrix} 0.105 \\ 0.368 \\ 0.779 \\ 1.000 \\ 0.779 \end{bmatrix},$$

$$\mathbf{m}_1^{(5)} = \begin{bmatrix} 0.018 \\ 0.105 \\ 0.368 \\ 0.779 \\ 1.000 \end{bmatrix}, \mathbf{m}_1^{(6)} = \begin{bmatrix} 1.000 \\ 1.000 \\ 1.000 \\ 1.000 \\ 1.000 \end{bmatrix},$$

$$h_1^{(i)} = \frac{\mathbf{m}_1^{(i)T} \mathbf{t}}{\mathbf{m}_1^{(i)T} \mathbf{m}_1^{(i)}},$$

$$h_1^{(1)} = -0.371 , h_1^{(2)} = -0.045 , h_1^{(3)} = 0.106 , h_1^{(4)} = -0.045 , h_1^{(5)} = -0.371 ,$$
$$h_1^{(6)} = -0.200 ,$$

$$o_1^{(i)} = \frac{(h_1^{(i)})^2 \mathbf{m}_1^{(i)T} \mathbf{m}_1^{(i)}}{\mathbf{t}^T \mathbf{t}},$$

$$o_1^{(1)} = 0.0804 , o_1^{(2)} = 0.0016 , o_1^{(3)} = 0.0094 , o_1^{(4)} = 0.0016 , o_1^{(5)} = 0.0804 ,$$
$$o_1^{(6)} = 0.0667 .$$

We see that the first and fifth centers would produce a 0.0804 reduction in the error. This means that the error would be reduced by 8.04%, if the first or fifth center were used in a single-neuron first layer. We would typically select the first center, since it has the smallest index.

If we stop at this point, we would add the first center to the hidden layer. Using Eq. (16.75), we would find that $w_{1,1}^2 = x_1 = h_1 = h_1^{(1)} = -0.371$. Also, $b_2 = 0$, since the bias center, $\mathbf{m}_1^{(6)}$, was not selected on the first iteration. Note that if we continue to add neurons in the hidden layer, the first weight will change. This can be seen from Eq. (16.75). This equation to find x_k is only used after all of the h_k are found. Only x_n will exactly equal h_n.

If we continue the algorithm, the first column would be removed from $\mathbf{U}$. We would then orthogonalize all remaining columns of $\mathbf{U}$ with respect to $\mathbf{m}_1$, which was chosen on the first iteration, using Eq. (16.54). It is interesting to note that the error reduction on the second iteration would be much higher than the reduction on the first iteration. The sequence of reductions would be 0.0804, 0.3526, 0.5074, 0.0448, 0.0147, 0, and the centers would be chosen in the following order: 1, 2, 5, 3, 4, 6. The reason that reductions in later iterations are higher is that it takes a combination of basis functions to produce the best approximation. This is why forward selection is not guaranteed to produce the optimal combination, which can be found with an exhaustive search. Also, notice that the bias is selected last, and it produces no reduction in the error.

P16.2 Figure P16.1 illustrates a classification problem, where Class I vectors are represented by dark circles, and Class II vectors are represented by light circles. These categories are not linearly separable. Design a radial basis function network to correctly classify these categories.

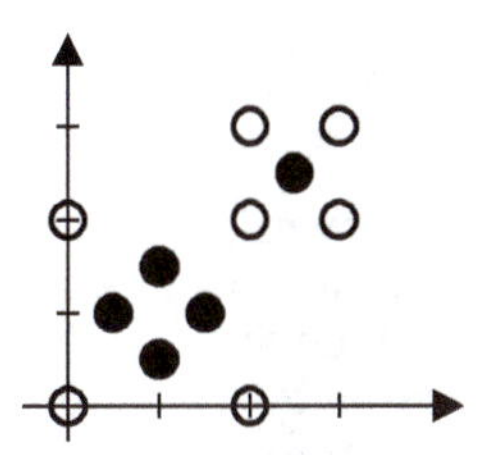

Figure P16.1 Classification Problem for Problem P16.2

From the problem statement, we know that the network will need to have two inputs, and we can use one output to distinguish the two classes. We will choose a positive output for Class I vectors, and a negative output for Class II vectors. The Class I region is made up of two simple subregions, and it appears that two neurons should be sufficient to perform the classification. The rows of the first-layer weight matrix will create centers for the two basis functions, and we will choose each center to be located in the middle of one subregion. By centering a basis function in each subregion, we can produce maximum network outputs there. The first layer weight matrix is then

$$\mathbf{W}^1 = \begin{bmatrix} 1 & 1 \\ 2.5 & 2.5 \end{bmatrix}.$$

The choice of the biases in the first layer depends on the width that we want for each basis function. For this problem, the first basis function should be wider than the second. Therefore, the first bias will be smaller than the second bias. The boundary formed by the first basis function should have a radius of approximately 1, while the second basis function boundary should have a radius of approximately $1/2$. We want the basis functions to drop significantly from their peaks in these distances. If we use a bias of 1 for the first neuron and a bias of 2 for the second neuron, we get the following reductions within one radius of the centers:

$$a = e^{-n^2} = e^{-(1 \cdot 1)^2} = e^{-1} = 0.3679, \; a = e^{-n^2} = e^{-(2 \cdot 0.5)^2} = e^{-1} = 0.3679$$

This will work for our problem, so we select the first layer bias vector to be

$$\mathbf{b}^1 = \begin{bmatrix} 1 \\ 2 \end{bmatrix}.$$

The original basis function response ranges from 0 to 1 (see Figure 16.1). We want the output to be negative for inputs outside the decision regions, so we will use a bias of -1 for the second layer, and we will use a value of 2 for the second layer weights, in order to bring the peaks back up to 1. The second layer weights and biases then become

$$\mathbf{W}^2 = \begin{bmatrix} 2 & 2 \end{bmatrix}, \; b^2 = \begin{bmatrix} -1 \end{bmatrix}.$$

For these network parameter values, the network response is shown on the right side of Figure P16.2. This figure also shows where the surface intersects the plane at $a^2 = 0$, which is where the decision takes place. This is also indicated by the contours shown underneath the surface. These are the function contours where $a^2 = 0$. These decision regions are shown more clearly on the left side of Figure P16.2.

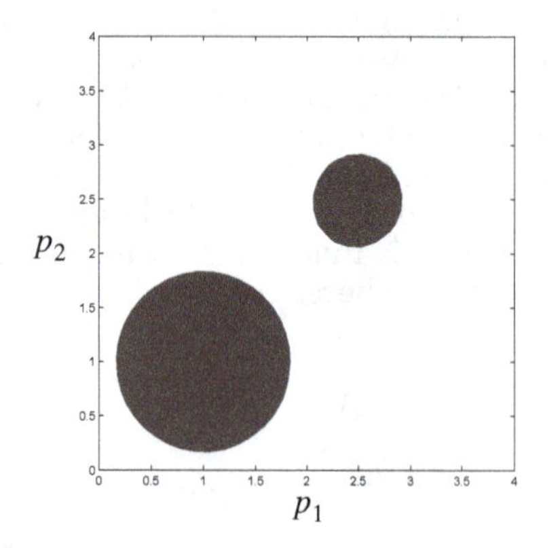

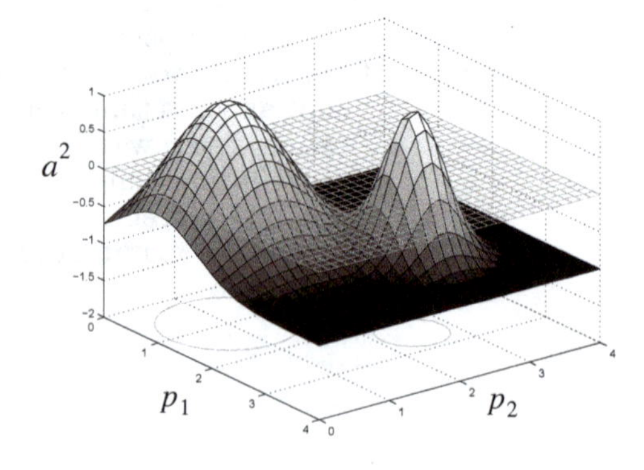

Figure P16.2 Decision Regions for Problem P16.2

P16.3 For an RBF network with one input and one neuron in the hidden layer, the initial weights and biases are chosen to be

$$w^1(0) = 0,\ b^1(0) = 1,\ w^2(0) = -2,\ b^2(0) = 1.$$

An input/target pair is given to be

$$\{p = -1, t = 1\}.$$

Perform one iteration of steepest descent backpropagation with $\alpha = 1$**.**

The first step is to propagate the input through the network.

$$n^1 = \|p - w^1\| b^1 = 1\sqrt{(-1-0)^2} = 1$$

$$a^1 = radbas(n^1) = e^{-(n^1)^2} = e^{-1} = 0.3679$$

$$n^2 = w^2 a^1 + b^2 = (-2)(0.3679) + 1 = 0.2642$$

$$a^2 = purelin(n^2) = n^2 = 0.2642$$

$$e = (t - a^2) = (1 - (0.2642)) = 0.7358$$

Now we backpropagate the sensitivities using Eq. (11.44) and Eq. (11.45).

$$\mathbf{s}^2 = -2\dot{\mathbf{F}}^2(\mathbf{n}^2)(\mathbf{t} - \mathbf{a}) = -2[1](e) = -2[1]0.7358 = -1.4716$$

$$\mathbf{s}^1 = \dot{\mathbf{F}}^1(\mathbf{n}^1)(\mathbf{W}^2)^T\mathbf{s}^2 = [-2n^1 e^{-(n^1)^2}]w^2\mathbf{s}^2 = [-2 \times 1 \times e^{-1}](-2)(-1.4716) = -2.1655$$

Finally, the weights and biases are updated using Eq. (11.46) and Eq. (11.47) for Layer 2, and Eq. (16.84) and Eq. (16.85) for Layer 1:

$$w^2(1) = w^2(0) - \alpha s^2 (a^1)^T = (-2) - 1(-1.4716)(0.3679) = -1.4586,$$

$$w^1(1) = w^1(0) - \alpha s^1 \left(\frac{b^1(w^1 - p)}{\|p - w^1\|}\right) = (0) - 1(-2.1655)\left(\frac{1(0-(-1))}{\|-1-0\|}\right) = 2.1655,$$

$$b^2(1) = b^2(0) - \alpha s^2 = 1 - 1(-1.4716) = 2.4716,$$

$$b^1(1) = b^1(0) - \alpha s^1 \|p - w^1\| = 1 - 1(-2.1655)\|-1-0\| = 3.1655.$$

Epilogue

The radial basis function network is an alternative to the multilayer perceptron network for problems of function approximation and pattern recognition. In this chapter we have demonstrated the operation of the RBF network, and we have described several techniques for training the network. Unlike MLP training, RBF training usually consists of two stages. In the first stage, the weights and biases in the first layer are found. In the second stage, which typically involves linear least squares, the second layer weights and biases are calculated.

Further Reading

[Bish91] C. M. Bishop, "Improving the generalization properties of radial basis function neural networks," *Neural Computation*, Vol. 3, No. 4, pp. 579-588, 1991.

First published use of the expectation-maximization algorithm for optimizing cluster centers for the radial basis network.

[BrLo88] D.S. Broomhead and D. Lowe, "Multivariable function interpolation and adaptive networks," *Complex Systems*, vol.2, pp. 321-355, 1988.

This seminal paper describes the first use of radial basis functions in the context of neural networks.

[ChCo91] S. Chen, C.F.N. Cowan, and P.M. Grant, "Orthogonal least squares learning algorithm for radial basis function networks," *IEEE Transactions on Neural Networks*, Vol.2, No.2, pp.302-309, 1991.

The first description of the use of the subset selection technique for selecting centers for radial basis function networks.

[ChCo92] S. Chen, P. M. Grant, and C. F. N. Cowan, "Orthogonal least squares algorithm for training multioutput radial basis function networks," *Proceedings of the Institute of Electrical Engineers*, Vol. 139, Pt. F, No. 6, pp. 378–384, 1992.

This paper extends the orthogonal least squares algorithm to the case of multiple outputs.

[ChCh96] S. Chen, E. S. Chng, and K. Alkadhimi, "Regularised orthogonal least squares algorithm for constructing radial basis function networks," *International Journal of Control*, Vol. 64, No. 5, pp. 829–837, 1996.

Modifies the orthogonal least squares algorithm to handle regularized performance indices.

[ChCo99] S. Chen, C.F.N. Cowan, and P.M. Grant, "Combined Genetic Algorithm Optimization and Regularized Orthogonal Least Squares Learning for Radial Basis Function Networks," *IEEE Transactions on Neural Networks*, Vol.10, No.5, pp.302-309, 1999.

Combines a genetic algorithm with orthogonal least squares to compute the regularization parameter and basis function spread, while also selecting centers and solving for the optimal second layer weights of radial basis function networks.

[Lowe89] D. Lowe, "Adaptive radial basis function nonlinearities, and the problem of generalization," *Proceedings of the First IEE International Conference on Artificial Neural Networks*, pp. 171 - 175, 1989.

This paper describes the use of gradient-based algorithms for training all of the parameters of an RBF network, including basis function centers and widths. It also provides a formula for setting the basis function widths, if the centers are randomly chosen from the training data set.

[Mill90] A.J. Miller, *Subset Selection in Regression*. Chapman and Hall, N.Y., 1990.

This book provides a very complete and clear discussion of the general problem of subset selection. This involves choosing an appropriate subset from a large set of independent input variables, in order to provide the most efficient prediction of some dependent variable.

[MoDa89] J. Moody and C.J. Darken, "Fast Learning in Networks of Locally-Tuned Processing Units," *Neural Computation*, Vol. 1, pp. 281–294, 1989.

The first published use of clustering methods to find radial basis function centers and variances.

[OrHa00] M. J. Orr, J. Hallam, A. Murray, and T. Leonard, "Assessing rbf networks using delve," IJNS, 2000.

This paper compares a variety of methods for training radial basis function networks. The methods include forward selection with regularization and also regression trees.

[PaSa93] J. Park and I.W. Sandberg, "Universal approximation using radial-basis-function networks," *Neural Computation*, vol. 5, pp. 305-316, 1993.

This paper proves the universal approximation capability of radial basis function networks.

[Powe87] M.J.D. Powell, "Radial basis functions for multivariable interpolation: a review," *Algorithms for Approximation*, pp. 143-167, Oxford, 1987.

This paper provides the definitive survey of the original work on radial basis functions. The original use of radial basis functions was in exact multivariable interpolation.

Exercises

E16.1 Design an RBF network to perform the classification illustrated in Figure E16.1. The network should produce a positive output whenever the input vector is in the shaded region and a negative output otherwise.

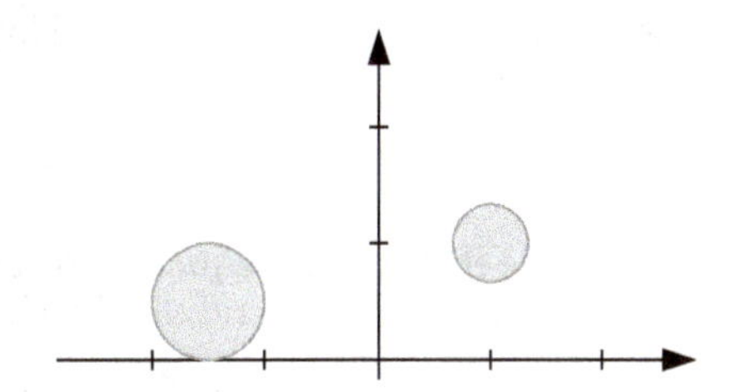

Figure E16.1 Pattern Classification Regions

E16.2 Choose the weights and biases for an RBF network with two neurons in the hidden layer and one output neuron, so that the network response passes through the points indicated by the blue circles in Figure E16.2.

Use the MATLAB® Neural Network Design Demonstration RBF Network Function (`nnd17nf`) *to check your result.*

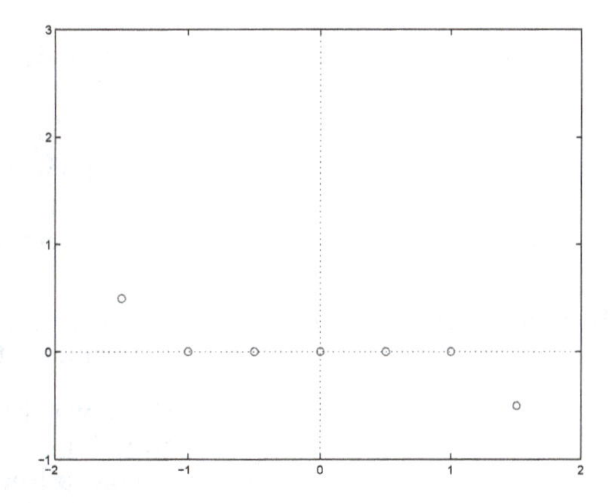

Figure E16.2 Function Approximation Exercise

E16.3 Consider a 1-2-1 RBF network (two neurons in the hidden layer and one output neuron). The first layer weights and biases are fixed as follows:

$$\mathbf{W}^1 = \begin{bmatrix} -1 \\ 1 \end{bmatrix}, \mathbf{b}^1 = \begin{bmatrix} 0.5 \\ 0.5 \end{bmatrix}.$$

Assume that the bias in the second layer is fixed at 0 ($b^2 = 0$). The training set has the following input/target pairs:

$$\{p_1 = 1, t_1 = -1\}, \{p_2 = 0, t_2 = 0\}, \{p_3 = -1, t_3 = 1\}.$$

i. Use linear least squares to solve for the second layer weights, assuming that the regularization parameter $\rho = 0$.

ii. Plot the contour plot for the sum squared error. Recall that it will be a quadratic function. (See Chapter 8.)

» 2 + 2
ans =
4

iii. Write a MATLAB® M-file to check your answers to parts i. and ii.

iv. Repeat parts i. to iii., with $\rho = 4$. Plot regularized squared error.

E16.4 The Hessian matrix for the performance index of the RBF network, given in Eq. (16.25), is

$$2[\mathbf{U}^T\mathbf{U} + \rho\mathbf{I}].$$

Show that this matrix is at least positive semidefinite for $\rho \geq 0$, and show that it is positive definite if $\rho > 0$.

E16.5 Consider an RBF network with the weights and biases in the first layer fixed. Show how the LMS algorithm of Chapter 10 could be modified for learning the second layer weights and biases.

E16.6 Suppose that a Gaussian transfer function in the first layer of the RBF network is replaced with a linear transfer function.

i. In Solved Problem P11.8, we showed that a multilayer perceptron with linear transfer functions in each layer is equivalent to a single-layer perceptron. If we use a linear transfer function in each layer of an RBF network, is that equivalent to a single-layer network? Explain.

» 2 + 2
ans =
4

ii. Work out an example, equivalent to Figure 16.4, to demonstrate the operation of the RBF network with linear transfer function in the first layer. Use MATLAB® to plot your figures. Do you think that the RBF network will be a universal approximator, if the first layer transfer function is linear? Explain your answer.

E16.7 Consider a Radial Basis Network, as in Figure 16.2, but assume that there is no bias in the second layer. There are two neurons in the first layer (two basis functions). The first layer weights (centers) and biases are fixed, and we have three inputs and targets. The outputs of the first layer and the targets are given by

$$\mathbf{a}^1 = \left\{ \begin{bmatrix} 2 \\ 1 \end{bmatrix}, \begin{bmatrix} 1 \\ 2 \end{bmatrix}, \begin{bmatrix} 0 \\ 1 \end{bmatrix} \right\}, t = \{0, 1, 2\}.$$

i. Use linear least squares to find the second layer weights of the network.

ii. Assume now that the basis function centers in the first layer are only potential centers. If orthogonal least squares is used to select potential centers, which center will be selected first, what will be its corresponding weight in the second layer, and how much will it reduce the squared error? Show all calculations clearly and in order.

iii. Is there a relationship between the two weights that you computed in part i. and the single weight that you computed in part ii? Explain.

E16.8 Repeat E16.7 for the following data:

i. $\mathbf{a}^1 = \left\{ \begin{bmatrix} 1 \\ 2 \end{bmatrix}, \begin{bmatrix} 2 \\ 1 \end{bmatrix}, \begin{bmatrix} -1 \\ 1 \end{bmatrix} \right\}, t = \{1, 2, -1\}.$

ii. $\mathbf{a}^1 = \left\{ \begin{bmatrix} 2 \\ 1 \end{bmatrix}, \begin{bmatrix} 1 \\ 1 \end{bmatrix}, \begin{bmatrix} 0 \\ 1 \end{bmatrix} \right\}, t = \{3, 1, 2\}.$

E16.9 Consider the variation of the radial basis network shown in Figure E16.3. The inputs and targets in the training set are $\{p_1 = -1, t_1 = -1\}$, $\{p_2 = 1, t_2 = 1\}$.

i. Find the linear least squares solution for the weight matrix $\mathbf{W}^2$.

ii. For the weight matrix $\mathbf{W}^2$ that you found in part i., sketch the network response as the input varies from -2 to 2.

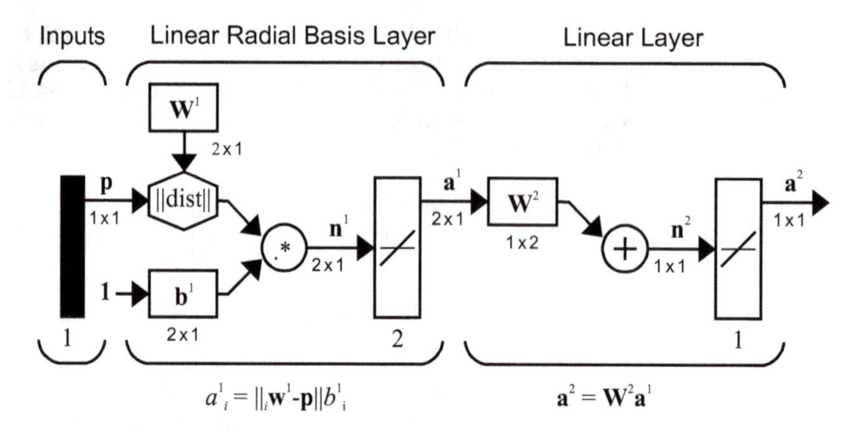

Figure E16.3 Radial Basis Network for Exercise E16.9

E16.10 Write a MATLAB® program to implement the linear least squares algorithm for the $1 - S^1 - 1$ RBF network with first layer weights and biases fixed. Train the network to approximate the function

$$g(p) = 1 + \sin\left(\frac{\pi}{8}p\right) \text{ for } -2 \le p \le 2 .$$

» 2 + 2
ans =
4

i. Select 10 training points at random from the interval $-2 \le p \le 2$.

ii. Select four basis function centers evenly spaced on the interval $-2 \le p \le 2$. Then, use Eq. (16.9) to set the bias. Finally, use linear least squares to find the second layer weights and biases, assuming that there is no regularization. Plot the network response for $-2 \le p \le 2$, and show the training points on the same plot. Compute the sum squared error over the training set.

iii. Double the bias from part ii and repeat.

iv. Decrease the bias by half from part ii, and repeat.

v. Compare the final sum squared errors for all cases and explain your results.

E16.11 Use the function described in Exercise E16.10, and use an RBF network with 10 neurons in the hidden layer.

i. Repeat Exercise E16.10 ii. with regularization parameter $\rho = 0.2$. Describe the changes in the RBF network response.

» 2 + 2
ans =
4

ii. Add uniform random noise in the range $[-0.1, 0.1]$ to the training targets. Repeat Exercise E16.10 ii. with no regularization and with regularization parameter $\rho = 0.2, 2, 20$. Which case produces the best results. Explain.

E16.12 Write a MATLAB® program to implement the orthogonal least squares algorithm. Repeat Exercise E16.10 using the orthogonal least squares algorithm. Use the 10 random training point inputs as the potential centers, and use Eq. (16.9) to set the bias. Use only the first four selected centers. Compare your final sum squared errors with the result from E16.10 part ii.

» 2 + 2
ans =
4

E16.13 Write a MATLAB® program to implement the steepest descent algorithm for the $1 - S^1 - 1$ RBF network. Train the network to approximate the function

$$g(p) = 1 + \sin\left(\frac{\pi}{8}p\right) \text{ for } -2 \le p \le 2 .$$

» 2 + 2
ans =
4

You should be able to use a slightly modified version of the program you wrote for Exercise E11.25.

i. Select 10 data points at random from the interval $-2 \le p \le 2$.

ii. Initialize all parameters (weights and biases in both layers) as small random numbers, and then train the network to convergence. (Experiment with the learning rate α, to determine a stable value.) Plot the network response for $-2 \le p \le 2$, and show the training points on the same plot. Compute the sum squared error over the training set. Use 2, 4 and 8 centers. Try different sets of initial weights.

iii. Repeat part ii., but use a different method for initializing the parameters. Start by setting the parameters as follows. First, select basis function centers evenly spaced on the interval $-2 \le p \le 2$. Then, use Eq. (16.9) to set the bias. Finally, use linear least squares to find the second layer weights and biases. Compute the squared error for these initial weights and biases. Starting from these initial conditions, train all parameters with steepest descent.

iv. Compare the final sum squared errors for all cases and explain your results.

E16.14 Suppose that a radial basis function layer (Layer 1 of the RBF network) were used in the second or third layer of a multilayer network. How could you modify the backpropagation equation, Eq. (11.35), to accommodate this change. (Recall that the weight update equations would be modified from Eq. (11.23) and Eq. (11.24) to Eq. (16.84) and Eq. (16.85).)

E16.15 Consider again Exercise E15.10, in which you trained a feature map to cluster the input space

$$0 \le p_1 \le 1 , \ 2 \le p_2 \le 3 .$$

Assume that over this input space, we wish to use an RBF network to approximate the following function:

$$t = \sin(2\pi p_1)\cos(2\pi p_2).$$

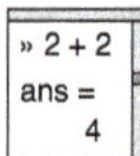

i. Use MATLAB to randomly generate 200 input vectors in the region shown above.

ii. Write a MATLAB M-file to implement a four-neuron by four-neuron (two-dimensional) feature map. Calculate the net input by finding the distance between the input and weight vectors directly, as is done by the LVQ network, so the vectors do not need to be normalized. Use the feature map to cluster the input vectors.

iii. Use the trained feature map weight matrix from part ii as the weight matrix of the first layer of an RBF network. Use Eq. (16.79) to determine the average distance between each cluster and its center, and then use Eq. (16.80) to set the bias for each neuron in the first layer of the RBF network.

iv. For each of the 200 input vectors in part i, compute the target response for the function above. Then use the resulting input/target pairs to determine the second-layer weights and bias for the RBF network.

v. Repeat parts ii to iv, using a two by two feature map. Compare your results.

17 Practical Training Issues

Objectives

Previous chapters have focused on particular neural network architectures and training rules, with an emphasis on fundamental understanding. In this chapter, we will discuss some practical training tips that apply to a variety of networks. No derivations are provided for the techniques that are presented here, but we have found these methods to be useful in practice.

There will be three basic sections in this chapter. The first section describes things that need to be done prior to training a network, such as collecting and preprocessing data and selecting the network architecture. The second section addresses network training itself. The final section considers post-training analysis.

Theory and Examples

In previous chapters, we have discussed a variety of neural network architectures and learning rules. Those chapters have placed special emphasis on the fundamental concepts behind each network. In this chapter, we will concentrate on practical aspects of training neural networks. Theoretical aspects and practical aspects are not mutually exclusive. It is only by combining a deep knowledge of network fundamentals with practical experience in using neural networks that we can get the most out of this technology.

Figure 17.1 illustrates the neural network training process. It is an iterative procedure that begins by collecting data and preprocessing it to make training more efficient. At this stage, the data also needs to be divided into training/validation/testing sets (see Chapter 13). After the data is selected, we need to choose the appropriate network type (multilayer, competitive, dynamic, etc.) and architecture (e.g., number of layers, number of neurons). Then we select a training algorithm that is appropriate for the network and the problem we are trying to solve. After the network is trained, we want to analyze the performance of the network. This analysis may lead us to discover problems with the data, the network architecture, or the training algorithm. The entire process is then iterated until the network performance is satisfactory.

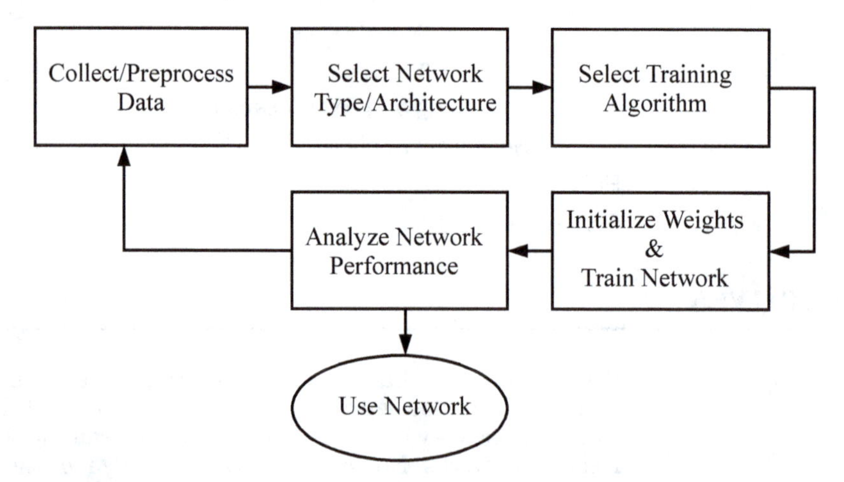

Figure 17.1 Flowchart of Neural Network Training Process

In the remainder of this chapter, we will discuss each part of the training process in some detail. We have divided this material into three major sections: Pre-Training Steps, Training the Network, and Post-Training Analysis.

Before we dig into the details of training, it is worth making a preliminary comment. Before beginning the neural network training process, you should first determine if a neural network is needed to solve your problem, or if some simpler linear technique might be adequate. For example, there is no need to use a neural network for a fitting problem, if standard linear regression will produce a satisfactory result. The neural network techniques provide additional power, but at the expense of more challenging training requirements. When linear methods will work, they are the first choice.

Pre-Training Steps

There are a number of steps that need to be performed before the network is trained. They can be grouped into three categories: Selection of Data, Data Preprocessing, and Choice of Network Type and Architecture.

Selection of Data

It is generally difficult to incorporate prior knowledge into a neural network, therefore the network will only be as good as the data that is used to train it. Neural networks represent a technology that is at the mercy of the data. The training data must span the full range of the input space for which the network will be used. As we discussed in Chapter 13, there are training methods we can use to insure that the network interpolates accurately throughout the range of the data provided (generalizes well). However, it is not possible to guarantee network performance when the inputs to the network are outside the range of the training set. Neural networks, like other nonlinear "black box" methods, do not extrapolate well.

It is not always easy to be sure that the input space is adequately sampled by the training data. For simple problems, in which the dimension of the input vector is small, and each element of the input vector can be chosen independently, we can sample the input space using a grid. However, these conditions are not often satisfied. For many problems, the dimension of the input space is large, which precludes the use of grid sampling. In addition, it is often the case that the input variables are dependent. For example, consider Figure 17.2. The shaded area represents the range over which the two inputs can vary. Even though each variable can range from -1 to 1, we would not need to create a grid in which both variables vary throughout their range, as shown by the dots in Figure 17.2. The network only needs to fit the function in the shaded area, since this is where the network will be used. It would be inefficient to fit the network outside the range of its use. This is especially true when the input dimension is large.

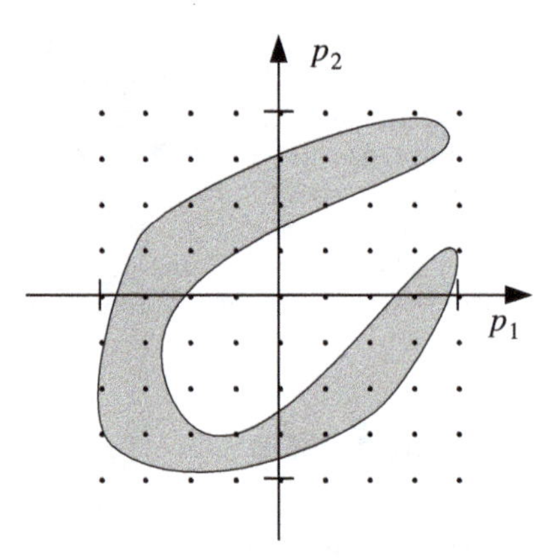

Figure 17.2 Input Range With Dependent Input Variables

It may not be possible to precisely define the active region of the input space. However, we can often collect data during standard operation of the system we are trying to model. In some cases, we have complete control over the design of the experiment during which data is collected. In these cases, we must be sure that the experimental setup drives the system through all conditions for which we plan to use the network.

How can we be sure that the input space has been adequately sampled by the training data? This is difficult to do prior to training, and there are many cases in which we have no control over the data collection process and must use whatever data is available. We will come back to this subject in the Post-Training Analysis section on page 22-18. By analyzing the trained network, we can often tell if the training data was sufficient. In addition, we can use techniques that indicate when a network is being used outside the range of the data with which it was trained. This will not improve the network performance, but it will prevent us from using a network in situations where it is not reliable.

After collecting the data, we generally divide it into three sets: training, validation and testing. As we discussed in Chapter 13, the training set will generally make up approximately 70% of the full data set, with validation and testing making up approximately 15% each. It is important that each of these sets be representative of the full data set — that the validation and test sets cover the same region of the input space as the training set. The simplest method for dividing the data is to select each set at random from the full data set. This usually produces a good result, but it is best to review the division to check for major differences between the sets. It is also possible in the post-training analysis to detect problems in the data division. We will have more to say about that later.

A final question we must ask about the selection of data is "Do we have enough data?" This is difficult to answer, especially before we train the network. The amount of data that is required depends on the complexity of the underlying function that we are trying to approximate (or the complexity of the decision boundaries that we are trying to implement). If the function to be approximated is very complex, with many inflection points, then this requires a large amount of data. If the function is very smooth, then the data requirements are significantly reduced (unless the data is very noisy). The choice of the data set size is closely related to the choice of the number of neurons in the neural network. This is discussed in the Choice of Network Architecture section on page 22-8. Of course, we generally don't know how complex the underlying function is before we begin training the network. For this reason, as we discussed earlier, the entire neural network training process is iterative. At the completion of training we will analyze the performance of the network. The results of that analysis can help us decide if we have enough data or not.

Data Preprocessing

The main purpose of the data preprocessing stage is to facilitate network training. Data preprocessing consists of such steps as normalization, nonlinear transformations, feature extraction, coding of discrete inputs/targets, handling of missing data, etc. The idea is to perform preliminary processing of the data to make it easier for the neural network training to extract the relevant information.

For example, in multilayer networks, sigmoid transfer functions are often used in the hidden layers. These functions become essentially saturated when the net input is greater than three ($\exp(-3) \cong 0.05$). We don't want this to happen at the beginning of the training process, because the gradient will then be very small. In the first layer, the net input is a product of the input times the weight plus the bias. If the input is very large, then the weight must be small in order to prevent the transfer function from becoming saturated. It is standard practice to normalize the inputs before applying them to the network. In this way, initializing the network weights to small random values guarantees that the weight-input product will be small. Also, when the input values are normalized, the magnitudes of the weights have a consistent meaning. This is especially important when using regularization, as described in Chapter 13. Regularization requires the weight values to be small. However, "small" is a relative term; if the input values are very small, we need large weights to produce a significant net input. Normalizing the inputs clarifies the meaning of "small" weights.

Normalization

There are two standard methods for *normalization*. The first method normalizes the data so that they fall into a standard range — typically -1 to 1. This can be done with

$$\mathbf{p}^n = 2(\mathbf{p} - \mathbf{p}^{min})./(\mathbf{p}^{max} - \mathbf{p}^{min}) - 1 , \tag{17.1}$$

where $\mathbf{p}^{min}$ is the vector containing the minimum values of each element of the input vectors in the data set, $\mathbf{p}^{max}$ contains the maximum values, ./

represents an element-by-element division of two vectors, and $\mathbf{p}^n$ is the resulting normalized input vector.

An alternative normalization procedure is to adjust the data so that they have a specified mean and variance — typically 0 and 1. This can be done with the transformation

$$\mathbf{p}^n = (\mathbf{p} - \mathbf{p}^{mean})./\mathbf{p}^{std}, \tag{17.2}$$

where $\mathbf{p}^{mean}$ is the average of the input vectors in the data set, and $\mathbf{p}^{std}$ is the vector containing the standard deviations of each element of the input vectors.

Generally, the normalization step is applied to both the input vectors and the target vectors in the data set.

Nonlinear Transformation

In addition to normalization, which involves a linear transformation, *nonlinear transformations* are sometimes also performed as part of the preprocessing stage. Unlike normalization, which is a standard process that can be applied to any data set, these nonlinear transformations are case-specific. For example, many economic variables show a logarithmic dependence [BoJe94]. In that case, it might be appropriate to take the logarithm of the input values before applying them to the neural network. Another example is molecular dynamics simulation [RaMa05], in which atomic forces are calculated as functions of distances between atoms. Since it is known that the forces are inversely related to the distances, we might perform the reciprocal transformation on the inputs, before applying them to the network. This represents one way of incorporating prior knowledge into neural network training. If the nonlinear transformation is cleverly chosen, it can make the network training more efficient. The preprocessing will off-load some of the work required of the neural network in finding the underlying transformation between inputs and targets.

Feature Extraction

Another data preprocessing step is called *feature extraction*. This generally applies to situations in which the dimension of the raw input vectors is very large and the components of the input vector are redundant. The idea of feature extraction is to reduce the dimension of the input space by calculating a small set of features from each input vector, and using the features as the input to the neural network. For example, neural networks can be used to analyze EKG (electrocardiogram) signals to identify heart problems [HeOh97]. The EKG might involve 12 or 15 signals (leads) measured over several minutes at a high sampling rate. This is too much data to apply directly to the neural network. Instead, we would extract certain features from the EKG signal, such as average time intervals between certain waveforms, average amplitudes of certain waves, etc. (See Chapter 20.)

Principal Components

There are also certain general-purpose feature extraction methods. One of these is the method of *principal component analysis* (PCA) [Joll02]. This method transforms the original input vectors so that the components of the transformed vectors are uncorrelated. In addition, the components of the transformed vector are ordered such that the first component has the greatest variance, the second component has the next greatest variance, etc. We generally keep only the first few components of the transformed vector, which account for most of the variance in the original vector. This results in a large reduction in the dimension of the input vector, if the original components are highly correlated. The drawback of using PCA is that it only considers linear relationships between the components of the input vector. When reducing the dimension using a linear transformation, we might lose some nonlinear information. Since the main purpose of using neural networks is to gain the power of their nonlinear mapping capabilities, we should be careful when using principal components to reduce the input dimension before applying the inputs to the neural network. There is a nonlinear version of PCA, called kernel PCA [ScSm99].

Coding the Targets

Another important preprocessing step is needed whenever the inputs or targets take on only discrete values. For example, in pattern recognition problems, each target will represent one of a finite number of classes. In these cases we need to have a procedure for *coding the inputs or targets*. If we have a pattern recognition problem in which there are four classes, there are at least three common ways in which we could code the targets. First, we can have scalar targets that take on four possible values (e.g., 1, 2, 3, 4). Second, we can have two-dimensional targets, which represent a binary code of the four classes (e.g., (0,0), (0,1), (1,0), (1,1)). Third, we can have four-dimensional targets, in which only one neuron at a time is active (e.g., (1,0,0,0), (0,1,0,0), (0,0,1,0), (0,0,0,1)). The third method tends to yield the best results in our experience. (Note that discrete inputs can be coded in the same ways as discrete targets.)

When coding the target values, we also need to consider the transfer function that is used in the output layer of the network. For pattern recognition problems, we would typically use sigmoid functions: log-sigmoid or tangent-sigmoid. If we use the tangent-sigmoid in the last layer, which is more common, then we might consider assigning target values to -1 or 1, which represent the asymptotes of the function. However, this tends to cause difficulties for the training algorithm, which tries to saturate the sigmoid function to meet the target value. It is better to assign target values at the point where the second derivative of the sigmoid function is maximum (see [LeCu98]). For the tangent-sigmoid function, this occurs when the net input is -1 and 1, which corresponds to output values of -0.76 and +0.76.

Softmax

Another transfer function that is used in the output layer of multilayer pattern recognition networks is the *softmax* function. This transfer function has the form

$$a_i = f(n_i) = \exp(n_i) \div \sum_{j=1}^{S} \exp(n_j) . \tag{17.3}$$

The outputs of the softmax transfer function can be interpreted as the probabilities associated with each class. Each output will fall between 0 and 1, and the sum of the outputs will equal 1. See Chapter 19 for an example application of the softmax transfer function.

Missing Data

Another practical issue to consider is *missing data*. It is often the case, especially when dealing with economic data, for example, that some of the data is missing. For instance, we might have an input vector containing 20 economic variables that are collected at monthly intervals. There may be some months in which one or two of the 20 variables were not collected properly. The simplest solution to this problem would be to throw out the data for any month in which any of the variables were missing. However, we might be very limited in the amount of data available, and it might be very expensive to collect additional data. In these cases, we would like to make full use of any data that we have, even if it is incomplete.

There are several strategies for dealing with missing data. If the missing data occurs in an input variable, one possibility is to replace the missing value with the average value for that particular input variable. At the same time, we could add an additional flag element to the input vector that would indicate that missing data for that input variable had been replaced with the average. This additional element of the input vector could be assigned the value 1 when the input variable was available, and 0 when the input variable was missing for that case. This would provide the neural network with information about which variables were missing. An additional flag element would be added to the input vector for every input variable that contained missing points.

If the missing data occurs in an element of the target, then the performance index can be modified so that errors associated with the missing target values are not included. All known target values will contribute to the performance index, but missing target values will not.

Choice of Network Architecture

The next step in the network training process is the choice of network architecture. The basic type of network architecture is determined by the type of problem we wish to solve. Once the basic architecture is chosen, we need to decide such specific details as how many neurons and layers we want to use, how many outputs the network should have, and what type of performance function we want to use for training.

Choice of Basic Architecture

The first step in choosing the architecture is to define the problem that we are trying to solve. For this chapter, we will limit our discussion to four types of problems: fitting, pattern recognition, clustering and prediction.

Fitting

Fitting is also referred to as function approximation or regression. In fitting problems, you want a neural network to map between a set of inputs and a corresponding set of targets. For example, a realtor might want to estimate home prices from such input variables as tax rate, pupil/teacher ratio in local schools and crime rate. An automotive engineer might want to estimate engine emission levels based on measurements of fuel consumption and speed. A physician might want to predict a patient's body fat level based on body measurements. For fitting problems, the target variable takes on continuous values. (For an example of training a neural network for a fitting problem, see Chapter 18.)

The standard neural network architecture for fitting problems is the multilayer perceptron, with tansig neurons in the hidden layers, and linear neurons in the output layer. The tansig transfer function is generally preferred over the logsig transfer function in the hidden layers for the same reason that inputs are normalized. It produces outputs (which are inputs to the next layer) that are centered near zero, whereas the logsig transfer function always produces positive outputs. For most fitting problems, a single hidden layer is sufficient. If the results with one hidden layer are not satisfactory, two layers are sometimes used. It would be rare in a standard fitting problem to use more than two hidden layers, although, for very difficult problems, deep networks, with many layers, have been used. Linear transfer functions are used in the output layer for fitting problems, because the target output is a continuous variable. As we saw in Chapter 11, a two layer network with sigmoid transfer functions in the hidden layer and linear transfer functions in the output layer is a universal approximator.

Radial basis networks can also be used for fitting problems. The Gaussian transfer function is most commonly used in the hidden layer for these networks, with linear transfer functions in the output layer.

Pattern Recognition

Pattern recognition is also referred to as pattern classification. In pattern recognition problems, you want a neural network to classify inputs into a set of target categories. For example, a wine dealer might want to recognize the vineyard that a particular bottle of wine came from, based on a chemical analysis of the wine. A physician might want to classify a tumor as benign or malignant, based on uniformity of cell size, clump thickness and mitosis.

In addition to fitting problems, multilayer perceptrons can be used for pattern recognition. The main difference between a fitting network and a pattern recognition network is the transfer function used in the output layer. For pattern recognition problems, we generally use a sigmoid function in the output layer. The radial basis function network can also be used for pattern recognition.

For an example of training a neural network for a pattern recognition problem, see Chapter 20.

Clustering

Clustering, or segmentation, is another use for neural networks. In clustering problems, you want a neural network to group data by similarity. For example, businesses may wish to perform market segmentation, which is done by grouping people according to their buying patterns. Computer scientists may want to perform data mining by partitioning data into related subsets. Biologists may wish to perform bioinformatic analyses, such as grouping genes with related expression patterns.

Any of the competitive networks described in Chapter 15 could be used for clustering. The self-organizing feature map (SOFM) is the most popular network for clustering. The main advantage of the SOFM is that it allows visualization of high-dimensional spaces.

For an example of training a neural network for a clustering problem, see Chapter 21.

Prediction

Prediction also falls under the categories of time series analysis, system identification, filtering or dynamic modeling. The idea is that we wish to predict the future value of some time series. An equities trader might want to predict the future value of some security. A control engineer might want to predict a future value of the concentration of some chemical, which is the output of a processing plant. A power systems engineer might want to predict outages on the electric grid.

Prediction requires the use of dynamic neural networks, as discussed in Chapter 14. The specific form of the network will depend on the particular application. The simplest network for nonlinear prediction is the focused time-delay neural network, which is shown in Figure 17.3. This is part of a general class of dynamic networks, called focused networks, in which the dynamics appear only at the input layer of a static multilayer feedforward network. This network has the advantage that it can be trained using static backpropagation algorithms, since the tapped-delay-line at the input of the network can be replaced with an extended vector of delayed values of the input.

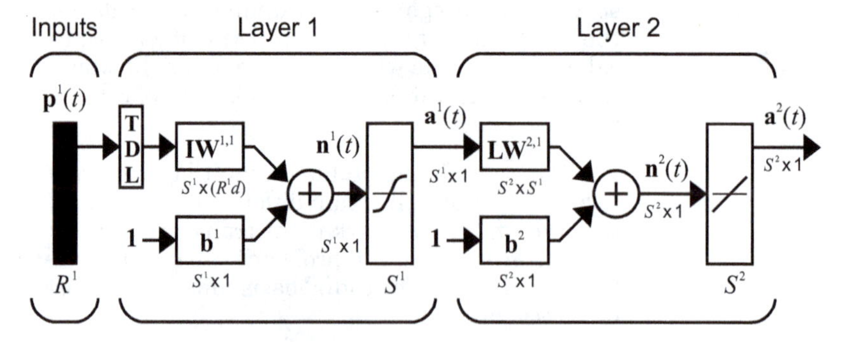

Figure 17.3 Focused Time-Delay Neural Network

For problems of dynamic modeling and control, the NARX network (Nonlinear AutoRegressive model with eXogenous input) is popular. This network is shown in Figure 17.4. The input signal could represent, for example, the voltage applied to a motor, and the output could represent the angular position of a robot arm. As with the focused time-delay neural network, the NARX network can be trained with static backpropagation. The two tapped-delay-lines can be replaced with extended vectors of delayed inputs and targets. We can use targets, instead of feeding back the network outputs (which would require dynamic backpropagation for training), because the output of the network should match the targets when training is complete.This is discussed in more detail in Chapter 22.

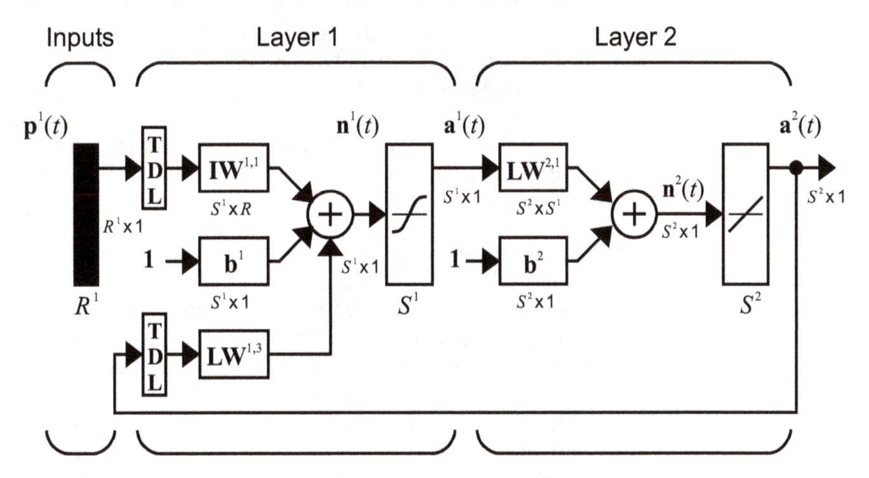

Figure 17.4 NARX Neural Network

There are many other types of dynamic networks that could be used for prediction, but the focused time delay network and the NARX network are the simplest of their type.

For an example of training a neural network for a prediction problem, see Chapter 22.

Selection of Architecture Specifics

After the basic network structure is chosen, we want to select the specifics of the architecture (e.g., the number of layers, the number of neurons, etc.). In some cases, the basic architecture choice will automatically determine the number of layers. For example, if the SOFM is used for clustering, then the network will have one layer. In the case of the multilayer network, which can be used for fitting or pattern recognition, the number of hidden layers is not determined by the problem, since any number of hidden layers is possible. The standard procedure is to begin with a network with one hidden layer. If the performance of the two-layer network is not satisfactory, then a three-layer network can be used. It would be unusual to use more than two hidden layers. The training becomes more difficult when multiple hidden layers are used. This is because each layer performs a squashing operation, as sigmoid functions are used in the hidden layers. This causes the derivatives of the performance function with respect to weights in the early layers to be quite small, which can cause slow convergence for steepest descent optimization. For very difficult problems, however, deep networks, with several hidden layers, can be used. Typically, parallel or GPU computing is required to train deep multilayer networks within a reasonable amount of time.

We also need to select the number of neurons in each layer. The number of neurons in the output layer is the same as the size of the target vector. The numbers of neurons in the hidden layers are determined by the complexities of the function that is being approximated or the decision boundaries that are being implemented. Unfortunately, we don't normally know how complex the problem is until we try to train the network. The standard procedure is to begin with more neurons than necessary and then to use early stopping or Bayesian regularization to prevent overfitting, as was described in Chapter 13.

The principal drawback to having too many neurons is that the network may overfit the data. If we use early stopping or Bayesian regularization, then we can prevent overfitting. However, there may be some situations in which we are concerned with the computation time or space required by the network (e.g., for real-time implementation on microcontrollers, VLSI or FPGAs). In these cases we want to find the simplest network that will fit the data. If you use Bayesian regularization, the effective number of parameters can be used to determine how many neurons to use. If, after training, the effective number of parameters is much less than the total number of parameters in the network, then the number of neurons can be reduced, and the network retrained. It is also possible to use "pruning" methods to eliminate neurons or weights in the network.

The number of neurons in the last layer is equal to the number of elements in the target vector. However, when there are multiple targets, we have a choice to make. We can have one network with multiple outputs, or we can have multiple networks, each with one output. For example, neural networks have been used to estimate LDL, VLDL and HDL cholesterol levels, based on a spectral analysis of the blood. It is possible to have one neural network with three neurons in the output layer to estimate all three cholesterol levels, or we could have three neural networks, with each one estimating only one of the three components. Theoretically, both methods should work, but in practice one method may work better than another. We generally start with one multi-output network, and then use multiple single-output networks if the original results are not satisfactory.

Another architectural choice is the size of the input vector. This is often a simple choice, which is determined by the training data. However, there are times when input vectors in the training data have redundant or irrelevant elements. When the dimension of the potential input vector is very large, it is sometimes advantageous to eliminate redundant or irrelevant elements. This can reduce the required computation and can assist in pre-

Input Selection

venting overfitting during training. The *input selection* process for nonlinear networks can be quite difficult, and there is no perfect solution. The Bayesian regularization method (Eq. (13.23)) can be modified to assist in input selection. It is possible to have different α parameters for different sets of weights. For example, we can let each column of the weight matrix in the first layer of a multilayer network have its own α parameter. If a given element of the input vector is irrelevant, then the corresponding α parameter would become large and force all elements of that column of the weight matrix to be small. That element could then be eliminated from the input vector.

Another technique that can assist in pruning the input vector is a sensitivity analysis of the trained network. In the Sensitivity Analysis section on page 22-28 we discuss this technique.

Training the Network

After the data has been prepared, and the network architecture has been selected, we are ready to train the network. In this section, we will discuss some of the decisions that need to be made as part of the training process. This includes the method for initializing the weights, the training algorithm, the performance index, and the criterion for stopping training.

Weight Initialization

Before training the network, we need to initialize the weights and biases. The method we use will depend on the type of network. For multilayer networks, the weights and biases are generally set to small random values (e.g., uniformly distributed between -0.5 and 0.5, if the inputs are normalized to fall between -1 and 1). As we discussed in Chapter 12, if we set the weights and biases to zero, the initial condition may fall on a saddle point of the performance surface. If we make the initial weights large, the initial condition can fall on a flat part of the performance surface, caused by saturation of the sigmoid transfer functions.

There is another approach to setting the initial weights and biases for a two-layer network. It was introduced by Widrow and Nguyen [WiNg90]. The idea is to set the magnitude of the weights in the first layer so that the linear region of each sigmoid function covers $1/S^1$ of the range of the input. The biases are then randomly set, so that the center of each sigmoid function falls randomly in the input space. The details of the method are as follows (assuming the inputs to the network have been normalized to values between -1 and 1).

Set row i of $\mathbf{W}^1$, ${}_i\mathbf{w}^1$, to have a random direction and a magnitude of

$$\|{}_i\mathbf{w}^1\| = 0.7(S^1)^{1/R}.$$

Set b_i to a uniform random value between $-\|{}_i\mathbf{w}^1\|$ and $\|{}_i\mathbf{w}^1\|$.

For competitive networks, the weights can also be set as small random numbers. Another possibility is to randomly select some of the input vectors in the training set to become initial rows of the weight matrix. In this way, we can be sure that the initial weights will fall within the range of the input vectors, so we will be less likely to have dead units, as described in Chapter 15. For the SOM, dead units are not a problem. The initial neighborhood size is set large enough so that all neurons will have the opportunity to learn during the initial stages of training. This will move all weight vectors into the appropriate region of the input space. Training can converge faster, however, if rows of the weight matrix are initially placed in the active input region.

Choice of Training Algorithm

For multilayer networks, we generally use gradient- or Jacobian-based algorithms, as described in Chapter 12. These algorithms can be implemented in either batch mode or sequential (also known as incremental, pattern or stochastic) mode. For example, in the sequential form of steepest descent (see Eq. (11.13)) we update the weights after each input is presented to the network. In batch mode (see page 12-7), all of the inputs are presented to the network, and the total gradient is computed by summing the gradients for each input, before the weights are updated. In some situations, the sequential form is preferred — for example, when on-line or adaptive operation is required. However, many of the more efficient optimization algorithms (e.g., conjugate gradient and Newton's methods) are inherently batch algorithms.

For multilayer networks with up to a few hundred weights and biases that are being used for function approximation, the Levenberg-Marquardt algorithm (see Eq. (12.31)) is usually the fastest training method. When the number of weights reaches a thousand or more, the Levenberg-Marquardt algorithm is not as efficient as some of the conjugate gradient algorithms. This is mainly because the matrix inverse calculation scales geometrically with the number of weights. For large networks, the Scaled Conjugate Gradient algorithm of [Moll93] is very efficient. This method is also attractive for pattern recognition problems. The Levenberg-Marquardt algorithm does not work as well for pattern recognition, in which the sigmoid transfer functions in the final layer are operating well outside the linear region.

Of the algorithms that can be implemented in sequential mode, the fastest are the extended Kalman filter algorithms. These algorithms are closely related to sequential implementations of the Gauss-Newton algorithm. Unlike the batch version of Gauss-Newton, they do not require an inversion of the approximate Hessian matrix. The decoupled extended Kalman filter implementation of [PuFe97] appears to be the most efficient of these types of algorithms.

Stopping Criteria

For most applications of neural networks, the training error never converges identically to zero. The error can reach zero for the perceptron network, assuming a linearly separable problem, as we showed in Chapter 4. However, it is unlikely to happen for multilayer networks. For this reason, we need to have other criteria for deciding when to stop the training.

We can stop the training when the error reaches some specified limit. However, it is usually difficult to know what an acceptable error level is. The simplest criterion is to stop the training after a fixed number of iterations. Because it is also difficult to know how many iterations will be required, the maximum iteration number is generally set reasonably high. If the weights have not converged after the maximum number of iterations has been reached, we can restart training, using the final weights from the first run as initial conditions for the restart. (We will talk more about how to tell if a network has converged in the Post-Training Analysis section on page 22-18.)

Another stopping criterion is the norm of the gradient of the performance index. If this norm reaches a sufficiently small threshold, then the training can be stopped. Since the gradient should be zero at a minimum of the performance index, this criterion will stop the algorithm when it gets close to the minimum. Unfortunately, as we have seen in Chapter 12, the performance surface for multilayer networks can have many flat regions, where the norm of the gradient will be small. For this reason, the threshold for the minimum norm should be set to a very small value (e.g., 10^{-6} for mean square error performance indices, with normalized targets), so that the training does not end prematurely.

We can also stop the training when the reduction in the performance index per iteration becomes small. As with the norm of the gradient, this criterion can stop the training too early. During the training of multilayer networks, the performance can remain almost constant for a number of iterations before dropping suddenly. When training is complete, it is useful to view the training performance curve on a log-log scale, as in Figure 17.5, to verify convergence.

If we are using early stopping, as discussed in Chapter 13, to prevent overfitting, then we will stop the training when the performance on the validation set increases for a set number of iterations. In addition to preventing overfitting, this stopping procedure also provides a significant reduction in computation; for most practical problems, the validation error will increase before any of the other stopping criteria are reached.

As shown in Figure 17.1, neural network training is an iterative process. Even after the training algorithm has converged, post-training analysis may suggest that the network be modified and retrained. In addition, several training runs should be made for each potential network to ensure that a global minimum has been reached.

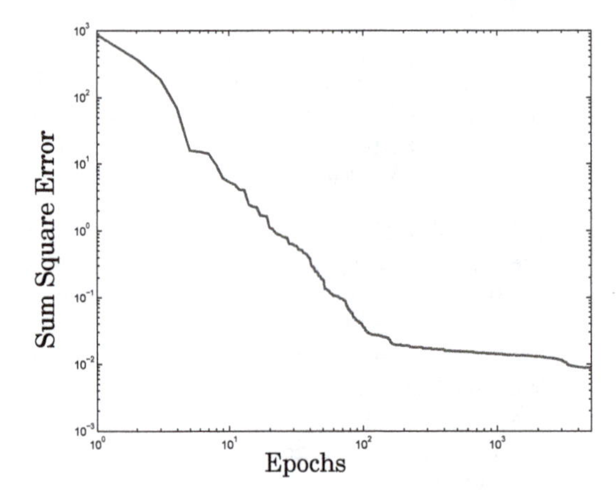

Figure 17.5 Typical Training Performance Curve

The previous stopping criteria apply mainly to gradient-based training. When training competitive networks, like the SOFM, there is no explicit performance index or gradient to monitor for convergence. The training stops only when the maximum number of iterations has been reached. For SOFMs, the learning rate and the neighborhood size are decreased over time. Typically, the neighborhood size is reduced to zero by the completion of training, so the maximum iteration number determines the end of training, as well as the rate of decrease in neighborhood size and learning rate. This is therefore a very important parameter. It is generally chosen to be more than ten times the number of neurons in the network. This is an approximate number, and the network needs to be analyzed at the completion of training to determine if the performance is satisfactory. (This will be discussed in the Post-Training Analysis section on page 22-18.) The network may need to be trained several times with different training lengths to achieve a satisfactory result.

Choice of Performance Function

For multilayer networks, the standard performance index is mean square error. When all inputs in the training set are equally likely to occur, then this can be written

$$F(\mathbf{x}) = \frac{1}{QS^M} \sum_{q=1}^{Q} (\mathbf{t}_q - \mathbf{a}_q)^T (\mathbf{t}_q - \mathbf{a}_q), \tag{17.4}$$

or

$$F(\mathbf{x}) = \frac{1}{QS^M}\sum_{q=1}^{Q}\sum_{i=1}^{S^M}(t_{i,q}-a_{i,q})^2 . \tag{17.5}$$

The scale factor that occurs outside the sum has no effect on the location of the optimum weights. Therefore, the sum square error performance index will produce the same weights as the mean square error performance index. However, the appropriate scaling can be useful when comparing errors on data sets of different size.

While mean square error is the most common performance index, there are others that have been used. For example, we could use mean absolute error. This would be similar to Eq. (17.5), except that the absolute value of the error would be used, instead of the square of the error. This performance index is generally less sensitive to one or two large errors in the data set, and is therefore somewhat more robust to outliers than is the mean square error algorithm. This concept can be extended to any power of the absolute error, as follows

$$F(\mathbf{x}) = \frac{1}{QS^M}\sum_{q=1}^{Q}\sum_{i=1}^{S^M}\left|t_{i,q}-a_{i,q}\right|^K , \tag{17.6}$$

where $K = 2$ corresponds to mean square error and $K = 1$ corresponds to mean absolute error. The general error given by Eq. (17.6) is referred to as the Minkowski error.

As we saw in Chapter 13, the mean square performance index can be augmented with the mean square weights, to produce a regularized performance index, which is used to prevent overfitting. The Bayesian regularization algorithm is an excellent training method for preventing overfitting. It uses a regularized performance index, and uses Bayesian methods to select the regularization parameter. See Chapter 13 for details.

Mean square error works well for function approximation problems, in which the target values are continuous. However, in pattern recognition problems, where the targets take on discrete values, other performance indices might be more appropriate. One performance index that has been proposed for classification problems is *cross-entropy* [Bish95]. Cross-entropy is defined as

Cross-Entropy

$$F(\mathbf{x}) = -\sum_{q=1}^{Q}\sum_{i=1}^{S^M} t_{i,q}\ln\frac{a_{i,q}}{t_{i,q}} . \tag{17.7}$$

Here we assume that the target values are 0 and 1, and they identify which of the two classes the input vector belongs to. The softmax transfer function is generally used in the last layer of the neural network, if the cross-entropy performance index is used.

As a closing note on the choice of performance index, recall from Chapter 11 that the backpropagation algorithm for computing training gradients will work for any differentiable performance index. If you change the performance index, you only need to change the initialization of the sensitivities in the last layer (see Eq. (11.37)).

Multiple Training Runs and Committees of Networks

A single training run may not produce optimal performance, because of the possibility of reaching a local minimum of the performance surface. It is best to restart the training at several different initial conditions and select the network that produces the best performance. Five to ten restarts will almost always produce a global optimum [HaBo07].

There is another way to perform multiple training runs and make use of all of the networks that have been trained. This is called the committee of networks [PeCo93]. For each training session, the validation set is randomly selected from the training data, and a random set of initial weights and biases is chosen. After N networks have been trained, all of the networks are used together to form a joint output. For function approximation networks, the joint output can be a simple average of the outputs of each network. For classification networks, the joint output can be the result of a vote, in which the class that is chosen by the majority of the networks is selected as the output of the committee. The performance of the committee will usually be better than even the best of the individual networks. In addition, the variation in the outputs of the individual networks can be used to provide error bars, or confidence levels, for the committee output.

Post-Training Analysis

Before using a trained neural network, we need to analyze it to determine if the training was successful. There are many techniques for post-training analysis. We will discuss some of the more common ones. Since these techniques vary, depending on the application, we will organize them according to these four application areas: fitting, pattern recognition, clustering and prediction.

Fitting

One useful tool for analyzing neural networks trained for fitting problems is a regression between the trained network outputs and the corresponding targets. We fit a linear function of the form

$$a_q = mt_q + c + \varepsilon_q , \tag{17.8}$$

where m and c are the slope and offset, respectively, of the linear function, t_q is a target value, a_q is a trained network output, and ε_q is the residual error of the regression.

The terms in the regression can be computed as follows:

$$\hat{m} = \frac{\sum_{q=1}^{Q} (t_q - \bar{t})(a_q - \bar{a})}{\sum_{q=1}^{Q} (t_q - \bar{t})^2}, \tag{17.9}$$

$$\hat{c} = \bar{a} - \hat{m}\bar{t}, \tag{17.10}$$

where

$$\bar{a} = \frac{1}{Q}\sum_{q=1}^{Q} a_q, \ \bar{t} = \frac{1}{Q}\sum_{q=1}^{Q} t_q. \tag{17.11}$$

Figure 17.6 shows an example regression analysis. The blue line represents the linear regression, the thin black line represents the perfect match $a_q = t_q$, and the circles represent the data points. In this example, we can see that the match is pretty good, although not perfect. The next step would be to investigate data points that fall far from the regression line. For example, there are two points around $t = 27$ and $a = 17$ that seem to be *outliers*. We would investigate these points to see if there was a problem with the data. This could be a bad data point, or it could be located far from other training points. In the latter case, we would need to collect more data in that region.

Outliers

In addition to computing the regression coefficients, we often also compute the correlation coefficient between the t_q and a_q, which is also known as the R value:

$$R = \frac{\sum_{q=1}^{Q} (t_q - \bar{t})(a_q - \bar{a})}{(Q-1)s_t s_a}, \tag{17.12}$$

where

$$s_t = \sqrt{\frac{1}{Q-1}\sum_{q=1}^{Q} (t_q - \bar{t})} \text{ and } s_a = \sqrt{\frac{1}{Q-1}\sum_{q=1}^{Q} (a_q - \bar{a})}. \tag{17.13}$$

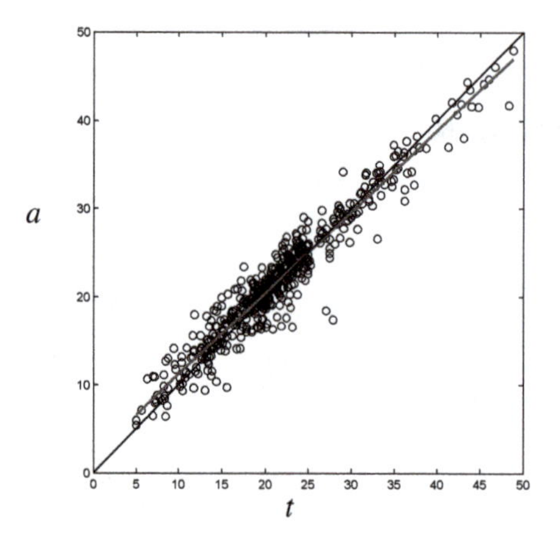

Figure 17.6 Regression Between Trained Network Outputs and Targets

The R value can generally range from -1 to 1, but we would expect it to be close to 1 for our neural network application. If $R = 1$, then all of the data points will fall exactly on the regression line. If $R = 0$, then the data will not be concentrated around the regression line, but will be randomly scattered. For the data of Figure 17.6, $R = 0.965$. We can see that the data does not fall exactly on the regression line, but the variation is relatively small.

The square of the correlation coefficient, R^2, is sometimes used instead of R. R^2 represents the proportion of the variability in a data set that is accounted for by the linear regression, and is also referred to as the coefficient of determination. For the data of Figure 17.6, $R^2 = 0.931$.

When the R or R^2 values are significantly less than 1, then the neural network has not done a good job of fitting the underlying function. A close analysis of the scatter plot may be helpful in determining problems in the fit. For example, we might find that when the targets are large there is more spread in the scatter plot. (This is not the case in Figure 17.6.) We might also notice that there are fewer data points with large targets. This would indicate that we need to have more data points in the training set for these target values.

Recall that the original data set was divided into training, validation (if early stopping is used) and testing subsets. The regression analysis should be performed on each subset individually, as well as the full data set. Differences between the subsets would indicate overfitting or extrapolation. For example, if the training set shows accurate fitting, but the validation and test results are poor, then this would indicate overfitting (which can sometimes happen, even when early-stopping is used). In this case, we might reduce the size of the neural network and retrain. If both the training and validation results are good, but the testing results are poor, then this could indicate extrapolation (where the testing data falls outside the training and validation data). In this case, we need to provide more data for training and validation. If the results for all three data sets are poor, then it might be necessary to increase the number of neurons in the network. Another choice is to increase the number of layers in the network. If you start with a single hidden layer, and the results are poor, then a second hidden layer could be helpful. First, try more neurons in the single hidden layer, and then increase the number of layers.

In addition to the regression/scatter plot, another tool that can identify outliers is a histogram of the errors, as shown in Figure 17.7. The y-axis represents the number of errors that falls within each interval on the x-axis. Here we can see that two errors are greater than 8. These represent the same two errors that we identified as outliers in Figure 17.6.

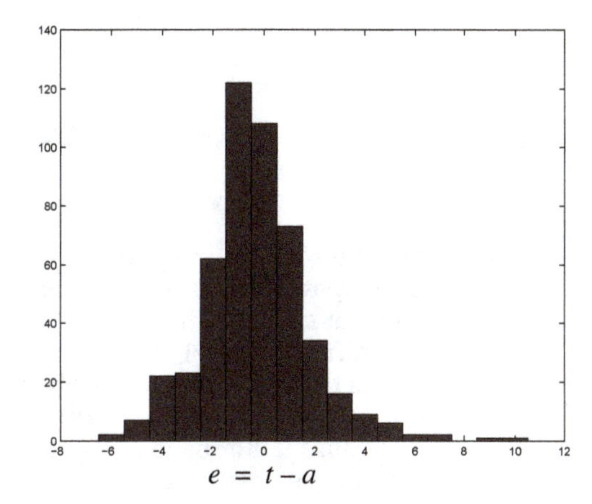

Figure 17.7 Histogram of Network Errors

Pattern Recognition

For pattern recognition problems, the regression analysis is not as useful as it is for fitting problems, since the target values are discrete. However, there is an analogous tool - the *confusion (or misclassification) matrix*. The confusion matrix is a table whose columns represent the target class and whose rows represent the output class. For example, Figure 17.8 shows a sample confusion matrix in which there were 214 data points. There were 41 input vectors that belonged to Class 1 and were correctly classified as Class 1. There were 162 input vectors that belonged to Class 2 and were correctly classified as Class 2. The correctly classified inputs show in the diagonal cells of the confusion matrix. The off-diagonal cells show misclassified inputs. The lower left cell shows that four inputs from Class 1 were misclassified by the network as Class 2. If Class 1 is considered a positive outcome, then the lower left cell represents *false negatives*, which are also

Confusion Matrix

False Negative

called Type II errors. The upper right cell shows that one input from Class 2 was misclassified by the network as Class 1. This would be considered a *false positive* or a Type I error.

False Positive

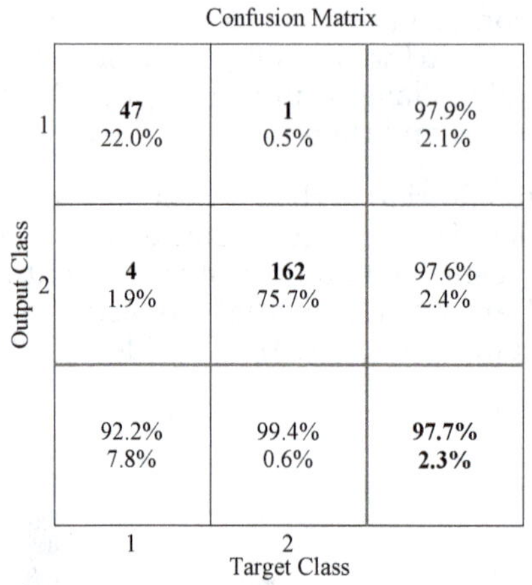

Confusion Matrix

Output Class \ Target Class	1	2	
1	**47** 22.0%	**1** 0.5%	97.9% 2.1%
2	**4** 1.9%	**162** 75.7%	97.6% 2.4%
	92.2% 7.8%	99.4% 0.6%	**97.7%** **2.3%**

Figure 17.8 Sample Confusion Matrix

ROC Curve

Another useful tool for analyzing a pattern recognition network is called the *Receiver Operating Characteristic (ROC) curve*. To create this curve, we take the output of the trained network and compare it against a threshold which ranges from -1 to +1 (assuming a tansig transfer function in the last layer). Inputs that produce values above the threshold are considered to belong to Class 1, and those with values below the threshold are considered to belong to Class 2. For each threshold value, we count the fraction of true positives and false positives in the data set. This pair of numbers produces one point on the ROC curve. As the threshold is varied, we trace the complete curve, as shown in Figure 17.9.

The ideal point for the ROC curve to pass through would be (0,1), which would correspond to no false positives and all true positives. A poor ROC curve would represent a random guess, which is represented by the diagonal line in Figure 17.9, which passes through the point (0.5,0.5).

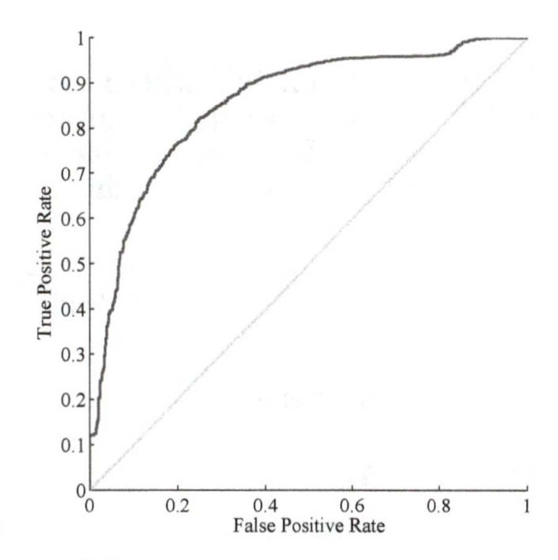

Figure 17.9 Receiver Operating Characteristic Curve

Clustering

Quantization Error

The SOM is the most commonly used network for clustering. There are several measures of SOM performance. One is *quantization error*. This is the average distance between each input vector and the closest prototype vector. It measures the map resolution. This can be made artificially small, if we use a large number of neurons. If there are as many neurons as input vectors in the data set, then the quantization error could be zero. This would represent overfitting. If the number of neurons is not significantly smaller than the number of input vectors, then the quantization error is not meaningful.

Topographic Error

Another measure of SOM performance is *topographic error*. This is the proportion of all input vectors for which the closest prototype vector and the next closest prototype vector are not neighbors in the feature map topology. Topographic error measures the preservation of the topology. In a well-trained SOM, prototypes that are neighbors in the topology should also be neighbors in the input space. In this case, the topographic error should be zero.

Distortion Measure

The performance of the SOM can also be assessed by the *distortion measure*:

$$E_d = \sum_{q=1}^{Q} \sum_{i=1}^{S} h_{ic_q} \| {}_i\mathbf{w} - \mathbf{p}_q \|^2 , \quad (17.14)$$

where h_{ij} is the neighborhood function, and c_q is the index of the prototype that is closest to the input vector $\mathbf{p}_q$:

$$c_q = \arg\min_j\{\|{}_j\mathbf{w} - \mathbf{p}_q\|\}. \quad (17.15)$$

For the simplest neighborhood function, h_{ij} is equal to 1 if prototype i is within some pre-specified neighborhood radius of prototype j, and equal to zero otherwise. It is also possible to have neighborhood functions that decrease continuously, such as the Gaussian function:

$$h_{ij} = \exp\left(\frac{-\|{}_i\mathbf{w} - {}_j\mathbf{w}\|^2}{2d^2}\right), \quad (17.16)$$

where d is the neighborhood radius.

Prediction

As we discussed earlier, one application of neural networks is the prediction of the future values of some time series. For prediction problems, we use dynamic networks, such as the focused time-delay neural network shown in Figure 17.3. There are two important concepts that are used when analyzing a trained prediction network:

1. the prediction errors should not be correlated in time, and
2. the prediction errors should not be correlated with the input sequence.

If the prediction errors were correlated in time, then we would be able to predict the prediction errors and, therefore, improve our original prediction. Also, if the prediction errors were correlated with the input sequence, then we would also be able to use this correlation to predict the errors.

Autocorrelation Function

In order to test the correlation of the prediction errors in time, we can use the sample *autocorrelation function*:

$$R_e(\tau) = \frac{1}{Q-\tau}\sum_{t=1}^{Q-\tau} e(t)e(t+\tau). \quad (17.17)$$

White Noise

If the prediction errors are uncorrelated (*white noise*), then we would expect $R_e(\tau)$ to be close to zero, except when $\tau = 0$. To determine if $R_e(\tau)$ is close to zero, we can set an approximate 95% confidence interval [BoJe96] using the range

$$-\frac{2R_e(0)}{\sqrt{Q}} < R_e(\tau) < \frac{2R_e(0)}{\sqrt{Q}}. \quad (17.18)$$

We can say that $e(t)$ is white, if $R_e(\tau)$ satisfies Eq. (17.18) for $\tau \neq 0$. This concept is illustrated in Figure 17.10 and Figure 17.11. Figure 17.10 shows a sample autocorrelation function for the prediction errors of a network that has not been adequately trained. We can see that the autocorrelation function does not fall totally within the bounds defined by Eq. (17.18), which are indicated by the dashed lines in the figure. Figure 17.11 shows the corresponding autocorrelation function when a network has been successfully trained. $R_e(\tau)$ falls within the bounds, except at $\tau = 0$.

Correlation in the prediction errors can indicate that the length of the tapped delay lines in the network should be increased.

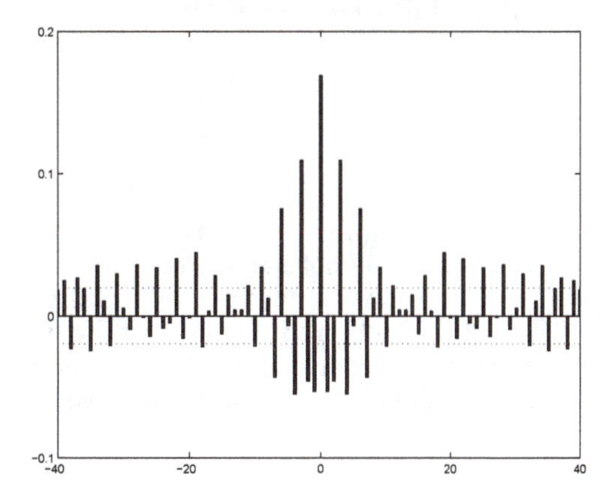

Figure 17.10 $R_e(\tau)$ for Inadequately Trained Network

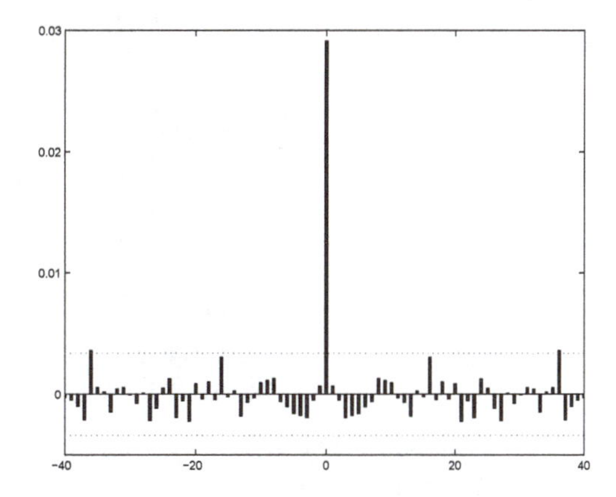

Figure 17.11 $R_e(\tau)$ for Successfully Trained Network

Cross-correlation

To test the correlation between the prediction errors and the input sequence, we can use the sample *cross-correlation function*:

$$R_{pe}(\tau) = \frac{1}{Q-\tau}\sum_{t=1}^{Q-\tau} p(t)e(t+\tau). \qquad (17.19)$$

If there is no correlation between the prediction errors and the input sequence, then we would expect $R_{pe}(\tau)$ to be close to zero for all τ. To determine if $R_{pe}(\tau)$ is close to zero, we can set an approximate 95% confidence interval [BoJe96] using the range

$$-\frac{2\sqrt{R_e(0)}\sqrt{R_p(0)}}{\sqrt{Q}} < R_{pe}(\tau) < \frac{2\sqrt{R_e(0)}\sqrt{R_p(0)}}{\sqrt{Q}}. \qquad (17.20)$$

This concept is illustrated in Figure 17.12 and Figure 17.13. Figure 17.12 shows a sample cross-correlation function for the prediction errors of a network that has not been adequately trained. We can see that the cross-correlation function does not fall totally within the bounds defined by Eq. (17.20), which are indicated by the dashed lines. Figure 17.13 shows the corresponding cross-correlation function when a network has been successfully trained. $R_{pe}(\tau)$ falls within the bounds for all τ.

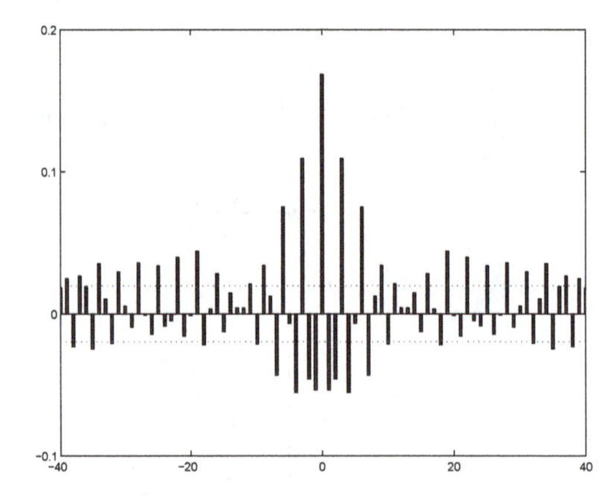

Figure 17.12 $R_{pe}(\tau)$ for Inadequately Trained Network

When using a NARX network, correlation between the prediction error and the input can suggest that the lengths of the tapped delay lines in the input and feedback paths should be increased.

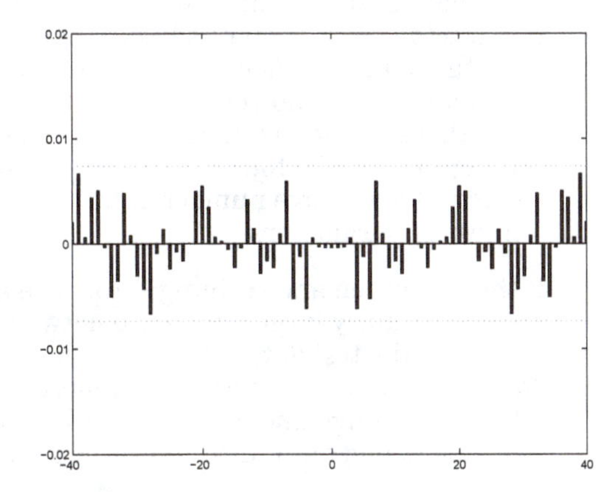

Figure 17.13 $R_{pe}(\tau)$ for Successfully Trained Network

Overfitting and Extrapolation

Recall from Chapter 13 that the total data set is divided into three parts: training, validation and testing. The training set is used to calculate gradients and to determine weight updates. The validation set is used to stop training before overfitting occurs. (If Bayesian regularization is used, then the validation set may be merged with the training set.) The test set is used to predict future performance of the network. The test set performance is the measure of network quality. If, after a network has been trained, the test set performance is not adequate, then there are usually four possible causes:

- the network has reached a local minimum,
- the network does not have enough neurons to fit the data,
- the network is overfitting, or
- the network is extrapolating.

The local minimum problem can almost always be overcome by retraining the network with five to ten random sets of initial weights. The network with minimum training error will generally represent a global minimum. The other three problems can generally be distinguished by analyzing the training, validation and test set errors. For example, if the validation error is much larger than the training error, then overfitting has probably occurred. Even though early stopping is used, it is possible to have some overfitting, if the training occurs too quickly. In this case, we can use a slower training algorithm to retrain the network.

If the validation, training and test errors are all similar in size, but the errors are too large, then it is likely that the network is not powerful enough to fit the data. In this case, we should increase the number of neurons in the hidden layer and retrain the network. If Bayesian regularization is used, this situation is indicated by the effective number of parameters becoming equal to the total number of parameters. When the network is large enough, the effective number of parameters should remain below the total number of parameters.

If the validation and training errors are similar in size, but the test errors are significantly larger, then the network may be extrapolating. This indicates that the test data fall outside the range of the training and validation data. In this case, we need to get more data. You can merge the test data into the training/validation data and then collect new test data. You should continue to add data until the results on all three data sets are similar.

If training, validation and test errors are similar, and the errors are small enough, then we can put the multilayer network to use. However, we still need to be careful about the possibility of extrapolation. If the multilayer network inputs are outside the range of the data with which it was trained, then extrapolation will occur. It is difficult to guarantee that training data will encompass all future uses of a neural network.

One method for detecting extrapolation is to train a companion competitive network to cluster the input vectors in the multilayer network training set. Then, when an input is applied to the multilayer network, the same input is applied to the companion competitive network. When the distance of the input vector to the nearest prototype vector of the competitive network is larger than the distance from the prototype to the most distant member of its cluster of inputs in the training set, we can suspect extrapolation. This technique is referred to as *novelty detection*.

Novelty Detection

Sensitivity Analysis

After a multilayer network has been trained, it is often useful to assess the importance of each element of the input vector. If we can determine that a given element of the input vector is unimportant, then we can eliminate it. This can simplify the network, reduce the amount of computation and help prevent overfitting. There is no one method that can absolutely determine the importance of each input, but a sensitivity analysis can be helpful in this regard. A sensitivity analysis computes the derivatives of the network response with respect to each element of the input vector. If the derivative with respect to a certain input element is small, then that element can be eliminated from the input vector.

Because the multilayer network is nonlinear, the derivative of the network output with respect to an input element will not be constant. For each input vector in the training set, the derivatives will be different. For this reason, we can't use a single derivative to determine sensitivity. One option would be to take the average of the absolute derivatives, or else the rms derivatives, over the entire training set. Another option would be to compute the derivative of the sum square error with respect to each element of the input vector. Each of these will compute a single derivative for each element of the input vector. The last computation can be performed with a simple variation of the backpropagation algorithm (see Eq. (11.44) to Eq. (11.47)). Recall from Eq. (11.32) that

$$s_i^m \equiv \frac{\partial \hat{F}}{\partial n_i^m}, \tag{17.21}$$

where $\hat{F}$ is a single error squared. We want to convert this to a derivative with respect to an element of the input vector, using the chain rule:

$$\frac{\partial \hat{F}}{\partial p_j} = \sum_{i=1}^{S^1} \frac{\partial \hat{F}}{\partial n_i^1} \times \frac{\partial n_i^1}{\partial p_j} = \sum_{i=1}^{S^1} s_i^1 \times \frac{\partial n_i^1}{\partial p_j}. \tag{17.22}$$

We know that

$$n_i^1 = \sum_{j=1}^{R} w_{i,j}^1 p_j + b_i^1, \tag{17.23}$$

therefore, Eq. (17.22) becomes

$$\frac{\partial \hat{F}}{\partial p_j} = \sum_{i=1}^{S^1} \frac{\partial \hat{F}}{\partial n_i^1} \times \frac{\partial n_i^1}{\partial p_j} = \sum_{i=1}^{S^1} s_i^1 \times w_{i,j}^1. \tag{17.24}$$

In matrix form, we can write this as

$$\frac{\partial \hat{F}}{\partial \mathbf{p}} = (\mathbf{W}^1)^T \mathbf{s}^1. \tag{17.25}$$

This will be the derivative for a single squared error. To get the derivative of the sum square error, we sum the individual derivatives for each single squared error. The resulting vector will contain the derivatives of the sum square error for each element of the input vector. If we find that some of these derivatives are much smaller than the maximum derivative, then we can consider removing those inputs. After removing the potentially irrelevant inputs, we retrain the network and compare the performance with the original network. If the performance is similar, then we accept the simplified network.

Epilogue

While previous chapters have focused on the fundamentals of particular network architectures and training rules, this chapter has discussed some practical aspects of neural network training. Neural network training is an iterative process involving data collection and preprocessing, network architecture selection, network training and post-training analysis.

The next five chapters will demonstrate some of these practical aspects, as we present some real-world case studies. The case studies will cover a variety of applications, including function fitting, density estimation, pattern recognition, clustering and prediction.

Further Reading

[Bish95] C.M. Bishop, *Neural Networks for Pattern Recognition*, Oxford University Press,1995.

This well-written and well-organized textbook presents neural networks from a statistical perspective.

[BoJe94] G.E.P. Box, G.M. Jenkins, and G.C. Reinsel, *Time Series Analysis: Forecasting and Control*, 4th Edition, John Wiley & Sons, 2008.

This is a classic text on time series analysis. It focuses on practical aspects, rather than theoretical derivations.

[HaBo07] L. Hamm, B. W. Brorsen and M. T. Hagan, "Comparison of Stochastic Global Optimization Methods to Estimate Neural Network Weights," *Neural Processing Letters*, Vol. 26, No. 3, December 2007.

This paper demonstrates that using multiple restarts of a local optimization procedure, like steepest descent or conjugate gradient, produces results that are comparable to global optimization methods, but with less computation.

[HeOh97] B. Hedén, H. Öhlin, R. Rittner, L. Edenbrandt, "Acute Myocardial Infarction Detected in the 12-Lead ECG by Artificial Neural Networks," *Circulation*, vol. 96, pp. 1798–1802, 1997.

Describes the use of neural networks in detecting myocardial infarctions, using the electrocardiogram.

[Joll02] I.T. Jolliffe, *Principal Component Analysis*, Springer Series in Statistics, 2nd ed., Springer, NY, 2002.

The most popular text on principal component analysis.

[LeCu98] Y. LeCun, L. Bottou, G. B. Orr, K.-R. Mueller, "Efficient BackProp," *Lecture Notes in Comp. Sci.*, vol. 1524, 1998.

This paper presents practical tips that improve the training of multilayer networks.

[Moll93] M. Moller, "A scaled conjugate gradient algorithm for fast supervised learning," *Neural Networks*, vol. 6, pp. 525-533, 1993.

The scaled conjugate gradient algorithm presented in this paper converges quickly, and with a minimum amount of memory requirements.

[NgWi90] D. Nguyen and B. Widrow, "Improving the learning speed of 2-layer neural networks by choosing initial values of the adaptive weights," *Proceedings of the IJCNN*, vol. 3, pp. 21–26, July 1990.

This paper describes a procedure for setting the initial weights and biases for the backpropagation algorithm. It uses the shape of the sigmoid transfer function and the range of the input variables to determine how large the weights should be, and then uses the biases to center the sigmoids in the operating region. The convergence of backpropagation is improved significantly by this procedure.

[PeCo93] M. P. Perrone and L. N. Cooper, "When networks disagree: Ensemble methods for hybrid neural networks," in *Neural Networks for Speech and Image Processing*, R. J. Mammone, Ed., Chapman-Hall, pp. 126-142, 1993.

This paper describes how you can combine the outputs of a committee of networks to produce results that are more accurate than any of the individual networks.

[PuFe97] G.V. Puskorius and L.A. Feldkamp, "Extensions and enhancements of decoupled extended Kalman filter training," *Proceedings of the 1997 International Conference on Neural Networks*, vol. 3, pp. 1879-1883, 1997.

The extended Kalman filter algorithm described in this paper is one of the faster sequential algorithms for neural network training.

[RaMa05] L.M. Raff, M. Malshe, M. Hagan, D.I. Doughan, M.G. Rockley, and R. Komanduri, "*Ab initio* potential-energy surfaces for complex, multi-channel systems using modified novelty sampling and feedforward neural networks," *The Journal of Chemical Physics*, vol. 122, 2005.

This paper describes how neural networks can be used to model molecular interactions.

[ScSm99] B. Schölkopf, A. Smola, K.-R. Muller, "Kernel Principal Component Analysis," in B. Schölkopf, C. J. C. Burges, A. J. Smola (Eds.), *Advances in Kernel Methods-Support Vector Learning*, MIT Press Cambridge, MA, USA, pp. 327-352, 1999.

This paper introduces a nonlinear version of principal component analysis using a kernel method.

18 Case Study 1: Function Approximation

Objectives

This chapter represents the first of a series of case studies with neural networks. Neural networks can be used for a wide variety of applications, and it would be impossible to provide case studies for each application. We will limit our presentations to five important application areas: function approximation (aka, nonlinear regression), density function estimation, pattern recognition (aka, pattern classification), clustering and prediction (aka, time series analysis, system identification, or dynamic modeling). For each case study, we will step through the neural network design/training process.

In this chapter, we present a function approximation problem. For function approximation problems, the training set consists of a set of dependent variables (response variables) and one or more independent variables (explanatory variables). The neural network learns to create a mapping between the explanatory variables and the response variables. In the case study we consider in this chapter, the system in question is a smart sensor. A smart sensor consists of one or more standard sensors that are coupled with a neural network to produce a calibrated measurement of a single parameter. In this chapter, we will consider a smart position sensor, which uses the voltages coming from two solar cells to produce an estimate of the location of an object in one dimension.

Theory and Examples

This chapter presents a case study in using neural networks for function approximation. Function approximation consists of defining a mapping between a set of input variables and a corresponding set of output variables. For example, we might want to estimate the price of a home, based on characteristics of the neighborhood, such as tax rate, pupil/teacher ratio in local schools and crime rate. Another example would be estimating octane number of a gasoline product at an oil refinery, based on measurements of reactor temperatures and pressures [FoGi07]. In the case study presented in this chapter, we will consider a smart position sensor system.

Description of the Smart Sensor System

Figure 18.1 illustrates the sensor arrangements for this case study. An object is suspended between a light source and two solar cells. The object casts a shadow on the solar cells, which causes the voltage out of the solar cells to decrease.

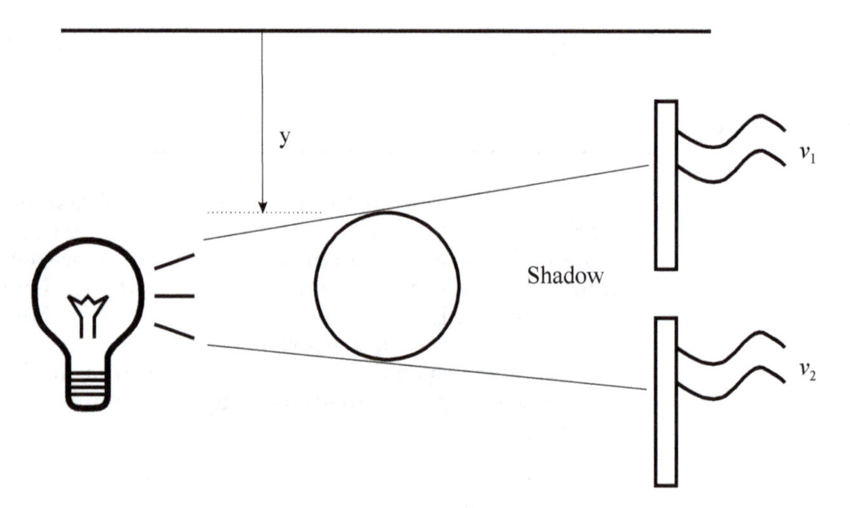

Figure 18.1 Position Sensor Arrangement

As the object position y increases, first the voltage v_1 decreases, then the voltage v_2 decreases, then v_1 increases, and finally v_2 increases. This is demonstrated in Figure 18.2. Our objective is to determine the object position from measurements of the two voltages. Clearly this is a very nonlinear relationship, so a multilayer network will be needed to learn the mapping. This is a classic type of function approximation problem, in which we are trying to learn the inverse of a function. The forward function is the mapping from y to v_1 and v_2. We want to learn the mapping from v_1 and v_2 to y.

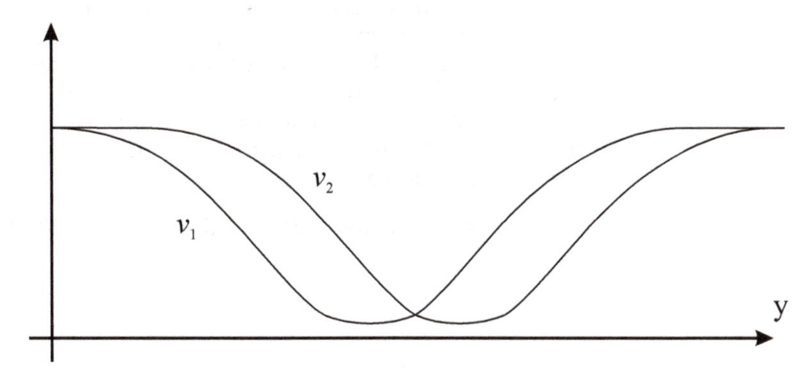

Figure 18.2 Example Solar Cell Outputs vs. Object Position

Data Collection and Preprocessing

In order to collect data for this process, we took measurements of the two solar cell voltages at a number of calibrated positions of the object. The object we used for these experiments was a table tennis ball. The data is displayed in Figure 18.3. There are a total of 67 sets of measurements. Each circle represents a voltage measurement at a calibrated position. The units of position are inches, and the units of voltage are volts. The flat regions at 0 volts for each curve occur where the shadow of the ball completely covers a sensor. If the shadow were large enough to cover both cells at the same time, we would not be able to recover ball position from cell voltages.

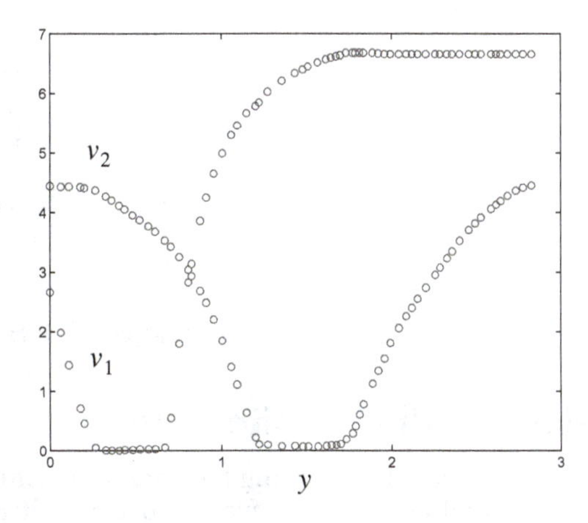

Figure 18.3 Data Collected from Solar Cells

The next step is to divide the data into training, validation and test sets. In this case, because we will be using the Bayesian regularization training technique, we do not need to have a validation set. We did set aside 15% of the data for testing purposes. To perform the division, we arranged the data in order, according to object position, and then selected every sixth or seventh point for testing. This resulted in 10 testing points. The testing points are not used in any way for training the network, but after the network has been completely trained, we will use the testing data as an indicator of future network performance.

The input vector for network training will consist of the solar cell voltages

$$\mathbf{p} = \begin{bmatrix} v_1 \\ v_2 \end{bmatrix}, \tag{23.1}$$

and the target will be ball position

$$t = y. \tag{23.2}$$

The data were scaled using Eq. (17.1), so that both the inputs and the targets were in the range [-1,1]. The resulting scaled data is shown in Figure 18.4

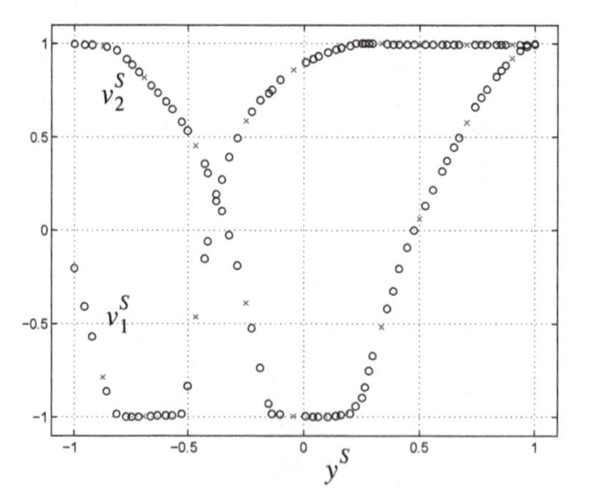

Figure 18.4 Scaled Data

Selecting the Architecture

Because the mapping between the solar cell voltages and the ball position is highly nonlinear, we will use a multilayer network architecture to learn the mapping. We know that there will be two elements in the input vector, which is defined in Eq. (23.1). The single target for the network is the ball position, given in Eq. (23.2).

Figure 18.5 shows the network architecture. We are using the tan-sigmoid transfer function in the hidden layer, and a linear output layer. This is the standard network for function approximation. As we discussed in Chapter 11, this network has been shown to be a universal approximator. There are cases in which two hidden layers are used, but we normally try first with one hidden layer. The number of neurons in the hidden layer, S^1, will depend on the function to be approximated. This is something that cannot generally be known before training. We will have more to say about this in the next section.

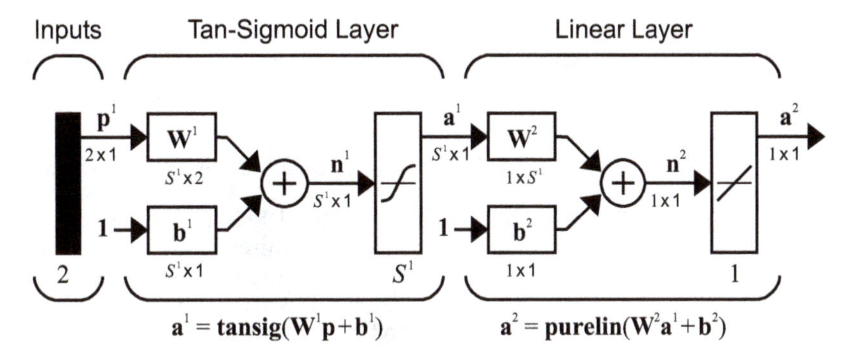

Figure 18.5 Network Architecture

Training the Network

Before beginning the training, we initialized the network weights using the method of Widrow and Nguyen described in Chapter 17. Then we used Bayesian regularization to train the network. Bayesian regularization, which we discussed in Chapter 13, is a very effective algorithm for training multilayer networks to perform function approximation. This algorithm is designed to train networks so that they generalize well, without the need for a validation set. Because the validation set can then be added to the training set, the performance is often better than that obtained with early stopping. (In the next chapter, we will give an example of using early stopping, with a validation set.)

Figure 18.6 illustrates the sum square error versus iteration number, while using the Bayesian regularization training algorithm. We used a network with 10 neurons in the hidden layer ($S^1 = 10$) for this case. The network was trained for 100 iterations, at which time the performance was changing very little.

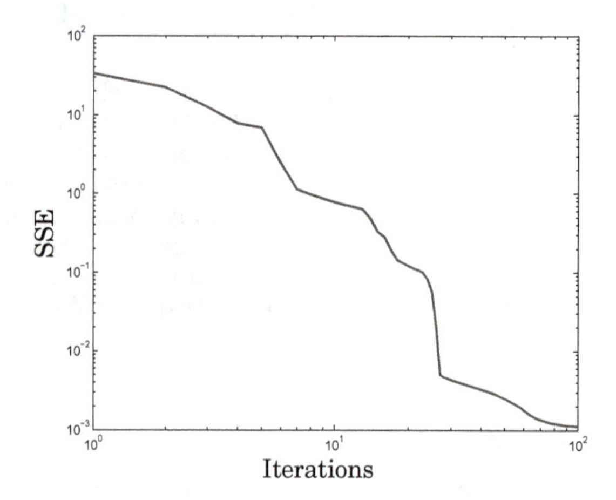

Figure 18.6 Sum Squared Error vs. Iteration Number ($S^1 = 10$)

Training has converged after 100 iterations, but we want to ensure that we have not fallen into a local minimum. For this reason, we want to retrain the network several times, using different initial weights and biases. (We use the Nguyen-Widrow initialization method described in Chapter 17.) Table 18.1. shows the final validation SSE for each of five different training runs. We can see that all of the errors are similar, although the errors are slightly smaller for runs 2, 4 and 5. Any of the weights from these five cases would produce a satisfactory network. We will discuss this in more detail in the next section.

1.121e-003	8.313e-004	1.068e-003	8.672e-004	8.271e-004

Table 18.1. Final Training SSE for Five Different Initial Conditions

Recall from Chapter 13 that the Bayesian regularization algorithm computes a parameter γ, which indicates the effective number of parameters that are being used by the network. In Figure 18.7, we can see the variation of γ during training. It eventually converges to 17.4. There are a total of 41 parameters in this 2-10-1 network, so we are only using about 40% of the weights and biases. For each of the five training runs discussed above, γ converged to values between 17 and 20. This indicates that we might be able to use a smaller network, if we are concerned about the amount of computation required to compute a network response.

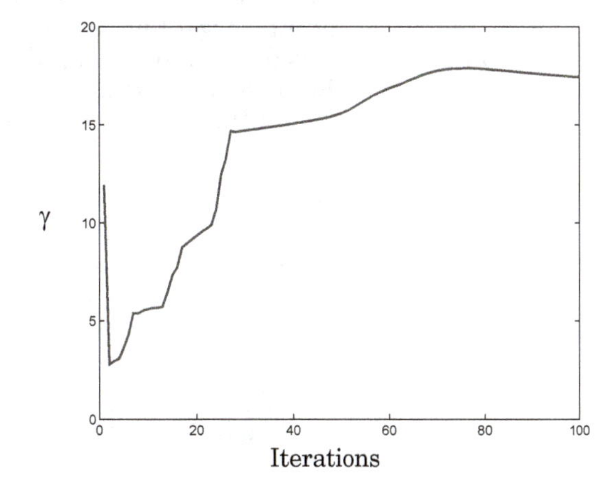

Figure 18.7 Effective Number of Parameters ($S^1 = 10$)

To see whether or not a smaller network would be satisfactory, we trained several networks with different numbers of hidden neurons. Since the effective number of parameters is near 20, we would expect that a network with five hidden neurons (21 weights and biases) might provide an adequate fit. Our experiments produced the results shown in Table 18.2. We can see that the performances of all five networks are roughly equivalent, except for the case where $S^1 = 3$, where the total number of parameters in the network is only 13.

$S^1 = 3$	$S^1 = 5$	$S^1 = 8$	$(S^1 = 10)$	$S^1 = 20$
4.406e-003	9.227e-004	8.088e-004	8.672e-004	8.096e-004

Table 18.2. Final Training SSE for Five Different Hidden Layer Sizes

The Bayesian regularization method allows us to train a network of almost arbitrary size, and yet insure that only the required number of parameters is effectively used. If we were concerned about the amount of time required to compute the network output (e.g., for real-time applications), then we would want to use the network with $S^1 = 5$. Otherwise, the original network with $S^1 = 10$ is satisfactory. We don't need to spend a lot of time finding the optimal number of neurons. The training algorithm will insure that we do not overfit.

Validation

An important tool for network validation is a scatter plot of network outputs versus targets, as shown in Figure 18.8 (in normalized units). We expect that for a well trained network the points in the scatter plot will fall close to the 45° output=target line. In this case, the fit is excellent. The figure on the left shows the training data, while the figure on the right shows the testing data. Because the testing data fit is as good as the training data fit, we can be confident that the network did not overfit.

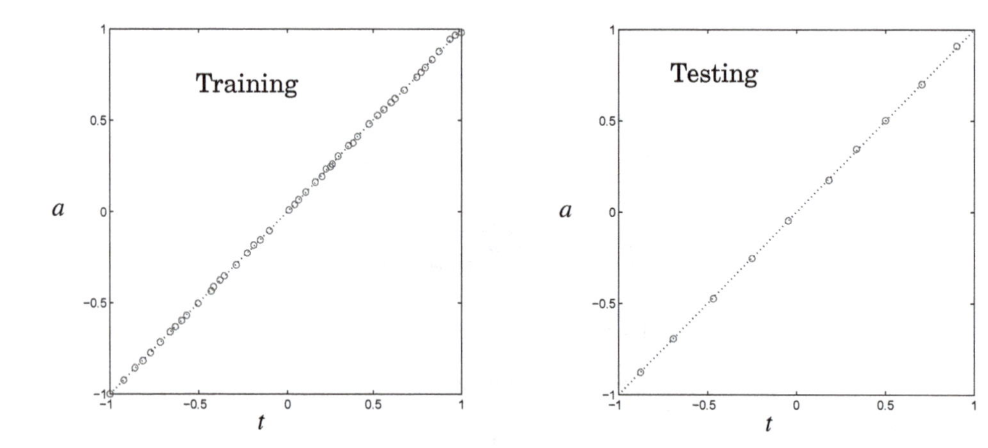

Figure 18.8 Scatter Plots of Network Outputs vs. Targets - Training and Testing Sets

Another useful plot is a histogram of network error, as shown in Figure 18.9. This gives us an idea of the accuracy of the network. For this histogram, we have converted the network output back into units of inches. This is done by applying the reverse of the target preprocessing function to the network output. The reverse of Eq. (17.1), for the targets, is given by

$$\mathbf{a} = (\mathbf{a}^n + 1).*\frac{(\mathbf{t}^{max} - \mathbf{t}^{min})}{2} + \mathbf{t}^{min} \qquad (23.3)$$

where $\mathbf{a}^n$ is the original network output, which was trained to match the normalized target, and .* represents an element-by-element multiplication of two vectors. After the postprocessing operation of Eq. (23.3), the resulting un-normalized output is subtracted from the raw targets, to produce an error in inches. Figure 18.9 shows the distribution of these errors for both training and testing sets. We can see that almost all errors are within one hundredth of an inch. This is within the accuracy of the original measurements, so we cannot expect to do better.

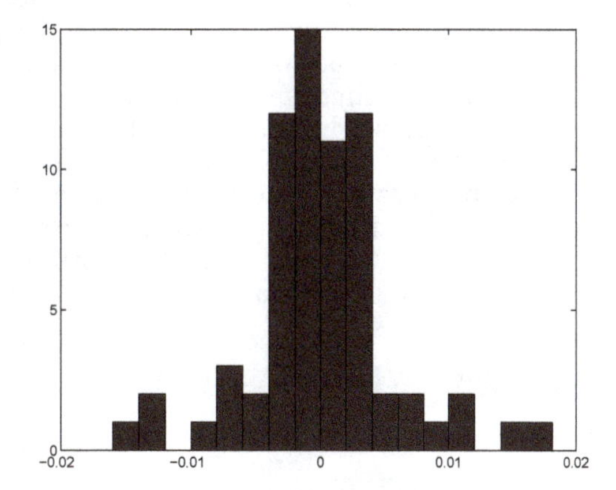

Figure 18.9 Histogram of Position Errors (in Inches)

Because this network has only two inputs, we can plot the trained network response, which is shown in Figure 18.10. (This figure shows the response from original unscaled inputs, in volts, to original unscaled output, in inches.) The blue circles indicate the path that is taken by the voltages as the ball is moved. Notice that the Bayesian regularization training has produced a smooth network response, even though the response is highly nonlinear.

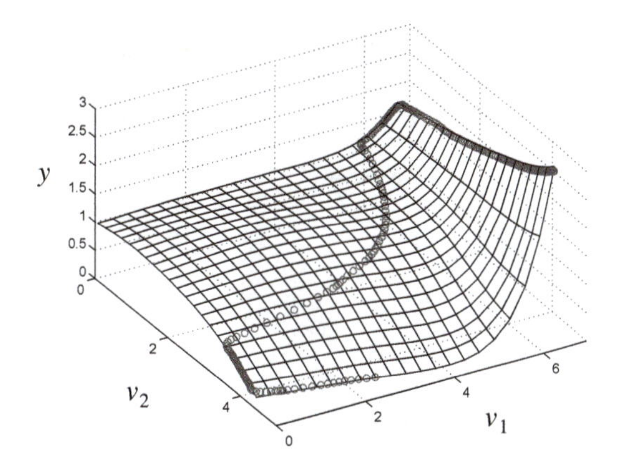

Figure 18.10 Network Response (Original Units)

Another thing to notice about Figure 18.10 is that training data only falls along the blue circles. The form of the network response in other regions is of no importance for the operation of the smart sensor system, since the network will never be used there. If the network were retrained, the shape of the response away from the blue circles might be very different, even though the response near the blue circles will always be the same. This concept is very important for many neural network applications. Often, during normal network operation, only a small portion of the input space will be accessed. The network only has to fit the underlying function in these regions where the network will be used. This means that the size of the data set can be modest, even when the input dimension is large. Of course, in these cases, it is critical that the training data span the full range of potential network operation.

Data Sets

There are two data files associated with this case study:

- ball_p.txt — contains the input vectors in the original data set
- ball_t.txt — contains the target vectors in the original data set

They can be found with the demonstration software, which is described in Appendix C.

Epilogue

This chapter has demonstrated the use of multilayer neural networks for function approximation. This case study is representative of a large class of neural network applications that could be termed "soft sensors" or "smart sensors." The idea is to use a neural network to fuse several raw sensor outputs into a calibrated measurement of some key variable of interest.

A multilayer network with sigmoid transfer functions in the hidden layers and linear transfer functions in the output layer is well suited to this type of application, and Bayesian regularization is an excellent training algorithm to use in this situation.

In the next chapter, we will look at another neural network application — probability estimation. We will also use multilayer neural networks for that application, but we will change the transfer function in the output layer.

Further Reading

[FoGi07] L. Fortuna, P. Giannone, S. Graziani, M. G. Xibilia, "Virtual Instruments Based on Stacked Neural Networks to Improve Product Quality Monitoring in a Refinery," *IEEE Transactions on Instrumentation and Measurement*, vol. 56, no. 1, pp. 95–101, 2007.

This paper describes the use of neural networks as soft sensors in a refinery. Measurements of reactor temperatures and pressures are used to predict octane number in a gasoline product.

19 Case Study 2: Probability Estimation

Objectives

This chapter represents the second of a series of case studies with neural networks. The previous chapter demonstrated the use of neural networks for function approximation. In this chapter we use a neural network to estimate a probability function.

Probability estimation is a special case of function approximation. In function approximation we want the neural network to map between a set of input variables and a set of response variables. However, in the case of probability estimation the response variables correspond to a set of probabilities. Since probabilities have certain special properties — they must always be positive, and they must sum to 1 — we want the neural network to enforce these conditions.

In the case study we consider in this chapter, the system in question is chemical vapor deposition of diamond. A carbon dimer (a bound pair of carbon atoms) is projected toward a diamond surface. We want to determine the probabilities for various reactions based on characteristics of the projected dimer. The input variables consist of such properties as translational energy and incidence angle, and the response variables consist of the probabilities of the potential reactions, such as chemisorption and scattering.

Theory and Examples

This chapter presents a case study in using neural networks for probability estimation. Probability estimation consists of determining the probabilities of certain events, based on a set of input variables. For example, we might want to know the probabilities associated with a patient having a certain disease, based on a set of laboratory tests. Another example would be determining the probability of a financial instrument going up in price, based on a set of market conditions.

For this probability estimation case study, we will train a neural network to estimate reaction rates in a chemical process. Chemical vapor deposition (CVD) of diamond is a process for making synthetic diamond. The idea is to cause carbon atoms in a gas to settle on a substrate in crystalline form. In order to study this process, scientists are often interested in reaction rates, which will determine how quickly the diamond can be created. In this case study, we will train a neural network to compute reaction rates as a carbon dimer (a bound pair of carbon atoms) interacts with the crystalline diamond substrate.

CVD

We will begin by describing the CVD process and how simulated data can be collected for this process. Then, we will show how a neural network can be trained to learn the reaction probabilities. The details of the procedure are described in [AgSa05].

Description of the CVD Process

During the CVD process, a carbon dimer is projected toward a diamond substrate. For the purpose of this study, we will assume that the dimer can react with the substrate in one of three ways: chemisorption (the atoms in the dimer become bound to the substrate), scattering (the atoms bounce off the substrate), or desorption (the atoms become bound to the substrate for a period of time, but are then released). There is another possible reaction that occurs with very small probability, but we will ignore it for this study. (See [AgSa05] for a full discussion.) We will train a neural network to estimate the probabilities of each of the reactions, based on various characteristics of the carbon dimer, which will be described below.

The notation we will use to define this interaction is illustrated in Figure 19.1. The black circle represents the carbon dimer, and the corresponding directed line represents the direction of the initial velocity vector. The blue star represents the location of the central carbon atom in the diamond substrate. The angle θ denotes the angle of incidence, i.e., the angle between the direction of the initial velocity vector of the carbon dimer and the perpendicular on the surface (the z direction). The impact parameter b is defined as the distance between the location of the central atom and the point of intersection of the initial velocity vector and the diamond surface (indicated by the origin of the axes in Figure 19.1). The angle ϕ represents the angle between the x axis and the line from the origin to the central atom.

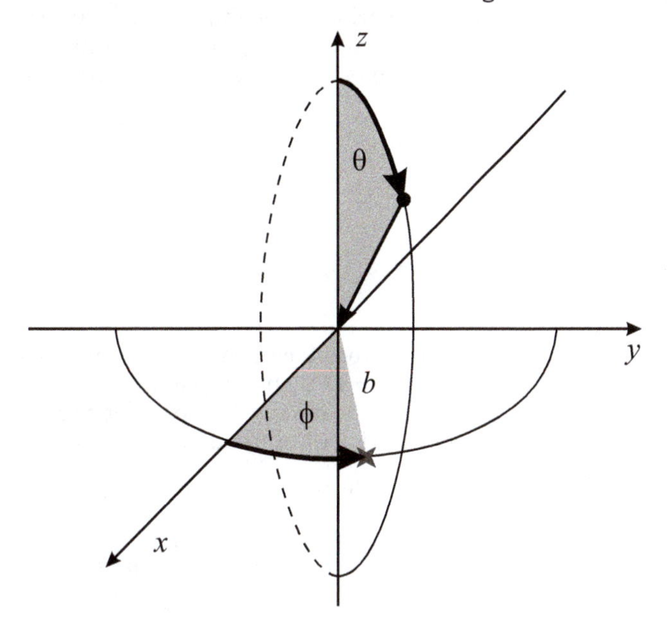

Figure 19.1 Notation for Carbon Dimer/Diamond Substrate Interaction

Data Collection and Preprocessing

Data for training the neural network are obtained by molecular dynamics (MD) simulations. In MD, the motion of atoms and molecules in a material under a given force are simulated, using known laws of physics to calculate the forces on individual atoms [RaMa05]. For this case study, we use a total of 324 atoms to model the CVD system. Out of these, 282 atoms of diamond substrate are used to model the crystalline face with 40 atoms of hydrogen on the top layer of the diamond surface, and 2 atoms in the C_2 dimer. In Figure 19.1, the (x,y) plane represents the location of the diamond substrate. Each of the carbon atoms on the top layer of the substrate, except the central atom and boundary atoms, is capped by a hydrogen atom. Any reactions will occur near the central atom.

For this study, we want to determine the dependence of probabilities for chemisorption, scattering, and desorption on b, θ, ϕ, rotational velocity (v_{rot}), and translational velocity (v_{trans}) of the C_2 dimer. The initial C_2 vibrational energy is set equal to the zero-point energy and the temperature of the lattice is maintained constant at 600 K [RaMa05].

This is an interesting function approximation problem, in that we don't have access to the true underlying reaction probabilities, which are unknown. We will obtain estimates of these probabilities by running Monte Carlo simulation experiments. We will need some notation to help us keep track of the various probabilities that we will work with. First, we will indicate the true underlying reaction probabilities by $P_X(\mathbf{p})$, where X refers to the reaction process, and $\mathbf{p}$ is the vector that characterizes the C_2 dimer:

$$\mathbf{p} = \begin{bmatrix} \theta \\ \phi \\ b \\ v_{trans} \\ v_{rot} \end{bmatrix} . \tag{19.1}$$

The reaction process can be chemisorption ($X = C$), scattering ($X = S$), or desorption ($X = D$). The probability estimates produced by the neural network will be indicated by $P_X^{NN}(\mathbf{p})$. The probability estimates obtained from the Monte Carlo simulations will be indicated by $P_X^{MC}(\mathbf{p})$.

The Monte Carlo estimates are obtained by

$$P_X^{MC}(\mathbf{p}) = \frac{N_X}{N_T}, \tag{19.2}$$

where N_X is the number of MD trajectories that resulted in reaction X and N_T is the total number of trajectories computed. The results of a given trajectory depend upon a multitude of input variables. These include the parameters included in $\mathbf{p}$, as well as the initial orientation of the C_2 dimer, the angle defining the C_2 rotational plane, the initial C_2 vibrational energy and its phase, the temperature of the system, and all of the variables that define the vibrational phases of the diamond surface. Because we are only interested in the effect of $\mathbf{p}$ on the reaction probabilities, the other variables are randomly set for each MD simulation, except that the initial C_2 vibrational energy is set equal to the zero-point energy and the temperature of the lattice is maintained constant at 600 K. Eq. (19.2) averages over the trajectories to estimate the underlying true probabilities $P_X(\mathbf{p})$. (As a note of clarification here, we use the term Monte Carlo to refer to the set of simulations that are obtained by setting a number of the variables to random values for each trajectory. We refer to the simulation of a single trajectory as an MD simulation, since the principles of molecular dynamics are used to perform the computations.)

This is a standard method used by chemists to estimate reaction probabilities. If they want to determine the effect of ϕ, for example, on the probabilities, they must run a series of Monte Carlo simulations at each value of ϕ that is of interest. This can be extremely time consuming. The required number of Monte Carlo trials can be quite large, if an accurate reaction probability is required. Our objective in this case study is to train a neural network to learn the true reaction probabilities as a function of the parameters in $\mathbf{p}$.

To train a neural network, we need a set of target outputs. Since we do not know the true underlying probabilities $P_X(\mathbf{p})$, we will use the estimates obtained from the Monte Carlo simulations $P_X^{MC}(\mathbf{p})$. We can think of these Monte Carlo probabilities as being noisy versions of the true probabilities. The neural network will need to interpolate these noisy values to produce an accurate estimate of $P_X(\mathbf{p})$ without overfitting. This is a good application for our generalization procedures, which were discussed in Chapter 13.

The data set consists of 2000 different $\{\mathbf{p}, P_X^{MC}(\mathbf{p})\}$ input/target pairs. Out of these 2000 data points, 1400 (70%) were randomly selected for training, 300 (15%) for validation, and 300 for testing. For each trajectory, the $\mathbf{p}$ were generated randomly, using physically-appropriate distributions for each variable [RaMa05]. A total of $N_T = 50$ different trajectories were run to obtain each $P_X^{MC}(\mathbf{p})$. This means that 2000x50 trajectories were run to create the entire data set.

The original units of the inputs are radians for ϕ and θ, angstroms for b, angstroms per picosecond for v_{trans} and radians per picosecond for v_{rot}. Before presenting the input data to the network for training, they are scaled using Eq. (17.1), so that each element of the input vector ranges from -1 to 1. The targets have values that are always in the range 0 to 1, since they represent probabilities. In the next section, we will describe a network architecture, in which the softmax transfer function of Eq. (17.3) is used in the final layer. This transfer function produces outputs that range from 0 to 1, so the original unscaled targets will work fine.

Selecting the Architecture

We will use a multilayer network for this application. We know that there will be five elements in the input vector, which is defined in Eq. (19.1). The target for the network can be a vector with three elements:

$$\mathbf{t} = \begin{bmatrix} P_C^{MC}(\mathbf{p}) \\ P_S^{MC}(\mathbf{p}) \\ P_D^{MC}(\mathbf{p}) \end{bmatrix}, \tag{19.3}$$

or we can use three different networks, each with a different $P_X^{MC}(\mathbf{p})$ as a target. We have tried both possibilities, and the results are similar.

In this case, there is an advantage to using the single network, with three elements in the output vector. The three targets represent probabilities. Therefore, they are always in the range 0 to 1, and they always sum to 1.

This is an ideal situation for using the softmax transfer function of Eq. (17.3), which is repeated here:

$$a_i = f(n_i) = \exp(n_i) \div \sum_{j=1}^{S} \exp(n_j) . \qquad (19.4)$$

This transfer function is different from others we have used, in that each neuron output a_i is affected by all of the net inputs n_j. (In the other transfer functions, the net input n_i affected only the neuron output a_i.) This does not cause any substantial difficulties in network training. The backpropagation algorithm of Eq. (11.44) and Eq. (11.45) can still be used to compute the gradient. However, the derivative of the transfer function is no longer a diagonal matrix. The derivative of the softmax function has the following form:

$$\dot{\mathbf{F}}^m(\mathbf{n}^m) = \begin{bmatrix} a_1^m\left(\sum_{i=1}^{S_m} a_i^m - a_1^m\right) & -a_1^m a_2^m & \cdots & -a_1^m a_{S_m}^m \\ -a_2^m a_1^m & a_2^m\left(\sum_{i=1}^{S_m} a_i^m - a_2^m\right) & \cdots & -a_2^m a_{S_m}^m \\ \vdots & \vdots & & \vdots \\ -a_{S_m}^m a_1^m & -a_{S_m}^m a_2^m & \cdots & a_{S_m}^m\left(\sum_{i=1}^{S_m} a_i^m - a_{S_m}^m\right) \end{bmatrix} \qquad (19.5)$$

The complete network architecture is shown in Figure 19.2. The input vector of Eq. (19.1) has 5 elements. The output vector has 3 elements, which is consistent with the target vector of Eq. (19.3). The transfer function in the hidden layer is the hyperbolic tangent sigmoid, and the softmax transfer function is used in the output layer. The number of neurons in the hidden layer, S^1, is yet to be determined. It depends on the complexity of the function that we are trying to approximate, but we do not know at this point how complex the function is. In general, the size of the hidden layer must be determined as part of the training process. We must choose S^1 so that the network provides an accurate fit to the training data, without overfitting. We will discuss this selection in the next section.

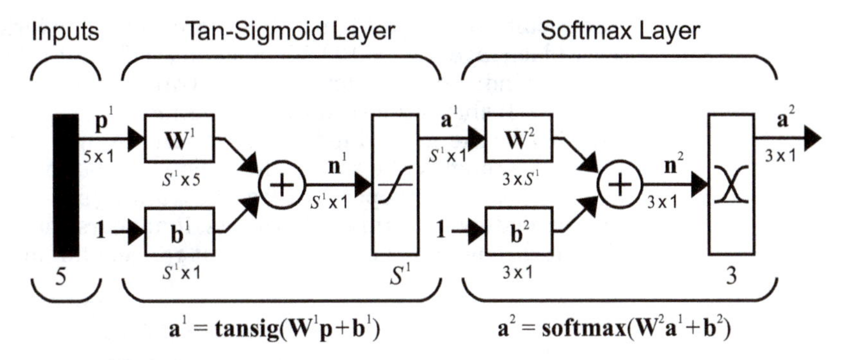

Figure 19.2 Network Architecture

Training the Network

We trained the network using the scaled conjugate gradient algorithm of [Mill93]. Many other conjugate gradient or Levenberg-Marquardt algorithms, such as those discussed in Chapter 12, would have also worked well. The targets for this problem have a significant amount of noise, so we are not expecting extreme accuracy in the final fit. We used early stopping, as described in Chapter 13, to prevent overfitting. We stopped the training if the error on the validation set failed to improve over 25 iterations. A typical training session is illustrated in Figure 19.3, which shows training and validation MSE. The minimum of the validation performance was reached at iteration 69. The algorithm continued for 25 more iterations, until iteration 94. Since the validation error was not reduced during those 25 iterations, the weights from iteration 69 were saved as the final trained values.

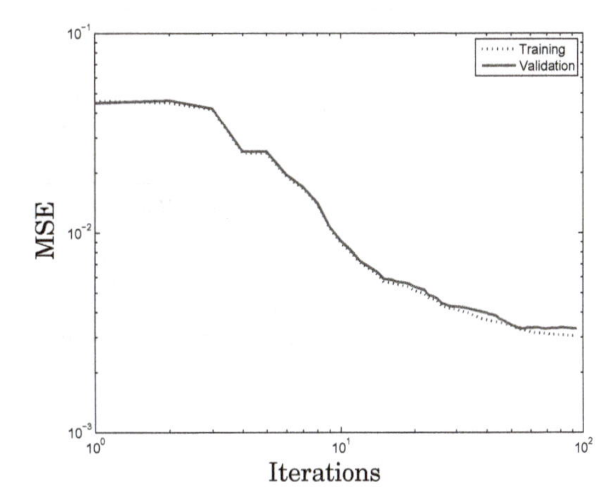

Figure 19.3 Training and Validation Mean Square Error ($S^1 = 10$)

The results shown in Figure 19.3 represent a network with 10 neurons in the hidden layer ($S^1 = 10$). We need to verify that this is a reasonable number. One indicator is a comparison of training and validation performance. Table 19.1. shows the training and validation root mean square error (RMSE) for the trained network. We can see that training and validation errors are roughly the same. The validation data was randomly selected and was selected independently of the training set. Because the errors were approximately the same on both sets, it appears that the network fit is consistent throughout the relevant input range, and no overfitting occurs.

	Training RMSE	Validation RMSE
$P_C(\mathbf{p})$	0.0496	0.0439
$P_S(\mathbf{p})$	0.0634	0.0659
$P_D(\mathbf{p})$	0.0586	0.0604

Table 19.1. Comparison of Training and Validation RMSE for $S^1 = 10$

It is also important to determine if the errors are as small as possible and if the fit is adequate. We will have more to say about that in the next section, but at this point we can try fitting networks with different numbers of hidden neurons. Table 19.2. shows the results of fitting a network with two hidden neurons. Again, the training and validation errors are consistent, which indicates lack of overfitting, but the errors are higher than those for $S^1 = 10$.

	Training RMSE	Validation RMSE
$P_C(\mathbf{p})$	0.0634	0.0627
$P_S(\mathbf{p})$	0.0669	0.0704
$P_D(\mathbf{p})$	0.0617	0.0618

Table 19.2. Comparison of Training and Validation RMSE for $S^1 = 2$

Table 19.3. shows the results for $S^1 = 20$. The validation error is slightly higher than the training error, which might indicate some overfitting. The main point is that neither training nor validation errors are significantly smaller for $S^1 = 20$ than for $S^1 = 10$. This indicates that ten hidden neurons are sufficient for this problem. We will investigate this further in the next section.

	Training RMSE	Validation RMSE
$P_C(\mathbf{p})$	0.0432	0.0444
$P_S(\mathbf{p})$	0.0603	0.0643
$P_D(\mathbf{p})$	0.0569	0.0595

Table 19.3. Comparison of Training and Validation RMSE for $S^1 = 20$

There is one further step that we want to make as part of the training process. We want to ensure that we have not fallen into a local minimum. For this reason, we want to retrain the network several times, using different initial weights and biases. (We use the Nguyen-Widrow initialization method described in Chapter 17.) Table 19.4. shows the final validation MSE for each of five different training runs. We can see that all of the errors are similar, so we have reached a global minimum at each run. If one error was significantly lower than the others, then we would use the weights that obtained the lowest error.

3.074e-003	2.953e-003	3.031e-003	3.105e-003	3.050e-003

Table 19.4. Final Validation MSE for Five Different Initial Conditions

We have determined that a neural network with ten neurons in the hidden layer produces a reasonable response without overfitting. The next step is to analyze the performance of the network. Depending on the results of that analysis, we might adjust the network architecture or training data and retrain the network.

Validation

An important tool for network validation is a scatter plot of network outputs versus targets, as shown in Figure 19.4. Here we can see that there is a strong linear relationship between the targets and the network outputs, but there appears to be quite a bit of variation. We might expect that for a well trained network the points in the scatter plot would fall exactly on the outputs=target line. Why do we have so much variation in this plot? The reason is that the targets of the network are not the true reaction probabilities, $P_X(\mathbf{p})$, but the Monte Carlo estimates $P_X^{MC}(\mathbf{p})$. There is noise in the targets.

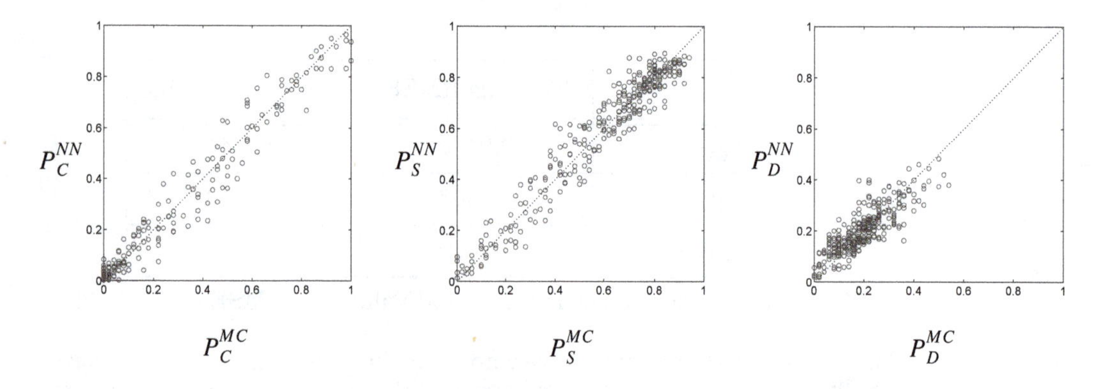

Figure 19.4 Scatter Plot of Network Outputs vs. Targets ($N_T = 50$)

The relationship between P_X^{MC} and P_X is such that ~95% of the time we expect to have

$$P_X - 2\Delta \le P_X^{MC} \le P_X + 2\Delta \,, \tag{19.6}$$

where

$$\Delta = \sqrt{\frac{P_X\{1 - P_X\}}{N_T}} \,. \tag{19.7}$$

This relationship is illustrated in Figure 19.5 for $N_T = 50$.

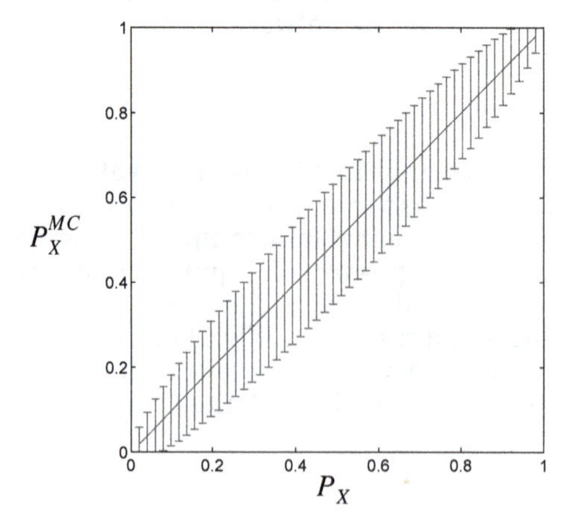

Figure 19.5 Expected Statistical Spread of P_X^{MC} for $N_T = 50$

By comparing Figure 19.4 with Figure 19.5, we can see that the spread in the data is explained by the statistical variations in P_X^{MC}. To further verify this observation, we generated additional testing data, in which 500 Monte Carlo trials were run to obtain each $P_X^{MC}(\mathbf{p})$ (i.e., $N_T = 500$). We applied this testing data to the network that was trained on the original data set with $N_T = 50$. The resulting scatter plots are shown in Figure 19.6. Here we can see that the spread has decreased dramatically from Figure 19.4, even though the network has not changed. This means that the neural network is fitting the true probabilities $P_X(\mathbf{p})$ and not the statistical fluctuations in $P_X^{MC}(\mathbf{p})$.

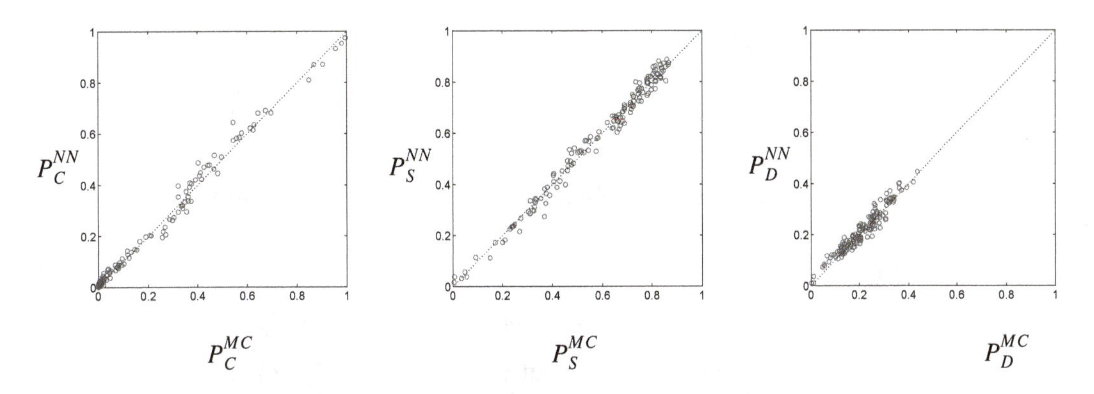

Figure 19.6 Scatter Plot of Network Outputs vs. Targets ($N_T = 500$)

After the neural network has been trained, it becomes a simple matter to investigate the effect of the input parameters on the reaction probabilities. In Figure 19.7, we see the effect of the impact parameter b on the reaction probabilities, as determined by the neural network. As the impact parameter is increased, the probability of chemisorption decreases, while the probabilities of scattering and desorption increase. (For this study, we have set θ to 5.4 radians, ϕ to 0.3 radians, v_{rot} to 0.004 radians per femtosecond, and v_{trans} to 0.004 angstroms per femtosecond.)

With standard methods, a study such as that shown in Figure 19.7 would take thousands of simulations. The trained neural network has fully captured the relationships between the parameters in $\mathbf{p}$ and the reaction probabilities. Therefore, we can perform arbitrary studies by simply computing the network responses at a varying set of input points. Note that the network interpolated smoothly through a noisy set of data points to capture the true underlying function. By using the early stopping technique, we prevented the network from overfitting the noise in the data.

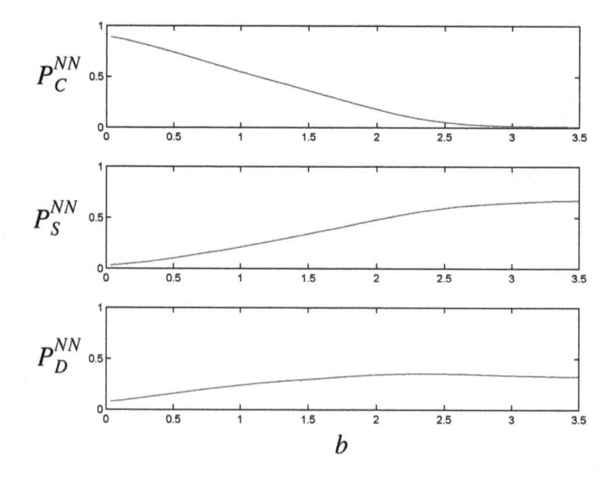

Figure 19.7 Reaction Probabilities vs. Impact Parameter

Data Sets

There are four data files associated with this case study:

- cvd_p.txt — contains the input vectors in the original data set
- cvd_t.txt — contains the target vectors in the original data set
- cvd_p500.txt — contains the input vectors in the $N_T = 500$ test set
- cvd_t500.txt — contains the target vectors in the $N_T = 500$ test set

They can be found with the demonstration software, which is described in Appendix C.

Epilogue

This chapter has illustrated the use of neural networks for probability estimation on a chemical vapor deposition problem. Monte Carlo simulations were used to provide estimates of the reaction probabilities. These estimates were used as targets for the neural network. The network was able to capture the true underlying probability function without overfitting to the errors in the Monte Carlo estimates. This was accomplished by using the early stopping procedure, which stops network training if the error on an independent validation set increases.

In the next chapter, we apply neural networks to a pattern recognition problem. We will also use multilayer neural networks for that application.

Further Reading

[AgSa05] P.M. Agrawal, A.N.A. Samadh, L.M. Raff, M. Hagan, S. T. Bukkapatnam, and R. Komanduri, "Prediction of molecular-dynamics simulation results using feedforward neural networks: Reaction of a C2 dimer with an activated diamond (100) surface," *The Journal of Chemical Physics* 123, 224711, 2005.

This paper describes the details of training a neural network to predict the reaction probabilities for chemical vapor deposition of diamond.

[Mill93] M.F. Miller, "A scaled conjugate gradient algorithm for fast supervised learning," Neural Networks, vol. 6, pp. 525-533, 1993.

The scaled conjugate gradient algorithm is a fast batch training algorithm for neural networks that requires a minimum of memory and computation at each iteration.

[RaMa05] L.M. Raff, M. Malshe, M. Hagan, D.I. Doughan, M.G. Rockley, and R. Komanduri, "*Ab initio* potential-energy surfaces for complex, multi-channel systems using modified novelty sampling and feedforward neural networks," The Journal of Chemical Physics, 122, 084104, 2005.

This paper explains how neural networks can be used for molecular dynamics simulations.

20 Case Study 3: Pattern Recognition

Objectives

This chapter presents a case study in using neural networks for pattern recognition. In pattern recognition problems, you want a neural network to classify inputs into a set of target categories, e.g., recognize the vineyard that a particular bottle of wine came from, based on a chemical analysis, or classify a tumor as benign or malignant, based on uniformity of cell size, clump thickness, mitosis.

In this chapter we will demonstrate the application of multilayer neural networks to the recognition of heart disease from a reading of the electrocardiogram. We will show each of the steps in the pattern recognition process: data collection, feature extraction, architecture selection, network training and network validation.

Theory and Examples

In pattern recognition (pattern classification) problems we are trying to categorize network inputs into their corresponding classes. Here are a few examples of pattern recognition problems:

- recognition of handwritten zip codes
- spoken word recognition
- disease recognition from a list of symptoms
- fingerprint recognition
- white blood cell classification

In the case study presented in this chapter, we will be looking for patterns in electrocardiogram signals that indicate the presence of a myocardial infarction (heart attack).

Description of Myocardial Infarction Recognition

An electrocardiogram (EKG) is a recording of the electrical activity of the heart over time. It generally consists of an array of different signals recorded at the same time. An EKG can consist of a single signal (also called a *lead*), although the standard EKG that is used for detailed interpretation consists of 12 leads. EKG's with as many as 15 leads are sometimes used. Each lead represents the electrical activity across two points on the body. The 12-lead EKG is determined from 10 electrodes that are placed on specific locations on the body. The calculation of the 12-lead potentials from the 10 electrodes is a somewhat complex calculation, and beyond the scope of this case study. The interested reader is referred to [Dubi00] for a more complete discussion of the EKG.

Through a careful analysis of the EKG, a physician can often determine the health of the heart. The shapes of the signals indicate the path of electrical flow in the heart as various muscles are contracted in a coordinated way to pump blood in and out. If a part of the heart muscle has been damaged because of a lack of blood flow through the coronary arteries (called a myocardial infarction (MI), or heart attack), then the path of electrical flow changes. A well-trained physician can discern from the changes in the EKG, if the heart has been damaged, and where the damage has occurred.

For this case study, we will train a neural network to recognize MI's, using information obtained from a 15 lead EKG.

Data Collection and Preprocessing

The EKG signals used for this case study were obtained from the PhysioNet database [MoMa01]. Data were extracted from the QT data set for healthy patients and patients with MI's. Each EKG consists of 15 leads. The leads are labeled I, II, III, aVR, aVL, aVF, V1, V2, V3, V4, V5, V6, VX, VY, VZ. Figure 20.1 shows a small portion of the lead I signal for one of the healthy patients.

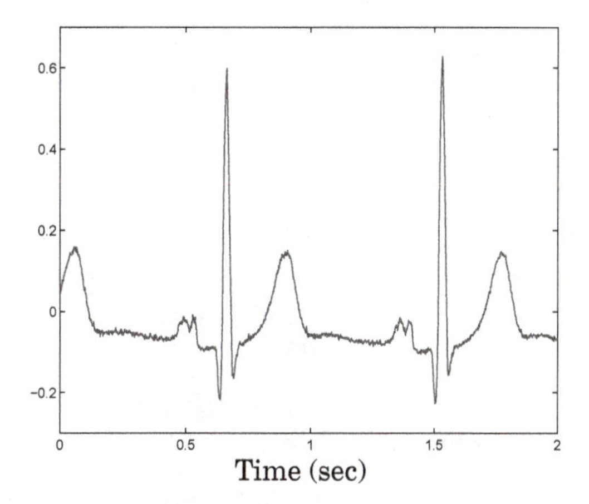

Figure 20.1 Example EKG Signal

Our data set has a total of 447 EKG records. Of these, 79 represent healthy patients, and the remaining 368 have an MI diagnosis. A diagnosis for each record was provided by a physician, but it is possible for the diagnosis to be in error. We will have more to say about this when we come to the validation of the network.

Each EKG consists of 15 leads measured at a rate of 1000 Hz for a period of several minutes. This is an enormous amount of data, and it would be impractical to use the entire EKG as an input to the neural network. As with many pattern recognition problems, we need to perform a *feature extraction* step before using the neural network to execute the pattern recognition step. Feature extraction involves mapping the high-dimensional input space into a space with fewer dimensions, in order to simplify and make more robust the pattern recognition step.

There are a number of general methods for dimensionality reduction. This includes linear methods, like the principal components method that we mentioned in Chapter 17, and nonlinear methods, like manifold learning [TeSi00]. For this case study, instead of using general methods to generate the low-dimensional feature space, we will extract features that are commonly used by physicians to detect abnormalities in the EKG. The first step is to consider a typical cycle of an EKG signal, as shown in Figure 20.2.

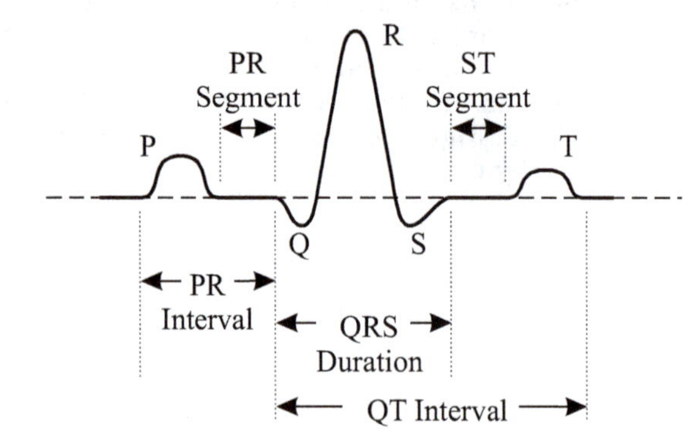

Figure 20.2 Prototype Cycle of an Electrocardiogram Signal

William Einthoven, in the early 1900's, was the first to carefully measure and analyze the EKG. He assigned the letters P, Q, R, S and T to the various deflections shown in the prototype cycle of Figure 20.2. He also described the electrocardiographic features of a number of cardiovascular disorders. He won the Nobel Prize in Medicine in 1924 for his discoveries. His features are still used to this day.

For this study, we have used some of the features that are in standard use by physicians, as well as other features that are related to the prototype cycle [Raff06]. The descriptions of the 47 features that we used are listed below. (If a description refers to the "amplitude" lead, this refers to the square root of the sum of squares of the VX, VY and VZ leads.)

Input Features to the Neural Network

1. age in years
2. gender, -1=female, 1=male
3. maximum heart rate in beats/min
4. minimum heart rate in beats/min
5. average time between heart beats in sec
6. rms deviation of the mean heart rate in beats/sec
7. full width at half maximum for the heart rate distribution
8. average qt interval for lead with max t wave
9. average qt interval for all leads
10. average corrected qt interval for lead with max t wave
11. average corrected qt interval for all leads
12. average qrs interval for all leads
13. average pr interval for lead with maximum p wave
14. rms deviation of pr intervals from average-max p lead

15. average pr interval for all leads
16. rms deviation for pr interval from average-all leads
17. percentage of negative p waves-max p lead
18. average percentage of negative p waves for all leads
19. maximum amplitude of any t wave
20. rms deviation of qt intervals
21. rms deviation of corrected qt intervals
22. average st segment length
23. rms deviation of st segment lengths
24. average heart rate in beats/min
25. rms deviation of heart rate distribution in beats/min
26. average rt angle averaged over all amplitude beats
27. number of missed r waves (beats)
28. % total qt intervals not analyzed or missing
29. % total pr intervals not analyzed or missing
30. % total st intervals not analyzed or missing
31. average number of maxima between t wave end and q
32. rms deviation of rt angle for all beats
33. ave qrs from amplitude lead
34. rms deviation of qrs from amplitude lead
35. ave st segment from amplitude lead
36. rms deviation of st segment from amplitude lead
37. ave qt interval from amplitude lead
38. rms deviation of qt interval from amplitude lead
39. ave bazetts corrected qt interval from amplitude lead
40. rms deviation of corrected qt interval from amplitude lead
41. ave r-r interval from amplitude lead
42. rms deviation of r-r interval from amplitude lead
43. average area under qrs complexes
44. average area under s-t wave end
45. average ratio of qrs area to s-t wave area
46. rms deviation of rt angle within each beat averaged over all beats in amplitude signal
47. st elevation at the start of the st interval for amplitude signal

To summarize, the data set contains 447 records. Each record has 47 input variables, and one target value. The target is 1 for a healthy diagnosis and -1 for an MI diagnosis.

One of the problems with the data set is that there are only 79 records for the healthy diagnosis, while there are 368 records for the MI diagnosis. If we train the network using the sum square error performance index, where all of the errors are weighted equally, the network will be biased to indicate the MI diagnosis. The ideal solution to this problem would be to collect more data from healthy patients. Let's say that this is not possible in this case, and we need to do what we can with the data available. One possibility is to use a weighted sum square error as the performance index, where errors for healthy patients would be weighted higher than errors for MI patients, so that overall healthy and MI contributions would be equal if each error were equal. Another simple approach is to repeat the healthy records in the data set, so that the total number of healthy records is equal to the number of MI records. This requires extra computation, but that is not a problem in this case study. Since it is the simplest solution, we will use it here.

After the data has been collected, the next step is to divide the data into training, validation and test sets. In this case, we randomly set aside 15% of the data for validation and 15% for testing. For the validation and testing sets, we did not include multiple entries of the healthy records. This was done only for the training set.

The data were normalized using Eq. (17.1), so that the inputs were in the range [-1,1]. Since the tangent-sigmoid transfer function will be used in the output layer of the neural network, the targets were set to values of -0.76 and +0.76, instead of -1 and 1, to prevent training difficulties caused by saturation of the sigmoid functions, as discussed in Chapter 17.

Selecting the Architecture

Figure 20.3 shows the network architecture. We are using the tangent-sigmoid transfer function in both layers. This is the standard network for pattern recognition. There are cases in which two hidden layers are used, but we normally try first with one hidden layer. The number of neurons in the hidden layer, S^1, will depend on the complexity of the decision boundaries needed for the pattern recognition task. This is something that cannot generally be known before training. We will start with 10 neurons in the hidden layer, and then test the network performance after training.

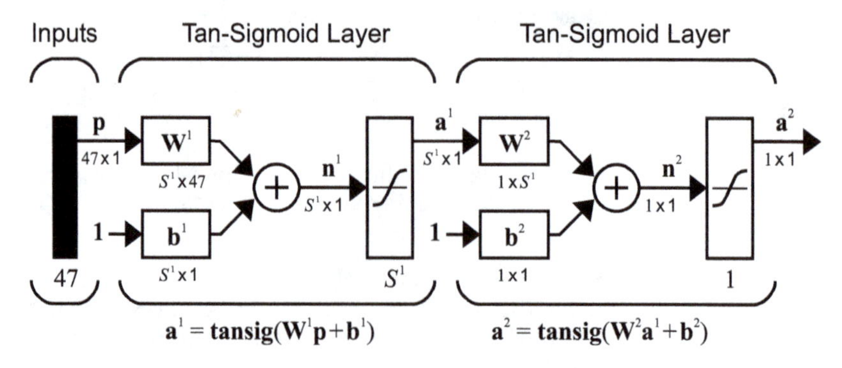

Figure 20.3 Network Architecture

Training the Network

We trained the network using the scaled conjugate gradient algorithm of [Mill93]. This algorithm is very efficient for pattern recognition problems. We used early stopping to prevent network overfitting.

Figure 20.4 illustrates the mean squared error versus iteration number. The blue line shows the validation error, and the black line shows the training error. We used a network with 10 neurons in the hidden layer ($S^1 = 10$). The minimum validation error occurred at iteration 16, as indicated by the circle in Figure 20.4, and the network parameters were saved at this point. Note that the validation error curve does not always fall at each iteration, and it may rise before falling to a lower value. We tested that the validation error was not reduced over 40 iterations before we finally stopped the training.

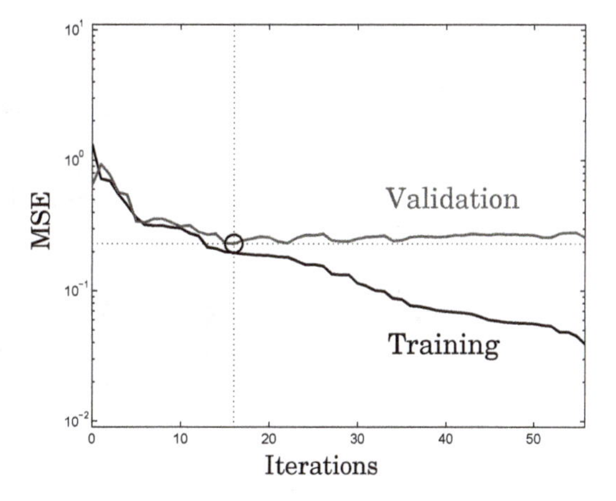

Figure 20.4 Mean Squared Error vs. Iteration Number ($S^1 = 10$)

Validation

As we discussed in previous chapters, an important tool for network validation in function approximation problems is a scatter plot of network outputs versus targets. For pattern recognition problems, the network outputs and targets are discrete variables, so a scatter plot is not particularly useful. Instead of the scatter plot, we use the confusion matrix, which was discussed in Chapter 17. Figure 20.5 shows the confusion matrix for our trained network on the test data. The upper left cell shows that 13 of the 14 healthy EKG's in the test set were classified correctly, while the 2,2 cell shows that 66 of the 71 MI EKG's were classified correctly. A total of 92.9% of the test data were classified correctly. The largest number of mistakes were for MI records that were classified as healthy (5), as shown in cell 1,2.

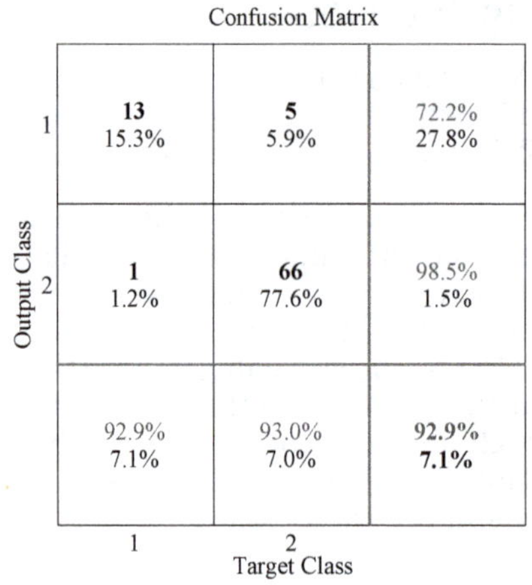

Confusion Matrix

Output Class \ Target Class	1	2	
1	**13** 15.3%	**5** 5.9%	72.2% 27.8%
2	**1** 1.2%	**66** 77.6%	98.5% 1.5%
	92.9% 7.1%	93.0% 7.0%	**92.9%** **7.1%**

Figure 20.5 Confusion Matrix for Test Data (One Data Division)

Another useful validation tool for pattern recognition problems is the Receive Operating Characteristic (ROC) curve, described in Chapter 17. Figure 20.6 shows the ROC curve (blue line) for the test data. The ideal curve would follow the path from 0,0 to 0,1 and then to 1,1. The curve for this test set is close to the ideal path.

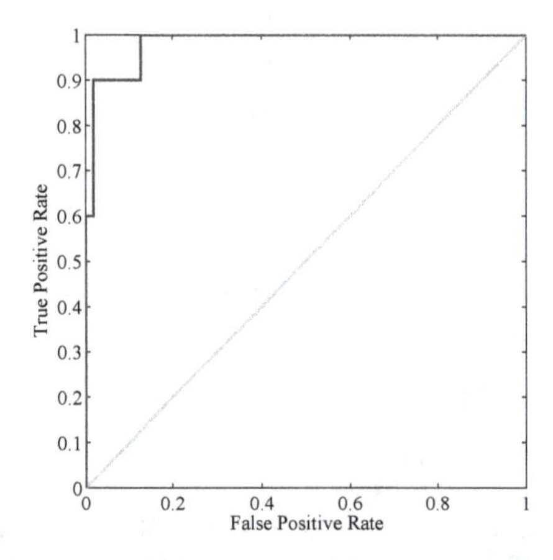

Figure 20.6 Receiver Operating Characteristic Curve (Test Set)

The results shown in Figure 20.5 and Figure 20.6 represent one division of the data into training/validation/testing sets. Because the data set is fairly small, especially in terms of healthy diagnoses, we might wonder how sensitive the results are to the data division. To investigate this sensitivity, we performed a Monte Carlo simulation. The data were divided 1,000 different times. For each division of the data, a neural network was trained with different random initial weights. The results of the 1,000 trials were averaged together, and the results are shown in Figure 20.7.

Confusion Matrix

Output Class \ Target Class	1	2	
1	**9.11** 14.0%	**3.69** 5.7%	71.2% 28.8%
2	**2.51** 3.9%	**49.83** 76.5%	95.2% 4.8%
	78.4% 21.6%	93.1% 6.9%	**90.5%** **9.5%**

Figure 20.7 Average Test Confusion Matrix for 1,000 Monte Carlo Runs

Figure 20.7 represents the average results over the 1,000 different networks and data divisions. There were approximately 12 healthy patients (on average) in each test set. Of these, more than 9 were correctly diagnosed. There were approximately 54 sick patients in each test set, and approximately 50 were correctly diagnosed. The average testing error was approximately 9.5%. Note that none of the patients in the test set were used to train the neural network, so these numbers should be conservative estimates of how the network should perform on new patients.

The average test results for the Monte Carlo simulation are similar to our original test results. However, in addition to knowing the average results, it is also helpful to look at the distribution of errors. Figure 20.8 shows a histogram of the percentage errors. The average percent error is 9.5%, but there is a significant spread in the distribution of errors. The standard deviation of the percent error is 3.5.

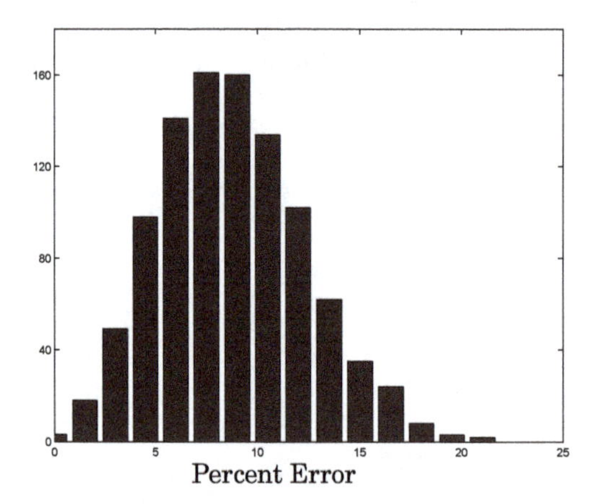

Figure 20.8 Percent Error Histogram (1000 Monte Carlo Trials)

This Monte Carlo process can be helpful in validating the data and the training process. For example, it is possible to identify which patients are misclassified in each Monte Carlo run. Those patients who are consistently misclassified (regardless of the division of the data) can be carefully investigated. These cases can be helpful in two areas. First, they can enable us to refine the data base. If, after reevaluation by clinicians, it is determined that a particular patient was mislabeled in the original data set, the data can be corrected. Second, if we find upon review that the patient was correctly labeled in the original data set, then we can use that patient to help improve the operation of the neural network classification. This may involve identifying new features which capture the key characteristics of the EKG, or it may involve obtaining more data with similar characteristics, in order to reinforce the training of the neural network.

The Monte Carlo process can be also used to help to improve the network performance. By combining the individual networks obtained from the Monte Carlo trials, we can often obtain a more accurate classification. The same input is applied to all of the networks, and the outputs can be combined through a "voting" procedure. We choose the class that is selected by the largest number of networks.

Data Sets

There are two data files associated with this case study:

- ekg_p.txt — contains the input vectors in the ekg data set
- ekg_t.txt — contains the targets in the ekg data set

They can be found with the demonstration software, which is described in Appendix C.

Epilogue

This chapter has demonstrated the use of multilayer neural networks for pattern recognition. In this case study, the pattern recognition network was used to classify EKG records into healthy and myocardial infarction diagnoses.

Most pattern recognition tasks involve a feature extraction step, in which the original data set is reduced in dimension. The features that were extracted from the EKG data consisted of characteristics of the prototype EKG cycle.

A Monte Carlo procedure was used as part of the network validation. The data were randomly divided a number of times into training/validation/test sets, and for each division a neural network was trained with random initial weights. The performances of all of the networks were analyzed to determine expected future performance. In addition, those records that were misclassified by most of the networks, regardless of the data division, were analyzed to assist in refining the data set and improving the pattern recognition.

In the next chapter, we apply neural networks to a clustering problem. We will use a self-organizing feature map network for that application.

Further Reading

[Dubi00] D. Dubin, *Rapid Interpretation of EKG's*, Sixth Edition, Tampa, FL: COVER, 2000.

This book describes the EKG in very clear terms, and leads you through the interpretation in a step-by-step way.

[MoMa01] G.B. Moody, R.G. Mark, and A.L. Goldberger, "PhysioNet: a Web-based resource for the study of physiologic signals," *IEEE Transactions on Engineering in Medicine and Biology*, vol. 20, no. 3, pp: 70-75, 2001.

This paper describes the PhysioNet data base that contains a large variety of recorded physiologic signals. The data base can be found at http://www.physionet.org/.

[Raff06] The features in the data set described in this chapter were designed and extracted by Dr. Lionel Raff, Regents Professor of Chemistry at Oklahoma State University.

[TeSi00] J. B. Tenenbaum, V. de Silva, J. C. Langford, "A Global Geometric Framework for Nonlinear Dimensionality Reduction," *Science*, vol. 290, pp. 2319-2323, 2000.

There are several different approaches to manifold learning, in which data is mapped from a high-dimensional space to a lower-dimensional manifold. This paper introduces a method called Isomap.

21 Case Study 4: Clustering

Objectives

This chapter presents a case study in using neural networks for clustering. In clustering problems, you want a neural network to group data by similarity. For example, market segmentation can be done by grouping people according to their buying patterns, data mining can be done by partitioning data into related subsets, and bioinformatic analysis can be done by grouping genes with related expression patterns.

In this chapter, we will apply clustering to a problem in forestry, in which we would like to analyze forest cover types. We will use the self-organizing feature map network of Chapter 15 to perform the clustering, and we will demonstrate a variety of visualization tools that can be used in conjunction with the SOFM.

Theory and Examples

This chapter presents a case study in using neural networks for clustering. In clustering problems, we generally don't have a set of network targets available, so clustering networks are trained by unsupervised training algorithms. Instead of training a network to produce a desired response, we want to analyze a data set to look for hidden patterns. There are many application areas for clustering. It is widely used in data mining, in which we analyze large data sets to identify similarities within subsets of the data. It is used in city planning, when town councils apportion regions of the city into areas of similar home type and land usage. It is used in image compression, in which a small set of prototype sub-images are identified and combined to represent a large collection of images. It is used in speech recognition systems, in which speakers are clustered into categories in order to simplify the problem of speaker-independent recognition. Clustering is used by marketers to identify distinct groups in their customer bases. It has also been used to organize large bibliographic data bases so that related material can be quickly accessed.

The neural network that we will use in this application is the self-organizing feature map (SOFM), which we introduced in Chapter 15. This clustering network has a unique attribute that enables us to visualize large data sets in many dimensions. We will focus on that visualization capability in this case study.

Description of the Forest Cover Problem

An important job of the forest service is to maintain accurate natural resource inventory information. One key characteristic that is recorded is the type of forest cover found in wilderness areas. This type of data can be expensive to collect, since it generally requires on-site inspection or estimation from remotely sensed data. [BlDe99] describes how forest cover type can be predicted from independent variables that can be more easily obtained. In this chapter, we will use the data described in that paper to perform a clustering analysis. We will demonstrate how an analysis of the data using an SOFM can allow us to visualize the high-dimensional space of independent variables and identify relationships between the forest cover types.

Ten independent variables that can indicate forest cover type are shown in Table 21.1. (There were 12 variables used in [BlDe99], but for ease of presentation, we selected only the first 10 for this case study.) These variables can be measured or estimated much more easily than forest cover type. We want to use the SOFM to find out whether these variables can be used to cluster the data in such a way as to separate regions with different forest cover types.

Variable Number	Description	Units
1	Elevation in meters	meters
2	Aspect in degrees azimuth	azimuth
3	Slope in degrees	degrees
4	Horz Dist to nearest surface water	meters
5	Vert Dist to nearest surface water	meters
6	Horz Dist to nearest roadway	meters
7	Hillshade index at 9am, summer solstice	0 to 255 index
8	Hillshade index at noon, summer solstice	0 to 255 index
9	Hillshade index at 3pm, summer solstice	0 to 255 index
10	Horz Dist to nearest wildfire ignition points	meters

Table 21.1 Description of Independent Variables

The forest cover types of interest in [BlDe99] are shown in Table 21.1. The data set we will use for this case study contains information on cover type, but we will not use this as part of the training process. We will use it to test the clustering ability of the SOM.

Label	Name
0	Krummholz
1	Spruce/Fir
2	Lodgepole Pine
3	Ponderosa Pine
4	Cottonwood/Willow
5	Aspen
6	Douglas-fir

Table 21.2 Forest Cover Types

Data Collection and Preprocessing

The data used in this study came originally from [HeBa99]. It contains the forest cover type for 30 x 30 meter cells obtained from US Forest Service (USFS) Region 2 Resource Information System (RIS) data. The original data set contained 581,012 observations of 12 independent variables and the forest cover type. We have used the first 20,000 observations, and we have used only the first 10 independent variables, which are described in Table 21.1. The forest cover types are given in Table 21.2. As mentioned previously, we did not use these for training.

For supervised learning, as demonstrated in the previous three chapters, after the data is collected, the next step is to divide the data into training, validation and test sets. For unsupervised learning, we don't generally divide the data in this way, because there is no need for a validation set to stop the training. Competitive training is typically performed for a fixed number of iterations. We use the entire data set for training.

The next step is to normalize the data. The data were scaled using Eq. (17.1), so that the inputs were in the range [-1,1]. (For the SOFM, data can also be scaled using Eq. (17.2), so that the input variables have a mean of 0 and a variance of 1.)

Before proceeding to train the network, it is often useful to view the input data. One convenient format for this is the scatter plot. Figure 21.1 illustrates a set of scatter plots among the input variables 7, 8 and 9. The diagonal plots in this figure are histograms for these three input variables, and the off-diagonal plots are the scatter plots. (We only show three of the variables in this figure because of the limits of the page size.)

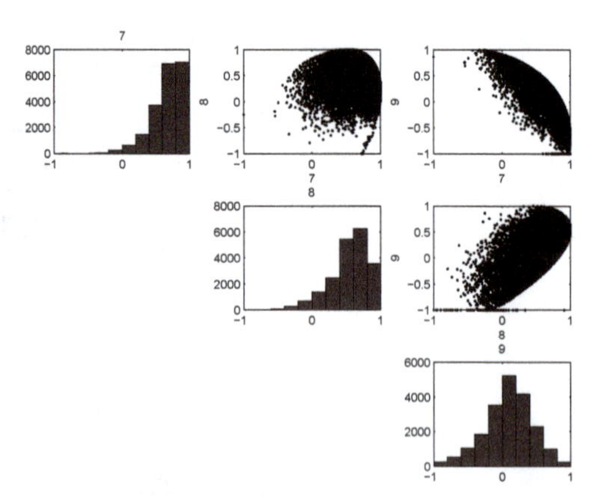

Figure 21.1 Scatter Plots for Input Variables 7, 8, 9.

There are several things we can look for in Figure 21.1. First, we want to see how well the data is scattered throughout the range. If some variables show little or no variation, then we would remove them from the analysis. We also look for correlation between the variables. For example, if the points in the scatter plot fell exactly along a line, then we would know that the two variables were linearly related. There would be no need to use both variables in the analysis. From Figure 21.1 we can see that there is some correlation between the variables, but they are not linearly dependent.

Selecting the Architecture

We will use the SOFM network, described in Chapter 15, to perform the clustering for this case study. The specific architecture is often selected based, in part, on the number of data points, so that there will be a reasonable amount of data associated with each prototype vector. (Recall that each row of the weight matrix represents a prototype vector. An input is associated with the prototype vector to which it is closest.) As the data set size increases, the number of neurons should increase as well, although not as rapidly. A rule of thumb is to have the number of neurons increase as the square root of the number of data points.

Figure 21.2 shows the architecture of the network we selected. We have 10 input variables (defined in Table 21.1), and we are using 150 neurons. The feature map is 15x10, and it uses an hexagonal arrangement of neurons. This means that each internal neuron will have six neighbors.

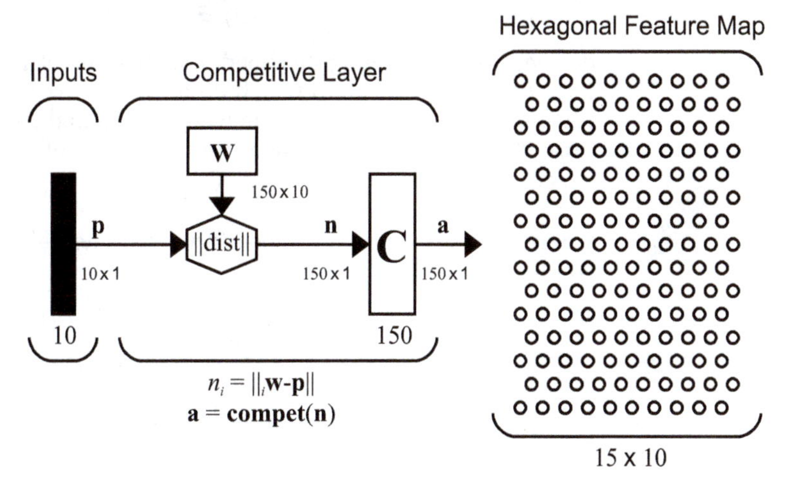

Figure 21.2 SOM Network Architecture

After the network has been trained, we will analyze the results to determine if the network architecture is satisfactory. In practical cases, we often try several different architectures. Unlike with supervised training, in which we have a clear performance measure (normally sum square error), there is no firm criterion for best performance in SOFM networks. Often what we are looking for is insight into the data set. It is somewhat of an art to selecting the best architecture and training regime for SOFM networks. This will become more clear when we analyze the results of the trained network in a later section.

Training the Network

Before beginning the training, the weight vectors (rows of **W**) were initialized using what is called *linear initialization* [Koho95]. First, a covariance matrix of the input vectors was computed. Then, the two eigenvectors of this matrix having the two largest eigenvalues were found. The rows of **W** were then assigned by taking the average of the input vectors and adding linear combinations of the two eigenvectors. This places all of the initial weight vectors in the space spanned by the two eigenvectors. This initialization process produces quicker training convergence than when using a purely random weight initialization. (It is also possible to randomly select input vectors from the training set to be the initial weight vectors.)

Recall from Eq. (15.21) the SOFM learning rule, which we repeat here:

$$\begin{aligned} {}_i\mathbf{w}(q) &= {}_i\mathbf{w}(q-1) + \alpha(\mathbf{p}(q) - {}_i\mathbf{w}(q-1)) \\ &= (1-\alpha){}_i\mathbf{w}(q-1) + \alpha\mathbf{p}(q) \end{aligned} \qquad i \in N_{i^*}(d) \qquad (21.1)$$

where i^* is the index of the winning neuron, and

$$N_i(d) = \{j, d_{ij} \le d\} \qquad (21.2)$$

defines the neuron neighborhood. For this case study, we have used a batch form of the algorithm, in which all of the inputs in the training set are applied to the network before the weights are updated. To develop the batch form, we can first modify the sequential form of Eq. (21.1) to

$${}_i\mathbf{w}(q) = {}_i\mathbf{w}(q-1) + h_{i^*,i}(\mathbf{p}(q) - {}_i\mathbf{w}(q-1)), \qquad (21.3)$$

where $h_{i^*,i}$ is the neighborhood function. The neighborhood function that would produce Eq. (21.1) is

$$h_{i^*,i} = \begin{cases} \alpha & i \in N_{i^*}(d) \\ 0 & i \notin N_{i^*}(d) \end{cases} \qquad (21.4)$$

Using this definition of neighborhood function, we can define a batch version of Eq. (21.1):

$${}_i\mathbf{w}(k) = \frac{\sum_{q=1}^{Q} h_{i^*(q),i}\mathbf{p}(q)}{\sum_{q=1}^{Q} h_{i^*(q),i}}, \qquad (21.5)$$

where k is the iteration number and $i^*(q)$ is the winning neuron for input $\mathbf{p}(q)$. Note that for the batch algorithm we have to distinguish between the iteration number and the input number, since all inputs are applied to the network at each iteration. This is in contrast to the sequential algorithm of Eq. (21.1), where there is one iteration for each input. Also, notice that the learning rate does not affect the batch algorithm, since it would appear in both the numerator and the denominator of Eq. (21.5).

For the neighborhood function of Eq. (21.4), this batch algorithm has the effect of assigning each weight to the average of the input vectors for which it is in the neighborhood of the winner. As with the sequential algorithm, the neighborhood size is decreased during training. The neighborhood size is set large at the beginning of training until all weights move into the region of the input space where the data lies. Then the neighborhood size is reduced, to fine-tune the position of the weights.

The batch algorithm requires many fewer iterations than the sequential algorithm, although each iteration requires much more computation. For this case study, we used two iterations of the batch algorithm. During the first iteration the neighborhood size was 4, and during the second iteration the neighborhood size was reduced to 1.

Validation

We will consider two numerical measures of the quality of a trained SOM: resolution and topology preservation (see page 22-23). One measure of SOM resolution is the *quantization error*, which is the average distance between each data vector and its winning neuron. If the average distance is too large, then there are many input vectors that are not adequately represented by any of the prototypes.

A measure of SOM topology preservation is the *topographic error*. This is the proportion of all input vectors for which the closest (winning) neuron and the next closest neuron are not adjacent to each other in the feature map topology. When this number is small, it means that the neurons that are neighbors in the topology are also neighbors in the input space. It is important that this topology be preserved, so that the visualization tools we will discuss later can provide valid insight into the data set.

For our trained SOM, the final quantization error was 0.535, and the final topographic error was 0.037. This means that for less than 4% of all input vectors, the winning neuron and the next closest neuron were not adjacent to each other. It appears that the SOM has achieved the correct topology by the completion of the training.

There are a number of visualization methods that can be used to assess the trained SOM network. One of the key tools is called the unified distance matrix, or *u-matrix*. This is a figure that shows the distance between neighboring neurons in the feature map. The u-matrix has a cell for each neuron in the feature map and an additional cell between each pair of neurons. The cells between neurons are color-coded with the distance between the corresponding weight vectors. The cells that represent the neurons are coded with the mean of the surrounding values. Figure 21.3 shows the u-matrix for our trained SOM.

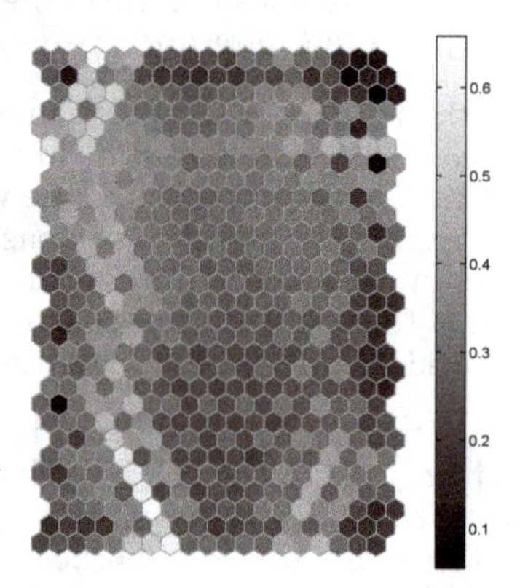

Figure 21.3 U-Matrix for Trained SOM

In Figure 21.3, the light-colored cells represent large distances between neurons. We can see that there is a string of light colored cells on the left side of the feature map. This indicates that the clusters associated with the neurons on the left side of the map are significantly different than those in the middle and right sides of the map. For this data set we actually know the forest cover types for each data point. We can label the feature map cells with the cover type that is associated with the closest input vector to that cluster center. The resulting labeled map is shown in Figure 21.4.

By comparing Figure 21.4 with Figure 21.3, we can see that the forest cover type 2 (see Table 21.2) is associated with the left edge of the feature map. As we move from left to right across the map, we see type 0 and 1 coded into the center section, followed by types 5, 3 and 4, with type 6 located mainly in the upper right section of the map. It is clear that the SOM has learned to cluster the data according to forest cover type.

To get more insight into how the SOM has clustered the data, we can produce a “hit histogram.” For this graph, we count how many times each neuron was the winning neuron for the entire data set. Since our data is labeled with forest cover type, we can also see where each type falls on the feature map. Such a graph is displayed in Figure 21.5. In each cell you can see a hexagram with a certain gray-scale. The size of the hexagrams indicate how many times the corresponding neuron was the winning neuron. The gray level of the hexagram indicates the forest cover type. The darkest hexagons correspond to type 0 forest covers, and the lightest hexagons correspond to type 6 forest covers. We can see that the various regions of the map have consistent colors. The left side has medium gray levels, corresponding to type 2 cover. The darkest levels are in the center-left region of the map, which corresponds to cover types 0 and 1. The lighter levels, which correspond to types 5 and 6, are in the center-right region, and the median levels of gray, corresponding to types 3 and 4, are on the right edge.

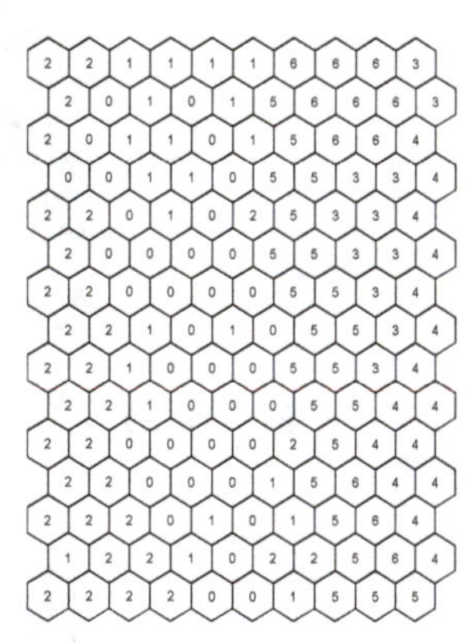

Figure 21.4 Labeled SOM

For many problems, we would not be able to label each input vector. The point here is that the SOM has been able to cluster the data into similar cover types, without knowing what the actual cover types were. This means that the 10 variables making up the input vectors have enough correlation with cover type to allow the SOM to make a useful clustering of the data.

Another tool that is useful in analyzing the trained SOM is the component plane. A component plane is a figure that represents a column of the weight matrix of the SOM. Each column corresponds to one element of the input vector; the jth element of column i represents the connection from input i to neuron j. In a component plane, each element of the weight is represented by a cell in the feature map topology at the location of the neuron to which it is connected. The gray level of the cell represents the magnitude of that element of the weight vector.

The ten component planes (one for each column of the weight matrix - each element of the input vector) for the trained SOM are shown in Figure 21.6. The first thing that we notice is that each of the columns is distinct. There are no two columns that have the same pattern. We can also see that input variables 1, 4, 5, 6 and 10 seem to be important in separating type 2 cover types from the rest of the data. They show patterns in which a boundary appears on the left edge of the feature map, where the type 2 cover types are clustered. By going back to Table 21.1, we can then locate the appropriate variables to see if we can deduce their connection to type 2 cover.

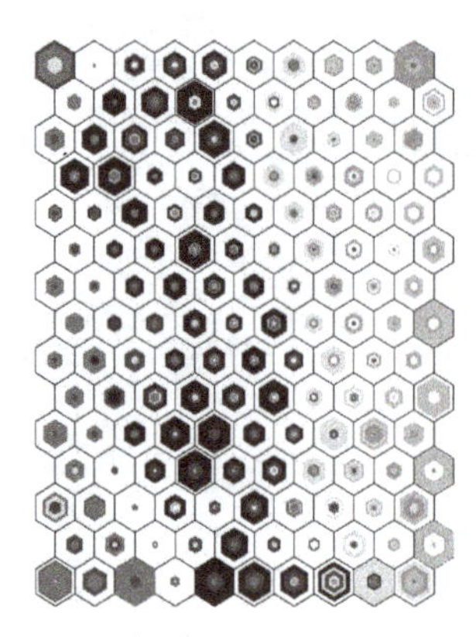

Figure 21.5 Hit Histogram for Trained SOM

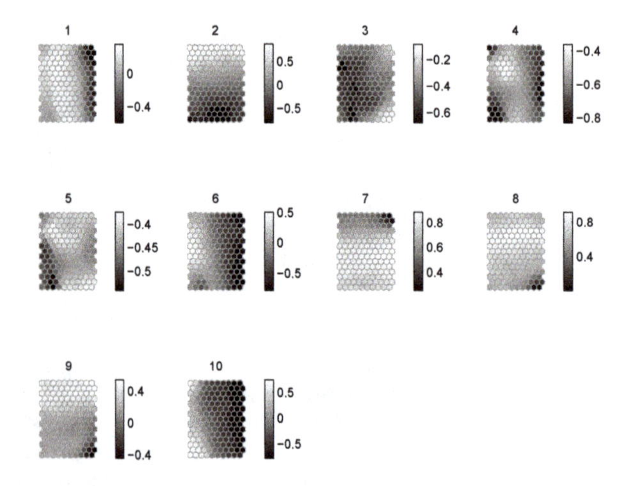

Figure 21.6 Component Planes for the Trained SOM

Data Sets

There are two data files associated with this case study:

- cover_p.txt — contains the input vectors in the data set
- cover_t.txt — contains the targets (labels) in the data set

They can be found with the demonstration software, which is described in Appendix C.

Epilogue

This chapter has demonstrated the use of SOM networks for clustering, in which input vectors in a data set are arranged so that similar vectors are placed in the same cluster. In this case study, the SOM was used to cluster forestry data. The idea was to cluster land into similar forest cover types.

One of the principal advantages of the SOM network, in addition to its ability to efficiently cluster a data set, is its ability to enable visualization of high dimensional data sets.

In the next chapter, we apply neural networks to a prediction problem. We will use a Nonlinear Autoregressive model with eXogenous inputs (NARX) network for that application.

Further Reading

[BlDe99] J. A. Blackard and D. J. Dean, "Comparative Accuracies of Artificial Neural Networks and Discriminant Analysis in Predicting Forest Cover Types from Cartographic Variables," Computers and Electronics in Agriculture, vol. 24, pp. 131-151, 1999.

This study compared neural networks and discriminant analysis for predicting forest cover types from cartographic variables. The study evaluated four wilderness areas in the Roosevelt National Forest, located in the Front Range of northern Colorado.

[HeBa99] S. Hettich and S. D. Bay, *The UCI KDD Archive* [http://kdd.ics.uci.edu], Irvine, CA: University of California, Department of Information and Computer Science, 1999.

The UCI Knowledge Discovery in Databases Archive. This is an online repository of large data sets which encompasses a wide variety of data types, analysis tasks, and application areas. It is maintained by the University of California, Irvine.

[Koho93] T. Kohonen, "Things you haven't heard about the Self-Organizing Map," *Proceedings of the International Conference on Neural Networks* (ICNN), San Francisco, pp. 1147-1156, 1993.

This paper describes the batch form of the SOM learning rule, as well as other variations on the SOM.

[Koho95] T. Kohonen, Self-Organizing Map, 2nd ed., Springer-Verlag, Berlin, 1995.

This text describes the theory and practical operation of the Self-Organizing Map in detail. It also has a chapter on the Learning Vector Quantization algorithms.

22 Case Study 5: Prediction

Objectives

This chapter presents a case study in using neural networks for prediction. Prediction is a kind of dynamic filtering, in which past values of one or more time series are used to predict future values. Dynamic networks, such as those described in Chapter 10 and Chapter 14, are used for filtering and prediction. Unlike the previous case studies, the input to these dynamic networks is a time sequence.

There are many applications for prediction. For example, a financial analyst might want to predict the future value of a stock, bond, or other financial instrument. An engineer might want to predict the impending failure of a jet engine. Predictive models are also used for system identification (or dynamic modeling), in which we build dynamic models of physical systems. These dynamic models are important for analysis, simulation, monitoring and control of a variety of systems, including manufacturing systems, chemical processes, robotics and aerospace systems. In this chapter we will demonstrate the development of predictive models for a magnetic levitation system.

Theory and Examples

This chapter presents a case study in using neural networks for prediction. In this case study, the predictor neural network is used to model a dynamic system. This dynamic modeling from data is referred to as system identification. System identification can be applied to a variety of systems: economic, aerospace, biological, transportation, communication, manufacturing, chemical process, etc. For this case study, we will consider a simple magnetic levitation system. Magnetic levitation has been used for many years in transportation systems. In our simple maglev system, we will suspend a magnet above an electromagnet. A maglev train works on a similar principle.

Description of the Magnetic Levitation System

The objective of this maglev system is to control the position of a magnet suspended above an electromagnet, where the magnet is constrained so that it can only move in the vertical direction, as shown in Figure 22.1.

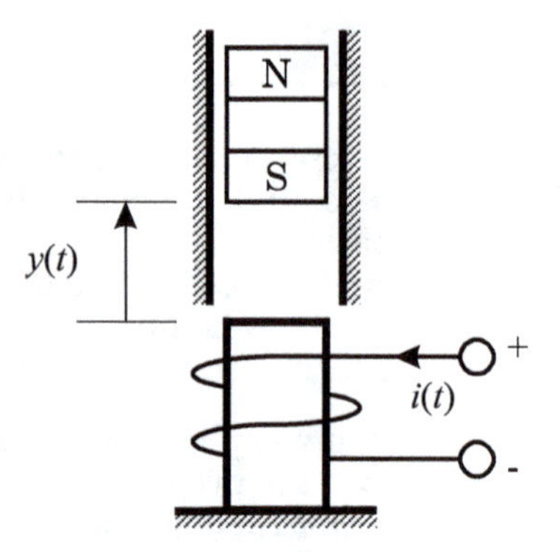

Figure 22.1 Magnetic Levitation System

The equation of motion for this system is

$$\frac{d^2y(t)}{dt^2} = -g + \frac{\alpha}{M}\frac{i^2(t)\operatorname{sgn}[i(t)]}{y(t)} - \frac{\beta}{M}\frac{dy(t)}{dt} \tag{22.1}$$

where $y(t)$ is the distance of the magnet above the electromagnet, $i(t)$ is the current flowing in the electromagnet, M is the mass of the magnet, and g is the gravitational constant. The parameter β is a viscous friction coefficient that is determined by the material in which the magnet moves, and α is a field strength constant that is determined by the number of turns of wire on the electromagnet and the strength of the magnet. For our case study, the parameter values are set to $\beta = 12$, $\alpha = 15$, $g = 9.8$, $M = 3$.

The objective of the case study will be to develop a dynamic neural network model that can predict the next value of the magnet position, based on previous values of the magnet position and the input current. Once the model has been developed, it can be used to find a controller that can determine the correct current to apply to the electromagnet, so as to move the magnet to some desired position. We won't go in to the control design in this case study, but the reader is referred to [HaDe02] and [NaMu97].

Data Collection and Preprocessing

For this case study, we did not build the maglev system of Figure 22.1. Instead, we created a computer simulation to implement Eq. (22.1). We used Simulink® to implement the simulation, but any simulation tool could be used. For our simulations, the current was allowed to range from -1 to 4 amps. The data were collected every 0.01 seconds.

To develop an accurate model, we need to be sure that the system inputs and outputs cover the operating range for which the system will be used. For system identification problems, we often collect training data while applying random inputs which consist of a series of pulses of random amplitude and duration. (This form of input is sometimes called a skyline function, because of its resemblance to a city skyline.) The duration and amplitude of the pulses must be chosen carefully to produce accurate identification. Figure 22.2 shows the input current, and the corresponding magnet position for our data set. A total of 4000 data points were collected.

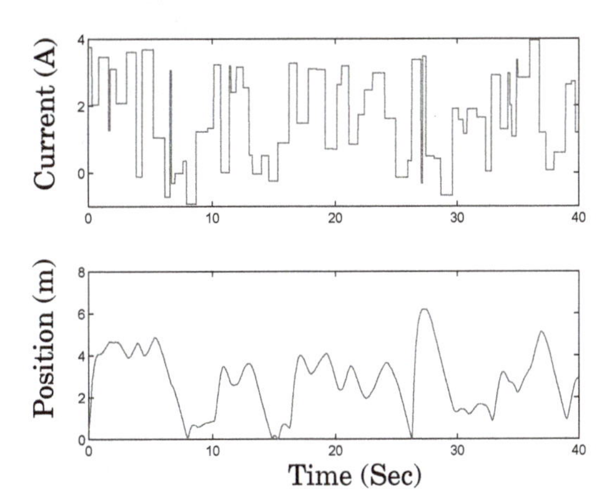

Figure 22.2 Magnetic Levitation Data

The skyline form of input function has the advantage that it can explore both transient and steady state operation of the system. Because some of the pulses are long, the system will approach steady state at the end of those pulses. Shorter width pulses explore transient operation of the system.

When steady state performance is poor, it is useful to increase the duration of the input pulses. Unfortunately, within a training data set, if we have too much data in steady state conditions, the training data may not be representative of typical plant behavior. This is due to the fact that the input and output signals do not adequately cover the region that is going to be controlled. This will result in poor transient performance. We need to choose the training data so that we produce adequate transient and steady state performance. This can be done by using an input sequence with a range of pulse widths and amplitudes.

After the data has been collected, the next step is to divide the data into training, validation and test sets. In this case, because we will be using the Bayesian regularization training technique, we do not need to have a validation set. We did set aside 15% of the data for testing purposes. When the input is a time sequence, it is useful to have the testing sequence consist of a contiguous segment of the original data set. For our tests, we used the last 15% of the data as the testing set.

The data were scaled using Eq. (17.1), so that both the inputs and the targets were in the range [-1,1]. The resulting scaled data is shown in Figure 22.3

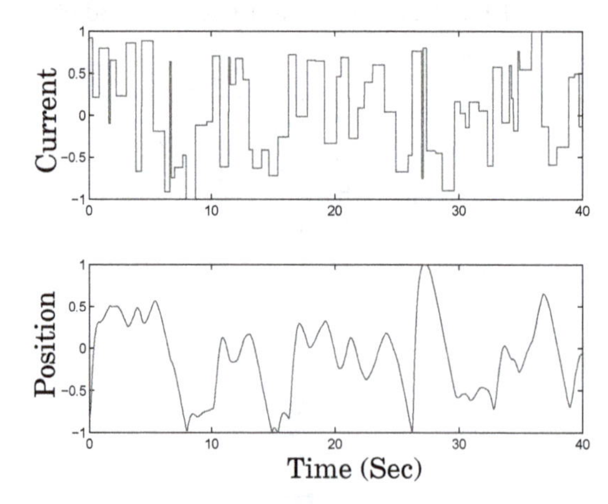

Figure 22.3 Scaled Data

Selecting the Architecture

There are many dynamic network architectures that can be used for prediction. A popular architecture is the nonlinear autoregressive network with exogenous inputs (NARX) network, which was discussed in Chapter 17. The NARX network is a recurrent dynamic network, with feedback connections enclosing several static layers of the network. The NARX model is based on the linear ARX model, which is commonly used in time series modeling.

The defining equation for the NARX model is

$$y(t) = f(y(t-1), y(t-2), \ldots, y(t-n_y), u(t-1), u(t-2), \ldots, u(t-n_u)), \quad (22.2)$$

where the next value of the dependent output signal $y(t)$ is regressed on previous values of the output signal and previous values of an independent (exogenous) input signal $u(t)$. (For our application, $y(t)$ is the position of the magnet, and $u(t)$ is the current going into the electromagnet.) We can implement the NARX model by using a feedforward neural network to approximate the function $f(\ )$. A diagram of the resulting network is shown in the Figure 22.4, where a two-layer feedforward network is used for the approximation. The output of the last layer of the network is the prediction of the next value of the magnet position. The network input is the current into the electromagnet.

We are using the tan-sigmoid transfer function in the hidden layer, and a linear output layer. As with the standard multilayer network, the number of neurons in the hidden layer, S^1, will depend on the complexity of the system being approximated. We will discuss this choice in the next section.

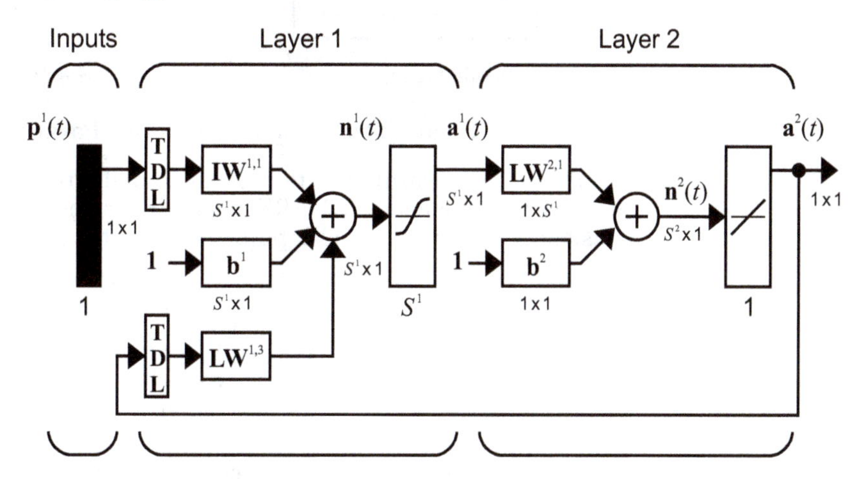

Figure 22.4 NARX Network Architecture

To define the architecture, we also need to set the length of the tapped-delay lines. The TDL for the inputs will contain the variables $u(t-1), \ldots, u(t-n_u)$, and the TDL for the outputs will contain the variables $y(t-1), \ldots, y(t-n_y)$. The TDL lengths n_u and n_y need to be defined. Because the defining differential equation in Eq. (22.1) is second order, we will start with $n_y = n_u = 2$. Later, we will investigate other possibilities.

Before demonstrating the training of the NARX network, we need to present an important configuration that is useful in training. We can consider the output of the NARX network to be an estimate of the output of the nonlinear dynamic system that we are trying to model. The output is fed back to the input of the feedforward neural network, as part of the standard NARX architecture, as shown on the left side of Figure 22.5. Since the true output is available during the training of the network, we could create a series-parallel architecture (see [NaPa90]), in which the true output is used instead of feeding back the estimated output, as shown on the right side of Figure 22.5. This has two advantages. The first is that the input to the feedforward network will be more accurate. The second is that the resulting network has a purely feedforward architecture, and static backpropagation can be used for training.

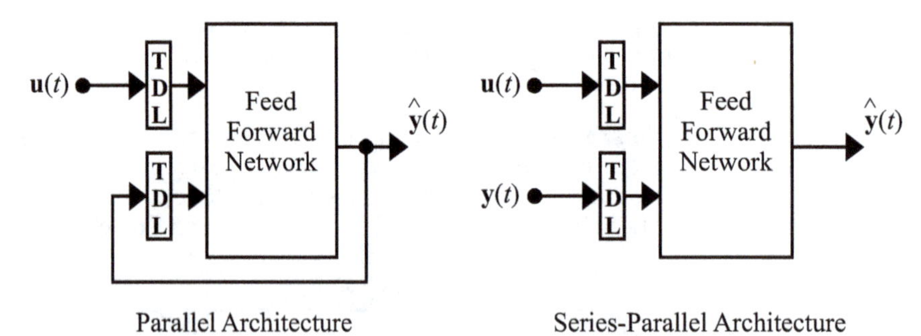

Figure 22.5 Parallel and Series-Parallel Forms

Using the series-parallel form, we can actually use a standard multilayer network to implement the NARX model. We can create an input vector that consists of previous system inputs and outputs:

$$\mathbf{p} = \begin{bmatrix} u(t-1) \\ u(t-2) \\ y(t-1) \\ y(t-2) \end{bmatrix} . \qquad (22.3)$$

The target is then the next value of the output:

$$\mathbf{t} = \left[y(t) \right] . \qquad (22.4)$$

Training the Network

We used the Bayesian regularization training algorithm, described in Chapter 13, to train the NARX network, after the weights were initialized using the Widrow/Nguyen method (see page 17-13). The prediction problem is similar to the function approximation problem, which was demonstrated in Chapter 18, and the Bayesian regularization method is effective for both applications.

Because we have 4000 data points, and the number of network weights and biases will be less than 100 (as we will see later), the chances of overfitting are very small. We do not need to use Bayesian regularization (or early stopping) in this case. However, because it can tell us the effective number of parameters, we like to use it whenever it is appropriate.

Figure 22.6 illustrates the sum square error versus iteration number, while using Bayesian regularization. We used a network with 10 neurons in the hidden layer ($S^1 = 10$) for this case. The network was trained for 1000 iterations, at which time the performance was changing very little. Several different networks were trained with different initial conditions, and the final SSE was similar for each, so we can be confident that we did not reach a local minimum.

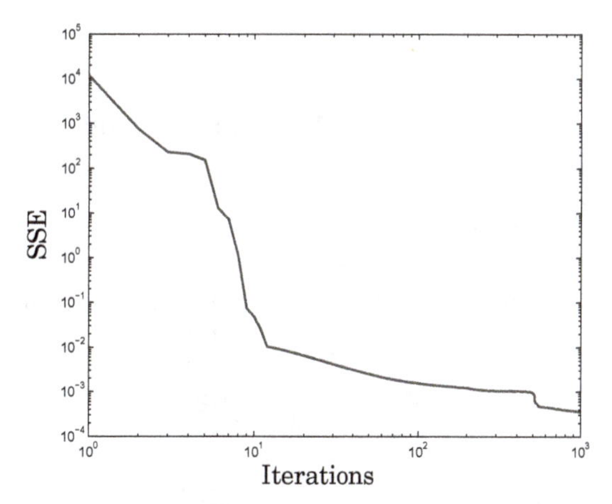

Figure 22.6 Sum Squared Error vs. Iteration Number ($S^1 = 10$)

In Figure 22.7, we can see the variation of the effective number of parameters γ during training. It eventually converges to 39. There are a total of 61 parameters in this 4-10-1 network, so we are effectively using less than 2/3 of the weights and biases. If the effective number of parameters was close to the total number of parameters, then we would increase the number of hidden neurons and retrain the network. That is not the case here.

There is no need to decrease the number of neurons, since the network computation time is not critical for this application. The only other reason for reducing the number of neurons would be to prevent overfitting. In terms of preventing overfitting, having 39 effective parameters is equivalent to having 39 total parameters. That is the beauty of using the Bayesian regularization technique. This method chooses the correct number of parameters for each problem, as long as we have a sufficient number of potential parameters in the network.

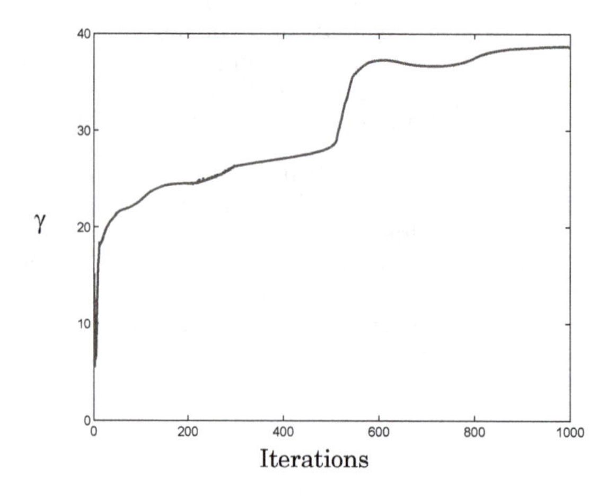

Figure 22.7 Effective Number of Parameters ($S^1 = 10$)

Validation

As we discussed in previous chapters, an important tool for network validation is a scatter plot of network outputs versus targets, as shown in Figure 22.8 (in normalized units). The figure on the left shows the training data, while the figure on the right shows the testing data. Because the testing data fit is as good as the training data fit, we can be confident that the network did not overfit.

For prediction problems, there is another set of tools for model validation. These tools are based on two basic properties of accurate prediction models. The first property is that the prediction errors,

$$e(t) = y(t) - \hat{y}(t) = y(t) - a^2(t), \tag{22.5}$$

should be uncorrelated with each other from one time step to the next. The second property is that prediction errors should be uncorrelated with the input sequence $u(t)$. (See [BoJe86].)

If the prediction errors were correlated with each other, then we could use that correlation to improve the predictions. For example, if prediction errors one time step apart had a positive correlation, then a large positive prediction error at the current time point would suggest that the prediction error at the next time point would also be positive. By lowering our next prediction, we could then reduce the next prediction error.

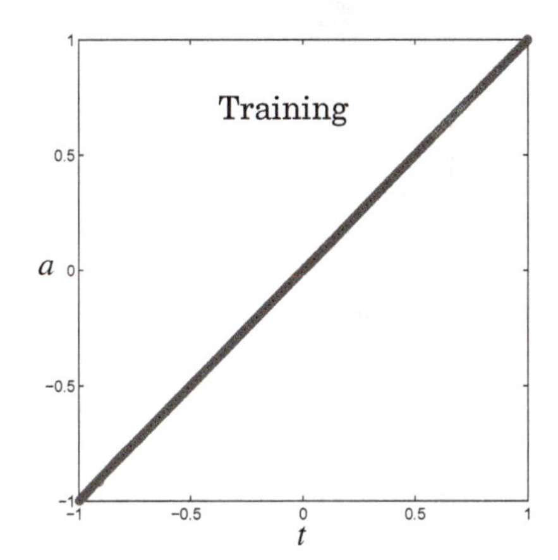

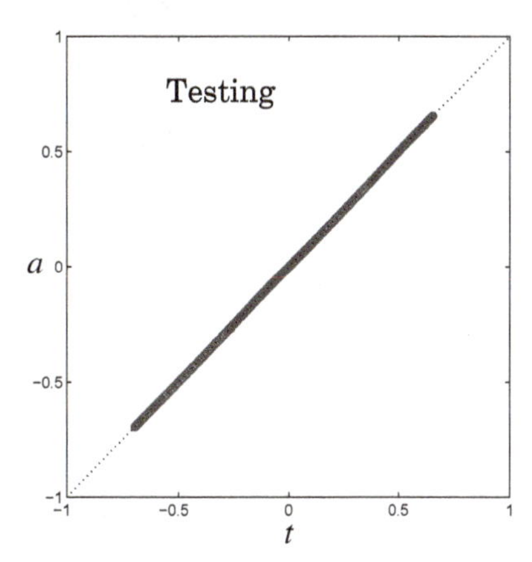

Figure 22.8 Scatter Plots of Network Outputs vs. Targets - Training and Testing Sets

The same argument holds for correlation between the input sequence and the prediction error. For accurate prediction models, there should be no correlation between the input and the prediction error. If there was correlation, then we could use this correlation to improve the predictor.

To measure the correlation in a time sequence, we use the autocorrelation function, which can be estimated by

$$R_e(\tau) = \frac{1}{Q-\tau} \sum_{t=1}^{Q-\tau} e(t)e(t+\tau) . \tag{22.6}$$

The autocorrelation function of the prediction error (in normalized units) of our trained network for the maglev problem is shown in Figure 22.9.

For the prediction error to be uncorrelated (termed "white" noise), the autocorrelation function should be an impulse at $\tau = 0$, with all other values equal to zero. Because Eq. (22.6) provides only an estimate of the true autocorrelation function, the values at $\tau \neq 0$ will never be exactly equal to zero. If the error sequence is white noise, then we can find confidence intervals for the $R_e(\tau)$ (see [BoJe86]) defined by

$$\pm 2 \frac{R_e(0)}{\sqrt{Q}} . \qquad (22.7)$$

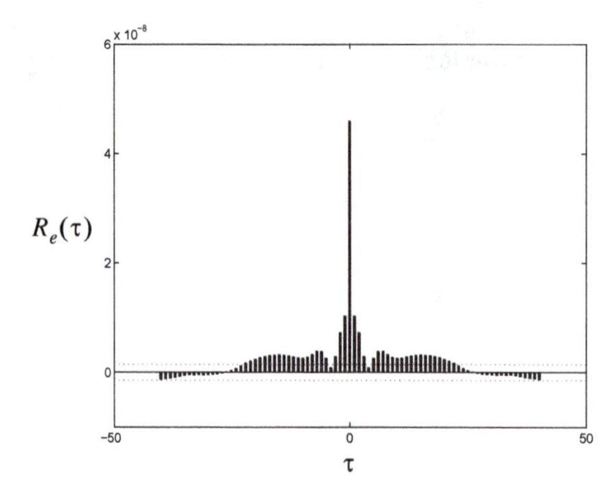

Figure 22.9 $R_e(\tau)$ ($n_y = n_u = 2$, $S^1 = 10$)

The dashed blue lines in Figure 22.9 indicate these confidence bounds. We can see that the estimated autocorrelation function for the prediction errors falls outside these bounds at a number of points. This indicates that we may need to increase n_y and n_u.

To measure correlation between the input sequence $u(t)$ and the prediction error $e(t)$, we use the cross-correlation function, which can be estimated by

$$R_{ue}(\tau) = \frac{1}{Q-\tau} \sum_{t=1}^{Q-\tau} u(t)e(t+\tau) . \qquad (22.8)$$

The cross-correlation function $R_{ue}(\tau)$ (in normalized units) of our trained network for the maglev problem is shown in Figure 22.9.

As with the estimated autocorrelation function, we can define confidence intervals to determine if the cross-correlation function is near zero

$$\pm 2 \frac{\sqrt{R_e(0)}\sqrt{R_u(0)}}{\sqrt{Q}} . \qquad (22.9)$$

The dashed blue lines in Figure 22.10 represent these confidence bounds. The cross-correlation function remains within these bounds, so it does not indicate a problem.

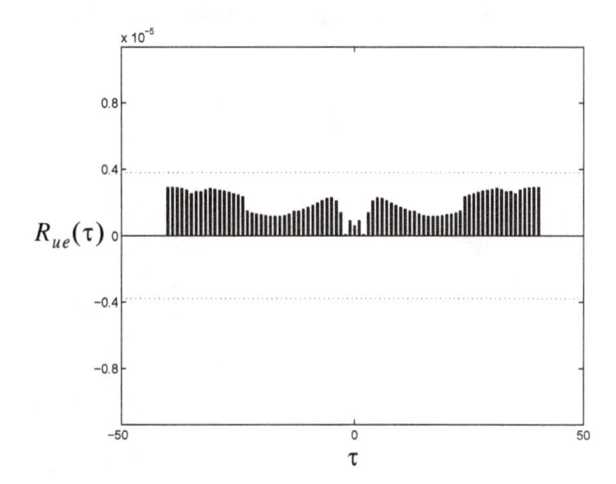

Figure 22.10 $R_{ue}(\tau)$ ($n_y = n_u = 2$, $S^1 = 10$)

Because the autocorrelation function of the prediction errors in Figure 22.9 indicated correlation in the errors, we increased the n_y and n_u values from 2 to 4 and retrained our neural network predictor. The resulting estimated autocorrelation function is shown in Figure 22.11. Here we can see that $R_e(\tau)$ falls within the confidence bounds, except at $\tau = 0$, which indicates that our model is performing correctly.

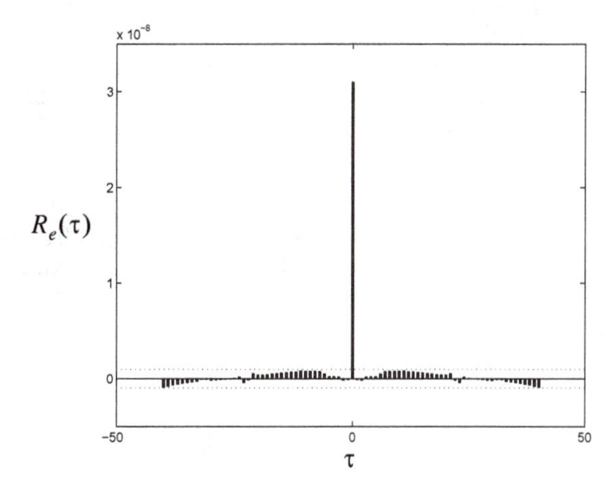

Figure 22.11 $R_e(\tau)$ ($n_y = n_u = 4$, $S^1 = 10$)

The estimated cross-correlation function, with the increased delay order, is shown in Figure 22.12. All of the points are well within the zero confidence interval. There is no significant correlation between the errors and the input.

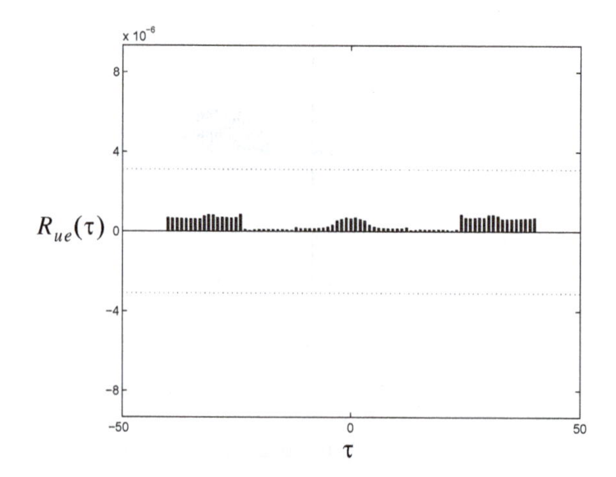

Figure 22.12 $R_{ue}(\tau)$ $(n_y = n_u = 4, S^1 = 10)$

With $n_y = n_u = 4$, we have white prediction errors, and there is no significant correlation between the prediction errors and the model input. It appears that we have an accurate prediction model.

The errors for the final prediction model are shown in Figure 22.13. We can see that the errors are very small. However, because of the series-parallel configuration, these are errors for only a one-step-ahead prediction. A more stringent test would be to rearrange the network into the original parallel form and then to perform an iterated prediction over many time steps. We will now demonstrate the parallel operation.

Figure 22.14 illustrates the iterated prediction. The solid line is the actual position of the magnet, and the dashed line is the position predicted by the NARX neural network. The network prediction is very accurate - even 600 time steps ahead.

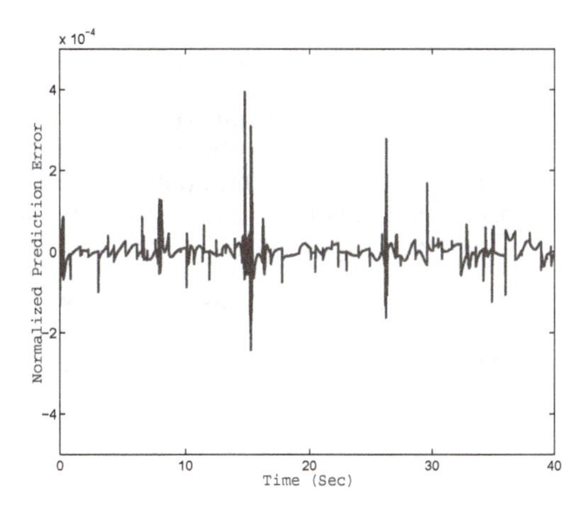

Figure 22.13 Prediction Errors vs. Time

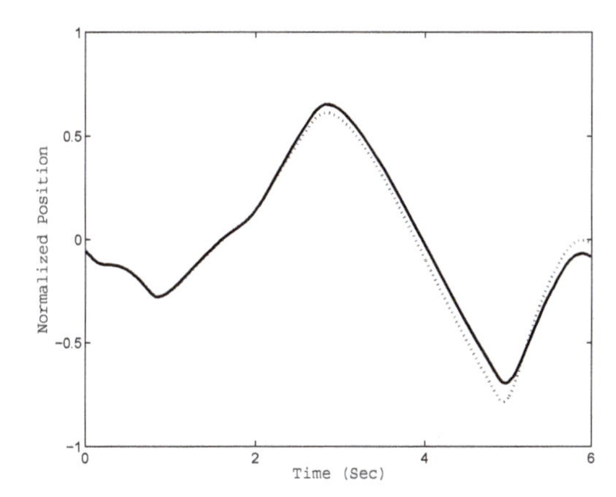

Figure 22.14 Iterated Prediction for the Maglev NARX Network

Data Sets

There are two data files associated with this case study:

- maglev_u.txt — contains the input sequence in the original data set
- maglev_y.txt — contains the output sequence in the original data set

They can be found with the demonstration software, which is described in Appendix C.

Epilogue

This chapter has demonstrated the use of multilayer neural networks for prediction, in which a future value of a time series is predicted from past values of that series and potentially other series. In this case study, the prediction network was used as a model of a magnetic levitation system. This modeling of dynamic systems is referred to as system identification.

A Nonlinear Autoregressive model with eXogenous inputs (NARX) network is well suited to this problem, and Bayesian regularization is an excellent training algorithm to use in this situation.

Further Reading

[BoJe94] G. E. P. Box, G. M. Jenkins, and G. C. Reinsel, *Time Series Analysis: Forecasting and Control*, Fourth Edition, Wiley, 2008.

A classic textbook on time series analysis and the development of prediction models.

[HaDe02] M. Hagan, H. Demuth, O. De Jesus, "An Introduction to the Use of Neural Networks in Control Systems," *International Journal of Robust and Nonlinear Control*, vol. 12, no. 11, pp. 959-985, 2002.

This survey paper describes some practical aspects of using neural networks for control systems. Three neural network controllers are demonstrated: model predictive control, NARMA-L2 control, and model reference control.

[NaMu97] Narendra, K.S.; Mukhopadhyay, S., "Adaptive control using neural networks and approximate models," *IEEE Transactions on Neural Networks*, vol. 8, no. 3, pp. 475 - 485, 1997.

This paper introduced the NARMA-L2 model and controller. Once the NARMA-L2 model is trained, it can be easily inverted to form a controller for the identified system.

[NaPa90] K. S. Narendra and K. Parthasarathy, "Identification and control of dynamical systems using neural networks," *IEEE Transactions on Neural Networks*, vol. 1, no. 1, pp. 4–27, 1990.

Classic early paper on the use of neural networks for the identification and control of dynamical systems.

A Bibliography

[AgSa05] P.M. Agrawal, A.N.A. Samadh, L.M. Raff, M. Hagan, S. T. Bukkapatnam, and R. Komanduri, "Prediction of molecular-dynamics simulation results using feedforward neural networks: Reaction of a C2 dimer with an activated diamond (100) surface," *The Journal of Chemical Physics* 123, 224711, 2005. (Chapter 19)

[Albe72] A. Albert, *Regression and the Moore-Penrose Pseudoinverse*, New York: Academic Press, 1972. (Chapter 7)

[AmMu97] S. Amari, N. Murata, K.-R. Muller, M. Finke, and H. H. Yang, "Asymptotic Statistical Theory of Overtraining and Cross-Validation," *IEEE Transactions on Neural Networks*, vol. 8, no. 5, 1997. (Chapter 13)

[Ande72] J. A. Anderson, "A simple neural network generating an interactive memory," *Mathematical Biosciences*, vol. 14, pp. 197–220, 1972. (Chapter 1)

[AnRo88] J. A. Anderson and E. Rosenfeld, *Neurocomputing: Foundations of Research*, Cambridge, MA: MIT Press, 1989. (Chapter 1, 10)

[Barn92] E. Barnard, "Optimization for training neural nets," *IEEE Transactions on Neural Networks*, vol. 3, no. 2, pp. 232–240, 1992. (Chapter 12)

[BaSu83] A. R. Barto, R. S. Sutton and C. W. Anderson, "Neuronlike adaptive elements that can solve difficult learning control problems," *IEEE Transactions on Systems, Man, and Cybernetics*, vol. 13, pp. 834–846, 1983. (Chapter 4)

[Batt92] R. Battiti, "First and second order methods for learning: Between steepest descent and Newton's method," *Neural Computation*, vol. 4, no. 2, pp. 141–166, 1992. (Chapter 9, 12)

[Bish91] C. M. Bishop, "Improving the generalization properties of radial basis function neural networks," *Neural Computation*, vol. 3, no. 4, pp. 579-588, 1991. (Chapter 16)

[Bish95] C.M. Bishop, *Neural Networks for Pattern Recognition*, Oxford University Press,1995. (Chapter 17)

[BlDe99] J. A. Blackard and D. J. Dean, "Comparative Accuracies of Artificial Neural Networks and Discriminant Analysis in Predicting Forest Cover Types from Cartographic Variables," Computers and Electronics in Agriculture, vol. 24, pp. 131-151, 1999. (Chapter 21)

[BoJe94] G.E.P. Box, G.M. Jenkins, and G.C. Reinsel, *Time Series Analysis: Forecasting and Control*, 4th Edition, John Wiley & Sons, 2008. (Chapter 17, 22)

[BrLo88] D.S. Broomhead and D. Lowe, "Multivariable function interpolation and adaptive networks," *Complex Systems*, vol.2, pp. 321-355, 1988. (Chapter 16)

[Brog91] W. L. Brogan, *Modern Control Theory*, 3rd Ed., Englewood Cliffs, NJ: Prentice-Hall, 1991. (Chapter 4, 5, 6, 8, 9)

[Char92] C. Charalambous, "Conjugate gradient algorithm for efficient training of artificial neural networks," *IEEE Proceedings*, vol. 139, no. 3, pp. 301–310, 1992. (Chapter 12)

[ChCo91] S. Chen, C.F.N. Cowan, and P.M. Grant, "Orthogonal least squares learning algorithm for radial basis function networks," *IEEE Transactions on Neural Networks*, vol.2, no.2, pp.302-309, 1991. (Chapter 16)

[ChCo92] S. Chen, P. M. Grant, and C. F. N. Cowan, "Orthogonal least squares algorithm for training multioutput radial basis function networks," *Proceedings of the Institute of Electrical Engineers*, vol. 139, Pt. F, no. 6, pp. 378–384, 1992. (Chapter 16)

[ChCh96] S. Chen, E. S. Chng, and K. Alkadhimi, "Regularised orthogonal least squares algorithm for constructing radial basis function networks," *International Journal of Control*, vol. 64, no. 5, pp. 829–837, 1996. (Chapter 16)

[ChCo99] S. Chen, C.F.N. Cowan, and P.M. Grant, "Combined Genetic Algorithm Optimization and Regularized Orthogonal Least Squares Learning for Radial Basis Function Networks," *IEEE Transactions on Neural Networks*, vol.10, no.5, pp.302-309, 1999. (Chapter 16)

[DARP88] *DARPA Neural Network Study*, Lexington, MA: MIT Lincoln Laboratory, 1988. (Chapter 1)

[DeHa07] O. De Jesús and M. Hagan, "Backpropagation Algorithms for a Broad Class of Dynamic Networks," *IEEE Transac-*

tions on Neural Networks, vol. 18, no. 1, pp., 2007. (Chapter 14)

[Dubi00] D. Dubin, *Rapid Interpretation of EKG's*, Sixth Edition, Tampa, FL: COVER, 2000. (Chapter 20)

[Fahl89] S. E. Fahlman, "Fast learning variations on back-propagation: An empirical study," in *Proceedings of the 1988 Connectionist Models Summer School*, D. Touretzky, G. Hinton and T. Sejnowski, eds., San Mateo, CA: Morgan Kaufmann, pp. 38–51, 1989. (Chapter 12)

[FoGi07] L. Fortuna, P. Giannone, S. Graziani, M. G. Xibilia, "Virtual Instruments Based on Stacked Neural Networks to Improve Product Quality Monitoring in a Refinery," *IEEE Transactions on Instrumentation and Measurement*, vol. 56, no. 1, pp. 95–101, 2007. (Chapter 18)

[FoHa97] D. Foresee and M. Hagan, "Gauss-Newton Approximation to Bayesian Learning," *Proceedings of the 1997 International Joint Conference on Neural Networks*, vol. 3, pp. 1930 - 1935, 1997. (Chapter 13)

[FrSk91] J. Freeman and D. Skapura, *Neural Networks: Algorithms, Applications, and Programming Techniques*, Reading, MA: Addison-Wesley, 1991. (Chapter 15)

[Gill81] P. E. Gill, W. Murray and M. H. Wright, *Practical Optimization*, New York: Academic Press, 1981. (Chapter 8, 9)

[GoLa98] C. Goutte and J. Larsen, "Adaptive Regularization of Neural Networks Using Conjugate Gradient," *Proceedings of the IEEE International Conference on Acoustics, Speech and Signal Processing*, vol. 2, pp. 1201-1204, 1998. (Chapter 13)

[Gros76] S. Grossberg, "Adaptive pattern classification and universal recoding: I. Parallel development and coding of neural feature detectors," *Biological Cybernetics*, vol. 23, pp. 121–134, 1976. (Chapter 1)

[Gros80] S. Grossberg, "How does the brain build a cognitive code?" *Psychological Review*, vol. 88, pp. 375–407, 1980. (Chapter 1)

[HaBo07] L. Hamm, B. W. Brorsen and M. T. Hagan, "Comparison of Stochastic Global Optimization Methods to Estimate Neural Network Weights," *Neural Processing Letters*, vol. 26, no. 3, December 2007. (Chapter 17)

[HaDe02] M. Hagan, H. Demuth, O. De Jesus, "An Introduction to the Use of Neural Networks in Control Systems," *International Journal of Robust and Nonlinear Control*, vol. 12, no. 11, pp. 959-985, 2002. (Chapter 22)

[HaMe94] M. T. Hagan and M. Menhaj, "Training feedforward networks with the Marquardt algorithm," *IEEE Transactions on Neural Networks*, vol. 5, no. 6, pp. 989–993, 1994. (Chapter 12)

[HeBa99] S. Hettich and S. D. Bay, *The UCI KDD Archive* [http://kdd.ics.uci.edu], Irvine, CA: University of California, Department of Information and Computer Science, 1999. (Chapter 21)

[Hebb 49] D. O. Hebb, *The Organization of Behavior*, New York: Wiley, 1949. (Chapter 1, 7)

[Hech90] R. Hecht-Nielsen, *Neurocomputing*, Reading, MA: Addison-Wesley, 1990. (Chapter 15)

[HeOh97] B. Hedén, H. Öhlin, R. Rittner, L. Edenbrandt, "Acute Myocardial Infarction Detected in the 12-Lead ECG by Artificial Neural Networks," *Circulation*, vol. 96, pp. 1798–1802, 1997. (Chapter 17)

[Himm72] D. M. Himmelblau, *Applied Nonlinear Programming*, New York: McGraw-Hill, 1972. (Chapter 8, 9)

[Hopf82] J. J. Hopfield, "Neural networks and physical systems with emergent collective computational properties," *Proceedings of the National Academy of Sciences*, vol. 79, pp. 2554–2558, 1982. (Chapter 1)

[HoSt89] K. M. Hornik, M. Stinchcombe and H. White, "Multilayer feedforward networks are universal approximators," *Neural Networks*, vol. 2, no. 5, pp. 359–366, 1989. (Chapter 11)

[Jaco88] R. A. Jacobs, "Increased rates of convergence through learning rate adaptation," *Neural Networks*, vol. 1, no. 4, pp 295–308, 1988. (Chapter 12)

[Joll02] I.T. Jolliffe, *Principal Component Analysis*, Springer Series in Statistics, 2nd ed., Springer, NY, 2002. (Chapter 17)

[Koho72] T. Kohonen, "Correlation matrix memories," *IEEE Transactions on Computers*, vol. 21, pp. 353–359, 1972. (Chapter 1)

[Koho87] T. Kohonen, *Self-Organization and Associative Memory*, 2nd Ed., Berlin: Springer-Verlag, 1987. (Chapter 15)

[Koho93] T. Kohonen, "Things you haven't heard about the Self-Organizing Map," *Proceedings of the International Conference on Neural Networks* (ICNN), San Francisco, pp. 1147-1156, 1993. (Chapter 21)

[Koho95] T. Kohonen, Self-Organizing Map, 2nd ed., Springer-Verlag, Berlin, 1995. (Chapter 21)

[LeCu85] Y. Le Cun, "Une procedure d'apprentissage pour reseau a seuil assymetrique," *Cognitiva*, vol. 85, pp. 599–604, 1985. (Chapter 11)

[LeCu98] Y. LeCun, L. Bottou, G. B. Orr, K.-R. Mueller, "Efficient BackProp," *Lecture Notes in Computer Science*, vol. 1524, 1998. (Chapter 17)

[Lowe89] D. Lowe, "Adaptive radial basis function nonlinearities, and the problem of generalization," *Proceedings of the First IEE International Conference on Artificial Neural Networks*, pp. 171 - 175, 1989. (Chapter 16)

[MacK92] D. J. C. MacKay, "Bayesian Interpolation," *Neural Computation*, vol. 4, pp. 415-447, 1992. (Chapter 13)

[MaNe99] J.R. Magnu and H. Neudecker, *Matrix Differential Calculus*, John Wiley & Sons, Ltd., Chichester, 1999. (Chapter 14)

[MaGa00] E. A. Maguire, D. G. Gadian, I. S. Johnsrude, C. D. Good, J. Ashburner, R. S. J. Frackowiak, and C. D. Frith, "Navigation-related structural change in the hippocampi of taxi drivers," Proceedings of the National Academy of Sciences, vol. 97, no. 8, pp. 4398-4403, 2000. (Chapter 1)

[McPi43] W. McCulloch and W. Pitts, "A logical calculus of the ideas immanent in nervous activity," *Bulletin of Mathematical Biophysics*, vol. 5, pp. 115–133, 1943. (Chapter 1, 4)

[Mill90] A.J. Miller, *Subset Selection in Regression*. Chapman and Hall, N.Y., 1990. (Chapter 16)

[Mill93] M.F. Miller, "A scaled conjugate gradient algorithm for fast supervised learning," Neural Networks, vol. 6, pp. 525-533, 1993. (Chapter 19)

[MoDa89] J. Moody and C.J. Darken, "Fast Learning in Networks of Locally-Tuned Processing Units," *Neural Computation*, vol. 1, pp. 281–294, 1989. (Chapter 16)

[Moll93] M. Moller, "A scaled conjugate gradient algorithm for fast supervised learning," *Neural Networks*, vol. 6, pp. 525-533, 1993. (Chapter 17)

[MoMa01] G.B. Moody, R.G. Mark, and A.L. Goldberger, "PhysioNet: a Web-based resource for the study of physiologic signals," *IEEE Transactions on Engineering in Medicine and Biology*, vol. 20, no. 3, pp: 70-75, 2001. (Chapter 20)

[MiPa69] M. Minsky and S. Papert, *Perceptrons*, Cambridge, MA: MIT Press, 1969. (Chapter 1, 4)

[NaMu97] Narendra, K.S.; Mukhopadhyay, S., "Adaptive control using neural networks and approximate models," *IEEE Transactions on Neural Networks*, vol. 8, no. 3, pp. 475 - 485, 1997. (Chapter 22)

[NaPa90] K. S. Narendra and K. Parthasarathy, "Identification and control of dynamical systems using neural networks," *IEEE Transactions on Neural Networks*, vol. 1, no. 1, pp. 4–27, 1990. (Chapter 22)

[NgWi90] D. Nguyen and B. Widrow, "Improving the learning speed of 2-layer neural networks by choosing initial values of the adaptive weights," *Proceedings of the IJCNN*, vol. 3, pp. 21–26, July 1990. (Chapter 12, 17)

[OrHa00] M. J. Orr, J. Hallam, A. Murray, and T. Leonard, "Assessing rbf networks using delve," IJNS, 2000. (Chapter 16)

[Park85] D. B. Parker, "Learning-logic: Casting the cortex of the human brain in silicon," Technical Report TR-47, Center for Computational Research in Economics and Management Science, MIT, Cambridge, MA, 1985. (Chapter 11)

[PaSa93] J. Park and I.W. Sandberg, "Universal approximation using radial-basis-function networks," *Neural Computation*, vol. 5, pp. 305-316, 1993. (Chapter 16)

[PeCo93] M. P. Perrone and L. N. Cooper, "When networks disagree: Ensemble methods for hybrid neural networks," in *Neural Networks for Speech and Image Processing*, R. J. Mammone, Ed., Chapman-Hall, pp. 126-142, 1993. (Chapter 17)

[PhHa13] M. Phan and M. Hagan, "Error Surface of Recurrent Networks," *IEEE Transactions on Neural Networks and Learning Systems*, vol. 24, no. 11, pp. 1709 - 1721, October, 2013. (Chapter 14)

[Powe87] M.J.D. Powell, "Radial basis functions for multivariable interpolation: a review," *Algorithms for Approximation*, pp. 143-167, Oxford, 1987. (Chapter 16)

[PuFe97] G.V. Puskorius and L.A. Feldkamp, "Extensions and enhancements of decoupled extended Kalman filter training," *Proceedings of the 1997 International Conference on Neural Networks*, vol. 3, pp. 1879-1883, 1997. (Chapter 17)

[RaMa05] L.M. Raff, M. Malshe, M. Hagan, D.I. Doughan, M.G. Rockley, and R. Komanduri, "*Ab initio* potential-energy surfaces for complex, multi-channel systems using modified novelty sampling and feedforward neural networks," *The Journal of Chemical Physics*, vol. 122, 2005. (Chapter 17, 19)

[RiIr90] A. K. Rigler, J. M. Irvine and T. P. Vogl, "Rescaling of variables in back propagation learning," *Neural Networks*, vol. 3, no. 5, pp 561–573, 1990. (Chapter 12)

[Rose58] F. Rosenblatt, "The perceptron: A probabilistic model for information storage and organization in the brain," *Psychological Review*, vol. 65, pp. 386–408, 1958. (Chapter 1, 4)

[Rose61] F. Rosenblatt, *Principles of Neurodynamics*, Washington DC: Spartan Press, 1961. (Chapter 4)

[RuHi86] D. E. Rumelhart, G. E. Hinton and R. J. Williams, "Learning representations by back-propagating errors," *Nature*, vol. 323, pp. 533–536, 1986. (Chapter 11)

[RuMc86] D. E. Rumelhart and J. L. McClelland, eds., *Parallel Distributed Processing: Explorations in the Microstructure of Cognition*, vol. 1, Cambridge, MA: MIT Press, 1986. (Chapter 1, 11, 15)

[Sarle95] W. S. Sarle, "Stopped training and other remedies for overfitting," In *Proceedings of the 27th Symposium on Interface*, 1995. (Chapter 13)

[Scal85] L. E. Scales, *Introduction to Non-Linear Optimization*, New York: Springer-Verlag, 1985. (Chapter 8, 9, 12)

[ScSm99] B. Schölkopf, A. Smola, K.-R. Muller, "Kernel Principal Component Analysis," in B. Schölkopf, C. J. C. Burges, A. J. Smola (Eds.), *Advances in Kernel Methods-Support Vector Learning*, MIT Press Cambridge, MA, USA, pp. 327-352, 1999. (Chapter 17)

[Shan90] D. F. Shanno, "Recent advances in numerical techniques for large-scale optimization," in *Neural Networks for Control*, Miller, Sutton and Werbos, eds., Cambridge, MA: MIT Press, 1990. (Chapter 12)

[SjLj94] J. Sjoberg and L. Ljung, "Overtraining, regularization and searching for minimum with application to neural networks," Linkoping University, Sweden, Tech. Rep. LiTH-ISY-R-1567, 1994. (Chapter 13)

[StDo84] W. D. Stanley, G. R. Dougherty and R. Dougherty, *Digital Signal Processing*, Reston VA: Reston Publishing Co., 1984. (Chapter 10)

[Stra76] G. Strang, *Linear Algebra and Its Applications*, New York: Academic Press, 1980. (Chapter 5, 6)

[TeSi00] J. B. Tenenbaum, V. de Silva, J. C. Langford, "A Global Geometric Framework for Nonlinear Dimensionality Reduction," *Science*, vol. 290, pp. 2319-2323, 2000. (Chapter 20)

[Tikh63] A. N. Tikhonov, "The solution of ill-posed problems and the regularization method," *Dokl. Acad. Nauk USSR*, vol. 151, no. 3, pp. 501-504, 1963. (Chapter 13)

[Toll90] T. Tollenaere, "SuperSAB: Fast adaptive back propagation with good scaling properties," *Neural Networks*, vol. 3, no. 5, pp. 561–573, 1990. (Chapter 12)

[VoMa88] T. P. Vogl, J. K. Mangis, A. K. Zigler, W. T. Zink and D. L. Alkon, "Accelerating the convergence of the backpropagation method," *Biological Cybernetics*, vol. 59, pp. 256–264, Sept. 1988. (Chapter 12)

[WaVe94] C. Wang, S. S. Venkatesh, and J. S. Judd, "Optimal Stopping and Effective Machine Complexity in Learning," *Advances in Neural Information Processing Systems*, J. D. Cowan, G. Tesauro, and J. Alspector, Eds., vol. 6, pp. 303-310, 1994. (Chapter 13)

[Werbo74] P. J. Werbos, "Beyond regression: New tools for prediction and analysis in the behavioral sciences," Ph.D. Thesis, Harvard University, Cambridge, MA, 1974. Also published as *The Roots of Backpropagation*, New York: John Wiley & Sons, 1994. (Chapter 11)

[Werb90] P. J. Werbos, "Backpropagation through time: What it is and how to do it," *Proceedings of the IEEE*, vol. 78, pp. 1550–1560, 1990. (Chapter 14)

[WeTe84] J. F. Werker and R. C. Tees, "Cross-language speech perception: Evidence for perceptual reorganization during the first year of life," Infant Behavior and Development, vol. 7, pp. 49-63, 1984. (Chapter 1)

[WhSo92] D. White and D. Sofge, eds., *Handbook of Intelligent Control*, New York:Van Nostrand Reinhold, 1992. (Chapter 4)

[WiHo60] B. Widrow, M. E. Hoff, "Adaptive switching circuits,"*1960 IRE WESCON Convention Record*, New York: IRE Part 4, pp. 96–104, 1960. (Chapter 1, 10)

[WiSt 85] B. Widrow and S. D. Stearns, *Adaptive Signal Processing*, Englewood Cliffs, NJ: Prentice-Hall, 1985. (Chapter 10)

[WiWi 88] B. Widrow and R. Winter, "Neural nets for adaptive filtering and adaptive pattern recognition," *IEEE Computer Magazine*, March 1988, pp. 25–39. (Chapter 10)

[WiZi89] R. J. Williams and D. Zipser, "A learning algorithm for continually running fully recurrent neural networks," *Neural Computation*, vol. 1, pp. 270–280, 1989. (Chapter 14)

B Notation

Basic Concepts

Scalars: small *italic* letters.....a,b,c

Vectors: small **bold** nonitalic letters.....**a,b,c**

Matrices: capital **BOLD** nonitalic letters.....**A,B,C**

Language

Vector means a column of numbers.

Row vector means a row of a matrix used as a vector (column).

General Vectors and Transformations (Chapters 5 and 6)

$x = A(y)$

Weight Matrices

Scalar Element

$w_{i,j}^{k}(t)$

i - row, j - column, k - layer, t - time or iteration

Matrix

$\mathbf{W}^{k}(t)$

Column Vector

$\mathbf{w}_{j}^{k}(t)$

Row Vector

${}_{i}\mathbf{w}^{k}(t)$

Bias Vector

Scalar Element

$b_{i}^{k}(t)$

Vector

$\mathbf{b}^{k}(t)$

Input Vector

Scalar Element

$p_i(t)$

As One of a Sequence of Input Vectors

$\mathbf{p}(t)$

As One of a Set of Input Vectors

$\mathbf{p}_q$

Net Input Vector

Scalar Element

$n_i^k(t)$ or $n_{i,q}^k$

Vector

$\mathbf{n}^k(t)$ or $\mathbf{n}_q^k$

Output Vector

Scalar Element

$a_i^k(t)$ or $a_{i,q}^k$

Vector

$\mathbf{a}^k(t)$ or $\mathbf{a}_q^k$

Transfer Function

Scalar Element

$a_i^k = f^k(n_i^k)$

Vector

$\mathbf{a}^k = \mathbf{f}^k(\mathbf{n}^k)$

Target Vector

Scalar Element

$t_i(t)$ or $t_{i,q}$

Vector

$\mathbf{t}(t)$ or $\mathbf{t}_q$

Set of Prototype Input/Target Vectors

$\{\mathbf{p}_1, \mathbf{t}_1\}, \{\mathbf{p}_2, \mathbf{t}_2\}, \ldots, \{\mathbf{p}_Q, \mathbf{t}_Q\}$

Error Vector

Scalar Element

$e_i(t) = t_i(t) - a_i(t)$ or $e_{i,q} = t_{i,q} - a_{i,q}$

Vector

$\mathbf{e}(t)$ or $\mathbf{e}_q$

Sizes and Dimensions

Number of Layers, Number of Neurons per Layer

M, S^k

Number of Input Vectors (and Targets), Dimension of Input Vector

Q, R

Parameter Vector (includes all weights and biases)

Vector

$\mathbf{x}$

At Iteration k

$\mathbf{x}(k)$ or $\mathbf{x}_k$

Norm

$\|\mathbf{x}\|$

Performance Index

$F(\mathbf{x})$

Gradient and Hessian

$\nabla F(\mathbf{x}_k) = \mathbf{g}_k$ and $\nabla^2 F(\mathbf{x}_k) = \mathbf{A}_k$

Parameter Vector Change

$$\Delta \mathbf{x}_k = \mathbf{x}_{k+1} - \mathbf{x}_k$$

Eigenvalue and Eigenvector

λ_i and $\mathbf{z}_i$

Approximate Performance Index (single time step)

$$\hat{F}(\mathbf{x})$$

Transfer Function Derivative

Scalar

$$\dot{f}(n) = \frac{d}{dn} f(n)$$

Matrix

$$\dot{\mathbf{F}}^m(\mathbf{n}^m) = \begin{bmatrix} \dot{f}^m(n_1^m) & 0 & \dots & 0 \\ 0 & \dot{f}^m(n_2^m) & \dots & 0 \\ \vdots & \vdots & & \vdots \\ 0 & 0 & \dots & \dot{f}^m(n_{S^m}^m) \end{bmatrix}$$

Jacobian Matrix

$$\mathbf{J}(\mathbf{x})$$

Approximate Hessian Matrix

$$\mathbf{H} = \mathbf{J}^T \mathbf{J}$$

Sensitivity Vector

Scalar Element

$$s_i^m \equiv \frac{\partial \hat{F}}{\partial n_i^m}$$

Vector

$$\mathbf{s}^m \equiv \frac{\partial \hat{F}}{\partial \mathbf{n}^m}$$

Marquardt Sensitivity Matrix

Scalar Element

$$\tilde{s}^m_{i,h} \equiv \frac{\partial v_h}{\partial n^m_{i,q}} = \frac{\partial e_{k,q}}{\partial n^m_{i,q}}$$

Partial Matrix (single input vector $\mathbf{p}_q$) and Full Matrix (all inputs)

$$\tilde{\mathbf{S}}^m_q \text{ and } \tilde{\mathbf{S}}^m = \left[\tilde{\mathbf{S}}^m_1 \; \tilde{\mathbf{S}}^m_2 \; \ldots \; \tilde{\mathbf{S}}^m_Q\right]$$

Dynamic Networks

Sensitivity

$$s^{u,m}_{k,i}(t) \equiv \frac{\partial^e a^u_k(t)}{\partial n^m_i(t)}$$

Weight Matrices

$\mathbf{IW}^{m,l}(d)$ - input weight between input l and layer m at delay d

$\mathbf{LW}^{m,l}(d)$ - layer weight between layer l and layer m at delay d

Index Sets

$DL_{m,l}$ - delays in the tapped delay line between Layer l and Layer m.

$DI_{m,l}$ - delays in the tapped delay line between Input l and Layer m.

I_m - indices of input vectors that connect to layer m.

L^f_m - indices of layers that directly connect *forward* to layer m.

L^b_m - indices of layers that are directly connected backwards to layer m (or to which layer m connects forward) and that contain no delays in the connection.

$$E^U_{LW}(x) = \{u \in U \ni \exists(\mathbf{LW}^{x,u}(d) \neq 0, d \neq 0)\}$$

$$E^X_S(u) = \{x \in X \ni \exists(\mathbf{S}^{u,x} \neq 0)\}$$

$$E_S(u) = \{x \ni \exists(\mathbf{S}^{u,x} \neq 0)\}$$

$$E^X_{LW}(u) = \{x \in X \ni \exists(\mathbf{LW}^{x,u}(d) \neq 0, d \neq 0)\}$$

$$E^U_S(x) = \{u \in U \ni \exists(\mathbf{S}^{u,x} \neq 0)\}$$

Definitions

Input Layer (*X*) - has an input weight, or contains any delays with any of its weight matrices

Output Layer (*U*) - its output will be compared to a target during training, or it is connected to an input layer through a matrix that has delays associated with it.

Parameters for Backpropagation and Variations

Learning Rate and Momentum

α and γ

Learning Rate Increase, Decrease and Percentage Change

η, ρ and ζ

Conjugate Gradient Direction Adjustment Parameter

β_k

Marquardt Parameters

μ and ϑ

Generalization

Regularization Parameters

α, β and $\rho = \frac{\alpha}{\beta}$

Effective Number of Parameters

γ

Selected Model

M

Sum Squared Error and Sum Squared Weights

E_D, E_W

Maximum Likelihood and Most Probable Weights

$\mathbf{x}^{ML}$, $\mathbf{x}^{MP}$

Feature Map Terms

Distance Between Neurons

d_{ij} - distance between neuron i and neuron j

Neighborhood

$N_i(d) = \{j, d_{ij} \le d\}$

C Software

Introduction

We have used MATLAB, a numeric computation and visualization software package, in this text. However, MATLAB is not essential for using this book. The computer exercises can performed with any available programming language, and the *Neural Network Design Demonstrations*, while helpful, are not critical to understanding the material covered in this book.

MATLAB is widely available and, because of its matrix/vector notation and graphics, is a convenient environment in which to experiment with neural networks. We use MATLAB in two different ways. First, we have included a number of exercises for the reader to perform in MATLAB. Many of the important features of neural networks become apparent only for large scale problems, which are computationally intensive and not feasible for hand calculations. With MATLAB, neural network algorithms can be quickly implemented, and large scale problems can be tested conveniently. If MATLAB is not available, any other programming language can be used to perform the exercises.

The second way in which we use MATLAB is through the *Neural Network Design Demonstrations*, which can be downloaded from the website hagan.okstate.edu/nnd.html. These interactive demonstrations illustrate important concepts in each chapter. The icon to the left identifies references to these demonstrations in the text.

MATLAB, or the student edition of MATLAB, version 2010a or later, should be installed on your computer in a a folder named MATLAB. To create this directory or folder and complete the MATLAB installation process, follow the instructions given in the MATLAB documentation. Take care to follow the guidelines given for setting the path.

After the Neural Network Design Demonstration software has been loaded into the MATLAB directory on your computer (or if the MATLAB path has been set to include the directory containing the demonstration software), the demonstrations can be invoked by typing **nnd** at the MATLAB prompt. All demonstrations are easily accessible from a master menu.

Overview of Demonstration Files

Running the Demonstrations

You can run the demonstrations directly by typing their names at the MATLAB prompt. Typing **help nndesign** brings up a list of all the demos you can choose from.

Alternatively, you can run the Neural Network Design splash window (**nnd**) and then click the Contents button. This will take you to a graphical Table of Contents. From there you can select chapters with buttons at the bottom of the window and individual demonstrations with popup menus.

Sound

Many of the demonstrations use sound. In many cases the sound adds to the understanding of a demonstration. In other cases it is there simply for fun. If you need to turn the sound off you can give MATLAB the following command and all demonstrations will run quietly:

```
nnsound off
```

To turn sound back on:

```
nnsound on
```

You may note that demonstrations that utilize sound often run faster when sound is off. In addition, on some machines which do not support sound errors can occur unless the sound is turned off.

List of Demonstrations

General

nnd - Splash screen.
nndtoc - Table of contents.
nnsound - Turn Neural Network Design sounds on and off.

Chapter 2, Neuron Model and Network Architectures

nnd2n1 - One-input neuron.
nnd2n2 - Two-input neuron.

Chapter 3, An Illustrative Example

nnd3pc - Perceptron classification.
nnd3hamc - Hamming classification.
nnd3hopc - Hopfield classification.

Chapter 4, Perceptron Learning Rule

nnd4db - Decision boundaries.
nnd4pr - Perceptron rule.

Chapter 5, Signal and Weight Vector Spaces

nnd5gs - Gram-Schmidt.
nnd5rb - Reciprocal basis.

Chapter 6, Linear Transformations for Neural Networks

nnd6lt - Linear transformations.
nnd6eg - Eigenvector game.

Chapter 7, Supervised Hebbian Learning

nnd7sh - Supervised Hebb.

Chapter 8, Performance Surfaces and Optimum Points

nnd8ts1 - Taylor series #1.
nnd8ts2 - Taylor series #2.
nnd8dd - Directional derivatives.
nnd8qf - Quadratic function.

Chapter 9, Performance Optimization

nnd9sdq - Steepest descent for quadratic function.
nnd9mc - Method comparison.
nnd9nm - Newton's method.
nnd9sd - Steepest descent.

Chapter 10, Widrow-Hoff Learning

nnd10nc - Adaptive noise cancellation.
nnd10eeg - Electroencephalogram noise cancellation.
nnd10lc - Linear pattern classification.

Chapter 11, Backpropagation

nnd11nf - Network function.
nnd11bc - Backpropagation calculation.
nnd11fa - Function approximation.
nnd11gn - Generalization.

Chapter 12, Variations on Backpropagation

nnd12sd1- Steepest descent backpropagation #1.
nnd12sd2 - Steepest descent backpropagation #2.
nnd12mo - Momentum backpropagation.
nnd12vl - Variable learning rate backpropagation.
nnd12ls - Conjugate gradient line search.
nnd12cg - Conjugate gradient backpropagation.
nnd12ms - Maquardt step.
nnd12m - Marquardt backpropagation.

Chapter 13, Generalization

nnd13es - Early stopping.
nnd13reg - Regularization.
nnd13breg - Bayesian regularization.
nnd13esr - Early stopping/regularization.

Chapter 14, Dynamic Networks

nnd14fir - Finite impulse response network.
nnd14iir - Infinite impulse response network.
nnd14dynd - Dynamic derivatives.
nnd14rnt - Recurrent network training.

Chapter 15, Associative Learning

nnd15uh - Unsupervised Hebb.
nnd15edr - Effect of decay rate.
nnd15hd - Hebb with decay.
nnd15gis - Graphical instar.
nnd15is - Instar.
nnd15os - Outstar.

Chapter 16, Competitive Networks

nnd16cc - Competitive classification.
nnd16cl - Competitive learning.
nnd16fm1 - 1-D feature map.
nnd16fm2 - 2-D feature map.
nnd16lv1 - LVQ1.
nnd16lv2 - LVQ2.

Chapter 17, Radial Basis Networks

nnd17nf - Network function.
nnd17pc - Pattern classification.
nnd17lls - Linear least squares.
nnd17ols - Orthogonal least squares.
nnd17no - Nonlinear optimization.

Chapter 18, Grossberg Network

nnd18li - Leaky integrator.
nnd18sn - Shunting network.
nnd18gl1 - Grossberg layer 1.
nnd18gl2 - Grossberg layer 2.
nnd18aw - Adaptive weights.

Chapter 19, Adaptive Resonance Theory

nnd19al1 - ART1 layer 1.
nnd19al2 - ART1 layer 2.
nnd19os - Orienting subsystem.
nnd19a1 - ART1 algorithm.

Chapter 20, Stability

nnd20ds - Dynamical system.

Chapter 21, Hopfield Network

nnd21hn - Hopfield network.

Index

W

CPSIA information can be obtained
at www.ICGtesting.com
Printed in the USA
LVHW020208190723
752608LV00011B/293